Era	Period		Age (millions of years)	Events
PALEOZOIC	Permian		225	Alleghenian (Appalachians)—Hercynian (Europe) orogenies
			280	Final assembly of Pangaea
	Carboniferous	Pennsylvanian	320	Early reptiles
				Extensive coal formation
		Mississippian	345	
	Devonian		395	Early forests, trees
				Acadian (Appalachians)—Caledonian (Europe) orogenies
	Silurian		430	Land plants
				Primitive fish
	Ordovician		500	Taconic orogeny (Appalachians)
	Cambrian		570	Evolution of multicellular and shelled organisms: 700–570
PRECAMBRIAN	Proterozoic		2300–2800	Grenville orogeny: ±1000
				Abundant iron formations
				Major gold deposits
	Archean		4600–4700	Earliest known bacteria and algae: ±3200
				Oldest rock (Southwest Greenland): ±3800
				Evolution of core-mantle-crust

EARTH

Mt. St. Helens in eruption in 1980 after 123 years of dormancy. Long before this most recent episode, geologists knew from the geologic record that Mt. St. Helens was the most active and explosive volcano in the contiguous United States. [U.S. Geological Survey.]

EARTH

THIRD EDITION

Frank Press
NATIONAL ACADEMY OF SCIENCES

Raymond Siever
HARVARD UNIVERSITY

W. H. FREEMAN AND COMPANY
San Francisco

Developmental and Manuscript Editor:
  Caroline Eastman
Project Editor: Betsy Dilernia
Designer: Robert Ishi
Production Coordinator: Bill Murdock
Illustration Coordinator: Richard Quiñones
Artists: Donna Salmon, John and Judy Waller,
  John and Jean Foster
Compositor: York Graphic Services
Printer and Binder: Arcata Book Group

Library of Congress Cataloging in Publication Data

Press, Frank.
  Earth.

  Includes bibliographies and index.
  1. Physical geology. I. Siever, Raymond. II. Title.
QE28.2.P7 1982   550   81-17451
ISBN 0-7167-1362-4   AACR2

1 2 3 4 5 6 7 8 9 0   KP   0 8 9 8 7 6 5 4 3 2

To Billie and Doris

Contents

Preface

TO THE STUDENT

Although it has been used by others, this book is written for beginning students who have had no previous college science courses and who may not necessarily intend to specialize in geology. For this reason, we have deliberately emphasized a broad view, one that stresses concepts and shows by many examples how science is actually done. Through the use of analogies to familiar processes, "kitchen physics" and "kitchen chemistry" in explanations, and many diagrammatic illustrations, we try to show the evidential bases of geological theories and the strong dependence of geology on the basic scientific disciplines of physics and chemistry. Because the organic world plays an important role in many geological processes, we introduce some notions of biological processes where appropriate. Yet first and foremost, the book is about geological processes.

As much as possible, we have tried to impart something about the motivations of contemporary geologists and their methods, both old and new. We want the student to share some of the excitement and exhilaration triggered by the many recent discoveries that have greatly increased our understanding of how this planet and the other planets work. Both of us have been active in research and teaching, and we believe the gap between what is new and what is taught to beginners should be narrowed. We take it as a challenge to integrate in a natural way the newest discoveries of plate tectonics, marine geology, geochemistry, geophysics, and lunar and Martian geology into the traditional discussions of such topics as geomorphology, sedimentation, petrology, volcanism, and structural geology.

We have not introduced the very new at the expense of eliminating the essential material, both traditional and modern, that a good course in geology should cover. By using the minimum vocabulary necessary for discussion of the concepts involved, we introduce the basic terminology that will suffice for more specialized courses that might follow. The coverage of subject matter is sufficiently complete that the book could, if used in its entirety, serve as a foundation on which to build a career in earth science. Nevertheless, our major aim is to reach the many students whose course in geology may be their sole college exposure to science and the study of the Earth. Ultimately, as responsible citizens they will participate in governmental policy decisions pertaining to geological questions.

The scientific study of the Earth has never been as important a concern for all peoples as it is today. The energy, food, and mineral resources of the Earth are now subjects for the daily newspapers and television; the peoples and governments of the world are coming to a new awareness of their importance for the well-being of society. More than at any time in the past century, today geologists are being asked such questions as how oil is distributed in the Earth and why we have no domestic supply of some valuable metals. Many of the questions relate to difficulties we encounter in managing our environment while attempting to exploit energy and mineral resources and produce the enormous quantities of industrial and other products required to support a huge and growing population. In various places in the book, therefore, we show how *knowledge* of the science is linked to its *uses,* both in detailed practical ways and in the making of policy decisions. As geology has become the focus of more attention, it has aroused the curiosity of young people about nature in general. Enrollment in introductory geology courses has increased greatly in the past decade. Today people travel more than ever before, and as they see the diversity of the Earth, they want to understand what they see.

Geology is in a golden age, as measured by the impact of the discoveries and theories of the past fifteen years that have given us profound new insights into the ways this planet works. For the first time in the modern history of the discipline, an all-encompassing synthesis of much of geological knowledge has been advanced and has gained acceptance across a broad spectrum of professionals in the field. That synthesis—plate-tectonic theory—born only fifteen years ago, has started to mature in the few years since the first edition of this book appeared. As with any major new theory, the work it has stimulated has enriched the subject and at the same time introduced complexities in a once-simple idea. Applications of the theory to new areas of the oceans and continents have bred new problems. Geologists are tackling these problems using a framework of thinking that is still only a decade old, one that has not yet settled into a well-worn path of familiar notions.

The plate-tectonic theory is not the only new development in geology. In the past few years the discoveries in planetary science—particularly the landing on Mars in 1976, the Pioneer mission to Venus in 1978, the Voyager missions to Jupiter and Saturn in 1980 and 1981—have added greatly to our knowledge of the evolution of the solar system and have given us better data for comparison with the planet we live on. This new knowledge has in turn stimulated a resurgence of work on the earliest rocks of Earth. But these are only a few of the high spots. The fact that every chapter of this book has had to be revised since the second edition is a measurement of our advancing knowledge on every aspect of the subject.

Our objective is to give an introduction to this broad field of knowledge that is as up-to-date as the results reported at the most recent meetings of the Geological Society of America and to present that new material in such a way that it is as understandable as today's newspaper yet not oversimplified to an extent that would destroy the richness of the discoveries.

We hope that this book will reach many students, and, in introducing them to geology, will impart something of the intellectual excitement, the growing relevance to societal problems, and the esthetics of the subject. If we succeed in this goal, the large investment we have made in time and energy will all have been worthwhile.

TO THE INSTRUCTOR

In preparing this new edition, we have benefited from the recommendations of hundreds of instructors. As a result many of the more complex sections have been simplified. The chapters on plate tectonics and deformation have been more fully integrated. We have made major revisions and have updated the chapters dealing with deformation, planets, and matter and energy. Without lengthening the book, we have added new material to most chapters, including such topics of current interest as the eruption of Mt. St. Helens, the discovery of hot springs on the sea floor, COCORP, the new significance of overthrust belts, pre-Mesozoic plate tectonics and continental drift, displaced terrains, and the exotic satellites of Jupiter and Saturn.

As in previous editions, each chapter is as self-sufficient as possible, so that chapters may be skipped in short courses or be taken up out of sequence according to the instructor's individual taste. (In the Instructor's Guide, Professor Roger Thomas makes several suggestions along these lines.) We have designed *repetition, review,* and *alternative restatement* into the book with a definite purpose: to enhance learning and to increase flexibility in the way the book is used.

We make extensive use of line drawings and photographs, plus boxed information and other pedagogical aids, for we believe such devices are of considerable value not only in simplifying difficult material but also in motivating the student. Explanatory sketches, diagrams, and photographs substitute for equations and help make up for lack of prior knowledge of other sciences. We also use illustrations as alternative statements of concepts presented in the text and as summaries of material covered earlier. To avoid interrupting the flow of the main text, occasional bits of parenthetical material appear as footnotes. These brief notes serve as slight amplifications of the text, as interesting sidelights, and as comments on some of the extraordinary personalities who have been part of the quest to understand the Earth. We have increased the number of boxes that expand in a more detailed way some materials of the text. These boxes are for the student who wants to understand more deeply some of the background of the subject. They are not necessary for understanding the text; some are pitched at a slightly higher level than the rest of the book.

The introductions at the beginning of the three parts of the book and the brief abstracts at the beginnings of chapters are designed to forecast in a general way the nature of the subject matter, how it fits together, and how it relates to other chapters. Summaries in list form appear at the ends of chapters to serve as systematic reviews of major conclusions. Questions were devised to help students test their comprehension of the materials either by essay or by solving concrete problems. The bibliographies include *Scientific American* Offprints on closely related subjects, elementary or slightly advanced paperbacks, government reports on specialized topics, and a few readily available technical articles from the geological literature.

The book is divided into three parts. Each part consists of chapters grouped together according to their relation to the major concepts of the Earth's dynamics. Part I groups topics relating to the Earth as an evolving planet and how we study it and its materials. In the first chapter we give a capsule history of the Earth and the first glimpse of the general theory about how it operates. A brief outline of plate tectonics is presented as a guide to succeeding chapters, where ramifications and implications of the theory are discussed with reference to the entire range of geological subjects considered in the book. The second chapter explores time in geology, the relationship between process and history, and emphasizes how field observations form the central basis of our knowledge of the geological cycle. The third chapter is concerned with the prime source of information on the Earth: rocks and minerals. The major concepts of mineralogy and petrology are introduced and linked to a brief but systematic discussion of the major rock-forming minerals and the three major rock groups.

Part II covers those aspects of the Earth dominated by the external solar heat machine, all of the surface processes that result from the Sun's radiant energy impinging on the surface of the planet, its atmosphere, and oceans. Erosion, transportation, and deposition of chemically altered and physically fragmented rocks, and the resulting sculpture of the surface are discussed in relation to tectonics and the dynamics of the atmosphere and oceans. Part II concludes with a chapter on the interactions of the biological world and Earth's inorganic materials, and how humanity as a geological agent has been profoundly changing the surface environment.

Part III explores the consequences of the internal heat machine of the Earth, and how it drives major movements of the interior and determines the structure of the whole planet. Internal heat, volcanism, and the kinds of igneous and metamorphic rocks that are produced by thermal processes are the subjects of the first group of chapters. The structure of the interior as deduced from seismology, gravity, and magnetism is then explored, in preparation for a detailed systematic explanation of plate tectonics. It is only at this point that we come to structural geology, which can then be treated in the context of the large-scale motions of lithospheric plates. A chapter on what we know of the nature and evolution of the other planets in the solar system follows, with major emphasis on lunar exploration. The book concludes with a chapter on Earth materials as resources, including an extended treatment of energy reserves and the central importance of energy costs in the recovery of all other resources.

We are indebted to Warren Hamilton (U.S. Geological Survey) for preparing a set of color slides to accompany the text. Norman Gilinsky helped us greatly in revising the glossary, preparing the index, and reading proof. Evelyn Stinchfield and Betsy Northrup helped with typing and other details of manuscript preparation. Caroline Eastman served admirably as developmental and manuscript editor.

October 1981

Frank Press
Raymond Siever

Acknowledgments

In preparing this edition we have been helped immeasurably by the many people who have pointed out corrections and given us thoughtful comments on the second edition or have provided new materials for tables and illustrations. Among these we are especially grateful to Clark Burchfiel (MIT), John Edmond (MIT), Ian Evans (University of Houston), Carol Gilchrest (Tulane University), Linda Meinke (MIT), Peter Molnar (MIT), Don R. Mullineaux (U.S. Geological Survey), Hallan C. Noltimier (Ohio State University), John Sclater (MIT), Edward C. Stone (Caltech), R. D. K. Thomas (Franklin and Marshall College), and M. Nafi Toksöz (MIT).

While we were planning this revision, W. H. Freeman and Company sent out a questionnaire to well-known teachers of introductory courses in geology asking for evaluation of the second edition and suggestions for the third. The responses—some of which included detailed comments—were invaluable as a guide to subject matter, level, and coverage, being based as they were on actual teaching experience using the book. Deserving special mention are the comments made by Wayne M. Ahr (Texas A&M University), John C. Butler (University of Houston), J. Allan Cain (University of Rhode Island), Richard L. Chase (University of British Columbia), Darrel S. Cowan (University of Washington), Tom Donnelly (State University of New York at Binghamton), Michael O. Garcia (University of Hawaii), Robert A. Gastaldo (Auburn University), Thomas L. Henyey (University of Southern California), Harry J. Klepser (University of Tennessee), Otto C. Kopp (University of Tennessee), Emlyn H. Koster (Alberta Research Council), Alexander R. McBirney (University of Oregon), Robert D. Merrill (California State University, Fresno), Hector J. Munn (George Lox College), Hermann Pfefferkorn (University of Pennsylvania), and Joseph J. Spigai (Office of Naval Research).

EARTH

THE EVOLUTION OF THE EARTH
AND HOW WE STUDY IT

This planet on which we live has undergone constant change throughout a long history. To understand what the Earth is and how it works today, we link direct observation of processes operating at the surface with indirect measurement of forces working in the interior. The fullest knowledge comes from deducing how the planet evolved—from its beginnings to its present state.

Formed almost five billion years ago from a mass of dust rotating around the infant Sun, the Earth grew into a medium-sized planet whose history has been dominated by two driving mechanisms. The first is the internal heat produced by radioactivity in the Earth. The second is the external heat supplied to the surface by the Sun. The internal heat melts rocks, makes volcanoes, and thrusts mountains upward. The external heat drives the atmosphere and the oceans and causes the erosion of mountains and the reduction of rock to sediment. New methods of studying how the internal and external forces drive the Earth have generated a wealth of new information and raised many exciting new questions. In the past two decades, geologists have gradually developed a new unifying theory that relates all of the dynamic Earth processes to the motions of large plates that constitute the outer shell of the planet. Called the theory of plate tectonics, it offers the most comprehensive model that geologists have ever had for explaining how the Earth works.

In the first part of this book we survey the ways in which we study this planet and what we have learned of its origin. These first chapters are a capsule of the book; in them we preview many subjects—particularly the nature of time and the materials of Earth—before they are explored in detail in later chapters.

1

History of the Earth and Solar System

*An introduction to the Earth, beginning with the cloud of dust and gas
from which the solar system formed a little less than 5 billion years ago,
to the birth of the Earth about 4.7 billion years ago, to the planet we know
today, with its hospitable atmosphere and rich resources, a planet still
active inside—as evidenced by earthquakes, volcanoes, ocean basins that
open and close, and continents that drift apart.*

THE UNIQUENESS OF PLANET EARTH

"Civilization exists by geological consent, subject to change without notice," said philosopher-historian Will Durant, reminding us of the remarkable circumstances that make this planet congenial to life as we know it. The Earth, after all, is a very special place—and not just because we humans inhabit it. More than a million life forms have developed on this unique spot in the solar system. *Homo sapiens*, the one species with the power of reason, is a rather recent arrival. In the study of geology, we not only explore the Earth as it exists today; we also seek answers to the questions of how it was formed, what it was like when first born, how it evolved to the planet of today, and, perhaps most exciting of all, what made it capable of supporting life.

No one knows precisely when the composition and state of the Earth's primitive atmosphere were just right for life to begin and evolve. We do know, however, that the large organic molecules that apparently preceded the evolution of the earliest forms of life could not have formed if the primitive atmosphere had contained as much oxygen as the one we now enjoy. Chemists tell us the oxygen would have destroyed them. The Earth's atmosphere and magnetic field acted as a shield against some of the biologically damaging radiation from space, just as they do today. Meteors, unbraked by the cushion of gases, would have bombarded Earth, leaving a crater-pocked, desolate surface that could never be softened by erosion. The atmosphere that exists today not only filters out most of the destructive ultraviolet radiation, but together with the ocean, it stores and redistributes solar energy, thus moderating climate. Without atmosphere and oceans, there would be much more extreme temperature differences between day and night, summer and winter, and equator and pole.

The list of conditions so favorable to life is long. Life as we know it is possible over a very narrow temperature interval—essentially within the limits set by the freezing and boiling points of water. This interval is perhaps 1 or 2 percent of the range between a temperature of absolute zero and the surface temperature of the Sun. How fortunate that Earth formed where it did in the solar system, neither too far from the Sun nor too close to it! Probably the conditions that would allow life to flourish anywhere in the universe do not differ much from those that have allowed life to evolve on Earth. This observation led astronomers to propose that other planets situated about as far from their suns as we are from ours might also

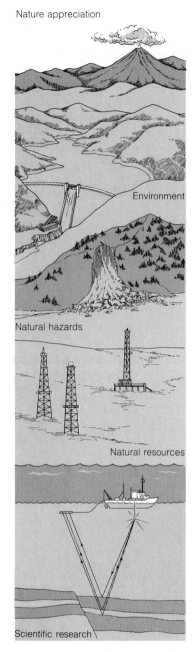

Nature appreciation

Environment

Natural hazards

Natural resources

Scientific research

Figure 1-1
Different aspects of geology—nature appreciation, environmental protection, hazard reduction, natural resources, scientific research.

the surface evolved. Were it running more slowly, all geological activity would have proceeded at a slower pace. The continents might not have evolved to their present form, and volcanoes might not have spewed out the water and gases that became the oceans and atmosphere. Iron might not have melted and sunk to form the liquid core, and the magnetic field would never have developed. The Earth would then have evolved as a cratered, dead planet similar to the Moon. Another scenario can be imagined. If there had been more radioactive fuel, and therefore a faster-running engine, volcanic gas and dust would have blotted out the Sun, the atmosphere would have been oppressively dense, and the surface would have been racked by daily earthquakes and volcanic explosions. Perhaps the Earth had such a fast-running era early in its history.

No wonder the Apollo astronauts were so profoundly affected by the view from the Moon of their home planet; its inviting blue atmosphere and white clouds stood in stark contrast to the desolate terrain of the Moon.

ASPECTS OF GEOLOGY

Although geology has ranked as a modern scholarly discipline for only two centuries, human beings have been curious about the Earth and its origins from the beginnings of prehistory. Stories of creation are found in the sagas and folktales of early civilizations. Perhaps the ancients derived some feeling of security by explaining their origin; perhaps they satisfied their own curiosity by reciting creation myths to their young, as if to put behind them the primeval chaos of an unknown creation. But a common theme was the creation, the bringing into existence, as contrasted with the idea of always being here.

Nature's Threat

The human need to understand and to be able to explain nature in order to gain protection from her vagaries survives to this day as a major motivation for the study of geology. Today we seek safeguards against nature's threats, earthquakes, landslides, volcanic eruptions, floods, droughts, and the destructive sea waves known as tsunamis. Perhaps even more dangerous are catastrophes of our own making, for our unique species has gained the power to trigger earthquakes, foul the atmosphere and oceans, and alter climates to the point of either initiating ice ages or melting the

have life. And Earth's size was just about right— not so small that its gravity was too weak to hold the atmosphere and not so large that its gravity would hold too much atmosphere, including harmful gases.

We will see that the Earth's interior is a gigantic but delicately balanced heat engine fueled by radioactivity, a fact that has much to do with how polar ice caps and flooding coastal cities. If civilization can do these things, so can it eventually predict or control most natural disasters.

Economic Geology

Were it not for the accessibility and diversity of minerals in the Earth's crust, our cultural level would never have progressed beyond the Stone Age. The discovery of the Earth's mineral wealth is the geologist's task. If we are to maintain our own present standard of living, let alone extend to the many the affluence enjoyed by the few, new mineral deposits must be found. Prospectors have long since found the easily accessible deposits of iron, copper, tin, uranium, oil, and other important minerals. It is a challenge to geologists to re-explore the world, using new tools and techniques to ferret out undiscovered deposits. New and pressing concerns of geologists are conservation and environmental protection. How can we most efficiently exploit nature's wealth without waste and without devastating the landscape? Somehow we must find answers to this question.

Scholarly Geology

Geology also has its "pure" aspects, in that it is interesting of and for itself. How a planet is born—the course of its evolution—and how it works today are only partially answered questions. Geologists are motivated to find answers because, like all scientists, they have unbounded curiosity and perhaps even a sense of uneasiness when important natural phenomena remain unexplained. Geologists will be hammering at outcrops, making geologic maps, exploring the sea floor, and scrutinizing Moon rocks as long as mountain-making, continental drift, sea-floor spreading, earthquakes, and other planetary features remain incompletely explained.

Geology for the Poets

Most students who enroll in geology courses have no intention of becoming professional geologists. They elect geology for many reasons. Perhaps they hope to heighten their appreciation of nature by gaining insight into her many ways. Perhaps concern about the environment has induced them to learn more about a key environmental science. Perhaps Norman Mailer expressed their motivation when he wrote of the moon flight of the

Apollo astronauts: "Yes, we might have to go out into space until the mystery of new discovery would force us to regard the world once again as poets."

ORIGIN OF THE SYSTEM OF PLANETS

Let us start at the beginning with the first and most difficult problem: How did the system of planets originate? This question has attracted the attention of great philosophers and scientists of the past two centuries. Yet it is a rare geological or astronomical congress that does not witness a fresh debate triggered by the latest experimental data or the newest theoretical advance pertaining to this question.

Explain the Observations

If anyone wishes to enter the lists with a pet hypothesis, the procedure is simple enough: On the basis of logical reasoning, develop a mechanism for the formation of the planetary system—a self-consistent mechanism that explains the mass and size distribution of the planets, the peculiarities of their orbits, and the relative abundances of elements in the planets and the Sun.

Whatever did happen, beginning about 4.7 billion years ago when the planets started to form, resulted in several amazing regularities and curious groupings in the solar system.

1. The planets all revolve around the Sun in the same direction, in **elliptical,** but almost circular, **orbits** that lie in nearly the same plane (Fig. 1-2); most of their moons also revolve in the same direction.

2. All the planets except Venus and Uranus rotate in the same direction as their revolution around the Sun—that is, counterclockwise as one looks from the North Pole to the South Pole of the Earth.

3. Each planet is roughly twice as far as the next inner one from the Sun, an ordering known as the **Titius-Bode rule.**

4. Although the Sun makes up about 99.9 percent of the mass of the solar system, 99 percent of the **angular momentum** is concentrated in the large planets (see Fig. 1-3 for an explanation of angular momentum).

5. The planets form two groups: the so-called **terrestrial planets**—Mercury, Venus, Earth, and Mars—an inner group of small, rocky, dense bodies (densities about 4 to 5.5 times that of water);

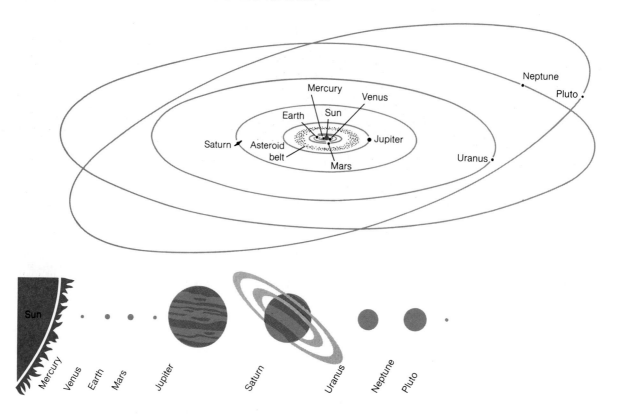

Figure 1-2
The solar system. Diagrammatic representations of the Sun and planets.

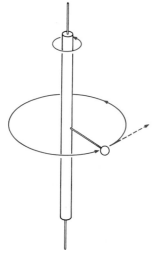

Figure 1-3
Angular momentum illustrated by a heavy steel ball fastened to the central shaft by a very light rod. Turning the shaft causes the ball to rotate around it. The ball's angular momentum is defined as the product of the mass of the ball, its velocity, and its distance from the rotation axis. [From *New Horizons in Astronomy* by J. C. Brandt and S. P. Maran. W. H. Freeman and Company. Copyright © 1972.]

and the **giant planets**—Jupiter, Saturn, Uranus, and Neptune—an outer group of large gaseous bodies with low densities (between 0.7 and 1.7 times that of water). In some respects—for example, their high gas content and low density—the giant planets are more like the Sun than like the terrestrial planets.

From chemical analysis of Earth rocks, Moon rocks, and meteorites that reach the Earth from interplanetary space, we surmise that the terrestrial planets are made up mostly (about 90 percent) of four elements: iron, oxygen, silicon, and magnesium. Spectroscopic studies* of the Sun show it to be composed almost entirely (99 percent) of hydrogen and helium. The high abundance of hydrogen and helium are features of the giant planets also.

---

*Light is emitted or absorbed in a characteristic way by different elements as they incandesce. Analysis of light into color components (more accurately, its spectral components) reveals the composition of its source. Thus the yellow color produced when common salt (sodium chloride, NaCl) is vaporized in a natural gas flame reveals the presence of the element sodium.

The Nebular Hypotheses

There has been no dearth of theories of creation over the centuries. The modern approach to the problem began in 1755 when the German philosopher Immanuel Kant hypothesized a primeval, slowly rotating cloud of gas, now called a nebula, which in some unspecified fashion condensed into a number of discrete, globular bodies.* By this **nebular hypothesis** Kant neatly explained the consistency of revolution and rotation directions, in that the rotation of the parent nebula is preserved in the rotation of the Sun, the revolution of the planets about the Sun, and the rotation of the planets about their axes—all in the same direction (Fig. 1-4).

The great French mathematician Laplace proposed essentially the same theory in 1796—surprisingly, without the mathematical formulation he was capable of providing. Historians of science will have to resolve the question of whether Laplace knew of Kant's work and why he chose not to subject his own nebular hypothesis to mathematical examination, for had he done so he might have discovered some serious flaws.

According to Kant and Laplace, the original mass of gas cooled and began to contract. As it did, the rotational speed increased (a consequence of the law of **conservation of angular momentum,** illustrated in Fig. 1-5) until successive rings of gaseous material were spun off from the central mass by centrifugal force. In the final stages the rings condensed into planets.

Not so, according to the great British physicists James Clerk Maxwell and Sir James Jeans, who showed about one hundred years later than there was not enough mass in the rings to provide the gravitational attraction for condensation into individual planets. The final blow was delivered at the close of the nineteenth century, when astronomer F. R. Moulton of Chicago showed that the nebular hypothesis violated item 4 above—namely, that the planets have most of the angular momentum. The conservation of angular momentum requires each part of a rotating, condensing nebula to keep its angular momentum; the Sun, which collected most of the mass, should have

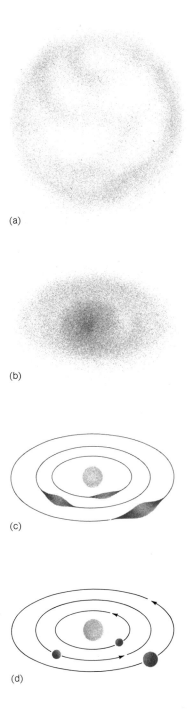

Figure 1-4
Schematic diagram of the nebular hypothesis. (a) A diffuse, roughly spherical, slowly rotating nebula begins to contract. (b) As a result of contraction and rotation, a flat, rapidly rotating disk forms with matter concentrated at the center. (c) Contraction continues, the proto-Sun is formed, and rings of material are left behind. (d) The material in the rings condenses into planets revolving in orbit around the Sun. [After *New Horizons in Astronomy* by J. C. Brandt and S. P. Maran. W. H. Freeman and Company. Copyright © 1972.]

*Kant's hypothesis, published anonymously, carried an impressive title: *Universal Natural History and Theory of the Heavens, or an Essay on the Constitution and Mechanical Origin of the Whole Universe, Treated According to Newtonian Principles.* The publisher went bankrupt, and the stock was seized by the creditors, so that very few copies reached public hands.

Figure 1-5
Conservation of angular momentum. When a skater pulls in her arms, she spins at a faster rate. Similarly, when a slowly rotating nebula contracts, its rotation speed increases.

gathered up most of the angular momentum of the system. Simply stated, the Sun doesn't rotate fast enough; it should have spun faster, just as the skater in Figure 1-5 spins faster when she pulls in her arms.

Collision Hypotheses

Wanted—a theory, now that Kant and Laplace were in disfavor. Geologist T. C. Chamberlin collaborated on one theory with Moulton, his fellow astronomer at Chicago, to revive an early proposal (1749) of Count Buffon of France—the **collision hypothesis.** This hypothesis holds that giant tongues of material were torn from the preexisting Sun by the gravitational attraction of a passing star. According to Chamberlin and Moulton, these broke into small chunks, or **planetesimals,** which went flying as cold bodies into orbits around the Sun in the plane of the passing star. By collision and gravitational attraction, the larger planetesimals swept up the smaller pieces and became the planets. Unfortunately, the several versions of collision theories have fatal weak-

nesses. According to astronomers, much of the material ejected from the Sun would have come from the interior and would have been so hot, perhaps 1,000,000°C, that the gases would have been dispersed throughout space with explosive violence rather than condensed into planets. Although more angular momentum would have been imparted to the planets by a passing star than by the rotation of a nebula, the amount predicted is still less than that observed. Finally, the vastness of space makes the probability of such a close approach of two stars extremely small.

Recent Theories

The thinking of the past few decades has been influenced by the discovery that space is not as empty as had been thought. Astronomical observations have detected, both in interstellar space and in nebulae, rarefied matter consisting of about 99 percent gas and 1 percent dust. The gases are mostly hydrogen and helium; the dust-size particles have compositions similar to those of terrestrial materials, such as silicon compounds, iron oxides, ice crystals, and a host of other small molecules, including organic ones. It is not surprising, then, that recent theories tend to be neo-Laplacian in that they revive the idea of a primordial, rotating cloud of gas and dust whose shape and internal motions were determined by gravitational forces and the forces of rotation. At some moment gravitational attraction became the dominant factor, contraction began, and the rotation speeded up (again, conservation of angular momentum). The cloud tended to flatten into a disk; matter began to drift toward the center, accumulating into the **proto-Sun.** The proto-Sun collapsed under its own gravitation, becoming dense and opaque as the material was compressed. There is growing evidence that the collapse of the nebula to form the Sun and planets was triggered by a nearby supernova—a cataclysmic explosion of a huge star with tremendous force. Some of the supernova debris has recently been found in meteorites. Whatever the cause of the collapse, it caused the internal temperature of the proto-Sun to rise to about 1,000,000°C, at which point "nuclear burning," or fusion, began. More precisely, the Sun began to shine with the initiation of the **thermonuclear reaction** (now unfortunately duplicated on Earth by H-bombs), in which hydrogen nuclei combine under intense pressure to form helium nuclei, releasing a huge amount of energy.

What about the disk of gas and dust enveloping the primitive Sun? How did it form planets? How did the planets pick up the necessary angular momentum, and why do the planets have different chemical compositions? There is little agreement among the experts about the answers to these questions.

A recent nebular-disk model, the *chemical-condensation-sequence* model, is receiving much attention lately because it seems to predict the observed variation in chemical composition and density of the planets. Initially, the disk was extremely hot, so that its materials were largely in gaseous form. As the disk cooled, various solid compounds and minerals condensed out of the gas, forming grains that gradually clumped together into small chunks, or planetesimals. The planetesimals coalesced, the bigger ones with stronger gravity pulling in all the nearly condensed matter. If a planet grew at a distance close enough to the Sun that it was too hot for certain materials to condense, those gases would be blown away by radiation and by matter streaming from the Sun. (The ejection of matter from the Sun during these early stages also accounts for the slowing of its rotation.) Near the Sun, where temperatures were highest, the first materials to condense were those with high boiling points, such as most metals and minerals. Thus Mercury, the planet closest to the Sun, is the densest (5.4 times the density of water) because it is richest in iron. Mercury's nearness to the Sun means that it formed at a temperature so high that mainly iron condensed. The lighter, rock-forming compounds, such as those made of magnesium, silicon, and oxygen, condensed more readily in the "cooler" environments of the terrestrial planets farther from the Sun. Easily evaporated materials, such as water, methane, ammonia, were mostly too volatile to remain on the terrestrial planets, but they could condense into ices in the cold outer reaches of the solar system, as, for example, on the satellites of the giant planets. Jupiter and Saturn were big enough, and had gravitational attractions strong enough, to hold on to all of their constituents and thereby retain the composition of the original nebula—mostly hydrogen and helium—much like that of the Sun.

The preceding should be taken for what it is—a hypothesis, a possible model. Perhaps some of these notions come close to what actually happened. We will know only after much more work is done, some of it now underway. Nebulae at different stages of development are being studied with the familiar optical telescope and with special devices that magnify x rays and radio waves. These invisible but detectable waves provide additional information about what goes on in the remote sections of the universe. Planetary probes have returned data on the nature and composition of the atmospheres and surfaces of Mercury, Venus, Mars, Jupiter, Saturn, and the Moon. All of this activity should in time give us a clearer picture of how it all started.

We have dwelt on the question of the origin of the solar system for several reasons. The evolutionary course followed by a planet is set by its initial state. The current state of Earth, some 4.7 billion years later, is reasonably well known to us. These two times in the course of **planetary evolution**—the beginning and now—are important constraints in developing models of how the Earth has changed throughout its history. The growth of ideas on the origin of the planets is an interesting story in the history of science. It illustrates how successive hypotheses are advanced, rejected, resurrected, and modified in the light of new observational data and theoretical concepts. Typical of all of science, this evolution of a theory is especially pertinent to modern geology, where a revolution in thought and concept is currently underway.

EARTH AS AN EVOLVING PLANET

This section serves as an introduction and preview to all that pertains, in this book, to the Earth's large-scale evolution. We will sketch the transformation of the Earth from an initially homogeneous body to a differentiated planet—that is, one in which the interior is divided into layers or zones that differ chemically and mineralogically. We will see how the process of differentiation is indirectly related to the formation of the atmosphere, oceans, continents, mountains, volcanoes, and magnetic field. The general sequence of events rather than details will be emphasized, for no other reason than that many of the details have yet to be worked out by future generations of Earth scientists.

Initial State—A Homogeneous Conglomeration

Think of the Earth as it was before the beginning of the geologic record—that is, before the formation of the oldest rocks now known, nearly 4 billion years ago. This period covers the first billion

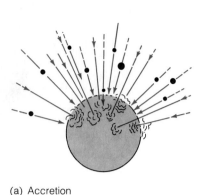

(a) Accretion

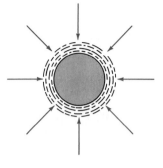

(b) Gravity compresses the original Earth into smaller volume

(c) Disintegration of radioactive elements and heat flowing away from points where particles are absorbed

Figure 1-6
Three mechanisms that would cause the early Earth to heat up. (a) In accretion, impacting bodies bombard the Earth and their energy of motion is converted to heat. (b) Gravitational compression of the Earth into a smaller volume causes its interior to heat up. (c) Disintegration of radioactive elements releases particles and radiation, which are absorbed by the surrounding rock, heating it.

years in the history of the Earth, the events of which we now attempt to reconstruct, even though we lack direct evidence in the form of rocks from these early times. The stage was set for the accumulation of the planet by the gathering up, or accretion, of planetesimals about 4.7 billion years ago. The new planet was probably an unsorted conglomeration, mostly of silicon compounds, iron and magnesium oxides, and smaller amounts of all the natural chemical elements. Although the planetesimals were relatively cold, three different effects began to heat up the growing planet (Fig. 1–6).

The Initial Temperature

Each infalling planetesimal carried much energy of motion—energy that was converted to heat upon impact. For example, a 4000-kilogram (5-ton) planetesimal hitting with a velocity of 30 km/sec or 19 miles/sec delivers as much energy as a 1-kiloton nuclear explosion. Although much of this heat was radiated back into space, a significant fraction was retained by the growing planet. Just how much is uncertain, since it depended on the mass, velocity, and temperature of the planetesimals and on the rate of accretion. At a high rate of accretion, the heated impact zone would have been covered by newly arrived material before the energy could radiate back into space, and the "buried" heat would have raised the temperature of the interior.

Compression also leads to a temperature rise. A common example is the heating of the barrel of an air pump, like those used to inflate bicycle tires. As we push down, the air is compressed quickly, too quickly for the heat to be diffused away, and the barrel heats up. The inner parts of the planet were squeezed under the growing weight of the accumulating outer parts. The energy expended in compressing the interior was converted to heat, which did not flow out because heat moves, or is conducted, very slowly in rock. As a result, the heat accumulated, and the temperature increased inside the Earth. Most geophysicists who have calculated the magnitude of heating think that accretion and compression resulted in an average internal temperature of about 1000°C for the newly organized planet.

Heat from Radioactive Disintegration

The heavy elements uranium and thorium, and the small fraction of potassium atoms that are heavier than ordinary potassium, are not very plentiful on Earth. Their occurence is measured in a few parts per million (one gram in a thousand kilograms of rock). Yet these elements have had a profound effect on the evolution of the Earth because of their **radioactivity.** Atoms of these elements spontaneously disintegrate by emitting atomic particles (helium nuclei and electrons) and are thus transformed into different elements. As the emitted particles are absorbed by the surrounding matter, their energy of motion is transformed into heat. The heat generated may seem inconsequential: about 20 calories of heat is emitted by a cubic centimeter of granite in the course of a million years. Although it would take some 500 million years to brew a cup of coffee using the radioactive heat released by a cubic centimeter of granite (assuming, of course, that the heat would be retained and not flow out), over the course of several billion years the temperature would rise to the melting point of granite—sooner if the temperature increase due to accretion and compression were also included. Thus the disintegration of the radioactive elements is a heat source that has persisted for billions of years. Radioactively generated heat, its outward flow slowed by the low thermal conductivity of rocks, warmed the newly formed Earth, initiating the process of planetary development.

HEATING, OVERTURN, AND FORMATION OF A DIFFERENTIATED EARTH

The Earth Heats Up

One can make reasonable guesses about the early temperature and radioactivity of the Earth and from them compute how the internal temperature may have changed in the years following the birth of the planet. Almost everyone who has done this finds the same important feature. Figure 1-7 shows one of the computational results. The curves illustrate how the internal temperature increased in the years following the planet's formation. The computer verifies what we might have guessed—that radioactive heat was generated more rapidly than it could flow away, so that the interior gradually warmed up. Figure 1-7 also shows the temperature at which metallic iron would melt within the Earth. The melting point increases with depth in the Earth because of the increasing pressure. The key feature of the model is that it indicates that about one billion years after the Earth was formed, the temperature at depths of 400 to 800 kilometers (250 to 500 miles) would have risen to the melting point of iron. Other models show that this state of affairs may have been reached only a few hundred million years after Earth's formation.

The Iron Catastrophe

Because iron is heavier than the other common elements of the Earth, when the iron in a layer began to melt, large drops would probably have formed and fallen toward the center, displacing the lighter materials there (Fig. 1-8).

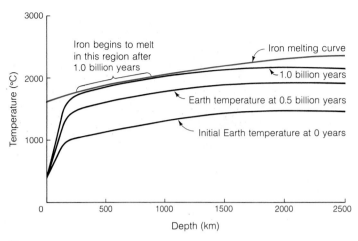

Figure 1-7
Temperature in the Earth's interior at different times in its history, according to a calculation by T. C. Hanks and D. L. Anderson. The lowest curve shows the initial temperature due to accretion and compression at 0 years. After 500 million years radioactivity warmed the Earth to the temperature shown by the next curve. After one billion years the interior heated to the melting point of iron at depths between 400 and 800 km, and iron began to melt in this region.

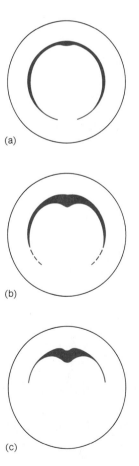

Figure 1-8
(a) The melting of iron leads to the formation of a heavy liquid layer. Drops begin to develop in later stages (b, c) and sink toward the center.

12

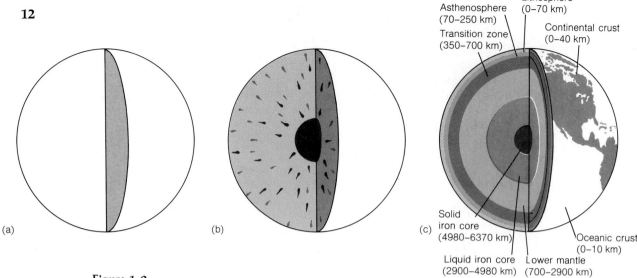

Figure 1-9
The early Earth (a) was probably a homogeneous mixture with no continents or oceans. In the process of differentiation, iron sank to the center and light material floated upward to form a crust (b). As a result, the Earth is a zoned planet (c) with a dense iron core, a surficial crust of light rock, and, between them, a residual mantle.

An abundant element, iron accounts for about one-third of the mass of the Earth. The melting and sinking of iron to form a liquid core at the center was therefore an event of catastrophic proportions. Iron "falling" toward the center released a huge amount of gravitational energy, which must eventually have been converted to heat. The process is basically the same as the one in which the gravitational energy of a waterfall is used to turn turbines and generate electrical energy. The heat released during the formation of the iron core was enough to raise the average temperature by some 2000°C, causing a large fraction of the Earth to melt.

Planetary Differentiation

The Earth underwent a profound reorganization after it warmed to the temperature at which iron melts. Approximately one-third of the primitive planet's material sank to the center, and in the process a large part of the body was converted to a partially molten state. The molten material, being lighter than the parent material from which it separated, floated upward to cool and form a primitive crust. Core formation was the beginning stage of the **differentiation** of the Earth, in which it was converted from a homogeneous body, with roughly the same kind of material at all depths, to a zoned, or layered, body with a dense iron **core**, a surficial **crust** composed of lighter materials with lower melting points, and between them the remaining **mantle** (Fig. 1-9). *Differentiation is per-*

haps the most significant event in the history of the Earth. It led to the formation of a crust and eventually the continents. Differentiation probably initiated the escape of gases from the interior, which eventually led to the formation of the atmosphere and oceans. It is as if our planet gave a "big burp" during this violent upheaval.

But what of the other planets? Did they go through the same early history? Information from the planetary probes indicates that all the terrestrial planets have undergone differentiation, but they have followed different evolutionary paths. For example, the Moon and Mercury evolved rapidly in the first 1 to 2 billion years, then apparently became inactive and geologically dead—that is, without the mountain-making, volcanism, earthquakes, erosion we observe on Earth. The giant planets will remain a puzzle for a long time. They are chemically so distinct from the terrestrial planets, and so much larger, that they must have followed an entirely different evolutionary course. Jupiter, for example, through some unexplained internal process, radiates two to three times as much energy as it receives from the Sun! Perhaps Jupiter and its 15 moons are akin to a small solar system whose sun never quite made it to the stage of shining.

The Earth Reborn

We take up the story of the evolution of the Earth following the catastrophic formation of the core. Early continents, if they existed, would have been engulfed and resorbed. In a sense the Earth was reborn without leaving a trace of its early history. According to the best estimates, all this took place between 3.7 and 4.5 billion years ago. We prefer to date the event at around 4 billion years ago, near the beginning of geological time, when the oldest rocks we can now find were first formed.

Convective Overturn

The flow of heat by conduction to the surface took place so slowly in the Earth that most of the heat accumulated within the Earth, causing the internal temperature to rise. When the interior became so hot that it became soft or molten and material could move, a more efficient mechanism took over the transfer of internal heat to the surface—namely, overturn, or **convection.** Convection takes place in liquids and gases that are hotter at the bottom than at the top. The hotter material expands, becomes lighter than the material above it, and floats upward, carrying its heat to the surface, where the hot matter cools and sinks again. Once convection began in the Earth, heat dissipated rapidly, and the planet quickly cooled. The mantle solidified, but the underlying iron core did not; it remains molten, for even 4 billion years is too short a time for it to have cooled off.

Although convection is most common in heated liquids and gases, surprisingly enough, it can occur also in solids. Under certain conditions, heated rock slowly moves upward, because its density decreases as it expands. We will see that convective flow in the Earth's solid interior has been proposed as the driving force for such large-scale geological processes as sea-floor spreading and continental drift. Convective overturn also produced a chemically zoned Earth.

Chemical Zonation

The eight most abundant elements on Earth are listed in Figure 1–10. Together these elements account for more than 99 percent of the mass of the Earth. About 90 percent of the Earth is made of the four elements iron, oxygen, silicon, and magnesium. Compare the abundance of elements in the crust with the values for the Earth as a whole. Because most of the iron sank to the core, that element drops to fourth place in the crust. Conversely, silicon, aluminum, calcium, potassium, and sodium are far more abundant in the crust than in the whole earth. This uneven distribution

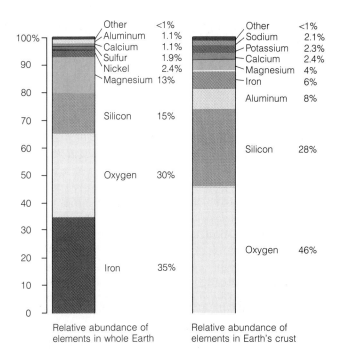

Figure 1–10
Relative abundance by weight of elements in the whole Earth and in the Earth's crust. Differentiation has created a light crust depleted in iron and enriched in oxygen, silicon, aluminum, calcium, potassium, and sodium.

of elements in the Earth is what we mean by *chemical zonation.*

It is interesting and significant that differentiation did not lead to a vertical arrangement of elements based entirely on their relative weights, as we might have supposed. The reason is that various elements formed compounds, and it was the chemical and physical properties of those compounds—properties such as melting points, chemical affinities, and densities—that governed the distribution of elements, rather than the properties of the elements themselves. For example, certain silicates of calcium, sodium, potassium, and aluminum—notably the *feldspars* ($CaAl_2Si_2O_8$, $NaAlSi_3O_8$, $KAlSi_3O_8$)—are easily melted. They begin to melt at temperatures as low as 700 to 1000°C, and when molten they are relatively light. One might speculate that compounds like these would melt early, rise to the surface by convection, and accumulate as crust. It should be no surprise that feldspars are the most common minerals in the Earth's crust.

The mantle, situated between the crust and the core, became the reservoir for magnesium-iron silicates, which melt less easily and are heavier than feldspars. From the abundances shown in Figure 1-10, we would expect that the dominant constituents are combinations of Fe, Mg, Si, and O. We should therefore not be surprised when we see later that most geophysicists and petrologists believe that the principal minerals in the mantle are olivine (Mg_2SiO_4—Fe_2SiO_4) and pyroxene ($MgSiO_3$—$FeSiO_3$).

Heavy elements like gold and platinum have little chemical attraction to oxygen and silicon, and most of these important metals probably sank into the Earth's core. But other heavy elements, like uranium and thorium, have strong tendencies to form oxides and silicates, which are light and could rise to accumulate in the crust. Gravity, however, is only secondarily responsible for the zonation of a planet; just as important are the relative abundances of the different elements and their chemistry, both of which determine the compounds they form. The properties of the compounds—densities and melting points—differ from those of the pure elements and govern the distribution of the elements.

Differentiation Slows the Engine

One important consequence of chemical zoning is the concentration of the heavy elements uranium and thorium in the Earth's crust as oxides and silicates. In the early stages, when these radioactive elements were evenly distributed throughout the Earth, they were probably highly effective in raising the temperature to the melting point of iron. As part of the process of differentiation, however, the radioactive "fuel" began to concentrate in the outermost layers, where the heat that was generated could be conducted through much shorter distances to the surface and so be lost more easily. In this way the differentiation of a planet may act to slow down the operation of the heat engine.

Formation of Continents, Oceans, and Atmosphere

We have only the most general notion of how the first continents formed. The story goes like this.

Box 1-1   THE RELATIVE ABUNDANCE OF THE ELEMENTS

We know about the chemical composition of the universe from spectroscopic examination of the light of the Sun and stars, from analysis of meteorites, and from the composition of rocks found near Earth's surface. The table shows our current knowledge of elemental abundances—the results of many measurements. The data support the notion that the relative abundance of elements in the universe is fairly constant throughout except for local variations, such as the dissipation into space of hydrogen and helium from the terrestrial planets or the internal differentiation of a planet.

It has been postulated that a large part of the Earth's interior consists of iron. But why iron rather than some other heavy metal? We have no direct way of determining that the core is iron. We do know, however, from physical calculations made on the basis of the mean density of the Earth and from seismological data that the innermost part of the Earth has a high density. The choice of iron seems appropriate because among all the candidates with high atomic weight, it is by far the most plentiful. Our confidence in this choice is reinforced when we note that 58 percent of all meteorites found are iron.

Abundance of the principal elements in the universe expressed as the numbers of atoms relative to a base of 10,000 atoms of silicon

Element	Atomic weight	Abundance in the universe
Hydrogen	1	266,000,000
Helium	4	18,000,000
Oxygen	16	184,000
Carbon	12	111,000
Neon	20	26,000
Nitrogen	14	23,100
Magnesium	24	10,600
Silicon	28	10,000
Iron	56	9,000
Sulfur	32	5,000
Argon	40	1,100
Aluminum	27	850
Calcium	40	625
Sodium	23	600
Nickel	59	478
Chromium	52	127
Manganese	55	93
Phosphorus	31	65
Chlorine	35	47
Potassium	39	35
Titanium	48	24
Cobalt	59	22
Fluorine	19	8

Source: After A. G. W. Cameron, 1981.

Lava flowing from the partially molten interior of the Earth spread over the surface and solidified to form a thin crust. This primeval crust melted and solidified repeatedly, and the lighter compounds gradually separated from the heavier ones and distributed at the top. **Weathering** by rainwater and other components of the atmosphere broke up and altered the rocks; **erosion** produced *sediments*—the residue of broken-down rock particles. As the sediments accumulated, they were in turn penetrated by hot gases and solutions from below, heated up and altered—"cooked," in a sense—into new rocks, or were entirely resorbed and recycled. The end product of such processes as these was the primitive nucleus of a continent, which grew as this process continued. As a rough check on this theory, we can take the amount of lava ejected each year by modern volcanoes and multiply it by a few billion years. The volume of rock produced corresponds fairly well with the volume of the continents.

If we start with the premise that the Earth accreted from cold planetismals, then the oceans came from the interior as a product of the processes of heating up and differentiation. Originally the water was locked up, chemically bound as oxygen and hydrogen in such minerals as potassium-aluminum mica, $KAl_3Si_3O_{10}(OH)_2$. As the Earth warmed and partial melting occurred, water was released and carried to the surface along with lava. As the lava reached the surface, much of the water escaped as hot vapor clouds. Even at today's rate of volcanic activity, the lavas that reached the surface in the past would have contained enough water vapor to fill the oceans in the course of geological time, though much of that water may have been recycled by the melting of water-rich oceanic sediments.

How and when the atmosphere began to develop is more difficult to determine. We can, however, give one self-consistent explanation that agrees with the known facts. There is little doubt that the earliest atmosphere was entirely different from the primarily nitrogen-oxygen one we live in now. The cold planetesimals, which aggregated to form the Earth, could have carried no atmosphere because they were too small to hold any gases by gravitation. Thus **outgassing** as part of the process of differentiation—that is, the release of gases from the interior due to internal heat and chemical reaction—is generally thought to be the primary source. From the chemical composition of lava and gases released by modern volcanoes, we surmise that volcanic gases consisted mainly of water vapor, hydrogen, hydrogen chlorides, carbon monoxide, carbon dioxide, and nitrogen (Fig.

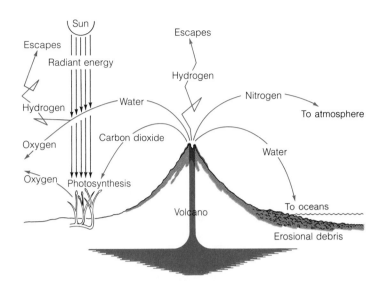

Figure 1-11
Volcanism has contributed enormous amounts of water, carbon dioxide, and other gases to the atmosphere and materials to the continents. Sunlight broke water molecules into its components, hydrogen and oxygen. Photosynthesis by plants removed carbon dioxide and added oxygen to the primitive atmosphere.

1–11). But the light hydrogen molecules could not have been held by the Earth's gravity, and the hydrogen would have escaped into space as it does steadily today. With energy supplied by sunlight, some water vapor in the upper atmosphere may have broken down to hydrogen and oxygen. Any oxygen so formed would not have remained free (uncombined) for long; it would have quickly combined with gases like methane and carbon monoxide to form water and carbon dioxide. It would also have combined with crustal materials—with metals like iron in olivines and pyroxenes to form iron oxides like **hematite** (Fe_2O_3).

The production of significant amounts of free oxygen and its persistence in the atmosphere probably came only after life had evolved at least to the complexity of green algae. Like all higher forms of green plants, these algae convert sunlight to organic matter by **photosynthesis,** using carbon dioxide and water to make organic matter and oxygen. Not until the production of oxygen exceeded its loss by chemical combination with other gases and metals could this by-product of photosynthesis have begun accumulating in the atmosphere.

The big question about the atmosphere concerns life itself: how did life develop in the "poisonous" primitive atmosphere and evolve the green plants that "purified" the air with free oxygen, clearing the way for higher forms of life to develop? Some answers are given in Chapter 12.

Much carbon dioxide was removed from the atmosphere by chemical combinations with calcium, hydrogen, and oxygen to form limestone, coal, and petroleum, the great bulk lying buried in the crust. The coal and petroleum are the great reserves of fossil fuels that we have relied on to power our industrial societies. The few hundredths of one percent of the atmosphere that remains as CO_2 is of course highly important to us in that it is an essential raw material for photosynthesis.

Earth is 140 million kilometers from the Sun. Some calculations show that if it were only 10 million kilometers closer, the higher temperature would have prevented water vapor from condensing into oceans, and carbon dioxide from being removed from the atmosphere and chemically locked up in rocks. It is doubtful that life would have formed under those circumstances.

The Account of the Earth's Evolution— Fact or Fiction?

The preceding story is, of course, speculation based on physical plausibility and extrapolation from experiment. *In all fairness, however, we should note that this is but one of several possible courses for our planet's evolution from among several existing hypotheses. There are other views for the nature of each stage, but about the general course of events there is fair agreement.* We are like children making a house out of a variety of blocks. The constraint is that each block has to fit with the ones under it and over it and that the whole structure has to stand up. But there may be more than one way to build the house.

Even if the details are obscure, we believe that the general sequence of events just described is a possible solution to the problem of the early history of the Earth because it explains several facts of paramount importance. For one, Earth has an iron core and a silicate mantle. In order for this separation to have taken place, the interior must have become hot enough for iron to melt. For another, the Earth's crust is chemically distinct from the mantle. The crust contains such high concentrations of certain elements (for example, uranium and thorium) that they must have been extracted from the whole of Earth's body. This implies a mobility that only extensive melting could provide. Finally, there is a source of thermal energy to which we can attribute the melting—radioactivity.

In the remainder of this book, we will be concerned mainly with what happened to the Earth following the stage of primary differentiation. The mantle solidified, a primitive crust and continents developed, oceans and atmospheres were produced by outgassing, and the processes that we know today were set in motion. Differentiation slowed the internal heat engine of Earth, leaving enough heat in the interior for the planet to continue to evolve to this very day.

THE EARTH MACHINE SET IN MOTION, MUCH AS WE KNOW IT TODAY

Plate Tectonics—A Unifying Theory

Geology books written in the 1980s have a distinct advantage over those written before the late 1960s. For almost 200 years, geologists have supported various theories of mountain-building, volcanism, and other major phenomena of Earth. None was general enough to explain well the whole range of geologic processes. Now, there exists a single, all-encompassing concept that interprets many of the major geologic features of the Earth. Such topics as the classification and distribution of rocks, the history of sedimentary rock sequences, the position and characteristics of volcanoes, earthquake belts, mountain systems, deep-sea trenches, and ocean basins were formerly described in more-or-less unrelated fashion. Today we have the advantage of being able to treat these and other topics in the context of a unifying theory, **plate tectonics,** in which the large-scale pattern, if not the underlying mechanism, is recognized. To be sure, some geologists have found flaws in certain parts of the theory, but few reject it entirely. Revolutions tend to be largely complete in twenty years.

The outermost shells of our concentrically zoned planet are illustrated in Figure 1–12, a representation of what we think we know of the Earth's zoning today. (Why we propose this picture will be explained in later chapters rather than here.) The **lithosphere** is depicted as the strong, solid outermost shell riding on the weak, partially molten **asthenosphere.** The continents are raftlike inclusions embedded in the lithosphere. Elsewhere only a thin crust tops the lithosphere.

The central idea of plate tectonics is remarka-

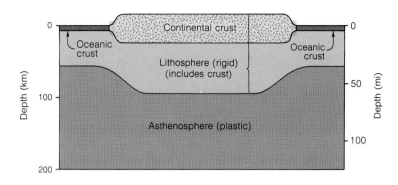

Figure 1-12
The outermost shell of the Earth is the strong, solid lithosphere, riding on the weak, partially molten asthenosphere. The lithosphere is topped by a thin crust under the oceans and a thicker continental crust.

bly simple. The lithosphere is broken into about ten large rigid plates, each plate moving as a distinct unit. The major plates and the directions in which they move are sketched inside the front cover. Many large-scale geological features are associated with the boundaries between the plates.

Plates spread apart along **divergent junctions** typified by the cracklike valley at the crest of the mid-Atlantic ridge. This particular feature is the contact between the American plate on the one hand and the Eurasian and African plates on the other (Fig. 1–13). The divergent junction is characterized by earthquake activity and volcanism. The void between the receding plates is filled by melted, mobile material that rises from below the lithosphere. The material solidifies in the crack, and the plates grow as they separate. Since new sea floor is created, this part of the process has been dubbed **sea-floor spreading.**

There seems to be little doubt that the Atlantic Ocean did not exist 150 million years ago. Instead, the large single hypothetical land mass reconstructed in Figure 1–14, named **Pangaea,** contained Eurasia, Africa, and the Americas. Pangaea was split by a crack that subsequently opened to form the Atlantic Ocean. As the continents drifted apart, the edges became new repositories for erosional debris brought to the newly formed sea from inland sources. The debris was laid down as wedge-shaped sedimentary deposits along submerged edges of the continents (Fig. 1–15).

If plates separate in one place, they must converge somewhere else, and they do. Plates grind together head-on along **convergent junctions.** It is no surprise that crumpled mountain ranges, deep-sea trenches, shallow and deep earthquakes, and volcanoes are all associated with convergence. Examples are the contact between the Nazca plate and the American plate (see inside of

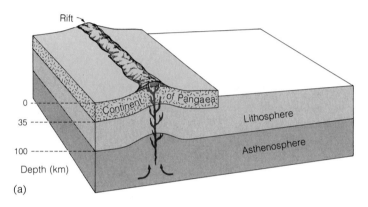

(a)

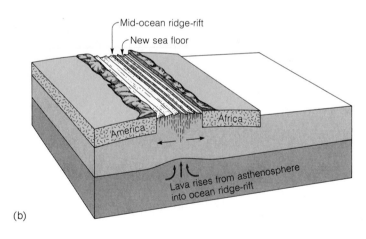

(b)

Figure 1-13
Stages in the development of a divergent junction and sea-floor spreading. (a) The lithosphere breaks, and a rift develops under a continent. Molten basalt from the asthenosphere spills out. (b) The rift continues to open, separating the two parts of the continent—in this example, America and Africa. A new ocean basin, the Atlantic, is created between the separating land masses. The active rift is marked by a mid-ocean ridge characterized by earthquakes and volcanism. [After "The Breakup of Pangaea" by R. S. Dietz and J. C. Holden. Copyright © 1970 by Scientific American, Inc. All rights reserved.]

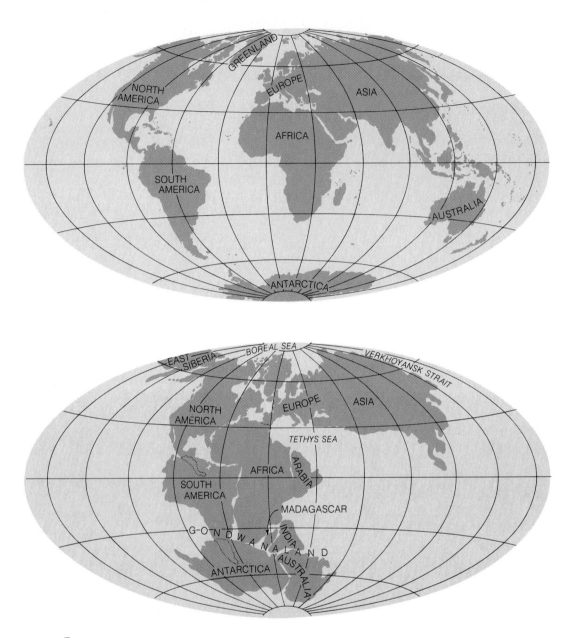

Figure 1-14
Single supercontinent of Pangaea, presumed to have existed 200 million years ago, may have looked like this.

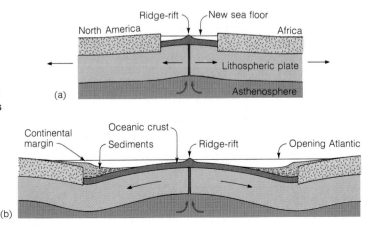

Figure 1-15
As a rift widens and a new sea floor is created (a), the trailing edges of the receding continents receive erosional debris from inland, deposited as thick wedges of sediment (b). [After "Geosynclines, Mountains, and Continent-Building" by R. S. Dietz. Copyright © 1972 by Scientific American, Inc. All rights reserved.]

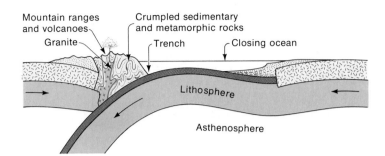

Mountain ranges and volcanoes
Granite
Crumpled sedimentary and metamorphic rocks
Trench
Closing ocean
Lithosphere
Asthenosphere

Figure 1-16
When plates collide, one plate usually buckles downward. The overriding plate is crumpled and uplifted. Deep-sea trenches (sites of the greatest of ocean depths), high mountain chains, volcanoes, and the greatest earthquakes are associated with such regions of plate convergence. [After "Geosynclines, Mountains, and Continent-Building" by R. S. Dietz. Copyright © 1972 by Scientific American, Inc. All rights reserved.]

back cover), where one finds the Andes Mountains, the Chilean deep-sea trench, and where some of the world's great earthquakes have been recorded. Another example is the Kuril-Kamchatka-Aleutian arc system, along which the Pacific plate runs into the Eurasian plate. We will see that convergent zones take on different manifestations, depending on spreading rates and whether the leading edges of plates are continental or oceanic.

Figure 1-16 shows the profusion of geologic activity associated with but one plate-collision scenario, in which the abutting edges are ocean floor and continent. The heavy, oceanic lithosphere descends into the mantle, beneath the lighter continental lithosphere. Downbuckling is marked by an offshore trench. Great earthquakes occur adjacent to the inclined contact between the two plates. The edge of the overriding plate is crumpled and uplifted to form a mountain chain parallel to the trench. Deep-sea sediments may be scraped off the descending slab and incorporated into the adjacent mountains. What a complicated mess! As the mountains are raised, erosion by wind, ice, and water wears them down. The debris eventually ends up in the oceans to repeat the cycle of rock formation and destruction over and over again.

Such regions of convergence, where lithosphere is consumed, have been named **subduction zones.** If the divergent zones are sources of new lithosphere, then the subduction zones are sinks in which material is consumed in equal amount. Otherwise the Earth would change in size, and neither geological evidence nor physical theory supports this conclusion.

Rocks caught up in a subduction zone are squeezed and heated. A new assemblage of minerals appears as a result of this "pressure-cooking" treatment. Just what minerals form depends on the pressure, temperature, and deformation to which the rocks are subjected by the subduction

process. **Metamorphism,** the cumulative effect of all of these processes, leads to the formation of new mineral grains and therefore new rocks. The sequence of sedimentary and metamorphic rocks in and near convergent zones may provide the key to unraveling the history of a collision between plates.

As the oceanic plate descends into the hot mantle, parts of it may begin to melt. The rock melt, or **magma,** thus formed floats upward, some of it reaching the surface as lava erupting from volcanic vents. In a manner not fully understood, the formation of magma in the subduction zone may be a key element in the creation of granitic rocks, the main rock of which continents are formed.

Plates can separate and collide, as we have just seen. They can also slip past one another along a **transform fault** (Fig. 1-17). The famed San Andreas fault of California is such a boundary.

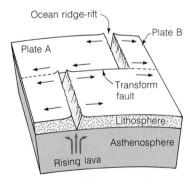

Ocean ridge-rift
Plate A
Plate B
Transform fault
Lithosphere
Asthenosphere
Rising lava

Figure 1-17
A transform fault is a plate boundary along which two plates slide past each other. In the example shown, Plates A and B are separating. The ridge crest, which marks the zone of spreading, is offset by the transform fault. The arrows indicate the opposite plate motions on the two sides of the transform fault between the ridges.

Figure 1-18
The San Andreas fault—the boundary between the American and Pacific plates.
The block in the background is being displaced to the right relative to the block
in the foreground, as evidenced by the stream offset. Carrizo Plains, California.
[Photo by R. E. Wallace, U.S. Geological Survey.]

There the Pacific plate slides past the American plate in a northwesterly direction (Fig. 1-18). This contact is characterized by contrasting geology on both sides and by large earthquakes originating near the surface, earthquakes of the kind that destroyed San Francisco in 1906.

What we have just described is the general pattern of the internal heat engine's work output as we can see it today. The process probably has been going on since the large-scale differentiation of the Earth ended some 4 billion years ago. A description, however, is not an explanation. We have unraveled the kinematics, or motions, of the process, but the dynamics, or forces, that are responsible have yet to be understood (Fig. 1-19).

So far we can do no more than make vague reference to convection serving as an internal heat engine. Furthermore, there is a wealth of major geologic features lying within the continents whose relations to plate tectonics, if any, have yet to be established. Nevertheless, a grand scheme of things has been recognized, and the course of geology has undoubtedly changed.

At the same time that the theory of plate tectonics has been taking shape, our understanding of the external heat engine of the Earth has also deepened. That engine, powered by the Sun's energy, drives the circulation of the atmosphere and oceans, providing the rain, wind, ice, and running water that erode the mountains of the Earth. As

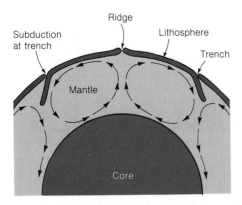

Figure 1-19
A simple model showing how convection currents in the deep interior might be the driving force of sea-floor spreading and continental drift. Hot matter rises under the ocean ridge and flows apart, carrying the plates along. Plates sink with cold matter at subduction zones.

soon as the atmosphere formed by outgassing, the external heat engine started to grind out the products of erosion and chemical decomposition. Rivers began to run to the ocean, delivering sediments to beaches and deltas; ocean currents carried the finest sediments farther from land, where in quiet waters they accumulated on the margins of continents and in the deep sea. The external heat engine is of immediate consequence to us, for aside from its importance in most of the geological processes we see operating at the surface of the Earth, we live in the midst of it and depend completely on its workings for food and energy.

The many ways in which the internal and external heat engines are linked provide exciting new territory for geologists and oceanographers to explore. As we find out more about each engine, we find out more about their interactions. Their interactions, too, can affect us drastically—as when an earthquake, that happens to occur during Peru's rainy season triggers a massive mudflow that kills thousands of people (1970), or when a volcano like Mt. St. Helens erupts sending so much volcanic ash into the atmosphere that the quantity of atmospheric dust over the whole world is significantly increased (1980).

In what follows we will deal with continents and ocean basins and interpret information contained in the different kinds of rocks in the conceptual framework of plate tectonics. Though that concept does not explain everything, and may not even relate to some kinds of phenomena, it gives us by all odds the best framework on which to lay the story of the Earth.

SUMMARY
1. The Sun and its family of planets formed when a primeval cloud of gas and dust condensed about 5 billion years ago. The planets differ in chemical composition according to their distance from the Sun and their size.

2. Earth probably grew to its present size by accretion of small chunks of matter. At its birth it was probably a homogeneous and relatively cool body.

3. Because of radioactivity, Earth began heating up, and within about a billion years or less the temperatures probably reached the melting point of iron. Drops of molten iron sank to Earth's center, and lighter matter floated up to form the outer layers that became the continents. Outgassing gave rise to the oceans and a primitive atmosphere. In this way the Earth was transformed to a differentiated planet with chemically distinct zones—an iron core, a magnesium-iron-silicate mantle, and a crust enriched in oxygen, silicon, aluminum, calcium, sodium, and radioactive elements.

4. The lithosphere, or outermost shell of the Earth, is broken into about ten large, rigid plates, which jostle each other because of their individual motions. The boundaries of these plates are zones of intense activity. Associated with them are many of the large-scale geological features produced by such processes as mountain-building, volcanism, creation and destruction of sea floor, and earthquake activity.

EXERCISES

1. What factors have made Earth a particularly congenial place for life to develop?

2. How does the chemical-condensation-sequence model explain the differences among the terrestrial planets? How and why do the terrestrial planets differ from the giant planets?

3. If you were an astronaut exploring another planet, what evidence would you look for to decide whether the planet was differentiated and whether it was still geologically active?

4. A recent hypothesis proposes that the Earth became chemically zoned not by differentiation but by accreting sequentially, first an iron core, then a rocky mantle, and finally a crust. How might this have happened? How can this idea be tested?

5. What are the advantages and disadvantages of living on a differentiated planet? On a geologically active planet?

6. Describe the central idea of plate tectonics and the large-scale geologic features associated with plate boundaries. How would you recognize an ancient plate boundary and a modern plate boundary?

7. Speculate on what life would be like today if the ancient continent of Pangaea had remained a single land mass instead of breaking up into Eurasia, Africa, and the Americas. How would the number and diversity of species be affected? Would climate be the same?

BIBLIOGRAPHY

Brandt, J. C., and S. P. Moran, *New Horizons in Astronomy.* San Francisco: W. H. Freeman and Company, 1972. (Chapters 7, 8, 15.)

Gass, I. G., P. J. Smith, and R. C. L. Wilson, eds., *Understanding the Earth.* Cambridge, Mass.: MIT Press, 1971.

Jastrow, Robert, and M. H. Thompson, *Astronomy: Fundamentals and Frontiers.* New York: John Wiley & Sons, 1972. (Part Three, The Planets.)

Laporte, L. F., *The Earth and Human Affairs.* San Francisco: Canfield Press, 1972.

Leveson, David, *A Sense of the Earth.* Garden City, N.Y.: Doubleday, 1972.

Press, F., and R. Siever, eds., *Planet Earth: Readings from Scientific American.* San Francisco: W. H. Freeman and Company, 1975.

Scientific American, *The Solar System.* September 1975.

Takeuchi, S., S. Uyeda, and H. Kanamori, *Debate About the Earth,* rev. ed. San Francisco: Freeman, Cooper & Company, 1970.

Trimble, Virginia, "Cosmology: Man's Place in the Universe," *American Scientist,* v. 65, pp. 76–86, 1977.

Wilson, J. Tuzo, ed., *Continents Adrift and Continents Aground: Readings from Scientific American.* San Francisco: W. H. Freeman and Company, 1976.

2

The Rock Record and the Geological Time Scale

Concepts of time are central to geology. Most geologic processes that shape the surface of the Earth and give structure to its interior operate over long times, up to millions and billions of years. The igneous, metamorphic, and sedimentary rocks exposed at the surface are the visible records of the geologic processes of the past. From the time and space relations that those rocks manifest, geologists have built the geologic time scale, which is used for placing the geologic events of Earth history in sequence according to relative age. Spontaneous decay of radioactive atoms in rocks gives absolute ages that date the geological periods and the origin of the Earth.

HOW LONG IS A LONG TIME?

The key difference between geologists and most other scientists is in their attitude toward time. Many chemical reactions and physical processes measured in the laboratory operate over time periods—or **time scales**—of seconds or fractions of seconds; geological processes that we can observe directly take place over a much broader time range (Fig. 2-1). Earthquake tremors may last seconds or minutes, and the seismic waves generated by that earthquake may take minutes or hours to travel through the Earth and along its surface. Erosion and transportation of an immense amount of boulders, sand, silt, and clay by a major river in flood operates over a few days. Sandbars move in and out from a beach in days or weeks. These processes we can see or feel happening. Others cannot be observed directly; for example, the Atlantic Ocean is spreading apart at the mid-Atlantic ridge on a much longer time scale—at a rate of a few centimeters per year. The gullying of overgrazed or carelessly farmed land by accelerated erosion takes place over a time scale of a few years. But when time scales are 50

years or more, our memories start to fail us, and we rely on historical records to measure how much a marshy tideland has filled in, how much a hillside has eroded, or how much a river has gradually changed its course (Fig. 2-2).

Despite an occasional earthquake or volcanic eruption, the Earth seems to provide us with a reasonably stable foundation upon which to build a civilization. This holds true on the time scale important to organized society—namely, hundreds or thousands of years. On a geological time scale of millions to hundreds of millions of years, Earth is far less stable. During that time continents, oceans, and mountain chains have moved horizontally and vertically through large distances. Although the evidence for long-term instability is all around us, only in recent years have scientists begun to recognize a worldwide pattern in these movements. Working out that pattern requires a means of timing the movements relative to each other and to a known time scale.

In Chapter 1 we discussed the time scales, millions of years, that elapsed during the formation of the Sun and the planets—times measured by radioactive clocks and inferred by reasoning from

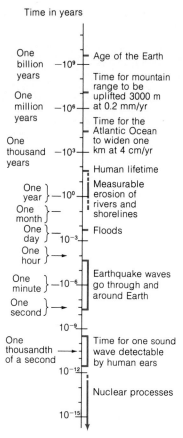

Figure 2-1
Orders of magnitude of times for some
common processes and events. The scale
is logarithmic; that is, it has equal divisions between successive powers of ten.

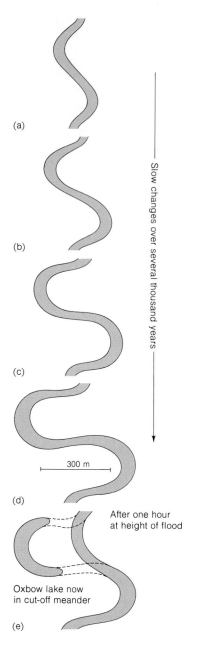

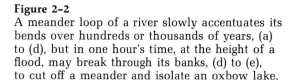

Figure 2-2
A meander loop of a river slowly accentuates its
bends over hundreds or thousands of years, (a)
to (d), but in one hour's time, at the height of a
flood, may break through its banks, (d) to (e),
to cut off a meander and isolate an oxbow lake.

the rates of the physical and chemical processes involved. We concluded that the Earth is about 4.7 billion years old. What has happened during all of the time between then and now? How long does it take geologic processes to make mountains and destroy them? What is the lifetime of a river? Time scales ranging from a few tens of years to a few billions of years are characteristic of most of the processes discussed in this book.

Why do we care about time-scales? One of the most important reasons comes from the heart of historical geology, the story of the evolution of the Earth as we now see it. We want to be able to reconstruct just what happened when. When were the Appalachians or the Rocky Mountains formed? When did the Atlantic Ocean start to open? Equally important in the value of time scales is the help they give us in figuring out *how* something happened. This comes from an old rule of physical science: If two things formed within different time scales, they are likely to have been made by different processes. Most of the time we use this rule of thumb in an unconscious way, but it is sometimes worthwhile to spell it out explicitly as a guide to thinking about a specific problem.

How We Estimate Very Slow Rates of Earth Processes

A few simple, rough calculations give surprisingly good estimates. No rocks older than about 200 million years have been found on the deep sea floor, even though many holes have been drilled there in the past 15 years. We may therefore take 200 million years* as some sort of an upper limit to the age of an ocean basin. If 10,000 kilometers is used as a representative width of an ocean—that is, the distance the plates have spread to form the basin—the spreading rate comes out to be 10,000 kilometers/200 million years = 5 centimeters/year (2 inches/year).

The well-known San Andreas fault in California is a transform fault along which the North Pacific plate slides past the North American plate (Fig. 2–3). At some places along the fault, survey measurements have been made for almost a century. During this period, slip along the fault, due both to earthquakes and to steady creep, has amounted to about 4 to 6 centimeters per year. On a longer time scale, the rate of movement can be determined by matching up distinctive geological formations that have been split by the fault, the separate parts moving away from each other. The average movement over the past several million years can be inferred in this way at about one centimeter per year in northern and central California. At that rate, about 25 million years ago the block containing the present San Francisco coast would have been at the latitude of the present Los Angeles! Later we shall describe more precise methods for obtaining rates of plate movements in different parts of the world, including dating of magnetic patterns on the sea floor and dating of sediment samples obtained in the Deep Sea Drilling Project.

Vertical movements can be evaluated by dating marine deposits that are now above sea level. Mountains made up of rocks that contain marine fossils have been uplifted 3000 meters (about 2 miles) in 15 million years, giving a rate of 0.2 millimeters per year. About 40,000 years ago, during the last ice age, Fennoscandia, the area including Norway, Sweden, and Finland, was covered by 2 to 3 kilometers of ice. The tremendous weight of the ice sheet loaded and depressed that part of the lithosphere. When a warming trend set in and the

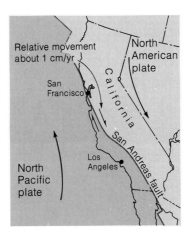

Figure 2–3
The North American plate is moving south relative to the North Pacific plate along the San Andreas transform fault at a rate of about 1 cm/yr.

ice melted away, the lithosphere there rebounded some 500 meters, as evidenced by ancient beaches that are at that elevation. Thus postglacial uplift of Fennoscandia took place at a rate of about 1 centimeter per year. This particular evidence gives us important data for determining the viscosity of the mantle.

Though they are slow, erosional processes are continuously wearing down the land surface (Fig. 2–4). Without mountain building and other vertical rejuvenation all land would eventually be reduced to sea level. Erosion rates can be estimated by adding up all of the disintegrated and dissolved products of erosion being carried away by rivers and wind. The rate for the North American continent has been estimated to be about 0.03 millimeter per year (about 1 foot per 10,000 years). Thus it takes hundreds of millions (10^8) of years to open an ocean basin, about 20 million (2×10^7) years to raise a mountain, and 10^8 years to cut it down to sea level.

Even these are relatively short periods on the geological time scale of the history of the Earth. The Earth has experienced many cycles of mountain building and erosion in the past four billion years. In the next sections we explore how we can date all of the events that are part of those cycles.

THE ROCK RECORD

The only record we have of things that happened on Earth in the geological past is the rocks that were preserved from erosional destruction.

---

*In many geological books and journals, m.y. is used for millions of years and b.y. for billions of years. AE (aeons) is also used by some for billions of years. And in a newer system of units, Ma (mega-ans) is used for millions and Ga (giga-ans) for billions of years.

Figure 2-4
Photographs of Bowknot Bend on the Green River in Canyonlands National Park, Utah. The upper one was taken by E. O. Beaman in 1871 on Major John Wesley Powell's second expedition down the Green and Colorado rivers; the lower one, by H. G. Stephens in 1968. Though taken almost 100 years apart, the two photographs show little change in rocks and formations. Millions of years of erosion were required to produce the river valleys of today. In contrast, the sandy bars and the banks of the river itself change more rapidly. [From U.S. Geological Survey.]

How to Use a Rock Section to Tell a Story: The Grand Canyon

There are many places where the bedrock, which everywhere underlies the surface, is exposed, not obscured by soil or loose boulders. Such exposures, called **outcrops,** range in size from small projections of weathered rock on a hillside (Fig. 2-5) and ledges in the beds of small streams to high cliffs that make the walls of canyons in mountainous terrain. Geologists also know how to use such engineering works as road cuts to study well-exposed sections of rock. No highway cut, however, could touch the dimensions of a place like the Grand Canyon of the Colorado River. And no cut has the beauty and grandeur of that fantastic gorge, which is more than a mile (1.6 kilometers) deep in places, 4 to 18 miles wide (6.5 to 29

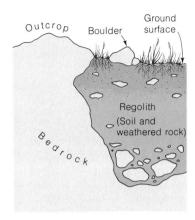

Figure 2-5
Cross section through surface
of the ground, showing relation of outcrop, boulders, and
soil.

kilometers), and 280 miles (450 kilometers) long.
The main part alone is 56 miles (90 kilometers)
long. It is still an adventure to travel down the
Canyon of the Colorado River in a small boat, repeating the first trip made in 1869 by geologist

Major John W. Powell, who later helped found the
U.S. Geological Survey. At every turn in the Canyon, from steep rims to inner gorge, one sees rock
in a multitude of forms large and small. Revealed
are patterns of rock characteristics that we use to
reconstruct geologic history.

The first thing we notice about the rocks of the
Grand Canyon is a pronounced horizontal **layer-
ing,** or **stratification** (Fig. 2–6). This **bedding,** as it
is most frequently called, is what we might expect
from the settling of particles from air or water to
form layers of **sediment,** (from the Latin *sedi-
mentum,* settled). This expectation is supported
by experience, for we can see horizontal layers of
sand being deposited on beaches or sandbars and
layers of mud and silt accumulating on floodplains of rivers (Fig. 2–7).

Once we make this generalization (which has
held up for the three hundred years since it was
first enunciated), we can go on to the obvious next
thought. Since it is absurd to think that a sedimentary layer can be deposited beneath a previously
deposited bed, we conclude that any new layer
added to a series must always be added on top.
Naturally we have to add the proviso that the

Figure 2-6
The Grand Canyon of the Colorado: a view south from the bordering Kaibab Plateau. Though the topography is rugged, the horizontal stratification of the upper
rocks in the canyon is clear. [From *Geology Illustrated* by John S. Shelton. W. H.
Freeman and Company. Copyright © 1966.]

Figure 2-7
Freshly deposited river sand and silt, showing original horizontality of sedimentary deposits. View is of a cut spaded through two layers of a sand bar in the Vermillion River, Illinois; the upper horizontal layer is made up of many inclined crossbeds and shows a ripplemarked surface. [Photo by Paul E. Potter.]

whole series has not been deformed and completely overturned at some later time. The time sequence of stratification is the simple basis for the **stratigraphic time scale,** a clock we can use to measure time and date events.

Simple as these generalizations are, they are a beautiful example of a well-known dictum: the truly great discoveries are the ones that are perfectly obvious *after* someone has pointed them out to us. It was Nicolaus Steno, a Danish court physician living in Italy, who in 1669 formulated the *principle of original horizontality* and the *principle of superposition,* which we have given above in modified form. He also stated the *principle of original continuity,* which holds that a sedimentary layer forms at the time of its deposition a continuous sheet that ends only by thinning to disappearance, by gradually changing to a bed of different composition, or by abutting against a wall or barrier such as a shoreline that confines the depositional area. From the law of continuity we intuitively grasp the idea that the face of a bed, as we see it in the excavation for a highway

cut or in the walls of the Grand Canyon, is the broken or eroded edge of a once-continuous sheet.

From these three principles we get the rudiments of our stratigraphic clock; that is, we establish a total length of time necessary for all of the rocks to be laid down and a time interval within the whole span for each layer. If we had some idea of how long each bed took to be laid down, and *if* all of the time span were accounted for by the time to lay down the sum of all of the beds, our clock would be constructed.*

---

*A special kind of sediment used to tell time, *varved clay,* forms in lakes that freeze over in the winter. The varves are layers of the clay that form couplets; one part of each couplet is a relatively thick, coarse-grained, light-gray silty clay that grades upward into a thin, fine-grained dark-gray clay. The light layers form in the summer and the dark ones in the winter, each pair marking one year. By counting the annual banding in lakes formed after the last ice age in northern Europe, Baron G. DeGeer of Sweden was able to determine that about 8,700 years had passed since the glacial ice had retreated from all of southern Europe. The accuracy of this method was confirmed by radioactive age determinations made much later, and only slight revision of the date had to be made.

Unfortunately, that last *if* is a big one. Based upon the observation of river floods and other kinds of sedimentation, we suspect that a certain amount of time is not represented by rock. The silts laid down on floodplains of rivers, such as the historic ones of the Nile River in ancient Egypt, do not accumulate steadily and uniformly. The time scale of flood deposits is about days long, but there is also a time scale for the times between floods, a time interval that may range from a few years to several decades. In other words, a hiatus, or interruption in sedimentation, may be two to three orders of magnitude (powers of 10) greater than the time for deposition of a layer of flood silt.

How Old Is a Fossil?

We have another, more powerful tool for establishing the time sequence of a series of layers of sedimentary rocks, and that tool is fossils, the remains of ancient organisms found in some of those rocks. **Limestones,** one class of sedimentary rock, are made up of calcium carbonate ($CaCO_3$), much of it in the form of fragments of fossil animal shells. **Shales,** rocks that are hardened and compacted clays and muds, and **sandstones,** rocks made of cemented sand grains, may also contain fossil materials such as shells and shell fragments. Some of the fossil shells found in such rocks can easily be identified by comparison with similar shells found today (Fig. 2-8). Many others look

Figure 2-9
Leaf-impressions of the extinct fernlike plant *Callipteris conferta* from the Hermit Shale. [From *Geology Illustrated* by John S. Shelton. W. H. Freeman and Company. Copyright © 1966.]

Figure 2-8
Ancient and modern cephalopods. The white shell is that of a modern chambered nautilus; the dark shell is that of a closely related form that lived about 200 million years ago. Because of such similarity of form, the significance of fossils as evidence of ancient life was recognized centuries ago. [Photo by R. Siever.]

vaguely like some animals that live today but are obviously different. And others are obviously some sort of organism's shell but not like anything alive today. Not all fossils are shells of invertebrates like clams, oysters, cowry shells, and periwinkles. Diggers in river beds millions of years old may uncover bones of vertebrates such as reptiles or mammals—sometimes even dinosaur remains. Fish skeletons and shark teeth are found in other rocks. Plant fossils are abundant in some rocks, particularly the rocks in and above coal beds, where one can recognize fernlike and other leaves, twigs, branches, and even whole tree trunks (Fig. 2-9).

What could be more natural than to jump to the conclusion that these fossils represent life of former times and that from them we could deduce the flow of evolution from the most primitive organisms to such a complicated one as modern *Homo sapiens.* The "jump" took some time to make, though. One of the first to make it in modern times (some Greeks had known it long before) was Leonardo da Vinci. He was followed later by Nicolaus Steno, who in the seventeenth century compared teeth from modern sharks with the so-called tongue-stones of Malta in the Mediterranean and concluded that they must have come from the same kind of shark (Fig. 2-10). Many objected to Steno's conclusion, but the similarities between the forms of modern animals, especially the hard parts, such as teeth, bones, and shells,

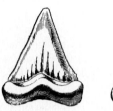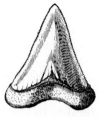

Figure 2-10
Fossil shark's teeth, as illustrated by Steno, 1667.

soon piled up in such numbers that the evidence was overwhelming and could not reasonably be dismissed either as "some accident of form" or as an obscure expression of God's wisdom in creating the Earth.

But what has all this to do with making a time scale? The beginning of an answer can be found in the rocks of the Grand Canyon, for there we can find fossils in many of the exposed rocks, particularly in the limestones. Each layer of limestone has a number of fossils of different species, but the assortment of fossils in one limestone may differ sufficiently from those in another as to be readily distinguishable. This vertical arrangement of different fossils is called the **faunal succession.** The sequence of life forms the fossils represent corresponds to the series of sedimentary rock layers bearing the fossils, the **stratigraphic sequence,** and the faunal and stratigraphic series have the same order (Fig. 2-11). For convenience in mapping, limestones and other rocks are grouped into **formations,** groupings of layers that are everywhere the same stratigraphic age and contain materials that for the most part have the same physical appearance and properties. This combination of appearance and properties is called **lithology.** Formations are a convenience for mapping. Some consist of a single rock type or lithology, such as limestone. Others may consist of interlayered thin beds of different lithologies, such as sandstone or shale. However they vary, each formation comprises a distinctive set of rock layers that can be recognized and thus mapped as a unit.

Once formations and stratigraphic sequences all over the world had been mapped by nineteenth-century geologists, it became apparent that everywhere the faunal successions matched the sequences. This is the rule in fossiliferous formations of all ages since the beginning of the Cambrian Period, when shelled animals evolved.*

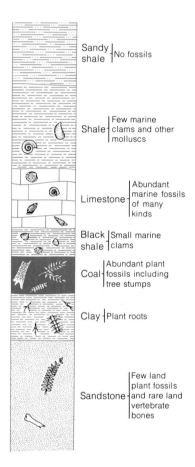

Figure 2-11
Faunal succession in a lithologic sequence: coal-bearing strata of Pennsylvanian age in Illinois. The assemblages of fossils are characteristic of each rock type.

Thus fossil assemblages can be used as "fingerprints" of formations; each assemblage has distinguishable characteristics, even though individual species may be present in several different formations.

It was this feature of fossil-bearing sediments that William Smith, an engineer and surveyor who had worked in coal mines and mapped along canals, began to see so clearly when he set about collecting fossils in southeastern England in 1793. Smith knew nothing of the idea of organic evolu-

---

*Up to the mid-twentieth century, geologists believed there were no fossils in rocks older than the Cambrian—called the Precambrian. We now know of the remains of single-celled organisms throughout most of Precambrian times (see Chapter 12). Paleontologists and paleobotanists are working now to erect faunal successions for these early times.

tion that Charles Darwin was to enunciate some decades later. He did note, however, that different formations contain different fossils, and he was able to tell one formation from another by the differences in the fossils. As Smith extended his mapping over much of southern England in the early part of the nineteenth century, he was able to draw up a stratigraphic succession of rocks that appeared in different places at different levels—a composite showing how the complete section would have looked if all of the rocks from those different places and levels were brought together in a single place.

Let us see how this idea works if we forget about names of fossils and formations but simply call them by arbitrary letters and numbers. At one place I a geologist sees the following horizontal succession with its characteristic fossil assemblage:

I

Rock	Fossil assemblage
A	a
F	f
K	k

In another place, miles away, he sees another horizontal succession:

II

Rock	Fossil assemblage
F	f
K	k
M	m

By simple use of the principle of superposition we infer that the composite sequence is:

A	a
F	f
K	k
M	m

But now at place III he discovers another sequence:

A	a
K	k
P	p

Here F is missing for some reason, but where does P belong—above or below M? We cannot tell from this information alone. This is a problem in making an ordered sequence with respect to time from different fragments or partial records. To resolve the problem, the **stratigrapher** will have to hunt for an outcrop that shows both P and M.

The missing F at III may be explained by erosion. Perhaps F was once deposited after K at place III, but it may have been eroded away before A was laid down. Or F may never have been laid down in the first place. In either case the boundary between A and K is an example of an **unconformity**; that is, a surface between two layers that were not laid down in unbroken sequence, where some intermediate deposit either never was deposited or was removed before the younger rock was deposited (Fig. 2–12). For example, there are places in the world today, typically in shallow marine waters, where there may be neither deposition nor erosion, but a constant balance between the two, so that no material accumulates.

The Grand Canyon Sequence Interpreted

Let us return to the Grand Canyon and work through the sequence from the base up. At the bottom, exposed in the inner gorge, are dark rocks that are not at all like the horizontally bedded sedimentary rocks above. They show no bedding and form bodies that seem to be inserted into and cut across structures in the surrounding rocks. Some of these rocks are made up of coarse crystals; others, of particles so small that they cannot be seen even with a hand lens. These characteristics are interpreted by geologists as evidence of **igneous** origin; that is, these rocks were formed by cooling and solidification of a hot molten liquid, or **magma.** The coarsely crystalline ones are inferred to have been emplaced in the rocks that surround them while those rocks were still buried deep in the earth; these **intrusives** originated as hot magma that pushed its way into cracks and other openings in the surrounding rocks. The large crystals are characteristic of intrusives and the result of the slow cooling of magma that takes place far below the surface. The fine-grained rocks, **extrusives,** were formed as lava flows and ash deposits from volcanic eruptions. Their characteristic fine texture indicates rapid cooling at the surface.

Another group of rocks exposed in the inner gorge are those that have a platy or leafy texture, called **foliation,** caused by the alignment of min-

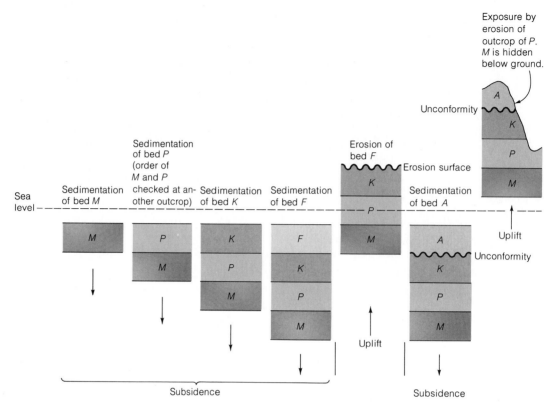

Figure 2-12
Sequence of events in the making of an uncomformity.

erals along straight or wavy planes. Foliation can be confused with bedding. In some of these foliated rocks true bedding can also be seen, but it is folded and crumpled. These **metamorphic** rocks (from the Greek, meaning to change form or to transform) were once sedimentary and igneous rocks, but they have been altered by the action of heat and pressure due to their deep burial in the Earth.

The lowest rocks of the inner gorge, the Vishnu formation (Fig. 2-13), are a complex mixture of these igneous and metamorphic rocks. Such rocks have no fossils, and there is no way to tell how old a rock is merely by looking at its minerals and their texture (geologists learned long ago that rocks formed by the same process at different times tend to look alike). Nevertheless, the Vishnu is at least the oldest rock we can see in the Grand Canyon inner gorge: it's at the bottom. Because of its position, it gives us our first glimpse of history there. The rocks of the Vishnu, originally formed as lava flows, ash deposits, and sediments, were deeply buried by the rocks that now overlie them, were metamorphosed by the resulting heat and pressure, and were later intruded by molten magma.

Above the Vishnu, and separated from it by a sharp line of discontinuity—an **angular unconformity**—is another series of rocks. An angular unconformity is a surface of erosion that separates two sets of beds whose bedding planes are not parallel. It thus signifies that the originally horizontal lower set was deformed and then eroded to a more-or-less even surface before the upper set was deposited horizontally upon it (Figs. 2-14 and 2-15). Rocks in many places show evidence of such physical deformation. Sedimentary layers, once horizontal, are in places **folded** (bent into a wavy structure) and **faulted** (broken and displaced along fractures) (Fig. 2-16). The same structural features, sometimes less easily recognized, are found in igneous and metamorphic rocks. We can therefore add an episode of deformation and uplift to our history of sedimentation, burial, and metamorphism.

The rocks above the Vishnu are interlayered sandstones, shales, and limestones. The Grand Canyon Series, as these rocks of the inner gorge are called, contains no fossils of shelled organisms such as those of Cambrian and younger age rocks and therefore cannot be tied to a standard faunal succession. All that we can say about the

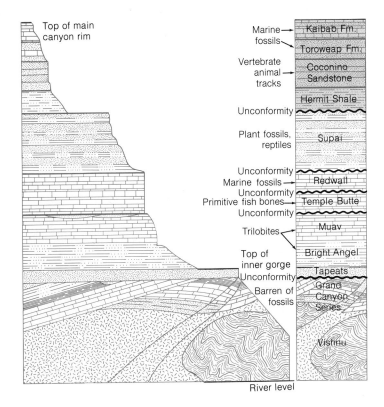

Figure 2–13
The Grand Canyon sequence. From the base up it consists of (1) the Vishnu (Precambrian), a complex group of metamorphic and igneous rocks; (2) the Grand Canyon series (Precambrian), tilted and faulted interlayered sandstones, shales, and limestones; (3) the Tapeats (Cambrian), a pebbly sandstone; (4) the Bright Angel Shale (Cambrian); (5) the Muav Limestone (Cambrian); (6) the Temple Butte Limestone (Devonian); (7) the Redwall Limestone (Mississippian); (8) the Supai Formation (Pennsylvanian), shales and sandstones; (9) the Hermit Shale (Permian); (10) the Coconino Sandstone (Permian); (11) the Toroweap Formation (Permian), mostly limestone; and (12) the Kaibab Limestone (Permian).

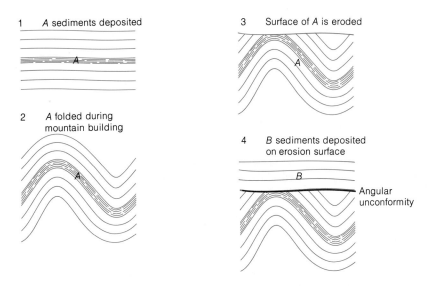

Figure 2–14
The order of geologic events in the making of an angular unconformity.

Figure 2-15
Uncomformable contact of Old Red Sandstone beds on vertical Silurian rocks at Siccar Point, Berwickshire, Scotland. It was from a study of this locality that James Hutton, in 1788, first realized the meaning of unconformity. The geologist is standing just above the contact. [Photo by R. Siever.]

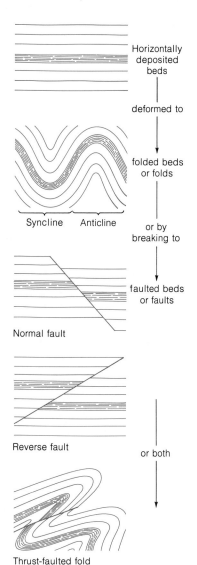

Horizontally deposited beds

deformed to

folded beds or folds

Syncline   Anticline

or by breaking to

faulted beds or faults

Normal fault

Reverse fault

or both

Thrust-faulted fold

Figure 2-16
The deformation of originally horizontal beds into folds and faults. Though easiest to see in sediments, folding and faulting can be found in all kinds of rocks.

age of this Series from this inspection is that it is younger than the Vishnu, older than the rocks above, and is tilted from its originally horizontal position. Sedimentary rocks like these are perfectly ordinary in every way except for two characteristics. They contain no shelled fossils—though they may have remains of microorganisms such as algae—and they may be associated with deformed and metamorphosed rocks like the Vishnu. In the nineteenth century rocks of this kind were set apart from younger fossiliferous strata and called the Precambrian. Though they were then thought always to be complexly folded and faulted as contrasted with much less deformed younger rocks, we now know many places where this distinction is invalid. From the rocks of the Grand Canyon Series we have a glimpse of a history after the erosion of an uplifted Vishnu terrain that includes sedimentation, moderate burial, and moderate deformation that led to tilting and gentle folding.

Another clearly discernible unconformity separates the Grand Canyon Series from the overlying pebbly brown Tapeats Sandstone. The Tapeats contains no fossils, but it can be dated by reference to the overlying formations because it blends into them without any break, thus forming a **con-**

formable succession. Farther up the canyon wall, the Tapeats gradually gives way to a formation consisting mainly of shale, a hardened equivalent of a muddy sediment, called the Bright Angel Shale.

The Bright Angel Shale contains a few fossils, most of which are **trilobites,** extinct arthopods related to modern crayfish (Fig. 2–17). The differences among the trilobites of different ages can be used by paleontologists to date these rocks. By matching fossil trilobite species in different stratigraphic sequences in different parts of the world, a composite succession has been worked out. As a result geologists learned that the part of the Bright Angel Shale lying just above the nonfossiliferous Tapeats Sandstone in the western part of the Canyon is older than the part of the Shale that occupies the same stratigraphic position in the eastern part of the Canyon (Fig. 2–18). This means that the sea in which the Bright Angel was deposited flooded the land in the east at a later date. This is evidence of a **transgression,** meaning that as the Bright Angel Shale was being deposited the sea gradually moved landward. Once again, simple geometric evidence leads to this conclusion: as the sea advanced slowly from west to east, it continuously laid down sand along beaches and mud in deeper water. The reverse, the withdrawal of a sea and the inverse distribution of sediments in relation to shorelines, is called a **regression.** The Bright Angel Shale grades upward into the overlying Muav Limestone, also a part of the transgressive sequence.

Even from a great distance, most of the formations of the canyon wall can be distinguished easily. For example, the Tapeats Sandstone forms an obvious rim at the top edge of the inner gorge. In contrast, most of the Bright Angel Shale, just above the rim, is hidden beneath rubble, which slopes gently upward from the top of the Tapeats to the base of a higher step, the base of the Muav Limestone (Fig. 2–19).

The next formation up, the Temple Butte Limestone, could easily be missed. It is thin, and in places along the Canyon wall it is missing entirely; in these places, the Redwall Limestone, which elsewhere overlies the Temple Butte, rests directly on the Muav. The importance of the Temple Butte is that it contains fossil skeletons of primitive fish. We know from the general succession of fossil animals that these fish lived at a much later time than the trilobites of the Muav. Fossils of many marine animals that lived between the times of deposition of the Muav and the Temple Butte are known from other formations in various parts of the world—but they are all miss-

Figure 2-17
A nearly complete trilobite and the tail of another from the Bright Angel Shale. [From *Geology Illustrated* by John S. Shelton. W. H. Freeman and Company. Copyright © 1966.]

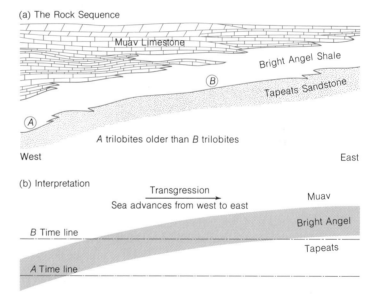

Figure 2-18
The stratigraphic relations of the Bright Angel Shale in the Grand Canyon region (a). The fact that trilobites at *A* are older than those at *B* leads to the interpretation shown in (b).

ing here. Thus there is evidence here of a great gap in the record, an unconformity between the Muav and the Temple Butte. If any sediment had been laid down during the time represented by this unconformity, it was later eroded without

Figure 2-19
Formations exposed in Diamond Creek, Grand Canyon. The Cambrian Bright
Angel Shale, which forms the rubbly slopes at the bottom, is overlain by the
cliff-forming Muav, Temple Butte, and Redwall limestones. Above the main cliff
is the thinner Supai Formation, which crops out with gentler slopes than the cliff-
forming units below. [Photo by Edwin D. McKee, U.S. Geological Survey.]

leaving a trace. The sequence implies a history of
Muav sedimentation and burial (but no deformation; it is still horizontal) before it was uplifted,
eroded, and later covered by sediments that
formed the Temple Butte Formation.

An unconformity between the Temple Butte
and the overlying Redwall Limestone* represents
another time gap, and an unconformity between
the Redwall and the Supai Formation, yet another. The Redwall's age we know from its sparse
content of fossil marine animals. The Supai contains no marine fossils, but it does contain fossils
of land plants like those found in the coal beds of
the United States and the Ruhr Valley in Europe.
Thus the Supai can be dated with reference to the
worldwide succession of plant fossils, a succession that has been worked out in much the same
way as the succession of animal fossils. Of even
greater interest are the fossil footprints of primitive reptiles found in the Supai. Thus it turns out
that we can tell something of time not only from

marine sediments but from those deposited on
land—the **terrestrial, or continental, deposits.**

Continuing up the canyon walls, we find another unconformity at the top of the Supai. Above
the Supai is a sandy red shale, the Hermit Shale,
which is succeeded by the Coconino Sandstone.
Not only does the Coconino contain more vertebrate animal tracks, but it has a distinctive form
of bedding, which is not uniform and horizontal
like that of many sediments but is composed of
many sets of interfering wedges of bedded material inclined at angles up to 35° from the horizontal. This form of bedding, called **cross-bedding,** is
characteristic of sand dunes on land and of dunes
formed by currents in rivers and under the sea
(Fig. 2–20; also see Fig. 2–7). On the basis of vertebrate animal tracks and the dune type of cross-
bedding, most geologists believe that the Coconino
Sandstone formed by wind.

The Coconino appears to be conformable with
the underlying Hermit and, with the overlying
Toroweap, a formation of limestone and red,
sandy shale. Next upward is a massive formation
of limestone and sandstone that forms the top of
the cliffs at the upper rim of the Canyon—the
Kaibab Formation.

If we were to inspect highlands above the canyon rim in the general region of the Grand Can-

---

*How the Redwall Limestone got its name is obvious from
any look at it in the Canyon. But the "red" in Redwall is unde-
served in at least one way, for when a piece of the Redwall is
cracked the fresh rock turns out to be gray. A close look at the
cliff shows that the limestone is stained, as if it had been rub-
bed with a reddish pastel chalk. The stain is washed down by
rain from the overlying sandy red shale of the Supai Formation.

Figure 2-20
Cross-bedding in the Coconino Sandstone. The cross-bedding of different layers interferes to produce such a complex
pattern that the true horizontal bedding
planes are not easily distinguished. [Photo
by Edwin D. McKee, U.S. Geological
Survey.]

yon, we would find formations younger than the
Kaibab. From their fragmentary successions we
could build a composite that would include red,
brown, yellow, and gray sandstones, conglomerates, and shales that contain the famous petrified
forest of tree trunks and, in some places, dinosaur
remains.

The rocks of the Grand Canyon have many stories to tell: of the advance and retreat of the seas
over the continent at this place, of the appearance
and disappearance of different kinds of organisms, and of the different kinds of marine and
terrestrial environments in which this remarkable
variety of sediments was deposited. But one of the
most important is the tale of time—the time that is
represented by the rocks of the Canyon and the
time that is recorded by the unconformities between so many of the formations. From the radioactive time scale, which is based upon the decay
of radioactive elements in minerals, we know that
the oldest formation, the Vishnu, is about 1400 to
1500 million years old, and the top of the Kaibab
about 225 million years old. An enormous amount
of time is represented by these rocks and unconformities.

ROCKS AS RECORDS
OF EARTH MOVEMENTS

Angular unconformities not only date erosion intervals, but they also record ancient earth movements. Beds below such unconformities were
folded, tilted, faulted, and uplifted before erosion

produced the more-or-less even unconformable
surfaces that we observe today. Erosion was, in
turn, followed by additional earth movement, for
only crustal subsidence could account for a
change from erosion to further sedimentation.
Unconformities, therefore, are records of periods
of mountain-building, even though the roots of
the mountains are all we can see today. **Discon-
formities**—time gaps between two units whose
bedding planes are parallel—are less dramatic,
but they too imply the same general sequence of
uplift, erosion, and subsidence.

There are other ways of telling a time sequence
too. Though igneous rocks are not layered, as sediments are, they too have characteristics that
place them in time. Igneous intrusions injected as
a mobile magma may show sharp contacts with
surrounding, or **country, rocks.** These contacts
cut across and interrupt original structures in the
country rock. Such cross-cutting intrusions are
thus **discordant.** They are typified by thin sheets,
called **dikes,** that may lie at any angle to bedding.
Intrusions may show **concordant** contacts, as in
the case of **sills,** which follow the bedding of the
sediments into which they were intruded. The
concordant and discordant field relations between igneous rock and adjacent sedimentary,
metamorphic, and other igneous rocks can be
used to date these formations in just the same way
that Steno's laws of original horizontality and
superposition can be used to figure out relative
ages of sediments. Similarly, folds and faults can
be fitted into time sequences as well.

Figure 2-21 is the record of a series of events

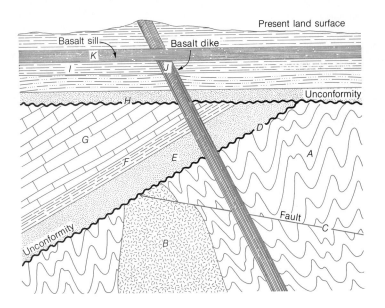

Figure 2-21
Relative time dating by field relations. The sequence of events shown must have been: (1) sediments *A* deposited; (2) sediments *A* deformed; (3) granite *B* intruded into *A*; (4) faulting along *C*; (5) erosion to produce unconformity at *D*; (6) sediments *E, F,* and *G* deposited; (7) tilting, uplift, and erosion to produce unconformity at *H*; (8) sediment *I* deposited; (9) basalt dike intruded at *J* and sill at *K*; (10) erosion to produce present land surface. Discordant contacts are illustrated by *B* and *J*, and a concordant contact by *K*.

that can be interpreted in only one way. Of course, once such a time-event jigsaw puzzle has been put together, as in this diagram, it looks easy. But sorting out the pieces and seeing how they fit can sorely tax the imagination of a field geologist. *To sum up, the ordering of geologic events with respect to a relative time scale is based on interpretations of sedimentary successions, igneous field relations, such as cross-cutting, and tectonic deformation, such as folding and faulting and angular unconformity.* By piecing together the information gleaned from the study of such field relations, geologists during the nineteenth century worked out the entire stratigraphic time scale.

Hutton and Uniformitarianism

Although the reasoning used here in interpreting the Grand Canyon sequence may seem obviously correct, it was not until the close of the eighteenth century that geologists were ready to believe that there had been any evolution of the Earth's surface. Up to then they had struggled to find an explanation for the field relationships of rock for-

mations consistent with the assumption that the Earth was created just as we see it today, with all of its river valleys, mountains, and plains placed where they are by God. The new way of looking at the Earth included the recognition that constant change takes place as geological forces modify the surface and the interior.

A Scottish gentleman farmer, James Hutton, led the way with a book carrying the bold title *Theory of the Earth with Proof and Illustration,* first presented to the Royal Society of Edinburgh in 1785. Hutton's greatness lies in his recognition of the cyclical nature of geological changes, and of the way in which ordinary processes, operating over long time intervals, can effect great changes. He reasoned from observation that rocks slowly decay and disintegrate under the action of water and air. This process—**weathering**—produces debris in the form of gravel, sand, and silt and furthers erosion of the land. Water and air also act to transport the debris, most of which ends up near or below sea level. The deposits are compacted, cemented, and ultimately become sedimentary rocks. At a later time, according to Hutton, subterranean heat and thermal expansion may produce an intrusion of igneous rock. The **plutonic** episode (named for Pluto, Greek god of the underworld) would be accompanied by upheaval of the sediments and deformation into folds and faults accompanying mountain-building, or **orogeny.** Marine sediments emerge as land, bringing deposition to a halt, and then erosion of the newly emerged highlands initiates the cycle all over again (Fig. 2-22).

Hutton observed and learned from the modern counterparts of each stage of his cycle: mountains erode, rivers carry debris to the sea, ocean waves pound rocks, sands and muds settle to the bottom and then are buried on the sea floor. Because the physical and chemical laws that govern geological behavior don't change with time, one can, by studying processes in the present, infer the behavior of those in the past. Hutton, followed by Charles Lyell (*Principles of Geology,* 1830), used and publicized this **principle of uniformitarianism.** Uniformitarianism, as we understand it today, does not hold that the rates of geological processes or their precise nature had to be the same. Volcanism may have been more frequent in the past than it is now. Nevertheless, ancient volcanoes surely released gases and deposited ash layers and lava flows, just as modern ones do when they erupt. One reason geologists were so intent on studying the eruption of Mt. St. Helens was to learn how to interpret the deposits of ancient volcanoes.

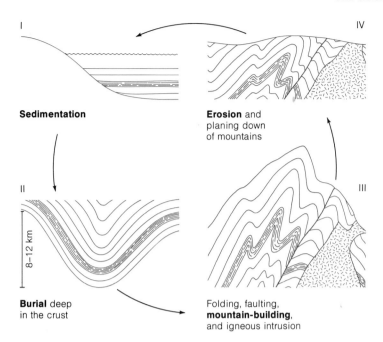

Sedimentation

Erosion and planing down of mountains

Burial deep in the crust

8–12 km

Folding, faulting, **mountain-building**, and igneous intrusion

Figure 2-22
The geologic cycle as deduced by Hutton.

Most of the concepts used in modern field interpretation go back to discoveries made by many late eighteenth-century and nineteenth-century geologists, but it was Hutton who first recognized that igneous bodies must be younger than the rocks they intrude. Hutton also pointed out that debris fragments in a sedimentary or igneous formation must have been derived from a parent rock older than the formation itself, and he was the first person to grasp the idea that a cycle of upheaval, erosion, subsidence, and sedimentation would show as an unconformity in the stratigraphic record.

With these principles established, nineteenth-century geologists opened a new era. The history contained in rock formations could at last be deciphered, and those who read it could travel backward in time to view ancient landscapes. It became possible to reconstruct the interrelations between mountains, oceans, climates, animals, and plants long since gone. By this time, geography and geology had historical counterparts, *paleogeography* and *paleogeology*. Geology entered a period of discovery and glamour.

EVOLUTION AND THE TIME SCALE

In 1859 Charles Darwin's *On the Origin of Species by Means of Natural Selection* was published, and the theory of organic evolution was launched. Along with it was launched one of the great controversies in the history of science that continues even today. The ideas of evolution were denounced by many as monstrous and antireligious, if not downright silly. But under the leadership of Lyell and Thomas Huxley, geologists and biologists moved in a few years to acceptance of the theory. The concept of evolution had enormous impact, for its theoretical framework gave support to the idea that the time-related changes in fossil species could be used to set up a stratigraphic time scale.*

Armed with a refined science of paleontology, based on Darwinian evolutionary theory, and with opportunities for travel afforded by the period of world exploration that accompanied European economic expansion and imperialism, geologists mapped the surface of the earth and fitted together what we now call the **Phanerozoic** (known) time scale (Fig. 2–23). The names of the time periods are taken either from the geographic

---

*Darwin's new theory was actually discovered by both Darwin and Alfred Russell Wallace, a young, unknown naturalist working in the East Indies (Indonesia). Darwin had in 1857 explained his theory in a now-famous letter to the great American botanist Asa Gray at Harvard; Darwin's theory had been germinating in preliminary drafts for almost 20 years. In 1858 Wallace sent Darwin a manuscript that he had written in the midst of a bout with intermittent fever in the Molucca Islands. Darwin's reaction was close to panic, for Wallace's manuscript outlined his theory in essential details. On the advice of two friends, Charles Lyell, then the most famous geologist in England, and Joseph Hooker, a leading English botanist, Darwin and Wallace both presented their views before the Linnaean Society of London on July 1, 1858. Independent simultaneous discovery of an idea whose time has come, and the rush to publish it, is not a new development in science!

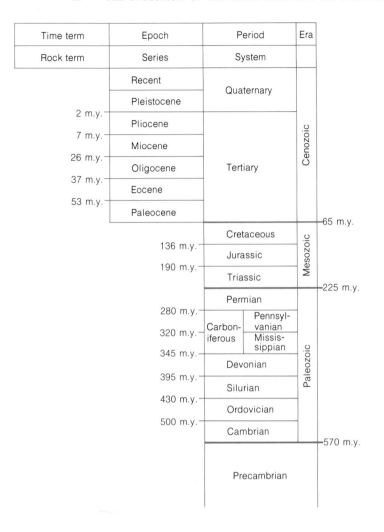

Time term	Epoch	Period	Era
Rock term	Series	System	
	Recent	Quaternary	Cenozoic
	Pleistocene		
2 m.y.	Pliocene	Tertiary	
7 m.y.	Miocene		
26 m.y.	Oligocene		
37 m.y.	Eocene		
53 m.y.	Paleocene		
65 m.y.		Cretaceous	Mesozoic
136 m.y.		Jurassic	
190 m.y.		Triassic	
225 m.y.		Permian	Paleozoic
280 m.y.		Carboniferous — Pennsylvanian	
320 m.y.		Carboniferous — Mississippian	
345 m.y.		Devonian	
395 m.y.		Silurian	
430 m.y.		Ordovician	
500 m.y.		Cambrian	
570 m.y.		Precambrian	

Figure 2-23
The geological time scale. Numbers at sides of columns are ages in millions of years before present.

locality where the formations were best displayed or first studied or from some characteristic of the formations. For example, the Jurassic is named from the Jura Mountains of France and Switzerland, and the Carboniferous is named from the coal-bearing sedimentary rocks of Europe and North America.

Each time period of the stratigraphic time scale is represented by its appropriate system of rocks, and we differentiate time units, the **periods,** from time-rock units (the rocks that represent time), the **systems.** Each major unit is divided: the period into **epochs,** and the system into **series.** Epochs and series have geographic names, except for the older names of many of the epochs, which are simply called Upper, Middle, or Lower. Thus the Upper Jurassic Series comprises the rocks of the Upper Jurassic Epoch of time.

ABSOLUTE TIME AND THE GEOLOGICAL TIME SCALE

The question of just how many years are represented by the rocks in the stratigraphic time scale has been debated for at least 2500 years. As far as we know, Xenophanes of Colophon (570–470 B.C.) first recognized the significance of fossils as remnants of former life and correctly inferred that sedimentary rocks originated as sediments on the sea bottom. Moreover, he concluded that such rocks and fossils must be of great age. Around 450 B.C., Herodotus, the great Greek historian, traveled through the lower Nile River valley. His observations led him to reason that the Nile delta must have been made from a series of floods. It then quickly followed, by his reasoning, that if a single flood were to lay down only a thin layer of sediment, it must have taken many thousands of years to build up the Nile delta. Herodotus also gave this deposit of sediment at the Nile's mouth the name by which we know it today. It's shape, roughly triangular, looked to him like the Greek letter *delta* Δ.

Aristotle and other Greek and Roman philosopher-naturalists strengthened and expanded this scholarly approach of observation combined with deduction. Early Christian scholars, such as St. Augustine, in all essentials continued the tradition. But the thread of that kind of thinking was temporarily lost in late medieval outgrowths of the revolution of Christian scholasticism and theological idealism, which found a solution to the problem of the age of the Earth and all things on it in one book: *Genesis.* The literal interpretation of this book of The Bible gained increasing devotion by churchmen and persisted in the beginnings of modern times in reaction against the scientific explorations of the Renaissance. Literalism, to many, was as explicit as the pronouncement by Archbishop Ussher of Ireland in 1664 that the Earth was created at 9:00 A.M., October 26, 4004 B.C. (presumably Greenwich mean time!).

As modern geology started gathering momentum, through the work of James Hutton and others, no attempts were made to guess at the age of the Earth; instead, there was an obvious return to the old Greek way of looking at things. The evidence demanded a long time for Earth processes to have had any effect in carving mountains and accumulating sediment. By looking at the time scale of processes and using the idea of uniformitarianism, geologists came to recognize that rocks are very old, and the Earth much older.

At the same time, physics enjoyed a new burst of activity. Applying the ideas of Galileo and of

Newton, who in 1687 established the basis for the theory of gravity, physicists could begin to calculate the time requirements for formation and orbiting of the members of the solar system. The times necessary seemed much too long compared with the shortness of time required by The Bible. Before the nineteenth century, however, physicists did not challenge religious orthodoxy: Sir Isaac Newton was a devoutly religious man—one we would call today a "fundamentalist." Even so, in the middle of the eighteenth century, Comte de Buffon of France analyzed melting and cooling rates of iron balls and put what he learned together with his guess that the interior of the Earth had to be iron (it was so dense) to calculate how long it would take the molten earth to cool. His result, 75,000 years, made no one happy; for the fundamentalists it was much too long, and for many geologists, much too short.

The plot thickened in 1854, when Herman von Helmholtz, one of the founders of the science of thermodynamics, seized on the problem of the Sun's luminosity. Not long before, Immanuel Kant had calculated that if the Sun's light came from ordinary combustion, it would have burned up in only 1000 years. Helmholtz realized that a way out of this bind was to infer that the Sun's light came from the heating required by the gravitational contraction of the immense mass of the Sun. His notion was that particles would literally fall toward the center and that the potential energy released in that fall would be converted to heat. Helmholtz estimated the age of the Sun and Earth at 20 to 40 million years.

Another great physicist of the nineteenth century, William Thomson, better known as Lord Kelvin, entered the debate. He expanded and revised Helmholtz's estimates, and using them plus Buffon's cooling-rate figures and other ingenious arguments, pronounced that the most probable value was something like 20 to 40 million years, probably closer to 20 million.

Many geologists found themselves unable to adjust their notions of the length of geologic time to accord with Kelvin's estimate. They argued vehemently that not everything was yet known about physics and that it was foolish to contradict the obvious evidence of the rocks in such composite sequences as those in the Grand Canyon, for they clearly must have taken many times 25 million years to accumulate. Others, such as Clarence King, one of the founders of the U.S. Geological Survey, who had done much of the geological exploration of the West, made his own calculations, compared them with Kelvin's, and concluded that Kelvin was about right.

What was, in retrospect, the best response came at the turn of the century from T. C. Chamberlin, head of the Department of Geology at what was then the brand new University of Chicago. Chamberlin, who held many agnostic opinions, added his speculation that there might yet be discovered new sources of energy (unknown to him then) within the particles of matter that would eliminate burning and gravitational contraction as the sole possible causes of the Sun's luminosity.

The discovery that was to be of world-shaking importance (both figuratively and literally, as it turned out!) came in 1895 when Henri Becquerel, a French physicist, discovered radioactivity in uranium salts almost at the same time that the German physicist Wilhelm Röntgen discovered x rays. Soon after, Marie Sklodowska-Curie made the crucial discovery and isolation of radium, a radioactive element. Thus in the space of a few months, the stage was set for the use of the clocks that are built into the nuclei of radioactive atoms—the clocks that were eventually to resolve the bitter argument about the age of the Earth. Yet neither Kelvin nor Chamberlin, to say nothing of their audiences, seemed to be aware of the extraordinary implications of these discoveries, though the journals of physics and chemistry and even the Sunday newspaper supplements were popping with the news.

It was not until 1905 that physicist Ernest Rutherford, who had been studying radioactive processes, suggested that radioactive minerals could be used to date rocks. He dated a uranium mineral in his laboratory at McGill University in Montreal the next year. In the same year B. B. Boltwood of Yale discovered "ionium," which turned out to be an isotope of thorium—the first isotope to be isolated* (see next section). Both Rutherford and Boltwood published ages of dated minerals, but it was not until F. Soddy, in 1913, clarified the nature of isotopes that the methods could be refined and made more accurate, for most of the early ages turned out to be too high. When the full series of decay products of radioactive disintegra-

---

*A typically long-delayed award of distinction came to Boltwood, who was credited with discovering the first isotope: a mineral was named after him, appropriately enough a uranium-bearing mineral. In this way science incorporates into its vocabulary the names of its great ones. Not all, of course, have minerals named after them. Kelvin's name was given to the absolute temperature scale (we refer to "degrees Kelvin"); Helmholtz's to a number of things, from electrical coils to equations and types of energy (Helmholtz free energy). Geologists have been immortalized not only by mineral names but by fossil names, like *Oryctocara geikiei*, a trilobite, whose specific name was given in honor of Archibald Geikie, a famous Scottish geologist of the nineteenth century.

tion was firmly established, it became clear that the Earth had to be not millions, but *billions,* of years old.

Within a few decades all of the apparent contradictions were cleared up. In the 1920s and 1930s, astronomers and physicists recognized that the immense energies liberated by nuclear processes must be responsible for the luminosity of the sun and that the heat given off by radioactive decay was the explanation for the heat trapped in the Earth's interior.

THE CLOCKS IN ROCKS: RADIOACTIVE ATOMS

What the pioneers of nuclear physics discovered at the turn of the century was that atoms of certain elements, the radioactive ones, spontaneously disintegrate to form atoms of a different element, liberating energy in the process. The important reason why radioactive decay offers a dependable means of keeping time is that *the average rate of disintegration is fixed and does not vary with any of the typical changes in chemical or physical conditions* that affect most chemical or physical processes. This means that once a quantity of a radioactive element is created somewhere in the universe, it starts to act like the balance wheel of a clock, steadily firing off one atom after another at a definite rate.

To tell time, we cannot simply look at a balance wheel or listen to ticking. We need some kind of reference, a face with hands and numbers on it or the light-emitting diode numbers of modern digital clocks. The numbers that we use to read the radioactive clock are supplied in the form of the new atoms, the **daughter elements,** that are formed from the old disintegrated ones, the **parent elements.** If we can identify and count the daughter element atoms, and if we know the rate of decay, we can work back to the time when there were no daughters but only parents. The idea is simple, but its practical implementation has required a major effort on the part of those geologists who combine their knowledge of nuclear physics with that of geology—the specialists in **geochronology.**

What Happens to a Radioactive Atom?

Every atom has a dense nucleus that contains practically all of the mass of the atom. Surrounding the nucleus is a cloud of **electrons** (Fig 2-24).

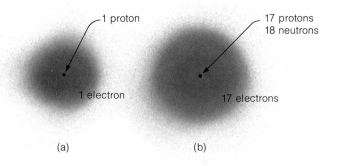

Figure 2-24
Graphical representation of the relative sizes of (a) hydrogen and (b) chlorine atoms. The radius of each atom is that of an electron cloud surrounding the nucleus, which contains the protons and neutrons. The sizes of the nuclei are greatly exaggerated; they are much too small to show on a true scale. The chlorine atom shown is of the isotope ^{35}Cl, which contains 18 neutrons.

The nucleus contains two kinds of particles, each with mass 1: a **proton** has a positive electrical charge of $+1$, and a **neutron** is electrically neutral. In a complete atom the number of protons in the nucleus is balanced by the same number of electrons in the cloud outside, each of which has a negative charge of -1. The number of protons (or electrons) is unique for each element and is called the **atomic number** (usually symbolized Z). The sum of the masses of protons and neutrons is the **atomic weight** of the atom.* All atoms of the same element have the same atomic number; for example, the atomic number of carbon is 6. Different isotopes of an element have the same number of protons but different numbers of neutrons. Carbon isotopes exist with 6, 7, and 8 neutrons, giving atomic masses of 12, 13, and 14 (Fig. 2-25). Of these isotopes, carbon-12 (^{12}C) and carbon-13 (^{13}C) are stable; that is, they do not change or spontaneously disintegrate. But ^{14}C is radioactive and spontaneously decays to the element nitrogen. Another element that decays spontaneously is rubidium-87 (^{87}Rb), which is transformed to strontium-87 (^{87}Sr). An important difference between the decay of ^{14}C and ^{87}Rb is the *rate* at which the atoms decay. That rate is commonly stated in terms of the **half-life**—the time required

---

*Atomic weights of naturally occurring elements, which chemists spent much of their time measuring accurately in the nineteenth century and early twentieth century, turn out not to be whole numbers. Thus the atomic weight of carbon given in tables is 12.011 relative to the atomic weight standard, carbon-12 as exactly 12. The reason for this is that the natural elements are mixtures of the various isotopes. The very small amount of ^{13}C and even smaller amounts of ^{14}C in most samples of carbon is enough to affect the average weight.

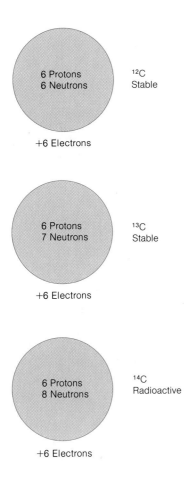

Figure 2–25
Carbon isotopes. All have the same number of protons, and thus atomic number 6, but each isotope has a different number of neutrons.

first major decay scheme to be understood and used for dating was the decay of the element uranium, which is still used extensively today. Uranium has two radioactive isotopes, each of which decays to an isotope of lead and helium. Another element, thorium, can also decay to lead. The half-lives of these decays amount to hundreds or thousands of millions of years, which makes them suitable for dating the oldest objects in our solar system.

One other radioactive isotope is of great importance in dating rocks, potassium-40. It decays by a scheme that has two paths. In one path, the ^{40}K decays to a calcium isotope, ^{40}Ca. About 89 percent of the ^{40}K atoms in any group of atoms follow this route. The remaining 11 percent of the ^{40}K atoms decay to form the inert gas argon, ^{40}Ar. The latter decay path is the one that is used for dating because the daughter, ^{40}Ar, can be easily distinguished from ordinary argon formed in other ways, whereas ^{40}Ca is ordinary calcium and atoms of radiogenic origin can not be distinguished from others.

for one-half of the original number of radioactive atoms to decay. At the end of the first half-life after a radioactive element is incorporated into a new mineral, one-half is left; at the end of a second half-life, one-quarter is left; at the end of a third half-life, an eighth is left; and so on, as shown in Figure 2–26.

We can now compare the rate of decay of ^{14}C, which has a half-life of 5570 years, with that of ^{87}Rb, which has a half-life of 47 billion years. It is the half-life that dictates that ^{14}C is commonly used for timekeeping for only the last 30,000 years or so of Earth history, a little more than 5 half-lives (at which point all except the last $\frac{1}{32}$ of the original amount is gone). In contrast, 3 billion years, about the age of some very old rocks dated on earth, is only about $\frac{1}{16}$ of a single half-life of ^{87}Rb. This fact makes rubidium-87 easily the choice for dating many old rocks.

Carbon-14 and rubidium-87 both undergo relatively simple decay processes (see Box 2–1). The

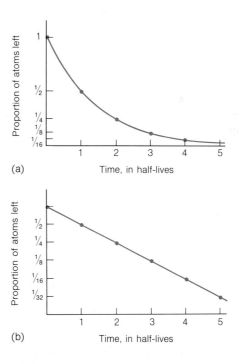

Figure 2-26
Graphs showing the decrease in the number of radioactive atoms with time. In each successive half-life, no matter how long or short, the number of atoms present at the beginning of that half-life period is halved. The relation between half-life and decrease in number of radioactive atoms with time is a straight line when shown with the vertical scale arranged logarithmically (b) rather than arithmetically (a).

Box 2-1 DATING ROCKS BY MEANS OF RADIOACTIVE MINERALS

The decay of radioactive atoms proceeds in different ways for various elements. Carbon-14 decays by emitting **β-particles** (electrons shot from the nucleus when a neutron is split into a proton and an electron). This changes its atomic number to 7, which is characteristic of the element nitrogen, but does not change its mass (because an electron has practically no mass); the new nucleus is thus nitrogen-14. We show this process by the following kind of equation:

$$^{14}C \rightarrow {}^{14}N + \beta$$

Rubidium-87 emits a β-particle in the same way as ^{14}C to produce strontium-87:

$$^{87}Rb \rightarrow {}^{87}Sr + \beta$$

Both radioactive isotopes of uranium decay by emitting **α-particles,** particles consisting of two protons and two neutrons (an α-particle is equivalent to the nucleus of a helium atom, ^{4}He); the decay reactions are written

$$^{235}U \rightarrow {}^{207}Pb + 7{}^{4}He$$
$$^{238}U \rightarrow {}^{206}Pb + 8{}^{4}He$$

Thorium also decays to lead by α-emission:

$$^{232}Th \rightarrow {}^{208}Pb + 6{}^{4}He$$

The two paths of radioactive potassium-40 decay are given by the following equations. For 11 percent of the atoms,

$$^{40}K + e \rightarrow {}^{40}A$$

where e is an orbital electron captured by the nucleus; for the remaining 89 percent,

$$^{40}K \rightarrow {}^{40}Ca + \beta$$

The equations for calculating radioactive ages are cast in terms of the number of atoms, n, that decay in a period of time (usually per second or per year, depending on which is more convenient) relative to the total number of atoms of that element (N) in any given amount of it. The proportion n/N is constant no matter what N is and is called the decay or **disintegration constant,** usually symbolized by the Greek letter lambda, λ. The other common measure of the rate of decay, the half-life, is usually represented by T. The half-life is simply related to λ: it is always

$$T = 0.693/\lambda$$

If we choose the unit of time for λ to be the half-life, T, of the element T, then $\lambda = 0.5$ by definition, and

$$\log \frac{N}{N_0} = -0.3y$$

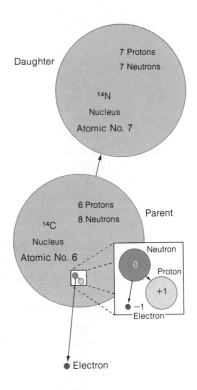

A radioactive carbon atom ^{14}C, spontaneously decays to a nitrogen atom, ^{14}N, by emitting a β-particle. This is the same process by which radioactive rubidium, ^{87}Rb, decays to strontium, Sr^{87}.

where N_0 is the original number of radioactive atoms in a sample of the parent mineral, and y is the number of half-lives that have elapsed since the mineral was formed. The age of the rock is determined by measuring N, deducing N_0 from amounts of daughter elements present, and solving for y. Then y multiplied by T, the half-life, gives the age in years.

The solution of the radioactive decay problem is generalized and shortened to one equation by using calculus:

$$dN/dt = -\lambda N$$

or, in integrated form (called the age equation),

$$N = N_0 e^{-\lambda t}$$

Reading the Clocks

Once isotopes were discovered and instruments were invented that could perform the chemical analysis for the ratios of parents to daughters, the business of dating rocks could begin. It was uranium-thorium decay that received first attention and which is, in a sense, the easiest to use, for it requires only an ordinary chemical analysis for uranium and lead. It is accurate enough to give approximate dates for many uranium minerals in rocks believed to have had little or no original lead. Unfortunately it cannot distinguish between the lead that originates from the different uranium isotopes and from thorium.

The next important development came in the 1920s and 1930s, when the **mass spectrometer** was invented. That instrument was designed to produce a beam of electrically charged atoms from the sample to be studied. The beam passes through electric and magnetic fields in such a way that the atoms are deflected by an amount that depends upon their masses. Thus isotopes of elements can be separated. The precision and sensitivity of these instruments has steadily improved, so that today even minute amounts of individual isotopes can be analyzed.

Not all radioactive decays are analyzed by mass spectrometry. Most ^{14}C ages are determined from the carbon in dead plant material. During growth, plants steadily incorporate a small amount of ^{14}C along with the other carbon isotopes contained in the carbon dioxide of the atmosphere. When a plant dies, photosynthesis stops, and no new ^{14}C is taken up. The relative amount of ^{14}C at this point is approximately the same as the ratio in the atmosphere (a ratio that we assume has remained constant for the last 100,000 years), but it steadily decreases with age as the ^{14}C radioactivity decays. The amount of ^{14}C remaining is measured indirectly by counting the decay particles being emitted by the ^{14}C still present in the sample. The count, called ^{14}**C activity,** can be used to calculate the age, since the number of decay particles produced is proportional to the number of ^{14}C atoms still present, and that count can be compared with the higher activity of a contemporary sample of ^{14}C.

Until a few years ago, it was difficult to extend this method to ages greater than 40,000 years. In the past five years, physicists have developed methods using particle accelerators to measure directly the ^{14}C atoms in the sample rather than count the decay particles. This has allowed the dating of materials as old as 70,000 years, dramatically increasing our ability to date events related to the recent ice ages and the growth of human culture.

When Were the Clocks Started?

Once the minerals in a rock are formed, any radioactive elements in those minerals keep ticking away all of the time. What we actually measure is the time elapsed since the radioactive parent elements became part of a rock from which their daughter elements could not escape. For example, when uranium becomes incorporated in rock-forming minerals that congeal from a molten state, it is separated by the crystallization process from the lead formed by previous decay. Once the decay process goes on in the solid rock, however, daughter elements are trapped, and quantities of lead are eventually produced. The amounts of parent and daughter present in a sample of rock are a measure of the time interval between now and the time the rock crystallized. Thus the methods based on the decay of uranium, rubidium, and potassium all give the date of crystallization of the rocks in which those minerals are found—and, by geologic inference, of any other rocks that bear a definite age relation to the rock analyzed. For example, when we date the crystallization of a granite, we know also from geology that the surrounding sedimentary rocks, into which the granite was intruded when it crystallized, can be no younger than the granite. If the geologic ages of the sedimentary rocks are known from fossils and those strata in turn overlie, and thus are younger than, other radioactively dated rocks, we can bracket the absolute age of the stratigraphically dated sedimentary rocks, even though they may contain no datable minerals.

Several factors can cause dates based directly on radiometric methods to be incorrect. First, one must be able to show that there has been no removal of a daughter element. For example, if groundwater solutions had dissolved some of the lead produced by uranium decay, the age would be underestimated. Second, one must recognize that a geological event, such as heating or partial melting of the rock in a later metamorphic episode, may have reset the clock to zero by allowing earlier-formed daughter elements to escape. The interpretation of potassium-argon ages is complicated by the fact that the daughter element is argon, a gas that can diffuse out of the solid mineral and so give a falsely young age for the rock. Since the argon diffusion rate depends strongly on the temperature, what is actually being dated may be interpreted as the time when the rock cooled enough to bring argon leakage to a stop. That time has been shown to be appreciably later than the original formation of the rock. In this case a difficulty in interpreting radioactive dating is turned to our advantage and used to learn more about the complex history of the rocks, such as cooling times or later metamorphic episodes.

AGES OF METEORITES

One of the triumphs of radioactive dating emerged only gradually as more and more workers dated meteorites. It became surprisingly apparent that all meteorites are of the same age, somewhere in the vicinity of 4.5 billion years old. This is true regardless of their composition—whether they are stony, made up of rock-forming minerals like those of the Earth's crust, or of the iron variety, made up mostly of alloys of iron and nickel. That there are no meteorites of any other age, regardless of when they fell to Earth, suggests strongly that all meteorites originated in other bodies of the solar system that formed at the same time that the Earth did. If this is true, and if there were no geologic processes to reset the clocks through geologic time, then the meteorites give the age of the Earth too.

HOW TO SYNCHRONIZE CLOCKS: THE STRATIGRAPHIC AND ABSOLUTE TIME SCALES

Geologists working with fossiliferous sediments have a fine watch by which to measure time. It is accurate enough to enable geologists to distinguish the relative ages of formations that are only a few meters thick and that may represent periods of time far less than a million years in duration. A million years, we have to remember, is only about $\frac{1}{5000}$ of the history of the Earth. The entire world rock record of fossiliferous sediments has been mapped and subdivided into the scheme of eras, epochs, and ages that is shown in Figure 2–23. A geologist studying fossiliferous rocks in the field needs only an average knowledge of paleontology in order to make a fairly accurate estimate of the epoch in which the rocks he or she is studying belong.

Using the stratigraphic time scale is like reading a watch that enables you to tell one time from another but gives you no idea of just how long a minute really is. It was not surprising, then, that

the discovery of radioactive age dating was immediately seen by some geologists as a way of making a combination clock that would function as an absolute as well as a relative timekeeper. Only ten years after the discovery of radioactivity, the first rocks were dated by the uranium-lead method. Eight years after that, Arthur Holmes, a young British geologist who had not yet received his doctor's degree, published the first edition of what was to become a classic work, *The Age of the Earth*. Holmes plotted radioactive age dates opposite the stratigraphic time scale by figuring out as closely as he could the age relations of the sediments, which were dated by fossils, and of the cross-cutting igneous rocks, which were dated by radioactivity. His estimates were remarkable, for most of them have held up in a general way. His first estimate of the beginning of the Cambrian placed it at about 600 million years before the present. His last estimate, published in 1959, a few years before his death, was the same as his first—and that in the face of thousands of more accurately determined dates made in the half century since (see Fig. 2–23 for a more recent summary of these dates).

Authorities still differ on precise details, but the general relation between the absolute and stratigraphic scales is undisputed. The Phanerozoic time scale of rocks containing fossils of higher organisms represents about 600 million years. It is divided into somewhat unequal divisions, the eras: the Paleozoic, about 350 to 400 million years long; the Mesozoic, about 150 million years long; and the Cenozoic, about 70 million years long (Fig. 2–27). The epochs too are unequal in length. The inequalities are the result of accidents of choice. Stratigraphers of the nineteenth century divided parts of the geologic column according to what seemed convenient or appropriate to the area they were studying. If the Chinese and Indians had done most of the stratigraphic work, the column would have been far different. The boundary between the Phanerozoic and Precambrian would still have been the same, however, for the appearance of shelled, higher forms of life occurred almost simultaneously (geologically speaking) all over the Earth; once those forms evolved, they spread rapidly.

Telling Precambrian Time

Because there are no fossils to rely on, the Precambrian had always been somewhat of a mystery to stratigraphers. Although they had been able to unravel complicated sequences of sedi-

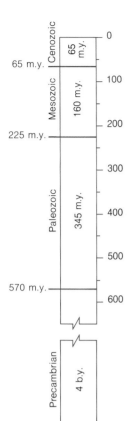

Figure 2-27
Unequal duration of the major eras of the geological time scale.

Phanerozoic. As we have noted before, stratigraphers can divide their columns into units that may be as little as a million years long, and they can usually estimate relative time to the nearest 10 million years. In contrast, the imprecision involved in sampling and analyzing Precambrian rocks for radioactive dating is such that even with the delicate and sensitive instruments now being used, old dates have to be given with large ranges of probable error, expressed as plus or minus as much as 100 million years. Such a range of possible uncertainty in age—200 million years from plus to minus—is equivalent to all the time of the Cambrian, the Ordovician, the Silurian, plus part of the Devonian periods of the Paleozoic. It is therefore not surprising that Precambrian stratigraphy still lacks in detail.

Figure 2-28 shows the kind of geologic map that has been worked out for Precambrian rocks in Canada. The provinces are drawn on the basis of radioactive ages. The oldest, the Superior province, is one in which most of the rocks have ages clustering about 2500 million years, corresponding to a period of mountain building (orogeny) called the Kenoran. The rocks of the Churchill

mentary and igneous and metamorphic rocks in small areas where it was possible to correlate from one outcrop to the next (just as we could get along without fossils in correlating from one wall of the Grand Canyon to the other), sheer guesswork was required in attempting to correlate one part of a continent with another. The advent of radioactive dating, however, changed that situation—so much that formerly well-established notions about the way the Precambrian fit together have been rethought.

There are two important differences between radioactive dating of the Precambrian and the stratigraphic dating of the Phanerozoic. First, the Precambrian events that can be dated are mainly episodes of igneous intrusion, metamorphism, or mountain-building; the Phanerozoic is dated by the ages of the sediments. Because of these differences in the two kinds of rock timekeepers, the Precambrian gives a much more discontinuous record, for the occurrence of intrusion, metamorphism, and mountain-building is spasmodic compared to the almost continuous record of sedimentation. Second, the resolution, or accuracy, of radioactive dating in the Precambrian, although it is steadily improving, is still lower than that of stratigraphic dating in well-known parts of the

Precambrian boundary of the Canadian shield

Province boundaries

Figure 2-28
Precambrian-age map of Canada, showing provinces drawn on the basis of radioactive age determinations. [After J. A. Lowdon et al., Geological Survey of Canada, 1963.]

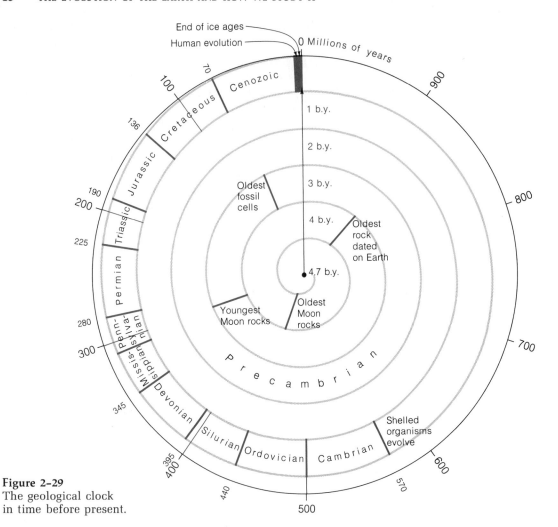

Figure 2-29
The geological clock
in time before present.

province, which partly surrounds the Superior, are about 1750 million years old, corresponding to the Hudsonian orogeny; and those of the Grenville province, are about 950 to 1000 million years old, corresponding to the Grenville orogeny. The same general kind of map can be made for other Precambrian areas of the world, such as those in Scandinavia or Africa. That such maps can be made on the basis of radioactive dating alone is a triumph of the method.*

We have not covered all of the ways of telling time from the rock record. Paleomagnetic stratigraphy is under steady development as a junior partner to the stratigraphic time scale (see Chapter 18). Geochemists have been using chemical reactions with known constant rates to date rocks that are too old for ^{14}C and too young for the other

radioactive methods. For example, many early human artifacts, such as arrowheads, are made of obsidian, a volcanic glass. Obsidian can be dated by the degree to which the glass, originally dry, has become hydrated, or chemically altered by absorbing water. Some rocks can also be dated by the microscopic tracks of decay particles ejected from the radioactive nucleus. The number of these fission tracks that are preserved in the rock can be translated into an age. Time is so central to the study of Earth that we continue to search for better clocks.

AN OVERVIEW OF
THE GEOLOGICAL CLOCK

We can now put together a clock for the whole history of the Earth. The clock is shown in the form of a spiral in Figure 2-29. Each revolution of the clock's hand corresponds to 1 billion years; each subdivision, the "hours," corresponds to 100

---

*The oldest rocks now known in the world are in Southwest Greenland, where an area of igneous and metamorphic rocks includes some as old as 3.8 billion years. These rocks are metamorphosed iron formations that look similar to those formed in much later times.

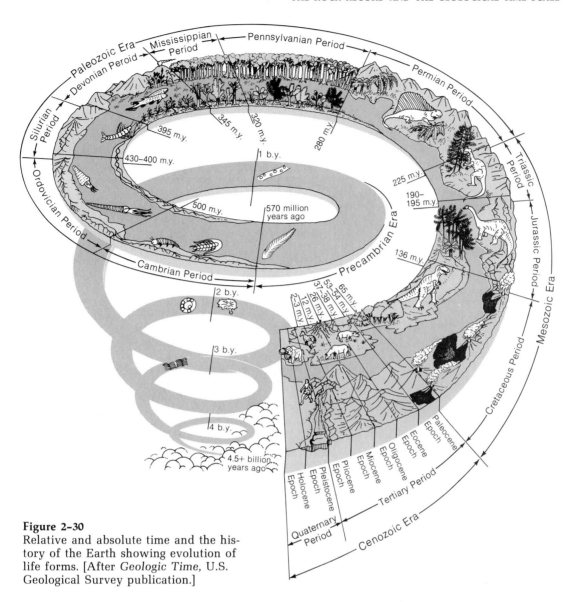

Figure 2–30
Relative and absolute time and the history of the Earth showing evolution of life forms. [After *Geologic Time,* U.S. Geological Survey publication.]

million years; and the "minutes" are 10 million years long. Geologists work backwards from the present as zero to the beginning at 4.7 billion years ago. From a look at this clock, we can see how short a period of the total of Earth history is taken up by the Phanerozoic eras, and what a minuscule amount of the time has elapsed since human beings evolved. History, as recorded by

artifacts, buildings, writings, and drawings, is only the last instant of all of this time (Fig. 2–30 and inside the back cover). One of the minor miracles of how people think is demonstrated by the way they can quickly shift mental gears and consider different time scales—from "Did the earth form 4.5 or 4.7 billion years ago?" to "Will this class hour never end?"

SUMMARY
1. Geologic time scales tend to be long; most are measured in thousands to millions of years. These slow rates of processes are measured by geologic change.

2. Rocks are made of minerals, which are chemical compounds, usually in crystalline form. Igneous rocks form by cooling of a magma. Sedimentary rocks show stratification

and form by settling of erosional debris and chemical precipitates. Metamorphic rocks form by the action of heat and pressure, which transforms preexisting rocks into new ones with different mineral compositions.

3. The field relations of rocks, such as stratigraphic sequences, concordant and discordant contacts with igneous rocks, and structural deformation, can be used to deduce a history of sedimentation, uplift, deformation, igneous activity, and erosion.

4. A stratigraphic time scale can be devised from sedimentary rocks by matching the evolutionary changes revealed by fossils in the strata with the stratigraphic sequence. The rocks of the Grand Canyon illustrate how field relations and fossils are used to infer a complex history of geological episodes.

5. Absolute time measurements can be made by using spontaneous decay of radioactive elements. Such dates of minerals containing radioactive substances provide an absolute age framework for the stratigraphic time scale and allow the dating of Precambrian rocks, which do not contain shelly fossils. Radioactive age dating of meteorites gives the time of Earth's origin.

EXERCISES

1. What are the time scales, and how might you measure the duration of the following man-made "geological" phenomena? (a) Digging a foundation for a small house; (b) excavating a highway road cut; (c) building a large dam on a major river; (d) building a pyramid in ancient Egypt. What does this tell you about the relationship of time scale to the geometric scale, or size, of the phenomenon?

2. Make a diagram similar to Figure 2–22 to show the following sequence of events: (a) deposition of sediments; (b) extrusion of thick lavas; (c) deformation of sediments and lavas; (d) erosion; (e) intrusion by igneous dikes and extrusive lavas; (f) erosion; and (g) deposition of sediments.

3. Many fine muds are laid down at a rate of 1 centimeter/1000 years; at this rate, how long would it take to accumulate a sequence 1 kilometer thick? If the sequence were interrupted every ten million years by a disconformity, during which there was no sedimentation for one million years, how long would it take to accumulate a kilometer?

4. Abraham Werner, an important figure in late eighteenth-century geology, believed that almost *all* kinds of rocks were deposited on the sea floor. What lines of evidence from the Grand Canyon or elsewhere could you use to show that this not so?

5. Which elements would be most appropriate for radioactive dating of formations of the following ages: (a) early Precambrian; (b) Mesozoic; (c) Miocene; (d) late Pleistocene.

6. What different kinds of geological event are dated by flakes of mica in (a) a granite, (b) a schist, and (c) a sandstone?

7. Over the past few hundred years, estimates of the age of the Earth have grown longer and longer. Might the present estimate of 4.7 billion years change too? If so, what geologic theories that pertain to geologic time might have to be re-evaluated?

BIBLIOGRAPHY

Berry, W. B. N., *Growth of a Prehistoric Time Scale.* San Francisco: W. H. Freeman and Company, 1968.

Deevey, E. S., Jr., "Radiocarbon Dating," *Scientific American,* February 1952. (Offprint 811.)

Dott, R. H., Jr., and R. L. Batten, *Evolution of the Earth,* 3rd ed., New York: McGraw-Hill, 1980.

Eicher, D. L. *Geologic Time,* 2nd ed. Englewood Cliffs, N.J.: Prentice-Hall, 1976.

Faul, H., "A History of Geologic Time," *American Scientist,* v. 66, pp. 159–165, 1977.

Hurley, P. M., *How Old Is the Earth?* New York: Doubleday and Company (Anchor Books), 1959.

Simpson, G. G., *The Life of the Past.* New Haven: Yale University Press, 1953.

3

Rocks and Minerals

Rocks and the minerals that make them up are the tangible record of geologic processes. The minerals of the Earth are understood in terms of their molecular architecture—the way their atoms are arranged in crystal structures. The kinds of atoms and their chemical bonding determine not only the crystal structures but the chemical and physical properties of minerals, all of which are used for their identification. Rocks are divided into the three major groups, igneous, metamorphic, and sedimentary, on the basis of origin. They are further subdivided within each group according to mineral composition and texture, which provide the data that allow us to interpret details of their origin.

Picking up pretty stones and showing or wearing them must go far back into human prehistory. The earliest records of practical use of stones, though, are of arrowheads and spear points made of flint (a sedimentary rock) or obsidian (volcanic glass), both of which are hard materials that break with sharp edges. From the practical use of individual stones as tools, weapons, and decorations, it was a big step to the wholesale mining or quarrying of rocks and minerals for building, for making clay for pottery, and then for the ores that contain metals. Today mining is done so expertly and intensively that geologists have come to concern themselves with the exhaustion of the world's valuable mineral resources; minerals of economic significance and their reserves are covered in Chapter 22.

THE MATERIALS OF EARTH

Because of the many uses of rocks and minerals, we have a practical curiosity about where they are found and how they were formed: we want to be able to find more. Yet there are other reasons, too, for rocks, as we have seen, are the only records of how the Earth evolved, and they are an important guide to how the Earth works today. For this reason, **mineralogy,** the study of minerals, and **petrology,** the study of rocks, are important subfields of geology. Finally, there is the intrinsic interest in the extraordinary range of the mineral kingdom, with its immense variety of color, form, and texture. Minerals and rocks, after all, give us the marble and alabaster of sculpture, the jade of Eastern carvings, and the pigments used by Rembrandt.

What Information Do We Want from Rocks?

If the nature of rocks is a clue to many of the things we want to know about the Earth, how do we go about interpreting it? We need a key, just as ancient historians needed the Rosetta stone to crack the "code" of Egyptian hieroglyphics before they could read that part of human history. First of all we want to find out just what the minerals are made up of and how the rock is put together from its constituent minerals. From its composition we should be able to say something about where the parent material came from and what it

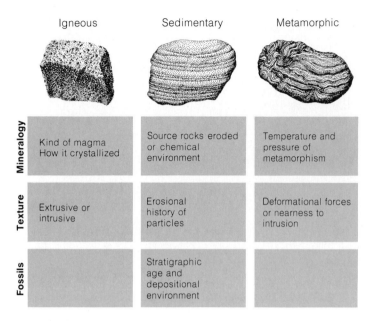

Figure 3-1
The mineralogy and texture of a rock are the keys to inferring its origin.

was like (Fig. 3-1). What was the magma like? Or, what were the source rocks of a sediment? Or, what were the preexisting rocks that were heated and compressed to make a metamorphic rock? From the composition and the texture of the rock we should also be able to tell something of the pressures and temperatures at which the rock was formed by comparing these properties with the artificial rocks and minerals made in the laboratory.

In this chapter we cover the nature of the rocks and minerals that make up the Earth's crust and mantle. We first explore the relation of rock to mineral. Next we show how the external appearance and properties of minerals are related to the way in which their fundamental building blocks—the atoms and ions of the chemical elements—are connected with each other in the internal architecture of crystals. With that picture in mind we can describe the mineralogy and textures of the three great classes of rocks.

Rocks Are Made of Minerals

A rock is many things. It is a collection of the particular chemical elements that make it up. Those elements are not found randomly mixed in a rock, but they are distributed among an assemblage of **minerals** (Fig. 3-2). A mineral is a solid chemical compound that is characterized by a definite composition or a restricted range of chemical compo-

sitions and by a specific, regular architecture of the atoms that make it up. Like all chemical compounds, minerals are homogeneous: a mineral cannot be separated mechanically into different substances. Minerals make up a rock just as bricks make up a brick wall, in a great variety of arrangements. In coarse-grained rocks the minerals are large enough to be seen with the naked eye. In some rocks the minerals can be seen to have **crystal faces,** smooth planes bounded by sharp edges; in others, such as a typical sandstone, the minerals are in the form of fragments without faces. In fine-grained rocks, the individual mineral grains

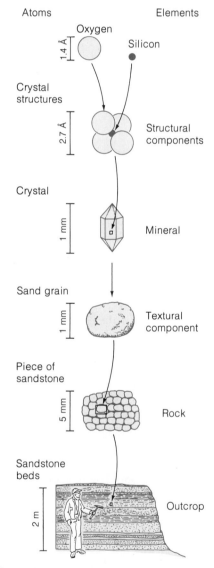

Figure 3-2
How atoms combine to form rocks. Atoms make up the small structural components that form minerals, which in turn combine to form rocks. Units of length are shown at left. Å stands for angstrom unit, $1 \text{ Å} = 10^{-8}$ cm.

are so small that they can be seen only with a powerful magnifying glass, the hand lens that the field geologist carries. Some are so small that a microscope is needed to make them out.

On the basis of certain characteristics, particularly physical and chemical properties, several thousand minerals can be distinguished, each defined by its unique set of properties. For thousands of years, people who have used minerals—whether miners looking for iron ore minerals or artists looking for minerals to grind into pigments—have used simple physical and chemical tests to distinguish one from another. Color is one obvious characteristic. Differences in hardness were found to make it easy to distinguish between minerals that look similar. How minerals break apart, some showing smooth cleavage planes and others rough irregular fractures, proved to be a reliable way to identify certain minerals. Simple chemical tests were found useful in the field, such as dropping acid on a mineral suspected of being calcite ($CaCO_3$) to see whether the mineral would fizz as it dissolved, releasing carbon dioxide bubbles.

Early in the study of minerals it was realized that all grains or crystals of a mineral, like quartz, have just about the same qualities regardless of the kind of rock in which they are found. Some minerals, particularly those that have a more complex mixture of atoms, vary slightly in their properties, depending on their precise composition. A mineral like garnet, for example, has a number of varieties. Each variety has its own range of composition, such as the proportions of iron and other elements, and hence, its own set of properties.

Rocks are not as uniquely defined by their properties as minerals are. Because of the immense number of ways in which the thousands of minerals can be combined, the geologist is faced with a bewildering array of rock types. The only way to make order out of this array is to classify like with like and to sort out by general type (Fig. 3-3). The major division of rocks into igneous, sedimentary, and metamorphic is just such an aid. Within each major division there are many groups and types. Using characteristic properties, we can divide the rock kingdom into several hundred general types, each with its own more-or-less distinctive earmarks.

Despite all of these numbers, a remarkable amount can be done by knowing even a small number of the most common minerals and rocks. In most parts of the world a field geologist can make an accurate geologic map by knowing only a few dozen major minerals and even fewer common rock types. This simplification is possible because most of the thousands of known minerals are either rare or unusual. In addition, many minerals can be lumped into groups. Thus the geologist who can recognize garnet will do well, even though a mineral sophisticate who can distinguish the many varieties of garnet by their slightly different chemical compositions might do better. Naturally, the more we can distinguish, the more the information gleaned, and the greater the power of our theories of explanation. That is why petrologists have to know a great deal about mineralogy.

Just how do we go about identifying minerals and explaining their origins from their characteristics? In the field we still use external form and other obvious physical properties. In modern laboratories, however, advanced instruments are used to learn the basic composition and atomic architecture of minerals. These are the underlying determinants of the other properties. From analysis of crystal structure and external form, we draw the best conclusions of origin. The next sections take us from the study of crystal faces to the explanation of physical properties in terms of the atomic arrangement.

CRYSTALS: FACES AND SYMMETRY

The regularity of crystal faces is the most striking feature of the external form of minerals, and for many years minerals were studied and identified mainly by analyzing their symmetry. In addition to his earlier contributions—enunciating the laws of stratigraphy and recognizing fossils—Steno wrote in 1669 that quartz crystals, wherever found, always show the same angle between similar crystal faces. By the late eighteenth century, his *constancy of interfacial angles* became accepted as a generality applicable to all minerals. By 1801 the major work of the great crystallographer René Haüy was accomplished, all in the midst of the great upheaval of the French Revolution. Haüy summarized the laws of *crystal symmetry*, the regularities of crystal faces. That symmetry, we now know, is a manifestation of the symmetry of the arrangement of the atoms that make up the crystal. As a consequence of the work of Haüy and others, the early part of the nineteenth century was a time of intense study of the relation between the external forms of minerals and their chemical composition, a development that paralleled the great geological exploration of the Earth and the growth of the geological time scale.

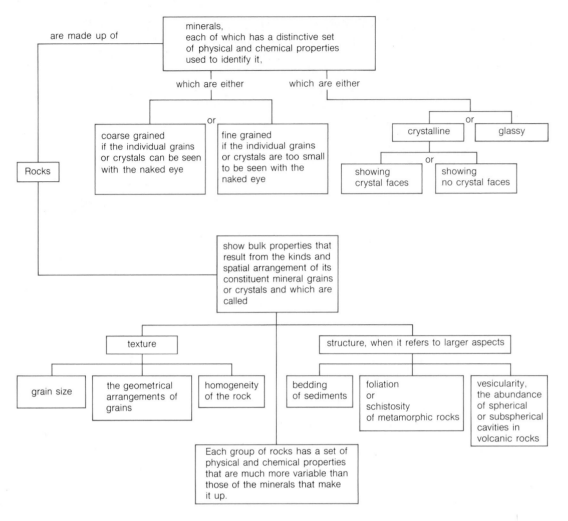

Figure 3–3
Flow chart of rock properties as determined by characteristics of the mineral constituents. The minerals ore classified both by their grain size and by crystal form. Rocks are grouped both by texture, the smaller-scale details of arrangement of mineral grains in the rock, and by structure, the larger-scale aspects of the mineral constituents.

How to Measure a Crystal

A crystal is a piece of matter whose boundaries are naturally formed plane surfaces. The geometry of a crystal may be relatively simple, as in the cubes of fluorite (Fig. 3–4) and of common salt, the mineral **halite;** or they may be beautifully complex, as in snowflakes, the crystals of ice (Fig. 3–5). Crystals are usually formed when a liquid solidifies or when a solution becomes **supersaturated**—that is, too enriched in dissolved material to hold it any longer—and the dissolved substances precipitate, or "drop," out of solution. When some liquids, such as molten silicates, congeal very quickly, the solid that forms is not crystalline but **glassy;** in this case crystals with plane faces do not form, but only masses with

Figure 3–4
Interlocking cubic crystals of fluorite. [Photo by Alfred A. Blaker.]

Figure 3-5
Photomicrographs of snowflake crystals. These were described as among
the choicest specimens of crystal architecture in a 25-year search by Wilson Bentley, pioneer photographer of snow crystals. [Photos by Wilson
Bentley, courtesy of Duncan Blanchard, State University of New York
at Albany.]

Figure 3-6
Obsidian, a volcanic glass. The curved,
sharply terminated fracture surfaces are
typical of *conchoidal fracture,* which
characterizes glasses and a number of
minerals. [Photo by R. Siever.]

Figure 3-7
Quartz exhibits a variety of crystalline forms ranging
from groups of crystals with well-developed crystal
faces, such as those on the left, to masses of crystals
intergrown in such a way that no crystal faces are
developed, such as in the broken pebble of vein
quartz on the right. [Photo by R. Siever.]

curved, irregular surfaces (Fig. 3-6). This absence
of crystallinity is typical of quickly cooled material in many lavas. The glassy texture is the result
of lack of regular, symmetric order in the arrangement of atoms.

The most useful measurements to be made on
crystals are those of the angles between faces (Fig.
3-7 and 3-8). From these angles the geometry of
all of the faces in relation to each other can be
constructed. Haüy and other mineralogists discovered that, in addition to constancy of interfacial angles, each kind of crystal exhibits other regularities: definite symmetrical relationships exist
among faces, and there are certain simple mathematical relations between the angles of all faces.
The marvel of finding such simplicity in the midst
of apparent complexity is what led Haüy and oth-

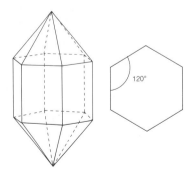

Figure 3-8
Drawing of a perfect quartz crystal.
A section at right angles to the long
axis shows a regular hexagon with
faces at 120° angles.

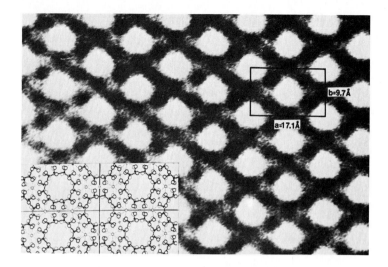

Figure 3-9
High-magnification electron micrograph of a section of the mineral cordierite, a magnesium-iron-aluminosilicate. This image was made perpendicular to one of the major axes of symmetry. The inset shows the idealized structure deduced from x-ray diffraction. [From Peter R. Buseck and Sumio Iijima, "High Resolution Electron Microscopy of Silicates," *American Mineralogist*, v. 59, 1974, Copyright © 1973.]

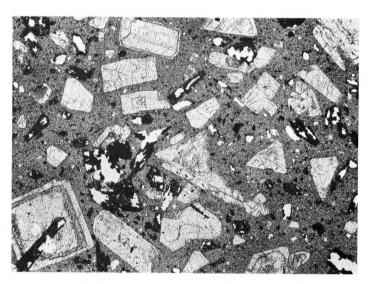

Figure 3-10
Microscopic view of a thin section (transparent slice) of diorite porphyry under polarizing light. The white crystals are feldspar; the large, dark, more-or-less rectangular ones are hornblende. The large crystals were all once floating in a melt, which is now represented by the fine-grained groundmass. Width of field 2.5 mm. [From *Geology of Soils* by Charles B. Hunt. W. H. Freeman and Company. Copyright © 1972.]

ers to infer that there must be an underlying order to the arrangement of the atoms in crystals—an idea that could not be verified until a century later when x rays were discovered and beamed through crystals.

Another element of order became evident after hundreds of different kinds of crystals had been measured. Mathematical analysis showed that there are only thirty-two different ways of arranging atoms about a point that will allow the building of a three-dimensional crystal that obeys symmetry rules. Measurement of angles between crystal faces and analysis of the symmetry of crystals led to a simple all-inclusive classification that consists of a limited number of major crystal systems and classes of those systems. The analysis of symmetry is still a part of modern mineralogy, though it is now studied by x-ray diffraction and electron microscopy rather than by the observation of angles between crystal faces (Fig. 3-9). Symmetry remains important because it is one of the basic clues to the pattern of regularity of the atoms in a crystal.

MINERALS AND THE MICROSCOPE

In 1858, at the end of the half century during which the modern laws of crystallography were worked out, the English geologist Henry Sorby published a memoir, *On the Microscopical Structure of Crystals*. It was Sorby's inspiration to use the knowledge that rocks become transparent if slices are ground thinly enough* and that the way crystals affect polarized light passing through them provides a means of identifying minerals and of studying the mineral compositions and textures of a wide variety of rocks. Using a polarizing microscope, Sorby could for the first time identify small grains or crystals that could be seen only with high-power magnification (Fig. 3-10).

From the various properties of minerals in polarized light, mineralogists have constructed detailed tables for the identification of crystals. Using these tables while studying thin sections of

---

*Thin sections are made by cutting off a slice of rock about 1 mm thick with a saw containing embedded diamonds. One side is then ground smooth with silicon carbide grinding powders and cemented to a glass slide. Next, the open side is ground down to the desired thinness, about 0.03 mm, the last gentle stages of polishing being monitored with a microscope. Some minerals are opaque no matter how thin they are ground. These are the metallic minerals, sometimes called the *opaque minerals*. An example is pyrite, FeS_2.

rock under the microscope, the geologist can tell just what kinds of minerals are put together in a particular pattern to make up the rock.

THE ATOMIC STRUCTURE OF MINERALS

Though many nineteenth-century mineralogists, following the lead of earlier scientists, had speculated about how the atoms that make up all matter might be arranged in crystals, there was at the time no way to confirm any relation between external form and internal structure. By the beginning of the twentieth century, many mineralogists were convinced that crystal form, chemical composition, and such physical properties as color and hardness, might be explained in terms of some hypothetical atomic pattern, but they still had no proof. That proof was provided in 1912 by the German physicist Max von Laue and two of his students when they irradiated a crystal of copper sulfate with x rays and produced an x-ray diffraction pattern on a film placed behind the crystal. Von Laue reasoned that if ordinary light could be **diffracted**—that is, deflected around corners to give a pattern of fringes when it is directed through very tiny openings such as pinholes or slits—so could x rays beamed through crystals. Von Laue guessed that if the mineralogists were right in their speculations about crystals being orderly arrangements of atoms, the pattern of x rays could reveal how the atoms are arranged in space. Only one year after that remarkable and crucial experiment, an English father-and-son team, William H. Bragg and William L. Bragg, published the first crystal-structure analysis of a mineral, halite. The Nobel Prize in physics was awarded to the Braggs for this work in 1915 when the younger Bragg, William L., was only twenty-five.

In the next dozen years a great many minerals were structurally mapped. This work paved the way for the Norwegian mineralogist and petrologist V. M. Goldschmidt to outline, in 1926, all of the general principles governing the ways that the atoms of different elements are assembled to form crystals. Finally, some explanation could be given for the long-known facts of the external appearance of crystals and their properties. The explanation lies in the structure of atoms, their systematic changes with increasing atomic number and atomic weight, their atomic size, and their ability to form various kinds of chemical bonds with other atoms.

The Gain and Loss of Electrons

As we noted in Chapter 2, an atom consists of a cloud of electrons around a relatively small nucleus of protons and neutrons. An atom may gain or lose one or more electrons, thus forming an **ion,** charged positively or negatively depending on the relative numbers of protons in the nucleus and electrons surrounding the nucleus. The number of protons remains the same, even though electrons may be lost or gained. A positively charged ion, a **cation,** has lost electrons; a negatively charged ion, an **anion,** has gained electrons.

Working out the rules that describe how electrons are tied to the nucleus was a major part of the business of atomic physics early in this century. Contemporary concepts of atomic structure picture the locations of electrons around the nucleus as orbitals whose shapes and sizes depend upon the energy level of the electron. The ease with which electrons are gained or lost is described in terms of the electronic structure of the elements, in which orbitals are represented as simple electron "shells"—spheres about the nucleus whose radii are proportional to their energy levels (see Fig. 2–24).

The elements can be classified on the basis of electronic structure (Table 3–1 and Fig. 3–11). One small set includes the elements whose outermost shell is populated by eight electrons. They are the *noble gases:* neon, argon, krypton, xenon, and radon. Because this is a stable configuration, these elements have little tendency either to gain or to lose electrons, and so do not form any important crystalline compounds. Helium, with only two electrons, is also stable and belongs with this group.

Another set of elements is made up of those whose outermost shells may gain or lose electrons. These elements include most of the common abundant ones, such as sodium, potassium, magnesium, calcium, and chlorine. These elements have a tendency either to gain or to lose electrons in order to assume the stable configuration of the inert gases. When a chlorine atom gains an electron, it becomes the chloride anion with eight electrons in the outer shell, thus achieving the same electronic structure as that of argon. Similarly, the element potassium has a strong tendency to lose an electron and form a cation with exactly the same structure as the chloride ion. The electrons that are gained and lost, called **valence electrons,** determine the chemical behavior of the elements.

Other elements have very weak tendencies to

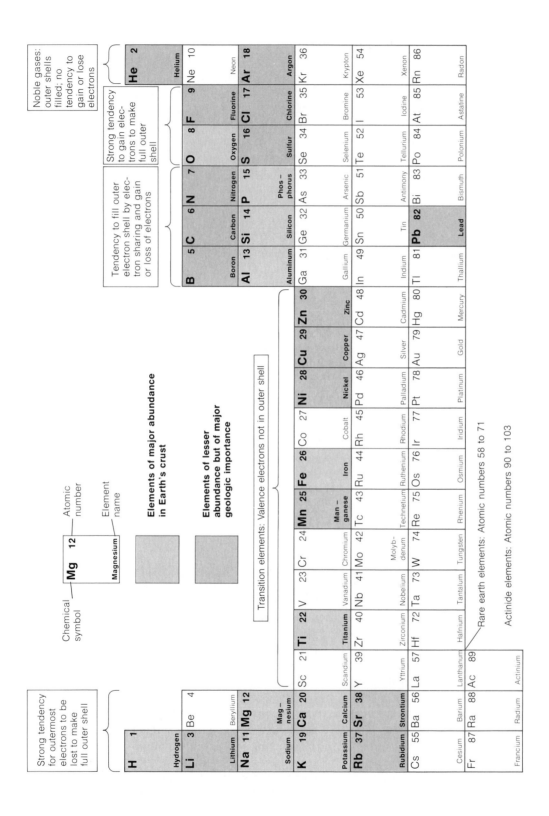

Figure 3-11
Periodic table of the elements (rare earth and actinide elements omitted). The elements of lesser abundance in the Earth's crust but of major geologic importance are those that are constituents of common rock-forming minerals or are of economic significance.

Table 3-1
Types of electronic structure of the elements (other than rare earths or actinides)

Type of element	Common elements	Characteristics
Noble gases	Helium, argon	Outer shells of elements have stable, filled configurations and no tendency to form chemical compounds.
Alkali metals Alkaline Earth metals	Sodium, potassium Calcium, magnesium	Outer shells have one or two valence electrons and strong tendency to lose them and form cations
Halogen group Oxygen group	Fluorine, chlorine Oxygen, sulfur	Outer shells need one or two electrons to attain stable configurations and have strong tendency to gain electrons and form anions
Boron group Carbon group Nitrogen group	Boron, aluminum Carbon, silicon Nitrogen, phosphorus	Outer shells tend to attain stable configurations by sharing electrons with other atoms and have weak tendency to gain or lose electrons
Transition elements	Iron, copper, zinc	Valence electrons tend to be gained or lost from an inner shell rather than the outer shell

gain or lose valence electrons but strong tendencies to share them with atoms of the same or different kind, thus achieving a stable configuration. In electron sharing, a shared electron cannot be considered to have been gained or lost. In a sense, both nuclei have "gained" the electron for whatever part of the time that it can be visualized as belonging to the one or the other nucleus. Carbon and silicon are two important elements in this group.

The transition elements are those whose valence electrons are in the next-to-outermost shells. These elements may form several different kinds of ions; iron, for example, forms two cations, Fe^{2+} and Fe^{3+} (see Chapter 4).

The importance of electronic structure, shells of valence electrons, and stability of the outer shell is that they all determine the nature of chemical bonding of one element to another in crystals. There are a number of bond types, and they are not mutually exclusive. A particular bond may be largely of one type but have some characteristics of another.

Ionic Bonds

The simplest form of chemical bond, in some ways, is the **ionic bond.** Bonds of this type are formed by electrostatic attraction* between ions

---

*Electrostatic attraction is described by *Coulomb's law:* the attraction (or repulsion) between two charged particles is directly proportional to their product and inversely proportional to the square of the distance between them.

$$E = k(q_1 \times q_2)/d^2$$

where E is the attractive force, the q's the number of charges on the ions or electrons, d the distance, and k a proportionality constant associated with the medium in which the attraction takes place. The value of k is 1 in a vacuum.

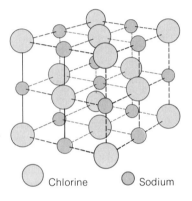

◯ Chlorine ⬤ Sodium

Figure 3-12
Cubic structure of sodium chloride.
The lines between ions are drawn only
to show the geometry of the cubic outlines.
Ions not drawn to scale.

of opposite charge. This attraction is of exactly the same nature as the attraction that makes hair stand up when some synthetic fabrics are drawn over it. Ionic bonds form most strongly between elements like sodium and chlorine, which have a strong tendency to gain or lose electrons and become cations or anions. In fact, the simplest of all ionically bonded crystalline substances is sodium chloride, whose crystal structure was the first to be worked out. From a look at Figure 3-12 one can see the symmetry of the ionic arrangement, that each ion of one kind is surrounded by six of the other. There is no "molecule" of NaCl as such. The six neighbors give rise to a new measure, the **coordination number.** Both sodium and chloride ions are in six-coordination. The six-sided figure defined by the lines joining the anions (chloride) that surround the cation (sodium) is an octahedron with the anions at the corners. In the same way, other simple geometric figures correspond to

Box 3-1 CRYSTAL SYMMETRY

One can see how a simple crystal is analyzed by taking a cube and noting, first, the obvious relation that all faces are always at 90° to each other and, second, that there are three mutually perpendicular imaginary planes such that each is a *plane of symmetry;* that is, each face on one side of the plane is mirrored on the other side. Or, one can hold opposite faces of the cube on an axis between thumb and forefinger and spin it around to find that there is a *fourfold axis of symmetry;* that is, in one complete rotation of 360° a face will be repeated four times. Another axis of rotation between opposite corners is a three-fold axis of symmetry. The essence of symmetry is this: a simple geometrical operation can be performed that will repeat a face in another position. To put it another way, if one performs an operation such as a rotation, a new face will occupy the same position that was occupied by another face before rotation—and one cannot distinguish the final appearance from the original one.

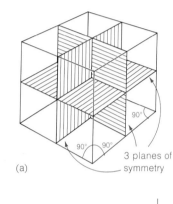

(a) 3 planes of symmetry

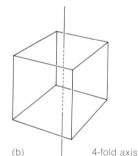

(b) 4-fold axis

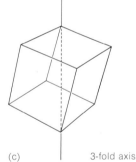

(c) 3-fold axis

Crystal systems

System	Minimum symmetry	Reference axes
Isometric	4 3-fold axes	3 mutually perpendicular; all of same length
Trigonal	1 3-fold axis	4 axes, 3 horizontal with 120° intersections, 1 vertical. Horizontal axes all of same length; vertical axis any length
Hexagonal	1 6-fold axis	Same as trigonal
Tetragonal	1 4-fold axis	3 mutually perpendicular; 2 of same length, 3rd of any length
Orthorhombic	3 2-fold axes or 3 symmetry planes	3 mutually perpendicular; each of any length
Monoclinic	1 2-fold axis and/or 1 symmetry plane	2 axes at oblique angle, 3rd axis perpendicular to plane of the other 2; each axis of any length
Triclinic	1-fold axis or center of symmetry	3 axes at oblique angles; each of any length

Note: In some classifications, hexagonal and trigonal are considered to be divisions of one crystal system, the hexagonal.

other coordination numbers: 3 for the anions at the corners of a triangle and 4 for the anions at the corners of a tetrahedron (Fig. 3–13). The coordination number can be estimated from the relative sizes of the ions. The **ionic radius** varies among the elements depending on atomic number and ionic charge of the ion. Comparisons between ions are then expressed as a **radius ratio,** the ratio of the cation radius to the anion radius. In general, the larger the radius ratio, the larger the coordination number. Though there are many additional complexities, mineralogists have been able to predict a large number of structures based on these criteria alone.

Cations are small, most of them less than 10^{-8} centimeters in radius. But most anions are large, as is the most common anion in Earth, oxygen. From this fact it is apparent that most of the space of a crystal is occupied by the anions and that the cations fit into spaces between them. Figure 3–14 shows the NaCl structure with ions of approximately the correct size.

Covalent Bonds

Compounds that achieve a stable electronic configuration by sharing electrons rather than gaining

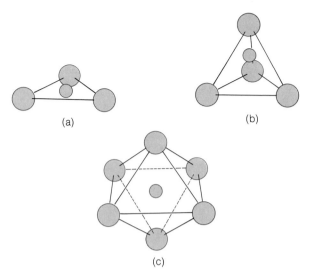

Figure 3-13
Coordination numbers correspond to regular geometric figures defined by lines joining the anions that surround a cation (small dark circle). (a) Coordination number 3 gives a plane triangle; (b) coordination number 4 gives a tetrahedron; (c) coordination number 6 gives an octahedron.

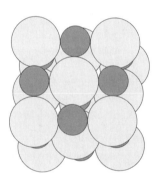

Figure 3-14
Cubic structure of sodium chloride, showing the ions in their correct relative sizes.

or losing them are held together by **covalent bonds.** The formation of such bonds depends on the number and distribution of shared electrons in the outer shells, and so the kinds of compounds and crystal structures formed are determined by more complex factors than the simple geometric ones determined by the ionic bonds in crystals. Elements that do not readily gain or lose electrons to form ions—for example, carbon—form bonds of this kind. The simplest covalent structure is that of diamond, in which every carbon atom (not an ion) is surrounded by four others (coordination number 4) arranged in a regular tetrahedron. When coordinated in this way, each carbon atom

shares an electron pair with each of its four neighbors and thus achieves a stable octet of electrons in its outer shell.

The van der Waals Bond

Much weaker bonds than ionic or covalent bonds exist between all ions and atoms in solids. They are named after the man who inferred their existence from weak attractive forces shown by atoms and molecules in gases. The **van der Waals bond** is a weak electrical attraction that is related to the asymmetry of certain atoms and ions. They play a role of moderate importance in some silicate minerals, where their presence is not masked by the more powerful ionic and covalent links.

The Structures of Some Common Minerals

Though the variety of known structure types in the universe is great, geologists usually encounter only a relatively small number, largely because most rocks are made up of silicate minerals, composed of the two most abundant elements in the Earth's crust, oxygen (O) and silicon (Si). Silicates are of a few main structure types. The basis for all silicate structures is the radius ratio of silicon to oxygen, about 0.30, which allows each silicon to be four-coordinated to four surrounding oxygens in a regular tetrahedron (Fig. 3-13). Silicate structures are made up of these tetrahedra arranged in different ways, with such cations as sodium (Na^+), potassium (K^+), calcium (Ca^{2+}), magnesium (Mg^{2+}), ferrous iron (Fe^{2+}) and ferric iron (Fe^{3+}) in the interstices (spaces) between. The bond between Si and O is about half ionic in character and half covalent, the Si sharing one of its outer electrons with each oxygen ion. Adjacent silicons may share oxygen ions, allowing networks of several kinds to be built up of tetrahedra. Many silicate minerals also contain aluminum, the third most abundant element in the Earth's crust. The radius ratio of aluminum to oxygen is 0.36, close enough to the silicon–oxygen ratio to allow aluminum to take the place of silicon ions in a tetrahedral structure. The ratio is large enough, however, to allow aluminum to be octahedrally coordinated too, like the ions in NaCl, and it may join adjacent tetrahedra by a largely ionic bond.

Cations that have similar coordination numbers and similar ionic radii tend to substitute for each other and make mixed compounds that we call *solid solutions,* which are analogous in every way to common liquid solutions. Natural olivines are

Table 3–2
Major silicate structures

Geometry of linkage of SiO$_4$ tetrahedra	Si/O ratio		Example mineral	Formula
Isolated tetrahedra: linked by bonds sharing oxygens only through cation	1:4		Olivine	$(Mg,Fe)_2SiO_4$
Rings of tetrahedra: joined by shared oxygens in 3-, 4- or 6-membered rings	1:3		Beryl	$BeAl_2(Si_6O_{18})$
Single chains: each tetrahedron linked to two others by shared oxygens. Chains bonded by cations	1:3		Pyroxene	$(Mg,Fe)SiO_3$
Double chains: 2 chains joined by shared oxygens as well as cations	4:11		Amphibole	$(Ca_2Mg_5)Si_8O_{22}(OH)_2$
Sheets: each tetrahedron linked to 3 others by shared oxygens. Sheets bonded by cations or alumina sheets	2:5		Kaolinite	$Al_2Si_2O_5(OH)_4$
Frameworks: each tetrahedron shares all its oxygens with other SiO$_4$ tetrahedra (in quartz) or AlO$_4$ tetrahedra	3:8		Feldspar (Albite)	$NaAlSi_3O_8$
	1:2		Quartz	SiO_2

solid solutions of variable amounts of iron and magnesium silicates. The pure magnesium olivine is Mg_2SiO_4, forsterite; the pure iron olivine is Fe_2SiO_4, fayalite. The composition of the natural solid-solution mineral is represented by the formula $(Mg, Fe)_2SiO_4$, which simply means that there are two magnesium or ferrous ions, in whatever combination, for every silicate group.

Silicates are classified and named according to the way the tetrahedra are linked, as shown in Table 3–2. Isolated tetrahedra are linked by mutual bonding to a cation between. Rings of tetrahedra are formed by bonding of two oxygens of each tetrahedron to adjacent tetrahedra in closed rings. Single chains form by the same linkage. In some minerals two single chains are combined to form double chains, in which the chains are linked by cations. Sheets are structures in which each tetrahedron shares three of its oxygens with adjacent tetrahedra to build planar lattices. In three-dimensional frameworks, the tetrahedra are linked by sharing oxygens with other tetrahedra. Aluminum may substitute for silicon in chains, sheets, and frameworks.

A different sort of building block is the carbonate ion. In this group of ions the carbon atom is surrounded by three oxygen atoms in a planar triangle. Groups of carbonate ions are arranged in sheets in a manner somewhat like that of the sheet silicates. The mineral calcite is made up of car-

Box 3-2 ELECTRONIC STRUCTURE OF THE ELEMENTS

The hydrogen atom is the smallest and simplest of all atoms, having just one proton in the nucleus and just one orbiting electron (atomic number Z-1). Electron orbits correspond to "shells" around the nucleus, for electrons do not follow a simple, single path as the planets do around the Sun but can be considered to occupy, at various instants, points on a sphere, or spheres, around the nucleus. As atomic number increases, so does the number of electrons and the number of shells. The table lists the shells and the maximum number of electrons each can hold. The single hydrogen electron occupies the **K-shell**, the innermost of the seven shells. The element of next higher atomic number is helium, which has two electrons, both in the K-shell (Z = 2). The K-shell can be occupied by only two electrons. The element of next higher atomic number, lithium, has three electrons (Z = 3), one of which occupies the next shell outward, the **L-shell.** This shell can hold eight electrons. The first two shells, K and L, fill up in order as the atomic number of the elements increases from hydrogen to neon. After that, from sodium (Z = 11) to argon (Z = 18), the next eight electrons start to fill up the **M-shell.** Then the regular order is interrupted and the **N-shell** starts to fill before the M-shell is completed. Then electrons are added to both M and N shells, until M is filled, whereupon with increasing atomic number, the **O-shell** starts filling and N continues being completed. This complication of shell filling helps to explain why the chemistry of the heavier elements is somewhat more varied and complex than that of the light elements, those with only two or three shells.

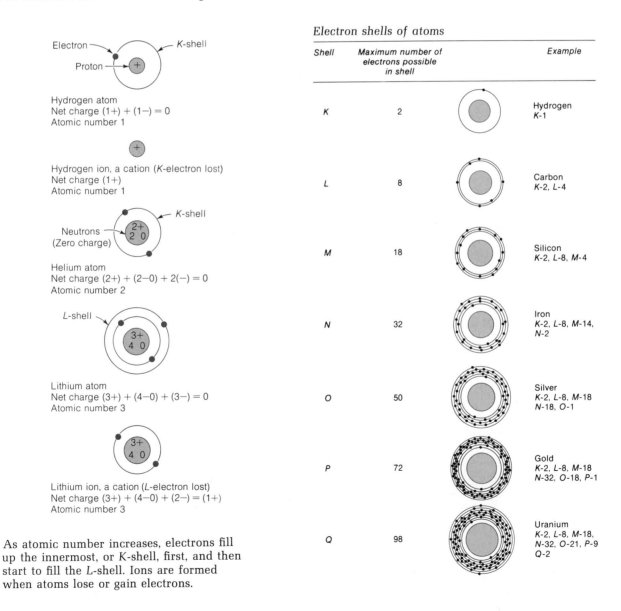

Electron shells of atoms

Shell	Maximum number of electrons possible in shell		Example
K	2		Hydrogen K-1
L	8		Carbon K-2, L-4
M	18		Silicon K-2, L-8, M-4
N	32		Iron K-2, L-8, M-14, N-2
O	50		Silver K-2, L-8, M-18 N-18, O-1
P	72		Gold K-2, L-8, M-18 N-32, O-18, P-1
Q	98		Uranium K-2, L-8, M-18, N-32, O-21, P-9 Q-2

Electron ——— K-shell
Proton ———

Hydrogen atom
Net charge (1+) + (1−) = 0
Atomic number 1

Hydrogen ion, a cation (K-electron lost)
Net charge (1+)
Atomic number 1

Neutrons ——— K-shell
(Zero charge)

Helium atom
Net charge (2+) + (2−0) + 2(−) = 0
Atomic number 2

L-shell

Lithium atom
Net charge (3+) + (4−0) + (3−) = 0
Atomic number 3

Lithium ion, a cation (L-electron lost)
Net charge (3+) + (4−0) + (2−) = (1+)
Atomic number 3

As atomic number increases, electrons fill up the innermost, or K-shell, first, and then start to fill the L-shell. Ions are formed when atoms lose or gain electrons.

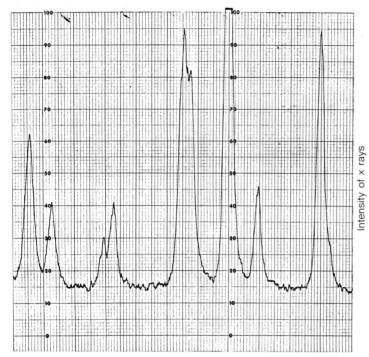

Intensity of x rays

Angle of beam to crystal

Figure 3–15
An x-ray diffraction pattern of a feldspar produced
by an x-ray beam from various planes of atoms in
the crystal as the crystal is rotated in the x-ray beam.
The peaks correspond to reflections from particular
planes, some more strongly reflecting than others.

bonate sheets and intervening planes of calcium
ions. The mineral dolomite is made up of the
same carbonate sheets separated by alternating
sheets of calcium and magnesium ions.

Glasses lack the regularity of structure that is
typical of the crystalline state. In silica glass, for
example, the atoms are arranged in silica tetrahedra, but instead of being linked in a regular repetitive manner, they are more or less randomly
arranged, as they would be in a liquid.

All natural crystal structures have a variety of
defects and imperfections. These are the result of
disturbances of the ordered arrangement of atoms
by dislocations, which are small "faults," by holes
where atoms are missing, or by places of misfit
around an impurity atom or point of disorder.
These defects, as well as the specific nature of
crystal faces, can be related to the conditions for
crystal growth and thus are clues to the environment in which the crystal grew. Since this is the
petrologist's goal in understanding how rocks
form, the study of the imperfections in crystal
structure is becoming almost as important as the
study of regularity and symmetry.

This brief tour through the world of crystal
structures is intended to give some idea of the
power of the principles of atomic physics, crystallography, and mineralogy, for it is from these disciplines that we have learned the extraordinary
details that enable scientists to explain how crystals are put together. Now that you have some
idea of what is known about structure, it is possible to show how x-ray diffraction was used to
learn it all.

X RAYS: THE DIAGNOSTIC TOOL OF MINERALOGY

Soon after von Laue discovered x-ray diffraction, W. L. Bragg found that a pattern of diffraction
corresponded to one that would be formed by reflection of the x-ray beam from planes within the
crystal, the reflection angles following certain
mathematical requirements. The many lines of
reflection that are found on x-ray patterns each
come from one of the planes (Fig. 3–15). The patterns of the lines can be used to find a probable
crystal structure. Today crystal structures are
analyzed by computer-controlled x-ray diffraction equipment. The results are fed into another
computer that spins out a complete structure
analysis, possibly even with crystal drawings.

X rays are used as a rapid means of identifying
minerals by their structures. They can be so used
because no two minerals have precisely the same
x-ray pattern, even though they may have the
same structure type. The precise pattern of sheet
silicates, such as the micas, muscovite, and biotite, differ. The spacings and the intensity of the
reflections depend on the kinds of atoms as well
as their arrangement, so each mineral is distinctive. Today the x-ray machine is used for simple
mineral identification almost as routinely as the
somewhat crude physical and chemical tests were
used in the last century. X-ray equipment, however, is not easily carried into the field, even
though some semiportable units have been invented, so it is still necessary to learn the simple
diagnostic features of the common minerals
(Appendix IV).

PHYSICAL AND CHEMICAL PROPERTIES OF MINERALS

A few years ago in Quincy, Massachusetts, a
scratched message was discovered on the window
of a house that John Hancock had owned in the
eighteenth century. Speculation was that Hancock had scratched his and a woman's initials on
the glass with his diamond ring. Whoever made

Table 3-3
Bond types and physical properties

Property	Ionic	Covalent	van der Waals
Structural types	Bond strength of any ion is uniform in all directions with high coordination numbers	Interatomic or ionic bonds are strong only in a few set directions, hence coordination numbers low	Bond strength of any ion is uniform in all directions with high coordination numbers
Hardness	Strong bonds give high hardness	Stronger bonds, in general, giving higher hardness than ionic bonds	Weak bonds give low hardness
Cleavage	Ease of cleavage, ranges poor to good; depends on distance between planes of atoms and individual bond strengths		Excellent cleavage
Melting points	Moderate to high; ions are present in the melt	Very high; molecules are present in melt	Low

Source: After R. C. Evans, *An Introduction to Crystal Chemistry*, 2nd ed., Cambridge University Press, Oxford, 1966.

the inscription perhaps knew that a diamond is the hardest mineral known and can easily scratch glass. What makes diamond so hard was then unknown. People also knew that mica, the sheety mineral commonly used for oven and lantern windows at that time, cleaves easily into paper-thin sheets as transparent as window glass, but they did not know why. They were aware of many distinctive and useful properties of minerals. The earliest and still the simplest ways of identifying many common minerals are based upon their hardness and cleavage and upon other physical properties.

Physical Properties, Bond Types, and Structures

We now know that the way atoms and ions are bound has a direct effect on their physical properties, as shown in Table 3-3. Few bonds are exclusively ionic or covalent; instead, most are hybrids. For this reason, the correlation with properties is only a weak generalization. Particular properties of a specific mineral can be explained on the basis of the kinds of atoms it is composed of and the bond types between atoms.

A good example is the mineral talc, of which talcum powder is made. Talc is composed of silicate layers weakly bonded to each other by van der Waals forces. Because they are so weakly bonded, the layers slip and break apart easily. It is that easy slippage of one layer over the other that gives talc its soapy feel and that makes it grindable to a fine, smooth, cosmetic powder. Beryl, whose clear, deep-green variety is emerald, is also a sheet silicate, but the bonds between the sheets in beryl are strong, predominantly ionic. As a result, the mineral is hard and breaks with difficulty in any direction.

Table 3-4
Mohs scale of hardness

Mineral	Scale number	Common objects
Talc	1	
Gypsum	2	Fingernail
Calcite	3	Copper wire or coin
Fluorite	4	
Apatite	5	Pocket knife
Orthoclase	6	Window glass
Quartz	7	Steel file
Topaz	8	
Corundum	9	
Diamond	10	

*Mohs arranged the scale so that the hardness difference between any two adjacent minerals is about the same, except for the difference between corundum and diamond, which is much greater. Mohs could not find a mineral of intermediate hardness, nor has anyone else.

Hardness

Just as a diamond scratches glass, so will a quartz crystal scratch a feldspar crystal because one is harder than the other. In 1822 Friedrich Mohs, an Austrian mineralogist, devised a scale of hardness based on the ability of one mineral to scratch another, that spanned the spectrum from the softest to the hardest mineral known (Table 3-4). Even with all of the elegant instruments now available to determine hardness by indentation or scratching, the **Mohs scale of hardness** remains the best practical way to identify an unknown mineral. A combination of common objects and a few of the minerals on the hardness scale are all one needs to bracket an unknown mineral between two points on the scale.

The hardness of any mineral depends on the strength of the bonds between ions or atoms; the stronger the bonds, the harder the mineral. Be-

Box 3-3 PROPERTIES OF ROCK-FORMING SILICATES

The major groups of silicates illustrate the influence of crystal structure and chemical composition on the properties of these important rock-forming minerals (See Table 3-3 and Appendix IV). Garnets and olivines are the two chief groups of silicates built of isolated tetrahedra linked by cations. Both groups are strongly bonded in all directions and moderately closely packed, so that they are hard and have no cleavage. The density of these minerals, relatively high, depends on both packing and chemical composition; in particular it depends upon the amount of iron present. Colors are also influenced by iron content. Olivines melt at very high temperatures; hence it is not surprising that they are found in high-temperature igneous rocks rich in iron and magnesium. Garnets are chemically complex; their composition reflects the bulk composition of the metamorphic rocks in which they are usually found.

Pyroxenes (single chains of tetrahedra) and amphiboles (double chains) are much more chemically varied than the olivines or garnets. Both structures are packed moderately closely, and the presence of iron makes both these groups fairly dense and moderately hard. But the cleavage of both is good parallel to the silicate chains, for across these surfaces bonds are weaker than the strong silicon-oxygen bonds of the tetrahedral chains. The two differ in cleavage angle because of the varying angles made by single and double chains. Pyroxenes have high melting points and are most likely to be found in igneous and high-temperature metamorphic rocks rich in iron and magnesium. The stability of amphiboles over large ranges of temperature and pres-sure is related to their chemical composition and its effect on the compactness of the structure and to the bond strengths between cations of different kinds.

Micas (sheet silicates) show the most striking dependence of physical properties on structure: their perfect cleavage is due to the weak bonds between sheets, where the weakly bonded cations, such as potassium, are held. The minerals are relatively low in density and light in color, except for biotite, the iron species. The micas are found in many different igneous, metamorphic, and sedimentary rocks, a fact related to the large range of temperatures and pressures over which they can be formed.

Feldspars, the most abundant of the silicates, have relatively simple compositions; they are made up of silicon-oxygen and aluminum-oxygen tetrahedra linked in three-dimensional frameworks, with the tetrahedra bonded to potassium (orthoclase, microcline, sanidine) and /or to sodium and calcium (plagioclase feldspars). Plagioclases range from albite, the pure sodium form, to anorthite, the pure calcium form. The strongly bonded silicon-oxygen and aluminum-oxygen framework gives these minerals their hardness and high melting temperatures. The cleavages are along planes of relatively weaker bonds between cations. There is sufficient variation in the way the three elements can combine to produce a wide range of melting temperatures and pressures; thus feldspars are characteristic of many kinds igneous, metamorphic, and sedimentary rocks, from the highest-temperature volcanics to the lowest-temperature sediments.

cause bond strengths may differ along the various crystallographic axes, hardness may also vary slightly in direction. Among the silicates hardness varies from 1 in talc to 8 in topaz. The strongest bonds are those of tetrahedrally linked Si and O. In quartz all bonds are of that type. In other silicates the Si and O atoms are also bonded to other elements, and it is those weaker bonds that determine the lower hardness. Metallic minerals may be difficult to scratch because they deform plastically; that character gives metals their malleability or ductility, properties that allow metals to be rolled into thin sheets or drawn into wires. In more brittle materials, hardness is small-scale breakage. The more ionic the bonding, the greater the brittleness, and most minerals tend to be somewhat brittle.

Cleavage

The term used for breakage along definite planar surfaces, typified by that of mica, is **cleavage.** The number of planes of cleavage varies among minerals. The cleavage planes of a mineral have some of the same characteristics as its crystal faces, though they are not to be confused with them. Cleavage planes always occupy definite and constant angles—as do interfacial angles of crystal faces—with respect to the symmetry or crystallographic axes of the crystal and are parallel to a possible crystal face.

Cleavage is the expression of differing bond strengths along the various planes of a crystal. The bonds across some of those planes may be very weak, like the van der Waals bonds that are

(a)

(b)

Figure 3–16
Calcite crystals. (a) A cleavage rhombohedron whose surfaces were made by breaking the crystalline material along zones of weak bonding strength in the crystal. (b) Faces formed by growth of the crystal from solution. [From *Dana's Manual of Mineralogy* (16th ed.) by C. S. Hurlbut, Jr., copyright © 1952, John Wiley & Sons.]

responsible for the easy breakage across the cleavage planes of mica. Calcite has three excellent cleavages parallel to the faces of a rhombohedron, the result of weak bonds (Fig. 3–16). Quartz is strongly bonded in all directions and has no good cleavage. Cleavage can be scaled on the basis of ease of cleaving with a chisel, from the most *perfect cleavages* of the micas to the *fair cleavage* of beryl (emerald). The most expert cleaving in the world is done in Amsterdam, where the center of the diamond "cutting" industry is; it is the excellent cleavage of diamond along many planes that makes possible the shaping of the hardest mineral known, for there is nothing harder with which to cut it!

Distinctive patterns of cleavage are identifying hallmarks of a number of common rock-forming minerals. Calcite and dolomite are easily recognized by their rhombohedral cleavage faces. Galena, lead sulfide, cleaves perfectly into cubes. Two important groups of silicates that frequently look alike otherwise, pyroxenes and amphiboles, can be distinguished on the basis of their cleavage patterns. Pyroxenes, single chains of silica tetrahedra, are bonded so that two good cleavage directions are almost at right angles (93°), whereas amphiboles, double chains, bond to give two cleavage directions at 56°.

Some minerals reveal the character of their bonding by bending rather than breaking. Bonds that are mainly covalent are stronger in one direction than another. In mica, the bonds along the sheets of the structure maintain themselves while the weaker bonds between sheets allow bending. Micas are elastic; they spring back after bending. Other sheet silicates, like chlorite, are flexible; once bent, they remain that way.

Fracture

The way in which minerals break other than along cleavage planes also serves to group them and help in identification. **Fracture** may be *conchoidal,* showing smooth, curved surfaces like those of a thick piece of broken glass (see Fig. 3–16); *fibrous* or *splintery, hackly* (from hackles, the stiff hairs or spines of some animals), and *uneven* or *irregular,* all of which are descriptive, though imprecise, terms. Geometric properties of such variability and irregularity as fracture still defy the attempts of scientists to devise simple, quantitative measures to make the description objective and uniform. The type and irregularity of fracture bears a complex relation to the breaking of bonds in directions that cut across crystallographic planes. In minerals whose bond strengths are about the same in all directions, fracture patterns depend on the abundance, kind, and distribution of submicroscopic fractures and crystal defects.

Streak

The name **streak** is given to the color of the fine powder that is produced when a mineral is scraped across a tile of unglazed porcelain, called a streak plate. The color is diagnostic for many minerals; for example, the iron oxide, hematite, always gives a reddish-brown streak regardless of the color of the particular mineral aggregate being streaked, which may be black, red, or brown.

Table 3-5
Mineral luster

Metallic: strong reflections produced by opaque substances.

Vitreous: bright, as in glass.

Resinous: characteristic of resins, such as amber.

Greasy: surface has the appearance of being coated with an oily substance.

Pearly: the whitish iridescence of materials like pearl.

Silky: the sheen of fibrous materials like silk.

Adamantine: the brilliant luster of diamond and similar minerals.

Luster

How the surface of a mineral reflects light gives it a characteristic **luster.** Mineral lusters are described by the terms in Table 3–5. The quality of light reflected from mineral surfaces is controlled by the index of refraction. That property, in turn, is related not only to the kinds of atoms present but to their bonding. For example, covalently bonded minerals tend toward adamantine luster, whereas ionically bonded minerals are more vitreous. Other lusters are related to irregularities of the surface.

Color

In July 1976 we found out from the Viking landing on Mars that the "red planet" really is red. The surface materials have a strong reddish hue that probably comes from hydrated ferric oxides. The same materials, thrown into the Martian atmosphere by strong winds, give the Martian sky a pinkish color. Many minerals show a characteristic color on freshly broken surfaces; others show characteristic colors on weathered or altered surfaces. We use not only the colors reflected from mineral surfaces but also the colors transmitted through minerals in microscopic thin sections. Color seen in thin sections is a good, quick guide to the identification of certain minerals. Some minerals—for example, precious opals—show a stunning play of colors on reflecting surfaces. Others change color slightly with a change in the angle of the light shining on the surface. Color may be a property of the pure substance or it may be the result of impurities. The color of pure substances is dependent on the presence of certain ions, such as iron or chromium, which strongly absorb certain colors of light. Most ionically bonded minerals with stable configurations of outer electron shells are colorless. Color is more characteristic of such minerals as olivine, which have bonds to transition elements (in this case iron) whose valence electrons are in an inner shell. Impurities, often too small to be seen except by the most powerful microscope, such as small dispersed flakes of hematite in a quartz crystal, impart a general color to an otherwise colorless mineral.

Specific Gravity and Density

Although the obvious difference in weight of a piece of hematite iron ore and a piece of sulfur of the same size is easily felt by hefting the pieces, a great many common rock-forming minerals have about the same range of **density** (= mass/ volume). Consequently, methods were needed that would make it easy to measure that property of minerals accurately. The standard measure of density is **specific gravity,** which is weight of the mineral in air divided by the weight of an equal volume of water at 4°C. The specific gravity is usually determined by measuring the weight of the mineral in air and in water. The difference between the two weights is equivalent to the weight of the equal volume of water. Because density not only is useful in identifying minerals but also depends strongly on the internal structure and composition of the substance, it has been measured accurately for most minerals.

Density is dependent on the atomic weight of the constituents and the tightness of packing of the atoms in the crystal structure. The iron minerals hematite and magnetite have high densities because of the high atomic weight of iron. Metals tend to have close-packed structures and therefore high densities. Covalently bonded structures tend to be more open and to have lower densities.

Density is also affected by pressure, and pressure-induced increases in density affect how rocks and minerals transmit light, heat, and elastic waves, such as those associated with earthquakes (see Chapter 17). Some chemical substances form more than one kind of crystal; the different kinds of crystals are called **polymorphs.** Calcium carbonate, for example, forms both calcite (density 2.71), and aragonite (density 2.93). The denser aragonite is known from experimentation to form and remain stable at high pressures, whereas calcite forms and is stable at low pressures.

Chemical Properties

The chemical compositions of minerals are the basis for the main classification of the mineral

Table 3-6
Chemical classes of minerals*

Class	Defining anions	Example
Native elements	None: no charged ions	*Copper,* Cu
Sulfides and similar compounds	Sulfide: S^{2-} and similar anions	*Pyrite,* FeS_2
Oxides and Hydroxides	O^{2-} OH^-	*Hematite,* Fe_2O_3 *Brucite,* $Mg(OH)_2$
Halides	Cl^-, F^-, Br^-, I^-	*Halite,* NaCl
Carbonates and similar compounds	CO_3^{2-}	*Calcite,* $CaCO_3$
Sulfates and similar compounds	SO_4^{2-} and similar anions	*Barite,* $BaSO_4$
Phosphates and similar compounds	PO_4^{3-} and similar anions	*Apatite,* $Ca_5F(PO_4)_3$
Silicates (see Table 3-2 for details of silicates)	SiO_4^{4-}	*Pyroxene,* $MgSiO_3$

* This classification, derived originally by Berzelius in the nineteenth century and used extensively by Dana, is a simplified form of the scheme used by Berry and Mason in *Elements of Mineralogy,* W. H. Freeman and Company, 1968.

kingdom. The chemical criterion used to classify the minerals is the anion of the mineral.* For example, halite, NaCl, is classed as a chloride, as is its close relative, sylvite, KCl. In this way, all minerals have been grouped into eight classes, as shown in Table 3-6.

Most of what we know about the chemical composition of minerals was gained through the use of ordinary wet chemical methods, by which materials are dissolved, separated into their constituent elements, and their weights or volumes then measured. Various other methods have been added over the past few decades as chemical instruments became more capable of measuring small quantities of elements in new ways. In the last twenty years the electron probe, a device that beams electrons at a sample mineral and analyzes the x rays generated, has been used to get good chemical analyses of very small crystals. Most recently it has been joined by the ion probe, which beams energetic ions at a mineral target and analyzes the resulting radiation from the constituent elements.

Mineralogists have also relied for many years on quick, simple chemical tests that can be made in the field. One such is the "acid test," in which dilute hydrochloric acid (HCl) is dropped on a mineral to see if it fizzes. If it does, it is likely to be calcite, a carbonate mineral.

All natural minerals contain impurities, enough of them so that it is frequently hard to decide whether an element is an integral part of the mineral or merely an extraneous contaminant. Elements that make up much less than 0.1 percent of the mineral are reported in analyses by the word "trace," and many elements are called **trace elements.** But most trace elements we rarely hear about, for they are neither abundant in the Earth (or the solar system) nor do they have any important industrial use. Typical of these are lanthanum (La), one of the rare earth elements, and scandium (Sc), an element that has affinities with aluminum and boron. Trace elements are analyzed by means of a number of instruments, most commonly by an emission spectrograph—an instrument that vaporizes the mineral with an extremely hot electric arc and then analyzes the light emitted by the burning substance. Also important is x-ray fluorescence, in which irradiation by x rays excites secondary radiation characteristic of an element. In another analytical method, neutron activation, an atomic reactor is used to bombard the mineral with fast-moving neutrons. Some of the originally nonradioactive elements in the mineral are changed into new, radioactive elements. The abundance of the new elements can then be analyzed by their radioactivity. One different atom among a billion others can be detected in this way.

Another part of chemical analysis is the determination of the relative proportions of different isotopes, the elements with the same atomic number (protons) but different atomic weights. In Chapter 2, we discussed the use of radioactive isotopes in dating rocks, but there are other isotopes

---

*The names "anion" for negative ion and "cation" for positive ion date back to early experiments with electrochemical cells, the ancestors of modern automobile batteries. Because the cathode, the negatively charged electrode, attracts positive ions, those ions became called *cations.* Similarly, the *anode,* the positively charged electrode, attracts negative ions, hence *anion.*

of some elements that are stable—that is, nonradioactive. Three of the most important are the heavy stable isotope of carbon, ^{13}C, the heavy stable isotope of oxygen, ^{18}O, and the heavy stable isotope of sulfur, ^{34}S. The study of these isotopes in minerals, in the atmosphere, and in natural waters has contributed greatly to geology. For example, the ratio of ^{18}O to ^{16}O has been used to determine ancient temperatures. The $^{18}O/^{16}O$ ratio in calcium carbonate ($CaCO_3$) depends on the temperature of the water from which the carbonate precipitated. Thus a fossil oyster shell can be analyzed using a mass spectrometer to learn the temperature of the sea in which the animal lived.

ROCKS AS MINERAL AGGREGATES

The architecture of a rock is not unlike that of a building—a design of building blocks assembled in certain ways. The building blocks of a rock, of course, are minerals. Without too much difficulty, we might be able to think of an enormous number of ways of putting minerals together, but most of the arrangements would bear little relation to anything we find in nature. The problem, then, is to find out the designs of nature and to learn why there are certain ways in which minerals are put together to form rocks and why there are many, many other ways in which they are not. Much of this book is devoted to questions such as these.

Rocks are first subdivided on the basis of origin. The three major categories of rock are igneous, metamorphic, and sedimentary. Origin is determined by a combination of rock characteristics, such as bedding, the layering that indicates a sedimentary origin; foliation, the preferred orientation of crystals that indicates a metamorphic origin; and a variety of other textures and structures. Although there are some differences in mineral composition among the three rock types, it is more useful to discuss mineralogical and textural attributes for each major group separately. Once the igneous, metamorphic, and sedimentary rocks have been described, it will be easier to see how the three groups can be distinguished from each other.

The Igneous Rocks

The igneous rocks can be subdivided on the basis of the minerals of which they are composed, the chemistry of the rocks serving as a clue to the composition of the magmas from which they so-

lidified. One of the first criteria used when the igneous rocks were first studied in the last century was the amount of silica, SiO_2, in the chemical analysis. During that premodern period of chemistry, the silica was thought to be derived from silicic acid, and so the more silica in the rock, the more "acidic" the magma was said to be. **Granite,** rich in silica, is the most abundant acidic rock. The rocks lower in silica were called basic. **Gabbro,** poor in silica, is the "basic" counterpart of granite. We now know that silica content is not a measure of acidity as that word is used in chemistry, but the terms persist even though what we really mean is "more or less **silicic.**" The amount of silica is not necessarily related to the amount of quartz for much of the silica may be combined in other silicate minerals. In the classification by silica content, the coarse-grained igneous rocks range in sequence from granite on the more silicic side, through **granodiorite** and **diorite,** to gabbro on the less silicic side.

The modern system of classifying the major groups, based on their chemistry and mineralogic composition, turns out to be much the same as the classification by silica content. The two terms most commonly used today, come from a broad division into light and dark minerals—and rocks—called, respectively, **felsic** and **mafic.** These terms were used because the dominant minerals of the light group are quartz and feldspars, both rich in silica (hence "felsic," from *fel*(s), feldspar, plus *ic*) and those of the dark group are pyroxenes, amphiboles, and olivines, all of which are rich in magnesium and iron (hence "mafic," from *ma*gnesium and *f*errous for iron, plus *ic*). The terms felsic and mafic are also used for rocks. Figure 3–17 shows graphically the division of igneous rocks according to their mineral content and their grain size. The varieties of feldspar are most important in the classification of the igneous rocks, both because they are abundant and because the proportions of different kinds of feldspar vary systematically from felsic to mafic rocks. Granite, at the left in Figure 3–17, is rich in potassium feldspar (mainly the mineral orthoclase), whereas the more mafic rocks, on the right, are dominated by sodium and calcium feldspars, the plagioclases. The less dark rocks are dominated by biotite mica and amphibole, and darker, more mafic rocks by pyroxene and olivine. Pyroxene and olivine are the major minerals of the **ultramafic** rocks, which are even lower in silica than the basalts and gabbros; **peridotite** is mainly olivine and pyroxene and **dunite,** olivine.

The other major basis for classification of the igneous rocks is, as already described in Chapter

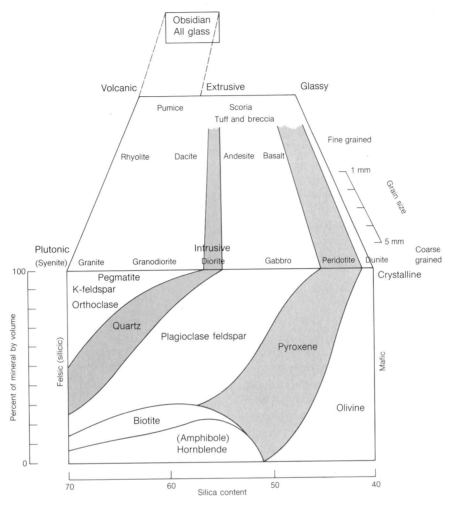

Figure 3-17
Igneous rock classification. On the front face of the cube the mineral composition
is plotted as percent by volume of a given mineral for a rock of given silica content
(horizontal axis). Thus a grandiorite of about 60% silica content (as determined by
chemical analysis and plotted at 60% on the horizontal scale) would contain about
15% hornblende, 12% biotite, 50% plagioclase feldspar, 18% quartz, and 5% K-feld-
spar. On the top of the block, textures are shown: front to back is grain size, corresponding to plutonic-intrusive-crystalline at front and to volcanic-extrusive-glassy
at back. Thus grandiorite is coarse grained, and dacite is the fine-grained extrusive
equivalent.

2, textural. We differentiate between the coarse-
grained granular rocks, or **phanerites** (from the
Greek, *phanero-*, meaning visible), and the fine-
grained **aphanites** (from the Greek *aphan-*, meaning invisible). A textural property that is in some
ways a crystallographic property is the degree of
glassiness, more commonly described by the opposite quality, the **crystallinity.** Both grain size
and crystallinity are products of the speed with
which a magma cools. The more rapid the cooling, the finer the grain size and the poorer the
crystallinity. The slowest-cooling magmas congeal in the last stages of intrusion, when magmas
become less viscous and contain more dissolved

gases. These give rise to **pegmatites,** very coarse-
grained rocks that may include crystals as much
as several meters across. The fastest-cooling magmatic materials, of course, are those that are
thrown high into the sky, where they instantly
freeze to glass. In Figure 3-17, the top surface of
the cube adds the dimension of texture to the
compositional classification.

The granular or coarse-grained igneous rocks
are subdivided on the basis of the abundance of
their characteristic minerals, which are easily
identifiable in coarse crystalline form in the field.
The fine-grained igneous rocks, the volcanics, are
more difficult to subdivide in the field because

their minerals and glasses are not so easily identified. Their textures, however, are a good basis for grouping. Some volcanic rocks (covered in more detail in Chapter 15) may show the texture of **pyroclastic** rocks. These result from volcanic explosions that expel magma that quickly cools to glass as in the May 1980 explosion of Mt. St. Helens. The finest fragments make **volcanic ash** and **dust.** Many larger particles are also formed. Mixtures of these components are agglomerations of pieces of glass, crystals that had started to form before the explosion, and fragments of previously cooled lava. The glass may be in the form of fragments of **pumice,** a frothy mass of glass with a great number of bubbles formed by gas escaping from the melt. The sharp, spiky glass fragments are called **shards.** The bubble holes in pumice and other extrusive rocks are **vesicles.** The solidified rocks that harden from ash falls and ash flows and all of the other varieties of ejected material are lumped under the term **tuff.**

The other major textural class of volcanics includes the lavas that flow from volcanoes or large fissures in the crust. The lavas have been given Hawaiian names: **pahoehoe** for the smoother, ropy kind, and **aa** for the sharp, spiky, more jagged kind (see Chapter 15). The differences in their appearance and structure are related to the lava's viscosity, which in turn is determined by the magma's composition. These rocks may be broken and recemented by fresh lava flows, forming a **breccia.**

The lavas and pyroclastics are named for their mineral and chemical compositions in the same way as the coarse-grained series, but fewer subdivision names are in common use. **Basalts,** the most abundant of volcanic types, are chemically the equivalents of gabbros but of finer texture, and **andesites** correspond to diabases and other rocks intermediate between granite and gabbro (Fig. 3–17). On the felsic side, **rhyolite** is the aphanitic counterpart of granite, and **dacite** the aphanitic counterpart of granodiorite. The chemical elements are distributed differently between glass and minerals in an aphanite than among minerals in the corresponding granular rocks. Thus the mineral compositions of the corresponding glassy and granular rocks are not exactly the same.

There are no minerals in the entirely glassy rock **obsidian,** which is an equivalent of granites and granodiorites. The glassy, sharp, broken edges of this highly silicic glass made it perfect for Indian arrowheads. The pumices, frothy glasses, are commonly silicic also.

Many igneous rocks are made up of a mixture of both large and small crystals. If some crystals are distinctively larger than the surrounding mass, or **matrix,** they are called **phenocrysts.** A rock with many phenocrysts is called a **porphyry.** The matrix of a porphyry may consist of coarse or fine crystals or, in volcanics, glass.

Putting together mineralogy and texture gives us a scheme of classification like the one shown in Figure 3–17. The mineralogical parameter measures the proportions and kinds of felsic and mafic minerals; and the textural parameter, grain size. As is true of most classifications, the rocks do not all fall neatly into pigeonholes, for composition and texture vary continuously, and we draw arbitrary dividing lines between them to preserve as nearly as possible the traditional meanings of names that have accumulated over two centuries of geological usage.

The Sedimentary Rocks

Mineralogy and texture are also useful in subdividing the sedimentary rocks (Fig. 3–18). They are used in combination to set apart two main groups, the detrital and the chemical. The **detrital sediments** are those that carry the earmarks of the mechanical transportation and deposition of the debris of erosion, **detritus,** by currents. The minerals are fragments of rocks or minerals broken and eroded from preexisting rocks, and so are called **clastic** (from the Greek *klastos,* to break). The rocks of which ancient mountains worn down by erosion were composed can be reconstructed from the minerals of detrital rocks. Quartz, feldspar, and the clay minerals make up the bulk of that contribution. The fragments tend to wear, and abrasion during transportation rounds the particles. During sedimentation, currents **sort** the minerals by size and weight with variable efficiency; the stronger the current, the larger the particle size carried. Size and sorting of clastic sedimentary particles are characteristic of the nature of the currents that carried them. As shown in Figure 3–18, these features also form the basis for subdividing the detrital sediments into (1) coarse-grained, the gravels and their hardened, or **lithified,** equivalents, the **conglomerates;** (2) medium-grained, the sands and **sandstones;** and (3) fine-grained, clays and muds and their lithified equivalents, the **shales. Mudstone** is a general term for rocks composed of more than 50 percent clay and silt. Shales are characterized by **fissility,** a splitting along bedding plane surfaces. Coarse sedimentary rocks composed of sharp, angular pieces of rocks and minerals are breccias, which contrast with the rounded pebbles and

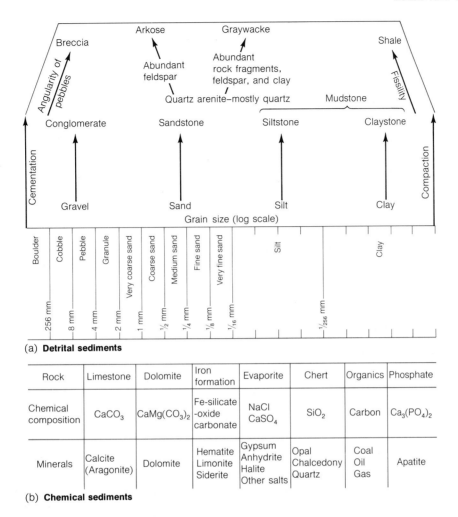

(a) **Detrital sediments**

Rock	Limestone	Dolomite	Iron formation	Evaporite	Chert	Organics	Phosphate
Chemical composition	$CaCO_3$	$CaMg(CO_3)_2$	Fe-silicate -oxide carbonate	NaCl $CaSO_4$	SiO_2	Carbon	$Ca_3(PO_4)_2$
Minerals	Calcite (Aragonite)	Dolomite	Hematite Limonite Siderite	Gypsum Anhydrite Halite Other salts	Opal Chalcedony Quartz	Coal Oil Gas	Apatite

(b) **Chemical sediments**

Figure 3-18
Sedimentary rock classification. (a) Detrital sediments. Along the front of the diagram the rocks are classified by grain size on the horizontal axis. Cementation is shown on the vertical scale. The names of the cemented, hard rocks are shown at the top and the corresponding loose, soft sediment terms below. On the top face are shown other textural characteristics and mineral composition of the sandstones. (b) Chemical sediments. These rocks are classified by chemical composition, with the major constituent minerals shown below. Thus chert is a rock composed largely of silica, SiO_2; the silica minerals of which it is composed include opal, chalcedony, and quartz.

cobbles of conglomerates. (Breccias may also be tectonically produced as rocks are sheared and broken along faults.)

The **chemical sediments** are precipitated from solution, mostly in the ocean, so that their minerals reflect the composition of the parent solution. The most abundant chemical rocks are **limestone** and **dolomite,** made up largely of calcium and magnesium carbonates, the minerals calcite and dolomite. Limestones may be made up in large part of fossils—shells formed by biochemical precipitation of calcium carbonate that animals extract from sea water. Other chemical sediments are also characterized by their chemical composi-

tion in relation to origin. The **evaporites** are composed largely of gypsum and halite, some including a complex group of other salts crystallized from evaporated sea water. The chemical rocks show a texture of crystal intergrowths resembling that of intrusive igneous rocks, on whose surfaces crystals can be seen to fit together like a carefully constructed flagstone walk.

Mineralogy is an important criterion for distinguishing varieties of detrital rocks, particularly the sandstones. **Quartzose sandstones** are mainly composed of quartz grains. This composition results from the erosion of feldspars and mafic minerals to leave only quartz, the most stable and resistant mineral component. **Arkose** is a sandstone that contains much feldspar in addition to quartz. **Graywackes** are poorly sorted dark sandstones that contain much feldspar and sand-sized rock fragments of metamorphic or volcanic rocks.

Shale, sandstone, and limestone, the three most abundant sedimentary rock types, account for more than 95 percent of the sedimentary rock in the crust. Of this fraction about 70 percent is shale, 20 percent sandstone, and 10 percent limestone.

The classification of sediments is based on mineralogy, texture, and chemical composition. The main criterion for subdivision of the detrital rocks is grain size, given as the diameter of the grain in millimeters. The boundaries are at 2 mm for gravel/sand, 1/16 mm for sand/silt, and 1/256 mm for silt/clay (Fig. 3–18a). The detritals are further grouped by their mineral content, mainly the relative amounts of quartz, feldspar, micas, and clay minerals they contain. The chemical sediments are grouped by chemical composition (Fig. 3–18b), which is reflected in their mineralogy, into the **carbonates,** limestone and dolomite; **evaporites,** including chlorides (halite) and sulfates (gypsum); **cherts,** or **siliceous rocks,** containing much silica either as quartz or other varieties of SiO_2; **iron formations,** containing much iron in silicate, carbonate, or iron oxide minerals; **organics,** the organic carbon sediments including coal, gas, and oil; and **phosphates,** rocks containing phosphate as a variety of the mineral apatite or in clay minerals.

Many limestones and dolomites—those made up of pieces of shells and grains of carbonate minerals moved by currents—look much like detrital rocks, and they are classed according to grain size in much the same way as the sands and muds.

The Metamorphic Rocks

Just as igneous rocks are divided into intrusive and extrusive and sediments into detrital and chemical, so are the metamorphic rocks divided into two broad genetic classes. They are the result of either **regional metamorphism** on the one hand or **contact metamorphism** on the other. Regional metamorphic rocks are produced by heat and pressure that transforms deeply buried rocks of all kinds, igneous, sedimentary and metamorphic. Contact metamorphics are made by the alteration of rocks near an igneous intrusion, largely from heat but also from pressure. Characteristic textures give clues to these two modes of origin. Most regional metamorphics show **foliation**—a platy,

wavy, or leafy structure imparted to the rock by the parallel alignment of minerals, particularly the sheety ones like mica. Some contact metamorphics are also foliated, but most tend to be granular, such as the **hornfelses,** which are very fine-grained silicate rocks of varied composition.

Foliation type and grain size are used, in combination, as the basis for subdividing the metamorphics into schist, slate, gneiss, and granulite (Fig. 3–19). The **schists** are characterized by partings along well-defined planes of medium-grained platy mica minerals. The **slates** have more perfect planar partings and are so fine grained that individual minerals cannot easily be seen. The **gneisses** are coarse grained and show much broader and less distinct foliation. They do not split or cleave as schists and slates do. The planes of foliation of slates and schists are called **slaty cleavage. Granulites** are like their textural equivalents, granular igneous rocks, in being mosaics of interlocking crystals of roughly equal size. They show only faint foliation, if any.

Within these textural groups, mineral assemblages form the basis for further dividing the rocks into small groups, or **facies.** The metamorphic facies originate in the mechanism by which the rock formed, for the minerals are determined by the temperatures and pressures required to form them. For example, albite-epidote-amphibole schists are the product of moderate temperature and pressure in regional metamorphism, whereas a **pyroxene hornfels** is the result of high temperature and moderate pressure in contact metamorphism (see Fig. 3–19). Because of the relations between texture and mineral facies—foliation and abundance of micas, for example—not all mineral facies are found in all textural types. Thus greenschist facies rocks are dominantly schistose whereas pyroxene hornfels rocks are granular. In normal field usage, the name of an abundant or prominent mineral constituent is used as a prefix to the metamorphic rock name—mica schist, for example is a schist that obviously contains much mica. Other metamorphic rocks are named for a mineral constituent that is greatly predominant, such as amphibolite. Marbles, metamorphosed limestones, are largely made of calcite. Quartzites, metamorphosed quartz arenites, are mainly quartz.

Some metamorphic rocks have characteristic textures produced by the crushing and mechanical deformation of grains as the rocks are folded and faulted. The broken, pulverized grain texture is called **cataclastic,** and fine-grained rocks produced by this kind of frictional action are called mylonites.

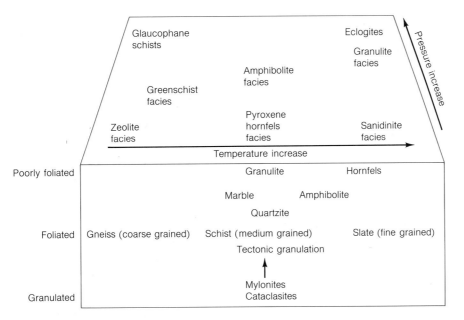

Figure 3-19
Metamorphic rock classification. The gradations in texture are shown at the front of the diagram. They range from the fine- to coarse-grained foliated rocks at the middle level to poorly foliated rocks at the top. The granulated rocks are typically fine grained. The textural properties are not necessarily correlated with mineral facies controlled by pressure and temperature. These are shown on the top of the diagram, with temperature increasing to the right and pressure to the back. Thus at moderate temperatures and pressures, the greenschist facies will be formed, but it may be a slate, a schist, or a quartzite, depending on its chemical composition.

The Information in Rocks

The characteristics of igneous, metamorphic, and sedimentary rocks can be used as a guide to identification (Fig. 3-20). Sedimentary rocks are identified as such mainly by their stratification, but also by their mineralogy and texture. The identification of a metamorphic rock depends largely on recognition of foliation and other textures. In the same way, the earmarks of igneous rocks are their mineralogy, textures, and structures. Within the major groups, identifications are made on the same dual basis of mineralogy and texture. Skeleton keys to the major rocks of each group are shown in Figures 3-21, 3-22, and 3-23.

From this brief summary of rock types, it should be apparent that mineralogy and texture are the major languages by which the information in rocks can be understood. The rounded quartz grains of a sandstone tell us about a history of erosion and transportation of fragments of a quartz-bearing igneous or metamorphic rock. The feldspar crystals in a granite inform us of the chemical composition of the magma and perhaps of the rate at which it cooled. The mica flakes of a schist are bits of information on the temperatures and pressures at which a shale was baked to make the schist. To read this information we need to know what the processes are by which they are made, where they operate in the Earth, and their relation to the larger architectural patterns of the Earth. It is these subjects that we move on to in the remainder of this book.

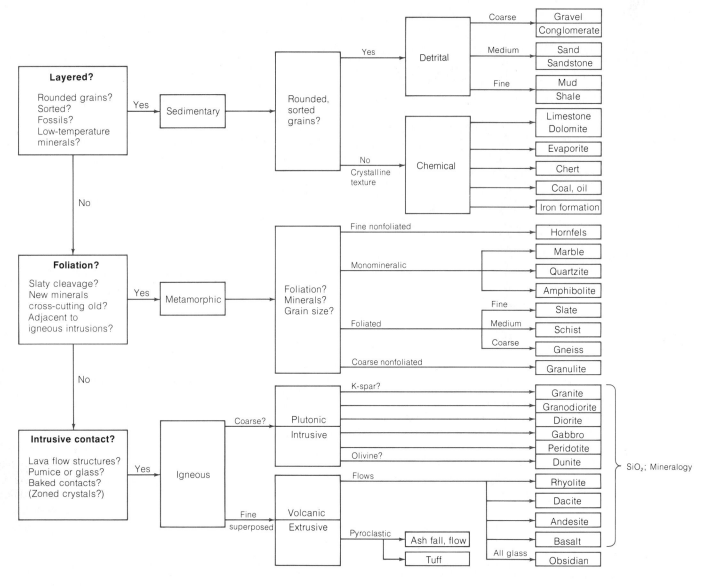

Figure 3–20
Rock identification flow chart.

SUMMARY

1. Rocks, and the minerals that make them up, are the record of the geological events of the past and contain information that is used to infer their origin.

2. Crystals exhibit external regularities of shape, constancy of interfacial angles, and symmetry relations that allow them to be divided into thirty-two crystal classes.

3. The optical properties of minerals—the distinctive ways in which they transmit light—can be analyzed by means of a polarizing microscope to identify them.

4. The atomic structure of minerals has become known largely as a result of x-ray diffraction studies, in which x rays are reflected by planes of atoms in the crystal.

5. The electronic structure of the elements explains the different kinds of chemical bonding of atoms in minerals. Bonding may be ionic or covalent, depending on whether valence electrons are transferred or shared. A weaker type of bonding, van der Waals, does not depend on the transfer or sharing of valence electrons.

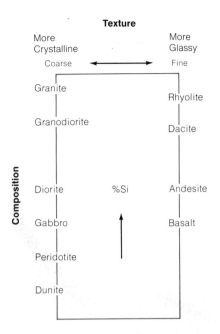

Figure 3-21
Skeleton chart of igneous rock classification.

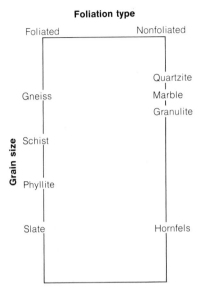

Figure 3-22
Skeleton chart of sedimentary rock classification.

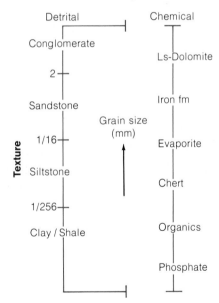

Figure 3-23
Skeleton chart of metamorphic rock classification.

6. The relative radii of ions in a substance determine the nature of the structures into which the ions can be organized.

7. Silicate mineral structures are made up of basic units of silica tetrahedrons bonded to each other and to other ions to form isolated tetrahedrons, rings, chains, sheets, and three-dimensional frameworks.

8. Physical properties of minerals—hardness, cleavage, fracture, streak, luster, color, and density—are determined by the kinds of elements making up the minerals, their bond types, and crystal structures.

9. Chemical properties of minerals are determined by the elements present and by their chemical bonding; minerals are classified on the basis of their anions.

10. Igneous rocks are divided on the basis of origin into intrusives and extrusives. The former are usually coarse grained; the latter, fine grained or glassy. They are also classified as felsic or mafic, depending on the kinds and relative amounts of light and dark minerals. Volcanic rocks are further subdivided by texture.

11. Sedimentary rocks fall into two main groups, the detrital, those formed from particles eroded from preexisting rocks, and the chemical, those formed by chemical precipitation of minerals from salt or fresh water. Detrital sediments are subdivided on the basis of grain size and mineralogy. The chemical sediments are subdivided primarily on the basis of chemical composition.

12. Metamorphic rocks are the products of either regional metamorphism, in which large volumes of rock are transformed by regional increases in pressure and temperature, or contact metamorphism, in which rocks close to igneous intrusions are transformed primarily by heat. Metamorphic rocks are classified according to the nature of foliation or cleavage and mineral assemblages or facies.

EXERCISES

1. If a given volume of rock is made up of one-third magnetite (specific gravity 5.2) and two-thirds quartz (specific gravity 2.7), would it be lighter or heavier than the same volume of a rock made up of two-thirds garnet (specific gravity 3.5) and one-third quartz?

2. Noteworthy for their absence from the mineral kingdom are crystals with 5-fold, or pentagonal, symmetry. Describe and draw some symmetries possible in a "pentagonal system."

3. Draw the electronic structures of magnesium ion (Mg^{2+}) and oxygen ion (O^{2-}). The atomic number of magnesium is 12 and that of oxygen 8. How would these ions differ from sodium ion (Na^+) and fluoride ion (F^-), which have atomic numbers 11 and 9, respectively?

4. What property of what common mineral is the basis for sandpaper? For what kind of jobs might you prefer to use emery paper? (Emery is a granular variety of the mineral corundum, Al_2O_3.)

5. Choose three minerals other than diamond from Appendix IV that you think might make good gemstones, and point out which physical properties might make them suitable for that purpose.

6. Name the properties of each of the following rocks, and describe the features that you would look for in order to identify them in outcrop: (a) a diabase dike, (b) a rhyolite flow, (c) a mica schist, (d) a conglomerate, (e) a hornfels, and (f) a marine shale.

7. No igneous rocks are composed of alternating beds of halite and gypsum, and no sedimentary rocks are made up largely of pyroxene. Give an example of other combinations of minerals, or groups of minerals, together with their outcrop appearance, that would be incompatible with (a) igneous, (b) sedimentary, and (c) metamorphic origins. Can you think of an example of a "forbidden" association of minerals and outcrop appearance—that is, a combination that could not be found in any natural rock?

BIBLIOGRAPHY

Bragg, Sir Lawrence, "X-Ray Crystallography," *Scientific American,* July 1968. (Offprint 325.)

Dietrich, R. V., and B. J. Skinner, *Rocks and Rock Minerals,* New York: John Wiley & Sons, 1980.

Ernst, W. G., *Earth Materials.* Englewood Cliffs, N.J.: Prentice-Hall, Inc., 1969.

Hurlbut, C. S., Jr., *Minerals and Man.* New York: Random House, 1969.

Loeffler, B. M., and R. G. Burns, "Shedding Light on the Color of Minerals," *American Scientist,* v. 64, pp. 636–647, 1976.

Mahan, B. H., *University Chemistry,* 3rd ed. Reading, Mass.: Addison-Wesley, 1975.

Simpson, B., *Rocks and Minerals.* New York: Pergamon Press, 1966.

Smith, C., "Materials." *Scientific American,* September 1967.

Williams, H., F. J. Turner, and C. M. Gilbert, *Petrography.* San Francisco: W. H. Freeman and Company, 1954.

THE SKIN OF THE EARTH:
SURFACE PROCESSES

The surface of the Earth is what we explore all of our lives. It is the interface between the solid planet and the gaseous envelope of air around it. Though we may go down into mines and fly through the air, it is the surface upon which we live and upon which we depend for our existence. The interface between the solid Earth and the vast bodies of water that cover much of it is being explored too; our curiosity about the sea floor has produced a crescendo of activity over the past two decades, but in satisfying our curiosity about this interface we must rely on instruments to do our seeing and feeling.

The dynamics of the surface are controlled by the Sun, which irradiates it. Solar energy drives the atmosphere in a complex pattern of winds that produces our climates and weather, and it drives the ocean's circulation in a pattern that is coupled to the atmosphere. The water and gases of the oceans and atmosphere react chemically with the solid surface and transport material physically from one place to another.

The processes that operate on the surface, then, are results of interactions between phenomena caused by the external heat engine and surface manifestations of the internal heat engine: mountains, volcanoes, and rocks brought up from the interior.

4

Weathering: The Decomposition of Rocks

Weathering is the chemical decay and physical fragmentation, under conditions at the Earth's surface, of minerals that formed for the most part at high pressures and temperatures in the interior of the Earth. Such minerals as feldspar dissolve partially after reacting with air and water, leaving a solid residue of clay. Other minerals, such as calcite and some mafic minerals, dissolve completely. The breakdown of rock masses to fragments ranging from boulders to clay is the result of chemical weathering combined with the forces of physical disintegration. The intensity of weathering depends upon climate, tectonics, original rock composition, and time. Given enough time, factors other than climate are relatively unimportant. The weathering process produces the clays of the world, all soils, and the dissolved substances that are carried by rivers to the ocean. The salt water of the ocean and the chemical sediments on the sea floors are themselves end-products of chemical weathering of the surface of the continents.

Mountains stand high, yet we know that they are worn away, slowly but inexorably, by erosion. Early geologists learned this by observing rivers carry sandy and muddy sediments from mountain tops to the plains below. Those sediments could have come only from the breakup of rocks. Rocks are hard, but we know they soften and disintegrate by noting the obliteration of markings on monuments and gravestones. Soils can be seen to consist partly of fine particles of rocks and minerals. Both the lowering of mountains by erosion and the origin of the raw materials of sediments can be explained in terms of weathering. Weathering is a twofold process: it is both fragmentation, or **mechanical weathering,** and decay, or **chemical weathering,** operating together, each helping and reinforcing the other. The smaller the pieces, the greater the surface area available for chemical attack and the faster the pieces decay; the faster the decay, the more the pieces become weakened and susceptible to breakage. Both chemical decay and physical breakup join with

rainfall, wind, ice and snow, and sliding and slumping to erode the surface and wear it away. At the beginning of this chapter we emphasize the chemical aspect of weathering, for it is in some ways the more fundamental driving force of the whole process; the effect of mechanical weathering, always important, is largely dependent on chemical decay, but it reinforces chemical action and is itself promoted by decay.

HOW LONG DOES A ROCK LAST?

The first person to make a tool of iron must have been the first to notice that it would rust, and perhaps that first toolsmith wondered why. An old nail in soil usually is so badly rusted that it can be snapped like a match, yet nails in wood in some of our colonial houses are still strong, covered with only a thin brownish film. A list of the kinds of places where nails rust quickly easily shows that exposure to air and moisture are important fac-

Figure 4-1
Two gravestones of the early nineteenth century in a cemetery in Wellfleet, Massachusetts. The lighter stone is limestone that has been ravaged by chemical weathering; the darker stone is slate, which remains practically untouched; even the delicate etching at the top of the stone is preserved. [Photo by R. Siever.]

tors. Rusting is a chemical process in which oxygen and water convert metallic iron to its oxidized form, ferric iron.

In contrast, some materials seem unchanged by exposure to air and water. Anyone who has hiked a trail to a lonely spot and discovered an old beer bottle knows that glass survives long exposure to the elements. Old bottles that have lain buried in moist soil do acquire a sometimes beautifully colored coating, but the main part of the glass itself seems immune to chemical attack. Glass itself does eventually decay but on a far longer time scale than metallic iron.

We can see the same range of response to wind, rain, snow, sun, and cold in old monuments and cemeteries (Fig. 4-1). Some gravestones seem to be made for the ages; others have lost almost all of their inscriptions. In a temperate climate, limestone goes fast, but granite lasts longer. In the aridity of desert lands, everything lasts. Today we can see well-preserved details on stone monuments left by early civilizations in the Middle East; that these details still remain is directly attributable to the dry climate. Alabaster (gypsum) sculpture would never have lasted outdoors in Wisconsin as it has in North Africa.

Though we usually think of organic materials as being quick to decay or putrefy, these substances, which are also found in ancient rocks and modern sediments, have as wide a range of responses to erosion as do inorganic minerals. Some tough resistant tissues of organisms, such as pollen grains, survive longer than many inorganic minerals. Human technology, which allows the synthesis of thousands of new chemicals, contributes an undue share to the mass of decay-resistant organic materials by producing huge quantities of nondegradable compounds, in particular the ever-present plastic products found scattered all over the land surface and floating in the middle of the ocean. We rely on the rapid decay of natural organic matter, primarily by bacteria, to return nutrients to soil and to produce fodder for farm animals from the fermentation of plant materials in silos. Some natural materials, however, just don't decay as rapidly as we would like—the millions of gallons of oil that get into the oceans as a result of spills of various kinds and leaks at offshore oil-drilling sites, for example. That oil originated as organic matter in sediments on the Earth's surface and was later preserved and altered by burial (see Chapter 22).

Why do some inorganic materials weather so quickly and others so slowly? From everyday observations we can contrast the differences among rocks of different composition, those in different climates, and those in different surroundings, from soil to bare mountainside. Mere age is of no consequence, for we can find fresh-looking Precambrian rocks and badly weathered Pleistocene ones. How long a rock has been exposed to erosion is more important, as anyone can testify by comparing a newly blasted road cut with one 20 or 30 years old. *Soil iself is both a factor in weath-*

ering and a result of it. The production of soil is evidently a *positive-feedback process,* one in which the product of the process works, by its presence, to increase the output of the process. Once soil starts to form, rock weathers more rapidly, and more soil is formed.

At the heart of the whole process are the ways in which different minerals react with water and air. We turn now to some important examples of the major rock-forming minerals, the first being the most abundant mineral in the Earth's crust, feldspar.

THE WEATHERING OF FELDSPAR

Feldspar is a key mineral in a great many igneous, metamorphic, and sedimentary rocks. Understanding its weathering behavior contributes much to our grasp of the weathering process in general. The place to start is in the field, making observations that give the clue to what the natural process is; then come the laboratory experiments, which enable us to quantify and identify the intermediate and final products of decomposition under controlled conditions.

Observing Decomposition in the Field

People in temperate regions are used to thinking of granite as the most permanent of rocks, but those in the humid tropics know that many granite boulders in soil can easily be kicked into a heap of mineral grains. Close examination of the particles and comparison with a fresh piece show that the crystals of feldspar are all punky and chalky with clay. Many of the crystals are so soft that they can be gouged with a fingernail, a condition conspicuously in contrast with that of the clear unaltered quartz crystals. The rock falls apart because the original interlocking crystal network of quartz and feldspar no longer holds together when the feldspar weathers to a loosely adhering clay. The white to cream-colored clay is the mineral kaolinite (sometimes just called kaolin), used in pure form as raw material for pottery and china.*

---

*The word kaolin is a French modification of the Chinese *kao-ling,* named for the place where the Chinese mined it for centuries before the finished product was exported as "china" to Europe. If Europeans had explored the state of Georgia before they visited China, they would have found one of the large kaolin deposits in the western hemisphere, and the name might have been different.

Figure 4-2
A weathered rock that contains much feldspar. The surface is corroded and pitted. In some of the enlarged cracks and joints, plants grow, helping further to weather the rock. Feldspar decomposition is an important part of this rock's weathering. As grains of feldspar alter to kaolinite, the rock is weakened. [Photo by R. Siever.]

Decay is not nearly so rapid in temperate climates, but any outcrop of a feldspar-bearing rock shows the beginnings of the same weathering to kaolin on some grains (Fig. 4-2). Only in the severely arid climates typical of some deserts does feldspar stay relatively untouched. From this we infer that water is essential to the chemical reaction by which feldspar becomes kaolin. The kind of reaction can be illustrated by a useful analogy to a familiar chemical reaction of the same kind: coffee-making. Freshly ground coffee beans plus hot water make a solution—the coffee—that is extracted from the solid, leaving behind spent coffee grounds as a residue (Fig. 4-3). How much can be extracted from the ground coffee depends on how much water is used, how hot it is, and how long it percolates. In the same way, the feldspar-kaolinite conversion depends on the amount of rainfall, the temperature, and the length of time water stays in contact with the feldspar-containing rock. These are the controlling factors, whether the rock is a boulder in soil or bedrock beneath a soil.

The water is of primary importance because it is such a good solvent for many natural materials. The reason is that the water molecule is polar—that is, asymmetric in charge distribution. The radius of the hydrogen atom is so small compared to the large oxygen atom that the hydrogens appear as small bumps on the surface of the oxy-

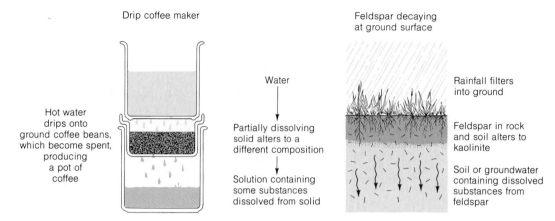

Figure 4-3
The process by which feldspar decays is analogous to the making of coffee. In both processes water dissolves some of the substance of the solid, leaving behind an altered material and producing a solution containing the substances drawn from the original solid.

gen in water. Partly as a result of this, the electrical charge is unevenly distributed and the molecule acts as a little magnet. This property gives it a weak electrical attraction for many solid surfaces, and this attraction tends to "pull" ions away from the surface of the solid minerals, thus dissolving the mineral.

We can write an equation for the weathering of the common feldspar of granite, orthoclase, which is made up of potassium (K), aluminum (Al), silicon (Si), and oxygen (O):

$$\text{feldspar} + \text{water} \rightarrow \text{kaolinite}$$
$$\text{KAlSi}_3\text{O}_8 \quad \text{H}_2\text{O} \quad \text{Al}_2\text{Si}_2\text{O}_5(\text{OH})_4{}^*$$

But chemical equations are like those of algebra: both sides have to balance. This requirement is imposed by the law of conservation of matter: the total number and kinds of atoms of the starting materials, the *reactants,* have to be the same as those of the *products* of the reaction. In what we have written, potassium appears on one side, as a component of the feldspar, but not on the other side. To make an equation of what we have written, we must add potassium to the right side. Since no potassium minerals are found in the weathered products, the potassium must have been carried away by a water solution (like the

coffee extracted from the grounds). To balance the equation completely, we must also add some silicon to the right side of the equation, in addition to that bound up in the solid kaolinite formed by reaction. Familiarity with these kinds of equations enables geochemists to guess intelligently at the form of the equation, but until the experiment has been carried out, the equation remains only a guess.

Dissolving Solids in the Laboratory

When trying to dissolve any of a great many materials rapidly, chemists often resort to an acid. Mixtures of strong acids are used as cleaning solutions for laboratory glassware because of their dissolving power. The experiment of choice, then, is to immerse feldspar in a readily available acid like hydrochloric acid, HCl. The feldspar is ground to a fine powder first, to speed dissolution by exposing more surface area to the solution—an effect we take advantage of when we use ground coffee instead of whole coffee beans. As the feldspar dissolves, the experimenter takes small samples from the solution and analyzes them for the expected dissolved potassium ion, K^+, and for dissolved silicon in some form. At first the feldspar dissolves rapidly, as measured by the steady increase in dissolved K^+ in the solution, but then the rate of dissolution decreases as the feldspar reacts more and more slowly and finally undergoes no further change (Fig. 4–4). If the solution continues to remain unchanged, it is presumed to be in **equilibrium** with the remaining feldspar powder. The analysis of the solution at this point shows the equation to be, in qualitative terms:

---

*Chemical formulas are written so that the number of atoms of a particular kind in the formula of any chemical compound is indicated by a subscript. The absence of a subscript indicates a single atom. Thus, feldspar, KAlSi_3O_8, consists of one atom of potassium, one of aluminum, three of silicon, and eight of oxygen. In a chemical equation, a number before the formula operates as a multiplier, the number of formulas needed to balance the equation. Thus $2\text{KAlSi}_3\text{O}_8$ indicates two atoms of potassium, two of aluminum, six of silicon, and sixteen of oxygen.

$$\text{feldspar} + \frac{\text{hydrochloric}}{\text{acid}} + \text{water} \rightarrow$$

$$\text{kaolinite} + \frac{\text{dissolved}}{\text{silica}} + \frac{\text{dissolved}}{\text{potassium}} + \frac{\text{dissolved}}{\text{chloride}}$$

Three major points about this equation are important: (1) *The potassium and silicon produced by dissolving the feldspar appear as dissolved material.* (2) *Water is used up in the reaction; it is absorbed into the kaolinite structure.* (3) *Hydrogen ion is used up in the reaction, and the solution becomes more basic as the reaction proceeds.*

This simple experiment includes the three main chemical effects of chemical weathering on silicates: it **leaches,** or dissolves away, cations and silica; it **hydrates,** or adds water to, the minerals; and it makes the solutions less acidic—more basic. If we were to dissolve feldspar in distilled water, the reaction would still proceed, though very much more slowly. It would produce a very dilute basic solution like that of a weak lye solution. The fact that hydrogen ion is used up makes this reaction look to a chemist like one in which an acid mixes with a base to produce a salt, as

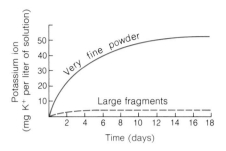

Figure 4-4
The rate at which potassium feldspar partially dissolves in water. The amount of feldspar that dissolves is measured by the amount of potassium ion liberated to the solution, which is measured as the number of milligrams of potassium per liter of solution. The feldspar dissolves at a fast rate in the first time intervals and then slows until finally no further change can be detected. Grinding the feldspar to a fine powder makes it dissolve faster and allows more of it to dissolve.

when hydrochloric acid mixes with potassium hydroxide, KOH, a strong base, to form potassium chloride, KCl, and water. In our experiment, the feldspar acted as the base.

Box 4-1 CHEMICAL WEATHERING OF FELDSPAR

The complete chemical equation for the weathering of feldspar to kaolinite is written

$$2KAlSi_3O_8 + 2(H^+ + Cl^-) + H_2O \rightarrow$$
$$Al_2Si_2O_5(OH)_4 + 4SiO_2 + 2K^+ + 2Cl^-$$

We write the acid HCl as $H^+ + Cl^-$ to emphasize that strong acids in solution are completely ionized—that is, they are present as individual ions, not molecules. Similarly, we write potassium chloride as K^+ and Cl^- rather than as KCl. The dissolved silica is written as SiO_2, for in nature silicon is not present as the free element. A chemically more realistic formula for dissolved silica is H_4SiO_4, silicic acid, which is simply $SiO_2 + 2H_2O$.

The equation for the weathering of feldspar can also be written with water alone considered as the acid, for it too can be a source of hydrogen ion (H^+). In that case, hydroxyl ion (OH^-) appears with the K^+ on the right side of the equation. Or one can consider carbonic acid (H_2CO_3), which is just carbon dioxide dissolved in water, the H^+ contributor. In that case bicarbonate ion (HCO_3^-) is written as the product (see Box 4-2).

It is as OH^- that water appears in the formula for kaolinite. The formula for kaolinite used to be written by some as $Al_2O_3 \cdot 2SiO_2 \cdot 2H_2O$ to show explicitly the presence of water.

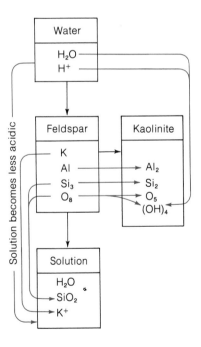

The flow of material in the alteration of feldspar to kaolinite by the dissolving action of water. Some water and some hydrogen ions (H^+) are absorbed into the kaolinite structure, but all of the potassium (K^+) and some of the silica (SiO_2) end up in solution.

But nature does not pour hydrochloric acid on granite outcrops. If we think again—now with the benefit of laboratory experience—about how feldspar dissolves in nature, we are led to the most common acid on the Earth's surface, carbonic acid, H_2CO_3. This weak acid is formed by the solution in rainwater of a small amount of carbon dioxide, CO_2, gas from the atmosphere.

Carbon Dioxide and Carbonic Acid

We are familiar with everyday solutions of CO_2 in water: soft drinks made by pumping pressurized CO_2 gas into the liquid. When a carbonated beverage is left to stand until bubbling has stopped, the dissolved CO_2 has decreased to a stable value, one we usually characterize as "flat" to the taste. The remaining dissolved CO_2 is in equilibrium with the small amount of CO_2 gas in the atmosphere. About 0.03 percent of molecules in the atmosphere are CO_2, a number that may seem small, but actually makes carbon dioxide one of the four most abundant gases, just behind argon (0.9 percent), oxygen (21 percent), and nitrogen (78 percent).

The planet Venus has an atmosphere that is almost entirely CO_2; so thick is the atmosphere at the planet's surface that the CO_2 concentration there is about 300,000 times that at the surface of the Earth. The Martian atmosphere is similarly made up mostly of CO_2 but is so thin that the CO_2 concentration is much less than that on Earth. In reading the following discussion of the role of CO_2 in weathering, think of the implications for weathering on those planets.

Carbon dioxide gas dissolves in water to form the compound carbonic acid.* Although carbonic acid is much weaker than hydrochloric acid, even such a weak acid can dissolve great quantities of rocks over a long time. When CO_2 dissolves in rainwater, the water becomes slightly acid, not enough to be noticed by plants and animals but enough to be slowly corrosive for feldspars. In areas where the air is greatly polluted with sulfurous gases from industries or power plants burn-

ing coal containing appreciable amounts of pyrite (iron sulfide), rainwaters may turn to "acid rain," for the sulfurous gases dissolve in rain in the same way as CO_2 to form sulfurous and sulfuric acids, both far stronger than carbonic acid. Although rainwater contaminated with sulfuric acid is still much too weak to sting our skins, it does noticeable damage to fabrics, paints, and metals. It also weathers our stone monuments and outdoor sculptures at such a rapid rate that public officials are beginning to take notice of their deterioration (Fig. 4–5). Acid rain is responsible for massive kills of fish in many lakes in the northeastern United States, Canada, and Scandinavia. The relation of coal-burning to acid rain is now deemed so important that a U.S. government project to track acid rain from its origin to downwind locations is underway.

How Does Weathering Proceed in Nature?

Now that we have a workable model for the major chemical reaction by which rainwater weathers feldspar, we have to return to field observations to see how the process works on outcrops and in the soil. Just as a nail rusts more quickly and deeply in the soil than it does lying on a rooftop, we find that feldspars on bare rock surfaces seem much better preserved than those in pebbles buried in wet soils, which may be crumbly and coated with soft rinds of kaolinite (Fig. 4–6). The chemical equation gives us a clue to this difference. First, the reaction requires water; during all of the dry periods, the bare rock is hardly touched. In moist soil the feldspar is constantly bathed in a corrosive acid-water solution. Second, there is more acid in the soil than in rainwater, and the stronger the acid the faster the decay. Some of the acids in soil are organic compounds produced by bacterial decay of plant and animal remains. Some bacteria seem able to grow even on bare rock particles and eat away at them. The acid content of soil water gets an extra boost from plant roots and bacterial decay, which produce CO_2 by **respiration** processes similar to the one upon which we depend. They oxidize organic matter and give off CO_2; we breathe oxygen and exhale CO_2. The amount of CO_2 in soil can be as much as ten times that in rainwater, making soil water an efficient dissolver of feldspars.

Weathering does not proceed at the same rate in all climates. One reason is that the speed of chemical reactions increases as temperature rises. Even more importantly, plants and bacteria grow and multiply much faster in warmer climates. An-

---

*Our carbonated beverages, such as artificially carbonated soda pop or naturally fermented beer and champagne, illustrate some important properties of CO_2 dissolved in water. Part of the "bite" of such beverages is the reaction of the weak carbonic acid with the tongue. The reverse of the reaction by which CO_2 is formed from carbonic acid is the one in which bubbles of CO_2 are released from a pressurized solution of a carbonated beverage when the pressure is released by uncorking. The amount of CO_2 that can be held in solution is dependent on the temperature; the colder the liquid, the more gas it can hold—a good reason for refrigerating such drinks!

(a)   (b)

Figure 4-5
Cleopatra's Needle, an obelisk of granite
(a) in Egypt and (b) after being moved to
Central Park, New York City. The obelisk,
and one like it in London, show evidence
of weathering much more extensive than
shown in the Egyptian photograph, taken
in 1880. The weathering is ascribed to the
humid climate of New York and London
acting on salts that infiltrated the obelisks
in their Egyptian history. [Photo (a) courtesy of the Metropolitan Museum of Art,
New York; photo (b) from Erhard Winkler, *Stone, Properties, Durability in Man's
Environment.* Copyright © Springer-
Verlag, New York, 1973.]

other reason is the contrast between wet and dry:
water is needed for the weathering reaction, and
vegetation grows more lushly in humid climates.
It is not surprising, then, that weathering is most
intense in tropical climates, which are both wet
and warm.

We will see later in this chapter how mechanical erosion is related to the speed of weathering
and to the topography. Just as the fresh feldspar of
the laboratory experiment reacted rapidly at first
and then slowed down, a fresh piece of feldspar
exposed at the surface of an outcrop by mechanical breakage will alter chemically faster than one
in an unbroken rock. Because the weathering
products in mountains are carried away quickly
by rains and melting snows, we cannot see the
effects there as easily as in lower, flatter areas,
where much of the clays and other products accumulate on gentle slopes and in valleys to form
soil.

Making the Clays of the Earth

We have described in detail how one kind of feldspar weathers to make one kind of clay. Some
other silicate minerals, such as mica and volcanic
glasses, weather in the same general manner,

Figure 4-6
Disintegrating boulder of granite partially embedded
in a glacial soil, shown partly excavated by digging.
The outer rim shows the original size of the boulder.
[Photo by F. E. Matthes, U.S. Geological Survey.]

some changing to kaolinite and others to different
clay minerals. Feldspar itself can alter to different
kinds of clay minerals under different weathering
conditions. **Smectite*** is an abundant clay formed
by weathering in many warm and semiarid climates. It is also the main clay product of the

---

*Until recently smectite was known as montmorillonite, but
clay specialists now prefer to use the new term for a more
general group of clays that absorb much water.

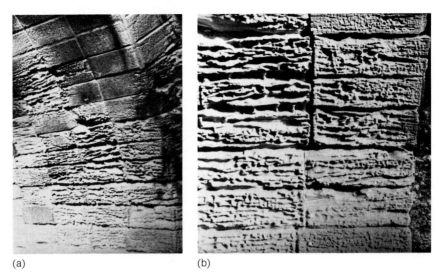

(a)   (b)

Figure 4-7
Weathered limestone blocks of a Roman aqueduct about 2000 years old
show the pitted, etched surface caused by chemical solution. (a) One of
the arches. (b) A closeup of a number of blocks. Pont du Gard, Provence,
France. [Photos by R. Siever.]

weathering of volcanic ash. Yet another abundant clay mineral, **illite,** typically develops from the weathering of sediments in temperate regions. Under the extreme weathering conditions of the tropics, kaolinite, itself a weathering product, can partially dissolve to give silica in solution, leaving a solid residue of the mineral gibbsite, aluminum hydroxide, $Al(OH)_3$. Gibbsite makes up the bulk of deposits of **bauxite,** the earthy ore that is the major source of economically extractable aluminum in the world.

All of the clay minerals are sheet silicates. Each clay type owes its distinctive character to the cations, such as sodium, potassium, magnesium, or calcium, that occupy positions in and between the sheets. Smectite, for example, has a structure and cation composition that gives it the ability to soak up large quantities of water. Kaolinite absorbs less water but is just right for making pottery and ceramics.

Because feldspar and other silicates that weather to form clays compose the large bulk of the igneous and metamorphic rocks of the crust, an enormous amount of clay is produced by the weathering of those rocks. Just how large is that figure? The total amount of sediment now existing over the whole Earth is about 3×10^{25} grams. Since about one-third of all sediments is made up of clay minerals, there must be about 1×10^{25} grams of clay in the Earth! Obviously the importance of chemical weathering as a source of sedimentary materals cannot be overemphasized.

Other kinds of weathering, however, result not in the formation of clay but in the complete disappearance of materials.

DISSOLVING WITHOUT A TRACE

Feldspar and other aluminosilicate minerals form clay by incomplete dissolution, but minerals of other chemical compositions dissolve completely, like salt in water. Quantitatively, the most important of these minerals are the carbonate minerals, calcite, $CaCO_3$, and dolomite, $CaMg(CO_3)_2$, which together make up the limestones of the world.

Limestone, the Soluble Sediment

The fast disappearance of limestone is evident to farmers and gardeners who spread finely ground limestone to reduce the acidity of the soil. The application has to be repeated every year or two because the limestone dissolves. Caves are characteristic of many limestone formations in humid climates. Their patterns of extensive passageways and the multitude of pits, holes, and irregular flutings on cave walls and ceilings are witness to the extensive dissolving power of groundwater. Caves and sinkholes are the indicators of underground water acting on limestones. Older limestone buildings show the same evidence (Fig. 4-7).

The dissolution of limestone brings us back to

the CO_2 system, for the carbonate ion, CO_3^{2-}, found in solid form in calcite is really CO_2 in another guise. Carbonate in solution is formed by dissolving CO_2 in water to form carbonic acid, as we have seen. That acid ionizes to form bicarbonate, HCO_3^-, and carbonate ions as well as hydrogen ions (Box 4-2). Carbonate ions are also found in solution after calcite dissolves, for that process is simply a separation of the calcium ions from the carbonate ions and of both from the solid. In the solution, the CO_3^{2-} tends to combine with H^+ ions from the water to form HCO_3^- (Fig. 4-8). The overall reaction by which calcite dissolves in a weak solution of carbonic acid, such as rainwater, is

$$\text{calcite} + \underset{\text{acid}}{\text{carbonic}} \rightarrow \underset{\text{ion}}{\text{calcium}} + \underset{\text{ion}}{\text{bicarbonate}}$$

$$CaCO_3 \qquad H_2CO_3 \qquad Ca^{2+} \qquad 2HCO_3^-$$

In nature, feldspar also weathers by reacting with carbonic acid. In both reactions, H^+ is used up as CO_2 reacts with water, and HCO_3^- is formed as a product. In this way CO_2 is extracted from the atmosphere and put into water. The more the weathering, the more CO_2 is removed from the atmosphere. Both reactions supply dissolved cations.

In a given quantity of water, limestone dissolves faster and in greater amounts than silicates. For both reasons, the chemical weathering of limestone accounts for more of the total chemical erosion of the land surface than any other rock, even though much larger areas are covered by silicate rocks. Some mafic silicate minerals also dissolve completely but more slowly. Although they by no means rival limestone in quantity eroded from the land surface, these minerals do account for a major fraction of the total amounts of iron, magnesium, and certain other elements weathered from rocks.

Box 4-2 CARBONIC ACID CHEMISTRY

We can write the following equation to describe the solution of carbon dioxide in water to form carbonic acid:

$$CO_{2(gas)} + H_2O_{(liquid)} \rightleftharpoons H_2CO_{3(solution)}$$

The double arrow stresses the fact that this reaction and the reactions given below are reversible; that is, they go both ways. Because the solubility of CO_2 in pure water is low, a liter of pure water contains only about 0.0006 gram of carbonic acid in equilibrium. Yet this is enough acid to weather rocks.

Once formed, carbonic acid spontaneously ionizes by another reversible reaction, in the same way that HCl or KCl does, to form a hydrogen ion, H^+, and a **bicarbonate ion, HCO_3^-**.

$$\underset{\text{acid}}{\text{carbonic}} \rightleftharpoons \underset{\text{ion}}{\text{hydrogen}} + \underset{\text{ion}}{\text{bicarbonate}}$$

$$H_2CO_3 \qquad H^+ \qquad HCO_3^-$$

This reaction is different from the ionization of the HCl and KCl because very few of the carbonic acid molecules are ionized; only one molecule out of a thousand is ionized at equilibrium. An even smaller fraction of HCO_3^-, only one ion out of 100,000 further ionizes to form another hydrogen ion and a **carbonate ion, CO_3^{2-}**

$$\underset{\text{ion}}{\text{bicarbonate}} \rightleftharpoons \underset{\text{ion}}{\text{hydrogen}} + \underset{\text{ion}}{\text{carbonate}}$$

$$HCO_3^- \qquad H^+ \qquad CO_3^{2-}$$

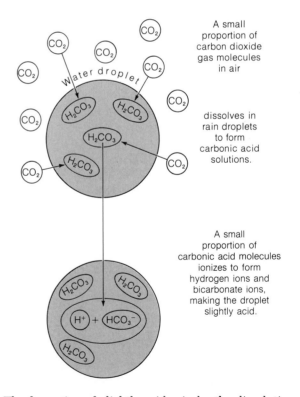

A small proportion of carbon dioxide gas molecules in air

dissolves in rain droplets to form carbonic acid solutions.

A small proportion of carbonic acid molecules ionizes to form hydrogen ions and bicarbonate ions, making the droplet slightly acid.

The formation of slightly acid rain by the dissolution of carbon dioxide gas molecules in water droplets in the atmosphere. Once dissolved, the carbon dioxide reacts with water to form carbonic acid, which then ionizes to form hydrogen and bicarbonate ions.

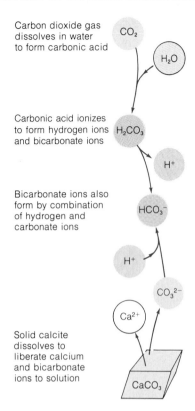

Figure 4-8
The general scheme by which calcite or other carbonate minerals dissolve in water containing dissolved carbon dioxide. The reactions are all reversible, and the system should be visualized with all reactions operating at the same time to produce the general net tendency shown by the arrows.

Mafic Minerals

The beating of ocean waves against rocky cliffs is one of nature's ways of hurling solvent at a rock. On the rocky coast of New England the chemical effects of the attack are seen when the geologist hunts for pyroxene crystals in a granodiorite. All that can be found are little cavities in the shape of pyroxene crystals. Olivine behaves the same way. Like limestone, both minerals wholly dissolve. A magnesium pyroxene, $MgSiO_3$, dissolves to give Mg^{2+} and dissolved silica. An iron pyroxene, $FeSiO_3$, dissolves to give Fe^{2+} and dissolved silica. The iron in pyroxene is **ferrous** Fe^{2+}, in contrast to the iron in hematite, Fe_2O_3, which is **ferric**, Fe^{3+}.* When iron pyroxene weathers, not only does the silicate structure dissolve, but the ferrous iron is oxidized by oxygen to ferric iron. Ferric

---

*Ferric iron is more oxidized than ferrous iron in the sense that it has fewer electrons. Oxidation can be defined as a loss of electrons in a chemical reaction, regardless of whether the element combines with oxygen itself. Reduction is the opposite of oxidation; it is a gain of electrons. In comparison with ferric iron, ferrous iron is more reduced.

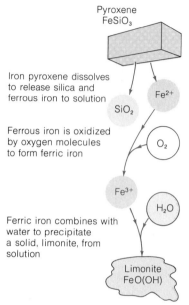

Figure 4-9
The general scheme by which an iron-rich silicate mineral, such as pyroxene, weathers. In the presence of oxygen and water, as at the Earth's surface, these reactions proceed sequentially and irreversibly. Limonite formed in this way is typically a poorly crystallized yellow–brown mass of the mineral geothite, sometimes mixed with another mineral, lepidocrocite, of the same composition but different crystal structure. Magnetic varieties of these minerals are used in magnetic recording tapes.

iron in the presence of oxygen is very insoluble in most natural surface waters, and it precipitates to form a poorly crystallized mineral that looks like rust, limonite, $FeO(OH)$, a ferric oxide that contains water in the form of hydroxyl (Fig. 4-9). The overall reaction, which includes oxidation and hydration of the iron and the dissolution of the silica, is

$$
\begin{array}{ccc}
\text{iron} & \text{oxygen} & \\
\text{pyroxene} + & \text{from} & + \text{water} \rightarrow \\
 & \text{atmosphere} & \\
4FeSiO_3 + & O_2 & + 2H_2O \rightarrow \\
\end{array}
$$

$$
\begin{array}{cc}
\text{limonite} & \\
\text{(hydrated iron} + & \text{dissolved} \\
\text{oxide)} & \text{silica} \\
4FeO(OH) & + \quad 4SiO_2 \\
\end{array}
$$

For every four iron atoms that are weathered after being brought to the surface of the Earth by mountain-building or volcanism, two atoms of oxygen are extracted from the atmosphere. Were it not for the rapid replenishment of oxygen by the photosynthesis of plants, the weathering of mafic rocks would in the long run—millions of years—have eliminated all oxygen from the air. Liberated by the weathering of mafic minerals, the iron precipitates as fine-grained, poorly crystalline limonite coatings and encrustations that color soils and stain rocks shades of red and brown. They also have served as the raw materials of artists' pigments—the ochres, siennas, and umbers. Moreover, they are the same minerals that gave the "red planet," Mars, its name.

One more dissolution process is important because so little of it happens. Quartz, SiO_2, dissolves completely leaving no residue, as does calcite, but the solubility of quartz is so extremely low and it dissolves so slowly that it is for practical purposes insoluble in water. As a result, most of the quartz in rocks remains chemically unaltered and survives as a solid residue of weathering. Quartz is thus a stable mineral under the conditions that prevail at the Earth's surface. But exactly what do we mean by stability of minerals?

CONCEPTS OF STABILITY

Stability can be demonstrated rather easily by trying to balance this book on one of its corners. You may succeed temporarily, but the slightest touch will topple it. The book is unstable in that position because the slightest change in its state—its position in this example—causes a larger change to another state: the book falls flat. When it is lying flat, lift it by one edge. As soon as you let it go, it immediately returns to its flat, or stable position. The book can be placed in an intermediate state by standing it upright; in this position it is **metastable,** because if pushed very slightly, it returns to its upright position, but given a bigger push, it falls to its stable state.

The chemical stability of minerals can be described in much the same way. For minerals the other states are the products of reactions, simple or complex. Iron pyroxene, for example, is unstable at surface conditions partly because it tends to dissolve in water. But that is not a good enough criterion, for salt dissolves even more easily in water, and it is stable. The difference is that after dissolving, the stable material, salt, reprecipitates or crystallizes as the same mineral, whereas the unstable one, iron pyroxene, has formed something else. The dissolved salt forms new salt crystals upon evaporation, but the dissolved pyroxene forms a mixture of crystals of hydrated iron oxide and silica. We can rank the relative resistance to decay of such stable minerals as salt and quartz under weathering conditions on the basis of their

Table 4-1
Stability of common minerals under weathering conditions at the Earth's surface

Most stable	Fe-oxides
	Al-oxides
	Quartz
	Clay minerals
	Muscovite
	K-feldspar (orthoclase)
	Biotite
	Na-feldspar (albite)
	Amphibole
	Pyroxene
	Ca-feldspar (anorthite)
Least stable	Olivine

solubility: the less soluble, the more resistant. Quartz wins hands down. Thus we see that stability, as it affects resistance to weathering, is also a relative property.

We can combine our knowledge of weathering in the field and chemical experimentation to rank minerals in the order of stability under weathering conditions, including alteration and solubility behavior, as in Table 4-1. The products of weathering, clays and oxides, are included in the table and are among the most stable minerals to further weathering.

These relative stabilities show in a general way how drastically different surface conditions on Earth are from the environments where the minerals were crystallized and remained stable. Muscovite, a mica, is formed at moderate temperatures and pressures, and hence has a small tendency to break down into clay minerals. Pyroxene, formed at higher temperatures and pressures, has a greater tendency to weather. The order of stability under weathering conditions is related to the stabilities of chemical bonds and crystal structures under different temperatures and pressure, subjects covered in the chapters on igneous rocks (Chapter 14) and metamorphism (Chapter 16). The order of stability of the minerals found in igneous rocks is roughly the reverse of the order of crystallization from a granitic melt as it cools; the first to crystallize, olivine, is also the first to weather.

The instability of silicates and the solubility of carbonates and other minerals are the driving forces of chemical weathering. That chemical decay is only part of the erosional process, however. Physical breakup, the fragmentation of rocks, is the other side of the weathering coin.

FRAGMENTATION

The physical process of fragmentation, the breakup of rock masses into boulders, pebbles, sand, and silt, is inextricably linked to the chemical processes of rock weathering. Chemical decay is not always apparent when one looks at a mass of broken rock at the foot of a cliff. Pieces may look freshly broken and unaltered. But however minutely, weathering has played some role in making the rock break. Fragmentation aids chemical weathering in that the physical breakup of the rocks opens channels for water and air to penetrate the rock farther and promote chemical decay. The rate at which chemical reactions take place increases when more surface area of the

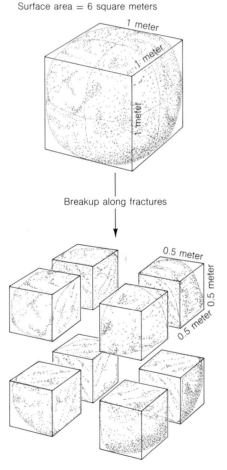

Single boulder, approximately 1 meter cubed
Volume = 1 cubic meter
Surface area = 6 square meters

Breakup along fractures

8 cobbles, each approximately 0.5 meter cubed
Volume = 1 cubic meter
Surface area = 12 square meters

Figure 4–10
As a large boulder breaks into smaller ones along fractures or joints, the surface area increases relative to volume, making more of the bulk's surface available for chemical decay.

solid is exposed to the gas or liquid reacting with it. Because the ratio of surface area to volume of rocks and minerals increases as the average particle size decreases, chemical weathering becomes more efficient as rocks are broken into small sand and silt particles (Fig. 4–10). Chemical and physical processes are both related to the action of organisms, from bacteria to tree roots, which play their own roles in the destruction of rock.

In studying such complexly intertwined processes, the geologist looks for special places where one or another effect is absent so that the others can be studied alone. Geologists have studied mechanical fragmentation in severely cold polar regions, where there is little active plant or bacterial life and little liquid water, factors that minimize chemical weathering. Even in such hostile regions, however, there was enough chemical decay to be significant. The study of the Moon gave a new and better opportunity to study fragmentation in the absence of ordinary chemical weathering as we understand it on Earth and so to compare mechanical erosion with and without chemical action.

We can be sure that there has been no chemical weathering of minerals in the basalts and other igneous rocks present in Moon rocks. These materials are as fresh as might be expected, having never been exposed to an atmosphere containing oxygen, carbon dioxide, and water. But, as many photographs show (see Figs. 21–4 and 21–7), the Moon's surface is composed of loose, fragmented pieces of rock, from boulders to fine dust. Fragmentation there must have resulted completely from physical mechanisms, such as the breakup of large parts of the surface rocks by the impact of meteorites (Chapter 21). The accumulation of fine dust particles is partly the product of large impacts, but much of it comes from the continual spray of small meteorites ranging in size down to the finest micrometeorites.

Other than fragmentation by these processes, Moon rocks show nothing of the kinds of fragmentation that we are so used to on Earth. Our "laboratory" investigation of another planet has shown us that fragmentation on Earth has no component that does not depend in some part on chemical reaction with the atmosphere. What, then, are the Earth's particles like, and how do they form?

Kinds of Fragmentation

A look at an outcrop of a well-bedded sandstone or a schist with pronounced cleavage shows how

Figure 4-11
Joint patterns developed in several directions in metamorphic rocks.
[Photos by R. Siever.]

important the original character of the rock is in determining the course of breakage. The sandstone breaks into slabs or plates along bedding planes, and the schist breaks into smaller, sometimes splintery pieces whose flat sides parallel the cleavage. The cleavage, schistosity, bedding, or other structural planes in rocks become zones of weakness, and cracks form along them (Fig. 4-11).

On a larger scale, unbedded rocks—or what we call **massive** rocks, like granite—tend to break up along regularly spaced planar cracks, called **joints,** at intervals of a meter to a few meters apart. Joints are produced by stresses associated with deep burial of sedimentary rocks and by structural deformation in the course of mountain building. To some degree they are characteristic of almost every rock type. In some igneous rocks the joints take the form of **sheeting,** one set of parallel, closely spaced, more-or-less plane surfaces, which in some exposures are curved.

All these cracks open when the rocks, once deeply buried within mountain ranges, are gradually brought to the surface and the weight of tons of overlying rock is removed. The process is probably not unlike the way in which poorly glued joints in a wooden chair separate as the clamps that held them together during gluing are released. Confirmation is given by the sometimes severe jointing that spontaneously occurs in deep mines as tunnels are opened and the pressure on the rock in those passageways is released.

Once the joints, bedding planes, or other cracks are opened ever so slightly, water, bacteria, and plant roots start to do their work in enlarging and wedging the openings apart (see Fig. 4-2). One of the most efficient physical mechanisms is freezing and thawing of ice. Water expands as it freezes, and the expansive force exerted during freezing is enough to crack cast iron engine blocks—and rock, too.* The process has the same effect as a wedge being pounded into the crack, enlarging it and finally splitting it open (Fig. 4-12). Crystallization of such minerals as salt or gypsum from salty solutions that infiltrated the cracks may also wedge fractures open. During warmer periods, chemical weathering by water and organisms helps the process along by weakening the miner-

_____

*Some of the properties of water, though familiar enough to us, are unusual in comparison with those of other liquids. One characteristic whose effects are far-reaching in our lives is water's **thermal expansion,** the way it expands or contracts when the temperature is changed. Like most substances, liquid water contracts when cooled. But between 4°C (about 39°F) and its freezing temperature, 0°C (32°F), it abruptly changes course and expands. When the liquid freezes, it again behaves contrary to most other materials and expands further as the rather open crystal structure of ice forms.

Figure 4-12
A disrupted glacial boulder in Wallace Canyon, Sequoia National Park, California. This boulder, originally one piece 2.5 m wide and 1.5 m high when it was deposited by a glacier, has been split into a series of parallel slabs by the freezing action of water that penetrated originally tight joints. [Photo by F. E. Matthes, U.S. Geological Survey.]

Figure 4-13
Exfoliation on a rocky slope, Yosemite National Park, California. The partings produced by weathering are nearly planar, and strongly resemble inclined joints. [Photo by F. E. Matthes, U.S. Geological Survey.]

als that hold the rock together. The role that plant roots play in opening joints is probably mostly in promoting chemical weathering, but they also exert physical pressure as they grow.

If freezing is so important, heating in the hot sun and cooling at night might be effective too—or so everyone thought for many years until someone tried it in the laboratory. Pieces of granite were alternately heated and cooled to temperatures far more extreme than those they would be subject to on any desert. The number of rapid alterations of hot to cold was equivalent to hundreds of years of days and nights, yet such rough treatment caused no visible fragmentation of the rock. Other experiments of the same kind have been tried with little success. In one experiment, water was sprayed on basalt to simulate dew at each cooling. The result was still negative, even though some chemical solution was demonstrated. Most geologists who have observed fragmentation in the desert still tend to believe that the stresses induced by thermal expansion and contraction in rocks may be significant. We know that the much more intense heat of fires can crack rocks open. Certainly the laboratory experiments so far are a poor copy of what happens in nature. We will have to wait for better experiments or observations to settle the matter.

Perhaps thermal expansion combines with frost action and chemical weathering to produce two commonly observed kinds of mechanical disinte-

Figure 4-14
Spheroidal weathering of a boulder of gabbro, Mesa Grande, San Diego County, California. [Photo by W. T. Schaller, U.S. Geological Survey.]

gration, exfoliation and spheroidal weathering (Figs. 4–13 and 4–14). **Exfoliation** is the peeling off of large, curved sheets or slabs of rock from the weathering surface of an outcrop. Exfoliating outcrops look rather like onion layers. **Spheroidal weathering** has the same general appearance but on a much smaller scale, in which rounded boulders spall (split) off layers or shells from the surface. Large-scale exfoliation may result in part from the relief of pressure. As each slab breaks off, it releases weight from the underlying mass, which then may expand enough, however slightly, to break the rock. In spheroidal weathering the pressure released by unloading is so minute that it cannot be important. Chemical weathering is thought to be of some importance in both these processes, producing mechanical stresses by uneven volume changes in the surface as minerals decay. This leads to the breakage of curved plates from the rock mass.

Sizes and Shapes of Fragments

The various processes that cause fragmentation produce a bewildering array of sizes and shapes. Careful study of the shapes of fragments that can be traced to their parent outcrop, even those kilometers distant, has shown that their shapes are largely inherited from the patterns of joints, bedding, cleavage, and other structures of the parent rock rather than being produced by the transporting medium. An experienced geologist can guess just from the shapes of pebbles on a beach what kinds of rock they came from. This inheritance extends even to many sand grains, whose shapes, modified only slightly by erosion and transportation, are derived from the original crystals or grains in the eroded source rocks (Fig. 4–15).

The sizes of fragments are a good clue to the intensity of mechanical erosion. A good general rule of thumb is that the higher or steeper the topography, the larger the fragments. Regardless of the rates of mechanical and chemical weathering that govern the break-up of rocks in a particular place, the fragments are largest close to their source. Once fragments have gradually worked downhill from their source, they enter new domains of weathering, and they undergo further fragmentation. Moreover, once boulders and pebbles fall into streams they break and abrade quickly, and the downstream decrease in size of pebbles becomes dramatic. Much of this size decrease is the result of the inability of most streams to carry large boulders very far except in rare floods, as we shall see in Chapter 7.

Figure 4–15
A microscopic view of a thin section of the Recluse Formation (Precambrian) in Canada. The grains are angular: they retain the original shapes they had when fragmented from their source rocks. Grains are poorly sorted by size, one of the characteristics of many graywacke sandstones transported by turbidity currents. (Magnification: ×44.) [From *Sand and Sandstone* by F. J. Pettijohn, P. E. Potter, and R. Siever, copyright © Springer-Verlag, New York, 1972.]

Sand grains, however, are not produced by mere wear—that is by the gradual abrasion of pebbles. Nor is any one sand grain the only remnant of what was once a pebble: this would require the disappearance, or wearing away, of too much material. The similarity in size, shape, and mineral composition of sand grains to the crystals that make up granites and other coarse-grained igneous and metamorphic rocks points to the origin of most sand grains. It is the mechanical fragmentation of rocks, not their gradual wearing away, that produces most sand. As the crystals of feldspars, micas, and other weatherable minerals are eaten away at the edges, the interlocked crystals loosen, and what was once whole rock starts to crumble into individual grains or groups of grains (Fig. 4–16) producing the "rotten" boulders found in the humid tropics. Where weathering is slower, rocks become smaller bit by bit as grains chip off.

Silt particles, smaller than sand grains, are crystals or broken fragments of crystals produced in the same way as sand grains. Clay minerals, on the other hand, are the chemical products of feldspar weathering. Clay-size particles are by defini-

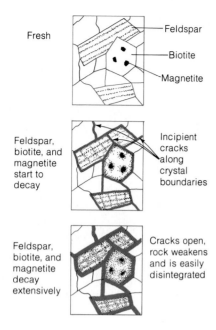

Figure 4-16
Microscopic views of stages in the disintegration of an igneous rock, such as granite. As chemical decay proceeds, grain boundaries weaken, and the rock begins disintegrating into fragments.

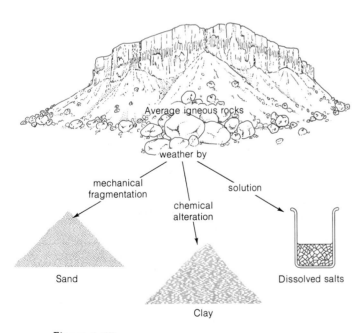

Figure 4-17
The chemical and physical processes of weathering transform the igneous rock into sand and clay particles and dissolved salts. The total volume of erosional debris produced is somewhat greater than the original, depending on the degree of chemical weathering, which adds carbon dioxide, water, and oxygen to the original composition.

tion those less than 1/256 millimeter in diameter (Chapter 3). Not all particles of clay size, however, are produced by chemical weathering; some clay-size particles of quartz, feldspar, and other minerals are formed by abrasion, disintegration, and physical fragmentation of larger particles.

Thus the different sizes and shapes of eroded particles—ranging from huge boulders to microscopic clay particles—are attributable to characteristics of their source rocks. The sizes and shapes of boulders, cobbles, and pebbles are largely inherited from patterns of jointing, bedding, and foliation, along which breakage occurs. Sand grains are largely the constituent crystals of disintegrated coarse-grained rocks; silts are broken and abraded finer crystals. Clays are mainly the chemical breakdown products of unstable minerals (Fig. 4-17).

SOIL: THE RESIDUE OF WEATHERING

The Soil Profile

Soil can be pictured as a thin layer that rests on bedrock, like a coating of rust on iron. The way to see the structure of a soil, its vertical profile, is to cut a trench through the soil down to bedrock, typically at a depth of about a meter (Fig. 4-18). The top section, or **A-horizon,** is usually the darkest for it contains most of the organic matter, or **humus**—tiny particles of decayed leaves, twigs, and animal remains. The A-horizon is a zone of intense biological activity, populated by everything from worms and insects to gophers, plant roots, and abundant microorganisms, all living in an ecological microcosm. In a thick soil that has developed over a long time, the minerals of the top layer are mostly clays and residual insoluble minerals like quartz. The soluble minerals are absent. Lower in the soil profile, in an intermediate layer called the **B-horizon,** there is relatively little organic matter, and soluble minerals and iron oxides are present. The lowest part, the **C-horizon,** is slightly altered bedrock, broken and decayed, mixed with clay. At its base it merges into solid bedrock. Soil scientists use many different classification schemes that are much more complex than this simple A-B-C scheme, but the general pattern is the same for most soils. The three zones may be difficult to distinguish in very young or poorly developed soils.

Geologists and soil scientists have much evidence that soils are thicker where they are older and where they receive warmth and moisture.

the more the solids are altered. The higher the temperature, the faster the reactions. All these effects combine to make the part of the solid closest to the fresh-water inflow react the most because it is there that the water carries the least solute. The water is less efficient a solvent as it passes on through the rest of the solids, for it is starting to aproach saturation, carrying the greatest amount of solute it can carry. As a result, the alteration proceeds along a "front" that moves slowly along the pipe as time goes by.

This is a simplified view of the process that occurs in soils. It explains why the tropical soils are thick and devoid of most unstable minerals and why arid soils tend to be thin and rich in unstable minerals. The complexities of clay composition also result in differences in soil mineralogy under different climates. Just as temperature speeds mineral alteration, it also stimulates the growth of vegetation and bacteria, which increase the rate of soil formation. In the extraordinarily dry, frigid environment of some bare ground areas of Antarctica, there are practically no microorganisms, much less larger organisms, and weathering is at a minimum.

A soil takes a long time to form. Many hundreds to several thousands of years may be required for

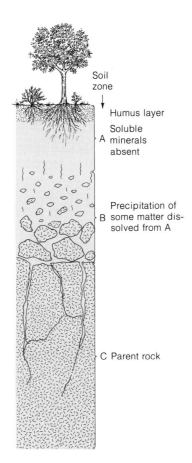

Figure 4-18
Typical soil profile developed on granite bedrock in temperate regions. The thickness of the soil profile depends on the age of the soil and the climate. Transitions from one zone to another are normally indistinct.

These are expected conclusions from what we know about the weathering of minerals. From them we can construct a model for soil formation.

Soil Formation in Relation to Climate

Rainfall and temperature both affect soil formation. We can compare a soil to a vertical pipe containing fixed solid chemicals that react with water. Water runs through soils just as it would the pipe, for soils are permeable, and rainwater runs through them before infiltrating the pore spaces of the bedrock below (Fig. 4–19). As rainwater passes over the solids in the pipe, it reacts chemically with them. Several factors affect the extent to which the solids and water react. The more fresh water runs through the pipe, the more the solids react with the fresh undersaturated solvent. The longer the time the water runs through,

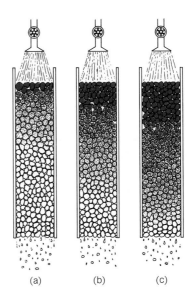

Figure 4-19
The extent of mineral alteration in a soil as illustrated by water flowing through a pipe packed with chemicals that react with the water. Columns (a), (b), and (c) represent successive stages in the downward progression of alteration zones under a steady flow of water. Alternatively, they can be thought to represent the results, at the same time stages, of three different rates of flow of water: (a) the lowest; (c) the highest.

an A-horizon to evolve to the point that there is a mat of decayed vegetation and organic matter with altered minerals and clays. The formation of a B-layer takes even longer, from ten thousand to a hundred thousand years. A soil that has evolved to maturity is in a steady state, in dynamic equilibrium with its climate. As soil is very slowly eroded from the top by natural processes—and by plowing and tilling—it is gradually deepened by chemical reaction, If erosion is rapid, chemical decay cannot keep up, and the soil is thinned, a serious consequence for agriculture. It is easy to see why farmers and governments promote use of those agricultural methods that do not accelerate soil erosion, for once gone, a soil takes a long time to form again.

The formation of soil can be diagrammed as a complex positive-feedback process involving the response of mineral decay and mechanical erosion to the variables of time, temperature, and rainfall and to the biological activity that is influenced by the first three factors (Fig. 4-20). If all this is true, then we should find a fair correspondence between soil type and thickness and climatic regions—and we do.

The Major Soil Groups

At least three major kinds of soil can be classified on the basis of the minerals present in their A- and B-horizons. In a high-rainfall area, the A- and B-zones of a soil are leached; that is, the soluble or quickly altered materials have been carried away by water. The resulting soil is rich in the insolubles, quartz, clay minerals, and iron oxide alteration products. Calcium carbonate is absent. From their abundance of aluminum and iron, the soils are called **pedalfers** (*ped* from the Greek word for soil; *al* and *fer* from the English and Latin names of aluminum and iron). Much of the eastern half of the United States and most of Canada are covered with fertile soils of this kind (Fig. 4-21a).

A second major group of soils are those that contain soluble minerals, indicating formation in dry, warm climates, like that of the western United States. Since calcium carbonate is an important constituent of these soils, they are called **pedocals** (Fig. 4-21b). Some of the calcium carbonate may be dissolved by occasional rainfall, but in a warm, dry climate, much of the soil water evaporates, leaving precipitated pellets and nodules of calcium carbonate, typically in the B-horizon of the soil. These nodules may form even where there is no limestone as a result of the reaction of calcium leached from silicates with bicar-

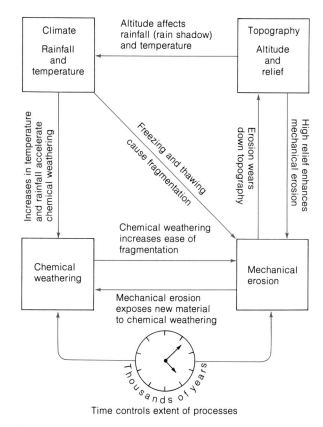

Figure 4-20
Interrelations among the major factors controlling weathering and erosion.

bonate formed by weathering reactions of carbonic acid. The pedocals contain less clay and more unaltered silicates than the pedalfers, again a result of the low rainfall. These soils are normally less fertile, as a result of the combination of mineralogy, texture of the material, and climate, which does not permit a rich population of organisms in the soil.

A third soil type is the **laterite,** a deep red soil of the tropics in which all silicates are completely altered, leaving mostly aluminum and iron oxides, from which not only calcium carbonate but also silica have been leached (Fig. 4-21c). So much water has passed through these soils that all but the most insoluble compounds are gone. If there is little iron in the parent bedrock, then bauxite, a special variety of this kind of soil, is formed. These tropical soils, though they may support lush vegetation in places, are not very fertile soils for crop plants, because clearing and tilling cause the humus to oxidize quickly under the warm climate rather than to accumulate to give a rich black soil. For this reason, many lateritic soils can be used for only a few years before they become barren and have to be abandoned.

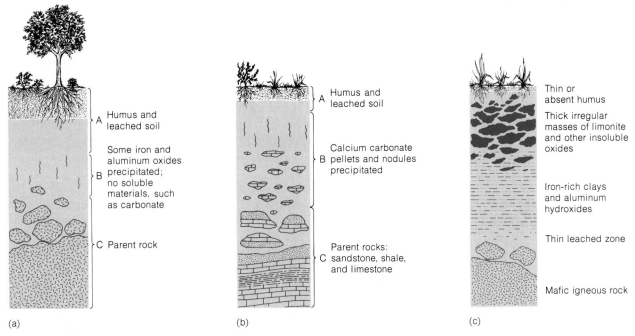

(a)   (b)   (c)

Figure 4-21
Major soil types. (a) Pedalfer soil profile developed on granite in a region of high rainfall. The only mineral materials in the upper parts of the soil profile are iron and aluminum oxides and silicates such as quartz and clay minerals, all of which are very insoluble. Calcium carbonate is absent. (b) Pedocal soil developed on sedimentary bedrock in a region of low rainfall. The A-zone is leached; the B-zone is enriched in calcium carbonate precipitated by evaporating soil waters. (c) Laterite soil profile developed on a mafic igneous rock in a tropical region. In the upper zone, only the most insoluble precipitated iron and similar oxides remain, plus occasional quartz. All soluble materials, even including relatively insoluble silica, are leached; thus the whole soil profile may be considered to be an A-zone directly overlying a C-zone.

WEATHERING: SOURCE OF SEDIMENT

Weathering is not only the main process by which rocks are destroyed; it is also the supplier of the raw material of sediment. By weakening rocks so that they are more easily fragmented, chemical weathering provides the rock and mineral particles of sedimentary detritus. The detritus bears the marks of the kind and extent of weathering in its composition. If chemical weathering is extensive, virtually the only products are clays, iron oxides, and quartz grains. Some remain as soil on the weathering surface; the rest are transported by rivers and the wind to sedimentary environments where muds and quartz sands are deposited. But such extensive weathering is an extreme. The intermediate stages of weathering are more common; therefore, it is important to learn the sedimentary products of such weathering.

Incomplete Weathering: Quartz and Feldspar

The survival rates of the two most abundant rock-forming minerals, feldspar and quartz, give us a rough index of weathering processes. The total amount of quartz stays relatively constant because it is so insoluble and hard, and therefore resistant to erosion. Feldspar, however, is gradually changed to clay; the longer and more intense the weathering, the less feldspar remains. Conversely, if chemical weathering is slight and mechanical weathering is the major process, feldspar grains are found in sediments in virtually the same proportions as in the source rock. Thus the ratio of feldspar to quartz is a general indicator as to which of the two weathering processes is more active.

For example, rapid mechanical wearing away of granitic mountains produces an accumulation of unweathered sand and gravel river deposits that have about the same mineral composition as the original granite (Fig 4-22). Arkoses, feldspar-rich sandstones that form in this way, can be distinguished from their source rocks only by the nature of the sorting and other sedimentary structures, such as bedding. As the highlands are worn down, perhaps a change in climate may take place. Chemical weathering may then play a dominant role, and the feldspar will be increasingly

(a) High granitic mountains: feldspar and quartz rapidly fragmented by mechanical weathering.

Transportation rounds fragments and sorts by size, producing arkose, a sandy sediment rich in feldspar.

(b) Low hills and plains: most feldspar weathered to clay before being washed away; quartz remains intact.

Transportation rounds and sorts quartz, producing sands composed almost entirely of quartz.

Figure 4-22
Two different kinds of topography on the same granitic bedrock weather to produce sandy sediments of differing feldspar content.

altered to clay. The sand transported and deposited by the rivers becomes largely quartz with much associated clay. The rocks formed from such deposits are quartz-rich sandstones, or quartz arenites. Unlike an arkose, which provides detailed information about the mineral composition of the source rock, the quartz arenite tells us mainly that the source rock contained quartz, even though an occasional grain of feldspar or some other silicate may give a bit more information on the source rocks. It is in this way that geologists working with sedimentary rocks can do detective work on the topography, climate, and rock composition of the past.

Implicit in the interpretation of climate and topography from weathering products is the control that parent rock composition has on soil composition and sedimentary detritus. It is again a

matter of time and intensity. Under intense, long-continued weathering, the soils and detritus reflect mainly climate and topography; the soil developed on a granite may look much like one on a shale. The shorter the time and the less intensive the weathering, the more the compositional differences of the source rocks show up in the soil and detritus. A young soil still shows some remains of unweathered minerals from which the parent rock composition can be inferred. But we gain one kind of information at the expense of another. The more we are able to deduce about the parent rock, the less likely it is that we will be able to get much information on climate because weathering has probably been too slight to leave a strong impression. As in an automobile junkyard, the older and rustier and more dismantled the car, the harder it is to tell its make, model, and year. The younger it is, the easier to estimate how rapidly it has rusted.

CHEMICAL WEATHERING AS THE REACTION OF ATMOSPHERE, OCEAN, AND CRUST

The materials that chemical weathering dissolves from rock eventually filter into the ground, move to the oceans in rivers, and make sea water the complex salty solution it is. The weathering processes that produce those solutions are chemical reactions in which atmospheric gases (carbon dioxide and oxygen) combine with water to attack exposed crustal materials of the earth. Thus by describing the overall effects of weathering, we can see how the atmosphere, the waters of the continents, the oceans, and the surface of the crust are all linked in a giant chemical system.

All the gases of the atmosphere and all the water on the Earth's surface originated deep within the Earth and escaped via volcanoes, hot springs, and other zones of connection between interior and surface. Probably most of this gas was released early in the Earth's history and has been added to slowly throughout geologic time. The water has accumulated primarily in the oceans, but much is stored as groundwater and in hydrated minerals of sedimentary origin. The great bulk of the carbon dioxide released has not accumulated in the atmosphere but has dissolved in water to form bicarbonate ion and then combined with calcium to make the limestones of the crust. Another fraction of the carbon dioxide was transformed to organic carbon by photosynthesis and was buried as organic matter in sediments, some of it ultimately altered to coal and oil. Oxygen accumulated in the atmosphere as the by-product of the photosynthesis that produced the organic carbon.

All of these components, the water, the gases, and even rocks now on the Earth's surface were once together deep in the Earth at high temperatures and pressures. The movements of solid lithosphere and magmas combine to allow the gases and water to separate from the rock-making materials from the interior and leak out into the atmosphere as the rocks are pushed to the surface in mountains, at mid-ocean ridges, and along volcanic island arcs. In the erosional processes that take place at the surface of the Earth, all those components come back into contact at much lower temperatures and pressures. A new set of low-temperature reactions—the weathering process—takes place between the rocks and gases: water and carbon dioxide combine to form carbonic acid, which weathers silicate minerals; oxygen oxidizes iron and other metals. Because the chemical stabilities of minerals that can be formed by these components are so different at low temperatures and pressures, an entirely new set of minerals, dominated by clays and carbonates, is formed instead of minerals like olivines and pyroxenes.

The ocean basins receive the dissolved products of the weathering. These materials do not simply accumulate there, for in the sea they come together to precipitate sedimentary minerals. The most abundant is limestone, formed in the sea by the combination of calcium and bicarbonate ions released by weathering. In this way the calcium is precipitated—mostly in the form of shells of organisms—at the same rate that it is brought in by rivers, and the ocean stays in a steady state, unchanging in the amount of calcium, even though ions are constantly coming in and going out. So it is with all of the other elements. Because of these steady-state reactions, the oceans have probably remained at about the same composition throughout much of the history of the Earth. The ocean

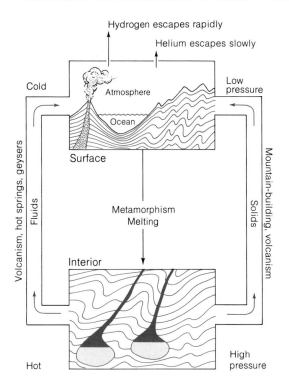

Figure 4-23
The chemical reaction system of the Earth is composed of two coupled subsystems, one in the interior, where igneous and metamorphic rocks are formed, and the other at the surface, where atmosphere, ocean, and crust react. Nothing escapes from this system except the lightest gases, hydrogen and helium, which have atoms light enough to escape the Earth's gravity field. [After R. Siever, "Sedimentological Consequences of a Steady-State Ocean Atmosphere," *Sedimentology,* v. 11, copyright © International Association of Sedimentology.]

and its sediments are the products of the last stage in the reactions between the atmosphere and the crust.

Thus the earth is like a giant chemical reaction chamber made up of two subchambers, a hot one in the interior and a cold one at the exterior (Fig. 4-23). The conveyor belt operating between the chambers is volcanism and tectonics.

SUMMARY
1. Chemical and mechanical weathering complement and reinforce each other in erosion.

2. Chemical reactions of weathering, such as oxidation of iron or decomposition of organic material, vary widely with climate and material.

3. Potassium feldspar weathers by partially dissolving to form solid kaolinite. Evidence for this reaction comes from field observation; its exact nature is known from laboratory experiment.

4. Feldspar weathers by reacting with water and hydrogen ion to form kaolinite, leaving potassium and silica in solution.

5. Carbon dioxide dissolves in water to form carbonic acid, an effective chemical weathering agent.

6. Chemical weathering is most active in warm, humid regions and least active in cold or arid regions.

7. Clay minerals—kaolinite, smectite, illite, and others—are the products of weathering of various materials under different climatic regimes.

8. Limestone, which is completely dissolved during chemical decay, is the rock type most rapidly weathered in humid climates.

9. Mafic minerals may dissolve completely. Iron contained in them is oxidized by the oxygen of the atmosphere and then precipitated as limonite in soils and sediments.

10. Relative chemical stabilities of various minerals at the Earth's surface are ranked on the basis of the kind and degree of solubility.

11. Fragmentation of rock into boulders, pebbles, sand, and silt is a mechanical process promoted by chemical weathering.

12. Studies of rock fragmentation on the moon show little of the normal mechanical erosion that we observe on Earth and give evidence of role of chemical weathering in fragmentation.

13. Zones of weakness open up as cracks and joints when rocks formed in the interior are exposed at the surface. Water seeps into the openings and causes weathering: when water freezes, it wedges the rocks apart.

14. Sizes and shapes of fragmented rock particles, from boulders to sand grains, are largely inherited from such characteristics of source rocks as jointing, bedding, and crystalline texture.

15. Soils, the residues of weathering over long periods of time, are of distinctive types as a result of climatic differences: pedalfers, pedocals, and laterites.

16. The quartz–feldspar ratio of a clastic sediment is a useful guide to the rate and kind of weathering in the eroding source area.

17. Chemical weathering is a reaction between the solid crust of the Earth and the atmosphere and oceans, all of which originated within the interior of the Earth.

18. The ocean keeps its chemical composition in a steady state by precipitation of chemical sediments as new dissolved substances are brought in by rivers.

EXERCISES

1. Rank the following rocks in order of their rapidity of weathering in a humid climate: limestone, quartz sandstone, amphibolite (a metamorphic rock composed of amphibole and feldspar), and arkose (a feldspar-rich sandstone). Give reasons for your choice.

2. Would you expect a basalt to weather more quickly than a granite in a semiarid climate? Would you expect a rhyolite to weather more quickly than a granite, its coarse-grained equivalent, in a semiarid climate? Why?

3. If you were a mountain climber and had to drive pitons (metal spikes used to hold ropes of climbers) into bare rocks at high altitudes, would you find it easier in desert mountains or mountains that receive high rainfall?

4. Describe how you might expect a calcium-iron pyroxene, $CaFe(Si_2O_6)$, to weather chemically and what the weathering products might be.

5. If we continue to burn oil and coal at an increasing rate, we may eventually double the carbon dioxide in our atmosphere. How might this affect weathering?

6. Suppose you are a conservation engineer assigned to design a plan for making soil form as quickly as possible on a fresh, broken rock rubble left by construction or mining operations. What would you do (without importing any soil from elsewhere)?

7. Which would you expect to fragment more rapidly into small pieces, a granite or a schist? What effects might climate have on the difference? How might the shapes of the pieces differ?

8. Of the various mechanisms of rock breakage described, which would you expect to be more important in the rainy tropics as opposed to cold polar regions?

9. What are the weathering processes that produce these three different sizes of material: blocky boulders up to several meters across, sand grains about 1 millimeter across, and clay particles a few thousandths of a millimeter in diameter? How far from the eroding outcrop might pieces of each size get in a given time period?

BIBLIOGRAPHY

Berner, R. A., and G. A. Holdren, Jr., "Mechanism of Feldspar Weathering," *Geology*, v. 5, no. 6, pp. 369–372, 1977.

Gauri, K. L., "The Preservation of Stone," *Scientific American*, June 1978. (Offprint 3012.)

Kellog, C. E., "Soil," *Scientific American*, July 1950 (Offprint 821.)

Loughnan, F. C., *Chemical Weathering of the Silicate Minerals*. New York: American Elsevier, 1969.

McNeil, M., "Lateritic Soils," *Scientific American*, November 1964. (Offprint 870.)

Millot, G., "Clay," *Scientific American*, April 1979. (Offprint 937.)

Siever, R., "The Steady State of the Earth's Crust, Atmosphere, and Oceans," *Scientific American*, June 1974.

5

Erosion and Landscape

Following the slow weathering that breaks down outcropping rocks to small fragments and clay, more rapid processes of active erosion and transportation take over to create the varied topography of Earth. Mass movements such as landslides erode mountainsides; together with other agents of erosion they help shape the surface. The sculpture of topography is illustrated by the contrast between the Grand Canyon and the Ozark Mountains. The forms and slopes of the valleys are clues to the agents of erosion that shaped them—the water, ice, and wind that act in opposition to the tectonic forces that elevate mountains. The evolution of landscape is a balance between uplift and erosion over the course of time.

Not a year passes without news of a flood, a mass of mud oozing down a hillside, or a catastrophic landside. On August 1, 1976, for example, the Big Thompson River in the Rocky Mountains of Colorado, suddenly swollen to a flood by heavy rain, swept through its narrow canyon, killing more than 60 people and washing away bridges, roads, and buildings. Left behind were new deposits of sand and gravel brought down from the eroding mountains. In 1972 Rapid City, situated at the foot of the Black Hills of South Dakota, was hit by a flash flood that shocked the country because of the suddenness with which it caused such vast destruction and tragic loss of life. Everywhere in the devastated city one could see rocks, mud, and silt brought in by the floodwaters (Fig. 5–1). Much of the debris of these floods began its downhill journey a long time before, having been produced at a normal rate by chemical and mechanical weathering in the nearby mountains. The prevailing erosional process was suddenly augmented by the rapid downhill movement of water-soaked soil and rock debris into the flooding streams. Rock fragments and clay that had taken tens to hundreds of years to produce and build up on the

slopes were caught up in fast downhill movements provoked by a storm that might occur only rarely.

This chapter is concerned with the active erosion and transportation that follow the slow, sometimes imperceptible, decay and fragmentation of chemical and mechanical weathering. The downslope transport by slow creep and by rapid slides and flows makes the debris of erosion available to the major long-distance agents of transport, rivers, wind, and glaciers, all of which shape the landscape of Earth into forms that tell of the processes acting on the face of the planet.

MASS MOVEMENTS

Downhill movement of the debris of erosion is not an obscure process to people who suddenly find themselves the victims of a disastrous mudflow or landslide. Earth movements are sudden, the accompanying volumes of sliding debris often enormous, and events seem unpredictable. Yet when geologists hear of some of these disastrous mass movements, they know that lack of foresight and

Figure 5-1
Jackson Boulevard in Rapid City, South Dakota, on June 10, 1972, after torrential rains caused a flash flood. [Photo by Perry H. Rahn.]

poor engineering planning are often responsible for them.

In Quebec the stage was set for a mudflow when too steep slopes were cut in soft formations of silt and clay as a highway was widened. After a period of heavy rain in 1955, the steepened cliffs became saturated with water and suddenly gave way in a flowing mass of debris that carried away buildings, roads, and people. Three deaths can be traced to poor highway engineering. In 1963 more than 200 million cubic meters of rock debris roared down steep slopes into the waters of a deep reservoir in the Italian Alps. The enormous volume of debris plunged into the reservoir and created a giant spillover of water at the dam. Downstream 2600 people lost their lives in the catastrophic torrent. The dam, 265 meters (870 feet) high, had been built in a steep-sided valley. The engineers who designed it ignored three warning factors: the weakness of the cracked and deformed layers of limestone and shale that walled the reservoir, an obvious geologic history of landslides, and a premonition of danger signalled by a small rockslide three years before. The landslide was not preventable and would have caused some damage to roads and buildings, and perhaps a few injuries, but the reservoir, whose waters caused the flood and multiplied the damage enormously, might have been placed in a geologically safer place. Occasionally geological foresight produces happier results. A few years ago a Japanese geologist observed conditions on a hillside that, he believed, threatened a landslide that would bury a nearby villge. Because he warned the villagers, predicting the imminence of the slide, he was instrumental in saving thousands of lives. The landslide came soon after and bore out his predictions. Fortunately, mass movements like this are infrequent, and most of them do not affect many people; many are known from the geological forms they leave behind rather than from first-hand observations.

Though downhill movement of debris is obviously induced by gravity, other factors determine why some movements are fast on some slopes and slow on others; and why some slopes are stable and others move if disturbed. First in importance is oversteepening of slopes, illustrated best by loose sand. Sandboxes have made nearly everyone familiar since childhood with the fact that a pile of sand assumes a characteristic slope; the angle is the same whether the pile is only a few centimeters high or three meters high. If some sand is scooped from the base of the pile very slowly and carefully, the angle of slope will steepen a little and hold. But if someone jumps on the ground near it, the slope cascades down and again assumes the original angle, which is called the **angle of repose** (Fig. 5-2). That angle varies with the size and shape of the particles; larger, flatter, and more angular pieces of loose material remain stable on steeper slopes.

Next in importance is the effect of a lubricant, usually water. Try to build a sand pile with dripping wet sand: it just flows until it becomes as flat as a pancake. Or to try to walk down a clay slope in the rain without sliding. Water filling the pores of permeable material allows the grains to slide past each other with little friction. The effects of slope and lubricant, of course, become more complex in other materials, depending on the nature of the mixture, the irregularities of slopes, the structurally stabilizing role of vegetation roots, and the amounts of water required to saturate the materials. Nevertheless, these effects are at the root of all downslope movements of boulders, sand, and soil. They work in different ways to produce the variety of what we call mass movements: landslides, mudflows, rockslides, slumps, and creep.

Landslides

Rockslides and landslides are rapid downhill movements of masses of rock or weathered erosional debris, soil, and anything else that may be in the way, such as houses, fences, and roads. **Rockslides** are movements of blocks of bedrock along joints or cracks, usually along the water-

Figure 5-3
The landslide in Madison Canyon that slid off the mountainside in the left background after an earthquake in August, 1959. The slide dammed the Madison River, forming the lake in the foreground. [Photo by J. R. Stacy, U.S. Geological Survey.]

lubricated bedding planes of tilted sediments. The preconditions for landslides are slopes too steep for the frictional forces to hold the heterogeneous mixtures of rock, soil, and unconsolidated sediment in place against the force of gravity. Typically, small landslides are formed where slopes are first steepened by erosion, either natural or artificial (such as excavations). When soaked by rain until saturated, the slope enters a critical stage of instability, ready to give. Once the stage of extreme instability is reached, a slide is inevitable. Vibrations of some kind trigger many slides. The more delicately poised something is, the smaller the jar needed to knock it over. Vibrations of strong earthquakes trigger slides in material that might otherwise have stayed fixed for a long time (Fig. 5-3). Most of the damage from the Alaskan earthquake of 1964 was caused by the slides it produced.

In the spring of 1925, melting snow and heavy rains made water abundant in the valley of the Gros Ventre River of the Jackson Hole area of Wyoming. That was the year of the Gros Ventre slide (Fig. 5-4), a classic among the landslides of the twentieth century. One eyewitness, a local rancher, barely escaped from the path of the slide on horseback. Warned by the roar of the slide, he looked up to see a huge mass of valleyside hurtling toward his ranch. From the gate to his prop-

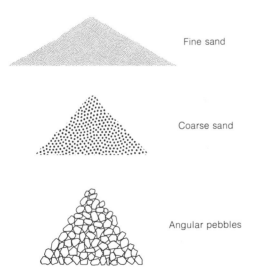

Fine sand

Coarse sand

Angular pebbles

Figure 5-2
The angles of repose of particle mounds increase as particle size increases and shapes become more angular.

Figure 5-4
The scar of the Gros Ventre slide. This photo, taken in 1966, more than forty years
after the slide, shows the still-unforested slopes of the scar left by the slide.
[Photo by R. Siever.]

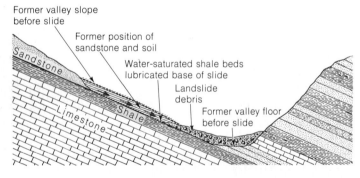

Figure 5-5
Cross section of the Gros Ventre slide, Wyoming.
Arrows show the direction of sliding of a mass of
soil-covered sandstone down the water-lubricated
surface of shale underlying it. The conditions for
this slide were set when the Gros Ventre River cut
its valley floor through the sandstone formation,
leaving unsupported the sandstone beds higher on
the valley slopes. [Modified from W. C. Alden, 1928,
Trans. Am. Inst. Mng. Met. Eng., v. 76, pp. 345–361.]

erty he watched the slide race past him at about
50 miles an hour and bury everything he owned.

It has been estimated that about 37 million
cubic meters of rock and soil slid down one side
of the valley, surged more than 30 meters (100
feet) up the opposite side, and then fell back to the
valley floor, damming the Gros Ventre River. Most
of the slide was a confused mass of blocks of
sandstone, shale, and soil, but one large block,
covered with soil and a forest of pine, slid down
as a unit. Two years later, the large lake that
formed behind the slide overflowed it, draining
the lake and causing a torrential flood in the valley below.

The forces that caused the Gros Ventre slide
were all natural. The stratigraphy and structure of
the valley (Fig. 5–5) made it almost inevitable. On
the side of the valley where the slide occurred, a
permeable, erosion-resistant sandstone formation
dipping about 20° toward the river rested on a
soft shale. The stage was set for the event to happen, sooner or later, after the channel of the Gros
Ventre River had cut through most of the sandstone and left it with virtually no support where it
intersected the river channel. Only friction along
the bedding planes between shale and sandstone
kept the layer of sandstone from sliding. Removal

Figure 5-6
Aerial view of slump in a hillslope underlain by weak shale. The slump is about
140 m (450 ft) wide. Stanley County, South Dakota. [Photo by D. R. Crandell,
U.S. Geological Survey.]

of this support by channel erosion was the equivalent of oversteepening. The heavy rains and melting snow saturated not only the sandstone but the surface of the underlying shale. Once the force of gravity overcame friction, almost all of the sandstone slid toward the bottom of the valley along the lubricated surface of the wet shale. This general pattern is characteristic of many other slides except that some slides take place along joints or other zones of structural weakness instead of along bedding planes.

In **unconsolidated** material—that is, in materials not held together by a cement or by a strong interlocking crystal structure—landslides start after a significant part of the whole mass is saturated with water and lubricated. In some slides, lower parts of the wet soil may undergo liquefaction, that is, transformation to a liquid state. In this state it flows easily and rapidly as a fluid, lubricating an overlying mass. Again, some shock, however small, can trigger the downslope movement of a whole unstable hillside of such material. The movement may take the form of a **slump,** in which the mass travels as a unit, in most places slipping along a surface that is concave upward (Fig. 5-6). Slumps and slides may be fostered by altering types of vegetation by herbicide spraying or burning, as is common on western U.S. wildlands. These practices increase the soil moisture and make slopes more prone to fail. Rockslides may move as slumps, but many landslides, rather than slumping as units, move as an incoherent jumbled mass flowing almost with the characteristics of a fluid. Ronald Shreve, a geomorphologist at the University of California, Los Angeles, believes that some landslides, once they start, race along on an undersurface of trapped, compressed air that provides an almost frictionless speedway.

Mudflows are particularly fluid forms of landslides that move as viscous tongues of mixed mud, soil, rock, and water (Fig. 5-7). Most common in hilly and semiarid regions, mudflows are generated by infrequent, sometimes prolonged, rains. The flows start down upper valley slopes and merge on valley floors. Where mudflows exit from confined valleys into broader, lower valley slopes and flats, they may splay out to cover large areas with a mixture of wet debris. Mudflows can carry large boulders, trees, and even whole houses. **Earthflows** are like mudflows but less fluid. They thus tend to be confined to slopes of weathered shale or clay, where they move short distances downslope. **Debris-avalanches** are high-speed flows usually associated with humid mountainous

Figure 5-7
A large mudflow that ran like a river down mountain slopes in Hinsdale County, Colorado. [Photo by W. Cross, U.S. Geological Survey.]

Figure 5-8
A mudflow deposit on the slopes of Mt. Rainier, Washington. The mudflow consists of a mixture of volcanic ash and rock fragments, including a large volcanic bomb (above the hammer). It was the mapping of these kinds of deposits that led the U.S. Geological Survey to predict the activity of the Mt. St. Helens volcano. [Photo by D. R. Crandell, U.S. Geological Survey.]

regions. Water-saturated soil eroded down to bedrock may move rapidly, carrying everything in its path with it. In 1962 a debris-avalanche in the Peruvian Andes traveled almost 15 kilometers (9 miles) in about 7 minutes, engulfing most parts of eight towns and killing 3500 people. Mudflows and debris-avalanches are easily triggered on slopes of volcanic cinder cones as their unconsolidated accumulations of ash and other erupted material become saturated during rains. Eruptions trigger some flows by associated earthquakes and tilting movements, cloudbursts, and sudden falls of volcanic debris. More than fifty-five such flows, many of them associated with eruptions, have occurred on the slopes of Mt. Rainier, Washington, in the past 10,000 years, according to U.S. Geological Survey geologists (Fig. 5-8). Ash and mudflows associated with the eruption of Mt. St. Helens in 1980 covered large areas and completely obliterated Spirit Lake at the foot of the volcano.

Rock Falls, Soil Creep, and Solifluction

Talus slopes are built up by the accumulation of many individual falls of talus blocks—angular rocks broken from the outcrop by physical and chemical forces (Fig. 5-9). Though each rock falls suddenly, the accumulation of talus is slow and steady, building up blocky slopes everywhere along the base of a cliff. The talus slopes themselves are eroded at their lower edges as the rocks are further broken down in size and carried away by mudflows and other mass movements or streams.

Soil creep, a downhill movement of soil, takes place so slowly that its rate of movement is difficult to measure over short time periods. Yet such slow movements are the reason why trees, telephone poles, fences, and other objects apparently fixed in the soil tend to lean or be displaced slightly downslope. By comparing the precise positions of marked stakes fixed in bedrock with those of others in the soil, geologists have learned that most creep ranges from a few millimeters per year to a few centimeters per year, depending on the kind of soil, steepness of slope, rainfall, and vegetation. The mechanics of creep are not fully understood, but it is related to minute movements caused by freezing and thawing, wetting and drying, thermal expansion, activities of burrowing organisms, and root movements of growing and decaying plants.

Depending on the degree of saturation, soil may move slowly or rapidly. In cold regions, where saturated ground becomes frozen, the movement of soil, called **solifluction,** takes place slowly, induced by freezing and thawing. Because ground in frigid areas may be frozen to great depths, the lower zones never thaw, but when the surface zones of the soil thaw, the water has no place to

go. The water-saturated upper layers then ooze downhill, carrying broken rocks and other debris. A related phenomenon is the movement of **rock glaciers.** These are like lobed rivers of rock that have moved down slopes and valleys in cold, mountainous regions. The pores between the rocks are partly filled with ice, and the motion is in many ways like that of true glaciers.

Though unspectacular, all of the motions of creep and related slow downslope travel of rock and soil account for a large part of the initial transportation of fragmented and weathered material from the primary sites of erosion to lower slopes, where they are further eroded and transported by streams.

Slope Wash

The prelude to organized river-channel transportation of detritus is **slope wash,** the generalize surface runoff of rainwater—the little rills and threads of water that gradually come together to form rivulets, then streams, then river. If you have the patience and don't mind the rain, you can observe how water running down a hillside moves small particles of silt and sand, gradually undermines clayey soil from beneath pebbles, and

Figure 5-9
Talus produced by a rockslide at the base of a rocky valley wall in Yosemite National Park, California. The trees in the foreground were defoliated by the blast of air that accompanied the slide. [Photo by F. E. Matthes, U.S. Geological Survey.]

Figure 5-10
Different aspects of continental topography are illustrated in this satellite photo of the Pacific Coast of South America at Antofagasta, northern Chile, looking across Chile toward Argentina. In the foreground, the irregular coast is dominated by mountainous terrain. Inland are plains and plateaus, and in the distance are the Andes Mountains. In the middle background are large basinal depressions filled with salt flats. [From National Aeronautics and Space Administration.]

makes them tilt and move. Heavy enough rain produces a **sheet wash** on even slopes—a continuous sheet of flowing water. Sheet wash does not normally persist very far down a slope before it organizes into tiny channels or rills, but a good place to see it is on sloping concrete embank-ments, such as those near bridges or culverts. The impact of raindrops and hailstones is sufficient to erode clay and silt and help lift them into the moving slope wash. Slope wash operates most effectively in such places as freshly graded highway shoulders and around new houses. The best way

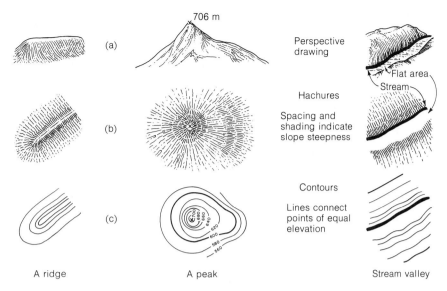

706 m

(a)

(b)

(c)

A ridge

A peak

Perspective
drawing

Hachures

Spacing and
shading indicate
slope steepness

Contours

Lines connect
points of equal
elevation

Flat area
Stream

Stream valley

Figure 5–11
Three topographic features, a ridge, a peak, and a stream valley, as represented (a) by perspective drawings as they might look from the ground; (b) by hachures, short straight lines that indicate the degree and direction of slopes as they might look from the air; and (c) by contours, lines that connect all points of equal elevation and thus more accurately represent the configuration of the surface. The more closely spaced the contour lines, the steeper the slope. Normally, every fifth line is heavier in weight and is numbered to indicate the elevation.

to avoid excessive erosion by slope wash is to plant grass and other vegetation. Leaves and stems shelter the surface from raindrops and retard the flow of slope wash; roots hold the soil together to resist erosion.

Though it is hard to get reliable figures on the quantity of material moved by slope wash and mass movements in general, almost all of the enormous load of erosional debris that eventually enters the rivers of the world must have first moved downhill by various combinations of those processes. The removal of earth materials, at different rates and by different processes, creates the forms of landscape. Fragmentation, weathering, and mass movements, in combination with rivers, the wind, and glacial ice, sculpture the land. We now turn to the landscape to see what we can deduce from it.

THE FORM OF THE SURFACE: HIGH AND LOW

What does the surface of the land look like from an orbiting satellite? The first impression is of a smooth sphere. A closer look shows prominent mountains and canyons, the surfaces departing most dramatically from the undulating contours of the plains or lowlands that characterize much of the Earth's surface (Fig. 5–10). Lakes and rivers fleck the continents, but it is the world-encircling ocean that makes Earth the blue planet. If that ocean were siphoned off, the sea floor would show some of its own kinds of mountains, plains, and valleys. Prominent mountains are also hidden by the South Polar ice cap on the continent of Antarctica and by the Greenland ice cap near the opposite pole.

How do we describe this varied surface of Earth? The heights and depths that give shape to the Earth's surface are called its **topography.** The standard level, or **datum plane,** to which we compare all heights or depths is usually **mean sea level,** the average between high and low tides. Vertical distances above mean sea level are called **elevations,** or **altitudes.** A **topographic map** shows the distribution of elevations in the area covered: in its two dimensions, the map gives a representation of the three-dimensional view we would get from looking down from the air. On most such maps this is done by means of **contours**—lines that connect points of equal elevation (Fig. 5–11). Less commonly used in topographic maps for professional use, but often used in atlases and road maps, are shadings. Hachures, short lines that indicate slope, give a much more realistic but less

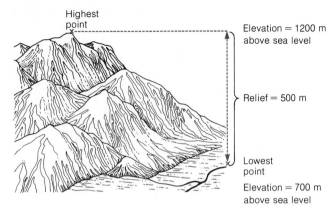

Figure 5-12
Relief is the difference between the highest and lowest elevations in a region.

precise idea of the land surface. Both contours and hachures bring out another measure of topography, its roughness, by illustrating the **relief**—the vertical distance between the highest and lowest points in the map area (Fig. 5-12). On a contour map relief is estimated by subtracting the elevation of the lowest contour, down in a river bottom, from the highest, on a hill or mountain top. Climbing a mountain is the best way to experience what relief means. It is one matter to climb to a 3600-meter (12,000-foot) peak from a high plateau of 3000 meters (10,000 feet) but quite another to climb a 2400-meter (8000-foot) peak starting from a few hundred meters above sea level.

A CONTRAST IN TOPOGRAPHY

There is much more to landscape than the grossest features seen from the great distance of a satellite. The infinite variety of form and detail that we look upon as scenery holds much of the information we need to deduce how it was made. Again we visit the Grand Canyon, this time to explore the natural architecture of its surface. By way of contrast, we then explore the Ozarks.

The Grand Canyon Landscape

The canyon is deep, broad, rocky, and ruggedly mountainous (Fig. 2-6). The Colorado River is rarely visible from the rim, though one can make out the edge of the gorge in which the river has been cutting its way downward for the past 15 to 20 million years. Steep cliffs compete with the more gentle rubbly slopes for our attention. What we see are the effects of **erosion**, the combination

of chemical and physical weathering acting together with the direct transporting power of the river in its channel and of the wind. In the Grand Canyon we can reconstruct the thought processes of James Hutton and other early geologists. We can see that the river has removed enormous quantities of rock by tracing such formations as the Kaibab Limestone from one rim to the other. Once a continous sheet, as it still is under the plateau back from the canyon, it has been breached, broken up, and taken away.

Merely by sampling its water, one can demonstrate the power of the Colorado River as an agent of erosion. Colorado River water, as engineers in charge of purifying it for drinking well know, is turbid with suspended fine clay. Sand in the river channel is constantly in movement. A tremendous volume of sand, silt, and clay is being transported all the time. One measure of this volume is the amount of sediment that is dropped by the river as it enters Lake Mead, created by the building of Hoover Dam. Since the completion of the dam in 1936, many millions of tons of sediment have accumulated on the lake bottom.* A close look at this sediment shows sand grains of the kind found in the Coconino and other sandstone formations, mica flakes that could have come from the Vishnu formation, and small pebbles of almost every kind of rock found in the canyon. It is clear that the river is a prime agent of removing weathered rock material from the canyon.

Now focus on the topography of the canyon. In summarizing the stratigraphic sequence in Chapter 2, we used the nature of slopes and cliffs to pick out formations. Steep cliffs are made of hard rocks that are resistant to erosion (Fig. 5-13). In the Grand Canyon these resistant rocks are limestones, and to a lesser extent sandstones. The gradual slopes, usually covered with talus, are underlain by shales. This differential response to erosion shown by more and less resistant rock types is one of the major controls on topography. Cliffs are formed as soft, less resistant, weak rocks erode and undercut overlying resistant rocks, which then break off in large blocks.

The valley of the Grand Canyon is much broader than the river itself. The river gorge is several times wider at the top than at the bottom, and a cut through it would show a simple

---

*This silting up is a common and unfortunate consequence of the erosional power of rivers. Sediment that would normally be carried downstream is deposited in the artificial lake formed by a dam. Some reservoirs are filling with sediment at so great a rate that the useful life of the dam for flood control and irrigation purposes is shortened to a few decades, hardly enough to have made such expensive projects worthwhile.

V-shaped profile. Above the inner gorge, in the main canyon valley, the profile is much more complex, but can be generalized to a much broader V-shaped valley. As the river cuts downward, the valley gradually widens. The metamorphic rocks of the inner gorge are much more coarsely crystalline and less porous than the rocks above and thus weather more slowly. Being much more resistant to erosion than the weaker Paleozoic rocks above, the Precambrian rocks support narrower canyon walls. Above, formations like the Bright Angel erode rapidly and broaden the valley, for shales fragment easily and disintegrate in water to form silt and mud.

The Grand Canyon teaches us about the nature of erosion of a river valley in terrain where most of the formations are horizontal sedimentary rocks. Another part of the erosion story emerges when we contrast the Grand Canyon with an area of different topography.

The Ozarks

The Grand Canyon has developed rather recently. Age is always such an important factor in geology that we ought to look at old mountains. One example in the middle of the continent is the Ozark Mountains. Situated in Missouri and Arkansas, the Ozarks, which are sometimes called a plateau, consist of a granite core surrounded by relatively flat-lying sedimentary rocks resting on the flanks of the granite. The sedimentary lithologies are the same as many of those in the Grand Canyon. The Ozarks are not in the same league with the Rockies in height or relief, but they do include some fairly hilly topography. The granite of the Ozarks shows the signs of moderate weathering. Outcrops are rounded and cracked, the joints enlarged by solution. The sedimentary rocks are more deeply weathered, and soils are well developed on them.

The slopes of the Ozarks are gentler and more rounded than those of the Grand Canyon. The valleys are broad V-shaped. Steep cliffs are few, and the streams meander along fairly wide valley bottoms. The dominant aspect is one of less vigorous erosion; the streams here carry much less detritus than high mountain streams.

Wherever we find older mountains, we see the low elevations, low relief, gentler slopes, and broader river valleys that typify the Ozarks. Older mountains are tectonically "quiet"; they are rarely the sites of earthquakes or volcanoes, nor do they show any evidence of recent rock deformation or metamorphism. The conclusion: once

Figure 5-13
The varied nature of the slopes of the Grand Canyon is an expression of the resistance to erosion of the kinds of rock; the steeper cliffs are formed in sandstones and limestones and the gentler, rubbley slopes formed on shale. [Photo by Edwin D. McKee, U.S. Geological Survey.]

the tectonic machine that created the mountains in a region has run down, erosion operating over a long time wears mountains down to low hills.

THE OVERALL PATTERN

In describing the formation and erosion of mountains, we might put it that "tectonics proposes and erosion disposes." The mechanisms that control the heights of mountains display an overall pattern that we can call a *negative-feedback process.* In this kind of process the results of a first (primary) action induce a proportionate secondary

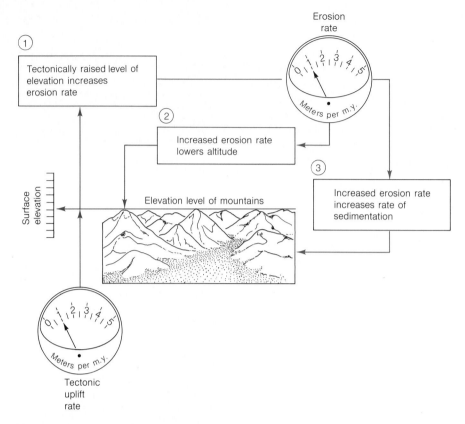

Figure 5-14
The negative-feedback loop that relates (1) uplift and erosion, (2) surface
elevation, and (3) sedimentation. Tectonic uplift causes an increase in erosion rate, which in turn lowers the surface elevation and increases the sedimentation rate. The elevation is thus a balance between tectonic uplift
and erosion rate.

action, and that secondary action reduces the first
action. A familiar example is what happens when
a thirsty person drinks a glass of water. The more
thirsty the person, the faster the gulping at first.
Just as the thirst induces the drinking, so the
drinking acts to decrease the thirst. As the thirst
decreases, the rate of drinking decreases; when
thirst is satisfied, drinking stops.

In the geological case, strong tectonic action
elevates mountains, provoking intense erosion
(Fig. 5-14).* The higher the mountains, the faster
erosion wears them down. But as long as mountain-building continues, tectonics prevails and
altitudes increase. As tectonic movement
slows—not because of any important effect of
erosion but because its own machine starts running down—the mountains rise at a slower pace.
For a time erosion keeps up with uplift, and the

mountains do not change in elevation. Then as
uplift slows further, erosion becomes dominant
and the elevations begin to lower. As the lowering
proceeds, the erosion slows too, the whole process
eventually tapering off. The relief is constantly
diminished by wearing away of the mountain
tops and filling in of valleys and low spots by sedimentation of the erosional debris. Sedimentation,
the consequence of erosion, acts to depress relief.

The World Distribution of Elevations

The relation of tectonics to erosion not only explains local and regional topography; at the largest scale, it also underlies the world distribution
of elevation. From accurate maps of Earth's topography, we can draw a **hypsometric diagram,** a
representation of the relative proportion of the
surface lying at each elevation. We make such a
diagram by plotting a graph of elevation against
the total area lying at that elevation (Fig. 5-15).
From these diagrams we can see that the maxi-

---

*No matter how extreme the uplift, there is a limit to moun-
tain heights. That limit is determined by the strength of the
rigid crust, which can support only so much weight before it
deforms plastically in response to any additional load.

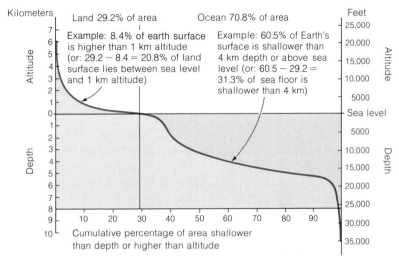

Figure 5-15
Hypsometric diagram of the surface of the Earth. The curve shows the
relative amounts of land and subsurface areas lying at various elevations
with respect to sea level by a plot of cumulative percent of area shallower
than a given depth or higher than a given altitude.

mum height of mountains above sea level, about
8900 meters (29,000 feet), is not much different
from the maximum depth of the deepest ocean
trenches below sea level, about 10,800 meters
(35,000 feet). Compared to the areas of land and
sea floor that are relatively close to sea level, such
extremes are few. The great bulk of the continental land surfaces lie within a few hundred meters
of sea level, and much of the ocean floor lies at
intermediate depths. Together with the relation of
erosion to uplift, this suggests strongly that on the
continents the land surface tends to be reduced to
a little above sea level. On the sea bottom the
tendency is to maintain some intermediate depth
below sea level; this tendency must represent
some other balance of forces than erosion coupled to uplift. Another inference that comes from
this diagram is that the recent tectonic activity
that has caused great changes in elevation on land
or in the sea, such as formation of mountain
ranges or trenches, is distributed in relatively narrow, restricted belts, a small proportion of the
total surface. Those belts are the edges of lithospheric plates of the present and recent past. The
face of the Earth is scarred by plate margins. The
older plate boundaries of the geologic past are
now healed and largely obliterated by later erosion and sedimentation, but here and there, in
places like the Appalachian Mountains, the old
scars show through.

CLIMATE AND TOPOGRAPHY

Freezing, thermal expansion, and the solvent
action of water, all strong factors in weathering,
are related to temperature and rainfall. So the relation between climate and erosion is clear. Moreover, since erosion shapes topography, climate
also influences topography—and is affected by it
as well. Logically then topography must have
some control on mechanical weathering. Here is
another example of how the results of one process
affect other related processes (see Fig. 4-20).

Rainfall and temperature indirectly affect mechanical weathering and erosion too, for rainfall
is the source of river water, which helps break up
and abrade rock and mineral particles. In addition to the obvious cooling effect of high altitudes,
topography also has various other effects on climate. For example, mountains cause rain shadows—dry areas on the lee slopes (See Chapter 8).
Chemical weathering is hastened in warm, humid
climates, and as that decay promotes mechanical
weathering, these climatic factors indirectly affect
mechanical weathering. In more rigorous climates, frost action and extreme daily variations in
temperature are direct causes of rock disintegration. One of the most important ways climate
works is by its control on vegetation, the direct
agent of much chemical alteration of rocks. Li-
chens—leafless, primitive plants—encrust rocks

in many areas even at high altitudes where other plants do not grow. They contribute in their small way to chemical weathering. A dry climate favors physical weathering over chemical weathering by slowing the decay produced by the solvent action of water and, equally important, by preventing the growth of much vegetation. Polar climates also inhibit the growth of plants and bacteria; chemical decay is minimized because in the extreme cold little moisture remains in the air.

Topography, Erosion, and Tectonics

If climate influences topography, and vice versa, it is also true that topography strongly controls weathering and erosion. High altitude and relief enhance mechanical weathering, partly by promoting freezing and thawing, but also because the fragmented debris on mountains quickly moves downhill, exposing new fresh rock to attack by the weather.

A climb up a typical high mountain shows evidence of high rates of mechanical erosion. At higher altitudes the soil thins. Above the timber line where nothing but shrubs and lower vegetation grow, soil becomes spotty, and much of the landscape is bare rock. Slopes of blocky talus alternate with steep cliffs and craggy points. The rocks lose the discolored, sometimes lichen-covered, rounded look that chemical weathering produces at lower elevations; instead the angular blocks have the appearance of freshly quarried stone and quickly wear out the boots of the climber. Talus slopes here may be unstable. A small disturbance, one careless step, may set off a roar of cascading blocks that can crush almost anything in its path. In the cold, hostile environment at the top, the wind drives snow and ice across bare rock.

Field geologists have long been well acquainted with the high rates of mechanical erosion at high elevations; from the evidence of these high rates and relatively little evidence of chemical decomposition, they conclude that the rate of mechanical weathering is high relative to the rate of chemical weathering in rugged mountains. This does not mean that chemical decay is inactive in high mountains, merely that its obvious effects are eclipsed by fragmentation. A comparative study of the Amazon River and the Andes Mountains found that the high Andes were the scene not only of accelerated mechanical weathering but of faster chemical decay as well—in spite of the cool, dry climate of those high altitudes. The extensive chemical decay was just not apparent in the high outcrops because the products were so quickly stripped away by mechanical erosion. The flatter lowlands of the Amazon jungle, where rotten granite cobbles and thick soils show the dominance of chemical weathering, are eroded much more slowly, both chemically and mechanically, than the uplands. In lowlands, where detritus is not removed so rapidly, the mechanical effects are masked by the chemical.

If topography is so important in weathering, a search for even more fundamental controls on weathering leads us from topographic effects to the deep-seated tectonic movements that determine the heights of mountains and depths of lowland basins. Ultimately it is those tectonic forces, which push up mountains, that influence the nature of erosion, and it is erosion that in turn affect's the kind of debris produced. Thus the rock and mineral fragments of ancient sediments produced by erosion give evidence of historical events. This chain of reasoning has already been touched on in Chapter 4. In the following section we explore more fully how tectonics can be deduced from sediments.

Ratio of Mechanical to Chemical Weathering

We can express the course of erosion in terms of a ratio of mechanical to chemical weathering. For example, in arid and very cold climates the ratio is high: mechanical breakup dominates. In high mountainous terrain the ratio is also high, for even though chemical weathering is active, mechanical fragmentation strongly predominates. In low plains, the ratio is low, for even though chemical decay is somewhat slower than in high mountains, mechanical erosion is so minor that chemical weathering is predominant.

We can gauge the ratio of past weathering processes by looking at the mineralogy and size of the detrital particles that make up the sedimentary rocks produced by erosion of lands that have long since disappeared. The more unstable the minerals and the larger the particle sizes, the higher the ratio must have been (Fig. 5-16). A coarse sandstone with abundant feldspar, biotite, and other minerals that are easily chemically weathered speaks to us of rapid erosion in mountains. Rocks containing large angular cobbles and pebbles are even more eloquent. Fine-grained quartz sandstones, siltstones, and clay-rich shales suggest low-lying source areas where chemical alteration was dominant.

But how do we sort out the effects of climate from those of topography? A conglomeratic sandstone might have come from a low-lying arid region instead of a high mountain, and shales and siltstones might have come from wet, warm hills. The problem is one that is not yet completely settled, but most evidence seems to support the notion that topography is more important than climate in determining the ratio of mechanical to chemical weathering. Moreover, it has become clearly recognized that topography influences climate much more strongly and directly than climate does topography. Finally, because topography strongly controls the rate at which debris is removed from eroding terrain, it thus promotes both chemical and mechanical weathering. No matter how wet the climate and how great the discharge of rainwater through rivers, mechanical erosion does not have much effect on a low-lying plain.

The logic of tectonic control of erosion, and thus sedimentation, is now apparent. Tectonic movements create mountainous topography and the kind of climates associated with mountains. Mountains are the places where the ratio of mechanical to chemical weathering is high, producing detritus rich in course, unstable rock and mineral fragments. The ancient sedimentary rocks made up of such debris are the key to the mountains of the past. The mountains of the present, though, are the ones we can observe. They and the myriad other forms of the land can tell us even more about how landscape is shaped and why it is different in one place than in another.

LANDFORMS:
THE PHYSIOGNOMY OF THE EARTH

The shapes that give variety to Earth's landscape have long been subjects of interest to artists and scientists, both of whom may look with analytical eyes. In the poem "New Hampshire," Robert Frost writes:

The Vermont mountains stretch extending straight;
New Hampshire mountains curl up in a coil.

Even though he did not use technical terms, Frost was being a **geomorphologist,** a student of the morphology (form) of the Earth. Poets, however, are free to use words in ways that scientists are not. Scientists must be able to communicate with as much precision as possible; hence the language of science is one of unambiguous description and

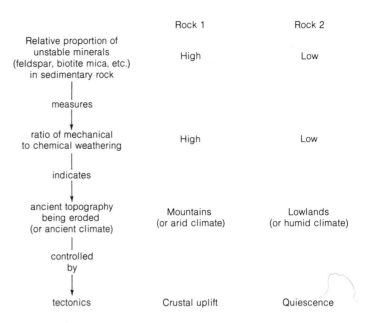

Figure 5–16
The relative proportion of stable and unstable detrital minerals in sediments can be used to infer the erosional origin of the detritus, and thus the topography and ultimately the tectonics of the source area.

analysis, and much of the modern vocabulary is of the scientist's invention. Most of the names of landforms, however, are older and come from the common language. There are hills and valleys, highlands and lowlands.

From Mountains to Molehills

As it turns out, even some of the most ordinary names of landforms are not easy to define precisely. We have used the word mountain many times in this book, yet we can define it no more precisely than to say that a mountain is a large mass of rock that projects well above its surroundings. A mountain can be a single peak, or it may be combined with others in mountain ranges, where it is hard to pick out distinct separate mountains but only peaks of various heights (see the figure in Box 20–2). How do we distinguish mountains from hills, other than by size? We do not. The gamut of size is continuous, from a 5-meter (16-foot) knoll good for playing "king-of-the-hill" to Mt. Everest, elevation 9524 meters (29,028 feet); precise limits are only arbitrary. We learn to accept that the Black Hills are higher than some Appalachian Mountains. In general, positive landforms more than a few hundred meters in elevation are called mountains.

Figure 5-17
Monument Valley, Utah. Mesas and pillars of sandstone rising above shale slopes.
Two structural terraces sustained by resistant sandstone beds are visible in the distance.
[Photo by Tad Nichols, Tucson, Arizona.]

A broad, flat area of appreciable elevation above its surroundings, at least on one side, is a **plateau.** Smaller plateaus are sometimes called **table lands,** and in the western part of the United States, some are called **mesas. Butte,** a French word for hill, is used, as is mesa, for smaller elevations with flat or rounded tops surrounded by steep slopes on all sides. Few plateaus have elevations much higher than 3000 meters (10,000 feet), but the Altiplano of Bolivia is at an elevation of 3600 meters (12,000 feet), and the high Tibetan plateau has an average height of almost 5000 meters (16,000 feet). Something in the nature of erosion produces peaked mountains when the terrain is higher than that. Many plateaus are relatively flat because they are floored with undeformed sediments or layers of lava flows (see Chapter 15 for a discussion of plateau basalts).

Though all hills and mountains are the result of upward tectonic or magmatic movements, their forms are dictated mainly by the erosional process, the rock of which they are formed, and its structure. As in the Grand Canyon, shales almost always form gentle slopes, much of the bedrock being completely covered either by a rock rubble or a soil, the difference depending upon climate and vegetation. Limestones are resistant to erosion in dry climates and tend to form steep cliffs.

But in the wet tropics, limestone erodes quickly, mainly by dissolving in water (see Chapter 4), and even small outcrops may be hard to find, much less steep cliffs. From these comparisons it is apparent that an important variable of landscape is the nature of the hillslopes. Are they steep or gradual, rock-covered or mantled with soil, convex or concave? A little further on in this chapter we will discuss how slopes can be used as important criteria for understanding landscape evolution. The relation of slopes to rock structure we discuss here.

Structural Control of Topography

The folds and faults produced by rock deformation in the course of mountain-building leave their mark on the Earth's face. These topographic expressions of deformation are often a guide to the structures that control them. Even the shapes of features of somewhat smaller scale, hills, valleys, and stream courses, are controlled by a complex interaction of erosion and structural elements.

Mesas, buttes, and the complex pattern of erosion of the Grand Canyon are forms that develop in many arid and semi-arid regions in horizontal, undeformed sedimentary rocks or lava beds with alternately resistant and easily eroded layers (Fig. 5-17). The resistant layers support the elevated flat-topped hills and make the cliffs; the softer layers below may erode to form gentler slopes or to undercut the capping strata. In horizontal beds of uniform resistance, the topography is much more homogeneous, all the hills and valleys having about the same slope.

In a tilted and eroded series of alternating resistant and soft beds, asymmetrical ridges called **cuestas** tend to form (Fig. 5-18). One side of a cuesta has a long, gentle slope determined by the angle of inclination of the resistant bed; the other side is a steep escarpment formed at the erosional edge of the resistant bed where it is undercut by erosion of a weaker bed beneath. Much more steeply inclined or vertical resistant beds erode to form **hogbacks**—narrow, steep ridges of similar but more accentuated shapes than cuestas (Fig. 5-19). Escarpments are also produced by near-vertical faults along which one side has been raised relative to the other. Faults can displace topography in such a way that the sides of hills appear sheared off as by a knife, river valleys show sudden detours, and ridges are offset (see also Fig. 20-53).

In the early stages of folding, the upfolds, **anticlines,** form ridges; the downfolds, **synclines,** form valleys (Fig. 5-20). Streams flow down the flanks of the anticlines and follow the course of the synclinal troughs. As continuing erosion bites deeper into the structure, the presence of resistant and soft beds becomes a controlling factor. For example, if the core of an anticline is soft, it may be eroded away to form an anticlinal valley (Fig. 5-21). If further erosion reachs a hard layer, an anticlinal mountain may once again form. In a region that has long been eroded, a pattern of linear anticlines and synclines produces a series of ridges and valleys, such as those of the Appalachian Mountains in Pennsylvania and adjacent states (Fig. 20-21). The positions of the ridges versus valleys are determined by the erosional resist-

Figure 5-18
Cuestas formed by gently inclined beds of hard sandstone, New Mexico. [Photo by W. T. Lee, U.S. Geological Survey.]

Figure 5-19
The eastern edge of the Front Range of the Colorado Rockies shows a prominent hogback of the steeply dipping Dakota Sandstone in the center and a lower hogback of the Niobrara Limestone to the right. [Photo by T. S. Lovering, U.S. Geological Survey.]

Figure 5-20
Valley and ridge topography formed on a folded terrain of sedimentary rock. The deformation is so recent (Pliocene) that erosion has not yet significantly modified the original structural forms of anticlines (ridges) and synclines (valleys). The circular dark areas in the foreground are salt domes formed when deeper salt formations were squeezed upward through overlying sediments as cylindrical masses. The scene is in Zagros Mountains, Iran, looking west over the Persian Gulf. [From National Aeronautics and Space Administration.]

ance of the rocks rather than the original anticlinal highs or synclinal lows formed by structural deformation.

The Low Spots

What is the opposite of mountain or upland? Is it a lowland plain or a valley? Those two different shapes, broad versus narrow in area, correspond to two ways in which negative elevations form. A plain is a relatively flat low area that is not being pushed up by tectonic activity. We can generalize further that lowlands are broad flat areas of two kinds. They are either stable, moving neither up nor down, or they are unstable, truly the opposite of mountains. Unstable lowlands are moving down, a tectonic process we call **subsidence.** This results in **basins,** depressions of various shapes, frequently found between mountain ranges, in which case they are **intermontane basins.** Valleys,

canyons, arroyos, gulches, and gullies all have the same general kind of geometry, and all are topographic lows that form by erosion by rivers in mountains and plains, highlands and lowlands, in stable and unstable areas. These low spots are less extensive than the large land areas of plains and lowlands. A closer look at lowlands shows that much of their large area is made up of smaller, broad, flat river valleys. Thus valleys and plains are distinguished on the bases of scale as well as process.

A special kind of valley—the tectonic valley—is long, narrow, relatively flat-floored, and bounded on both rims by faults (Fig. 20-53). Some of these valleys are occupied by major rivers, the upper Rhine being one of the most famous. The Great Valley of California, the route of the San Joaquin and Sacramento rivers, is also a tectonic valley. A distinctive variant of the tectonic valley is the **rift valley,** formed by incipient or active spreading apart of lithospheric plates. The great African rift

(a) Early stage of folding:
ridges over anticlines,
streams in synclines

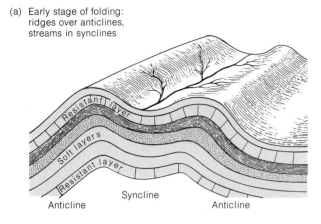

(b) Later stages of erosion:
ridges may overlie synclinal axes
if capped by resistant beds

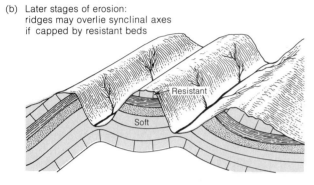

Figure 5–21
Two stages of the development of ridges and valleys
in folded mountains. The original positions of ridges
can be determined by the shape of the structure, as
in (a). At later stages (b), however, anticlines may
be breached and eroded more rapidly through nonresistant rocks, and ridges may be held up by caps
of resistant rocks.

Figure 5–22
Large sinkhole, about 130 × 100 × 45 m (425 × 350 ×
150 ft) that formed in 1972 in central Alabama as a
result of collapse of cavernous limestone below the
surface. This may be the largest sinkhole to have
formed in recent years. [From U.S. Geological
Survey.]

valleys are occupied by large lakes, such as Lake
Tanganyika. The River Jordan and the Dead Sea
occupy another rift valley. Beneath the ocean,
great rift valleys form the axes of mid-ocean
ridges. In fact, it was the discovery of the mid-
ocean rift system in the late 1950s that sparked
much of the exploration of sea-floor ridges that
led to the concepts of sea-floor spreading and
plate tectonics. It was topography that gave the
first clue.

A walk around the Mammoth Cave district in
Kentucky reveals an entirely different kind of
negative topography, for in walking across a
meadow you might come upon a deep, round
hole, a **sinkhole** (Fig. 5–22). Some of such holes
are filled with water; others quickly "drink up" all
the water from a rainfall and remain open. Sinkholes may suddenly appear in urban areas too, as
happened in May 1981 in several cities in Florida,
where gaping holes swallowed cars, pavements,
and entire houses. Streams disappear into the

ground at some sinkholes, and great gushing
springs well up from others. This kind of land surface, called **karst topography,** is produced in
areas of plentiful rainfall where the bedrock is
limestone. As in Kentucky, it is associated with
caves in the limestone formations. The irregular
narrowing of openings in all directions and the
signs of dissolution on both ceilings and floors in
many caves serve as evidence that they were
formed while completely filled with water and
thus well below the surface, at depths where
water fills all pores and cavities (see Chapter 6).
We can now walk around in these caves because
the water has since dropped as a consequence of
regional uplift or changes in climate or topography. **Stalactites,** icicle-like forms of newly precipitated limestone hanging from the cave roof, are
formed where groundwater drips down through
the cave. In the same way, the drip builds up a
stalagmite on the cave floor below the stalactite.
All these and other forms of **dripstone** are the
hallmarks of later stages of cave formation above
the water table. The conclusion is inescapable
that the main erosive process in karst regions is
solution of limestones by water. It is a chemical
topography. How karst and caves are formed by
groundwater is described in Chapter 6.

Low areas, whether stable or subsiding, rarely
persist for long because they are the natural
dumping grounds for sediment. Many intermontane basins of the Rocky Mountains would

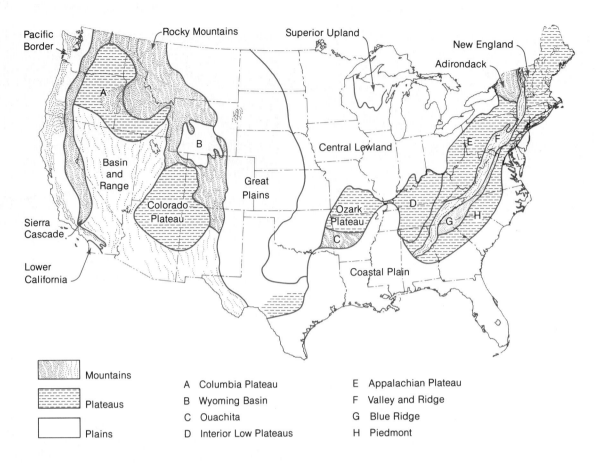

Figure 5-23
The main landforms of the United States. About one-quarter of the land is mountains, one-quarter plateaus, and about half plains. A more detailed map of these regions is shown in Figure 20–42.

be far deeper than they now appear if all of the gravels, sands, and clays eroded from the surrounding mountains were removed. The Gulf Coast country of Mississippi, Louisiana, and eastern Texas has been subsiding steadily for many millions of years, but since sedimentation there keeps pace with subsidence, the land stays above sea level. Other forces help shape landscape into distinctive landforms. The landscapes sculpted by glacial ice are among the most varied and complex (see Chapter 9). Desert topography owes its character in large part to wind action (Chapter 8).

The Face of North America

The landform map of North America shows the main relief patterns of the large topographic units and how clearly the erosional forms relate to tectonics and sedimentation (see Figs. 5–23 and 20–42). The major mountain chains are in the East and in the West; the older Appalachians are much lower and less rugged than the Rockies. Between

are the prairies and plains; their gentle southward slope is well indicated by the drainage pattern. All rivers run to the Gulf of Mexico. A low **coastal plain** lies along much of the central and southern Atlantic coasts and the Gulf coast. North of the Great Lakes are the low-relief plains and lakes of Ontario and Quebec. West of the Rocky Mountains is the Basin and Range province, a region of many smaller chains of mountains alternating with elongate basins. This is the area some geologists see as a potential huge rift valley that may one day separate the continent into two parts. West of the Basin and Range come the Sierra Nevada, the Cascades, and, to the north, the British Columbian Rockies. The western edge of the continent is marked by the Coast Ranges on the far side of the Great Valley of California, the valley of the Willamette River in Oregon, and its northward extensions. Alaska and the Yukon of Canada are extensions of the Rockies, or, as the whole western mountain belt of the Americas is called, the **Cordillera** (from the Spanish for "rope," to describe long narrow mountain ranges).

Figure 5-24
The eastern slope of the Sierra Nevada, a great fault escarpment nearly 2 miles (3.2 kilometers) high. The view is from the Owens Valley, Inyo County, California. These young mountains are high, steep, and rugged. [Photo by W. C. Mendenhall, U.S. Geological Survey.]

The history of this whole continent is dimly shown by its geomorphology, each event being seen through the veil of later episodes of mountain-building, erosion, and sedimentation. If geomorphology is to be of historical value, however, its analysis has to be based on a theory of landscape evolution. Is there some orderly sequence by which mountains are worn down to plains?

EVOLUTION OF LANDSCAPE

What is the evolution of the topography of a mountain range or a low plain? If it hasn't always existed in its present form, then it must have existed in some preceding state or stage, and it will necessarily give way to a succeeding one. Perhaps the best way to answer the question is to study some young mountains, such as the Sierra Nevada of California (Fig. 5-24); their characteristics can then be compared with those of older mountains that have been exposed to erosion for a long time (Fig. 5-25). The youngest individual mountains are recently erupted volcanoes, but the youngest large mountainous region is the Himalayas, pushed up by the collision of the Indian subcontinental plate with the Eurasian plate. We can date the major uplift as Pleistocene because that is the age of the youngest sedimentary layers uplifted, but uplift is undoubtedly still going on. The mountains are the highest in the world, the relief is great, and the slopes and peaks are steep and rugged. Though the height of a volcano like Mt. Pelée may not be so great, it is similar in ruggedness and relief, and its slopes are as steep as can be held up by the materials (see Chapter 15).

Figure 5-25
The Great Smoky Mountains National Park in the Appalachian Mountains of Tennessee. The generally gentle slopes are those of a tectonically quiet mountain belt in an advanced stage of erosion. [Photo by W. B. Hamilton, U.S. Geological Survey.]

When we study old mountains like the Appalachians, we find that their heights and relief are much lower. Although there may be many cliffs that are tricky to climb, most slopes are more gradual, and the mountain tops have a softer, rounded appearance.

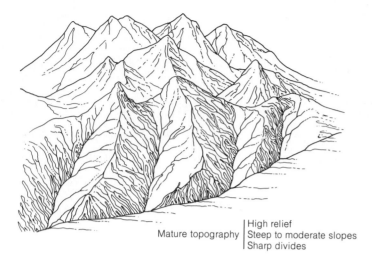

Mature topography | High relief
Steep to moderate slopes
Sharp divides

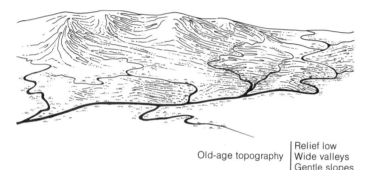

Old-age topography | Relief low
Wide valleys
Gentle slopes

Figure 5–26
Two types of topography thought by W. M. Davis to be mature and old-age stages of an evolving landscape as it is worn down by rivers.

Here, then, is the germ of an idea for a time ordering of the sequence of stages in wearing down mountains. William Morris Davis, a Harvard geologist, brought this kind of study of topography to its fullest expression at the end of the nineteenth century when he studied mountains and plains all over the world. He characterized the **geomorphic cycle** of topography as progressing from the high, rugged mountains of youth to the rounded forms of maturity to the worn down plains of old age (Fig. 5–26). Perfect plains, the theoretical end-products, are never realized, Davis said; **peneplains,** relatively flat erosion surfaces that approach perfect plains do result from very long periods of erosion. Flat surfaces of unconformities, such as the one at the top of the Precambrian in the Grand Canyon, must represent something like a peneplain. Here and there, standing uneroded above such a plain, would be erosional remnants, perhaps of more resistant rocks. Davis believed Mt. Monadnock in Southern New Hampshire to be such a hill and named all such **monadnocks** after it.

The geomorphic sequence has proved useful in deciphering geomorphic histories; the assumptions underlying this concept do not stand up quite so well as the sequence itself. The process is now looked upon as one that a chemical engineer might describe as a "continuous-flow" process, in which all parts of the operation are going on simultaneously rather than a "batch" process in which the operation stops and starts periodically.* Davis thought mountains to have been elevated over a very short time period, almost instantaneously from a geological point of view (a batch), and then to have stayed fixed tectonically as erosion slowly wore them down through the stages of the cycle. Today our knowledge of the long, drawn-out, intermittent, and uneven nature of mountain-building makes that assumption untenable. It is more of a continuous process by which mountains keep on moving up while they erode, the net effect being determined by shifts in the balance between the two (Fig. 5–14). Another assumption that flawed Davis' theory was that the sequence is characteristic of all rocks in all climates; it is now generally agreed that progression from youth to old age can produce a variety of other forms, particularly in arid climates.

Current views of geomorphic evolution emphasize the **steady state,** in which a balance of forces results in a topography unchanging with time, though rocks are steadily being uplifted and eroded. The progression from youth to old age occurs only after tectonic activity dies down, and it may follow different paths in different regions. Such a concept tends to deny the possibility of reconstructing much of the geomorphic history of the region, for it says that what a region is like is the product of currently active operations, not what went before. What it will be next has little to do with what it is now. Still more complex models incorporate all of the operating variables: tectonics, rock type, extent of rock weathering, and drainage in relation to climate and vegetation, together with the possibility of some randomness in the course of events. This leads to a low degree of predictability and some sense of disillusion for those who had hopes for a thoroughly orderly and predictable pattern.

---

*One way to illustrate the difference between a batch process and a continuous-flow process is to contrast two ways of baking bread. In the home (or in an old-fashioned bakery), a batch of dough is prepared, formed into loaves, and then baked. In a modern, computerized bread "factory," the flour and other ingredients are continuously mixed in vessels that allow the finished dough to flow out at one end while the raw ingredients enter at the front. The dough is on a continuous assembly line while being formed into loaves and baked.

Isn't it strange that the part of the Earth that we see and walk over all the time is so difficult to make a simple order of? Perhaps the very wealth of information and the complexities of all of the variables that we know have an effect on landscape are what make it so resistant to analysis.

Perhaps the subject waits for those who would do as Einstein did: in describing his thought patterns about relativity, he wrote that "I had to divest myself of my intuition—then I could think clearly."

SUMMARY

1. Mass movements, including rock falls, landslides, slump, and creep are slow to rapid downhill movements induced by gravity when a slope is oversteepened by natural erosion or by construction.

2. Mudflows, debris-avalanches, and earthflows are water-saturated, nearly fluid flows.

3. Slope wash is the generalized unchanneled flow of rainwater on slopes—the prelude to organized rills, streams, and rivers.

4. Topography is the areal distribution of the heights and depths of the Earth's surface, usually represented by contour maps.

5. The Grand Canyon is a landscape developed by river erosion of a series of flat-lying sediments of varying resistance to erosion. The valley is V-shaped.

6. The Ozark Mountains are a landscape of gentler forms and low relief on a very old granite uplift that has been eroded for long geologic times.

7. In the overall pattern of landscape development, tectonic uplift, which tends to produce high altitudes, is countered by the erosional wearing down by water and ice.

8. Tectonics creates mountains, and topography strongly affects erosion. Mechanical weathering dominates in areas of high topographic relief, but is relatively less important in low-lying regions. The ratio of mechanical to chemical weathering is a function of both climate and topography, and ultimately of tectonic activity.

9. Landforms, such as hills, plateaus, and valleys, are controlled by the erosional process acting over geologic times on a variety of resistant and nonresistant rock types arranged in different structural patterns.

10. Karst topography is formed by chemical solution of limestones in areas of moderate to high rainfall.

11. The landform map of North America reveals the diverse effects of tectonic and erosional activity, modified by sedimentation.

12. Evolution of landscape, formerly perceived as simple erosional stages of youth, maturity, and old age following a single uplift, are today understood as series of states of balance between structural uplift and erosion.

EXERCISES

1. Would prolonged drought have any effect on the frequency of landslides? Why?

2. What geologic conditions would you want to check on before you built or bought a house at the base of a steep, soil-covered hill?

3. What role do earthquakes play in landslides and related phenomena? Would you advise people living in areas where there are many earthquakes to show more, less, or the same awareness of these mass movement hazards than those in areas where earthquakes are infrequent?

4. In some areas the relief of the natural topography is less than that of human architecture or construction. What kinds of topographic areas are they, and what is the magnitude of relief?

5. If you were to see the topographies of the Ozarks and the Grand Canyon 20 million years from now, assuming no great change in climate and no tectonic activity, would you expect greater or lesser differences between them than exists today? Describe what each might look like at that time in the future.

6. Describe the changes in topography that might take place if the Ozark Mountain region were to be uplifted by 5000 meters (a) very rapidly, and (b) very slowly.

7. In flying across the country, what topographic features would you look for to tell (a) whether you were flying over structurally deformed or flat-lying sedimentary rocks, and (b) whether you were flying over beds or rock bodies of uniform resistance to erosion or varying resistance to erosion.

8. Does a mountain peak on land erode at the same rate as a seamount rising from the sea floor?

9. Describe three different geologic processes by which deep valleys can form on land. Which one(s) of these may also be responsible for deep valleys on the sea floor?

10. Consider two mountain ranges of the same age, elevated by the same plate motions, one composed entirely of limestone, the other of basalt, in a humid tropical region. Which mountain range would you expect to be higher? Would their relative heights differ in an arid regime?

BIBLIOGRAPHY

Bloom, A. L., *The Surface of the Earth.* Englewood Cliffs, N.J.: Prentice-Hall, 1969.

Crandell, D. R., and H. H. Waldron, "A Recent Volcanic Mudflow of Exceptional Dimensions from Mt. Rainier, Washington," *American Journal of Science,* v. 254, pp. 349–362, 1956.

Davis, W. M., *Geographical Essays.* Boston: Ginn and Company, 1909. (Reprinted 1954, Dover, New York.)

Hunt, C. B., *Natural Regions of the United States and Canada.* San Francisco: W. H. Freeman and Company, 1973.

Jennings, J. N., *Karst.* Cambridge, Mass.: MIT Press, 1971.

Legget, R. F., *Geology and Engineering.* New York: McGraw-Hill, 1962.

Schuster, R. L., and R. J. Krizek, eds., *Landslides: Analysis and Control.* Washington, D.C.: U.S. Transportation Research Board Special Report 176, 1979.

Wahrhaftig, C., and A. Cox, "Rock Glaciers in the Alaska Range," *Geological Society of America Bulletin,* v. 70, pp. 383–436, 1959.

Young, A., *Slopes.* Edinburgh: Oliver & Boyd, 1972.

6

The Natural Water Cycle and Groundwater

Water dissolves during weathering, then carries the dissolved material away—into the ground or into rivers, most of which ultimately empty into the ocean. The movement of the Earth's waters from one place to another and the dissolved loads carried by them are parts of a continuous overall pattern: the hydrologic cycle. Groundwater accumulates by infiltration of water into soils and bedrock and reappears at the surface in springs and stream beds. Groundwater levels, and thus water supplies, are a balance between the rate of infiltration and the rate of loss by springs, streams, and pumping from wells. The evolution of surface waters and the ocean are related to the escape of gases from the interior.

WATER, WATER, EVERYWHERE . . .

In an age of rocket exploration of the planets and nightly weather satellite photographs on television, all of us quickly recognize the color, the expanse of water that has given Earth the name the "blue planet." The dry statistic that about two-thirds of the Earth's surface is covered with water does not convey the same sense of the immense amount of water as do the images from space. Geologists have long been aware of the importance of our oceans of water and all the clouds of water vapor in the sky, for that water sustains life and takes part in almost every process on the surface of Earth (Fig. 6–1). Thus it is not surprising that the study of water, the science of **hydrology,** is an important part of geology. It is also central to oceanography and meteorology.

In this chapter we shall concentrate on how water moves from one place to another on and in the crust and, in so doing, acts as the great transporting agent of dissolved material weathered from rocks. Ice and snow, a different form of water, also play their role in hydrology; for this reason we include them in our inventory of water

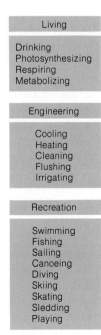

Figure 6–1
A few of the many ways in which we use water. It is so taken for granted that we are frequently unaware of water's pervasive presence.

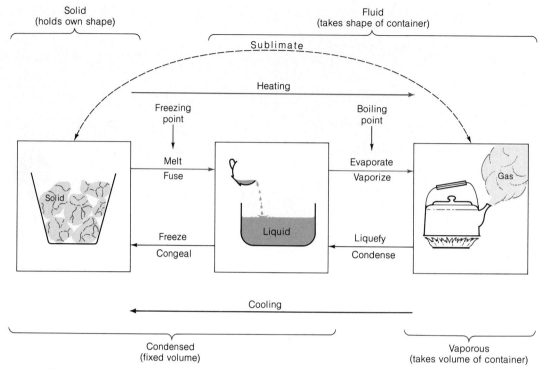

Figure 6-2
The three states of matter, as exemplified by water: ice (solid), water (liquid), and water vapor or steam (gas).

on the Earth although we consider the geologic work of glaciers later, in Chapter 9.

The compound water is so well known to us that we will use it here to exemplify the three states of matter, solid (ice and snow), liquid (water), and gas (water vapor) (Fig. 6-2). Ice is a crystalline mineral whose individual crystal forms are displayed in the enormous variety of snowflakes. A glacier is, in a sense, a "rock" body made up of an enormous quantity of interlocking crystals of the mineral ice. As a liquid, water is a good solvent and, when in motion as strong currents, able to carry mud, silt, sand, and boulders. It is in the range of temperatures at which H_2O is liquid that life on Earth exists, for water is the carrier of dissolved food that is vital for every form of life, from bacterial to human.

Water can slowly evaporate to form an invisible vapor at any temperature as well as during boiling. Even ice can "evaporate" to vapor—a process we call **sublimation.** Water vapor in air is not pure but is mixed with other gases, nitrogen and oxygen. The **relative humidity** is the amount of water vapor that is in the air compared to the maximum amount the air could hold at its present temperature. Because the amount of vapor air can hold changes with temperature, air saturated with

water vapor at one temperature will condense some of its vapor to water droplets when it is cooled to a lower temperature.

The white clouds of the sky or the billows of steam that we associate with vapor are actually composed of tiny droplets of liquid water formed when the air and vapor cool. It is as vapor that water is transported through the Earth's atmosphere—from the oceans to the continents, where it condenses and is precipitated as rain or snow.

WATER TRANSPORT: THE HYDROLOGIC CYCLE

Not only is our planet very watery, but gravity ensures that it will stay that way. Although much water is moved around from one place to another, the total is conserved.* The **hydrologic cycle** is a

---

*An occasional molecule of H_2O in the upper atmosphere splits into separate hydrogen and oxygen atoms under the effect of strong radiation from the Sun. Hydrogen is the lightest element, and some of its atoms can move fast enough to escape permanently from Earth's gravity—just like a rocket. Thus there is actually a tiny loss of water to outer space as a result of the steady loss of hydrogen from the atmosphere.

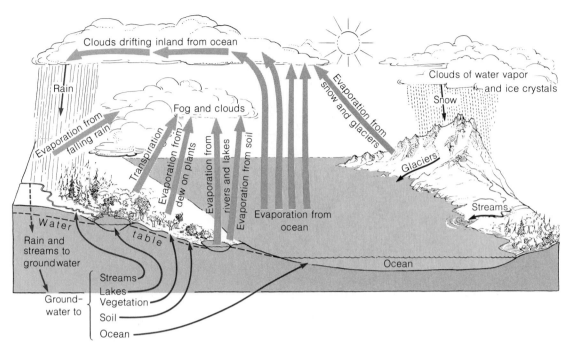

Figure 6-3
The hydrologic cycle. Movement into the atmosphere by evaporation is matched by precipitation as rain and snow. [After *Principles of Geology,* 4th ed., by J. Gilluly, A. C. Waters, and A. O. Woodford. W. H. Freeman and Company. Copyright © 1975.]

simplified description of the ways in which waters move from one place to another and of the amounts transported (Fig. 6–3).

The external heat engine of Earth, powered by the sun, drives the hydrologic cycle at the surface. It does so mainly by evaporating water from the surface of the warm oceans of the tropics and transporting it by winds, themselves driven by the temperature differences between the hot and cold parts of the globe. The water in the atmosphere condenses to clouds and eventually falls as rain or snow. Much of the rain soaks into the ground by **infiltration** to form **groundwater.** What does not soak in collects as **runoff,** which finds its way into streams and rivers and runs back into the oceans. Some of the water in the ground may return directly to the atmosphere by evaporating through the soil surface. Another part may be absorbed by plant roots, carried up to leaves, and returned to the atmosphere by **transpiration**—the release of water vapor by plant foliage. A major fraction of the groundwater stays in the ground, most of the shallow subterranean waters slowly moving through near-surface formations and ultimately exiting into stream beds, springs, or the oceans.

How the total flow from one place to another adds up—what we call a **mass balance**—is shown in Figure 6-4. From this elementary "water-flow budget" for land and sea, we can see that more

water evaporates from oceans than falls on them as rain. This discrepancy is exactly balanced by the return of water via runoff from the continents, which itself exactly balances the excess of precipitation over evaporation on land.

Knowledge of the cycle informs the judgment of water conservationists who tell us that there is a limit to how much natural fresh water we can ever get to satisfy the insatiable thirst of our industrial society. That limit is determined by the total amount of rainfall that reaches the continents, for there is no other source. The only way we could increase the total supply would be to "manufacture" fresh water by modern chemical methods of desalting sea water—accomplishing what nature does by evaporation and precipitation.

The United States and many other developed countries of the world must soon begin carefully planning the use of these large, but not infinite, water resources for the ever-growing demands of industry, energy production, urban population concentrations, and agriculture. A more detailed breakdown of the water budget for the United States (Fig. 6-5) gives the "road map" for the decisions that will have to be made. Though only about 7 percent of the total rainfall is used for agriculture, industry, and cities and towns, another 71 percent is unavailable because of loss

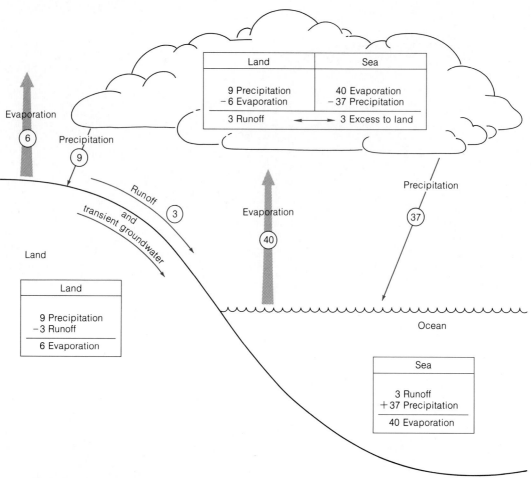

Figure 6-4
The mass balance, or water-flow budget, of land, seas, and atmosphere. All figures are in 10^{13} m³/year. 10^{13} m³ $= 2.6 \times 10^{15}$ gallons $= 8.1 \times 10^9$ acre-feet. An acre-foot is the volume of water that would cover one acre to a depth of one foot.

by evaporation, transpiration, and infiltration (though much of this may go to usable groundwater supplies). Much of the water used for manufacturing and the production and processing of mineral products is reused as effluents reenter the cycle. The 22 percent of untapped stream water represents one of the biggest potentials. That is why the pollution of rivers poses a much greater threat than the loss of their esthetic and recreational values: we are likely to need all the good water we can get as the demands of our population increase. As of 1975, the United States was using 1.7 billion cubic meters of water per day (5.1×10^{11} cubic meters per year) from surface and groundwater sources, and each year we use more. Total potential water supplies are only about three times this present use. Unfortunately, abundant river-water supplies are not always near the areas of high demand, and piping water for long distances is expensive. Thus, in areas with limited supplies of river water, it will become imperative to maintain—or restore—a flow of unpolluted, usable water.

Surface Runoff

The direct relation between rain and stream runoff is obvious—especially to anyone who has ever experienced a flash flood following a torrential rainstorm. The yearly amount of water flowing in streams in any region is directly related to the annual rainfall. Most stream beds in arid or semiarid regions are dry except right after the rare rains. The difference between wet and dry is particularly conspicuous when crossing a mountain range into a "rain shadow"—an area of low rainfall on the leeward slope of a mountain range.

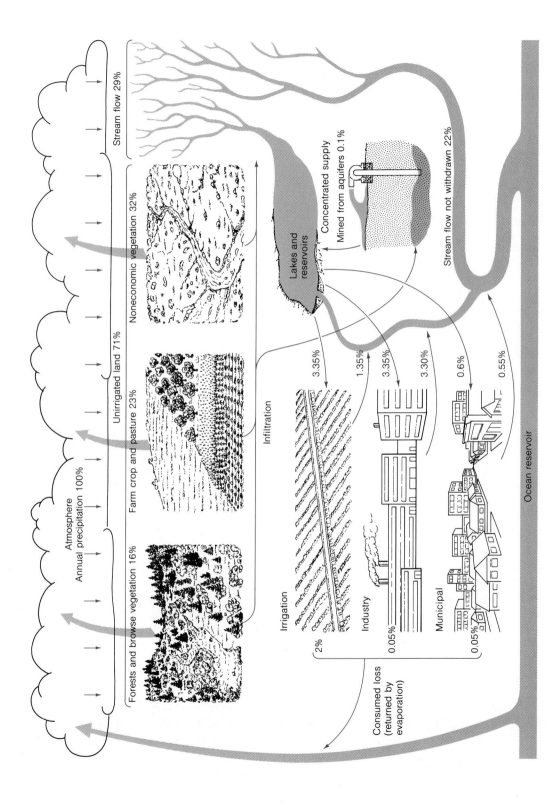

Figure 6-5

The hydrologic cycle for the United States shows the fraction of annual precipitation used in a highly developed nation. Twenty-nine percent of the rainfall arrives at the oceans (bottom) via stream flow; 71 percent falls on various types of unirrigated land, returning directly to the atmosphere (top) by transpiration and evaporation. Water withdrawn for irrigation, industry, and municipal use is shown at left to constitute only 7.3 percent of the total.

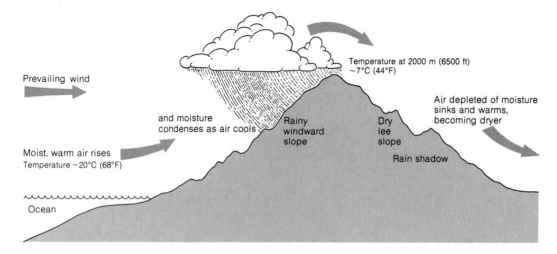

Figure 6-6
A mountain range may produce a rain shadow on its lee slope by forcing warm, moist air to rise with a prevailing wind from the ocean. The rising air cools, causing precipitation on the windward slope and leaving the lee slope dry.

Moisture-laden air rises as it crosses high mountains, cooling and precipitating much of its water vapor on the windward slope (Fig. 6-6). By the time the air reaches the opposite, leeward slope, it has lost much of its moisture. As it drops and warms, its relative humidity decreases, and little rain falls. The Cascade Mountains of Oregon offer a striking example of the contrast. The western, windward slopes have high rainfall, the abundant water draining into the Willamette River valley with its lush natural and agricultural vegetation. But just a few tens of miles to the east, over the crest of the range, the countryside is dry and relatively barren. The Sierra Nevada, an even broader rain-shadowing mountain range, keeps the Great Basin to the east a semiarid region swept by dry winds.

In this way, we arrive at a perhaps unexpected relation between climate, especially rainfall, and the tectonic forces responsible for the mountains. The Cascades, a range of extinct and active volcanoes like Mt. St. Helens, are the surface expression of a zone along which the Gorda plate has been subducted beneath the western edge of the North American plate, throwing up a belt of basaltic and andesitic volcanoes above the descending oceanic plate (Fig. 1-16). The Sierra Nevada is made up of granitic rocks that were intruded and thrust up as a large mass during a major plate-collisional mountain-building episode of the Cenozoic (see Chapter 20).

Most of the surface runoff is transported by the large stream networks of major river systems. Despite the enormous number of streams on the continents, about half of the entire runoff from

Table 6-1
Water flows of some great rivers
(in m³/sec)

Amazon, South America	113,330
La Plata, South America	79,300
Congo, Africa	39,600
Yangtze, Asia	21,800
Brahmaputra, Asia	19,800
Ganges, Asia	18,700
Mississippi, North America	17,500

the land areas of the world is carried by only about seventy major systems. Even more dominant are a few major rivers, such as the Mississippi and the Amazon (Table 6-1).

Rivers ebb and flow in discharge from season to season, mainly in response to seasonal precipitation patterns (Fig. 6-7). In some tropical regions it rains heavily only for a few months of the year. In some maritime climates, like parts of the Pacific Northwest of North America and most of the British Isles, the rain is abundant and more constant throughout the year. A different pattern of streamflow is shown by rivers draining snow or ice-covered country, where spring meltwaters can swell the streams to flood stage. Here the response of runoff to precipitation lags over the winter months while the precipitation that fell as snow is stored in snow and ice fields. The runoff jumps when the store is released by warming in the spring. Similarly, lakes and large areas of swamp and marshland, sometimes called wetlands, act as storage depots for runoffs (Fig. 6-8). These reservoirs, by their large volume, smooth out the big

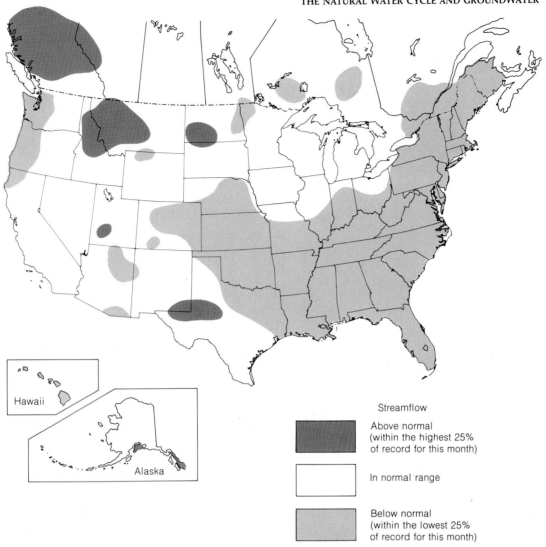

Streamflow

Above normal
(within the highest 25%
of record for this month)

In normal range

Below normal
(within the lowest 25%
of record for this month)

Hawaii

Alaska

Figure 6-7
Streamflow and groundwater conditions are strongly affected by seasonal and yearly climatic variations. This map of streamflow for January 1981 shows a continuing drought in the East. The lower streamflow depleted many surface and groundwater reservoirs to levels far below average. Each month Canada and the United States jointly release such evaluations of streamflow and groundwater conditions. [From U.S. Department of Interior and Department of the Environment, Water Resources Branch, Canada.]

variations in discharge of rivers and release a steady outflow downstream.* (Some flood control dams serve a similar purpose.) It is for this reason that some geologists have worked to stop the artificial draining of wetlands by real estate developers; if unchecked, this practice could wreck much of the natural flood regulation capacity in some areas.

But all of these storage areas for precipitation and runoff are minor compared to the water stored in the ground. Why do rivers continue to run day after day after several weeks of no rain—and no melting snow to feed them? How do springs continue to flow day after day to feed the rivers and lakes? Groundwater flow is the answer.

WATER IN THE GROUND

How do we know about the water in the ground? We can literally see the rain disappear into the ground in many places. Watch rain falling on a beach sometime. Some soils act like porous sand in sponging up rainwater; others have dense, tough clay layers, or "hardpans," that impede the

---

*Wetlands may soon find new use as natural water purifiers for sewage from residences. In experiments in Florida, a cypress swamp has been shown to work well as a secondary sewage treatment "plant" when effluent from a primary treatment plant is piped directly into it.

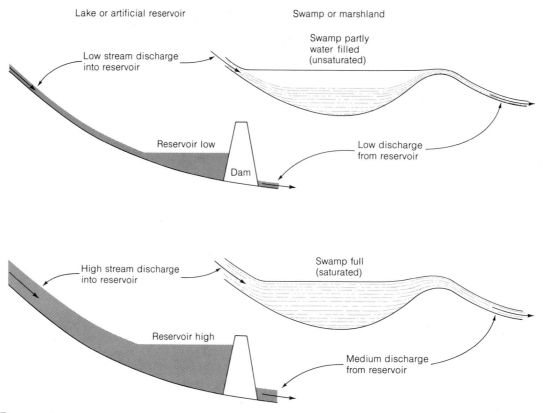

Lake or artificial reservoir

Swamp or marshland

Swamp partly
water filled
(unsaturated)

Low stream discharge
into reservoir

Reservoir low

Dam

Low discharge
from reservoir

High stream discharge
into reservoir

Swamp full
(saturated)

Reservoir high

Medium discharge
from reservoir

Figure 6-8
A swamp or marshland acts much like a natural lake or an artificial reservoir behind a dam in
storing water during times of rapid runoff and slowly releasing it during periods of little runoff.

infiltration of water and are poor absorbers of
water. Poorest of all are the rocky surfaces of
mountains, which have little or no soil. There, little water runs into the ground; what does, goes
through joints and cracks.

Bubbling springs show us where some of the
water that soaks into the ground emerges; good
places to look are on the lower slopes or the base
of a hill or valley wall. If you drill a well, and
have picked a good place to do it—for example,
where porous sandstone beds are near the surface
in a temperate climate—water will fill the bottom
of the well. These sandy or other kinds of beds
that carry waters are **aquifers.** If you are lucky,
you might even find water flowing up and out of
the hole: in some way the water is under sufficient
pressure to force it up the well. A few more facts
and you will have enough to get the general picture. After long periods of dryness, wells run dry,
the ground becomes dried out well below the surface, and smaller rivers shrink or even dry up entirely. There has to be some underground connection between the water that sinks in from rain on
the one hand and the rivers, springs, and wells on
the other.

The underground connection has to be through
soil and rock. Except in cavernous limestone networks and some open lava tubes in volcanic terrains, there are no open spaces under the ground
where rivers run as they do on the surface. Suppose you fill a short length of pipe with clean
sand, put a screen at the bottom, and pour water
in at the top. In a short time you will see the water
dripping out at the bottom. The water travels by
winding through the pore spaces between the
sand grains. The smaller the pore spaces and the
more tortuous the journey, the more slowly the
water flows through the sand. This property by
which solids allow fluids to pass through them is
permeability. Sands are permeable; hard, compacted muds and shales are not. The permeability
depends largely on the amount of pore space between the grains or crystals of the rock, its **porosity.** Rocks vary widely in porosity. Most igneous
and metamorphic rocks are not very porous; only
1 or 2 percent of many of those rocks consists of
pores. Unless they are fractured, these low-poros-
ity rocks are quite impermeable. Many sedimentary rocks, particularly sandstones, are very porous, as much as 30 to 40 percent porosity.

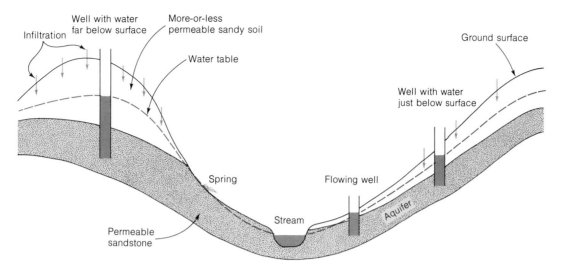

Figure 6-9
The groundwater table lies at varying depths beneath the surface, as shown by the water level in wells. The water table intersects the ground surface at springs and river beds.

Because in some rocks the pores are all connected to give an easy path and in others the pores are disconnected and do not allow any exit, not all porous rocks are equally permeable. Clay particles obstruct paths of flow between many pore spaces in some sandstones and so may significantly decrease their permeability. This is an important factor in obtaining a good flow of oil: many such sandstone beds are saturated with oil, but their permeability is too low to allow economic pumping.

Thus the ground acts like a sponge, soaking up rain in some places and leaking it out at others. Adding geological information about the structure of formations and their depth below the surface to data on the depth of water in wells dug at various places gives us the picture in Figure 6-9.

The Groundwater Table

If we connect the levels of water in the wells shown in Figure 6-9, we get a surface that we call the **groundwater table,** or just **water table.** That surface is the boundary between two zones of rock or soil: below the water table, the pores are completely filled with water, and the zone is called the **saturated,** or **phreatic,** zone; above the water table is an **unsaturated,** or **vadose,** zone. The unsaturated zone is also called the **zone of aeration** because the pore spaces are filled partly with water and partly with air. Some of the water in this upper, unsaturated zone is at times in transit down to the water table, but much water is held in smaller pore spaces in the same way that sand in a sandbox or beach stays moist long after it has been wetted. Evaporation of these little drops buried in pores is slowed almost to nothing because of the effects of surface tension. Surface tension is shown by a water surface that will support a needle or razor blade gently placed on it. This effect—whereby a skinlike surface layer of the fluid a few molecules thick is in a state of tension that gives it supporting strength—greatly reduces evaporation in small droplets.

Knowing the depth and shape of the water table is one of the important aims of the groundwater hydrologist and the driller of water wells. Predicting it from a combination of general understanding and detailed information on a specific area is becoming increasingly important in North America, particularly in the broad, thirsty areas of the Great Plains but also in many other parts of the West and Southwest. Even in regions of relatively abundant rainfall, it is no longer easy to put down a well almost anywhere and get water. Dowsers claim that they can somehow divine where water is underground with a forked stick or other inanimate object rather than with geological knowledge. But scientists remain firm in their conviction that dowsers have offered neither any mechanism by which their method might work nor any convincing proof that its rate of success is better than chance.

The water table generally follows, with a more subdued form, the contour of the surface topography. The "outcrops" of the water table are springs and the beds of rivers; it is there that water drains

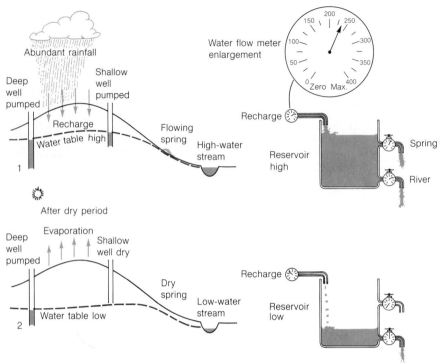

Figure 6-10
The depth of the water table fluctuates in response to the balance between what
is added by precipitation and what is lost by evaporation plus discharge from wells,
springs, and streams. The changing height of the water table is a reflection of the volume
of water stored in the ground and can be analyzed as a reservoir with inputs and outputs. The flow meters indicate the relative changes typical from spring to late summer.

out of the land. That drainage gives the clue to the
fact that most groundwaters near the surface are
not static but move slowly through the ground
while the water table stays in the same place. The
depth of the water table is another example of a
steady-state process: it reflects a balance between
the rate of infiltration, usually called **recharge,**
and the rate of discharge at rivers and springs—or
pumped water wells. The water table stays at the
same place when rain falls frequently enough
to balance the river, spring, and well outflow.
Any imbalance, typically by seasonal fluctuation
of rainfall, raises or lowers the water table (Fig.
6-10). A long-term imbalance results if there is either a decrease in recharge, such as from a prolonged drought, or an increase in the discharge.
This second factor accounts for the slow drop in
water tables in many regions. Where discharge
has steadily climbed as a result of constantly increasing pumping of water wells the imbalance
has depleted the reservoir—for an aquifer is just
as much a reservoir as any surface lake or pond
that can be filled up or depleted.

The flow of groundwater and the position of the
water table may be complicated by alternating
permeable aquifers and relatively impermeable

beds, called **aquicludes,** that hinder or prevent
water movement. When aquicludes lie both over
and under an aquifer, they produce a **confined**
water reservoir (Fig. 6-11). The water pressure in
such an aquifer depends on the difference in
height between it and the recharge area. If the
water table in the recharge area is higher than the
surface where a new well is drilled, water will
flow spontaneously out of the well; such wells are
called **artesian** wells. If enough artesian wells
flow long enough to deplete the reservoir and
lower the water table in the recharge area, the
pressure will drop. Eventually the wells stop
flowing and have to be pumped. An aquiclude
underlying a discontinuous aquifer may support a
perched water table above the main water table,
as in Figure 6-12. Some perched water tables extend over hundreds of square kilometers.

Yet another complication is faced by the people
who live at the edge of the ocean. Residents of
some shoreline areas have found a salty taste in
the water from wells that once held fresh water.
The extensive pumping of wells in the area has
caused an incursion of salt water. There is a sloping underground boundary between the fresh
water under land adjacent to the sea—or under an

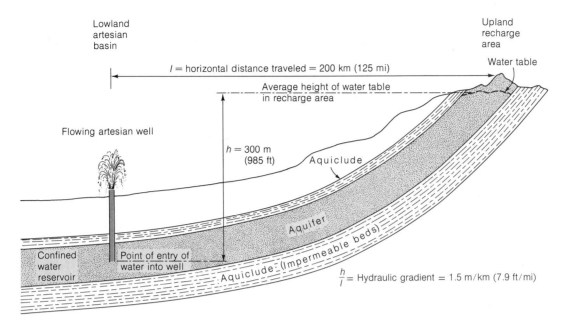

Figure 6-11
A confined water reservoir is created where water enters an aquifer situated between two confining aquicludes. The artesian well flows in response to the natural (before the well was drilled) pressure difference between the height of the water table in the recharge area and the bottom of the well, equivalent to the pressure of a water column 300 m high in this example. The actual pressure difference that governs the flow from the top of the well is the difference between the elevation of the water table and the top of the well. If the well head were as high as the water table in the recharge area, there would be no pressure difference and thus no flow.

island—and the salt water in the sea, as shown in Figure 6-13. If withdrawals by pumping are not excessive, the recharge of the fresh water on the land is enough to keep the water table elevated above sea level and maintain a reservoir of fresh water bulged into the sea water at depth. But if withdrawals exceed recharge, the water table is lowered and the reservoir depleted. The pressure of fresh water on the bulge decreases, and sea water moves in. Those closest to the shore are the first affected. Cape Cod, Long Island, Bermuda, and many other near-shore communities and islands have this problem to contend with. Systems of monitoring wells have been set up in some areas to provide early warning of salt-water contamination. There is no solution to the inexorable incursion of sea water except to reduce the pumping or to increase the recharge artificially, as will be discussed later in this chapter.

How Fast Does Groundwater Move?

From studies of the response of wells and springs to rainfall recharge, we have learned that the rate of groundwater movement is fairly slow, a fact of

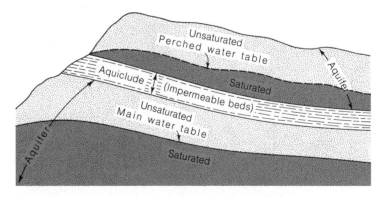

Figure 6-12
A perched water table is separated from the main water table by an aquiclude composed of shale or other impermeable beds.

nature that is lucky for us. It is that slowness that keeps the underground reservoirs full, for fast discharge would mean that water wells would run dry after just a few days without rain. What makes some groundwaters travel rapidly and others slowly? The details of the answer were worked out in the middle of the nineteenth century by Henry Darcy, the town engineer of Dijon,

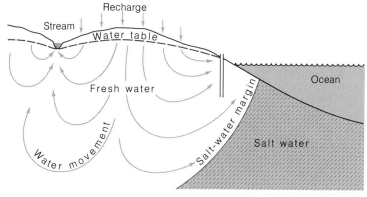

(a) Before extensive pumping

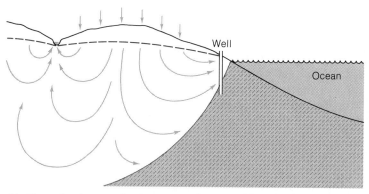

(b) After extensive pumping by many wells

Figure 6-13
The boundary between fresh groundwater and salt water along shorelines is determined by the balance between recharge and discharge to the ocean or wells. Normally, as in (a), the pressure of fresh water keeps the salt-water margin slightly offshore. Extensive pumping, as in (b), lowers the pressure of the fresh water, allowing the salt-water margin to move inland.

France. From observations of the heights of water in wells, the distance traveled, and the permeability of the rocks through which the water flowed came what we now call Darcy's law.

Darcy found that, for a given aquifer, the rate of water flow from one place to another is directly proportional to the drop in vertical elevation between the two places and inversely proportional to the horizontal distance the water travels. The relation between the horizontal and vertical distances can be seen to be akin to the angle of an inclined plane; it is, in fact, the force of gravity that makes water run "downhill" underground just as it makes a ball roll with different speeds down inclined planes of different slopes. Darcy also found that the greater the permeability, the faster the flow.

Velocities have been found by using Darcy's law and dating groundwater by radioactivity.[*] Velocities can also be determined experimentally by introducing dye into groundwater at a well used as an artificial recharge point and noting the time lapse before it shows up at a nearby pumping well. Movement in deeper formations is rarely faster than 100 centimeters per day or slower than 0.5 centimeters per day. In most aquifers, groundwater moves at rates of a few centimeters per day. In very permeable gravel beds near the surface, groundwater may travel as much as 15 centimeters per day.

How Far Does Water Travel Underground?

The length of time that groundwater spends underground from recharge to discharge depends not only on its velocity but on how long its path is. Routes of travel vary widely, and in a typical flow, such as shown in Figure 6–13, some routes may be many times as long as others.

So far we have concentrated on flow over relatively small areas, such as a hillside or valley. Flow is also demonstrable over regions the size of counties or groups of counties hundreds of square kilometers in area. Many areas in the Great Plains and parts of the central Midwest are underlain by sandstones of various ages that transport waters for hundreds of kilometers. Over such broad areas, flow is dominated by the geologic position of permeable formations, the aquifers. Typically the pattern is like that shown in Figure 6–11, the essential condition being a continuous aquifer with a recharge area in an upland area so that the water can run "downhill" underground. Recall from Darcy's law that the discharge and velocity are proportional to the change in height over the distance. In the Midwest the vertical height is rarely more that a few hundred meters (in most areas it is much less), but the horizontal distances are measured in hundreds of kilometers; hence the rates of water movement are slow, and the recharge takes a long time. What are the conse-

---

[*]Studies on the rate of groundwater movement have been confirmed and made more precise by application of radioactive dating to groundwater, using radioactive carbon, ^{14}C, to date the dissolved carbon dioxide in the waters. On entering the ground the water contained in its dissolved CO_2 the minute amount of ^{14}C characteristic of rainwater, but as it slowly traveled through the ground the ^{14}C decayed radioactively. The velocity is obtained by measuring the distance from the well in which the water was sampled to the recharge area and dividing this by the age as determined by ^{14}C.

Box 6-1 Darcy's Law of Groundwater Motion

The law formulated by Darcy is given in terms of the volume of water moving through any opening in a given amount of time, essentially a velocity term, and the geometry of the general flow, or the ratio of the vertical to the horizontal distance.

equation by multiplying the right-hand side by a proportionality factor, K. Darcy identified K as a measure of the permeability of the rock, or in other words, how easily it transmits water.

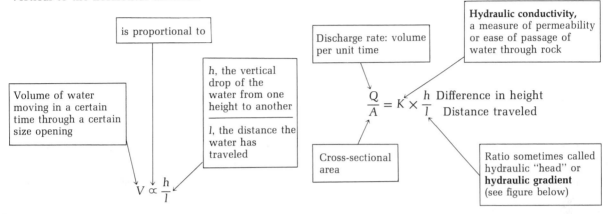

Darcy reasoned that the permeability of a rock is what slows down the flow for a given drop of height h in a certain distance l. So he made this proportion into an

From this equation, we can either determine the velocity of flow or, knowing the velocity, the hydraulic conductivity.

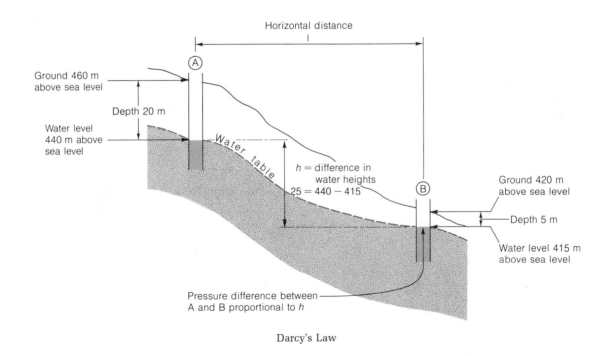

Darcy's Law

quences for us? If extensive water-well pumping removes the water faster than the slow recharge can fill it, the reservoir will gradually become smaller.

Mining Groundwater

A baby born when the first wells were drilled into the Ogallala, a formation of sands and gravels of Tertiary age in western Texas and New Mexico, would be about a hundred years old now. He or she would have lived to see the water pressure of those wells steadily diminish and the water table drop drastically, by as much as 30 meters (100 feet). The Ogallala aquifer has been the water support of a population that has gradually grown to more than half a million people in this century; it has provided ample water for irrigation for the agriculture that is the economic base of the region. This aquifer is very slow to recharge naturally because of evaporation and some impermeable soils. Water has been pumped from it so extensively that at present rates of withdrawal and recharge the water table would take several thousand years to recover its original position. Other aquifers in the high plains tell the same story in varying degrees. The size of the problem is revealed by the statistics. More than 314 billion liters (83 billion gallons) of groundwater are pumped daily from aquifers in the United States. Deeper aquifers are being explored in hopes of increasing production.

The amount of usable fresh water in the ground is staggering (Fig. 6–14). Certain areas, such as the Pacific Northwest, have great reserves, estimated to be about 1.2 quadrillion gallons in the upper 330 meters (1000 feet) of the states of Idaho, Oregon, and Washington. Recent estimates are that usable groundwater amounts to much more than 90 percent of all of the fresh water on Earth. Rivers, lakes, reservoirs, and soil moisture account for the remainder. Glacial ice is the equivalent of seven times the groundwater supply. Yet, because of the extreme slowness of recharge and the slow rates of water movement, the groundwater in such areas as western Texas is in every sense being mined, just like coal, so that the amount left in the ground steadily diminishes as production continues. For all practical purposes, this water is an exhaustible resource that, once gone, cannot be replenished. We are not yet in a crisis situation nationally, but certain communities, in areas where depletion is in an advanced stage, are in for tough times.

The problem is to find untapped supplies and plan for the recharging of aquifers that can be replenished, particularly those that can be recharged in a short time. Long Island, in New York, offers one example of success at groundwater recharge. Five million people there depend on wells for almost half their water supply. By the early 1940s, extensive pumping had depleted the reservoir to the extent that salt water was encroaching on the aquifer, and a state-regulated program of artificial recharge was started to reverse the deterioration of the water supply. By drilling an extensive system of recharge wells so that used water could be put back into the aquifer, and by constructing large basins to help accelerate infiltration from surface waters, including storm and industrial waste drainage, the aquifer was built back up to usable levels. A measure of the program's effectiveness came in 1972 and 1973. After unusually wet seasons on Long Island, the water table was so high that water seeped into the basements of many homes unwisely built in low, poorly drained areas in times when the water table was low.

The Long Island story is one of success up to now and is an encouraging example of a stitch in time. But the situation remains precarious, for population and water demands grow daily. One of the unlooked-for results of rapid growth in urban and suburban areas such as Long Island is that natural recharge of the aquifer through seepage into soil decreases as more and more buildings, paved streets, sidewalks, driveways, and parking lots appear. Our expanding urban areas not only deplete the underground reservoirs by pumping—they interfere with its natural refilling.*

Karst: The Erosive Power of Groundwater

Along the northeastern shore of the Adriatic Sea, where Yugoslavia and Italy meet, lies the Karst region. An irregular terrain of hills and low mountains with many small depressions, it has evolved on limestone bedrock that has been eroded and shaped by the dissolving power of groundwater in a humid climate (Chapter 5). In such karst topography, sinkholes (see Fig. 5–22)

---

*Pumping of groundwater can also lead to surface disturbances such as subsidence of areas under which much groundwater has been withdrawn. A 4300-mi² area of the San Joaquin Valley of California has subsided a foot or more since 1920. Pumping has even triggered slow movements along a fault between Tucson and Phoenix, Arizona.

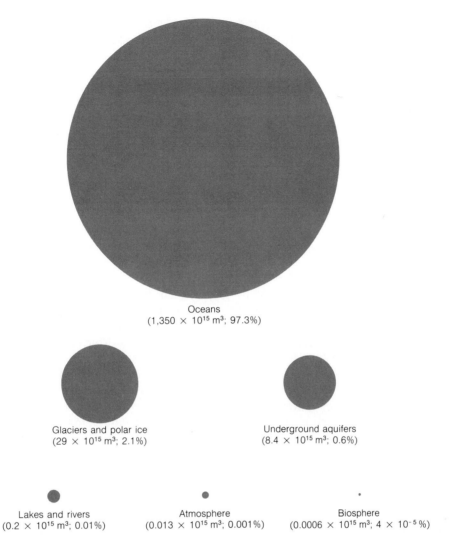

Oceans
(1,350 × 10¹⁵ m³; 97.3%)

Glaciers and polar ice
(29 × 10¹⁵ m³; 2.1%)

Underground aquifers
(8.4 × 10¹⁵ m³; 0.6%)

Lakes and rivers
(0.2 × 10¹⁵ m³; 0.01%)

Atmosphere
(0.013 × 10¹⁵ m³; 0.001%)

Biosphere
(0.0006 × 10¹⁵ m³; 4 × 10⁻⁵ %)

Figure 6-14
Distribution of water on the Earth. The amount of water present in various natural
reservoirs is represented in terms of comparative spherical volumes. The content of
each reservoir is given in cubic meters and as a percentage of the whole. Although
there is an enormous quantity of groundwater, much of it may be unusable because
of its high concentration of dissolved solids. [From "The Control of the Water
Cycle," by José P. Peixoto and M. Ali Kettani. Copyright © 1973 by Scientific
American, Inc. All rights reserved.]

are combined with underground caverns and a
surface drainage of short and scarce streams that
end abruptly in sinkholes. The amounts of limestone that have dissolved to make caverns may be
huge. Mammoth Cave, Kentucky, has tens of kilometers of interconnected caves; the large room at
Carlsbad Caverns, New Mexico, is more than 1200
meters (4000 feet) long, 200 meters (600 feet) wide,
and 100 meters (350 feet) high.

The groundwater in karst regions is made
"hard" by the calcium carbonate dissolved in it.
Hardness and other chemical characteristic properties of water—the result of its geologic origins—are important to all of us, for the quality of
the water that we drink is as vital to our well
being as the quantity.

THE QUALITY OF WATER

When Portia speaks of mercy in the Shakespeare
that so many have learned, she uses a common
metaphor:

> *The quality of mercy is not strain'd.*
> *It droppeth as the gentle rain from heaven.*
> MERCHANT OF VENICE, ACT 4, SCENE I

Rainwater is our usual standard for purity, and in parts of the Earth where the air is still clean, so is the water. A great many, if not most, water wells produce fine water of excellent quality, just as good as rainwater to drink. Yet centuries before anyone thought of chemical contamination of our air and water, people who drank some well waters made wry faces at the taste and tended to prefer wine or beer. Some groundwaters taste of "iron"; others may have a disagreeable odor. Every few months a newspaper somewhere features a story about water containing dissolved chemicals that appear to be pollutants. What makes the difference?

Rainwater to Groundwater

One place where we can sometimes find groundwater about as pure as rainwater is under a bare dune made of pure quartz sand. Analyses of such waters show that they have about the same amount of carbon dioxide dissolved in the water as the rain—and little else. Compare this with groundwater from a karst terrain, a **hard water**—one that contains dissolved calcium and usually some magnesium, which make it difficult to make a lather or soap suds. Such water may also have a slight taste given by other ions in solution.

These comparisons suggest the same conclusions that we drew from observations of rock weathering (Chapter 4)—rainwater infiltrating the ground chemically alters rock and soil and in so doing becomes a groundwater containing dissolved ions and other components contributed by the plant and animal life on and in the soil. The water under the quartz sand dune is like rainwater because quartz is so insoluble that for practical purposes it is inert, and neither soil nor vegetation contributes dissolved substances to the water. Unlike quartz sand, surface materials of soil, limestone, and shale react with rain to produce dissolved materials that contribute to the taste and hardness of the groundwater.

How Good Is the Water?

The amounts of dissolved material in **potable** groundwaters—that is, waters usable for drinking—are very small, usually measured by weight in parts per million (ppm), a unit made up in the same way as the more familiar percent, which is just parts per hundred. For example, 1 ppm is the same as 0.0001 percent, but it is much easier to use parts per million because we can almost always avoid decimals and all those zeroes. One hundred and fifty ppm is a typical figure for the total dissolved material in a good water, for no natural water is as pure as a carefully distilled water. The upper limit for potable water for human consumption is usually 500 to 1000 ppm; for watering livestock, it is usually 2000, though both humans and animals can get used to more in some circumstances. A water can measure in the upper range of permissible dissolved ions and yet still taste good if its dissolved ions consist largely of a mixture of calcium and bicarbonate ions. Yet a water containing only several hundred ppm of sodium and chloride ions will taste slightly salty, and water with 70 or 80 ppm sodium may taste all right but be a hazard to those on low-sodium diets.

The worst offenders to water taste are the dissolved organic compounds, many introduced by human activities. Some of these are harmless; others may be toxic. Some dissolved components have no taste but beneficial qualities. For example, the movement for fluoridation of water supplies was sparked by the discovery that in some regions small amounts of naturally occurring fluoride in groundwater are responsible for great resistance to tooth decay. Conversely, however, some toxic elements have no taste but in more than minute quantities are dangerous to health. Concentrations of such poisonous elements as lead and arsenic must be watched carefully. No matter how good the water may taste, it is unsafe to drink it if these elements exceed a tenth to a twentieth of a part per million.

A few groundwaters are naturally softened as they pass through and react with formations containing zeolite minerals. Zeolites, hydrous silicates, have a strong chemical tendency to absorb certain ions from water in exchange for other ions bound in the solid. If, for example, a water carrying calcium ion travels through a zeolite formation that exchanges sodium ion for the calcium, the water comes out with sodium instead of calcium, and the zeolite becomes richer in calcium and poorer in sodium. This process has been put to commercial use by attaching tanks of zeolite to the water inlet of homes or entire cities as a water softener (Fig. 6–15). The commercial zeolite eventually gets used up—saturated—and has to be changed.

Groundwaters are always clear of solid materials as they come from the aquifer. The intricate passageways of the aquifer act as a fine filter and remove small particles of clay or any other solids. They even strain out bacteria and large viruses. The dirtiest sewage will be clear of suspended

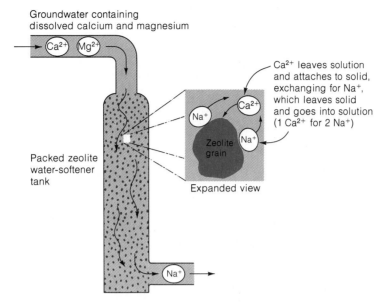

Figure 6-15
A zeolite water softener operates by exchanging sodium
(Na^+) or other ions in the solid zeolite for the ions that make
the water hard, calcium (Ca^{2+}) and magnesium (Mg^{2+}). The
exchanging ions always balance total electrical charge; thus
one Ca^{2+} will exchange for two Na^+.

solid matter once it has gone through a thick bed
of sand. The dissolved matter is another story. It is
not strained out.

Contamination of Water Supplies

In the past few decades the tremendous increase
in the use of septic tanks for home sewage disposal has contributed a new source of dissolved
material to groundwater. The septic tank waters
seep into the soil and, where water-supply aquifers are shallow, contaminate groundwater with
phosphate and foaming agents from detergents
and a variety of other substances that are undesirable or potentially harmful to health, such as ni-
trates. Water filtering into the ground through city
dumps, chemical industry waste lagoons, or sanitary landfill operations may introduce a number
of undesirable compounds into groundwaters.
Extensive salting of icy roads in the North has led
to sodium chloride contamination of some aquifers. Groundwaters in the vicinity of some storage
areas for road salt contain concentrations of cadmium chloride, chromium, and sodium, all in excess of limits for potable water. Still another
source of groundwater contamination is nitrate
fertilizer. The nitrate is very soluble, and although some is used by plants—the purpose of the
spreading—much dissolved nitrate escapes unused into deeper parts of the soil and into ground-

water. In combination, these sources (sewage and
fertilizer) have increased nitrate levels in some
aquifers to the worry point, for nitrate is toxic to
humans even in amounts as small as 10 to 15 ppm.

It must be obvious that the faster the recharge
of the aquifer, the more quickly it might become
contaminated from these sources. But the fast recharge also means that the aquifer can recover in
a short time, once the sources of contamination
are closed off. Contamination of slow-recharge
reservoirs is more serious, for by the time the results of many years of contamination start showing up, it is too late for quick recovery. Because of
the definite links between surface waters and
groundwaters, we cannot afford to view the surface and the subsurface as separate entities. Integrated planning for groundwater reservoirs in relation to pollution of and from surface sources
will become increasingly necessary in order to
maintain the quality of our water supplies.

Deeply Buried Waters

Below the groundwater table, rocks are almost
everywhere saturated with water, especially
below depths of a few hundred meters (Fig. 6–16).
No matter how deeply oil wells are drilled, geologists always find water in permeable formations.
In most places the water below the upper few
hundred meters becomes increasingly concen-

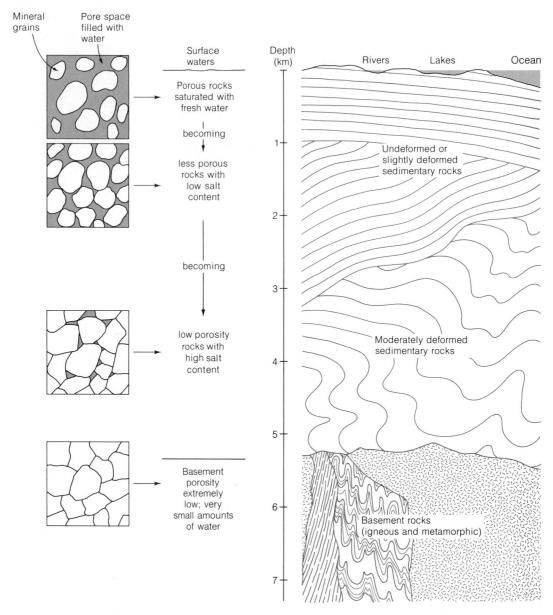

Figure 6–16
A typical section through continental crust, showing the distribution of water. Most water is at the surface or in sedimentary rocks buried at shallow depths. Depth to basement varies greatly, but it is rarely more than 10 km. Porosity and water content generally decrease with increasing depth and greater structural deformation.

trated in dissolved materials; for this reason they are frequently called **brines.** They vary widely in chemical composition, just as surface waters do, but they are much older than the near-surface waters and presumably have had a much more complex history. These waters move very slowly, probably at speeds of less than a centimeter per year. For example, in southern Illinois there is a series of (Paleozoic) sedimentary formations arranged in a spoon-shaped basin a few kilometers thick; the deep groundwaters in this area are

much more salty and concentrated than shallower waters. In other areas deep groundwaters are rich in ordinary salt, NaCl, particularly where a water has been steeped in buried salt beds. Because many deep waters are in some ways similar in composition to sea water, they used to be considered remnants of original sea water trapped in the pores of sediments at the time of deposition. But detailed studies of the chemistry of the waters of the Illinois basin have shown that all of these waters have become salty by reacting with various rocks over a long time interval. The deeper and the older the waters, and the more time they have had to react with rocks, the greater is the content of dissolved solids.

How Deep Does Water Go?

A deep well may penetrate an unconformity under a thick sequence of sedimentary formations and enter what is called **basement**—by definition, a highly complex mixture of igneous and metamorphic rocks that everywhere underlies the sedimentary rocks of the continents. Basement may occur from a few meters to ten kilometers or more below the surface. In the Grand Canyon, the basement is Precambrian, including the Grand Canyon Series and the Vishnu formation. Elsewhere the basement may be Paleozoic, as it is along parts of the Eastern Seaboard of the United States. There flat-lying Mesozoic and Cenozoic sedimentary rocks are separated by a widespread unconformity from deformed and metamorphosed rocks of Devonian age and older. Because igneous and metamorphic rocks have low porosities and permeabilities, the basement is relatively dry, though there are always minute amounts of water in small cracks and along boundaries between crystals. Some sedimentary sequences, notably those along the Gulf Coasts in eastern Texas, Louisiana, and Mississippi, are so thick that the depth of basement is known only from seismological evidence. Here the sedimentary formations show a decrease in permeability with depth, caused partly by compaction due to the immense weight of overlying rock. Another cause of the decrease in permeability is the strong tendency for pore spaces to be filled by such minerals as quartz or calcite, which precipitate under the chemical conditions of deep burial. Though the deepest wells drilled so far—more than 9 kilometers (29,500 feet)—still penetrate permeable rocks, we can nevertheless assume that at much greater depths these sedimentary rocks will be as dense and dry as basement rocks. We now know enough about the problems of deep drilling technology to make it likely that we will be able to drill to these depths in the near future.

The conclusion is inescapable: the accessible waters of the Earth are limited to the surface and the near-surface part of the crust. Most of it is surface water—in oceans, rivers, and lakes; the remainder is stored in rocks, to depths of not much more than 10 kilometers. The overwhelming majority of the world's *fresh* waters are in the ground rather than on the surface. Although many deeply buried waters are salty (brines), most near-surface groundwater is fresh. Rivers and lakes account for the only a small fraction of the total fresh water.

Thermal Waters

In such places as Marienbad, Germany; Aix-les-Bains, France; Karlsbad, Czechoslovakia; Hot Springs, Arkansas; the restorative properties of mineral baths in hot spring pools have been known (or assumed) from the earliest days of civilization. They are also a part of our story, for the activity of these mineral waters, or **hydrothermal** waters ("hot water"), is responsible for extensive deposits of travertine and other minerals that encrust rocks around the pools. Those waters that do not reach the surface form, at great depth, many of the richest metallic ore deposits of the world (see Chapter 22). Thermal waters are found in areas of current or recent igneous activity—both deep-seated, intrusive, activity and surface, extrusive, activity. In the oceans thermal waters are associated with sea water circulating through mid-ocean ridges, giving rise to submarine hot spring vents (see Chapters 10 and 22). On the continents the water itself is mostly **meteoric** (meaning derived from rainwater, and having nothing to do with meteors except for the derivation of the name). Such waters may be very old; the water at Hot Springs, Arkansas, has been determined to be derived from rain and snow that fell over 4000 years ago and slowly infiltrated the ground. Meteoric water may be mixed with water originally dissolved in the magma itself, called **magmatic** water. In areas of igneous activity, surface waters moving downward encounter hot masses of rock, become heated, and mix with the magmatic waters at depth. Return circulation takes place along fracture zones or other channels that intersect the surface at continuously fed hot springs or geysers, some of which spout intermittently like the spectacular Old Faithful in Yellowstone Park (see Fig. 6-17 and 15-33).

Hot water is a much more efficient dissolver of rocks than cool rainwater, and so these hydrothermal solutions are more concentrated in many ions than ordinary groundwaters. As the hot solutions rise to the surface, they cool. If these hot solutions are saturated at the higher temperature, they become supersaturated as they cool and deposit the excess as mineral precipitates. The extensive whitish to grayish, muddy flats around some of the hot springs in Yellowstone Park are made up of fine particles of silica precipitated

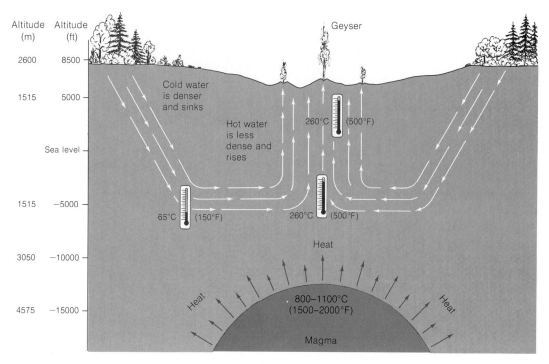

Figure 6-17
Circulation of water over a magma produces geysers or hot springs. Cold rainwater soaks into the soil and filters down through permeable rocks. As it approaches the magma, it heats up, becoming less dense, and thus sets up a circulation system that returns it to the surface.

from the solutions that had originally, when hotter, dissolved it from silicate rocks at depth, near the hot igneous rock. **Travertine** is a calcium carbonate deposit commonly precipitated at cooling hot springs from solutions that passed through limestone formations at depth. Other hot waters may contain large amounts of sulfur. Our ancestors were right in their belief that such smelly, sulfurous waters escaped from the "hellish" interior of Earth.

In the search for new and clean sources of energy, scientists have turned their attention to hot water areas as natural sites for the development of steam turbines and other devices to convert the heat into electricity. Though this is not a generally applicable solution to the problem of our rapidly climbing energy needs, the potential for efficiently harnessing this source of energy is good in areas that do have hot springs or geysers. California, Iceland, Italy, and New Zealand are already exploiting hot spring areas for power in a practical way (see Chapter 22).

WHERE DID ALL THE WATER COME FROM?

Sooner or later, after geologists began to question seriously how the Earth evolved as a planet,

someone had to tackle the origin of all the water on Earth, for if it were not always here, it must have come from somewhere. In the late 1940s a remarkably versatile geologist who was equally at home doing field mapping, calculating chemical equations, and studying the physics of how water currents carry sediment, found himself attracted to this problem. William W. Rubey, a researcher for the U.S. Geological Survey, started wondering whether the saltiness of the ocean had changed much in the past. As frequently happens to people of insatiable curiosity, that question led to others: why was the ocean salty in the first place, and where did all the water and salt come from?

Rubey compared the average chemical compositions of igneous, metamorphic, and sedimentary rocks and reconfirmed what geologists had long known—that sedimentary rocks are much richer in water and carbon dioxide than the others. He went on to add up the amounts of all of the water, carbon dioxide, chlorine, and a few other substances that are now present in sedimentary rocks, atmosphere, and sea waters (Table 6-2). He then compared those totals with estimates of the amounts of these substances that could have been derived by the weathering of igneous rocks. This attempt at a mass balance was reasonable; after all, we know that the process of chemical weathering of igneous rocks produces sediments and

Table 6-2
Some volatile substances found at or near
the Earth's surface

Compound or element	Where found
Water (H_2O)	As liquid in oceans; as vapor in atmosphere; chemically bound in hydrous minerals, such as clays, in sedimentary rocks.
Carbon dioxide (CO_2)	Dissolved in oceans; as gas in atmosphere; as carbonate minerals, such as calcite ($CaCO_3$), in limestones and other sedimentary rocks; in reduced form as organic carbon in oil, gas, coal, and other organic matter.
Chlorine (Cl)	As chloride ion (Cl^-) dissolved in oceans; as component of saline minerals, such as halite (NaCl), in evaporite deposits.
Sulfur (S)	As sulfate ion ($SO_4{}^{2-}$) dissolved in oceans; as calcium sulfate ($CaSO_4$) in the form of gypsum and anhydrite in evaporite deposits; as sulfide ion (S^-) in iron sulfide (FeS_2) in the form of pyrite in sedimentary rocks.

the dissolved ions in ground and surface waters that eventually end up in the oceans.

Rubey found an enormous discrepancy. There was far too much water, carbon dioxide, and chloride in the atmosphere, oceans, and sedimentary rocks to be accounted for by the weathering of igneous rocks, which contain little of these substances. The only reasonable way to make the budget balance was to revive an idea first proposed in the nineteenth century—that all of the gases released by volcanoes, hot springs, and geysers, lumped together as **volcanic emanations,** could make up the difference. These gases (water vapor, carbon dioxide and other carbon gases, hydrogen chloride, chlorine, hydrogen sulfide, sulfur dioxide, and others) are present in abundance on or near the surface or buried in different forms (carbon dioxide as carbonate in limestones, for example) in sedimentary rocks. Rubey called them **excess volatiles.** He concluded that almost all of the water vapor and other gases were released from deep in the earth by volcanism. Since chemists use the term **outgassing** for the release of gases from a solid or liquid as a result of heating or some other cause, it was natural to refer to release of these gases as *outgassing of the Earth.*

Outgassing does, indeed, seem to be the process responsible for the formation of the oceans and the atmosphere. But the question remains, when did it happen? Did it happen all at once or over a long period of time? Rubey looked first at the geologic record for evidence that might indicate whether any change in the salinity of sea water

had ever taken place. Knowing that marine organisms have little tolerance for such changes, Rubey argued that, based on fossil evidence dating back to the Cambrian, marine life has never been drastically affected by variations in the salt content of the oceans. He concluded that neither sea water nor the atmosphere has changed much in composition over most of geologic time and that the volcanic gases from the interior accumulated slowly and steadily.

THE CONSTANCY OF SEA-WATER COMPOSITION

Rubey's paper opened up a whole new line of thought. In 1959 the world's first international oceanographic congress was held in the U.N. Building in New York. For a week the delegates' lounges were alive with talk about the outpouring of new information on the oceans instead of about the diplomatic activities of ambassadors. At one session a brilliant Swedish physical chemist, the late Lars Gunnar Sillén, brought his deep knowledge of the quantitative data on inorganic chemical reactions to bear on the problem of explaining the chemical composition of the oceans. Sillén started out with pure water mixed with dissolved carbon dioxide and oxygen in equilibrium with the amounts present in the atmosphere. He then calculated how that water could chemically react to equilibrium with all of the clay minerals, carbonates, and other minerals characteristic of the sediments flooring the oceans. He further calculated the same kinds of reaction between water and all of the unweathered igneous minerals that are transported to the oceans as detritus eroded from continental rocks. He concluded that the final product would be a solution much like sea water. This analysis made even more probable Rubey's estimate that sea water was always of the same composition and emphasized even more strongly that sea water is the expectable ordinary consequence of weathering and sedimentation.

At about the same time, other geochemists showed that the chemical elements dissolved in sea water are being precipitated onto the bottom as sediments at about the same rate as they are being supplied to the ocean in solution by the world's rivers. Thus we now see the oceans as the product of a steady-state process rather than the result of some early, single, cataclysmic event. Our view of Earth history has been sharply altered by the gradual recognition that steady-state processes are the rule rather than the exception.

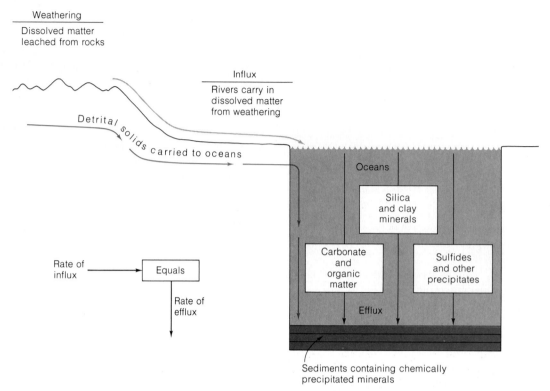

Figure 6-18
A model of the steady-state ocean. The rate of influx of dissolved substances by rivers matches the rate of efflux by chemical precipitation of sedimentary minerals. Detrital solids formed by erosion accompany the dissolved material and are deposited as sands and muds in various mixtures with the chemical precipitates.

How Long an Atom Stays in the Ocean

The steady state of the oceans leads to another concept that relates the inflow or outflow of an element to its total amount in the sea. A good example might be a crowded party, where many more people have been invited than fit comfortably in the rooms. As people come in, the rooms quickly fill up, and then as more arrive, the crush gets so much that people start to leave. At its most active, the party is in a steady state, arrivals and departures balanced, and the rooms saturated. Now, even though some come early and stay late, and others arrive, see the crowd, and leave, there is an average length of time between each person's arrival and departure that we can call **residence time.** The length of stay at the party is determined by the number of people that can be jammed into the rooms, which we will call the capacity, and the rate of arrivals, which we will call the rate of influx (this, of course, is equal to the rate of departures, or efflux).

$$\text{Residence time} = \frac{\text{capacity}}{\text{rate of influx}}$$

A simple numerical example: if a room will hold thirty people and a new person arrives every two minutes (1/2 person per minute), you can verify that the average residence time has to be one hour.

Applying all this to the oceans allows us to calculate residence times for the dissolved elements in sea water (Fig. 6–18). You can visualize this as the average length of time that elapses between the entry of an atom into the sea and its removal by sedimentation. Calcium is carried in abundance by rivers, and even though it is readily taken up by many kinds of organisms that build shells composed of calcium carbonate, it has a long residence time, 8 million years. Sodium has one of the longest residence times, 260 million years. Iron, on the other hand, stays in the ocean only 140 years. Knowing the residence time of elements is important not only for working out the general scheme of things but also for predicting the behavior of toxic or radioactive elements in the ocean. It is important today, for example, to know the residence time of crude oil spilled on the sea by tankers before it is decomposed or sedimented out on the sea bottom or its beaches. Likewise, as we will discuss in Chapter 12, it is crucial that we know the residence time of carbon dioxide if we are to be able to predict the effects of excess CO_2 in the atmosphere and oceans. This whole field of study is most active, and we are now beginning to see how plate tectonics and geochemical connections between the surface and the interior of the Earth may have a bearing on the compositions of sea water and sediments.

When All the Water Came Out

Rubey thought that the outgassing of the interior took place gradually, but information that has accumulated during the years since the late 1940s has persuaded some scientists that the great bulk of the Earth's atmosphere and oceans may have been outgassed early in its history and have been partially recycled ever since. Part of this evidence is in the isotopic composition of the gases from modern volcanoes. These data indicate that the volcanic emanations originate in the surface or crust and are recycled only through the outer 25 to 30 kilometers of the Earth rather than outgassed from the deep interior. Current ideas on the timing of the astronomical and geological events in the early history of the Earth also point to early outgassing. For instance, the Moon and the Earth were much closer during the early part of their joint history, and the Earth would have been stressed by the strong gravitational pull of the Moon. Another factor is the early timing of differentiation, for the melting of iron and the formation of an iron core would have caused a profound disturbance that would also have promoted outgassing (see Chapter 1). Finally, the theory of plate tectonics holds that there is continuous mixing and recycling of material between the crust and the mantle; it also provides a rationale for the distribution and kinds of volcanic activity that contribute gases to the atmosphere and oceans today. Thus we have evidence that forces promoting extensive outgassing would have been strongest in the earliest stage of Earth history, and we have a mechanism—plate tectonics—that would allow for continuous recycling of some of the gases through the crust and mantle and also account for the gaseous volcanic emanations of today.

Putting all of these lines of evidence together, we refer again to what was called in Chapter 1 the "big burp." It is likely that there was a huge outgassing of the excess volatiles at the time when much of the earth became molten as a result of the accumulation of radioactively produced heat. At the same time, the body of the Earth would have been distorted by the large Earth tides caused by the nearby Moon. That too would have promoted the escape of gases from the interior. Since that early time of outgassing, the excess volatiles have been cycling through the crust and mantle. The temporarily dense early atmosphere and the newly formed oceans would have produced a large volume of primitive sediments from the reaction of the gases with primitive igneous rocks. The primitive igneous and sedimentary rocks would long since have been deformed or subducted by plate and other tectonic movements.*

Thus our concept of the history of the oceans and atmosphere includes a stage of rapid evolution early in the Earth's history, during which the steady-state "machine" was set up, followed by a second stage, almost all of geologic time, during which the machine has been steadily operating at more or less the same rate.

Knowing the importance of recycling of excess volatiles, we can integrate the movements of water in the hydrologic cycle with the routes of the dissolved ions and gases, such as carbon dioxide, that it carries (Fig. 6–19). The water is recycled at the surface of Earth by evaporation from land and sea, circulation as vapor and clouds in the atmosphere, and precipitation on land and sea. On land the snow may form glaciers, whose meltwaters join ordinary rainwater runoff in bringing land precipitation back to the sea. Surface runoff is connected to shallow, subsurface groundwater flow, much of which eventually returns to the surface in the beds of streams and lakes. Some surface water is incorporated into hydrous minerals and reaches the interior, where it is recycled along with solids by the tectonic, magmatic, and metamorphic processes of exchange that take place between the crust and upper mantle of the Earth. In this way the surface hydrologic cycle, powered by the external heat engine, the Sun, is linked to the Earth's internal heat engine, powered by radioactivity. Water from the interior eventually returns to the surface in volcanic emanations, which may escape either at the surface or in the deep sea. Volcanism also helps recirculate carbon dioxide, the second most abundant volcanic gas. Carbon dioxide dissolved in rain and surface waters is important in the

---

*Certainly not all excess volatiles came out very early in Earth history; additional outgassing of primordial material continued at a much slower pace throughout geologic time. The relation of outgassing to the formation of continental landmasses early in the Precambrian is currently an active subject, with much attention being given to the oldest rocks in the world in Southwest Greenland.

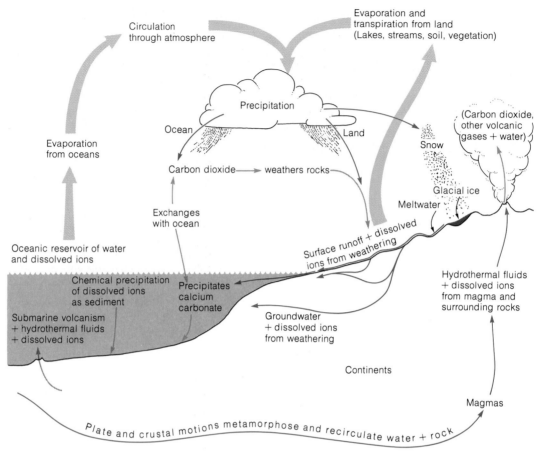

Figure 6-19
Overall hydrologic cycle. Routes of carbon dioxide are shown in black, water in brown.

weathering of rocks, for it releases ions from decomposed rock. Surface water and groundwater constantly carry the dissolved substances to the oceans, where they steadily recombine as the chemically precipitated fraction of sediments. Calcium carbonate formed in this way effectively buries some of the carbon dioxide from the atmosphere. The carbon dioxide of the oceans and atmosphere are in equilibrium with each other by exchange across the surface of the ocean. All of the operations are interrelated and combine to keep the ocean constant in composition as water evaporates from the surface and as precipitates settle to the bottom.

Thus we can again visualize the operation of this chemical processing plant as a coupling of two subunits powered by the two heat engines of Earth (Fig. 4-23). In the surface subunit, water is the pipeline carrying the products of weathering to the sea, where they are removed by sedimentation. The water is purified by evaporation and recycled to continue its weathering work. In the subunit operating within the Earth, the water is recycled with solid rock as part of interior movements of crust and mantle, ultimately to surface again as volcanic emanations. In the surface subunit—that is, in what geologists call the hydrologic cycle—a molecule of water may be recycled in a few years, but in the interior subunit recycling may take millions of years.

SUMMARY

1. The properties of water are unique indeed; existing as gas, liquid, and solid on the Earth's surface, water participates in all geological processes.

2. The hydrologic cycle is a description of the pathways by which water evaporates from land and sea surface and returns by precipitation as rain, ice, and snow.

3. Surface runoff is directly related to the amount of precipitation and thus reflects climate.

4. Groundwater forms by infiltration of precipitation through soil into permeable aquifers.

5. The groundwater table generally follows the contour of the land surface and changes in level in response to precipitation.

6. Groundwater moves at slow to moderate rates through aquifers, the rates depending on the permeability and the hydraulic gradient. Water may travel long distances underground.

7. Groundwater supplies may be depleted by excessive pumping and insufficiently rapid recharge. The water supply in some places can be replenished by increasing recharge.

8. The quality of water depends on the kinds of materials dissolved in the water during its passage through soils and aquifers. Hard waters form when abundant calcium dissolves. Potable waters contain only small amounts of dissolved material.

9. Groundwaters may be contaminated by improper sewage disposal and by seepage of chemical nutrients applied to soils.

10. Groundwaters become more concentrated in dissolved solids at greater depths. Waters fill pore spaces of rocks down to depths where porosity decreases to the vanishing point.

11. Thermal waters occur in the presence of igneous intrusions. Thermal waters form a variety of deposits around hot springs and geysers.

12. The waters of the surface evolved from the escape of volcanic gases from the Earth's interior.

13. The composition of sea water has stayed constant throughout most of geologic time because of the balance between inflow of dissolved material from rivers and precipitation of sediments. Residence time is the average length of time an atom stays in the ocean before precipitating.

14. Much of the gases and water at the surface escaped from the interior early, setting in motion the hydrologic cycle much as we know it today.

EXERCISES

1. If the polar ice caps were to melt completely as a result of slight warming of Earth, how might the quantities in the hydrologic mass balance of Figure 6-4 change?

2. If there were a vast increase in the number of desalination plants making fresh water out of sea water all over the world, so that a significant volume of fresh water were produced, would it affect the workings of the hydrologic cycle? If so, how?

3. Assume that in the future, motions along the San Andreas fault, a plate boundary, were to change to divergence, separating the San Joaquin Valley from the coastal mountains to the west and changing it to a low coastal plain bordering a new arm of the Pacific. What changes in the hydrologic budget of the valley might follow?

4. Several hundred meters under the city of Chicago are permeable aquifers of sandstone, which crop out in southern Wisconsin, 100 to 200 miles north of the city. One-hundred years ago, Chicago got much of its water from flowing wells drilled into these aquifers, but water is now difficult to pump from the wells, and the city takes its water from Lake Michigan. What do you think might have been the history of Chicago's water supply?

5. The recharge area of an important aquifer is slowly being paved over as a result of urbanization. Can you devise ways of increasing infiltration in this area?

6. If, as recently happened in one state, it were discovered that radioactive waste from a nuclear processing plant had seeped into groundwater in the immediate vicinity of the plant, what kind of information

would you want in order to predict the length of time before that waste might appear in a groundwater aquifer many kilometers away?

7. Why do most well drillers not recommend drilling too deeply, many hundred of meters, to get pure water supplies?

8. A water reservoir for a city of 100,000 people supplies 10 million cubic meters of water per year and has a capacity of 20 million cubic meters. What is the residence time for water in the reservoir if it stays full all the time?

BIBLIOGRAPHY

Davis, S. N., and R. J. M. DeWiest, *Hydrogeology,* New York: John Wiley & Sons, 1966.

Dunne, T., and L. B. Leopold, *Water in Environmental Planning.* San Francisco: W. H. Freeman and Company, 1978.

Garrels, R. M., and F. T. Mackenzie, *Evolution of Sedimentary Rocks.* New York: W. W. Norton, 1971. (Chapter 4.)

Leopold, L. B., *Water: A Primer.* San Francisco: W. H. Freeman and Company, 1974.

Peixit, J. P., and M. A. Kettani, "The Control of the Water Cycle," *Scientific American,* April 1973. (Offprint 907.)

Revelle, R., "Water," *Scientific American,* September 1963. (Offprint 878.)

Sayre, A. N., "Ground Water," *Scientific American,* November 1950. (Offprint 818.)

7

Rivers: Currents, Channels, and Networks

Most of the debris of erosion is carried downhill by running water in the streams of the Earth. Turbulent currents excavate river valleys by eroding channels and moving particles downstream. The ability of a river to transport particles in different amounts and sizes depends on current velocity, the total amount of water carried, and the downhill slope of the channel. Currents form ripples, dunes, and bars in the channel sediment. A river can be analyzed as a complex system of drainage patterns that form in response to the particular conditions of climate, bedrock, and particles produced by weathering.

Rivers grip our imaginations. They bring out clichés like "mighty waters" and "raging torrents" and, at more serene moments, others like "majestic flow" and "babbling brook." The variety of clichés, most of which center on the idea of constant flow, is matched only by the variety of forms themselves. Rivers may be straight, but they may bend and loop around too. A river may flow in a broad low swale that barely has the form of a valley, but in another place that same river may rush between the nearly vertical walls of a narrow gorge. A high mountain stream in the Rockies may be a hazard to cross because of the jumble of rocks through which the water rushes, but at times of drought you can walk almost all the way across the fine sand and mud on the bed of the Missouri River at Omaha. Rivers are clear, and rivers are turbid. Some are fast, and others slow.

The factors responsible for these differences in how water moves, what it carries, and its effects on the landscape make up the subject of this chapter. We start with some notions about how water moves in currents and how the movement enables it to carry things of all sizes and shapes, from mud, sand, and pebbles to whole houses ripped from their foundations by floods.

HOW CURRENTS MOVE PARTICLES

If you slowly pour cold syrup over melting butter on pancakes, you can observe a most basic form of fluid flow: a pattern of **streamlines**—lines of flow—is formed by thin strands of melted butter flowing along with the syrup. As long as the syrup is cold, and therefore thick and sluggish, the streamlines show the two fluids moving steadily without mixing. This is the essence of **laminar flow,** flow in which the particles of the fluid move in parallel layers with no mixing of material across the boundaries between layers (Fig. 7-1). Some, however, like their syrup hot, and that

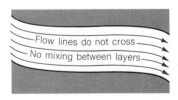

Figure 7-1
Laminar flow of a fluid between two solid channel walls. In this type of flow, flow lines do not cross, and there is no mixing of fluid between layers.

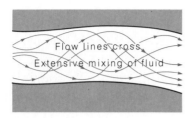

Figure 7-2
Turbulent flow of a fluid between two solid channel walls. Flow lines in this type of flow cross in a complex pattern; unlike laminar flow, no definite layers of fluid can be discerned, and extensive mixing of the entire fluid mass occurs.

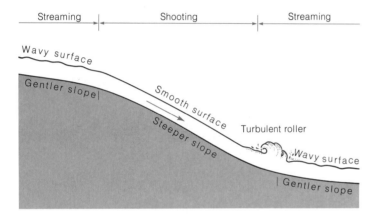

Figure 7-3
Streaming (tranquil) and shooting (rapid) flow are determined by slope angle. Beyond a certain steepness of slope, slower, wavy-surfaced streaming flows abruptly change to faster, smooth-surfaced shooting flows. Where the slope again becomes gentle, the flow returns to streaming, the transition being marked by a turbulent roller, or eddy.

changes the picture. The property of fluids that makes them move slowly or rapidly when poured is viscosity; the viscosity of such fluids as automobile lubricating oils is usually determined by measuring the time that a certain quantity of fluid takes to flow through a small opening, like the bottom of a funnel. Viscosity is a function of temperature, and the viscosity of most fluids is lower—that is, they flow faster—at higher temperatures. Thus hot syrup, which is less viscous than cold syrup, may not move in obvious, slow laminar flow but in a faster, more confused pattern, one of **turbulent flow** (Fig. 7–2). In a quiet room, tobacco smoke, another fluid, will at first rise from a cigarette in a straight, vertical plume of laminar flow and then break up into a turbulent, eddying, complex flow in which the streamlines cross, interfere, and eventually dissipate by completely mixing with air. The essence of turbulence

is in the mixing of the flow in complex patterns. The general behavior is the same whether the fluid is a gas, like smoke in air, or a liquid, like syrup or water.

The most important factor that determines whether the flow of water is laminar or turbulent is the velocity. Only very slow flows are laminar; flow in practically all of the streams we see is turbulent. Secondary factors are the roughness of the stream bottom and the depth of the stream. The shallower the stream and the smoother the bottom, the more likely the flow is laminar. Laminar flow in nature can be seen in thin sheets on nearly flat slopes and occasionally in very cold water in streams running over smooth bottoms. Turbulent flow varies in the intensity of eddying and swirling movements, again mainly a matter of stream velocity.

Strange behavior is shown by a fast-flowing turbulent stream where it suddenly steepens and starts to flow even faster than before. The rush of water smooths to what looks like straight, uneddied movement, even though it is still basically turbulent. Its surface smoothness is deceptive, for the flow is actually so fast that it is commonly called **shooting flow,** as opposed to the more tranquil, or **streaming flow,** at lower velocities (Fig. 7–3). Shooting flow can be seen typically in rocky rapids or where water flows down steep narrow chutes.

Movement of Particles

These three forms of flow—laminar, turbulent, and shooting—differ enough in velocity and in other characteristics that they may be likened to fine, medium, and coarse sandpaper in their ability to erode stream bottoms. Laminar flows are gentle; they erode the bottom slightly if at all and carry only the smallest of particles—those of clay size. Turbulent flows, depending on their speed, can move particles from clay size up to pebbles and cobbles. The fastest streams, including all shooting flows, wear away solid rock and carry along coarse particles at high speeds. The debris of erosion that rivers deposit as sediment, **alluvium,** covers the valley floors of rivers. Many ancient sedimentary rocks originated in this way.

Turbulent currents move particles downstream mainly by lifting them into the flow or by rolling or sliding them along the bottom (Fig. 7–4). Smaller particles, clay and silt, are easily pulled up into the flow by upward-moving eddies. Once in the stream, the particles are carried along in suspension; all particles so carried by a stream at

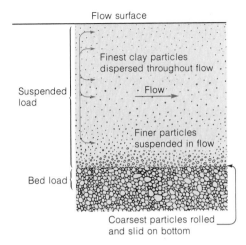

Figure 7-4
A current flowing over a bed of sand, silt, and clay particles transports particles in two ways: (1) as bed load, which moves by sliding and rolling along the bottom, and (2) as suspended load, which moves by being temporarily or permanently suspended in the flow itself. The finest clay particles remain permanently suspended in all but extremely sluggish flows.

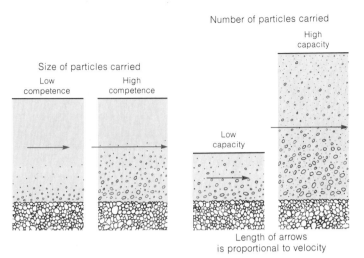

Figure 7-5
The competence of a current to carry particles is a measure of the largest size particles it can carry. The capacity of a current is a measure of the number of particles it can carry. Although both depend on velocity to some extent, competence depends much more directly on velocity, and capacity depends largely upon discharge, the volume of water flowing past a point in a given time.

any one time are its **suspension load.** The forward force of the moving current acts more directly on the larger grains at the bottom, pushing, rolling, and sliding them along. Particles moved in this way constitute the **bed load** of a stream. Intuitively we grasp the idea that the stronger the current, the larger the particles it can carry in suspension or as bed load. The measure of this ability to carry particles of different size we call **competence** (Fig. 7-5). The greater a stream's competence, the larger the particle size it can carry; the straightforward way to measure a stream's competence is by the largest particle sizes it can carry.

Just as dump trucks of various sizes can all carry sand, but some more than others, so streams carry similar loads, but in different amounts depending on their size. The load of particles of all sizes that a stream can transport is its **capacity.** Although the capacity of a stream is related to its size, more important than the width and depth of a stream is the amount of water that flows in it—the **discharge,** or the volume of water that flows past a given point in a specified unit of time. Discharge is commonly measured in cubic meters per second or cubic feet per second.* A typical small

stream may vary in discharge from 10 to 10,000 cubic feet per second (0.28 to 280 cubic meters per second). The Mississippi River discharge is at times as low as about 50,000 cubic feet per second (1400 cubic meters per second), but in flood may amount to more than 2,000,000 cubic feet per second (57,000 cubic meters per second). The Mississippi, though it can ordinarily carry only clay, silt, and sand, carries enormous quantities of those small particles; in contrast, a fast mountain stream may move boulders, but only a few of them.

A stream's competence for suspended load is a balance between the uplifting forces that turbulence exerts on particles and the force of gravity, which pulls the grains back down to the stream bed. Particles of silt and clay are so easily lifted and drop back so slowly that they tend to remain in suspension in all but the most still waters. The **settling velocity**—the speed with which particles fall through the fluid to the bed—is much greater for sand grains than for silt and clay particles. The typical movement of sand grains is an intermittent jumping, called **saltation,** in which they are sucked up by eddies into the flow, travel for a while, and then fall back to the bottom (Fig. 7-6). The bigger or denser the grain, the shorter its travel time and the longer it rests on the bottom before being lifted again by a particularly strong eddy. The lighter the grain, the more frequently it is picked up and moved along above the stream bed.

---

*Discharge is measured in practice at a gauging station, where the height, or *stage,* of a river is measured together with the velocity of the flow, from which the discharge can easily be calculated. There are now more than 7000 gauging stations in the United States.

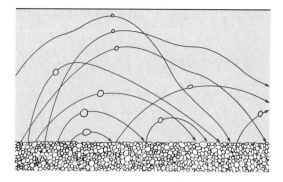

Figure 7-6
Saltation is an intermittent "jumping" motion of
grains. Turbulent eddies pull grains up into the flow,
where they travel with the flow for a distance before
falling back to the bed. In general, the smaller the
particle, the higher it will jump and the farther it
will travel. Turbulence keeps some very small particles suspended throughout a long travel path before
they fall back to the bed.

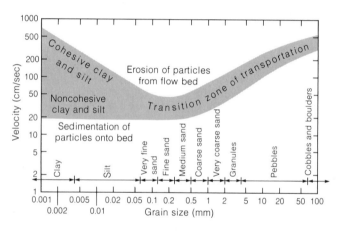

Figure 7-7
Experimental data on the relation between grain size
of particles and velocity of a flow one meter deep
over a flat, granular bed. For a specific grain size,
the lower line is the velocity at which all particles of
that size fall to the bed; the upper line is the velocity at which all particles of that size continue to be
picked up from the bed. The transportation zone is a
transition zone in which particles already suspended
remain in the flow and in which there is no tendency for either erosion or sedimentation to take
place. Cohesive behavior of most fine-grained natural minerals makes them more resistant to erosion
than noncohesive clays and silts. [After F. Hjulstrom,
as modified by A. Sundborg, 1956, "The River
Klarälven," *Geografisk Annaler*.]

The measurement of the exact relations between current velocity and competence has been
a major concern of geologists and hydraulic engineers who study transport of sediments by
streams. The geological objective is to use these
measurements to obtain some idea of the nature
of the sediment that is produced by a particular
current. Geologists could then work backwards to
infer the nature of ancient currents from the sizes
of grains in sedimentary rocks. Engineers, on the
other hand, frequently want to predict what kind
of sediment, and how much of it, will be brought
in by a river to a lake behind a dam. Figure 7-7
shows the result of experimental measurements
of competence in relation to velocity. From such a
graph one can quickly read off the size of a particle that can be carried at a given velocity, or vice
versa. The band indicates the lack of precision of
the data from experiments in which it is impossible to control all factors completely and rigidly.
The unexpected upturn that the curve shows for
clay-size particles does not mean that they are
harder to carry but that they are harder to lift
from a smooth clay bottom because the particles
cling together with strong cohesive forces. This
graph is applicable only to flows of one meter
depth. Other, more complex graphs are more generally applicable to flows of any depth.

Bedding and Sedimentary Structures

The movement of sand grains in current creates
ripples and dunes on the stream bed as well as the
familiar horizontal bedding planes (Fig. 2-7). **Ripples** in sand are the low narrow ridges that are
separated by wider troughs, both of which can be
seen on the surfaces of windswept sand dunes, on
underwater sandbars in shallow streams, and
under the waves at beaches (Fig. 7-8). They come
in an assortment of shapes, sizes, and patterns
that are characteristic of the currents that form
them. Ripples and a variety of other forms produced in the process of sedimentation are called
sedimentary structures. Information on the connection between current and sedimentary structures comes mainly from laboratory experiments
in which the streams of nature are simulated in
flumes—long tilted channels in which the flow of
sand and water can be controlled. Flumes allow
careful measurement of velocity and other variables plus direct observation, through transparent
side walls, of the ways in which grains move and
accumulate on the bed to form ripples and stream
dunes.

One of the best ways to obtain good data from a

Figure 7-8
Ripples in sand form in response to current action of wind or water. Their size and spacing increase as the current velocity increases. The shadows show that the ripples here are steeper on the slope toward the top of the page; that is the down-current direction. [Photo by P. E. Potter.]

dynamic experiment is to hold all variables constant except one and to change that one variable systematically and measure the change in the result. In 1961, members of a U.S. Geological Survey group reported the results of their experiments on the mechanics of river flow. Using sensitive equipment and interpreting their results in terms of modern theory, they repeated a simple, classic flume experiment done in 1914 by one of the great geologists of that time, G. K. Gilbert. Imagine that you are redoing this experiment. Beginning with a flume containing a preformed, flat bed of sand grains, all of the same size, you open a faucet just a crack to start the water flowing as a very weak current. At this low velocity of flow, less than 5 centimeters per second ($\frac{1}{8}$ mile per hour), the current is too weak to move any grains, and the bed remains flat and undisturbed (Fig. 7–9a). Upon turning the water up a bit more to make the current a little stronger, a little more than 10 centimeters per second ($\frac{1}{4}$ mile per hour), you begin to see

a few grains here and there roll or slide short distances, and a rare one saltate briefly. The bed stays flat. After another increase in flow, however, you notice that many grains are saltating, and after a little while at this flow velocity, about 20 centimeters per second ($\frac{1}{2}$ mile per hour), small ripples a few centimeters high start to form all over the bed. Soon you notice that the ripples themselves are moving like waves almost imperceptibly downstream whereas the particles themselves move much more rapidly.

The ripples have a gentle slope upstream and a steep slope downstream: they are **asymmetrical ripples.** At the transparent sides of the flume, you can see the ripples in cross section and observe the inclined bedding that we call **cross-bedding** (Fig. 7–9b). Here, of course, the cross-bedding is on a very small scale compared to that of the Grand Canyon (Fig. 2–20). The angle of the cross-bedding is the angle of the downstream, or lee, slope of the ripple.

The next higher velocities make the ripples move faster and grow larger, until they are many centimeters high, large enough to be called **dunes** (Fig. 7–9c). Dunes have the same general form and structure as ripples but are larger; they form under water as commonly as those that people more frequently observe forming under air. As the dunes in the flume grow larger, smaller ripples form and climb up their backs and disappear over the lee slope. Increasing the velocity still further, to more than 50 centimeters per second. (1.25 miles per hour), changes the character of flow from streaming to shooting flow, and as it does the bed responds quickly: ripples and dunes disappear, and the bed becomes flat again. But this time all the grains in it appear to be in rapid motion. Increasing the flow still more, to a velocity seldom met in rivers, produces a more irregular bed with some dune-like forms that move upstream (Fig. 7–9d) and, finally, a complete washout of the whole bed of sand, which becomes suspended in the flow.

Sedimentologists—geologists who specialize in the study of sediments—have worked hard to infer from the ripples and cross-bedding preserved in rocks the kind and intensity of current that formed them. Their study of sedimentary structures produced in flumes has shown that the geometry of the bed is determined by three main variables: the velocity, the grain size of the sediment, and the depth of the flow. Observations of modern sediments of rivers and other environments and the currents that formed them have correlated with the laboratory studies. For example, we have learned to recognize ripples formed

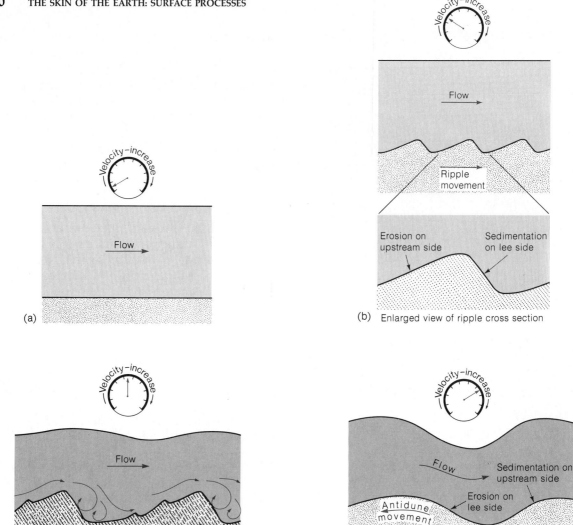

Figure 7–9
A flume experiment showing the change in sand bed with increasing stream velocity. (a) At this flow stage, sediment transport takes place by sliding and rolling motions of single grains. (b) At slightly higher velocity, a rippled bed forms. Major transport is by bed load and saltation. Ripples migrate downstream and have cross-bedded structure. (c) At moderate velocity transport is by rippled dunes. Dunes have the same cross-bedded structure as ripples. Because ripples migrate downstream faster, they tend to climb over the backs of the dunes. Strong reverse eddies form in the lee of dunes. (d) At high velocity antidunes migrate upstream, while there is extensive sand-grain movement downstream by suspension and bed load, almost the entire bed surface being in motion. [After D. A. Simons and E. V. Richardson, 1961, "Forms of Bed Roughness in Alluvial Channels," *Am. Soc. Civil Eng. Proc.* v. 87. Hy3, pp. 87–105.]

by wave action, for they are symmetrical and have much sharper crests than current ripples. Formed by the back-and-forth movement of waves, they are called **oscillation ripples** (Fig. 7–10). Different cross-bedding forms are diagnostic of other environments. In river bars, for example, the current scoops out long, spoon-shaped troughs that are quickly filled by cross-bedded sand, as in Figure 7–11; geologists can recognize this **trough cross-bedding** pattern in ancient rocks.

EROSION BY CURRENTS

Though it is obvious from observations of sediment transport that currents can erode a bed of soft sediment by picking up particles and carrying them downstream, one cannot so easily observe the erosion of solid rock by a stream, because it happens much more slowly. Here the analogy to the action of sandpaper (or sandblasting) is apt, for much of the erosive power of a stream comes from the abrasion of the bottom by the sand and

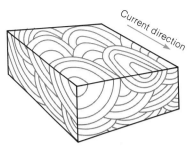

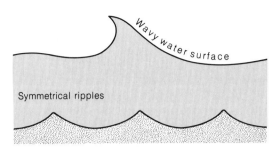

Figure 7-10
Oscillation ripples are symmetrical forms produced by waves. They commonly have much sharper crests than current ripples.

Figure 7-11
Trough cross-bedding is produced on river bars as troughs are scoured and then filled with sand, building up a series of interfering structures as earlier ones are partly eroded. The open end of the trough always faces downcurrent, but the direction of cross-bedding inclination may be variable, depending on the orientation of the exposure with respect to the troughs.

gravel it carries. On some river bottoms sand grains and pebbles swirling in fast eddies wear **potholes**—rounded depressions and deep holes—into the solid rock of the riverbed (Fig. 7-12).

We rarely see the bedrock floor of a river except in rocky streams in mountains or where a stream crosses rapids, and even there most of it is covered by boulders of all sizes. The floors of most rivers that flow in broad valleys in plains or lowlands are covered with the sand, gravel, and mud that they carry. These rivers usually scour their bedrock bottoms only at high flood stages. The bedrock channels and walls of a river are eroded physically by a variety of other mechanisms. Just as in weathering on the outcrop, chemical decay along joints and cracks helps the breakup of bedrock, but here it is aided by the sledgehammer action of boulders slamming against solid rock. Chemical action is also helped by a variety of organisms, among them the aquatic algae that attach themselves to rock surfaces (and make them slippery to cross). Once large blocks have been loosened by decay along joints, they are dislodged by strong upward eddies that literally pull them up and out by what amounts to violent, sudden suction, a plucking action. Channels at the bottoms of waterfalls are eroded at an enormously rapid rate by the plunging action of the water and transported rocks, which fall with tremendous impact. Waterfalls are eroded and move upstream as the cliff that forms the falls is undercut and the upper beds collapse. This situation occurs where a nearly horizontal, erosion-resistant bed overlies more easily erodible formations. A well-known example, Niagara Falls, has been studied to learn the rate of its headward erosion. Historical rec-

Figure 7-12
Potholes in granite ledges exposed at low water in the James River, Virginia. The pothole in the right foreground shows gravel at the bottom, which is the scouring agent that erodes the holes in solid rock. [Photo by C. K. Wentworth, U.S. Geological Survey.]

ords show that the main (horseshoe) part of the falls has been moving upstream at a rate of more than one meter per year.

Erosion of soft unconsolidated material or of weak sandstone and shales by a river is much more obvious than that of bedrock channels and falls. Scouring and slumping of soft river banks at high water is common. Gullies in soils erode headward at a rapid pace. On a small scale, one can observe this kind of erosion happening during a strong rainstorm. Evidence of intense erosion of

Figure 7-13
The Badlands of South Dakota, a region underlain by easily eroded horizontal sandstones and shales, which have been heavily gullied by slope wash, rills, and small streams. [Photo by N. H. Darton, U.S. Geological Survey.]

river banks, channels, and surrounding areas is widespread after a major flood. In one of the fastest eroding regions in the United States, the basin of the Eel River in northern California, 310 million tons of rock and soil was removed from 8000 square kilometers (3100 square miles) in a recent 10 ten-year period. Some accelerated erosion may be the result of human activities, such as logging. In dryer climates erosion of soft sedimentary rocks can produce badland topography, in which erosion is so rapid that little vegetation can grow to stabilize slopes and resist erosion (Fig. 7-13).

Strong as the direct evidence of erosion is to us, it was not the only thing that convinced Hutton almost two hundred years ago that valleys are eroded by the streams that flow in them. For Hutton, the match between the rocks and debris carried by a stream and the bedrock of the valley could be interpreted reasonably only as an indication of erosion. As we have seen (Chapter 5), the stratigraphic matching of sedimentary beds on opposite valley walls showed that discontinuity represented by the valley had to come later by erosion. The many lines of evidence, all supporting each other, lead to only one conclusion: valleys, whether deep and narrow or broad and

wide, are cut into rock by streams. This does not always mean that the stream now in the valley is the one that carved it. As we shall see in Chapter 9, mountain glaciers—rivers of ice—cut U-shaped valleys, and these are usually occupied by streams that inherited the valley after the ice melted.

THE RIVER AS A SYSTEM

Rivers have been personified as living things by various writers, who, like most of us, are affected by their constant motion and change. Rivers are dynamic systems that exhibit a balance between input and output. The input is the total of all the water that reaches a river as surface runoff and groundwater plus the erosional debris from the entire area drained by it. The output is water and sediment, which are both ultimately carried to the ocean. In detail, any unchanging stretch of river has adapted itself in shape and size to the input, so that output matches input. Changes in a river are the result of imbalance. Most changes are adjustments of its channel during normal flows, not floods. Floods are extreme examples of a river channel's inability to carry off a suddenly in-

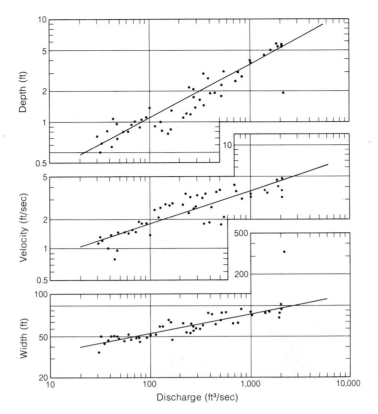

Figure 7-14
Changes of width, depth, and velocity with discharge at a specific river station: Seneca Creek at Dawsonville, Maryland. The drainage area of this stream is about 260 km². Discharge is given in cubic feet per second. [After *Fluvial Processes in Geomorphology* by L. B. Leopold, M. G. Wolman, and J. P. Miller. W. H. Freeman and Company. Copyright © 1964.]

creased input of water and sediment; the river responds by overflowing and "inventing" new channels or flow paths to carry the water. When input decreases to normal, the river shrinks to its normal size.

As we noted earlier, the amount of water flowing past a given point per unit of time is the discharge, which is by definition equal to the cross-sectional area of the river multiplied by the velocity:

$$\begin{matrix} \text{Discharge} \\ \text{volume} \\ (\text{ft}^3 \text{ or } \text{m}^3/\text{sec}) \end{matrix} = \begin{matrix} \text{Cross-sectional} \\ \text{area of stream} \\ (\text{usually width times} \\ \text{depth in ft}^2 \text{ or m}^2) \end{matrix} \times \begin{matrix} \text{Velocity} \\ (\text{ft or m/sec}) \end{matrix}$$

Because rivers are generally wide in relation to their depth, and are fairly easily describable as simple rectangles in cross section, the width-depth relationship is more useful than cross-sectional area. Now we ask: what happens to width and depth for normal flows as discharge increases—for example, as a response to increased rainfall runoff? Just looking at this relationship, one can see that the water will have to run faster, or the width or depth will have to increase by the

river's erosion of its channel, or some other combination of factors will have to change. We could think about what *might* happen for a long time, but the way to find out is to measure what actually does happen. Figure 7–14 shows typical graphs made at a measuring station; the graphs show that there is a systematic increase in all three factors, width, depth, and velocity, as discharge increases. The proportion of increase varies from one stream to another, depending on the debris load of the river and the ease with which it erodes its banks and channel bed, for it is erosion that widens and deepens a channel.

How much sediment the stream carries—its transport capacity—is also related to the discharge. The higher the discharge, the larger the load of sediment, as is shown in Figure 7–15. There is also a relationship between discharge and the size of the material eroded or deposited, for the flow picks up or deposits material in accord with the graph in Figure 7–7.

Discharge in most rivers increases downstream as more and more water is collected from tributaries. Inevitably, then, the width, depth, and ve-

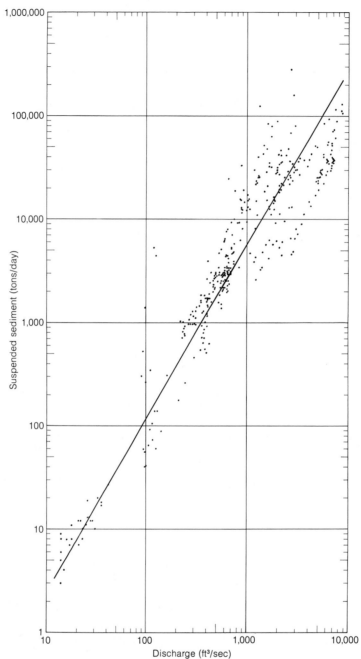

Figure 7-15
Increase of suspended sediment carried by a river in relation to increased discharge at a specific river station: Rio Grande near Bernalillo, New Mexico. [After *Fluvial Processes in Geomorphology* by L. B. Leopold, M. G. Wolman, and J. P. Miller. W. H. Freeman and Company, Copyright © 1964.]

locity must change too. The progressive downstream change in these factors has been observed to follow a pattern similar to that shown by an increase in discharge at one place: width, depth, and velocity all increase systematically, width far more than depth. Velocity does not increase downstream as much as one might expect from the increase in discharge because the decrease in the downhill slope of a channel is accompanied by a decrease in velocity. After all, as Galileo showed the world 350 years ago, the steeper the slope, the faster the motion. The **longitudinal profile** of a river—that is, a plot of the relative elevation of its stream bed from headwaters to mouth—is concave upward, which tallies with the steady downstream decrease in slope (Fig. 7-16). Although decreasing slope tends to slow the current, its influence may be overcome by an increase in discharge that makes the stream flow faster. In many rivers the discharge does not increase greatly downstream, and the lower downstream slope reduces the velocity. Because the slope is itself a product of the river's capability of downward erosion and the opposing capability of depositing sediment in its channel, the longitudinal profile is a response the river makes to the balance between input and output of water and sediments.

The concave upward longitudinal profile is remarkably general; it evolves in small rills and large rivers, in natural streams, and in artificial water courses. A feature so general must be bound up with some basic dynamics common to all streams. The downstream increase in discharge gives one clue. As the discharge increases, we have seen, so in general do the width and depth of the stream. But as the width and depth increase, so does the ratio of cross-sectional area to perimeter, for the cross-sectional area increases in proportion to the square of the radius (πr^2), whereas the circumference increases in proportion to the first power of the radius ($2\pi r$). The effect of this increase is the same as that produced by increasing the diameter of a pipe: it allows the volume of flow to increase. The key here is that the circumference of a pipe or the perimeter of a stream is what controls the friction between the flowing water and the walls of the pipe or the banks and floor of a stream. As the volume of water increases, proportionally less of it is in contact with the sides, and so friction, which retards the flow, is relatively decreased. The less the friction, the more rapid the flow velocity.

By similar analyses we can list all of the factors that change in the downstream direction that in-

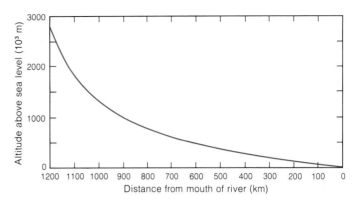

Figure 7-16
Longitudinal profile of the Platte and South Platte rivers from the mouth of the Platte at the Missouri River in Nebraska to the headwaters of the South Platte in central Colorado. This profile shows the typical concave-upward slope of rivers, much steeper at the headwaters than near the mouth. [Generalized from data of H. Gannett, in *Profiles of Rivers in the U.S.*, U.S. Geological Survey Water Supply Paper 44, 1901.]

fluence a river's longitudinal profile (Table 7-1). Putting all of these interrelated factors together in a single formula that would express quantitatively how a concave profile develops is a formidable task—one that has been only partially done. What is already clear, however, is that as they move downstream, rivers are able to transport more material of finer size at a higher discharge but at a lower slope. The concave upward profile seems to be another one of the steady-state equilibrium patterns that characterize so many Earth processes.

The longitudinal profile is also responsive to **base level**—the level or elevation at which the

Table 7-1
Normal changes in flow properties in the downstream direction and effects on river flow

Property of flow	Normal change in downstream direction	Effect on river downstream
Discharge	Increases	Velocity increases
Width	Increases	Ratio changes so that friction is decreased
Depth	Increases	
Velocity	Increases or decreases depending on changes in discharge and slope	Increases capacity to erode and transport
Frictional resistance to flow	Increases or decreases depending on perimeter and load	An increase causes decrease in velocity; a decrease causes increase in velocity.
Sediment load (suspension + bed load)	Increases	Increases friction
Size of sediment particles (gravel, sand, silt, clay)	Decreases	Decreases flow resistance
Slope of river bed	Decreases	Decreases velocity

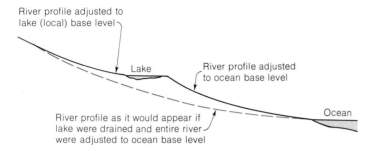

Figure 7–17
Regional and local base level as illustrated by a river flowing into a lake and from the lake into the ocean. In each river segment, the longitudinal profile adjusts to the lowest level it can reach. The lake may be a natural one or an artificial reservoir behind a dam. If the lake were drained and the river allowed to erode downward, the river bed would adjust eventually to a single concave-upward profile.

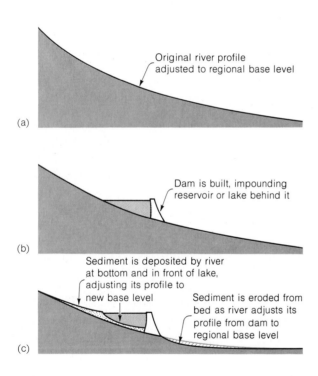

Figure 7–18
Change of base level and its consequences on a river profile. The original profile, (a), adjusted to equilibrium with the regional base level, is altered when a dam is built. The dam impounds a lake behind it (b) and raises the local base level. The river responds upstream by adjusting its bed to a new profile as sediment is deposited in and in front of the lake, and it responds downstream by eroding its bed to a new profile with respect to the original regional base level (c).

mouth of a stream enters a large standing body of water, a lake or the ocean, and so disappears as a river (Fig. 7–17). Streams cannot cut below base level, for base level is "the bottom of the hill"— the lower limit of the longitudinal profile. A river may have a number of local base levels; for example, lakes or waterfalls along its course form limiting levels that affect the flow of stretches immediately upstream. Changes of base level cause a stream to change its characteristics, so that the water, sediment, and channel geometry come to a new balance. The most common adjustment to a raising of base level is predictable from the variables involved (Fig. 7–18). The slope decreases, causing a reduction of velocity, which in turn decreases the stream's sediment-transporting ability. The stream then responds by dropping some of the sediment on the bed, which makes the concavity somewhat shallower than before. The opposite change, lowering the base level, increases the slope, the velocity, and the capacity to transport and erode the stream bed. The precise course of the change is predictable only when all of the flow-sediment relationships are known. For example, a stream greatly choked with sediment might be practically unaffected by a slight steepening of profile caused by a lowered base level because the steepening would not be enough to make a significant increase in the load-carrying capacity in relation to the sediment supply. The effects of changing base levels of streams by sea-level rises and falls during geologic history is receiving new attention as geologists seek to fit worldwide sea-level fluctuations with plate tectonics, continental drift, and worldwide glaciations.

The ways that flow, sediment transport, and channel geometry adjust to changes in base level, to keep a stream in balance lead to the concept of the **graded stream.** J. Hoover Mackin, a leading American geomorphologist of the mid-twentieth century described a graded system as "one in which, over a period of years, slope is delicately adjusted to provide, with available discharge and with prevailing channel characteristics, just the velocity required for the transportation of the load supplied from the drainage basin. The graded stream is a system in equilibrium; its diagnostic characteristic is that any change in any of the controlling factors will cause a displacement of the equilibrium in a direction that will tend to absorb the effect of the change." Rivers adapt by a variety of depositional and erosional patterns and by altering the shape of the channel.

Figure 7-19
Alluvial fans in the Mohave Desert, California. Each cone-shaped fan
has been deposited where the channel changes abruptly at the base of the
mountains from a narrow, confined valley upstream to a broad, unconfined
plain. The fans have grown together at their feet, completely obscuring the
original mountain slopes. [Photo by J. R. Balsley, U.S. Geological Survey.]

River Channels and Their Deposits

If you wanted to test a river severely to see how it
would adjust to a change, you might choose a
place where a stream leaves a narrow valley in a
mountainous area and suddenly enters a broad,
relatively flat valley or plain at a fault scarp. The
change from the narrow channel of the mountain
stream to the unconfined slopes of the plain is
abrupt, and so is the river's response—a sudden
slowing of velocity and an immediate dumping of
sediment in a cone or fan-shaped accumulation
called an **alluvial fan** (Fig. 7-19). The fan itself
normally shows the characteristic concave upward profile (Fig. 7-20). On the steep upper slopes,
alluvial fans are typically dominated by coarse
materials, boulders and cobbles, gravel and sand;
on their lower, gentler slopes, by finer sands, silts,
and muds. Fans from many adjacent streams
along a mountain merge to form a long wedge of
sediment whose external appearance may mask
the outlines of the individual fans that make it up.
As these compound fans grow upward and outward from a mountain front over thousands to
millions of years, they build up extensive sedimentary accumulations many cubic kilometers in
volume.

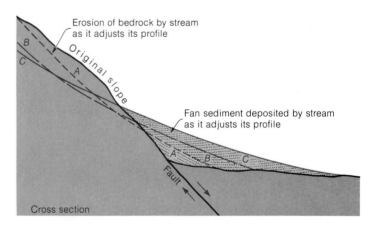

Figure 7-20
Formation of an alluvial fan as a stream adjusts its
profile to a fault scarp. The original fault slope is
eroded above the fan as the fan is built upward and
outward from the base of the scarp in successive
positions A, B, and C.

Figure 7–21
Braided stream choked with erosional debris, near the edge of a melting glacier.
Muddy River, near junction with McKinley River, Alaska. [Photo by B. Washburn.]

Channel Patterns

Streams running down alluvial fans frequently break up into small multiple channels that branch into distributary networks. This too is a consequence of the sediment-flow relationship. Because the sediment load is large in relation to discharge, slope, and channel depth, the shallow channels tend to fill, and the streams then break through their walls to form new channels. Many of the subsidiary channels rejoin farther downstream, giving the stream as a whole a familiar pattern from which it gets the appropriate name **braided stream** (Fig. 7–21). Braided streams are not restricted to alluvial fans by any means; they are also typical of rivers running in plains and low valleys—anywhere, in fact, where discharge is low in relation to sediment load and the stream cannot transport enough to maintain a single, deep channel. The braided pattern of multiple channels develops within the confines of a single larger channel, and there may be several scales of braiding, or braids within braids.

The word **meander** comes from the name of a river in Turkey fabled in ancient times for its twisting, winding course of looping bends. In geology the word refers to the more-or-less regular bends that may form in any river (Fig. 7–22). Meanders are normal for rivers flowing on low slopes in plains or lowlands where the sediment is fine sand, silt, or mud. In terrains of higher slope and harder bedrock they are less abundant but still more common than long straight stretches of river. Meanders are not restricted to rivers that carry sediment. They are common in meltwater streams on glaciers that carry no sediment. Meanders are even found in flows other than rivers. The Gulf Stream of the Atlantic Ocean meanders, and there appears to be a kind of meander pattern to jet streams in the stratosphere. Some meander-

Figure 7-22
Meander bed and floodplain of the Animas River a few miles above Durango,
Colorado. The river flows toward the lower left. [From *Geology Illustrated* by
John S. Shelton. W. H. Freeman and Company. Copyright © 1966.]

Figure 7-23
Incised meanders of the San Juan River, San Juan County, Utah.
[Photo by E. C. La Rue, U.S. Geological Survey.]

ing rivers have eroded deeply into bedrock (Fig.
7-23). Meanders are common in streams flowing
in limestone caverns, where they erode mainly by
dissolving the limestone. To understand this most
general of forms we compare it to other channel
patterns.

The three kinds of channel pattern—braided,
meandering, and straight—each blend from one to
the other without sharp division. Few long
stretches of any river can be counted as truly
straight; there are always some curves and bends.
We call the measure of how straight or curved a

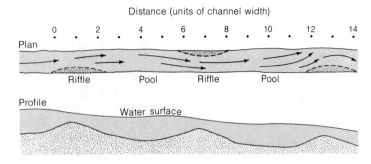

Figure 7-24
The straight reach of a river has a more-or-less un-even bed that consists of alternating deeps and shallows, known to trout fishermen as riffles and pools. The humps in the stream bed that give rise to the riffles tend to be situated alternately on each side of the stream at intervals roughly equal to five to seven times the local stream width. As a consequence the stream at low flow seems to follow a course that wanders from one side of the channel to the other, in a manner having an obvious similarity to meandering. [After "River Meanders" by L. B. Leopold and Langbein. Copyright © 1966 by Scientific American, Inc. All rights reserved.]

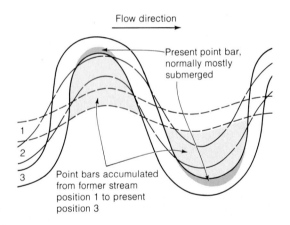

Figure 7-25
Lateral movement of river meanders gradually enlarges point bars: 1, 2, and 3 are three stages in the meander movement.

river is its **sinuosity,** the ratio of its actual channel length to the straight-line distance down valley. A sinuosity of 1 corresponds to the perfect straight stretch. Meandering streams have sinuosities of 4 or more.

Rivers in straight channels do not flow precisely straight. The **thalweg**—the line of maximum depth of a river (usually close to the line of strongest current)—winds from one side to the other of straight stretches. Where the thalweg approaches one bank, sand bars or mud banks tend to accumulate on the opposite bank, and a series of such

bars may be found alternating from one side to the other along the channel. Straight streams also tend to have undulating beds in which shallows, called **riffles,** regularly alternate with deep spots, **pools,** usually at a repeating distance of five to seven times the channel width (Fig. 7-24). Pools and riffles tend to form where there are many different sizes of sediment in the stream; the coarser sizes collect on riffles, where they may become gravel bars, and the finer sizes floor the pools. Riffles and pools, bars and mudbanks along straight rivers, and meanders all show a wavelike regularity in their spacing, and seem to be different forms of the same general behavior pattern. Like waves, these forms travel, migrating downstream in much the same way that waves travel down a whipped rope. The patterns are stable. A river does not tend to meander one week and straighten the next. All of these facts support the idea proposed by Luna B. Leopold and W. B. Langbein, two of our most acute analysts of streams, that a river tends to take the course of least resistance—to minimize the work it must do in running downhill with a certain sediment load along a certain slope. The characteristic channel patterns are appropriate "solutions" that rivers find for different sets of prevailing conditions.*

Meanders, because they are so easily seen, have been studied the most. They move both downstream and to the side, and we have the results of many river surveys to show just how they do. The sideways movement is accomplished by erosion along the outside of the bend, where the current is strongest, and by deposition of a curved bar on the inside of the bend, called a **point bar,** where the current is weakest (Fig. 7-25). Some meanders on the Mississippi shift as much as 20 meters per year. As the meanders move sideways and downstream, so do the point bars, building up an accumulation of sand and silt that covers the part of the valley floor over which the channel has migrated. Characteristic patterns of cross-bedding in these deposits have been recognized to be the same as those in ancient sandstones, and the downcurrent direction of cross-bedding has proved to be a major clue to the direction of flow of river systems that date back hundreds of millions of years.

---

*"Channelization," the artificial straightening of a sinuous channel to one with uniform gradient and cross section, is sometimes done in an attempt to improve drainage and flood control and to reclaim wetlands of the floodplain for farming or urban development. Many channelization projects have led to adverse environments effects, such as deterioration of aquatic life, accelerated erosion and silt deposition, and even poorer flood control.

As meanders move, they may progress unevenly, so that the relatively straight parts between bends get closer and closer to one another, as is shown in Figure 2–2. When they get close enough, the river may cut across the next loop, shortening its course and leaving behind the abandoned bend as an **oxbow lake.** These lakes soon fill with mud and silt, which support the growth of reeds and other vegetation. Although the cut-off may take place anytime, it is most likely to happen during a flood stage, when the stress on the banks is high. The shortening of the river's course is only temporary in comparison with the many thousands or millions of years of a river's lifetime, for new meanders are constantly evolving and enlarging, lengthening the course. Over a long time period on a given slope, the length of the channel tends to remain constant. This is another way of stating that the river tends to maintain the same concave longitudinal profile.*

The River Floodplain

As meanders migrate back and forth across a river valley, laying down deposits of gravel, sand, and silt in the channels as they go, they create a wide belt of almost flat plain (Fig. 7–26), a **floodplain.** It is this major part of the stream valley that is covered with water when a river overflows its banks at flood stages. Thus floodplain deposits are formed both by lateral migration of the channel at nonflood stages and by overbank deposits in floods. The areal extent of a floodplain may be large compared to that of the river channel itself. For example, that part of the floodplain of the Mississippi River between the junction of the Ohio and Mississippi rivers at Cairo, Illinois, and New Orleans at the delta, embraces about 80,000 square kilometers (about 30,000 square miles).

---

*In *Life on the Mississippi,* Mark Twain humorously expressed his version of a common misunderstanding of scientific extrapolation: "In the space of one hundred and seventy-six years the Lower Mississippi has shortened itself two hundred and forty-two miles. That is an average of a trifle over one mile and a third per year. Therefore, any calm person, who is not blind or idiotic, can see that in the Old Oolitic Silurian Period, just a million years ago next November, the Lowest Mississippi River was upward of one million three hundred thousand miles long, and stuck out over the Gulf of Mexico like a fishing rod. And by the same token any person can see that seven hundred and forty-two years from now the Lower Mississippi will be only a mile and three-quarters long, and Cairo and New Orleans will have joined their streets together and be plodding comfortably along under a single mayor and a mutual board of aldermen. There is something fascinating about science. One gets such wholesale returns of conjecture out of such a trifling investment of fact."

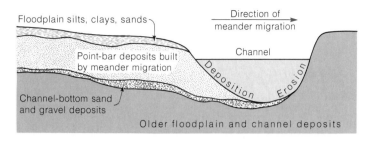

Figure 7–26
Formation of river-valley deposits by a meandering channel. As the meanders swing from one side of the valley to the other, they leave a train of point-bar and channel deposits from former channel positions, overlain by finer-grained silts, clays, and occasional sands deposited by floods. Point bars are usually cross-bedded, and floodplain deposits are horizontally bedded and rippled.

Lateral migration floodplain deposits are both coarse and fine. The coarse material, usually gravel and coarse sand, is deposited as channel beds or bars, which tend to be smeared out by lateral migration of the river. The fine material, silt and clay, is deposited from the lateral migration of muddy parts of banks and bars of the river channel. Some of the fine material comes from the fill of oxbow lakes.

When a river overflows its banks, the flow of sediment-laden water rapidly decreases as it spreads out over the floodplain. Along the strip bordering the river banks, where the decrease is is particularly rapid, much coarse sediment—usually gravel and sand—is deposited. Lesser amounts of finer sediment—silt and clay—are distributed more widely over the floodplain. In this way, successive floods build up ridges on both sides of a river channel, to form **natural levees** that confine the river within its banks between flood stages (Fig. 7–27). Where a flooding stream breaks through the banks, it drops fans or linear accumulations of sand and gravel that extend away from the channel in **splay** deposits. Some levees have been built up many meters above the surrounding plain, so that the plain may be lower than the river surface.

Rivers that carry mostly fine-grained sediment in suspension do not build natural levees, or at most deposit very low ones. Though perhaps less dramatic than the deposits of coarse sediments, the fine deposits may be more important for society; the floodplains of the Nile, the Tigris and the Euphrates were important in history and prehistory for their rich agriculture on fertile silt and clay deposited at times of flooding. The floodplains of the Ganges still play an important role in Indian life and agriculture today.

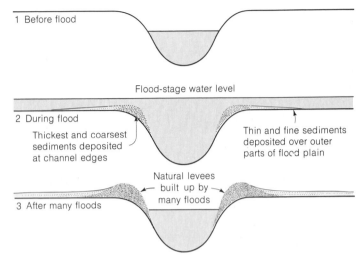

Figure 7-27
The formation of natural levees by river floods. As the river overflows its banks, it rapidly decreases in velocity away from the channel and drops most of its sediment, the coarser fraction near the channel and the finer fraction as a thinner layer of silt and clay over most of the floodplain. Successive floods build up the natural levees to ridges many meters high.

Figure 7-28
Floodwaters of the Feather River at Nicolaus, Sutter County, California, after a break in the river's levee during a flood in December 1955. The river channel is to the left (note broken bridge), and the break in the levee is in the left foreground. [Photo by W. Hofmann, U.S. Geological Survey.]

Rivers regularly overflow onto their floodplains, as any river-bottom native can tell you (Fig. 7-28). Rivers reach the **bankfull** stage, that level at which the water completely fills the channel, every year or two on the average. They break their banks and flood every two or three years, again on the average. Most of these frequent floods are small. The larger ones come much less frequently. We speak of a ten-year flood as flood of a certain discharge that has a probability of recurring every ten years. A fifty-year flood might be twice as big as the ten-year flood but will recur on the average only once every fifty years. These are probability statements, not predictions, and those who live in river towns have learned not to relax after one large flood in the false security that the next one will not recur for another so many years. The U.S. Geological Survey has published maps of the potential for recurrence of floods of given discharges in the United States.

A ten-year flood is a moderate event, normally only one and a half to two times the low flood that occurs on the average every year, the "mean annual flood." The fifty-year flood is more likely to be a catastrophic event and may be as much as three times the mean annual flood. The floodplain is overflowed by a depth of water proportional to the magnitude of the flood. The fifty-year flood, for example, may cover the floodplain with a depth of water equal to about eight-tenths the bankfull depth, whereas the ten-year flood will be only about half the bankfull depth. The depths vary widely with the width of the floodplain and the details of its topography.

The flood of 1976 on the Big Thompson River of the Colorado Front Range is a well-studied example of an unusually damaging flood in a mountainous river canyon (Fig. 7-29). On the night of July 31, an unusual weather pattern brought intense rainfall: between 6:30 and 11:30 P.M. nearly one year's total precipitation—over 30 centimeters (about 12 inches)—fell in the river drainage area. The intense rainfall on a steep terrain quickly transformed the Big Thompson into a torrent with extremely high velocities, averaging over 750 centimeters per second on many steep tributaries, ten times normal flows. Discharge reached a peak flow of 883 cubic meters (31,200 cubic feet) per second, more than four times the previous record high flow reached 31 years earlier. The flood claimed 139 lives; the property losses of 35.5 million dollars included the destruction of 250 structures and over 400 automobiles.

Historic, climatic, hydrologic, and geologic evidence indicate that other streams along the east

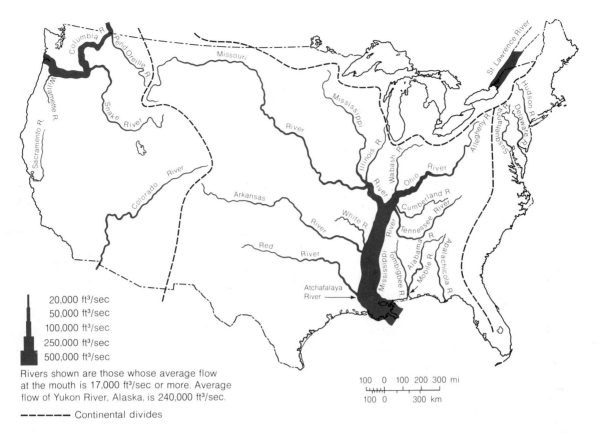

Figure 7-30
Large rivers of the United States. The width of rivers is proportional to their discharge. The major drainage basins that empty to the Pacific Ocean, the Gulf of Mexico, the northern and southern parts of the Atlantic Ocean are separated by the continental divides. [After U.S. Geological Survey.]

slope of the Front Range are susceptible to the same intense flooding. Geologists have mapped deep channel scours, older flood debris of extremely large size—the 1976 flood moved a boulder of more than 275 tons—and large fans of flood origin. This kind of evidence suggested to one specialist that a flood of this magnitude had not occurred in the main channel for several thousand years; a U.S. Geological Survey report based on historical figures and statistical analysis found it to be an unusual but not unprecedented event with a probability of recurring every 200 years. Though we need to learn much more about flood forecasting, our estimates are being improved by combining geological evidence with historic records.

Does a river do most of its work of erosion and transport during everyday stages, during small or moderate floods, or during the major events that come only once in a generation? It appears that the great bulk of the sediment in the few rivers studied is carried by floods that recur at least every five years—events of moderate intensity that happen often. Though the most infrequent events are very intense, they do not recur often enough to add much to the total. The amount transported by rivers in their everyday stages is too small to contribute significantly.

DRAINAGE NETWORKS

About fifty miles west of Denver, U.S. Highway 40 crosses Berthoud Pass at an altitude of 3450 meters (11,314 feet) above sea level. At the crest of this pass is the **continental divide,** the line west of which all water flows eventually to the Pacific Ocean and east of which to the Atlantic (Fig. 7-30). The continental divide is a continuous line that runs north and south the length of North America, dividing the continent into two enormous areas that contribute water to the major

(a)

(b)

Figure 7-29
The small community of Waltonia in the Big Thompson River in Colorado is shown
in these photographs taken before and after the devastating flood of July 31 and
August 1, 1976, on the Big Thompson. Two large motels, several other buildings,
and U.S. Highway 34, shown in (a), were washed away, as shown in (b), taken two
days after the flood on August 3, 1976. The buildings on higher ground in the upper
right remained intact. [U.S. Geological Survey.]

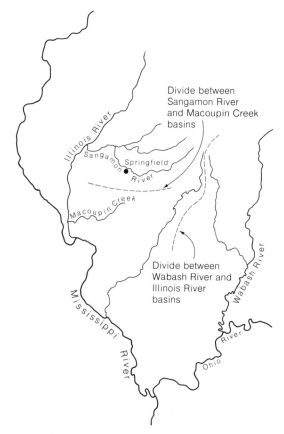

Figure 7–31
River drainage basins of central and southern Illinois, showing divides between the rivers discussed in the text and the hierarchical nature of drainage networks.

river systems on either side. In northern Wisconsin, a fairly low east–west rise at an elevation only a few hundred meters above sea level is another continental divide, separating all of the water that flows to the north, to the Great Lakes and the St. Lawrence River, from the water that flows to the south via the Mississippi and its tributaries. These divides and all of the smaller ones, down to the ridges that separate small ravines, separate **drainage basins,** the areas that funnel all their water into the streams draining them.

A good example of the hierarchy of drainage basins and divides can be found at Springfield, Illinois, in the drainage basin of the Sangamon River (Fig. 7–31). To the south is the divide between the Sangamon and the smaller drainage basin of Macoupin Creek. The rivers of both drainage basins empty into the Illinois River, and so Springfield is also in the larger Illinois River drainage basin. This larger basin is separated by a divide some sixty miles east of Springfield from

the drainage basin of the Wabash River, which drains eastern Illinois and western Indiana to the Ohio River near Shawneetown, Illinois. And so it goes. The Ohio and Illinois both drain to the Mississippi, which is the master river of the whole drainage basin east of the north–south continental divide and south of the east–west continental divide. Figure 7–32 shows the Colorado River basin, a large region of the southwestern United States that is drained by a single river system.

Divides are not immune from change. By virtue of discharge, slope, or other factors, the stream on one side of a divide may be able to erode and transport much more rapidly than the stream on the opposite side. It may, as a result, break down the divide between the two at some place along it and take part or all of the drainage of the slower stream. Victorian geologists of the last century used a moralistic word for this kind of "capture" and called it **stream piracy.** The breaching of divides and capture of streams is common in small rills, becomes less common as streams grow larger, and is rare in large rivers. Piracy explains such oddities as narrow valleys or gorges that have no active streams running in them: the part of the stream that cut the valley was drained away, its course having been changed by the pirate stream.

The streams in all drainage basins follow certain rules. All are connected in a one-way network by which smaller tributaries drain into larger ones with a definite pattern. The number of streams and their distance apart both follow a fairly orderly distribution. Most tributaries of about the same size are about the same length, and the intervals between the mouths of tributaries are fairly uniform. The larger the drainage area, the longer the stream, the ratio between the two being constant for similar terrains. Robert Horton, an American hydraulic engineer, was the first to use as a measure of the hierarchy of streams their **order**—that is, their position in the tributary network (Fig. 7–33). Though a number of systems of ordering streams have been devised over the years, Horton's original scheme is simple and gives the idea: A stream of order 1 has no tributaries; a stream of order 2 has tributaries of order 1; a stream of order 3 has tributaries of order 2; and so on. Thus the order is defined by the order of tributaries; we count up as we move downstream. As the order of streams increases, the following changes are systematic (see Fig. 7–34): the length of main streams increases, the number of main streams decreases, and the drainage area increases. These general characteristics

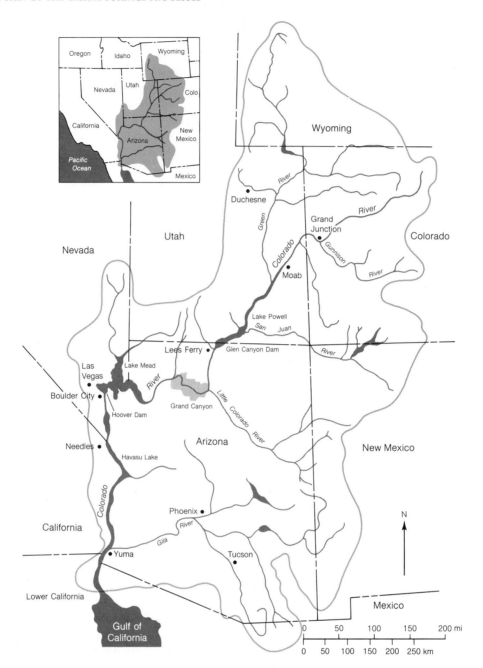

Figure 7-32
The natural drainage basin of the Colorado River covers about 630,000 km²
(243,000 mi²), a large part of the southwestern United States. [After U.S.
Geological Survey.]

of drainage networks are typical of many kinds of systems. For example, even the blood circulatory systems of mammals seem to have some of the same characteristics. Here, as in the meander story, there are some underlying laws that govern a wide range of phenomena that at first sight seem unrelated. Many believe these general patterns reflect the presence of a random element in the geometrical arrangement of all these networks; others continue to search for specific combina-

tions of rock type, channel shape, climate, structure, and other factors that determine the network pattern.

Though all drainage networks branch in the same way, the shape of their patterns varies greatly from one kind of terrain to another, mainly as a response to the rock type or structural pattern of folds and faults. The most common is **dendritic** drainage, named for the characteristic pattern shown by most branching deciduous trees

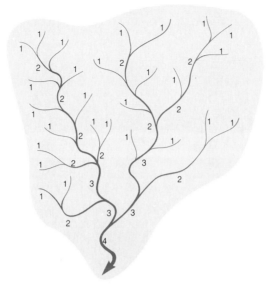

Figure 7-33
Order numbers of streams are designated by their position in the hierarchical network pattern according to the simple system of Robert Horton. Streams of order 1 have no tributaries, those of order 2 have tributaries of order 1, those of order 3 have tributaries of order 2, and so on. The number of streams of a given order in a drainage basin decreases strongly as order number increases. Thus, in the illustration above, there are twenty-seven streams of order 1, seven of order 2, two of order 3, and one of order 4.

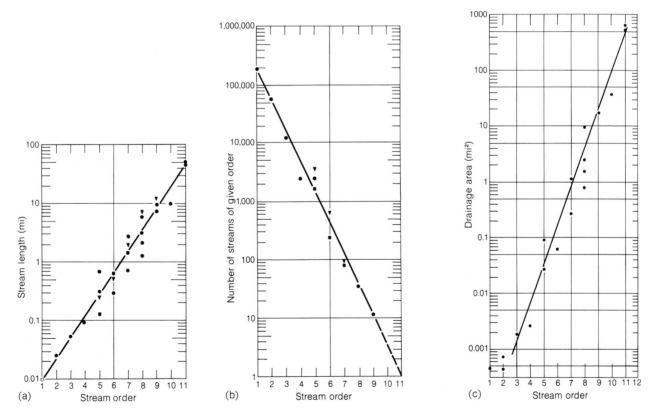

Figure 7-34
Relation of stream characteristics to stream order in drainage network of arroyos near Santa Fe, New Mexico. (a) Relation of stream length to stream order. (b) Relation of number of streams to stream order. (c) Relation of drainage area to stream order. [After *Fluvial Processes in Geomorphology* by L. B. Leopold, M. G. Wolman, and J. P. Miller. W. H. Freeman and Company. Copyright © 1964.]

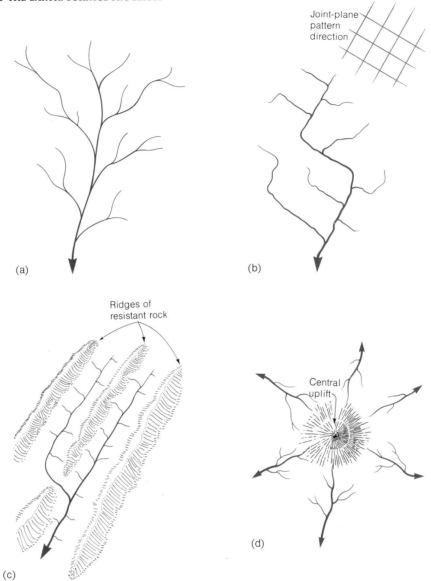

Figure 7–35
Typical drainage patterns. (a) A dendritic pattern is characterized by branching similar to that of the limbs or roots of trees. (b) In a typical rectangular drainage pattern developed on a strongly jointed rocky terrain, drainage tends to follow the joint pattern. (c) Typical trellis drainage develops in valley and ridge terrain, where rocks of varying resistance to erosion are folded into anticlines and synclines. (d) Radial drainage patterns develop on a large single peak, such as a large dormant volcano or a domal uplift.

(Fig. 7–35a). This fairly random pattern is typical of terrain floored by uniform rock types, such as horizontal sedimentary rocks or massive igneous or metamorphic rocks. Differential weathering of fractures or joint systems in bedrock localizes stream flow, producing a more ordered and geometrical **rectangular** drainage pattern (Fig. 7–35b). A special variety of rectangular drainage, the **trellis** pattern, is formed where bands of rock resistant to weathering alternate with bands that erode more rapidly, typically in a region where rocks

have been deformed into a series of parallel folds (Fig. 7–35c). Drainage from a central high point, such as a volcano or a domal uplift is **radial** (Fig. 7–35d). From this series of examples, one can see that drainage patterns reflect mainly the underlying structure of the rocks and their varying resistance to erosion, though the density of drainage depends more on climate. Areas of high rainfall tend to have many more streams than arid regions. One rock type, limestone, may virtually eliminate surface drainage in areas of high rainfall because it dissolves to form caves and underground drainage. The northern part of the humid, rainy Yucatan Peninsula of Mexico has no streams because all of the water sinks into porous, cavernous limestone.

Geologic events may strongly control the course of a river. Streams flowing in a dendritic pattern on flat-lying sediments may erode through the sediments to expose underlying, strongly folded and faulted rocks of varying resistance to erosion. Because these **superposed** streams, preordained in their courses by their former bed of sediments, cannot break out easily to form a wholly new drainage pattern appropriate to the new bedrock, they adjust by cutting relatively narrow valleys or gorges through ridges of resistant rock, and elsewhere meandering through broad valleys in softer rock (Fig. 7–36).

THE END OF THE LINE: DELTAS

Sooner or later, the end comes to all rivers—they empty into large standing bodies of water. Although this is so well known as to appear intuitively obvious, a real understanding of the way it happens comes from watching the behavior of a river as it enters the sea and mixes turbulently with the surrounding water (Fig. 7–37). The mixing gradually transfers the powerful forward momentum of the flowing river to the surrounding water and dissipates the current. It does so in

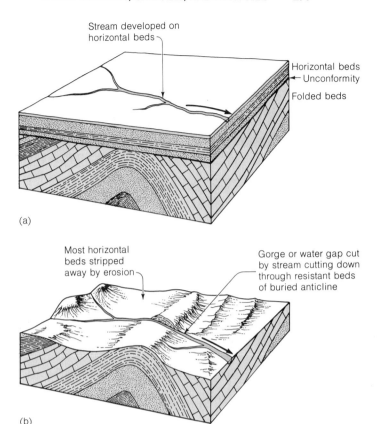

(a)

(b)

Figure 7–36
The development of a superposed stream by erosion of horizontal beds unconformably overlying folded beds of varying resistance to erosion. As the downcutting stream encounters the buried anticline, it erodes a narrow gorge, or water gap, in the resistant beds of the anticline.

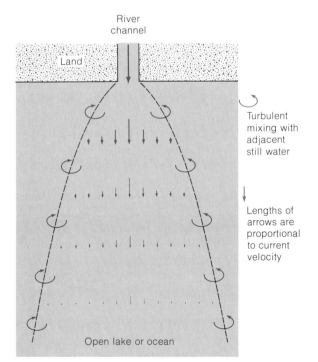

Figure 7–37
Decrease in the velocity of flow as a river enters an open body of water, such as a lake or an ocean. The current persists for the longest distance along the main line of flow and drastically decreases to either side. The current mixes with still water around the edges of the flow, expanding and losing momentum as it does so. [Data in part from C. C. Bates, "Rational Theory of Delta Formation," *Bull. A.A.P.G.*, v. 37, pp. 2119–2162, 1953.]

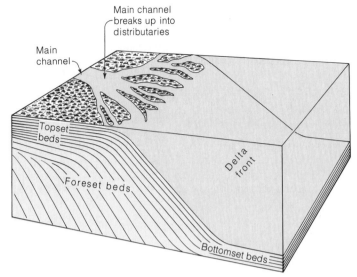

Figure 7–38
A typical fresh-water delta, characterized by well-defined topset, foreset, and bottomset beds. The slope of the delta front is fairly steep, up to 25°. The main river channel may or may not break up into distributaries, depending on the stream discharge and the sediment load: the greater the load in relation to discharge, the greater the tendency to form distributaries.

much the same way that a billiard ball shot into a large cluster of balls on a small pool table loses its forward motion by transferring it to movement of all of the balls as they break apart and rebound from the cushions and from each other, gradually coming to a stop as friction overcomes their motion.

The current decays more or less gradually, depending on its volume and velocity. The discharge of the Amazon River, the world's largest, was recently measured at 4 billion gallons per minute (15 million cubic meters per second) at its mouth by a team of U.S. and Brazilian hydrologists. That flow from the river maintains its integrity many kilometers out to sea, though with decreasing velocity. In contrast, small streams entering a turbulent, wave-swept coast may mix so rapidly that the current disappears almost immediately beyond the mouth.

Delta Sedimentation

The dissipation of the river currents sets in motion a train of events that accounts for some major sedimentary deposits: deltas. The dying current loses its competence for transporting particles and so deposits the series of sedimentary layers that make up the delta. The simplest deltas are formed by modest streams entering fresh-water lakes. Because the density of the entering river water is the same as that of the surrounding lake water, the current mixes in all directions in a conelike pattern and rapidly slows to a halt. The coarsest material is dropped first, followed by medium and fine materials farther out. In the usual situation, where the lake bed slopes away from the floor of the river channel, the coarse material builds up a depositional platform like that shown in Figure 7–38. **Foreset** beds, inclined downcurrent from the delta front in every way analogous to cross-bedding, are covered by thin, horizontal **topset** beds and preceded on the lake bottom by thin, horizontal **bottomset** beds. This sequence has been reproduced, on a small scale, in laboratory flumes.

Deltas formed in the sea are generally the same in form but tend to be stretched out more in the horizontal dimension. The reason for this is that the fresh water of the river, (density about 1 gram per cubic centimeter) enters sea water (density 1.02 grams per cubic centimeter). The lighter fresh water tends to float on the sea water and consequently mixes with it only in two dimensions—horizontally but not vertically as in a fresh-water lake. A lower mixing rate means that the current dissipates more slowly, so that coarse, medium, and fine materials spread out along a much longer path. The slope of foreset beds on large marine deltas may be only a few degrees.

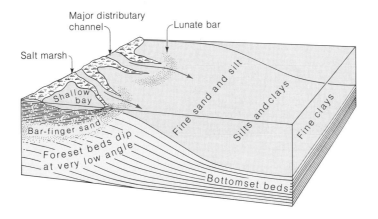

Figure 7-39
A typical marine delta, in which the foreset beds are fine-grained and deposited at a very low angle, normally only 4° or 5° or less. Lunate bars—
crescent-shaped sand accumulations—form at the distributary mouths,
where the current velocity suddenly decreases. As the delta builds forward,
the bar-finger sand is built up by the advance of the lunate bar and the
distributary channel. Between channels, shallow bays fill with fine-grained
sediment and become salt marshes. This general structure is found on the
Mississippi River delta.

Major rivers, such as the Mississippi and the
Nile, form deltas thousands of square kilometers
in area with complex structures. The Mississippi
has been studied for generations by hydraulic
engineers responsible for the operations of the
waterway and by geologists trying to work out the
relations between river, sediment, and sea. The
Mississippi delta, like most others, reverses the
normal dendritic drainage pattern in which
smaller tributaries join to form bigger ones. As it
approaches the sea and lowers its slope, the river
branches to form **distributaries** that funnel off the
water in various paths to the sea, many small ones
and three large ones near the mouth (Fig. 7–39). At
the mouth of main distributaries the coarsest sediment, sand, is dropped as crescent-shaped **lunate
bars;** silt and mud are carried farther out. The
sediments at these distributaries grow seaward as
long sand fingers, **bar-finger sands;** the finer muds
and silts are washed by shore currents and waves
into the bays between, gradually filling them with
mud and swamp deposits. In this way the whole
delta is built out into the sea (Fig. 7–40).

Because marine deltas may be strongly affected
by waves and tides as well as by the river flow
itself, there is a three-way classification of deltas:
river-, wave-, and tide-dominated deltas. The Mississippi has a river-dominated delta. High river
discharge and sediment load are coupled with
weak waves and tides to produce a delta plain

much like ordinary alluvial plains except for the
distributaries. Frequent flooding causes breakouts
to form new channels and deposits of river sediment that fill in the quiet shallow marine areas
between the distributaries. The delta front, the
area where the river enters the sea, is controlled
by the river outflow with little interference from
waves and tides.

Other deltas are dominated by waves and their
resultant longshore currents (Chapter 10), which
move the sediment along the shore as rapidly as it
is dropped by the river. There the delta front is a
relatively even beach shoreline with only a slight
protuberance at the river mouth. In tide-dominated deltas, the lower distributaries are invaded
by tidal flows which spill over channels and produce extensive tidal flats on the lower delta plain.
On the delta front, tidal currents shape many
elongate sand ridges parallel to the direction of
the currents. Varying strengths of tides, waves,
and river flows produce intermediate delta types.
If the combination of waves and tides is strong
enough, the river's detritus is spread along the
coast, in some places for hundreds of kilometers,
and no delta is formed.

The delta of the Mississippi, like many other
large deltas, has been growing for many millions
of years. It started out around Cairo, Illinois, in
Cretaceous times and has advanced about 1600
kilometers since then. The pattern of its growth

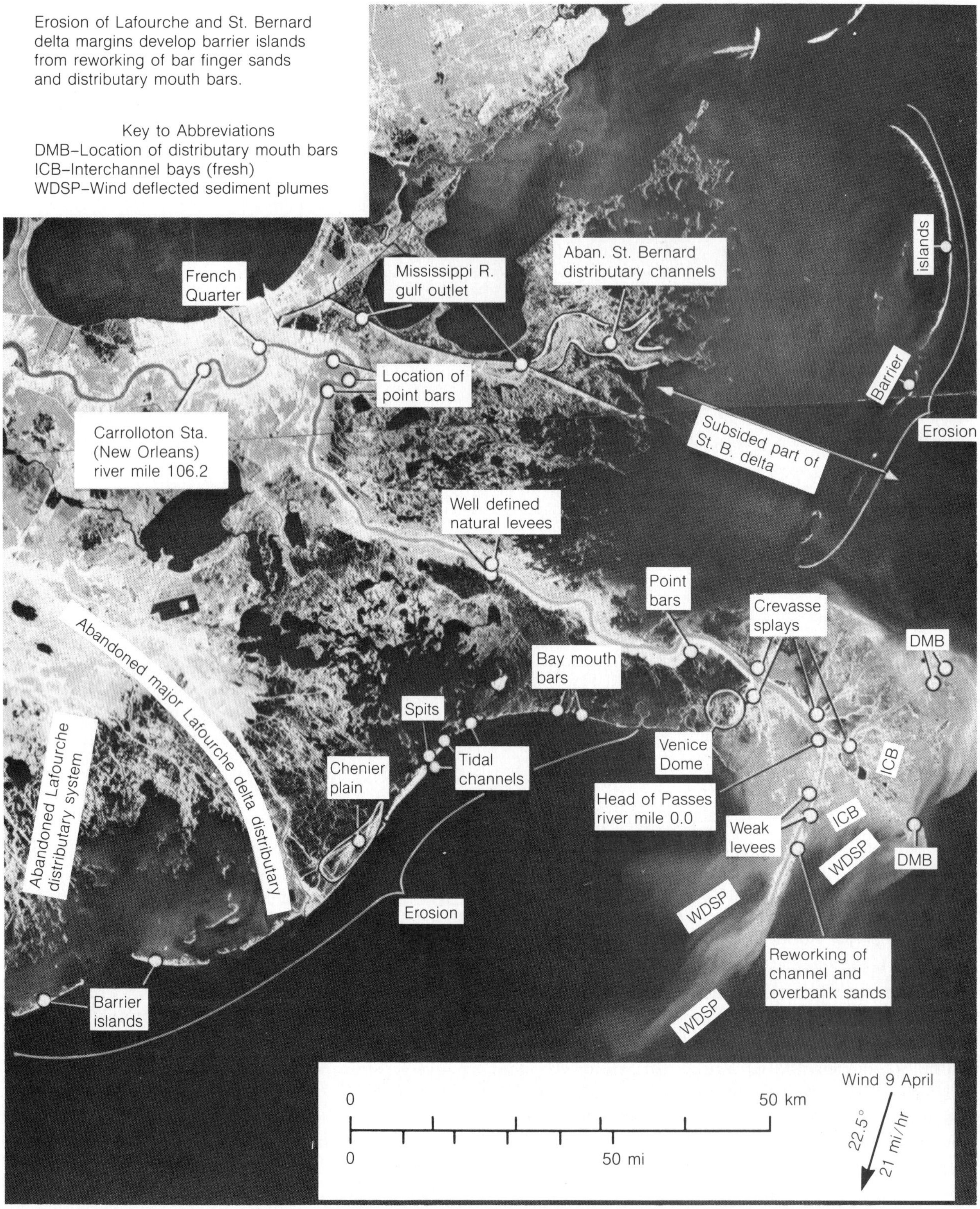

Erosion of Lafourche and St. Bernard delta margins develop barrier islands from reworking of bar finger sands and distributary mouth bars.
Key to Abbreviations
DMB–Location of distributary mouth bars
ICB–Interchannel bays (fresh)
WDSP–Wind deflected sediment plumes
French Quarter
Mississippi R. gulf outlet
Aban. St. Bernard distributary channels
Barrier islands
Location of point bars
Carrolloton Sta. (New Orleans) river mile 106.2
Subsided part of St. B. delta
Erosion
Well defined natural levees
Point bars
Crevasse splays
DMB
Bay mouth bars
Spits
Tidal channels
Chenier plain
Venice Dome
Head of Passes river mile 0.0
ICB
Weak levees
ICB
WDSP
DMB
Abandoned major Lafourche delta distributary
Abandoned Lafourche distributary system
Erosion
WDSP
Reworking of channel and overbank sands
WDSP
Barrier islands
0
50 km
0
50 mi
Wind 9 April
22.5°
21 mi/hr

Figure 7-40
Landsat 2 orbiting satellite photograph of the Mississippi delta. The sensitivity of this film to infrared radiation reveals much geological detail of the exposed and shallow-water parts of the delta, as indexed on the photograph. [From G. T. Moore, "Mississippi River Delta from Landsat 2," *Bull. Amer. Assn. Petr. Geol.* Copyright © American Association of Petroleum Geologists, 1979.]

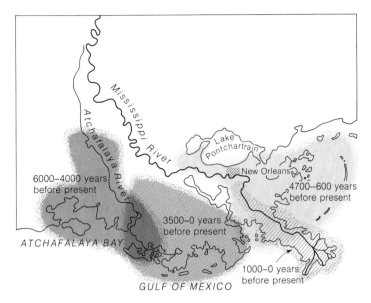

Figure 7-41
The Mississippi delta complexes of the past 6000 years are a series of major distributary centers that shifted with time as the river built its delta first in one direction and then in another. The oldest complex that has been mapped underlies the present area of Atchafalaya Bay and the mouth of the Atchafalaya River. It was succeeded by a series of deltas deposited to the east and west of the modern birdfoot delta. For more than a decade before 1955, the Mississippi River was diverting an increasing amount of its discharge to the Atchafalaya, an indication that the river was again about to shift its major site of delta building. At that time, the Army Corps of Engineers began construction to prevent major diversion into the Atchafalaya. [Map compiled from various sources, including C. R. Kolb, and J. R. Van Lopik, 1966 in M. L. Shirley (ed.) *Deltas in their Geological Framework,* Houston Geol. Soc.; D. E. Frazier, "Recent Deltaic Deposits of the Mississippi River: Their Development and Chronology," *Gulf Coast Assn. Geol. Soc. Trans.,* v. 17, pp. 287–315, 1967.]

for the past few thousand years has been mapped by carbon-14 dating of wood found by drilling into the different kinds of deposits at various depths (Fig. 7-41). Deltas grow in one direction for a while and then shift to another, seeking a shorter path to the sea. The individual subdeltas interfinger and partially pile up on top of each other to form a rather complex mass of sediment. Because of this complexity, it is no mean task to sort out the rock record of major ancient deltas formed hundreds of millions of years ago, but the geologist has one good clue: comparison of coarse sandy beds containing fossils of land plants or animals with nearby finer beds of the same age containing marine fossils. Just such evidence has shown parts of the Catskill Mountains of New York State to be an ancient delta formed in Devonian times. Coarse, land-deposited sandstone and conglomerate in the east, near the Hudson River,

interfinger with finer marine sandstones and shales farther west. The river forming that delta came from a belt of high mountains roughly paralleling the Taconic and Green Mountains and ancestral to them.

As rivers end their journey, they drop not only solid particles of rock detritus but also the dissolved salts coming from weathering, which mix into ocean water. The rivers of the Earth constitute the major transport system for the products of weathering, most of which end up in the sea. The sediment load of the world's rivers that is discharged to the oceans is estimated to be about 18 trillion tons per year. The material carried in solution in river water adds another 4 trillion tons per year. Water, however, is not the only agent that erodes, transports, and deposits. In the next chapter, we discuss another, the wind.

SUMMARY

1. Water flows in laminar and turbulent currents. Most natural streams flow turbulently and erode, transport, and deposit solid materials. Depending on turbulence, velocity, and slope, streams have competence to carry particles of a certain size and a certain capacity of total sediment load. Particles move by suspension, saltation, and rolling.

2. The sandy sediment deposited by stream currents typically shows cross-bedding and ripples. Ripples, dunes, and flat surfaces form in an orderly sequence on sand beds as the velocity is increased.

3. Currents effectively erode both unconsolidated material and hard rock by undercutting, scouring, sandblasting, and plucking.

4. The river is a dynamic system characterized by discharge, velocity, and slope. Changes in one are accompanied by mutual changes in the others as the river adjusts its longitudinal profile.

5. Longitudinal profiles tend to be graded in adjustment to base level.

6. Alluvial fans are river deposits formed in response to a sudden change in slope. Channels form meanders and pools and riffles with regular patterns of point bars and other channel deposits.

7. Most of the buildup on floodplains happens during the moderately frequent, moderately high floods that come every few years.

8. Drainage networks show hierarchical branching patterns that follow rules of distribution. The pattern of drainage may be dendritic, rectangular, trellis, or radial, depending on topography, rock type, and structure.

9. Deltas are the major sites of deposition of river sediment. They show a structure of topset, foreset, and bottomset beds that differs from lake delta to ocean delta. Not all rivers have deltas; the sediment of these rivers is dispersed by strong shoreline currents or tides.

EXERCISES

1. Would you expect the flow of small rills flowing from the melting of snow and ice in early spring to be laminar or turbulent? Why?

2. What could you infer about the changing current conditions that produced a series of sand beds in which the lowest beds are cross-bedded in a series of large interfering troughs, the middle beds finely rippled, and the highest beds horizontal with no ripples or any other structures?

3. Would you expect the ratio of first- to second-order streams in a trellis drainage system to be much greater than, about the same as, or much less than the ratio of first- to second-order streams in a dendritic drainage system? Why?

4. In the middle part of its length, a major stream runs through a belt of cavernous limestone with Karst topography. How would its discharge-velocity relations be affected, and how might the drainage network pattern change?

5. You are the city engineer of a city of 100,000 population, built partly on the floodplain of a major river. Two years ago the city suffered the most disastrous flood in its 200-year history, and the city council now says that you need not worry about contingency planning for another major flood for the next few years. What would be your response to the council?

6. One popular engineering project used to control rivers is channelization—the artificial straightening and widening of the river channel. What natural changes might you expect to take place over a number of years after the river is left free to adjust its course naturally after channelization?

7. Both alluvial fans and deltas have drainage patterns in which a channel breaks up into a number of distributaries. Is there a common cause for this behavior? If so, what is it? If not, what are the causes in each case?

8. Given a simple drainage network of four first-order streams, each with a discharge of 250 cubic meters per second, that drain into two second-order streams and finally into a single third-order stream, what will the discharge become in the second- and third-order streams?

9. Only a few years after a dam was built, the reservoir behind it became silted up. Could this have been predicted beforehand? What quantity would you have measured to predict?

BIBLIOGRAPHY

Leopold, L. B., and W. B. Langbein, "River Meanders," *Scientific American,* June 1966. (Offprint 869.)

Leopold, L. B., M. G. Wolman, and J. P. Miller, *Fluvial Processes in Geomorphology.* San Francisco: W. H. Freeman and Company, 1964.

Matthes, G. H., "Paradoxes of the Mississippi," *Scientific American,* April 1951. (Offprint 836.)

Morgan, J. P., ed., *Deltaic Sedimentation, Modern and Ancient.* Tulsa, Ok.: Society of Economic Paleontologists and Mineralogists Special Paper 15, 1970.

Morisawa, M., *Streams—Their Dynamics and Morphology.* New York: McGraw-Hill, 1968.

Schumm, S. A., *The Fluvial System.* New York: John Wiley & Sons, 1977.

8

Wind, Dust, and Deserts

Wind, though less powerful than water currents, can erode sand and silt effectively, particularly in arid regions. The deserts of the world, sites of intensive wind action, display special topographies and erosional and depositional processes more intense than those of humid regions. Sand dunes, depositional landforms created by wind, are heaps of sand that accumulate in different shapes and move in response to the abundance of the sand supply, the strength and direction of the wind, and the nature of the bedrock surface.

All of us have been caught, at one time or another, in a wind so strong that it could have blown us over if we hadn't leaned into it or held on to something solid. A wind strong enough to move a body weighing more than 50 kilograms (110 pounds) is easily capable of blowing sand grains into the air, as anyone who has ever been in a sandstorm can attest. Wind is a turbulent stream of air with an ability to erode, transport, and deposit sediment much like that of water, for the same general laws of fluid motion that govern liquids govern gases as well. There are differences, of course, and they are traceable to two properties of wind: its low density and the fact that its flow is unconfined, not restricted to channels. The low density of air limits its competence to move larger particles, and its unconfined flow enables it to spread over wide areas and high into the atmosphere. In contrast to rivers, whose discharge is dependent upon rainfall, it is a lack of rain that allows wind to work most effectively. Most rivers flow uninterruptedly for long periods of time; winds, on the other hand, can spring up and die down in a short time. In this chapter, we shall discuss arid environments, the deserts of the Earth, because so many of the geologic processes of the desert are related to the work of the wind. Winds also are coupled to the

waters of the ocean as they blow waves into being and influence the general circulation of the oceans. A common term used to describe the activity and deposits of the wind is **eolian**, from Aeolus, Greek god of the winds.

HOW MUCH CAN THE WIND CARRY?

Winds are highly variable in direction and power (see Table 8–1). In temperate climates we are used to winds that come mainly from one direction, the prevailing westerlies. (Fig. 8–1). Those in the tropics are familiar with the equatorial easterlies. Yet within these belts the winds are variable in direction and power over time intervals from minutes to days, depending on the movement of air masses and storms. From minute to minute, winds ranging from a breeze to a gale can be highly changeable.

Many of us live with wind fluctuations during the day, such as the sea breeze that blows during warm summer days and dies down in the evening, or the daytime valley breeze and nighttime mountain breeze in high-relief terrain. More constant are the steady, strong winds that blow for days without letup, like the dry Chinook winds of the

Table 8-1
Terminology of wind speeds, according to 1939 international agreement and U.S. Weather Bureau

Beaufort number*	Wind speed (km/hr)†	International description	U.S. Weather Bureau description	Effect of wind on the sea
0	<1	Calm		
1	1–5	Light air	Light wind	Mirror surface to small wavelets
2	6–11	Light breeze		
3	12–19	Gentle breeze	Gentle–moderate	Large wavelets to small waves
4	20–28	Moderate breeze		
5	29–38	Fresh breeze	Fresh wind	Moderate waves, many whitecaps
6	39–49	Strong breeze	Strong wind	Large waves, many whitecaps, foamy sea
7	50–61	Moderate gale		
8	62–74	Fresh gale	Gale	High waves, foam streaks, and spray
9	75–88	Strong gale		
10	89–102	Whole gale	Whole gale	Very high waves, rolling sea, reduced visibility
11	103–117	Storm		
12–17	>117	Hurricane	Hurricane	Sea white with spray and foam, low visibility

* The Beaufort scale was devised near the beginning of the nineteenth century as a way of measuring winds at sea by noting types and heights of waves, spray, and visibility.
† Today, wind speed is usually measured in knots (nautical miles per hour; 1 nautical mile = 1.15 statute miles or 1.85 kilometers).
Source: Based in part on N. Bowditch, 1958, "American Practical Navigator," U.S. Navy Hydrographic Office Publ. 9.

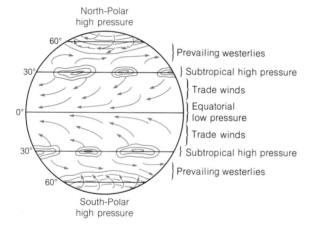

Figure 8-1
The global pattern of belts of surface winds and barometric-pressure cells. The subtropical high-pressure cells move somewhat and change size from summer to winter. Arrows show generalized wind directions.

eastern slopes of the Rocky Mountains and the Mistral that howls down the Rhone Valley of France in the winter.

The distribution and intensity of winds in combination with climate has much to do with the location of wind erosion and windblown deposits on Earth. The prevalence of sea breezes, for example, is responsible for the sand dunes blown back from sandy beaches along coasts. Most of the geologic work of the wind is done by the moderately infrequent strong winds of long duration, just as the major part of a river's geologic work is done by floods. Neither the everyday mild wind

nor the rare tornado, whose winds may exceed 160 kilometers (100 miles) per hour but which covers only a narrow strip on the Earth's surface for a short time, is responsible for much geologic change. Hurricanes and typhoons are important agents because of their frequency in certain maritime regions, but, because of the rain they bring, they do their work by causing floods and stirring up waves rather than by blowing sand or dust: their rain washes dust particles out of the air and by wetting down the ground surface, prevents pickup of further particles. This points to dryness as a crucial prerequisite for erosion and transport by the wind.

Sand Transport

The wind exerts the same kind of force on particles on the land surface as a river current exerts on its bed, the turbulence and forward motion combining to lift particles into the windstream at least temporarily.

The force of a wind on a surface depends not only on the wind speed but on the roughness of the surface which induces greater turbulence. A wind of 10 meters per second (22 miles per hour) exerts a force of 30 tons per square kilometer on a smooth mown grass field but about 70 tons per square kilometer on a rough ripe wheat field. Thus as wind over a coarse sandy surface first ripples the surface and then builds dunes, the roughness increases, producing a positive feedback and increasing the wind force.

The amount of sand that can be moved by the winds of various strengths is shown in Figure 8-2, taken from a famous book on sand and sand dunes by R. A. Bagnold, a British officer in the desert forces in Egypt and Libya. Before and during World War II, Bagnold pioneered the study of the forms and movements of desert sand. As can be seen from the graph, half a ton of sand per day can be moved over a meter-wide strip of ground by a strong wind of 48 kilometers (30 miles) per hour. As wind increases to gale force, about 80 kilometers (50 miles) per hour, the rate of sand movement increases more rapidly. No wonder that a whole house can be buried in a long sandstorm driven by strong winds.

Saltation, the bounding and jumping movement of grains that we explored in the previous chapter, operates in the air in the same way that it does under water (Fig. 8-3). In the air, however, it is much more effective: partly as a result of the lower frictional and retarding force of the air, saltating grains frequently rise to heights of 50 centimeters over a sand bed and up to 2 meters over a pebbly surface. In a strong wind, there may be enough saltating grains to form a cloudy layer near the ground dense with blowing particles and capable of sandblasting any object in its path (Fig. 8-4). The saltating grains falling back to the ground hit the surface with all the force of the wind combined with gravity, cushioned hardly at all by the air. This strong impact induces saltation of some of the grains on the surface as they are struck. Its much wider effect is the pushing forward of struck grains, not enough to throw them up into the air but enough to cause a general forward creep of sand particles along the surface as the rain of saltating grains falls on it. A sand grain striking the surface can move another grain up to six times its own diameter. Because saltation blows smaller grains more quickly and surface creep moves larger particles more slowly, the two will sometimes separate: the fine sand blows away, leaving behind a pavement of coarser sand and gravel. The fine sand, up to 0.1 millimeter in diameter, accumulates in dunes and sheets downwind.

Dust Transport

Turbulent winds can easily sweep small dust particles high into the atmosphere, once they are lifted off the ground surface (Fig. 8-5). Picking up small particles of clay or dust from the ground in the first place, however, is not so easy mainly because, in a thin layer less than one millimeter

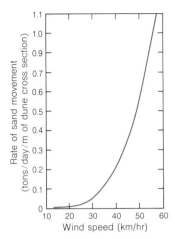

Figure 8-2
The amount of sand moved across each meter of width of a dune cross section in relation to wind speed. High-speed winds blowing for several days can move enormous quantities of sand and change dune positions markedly. [After R. A. Bagnold, *The Physics of Blown Sand and Desert Dunes*, London, Methuen, 1941.]

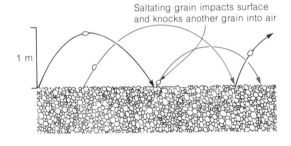

Figure 8-3
Saltation transport of sand grains in wind. As a saltating grain falls to the ground, it may strike another grain lying on the surface with sufficient force to throw the struck grain into a saltation trajectory.

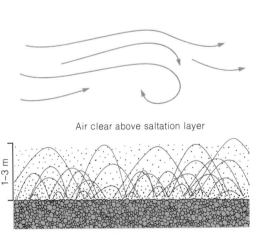

Figure 8-4
Saltation layers formed during many windstorms are so dense that the ground cannot be seen, yet the top of the layer may be sharp and the air above it clear.

Figure 8-5
Dust storm raised by a northeast wind over the vineyard district of southern California.
[From *Geology Illustrated* by John S. Shelton. W. H. Freeman and Company. Copyright © 1966.]

thick right next to the ground surface, the air is practically still. Sand grains protrude through this layer into the windstream above, and thus are caused to roll or saltate; smaller particles, however, are directly affected only at high wind speeds. But once a saltating sand grain or other particle hits the ground and kicks up a bit of dust, the wind catches it. Smaller dust particles, many of which are only a few thousandths of a millimeter in diameter, settle so slowly that the slightest air currents keep them suspended. Violent volcanic explosions can inject pyroclastic dust high into the stratosphere, as Mt. St. Helens did, and such dust may take years to settle.

The capacity of air to hold dust is enormous. In large dust storms, one cubic kilometer of air may carry up to 1000 tons; if such a storm covers hundreds of square kilometers, it may carry more than 100 million tons of dust. In the worst dust storms of the 1930s in the "dust bowl," the western plains of the United States, whole farmhouses were buried in the dust. Wind erosion of dry plains plays an important role in denuding arable soils, a major reason why soil conservationists are worried about the preservation of our agricultural lands.

Detritus Carried by the Wind

Sand grains carried by the wind may be of almost any mineral, but most are quartz, which reflects the dominance of that mineral in most sands and sandstones. The grains typically look frosted, like the ground-glass focusing screens of many cameras. It has been shown by experiment that much of this frosting results from the long-continued action of dew. This moisture, even the tiny amount found in arid climates, is enough to dissolve away little pits and holes on the grain, creating a matte surface. The frosting, which is limited to the grains lying on the surface of the ground, is quickly smoothed and polished when the grains are blown into a river and transported by water.

Thus frosted grains are good evidence of eolian origin. Experiments and observations show that wind alone could not produce the frosted surface. This evidence contradicts a long-held inference that the matte appearance of small grains is the result of sandblasting, which is known to produce frosting on large glass objects.

Given enough sand-size rock or mineral fragments of any variety, the wind will blow them into dunes. In Bermuda, for example, winds from the ocean have produced dunes made up of the calcium carbonate fragments of shells and coral. The White Sands National Monument area in New Mexico is largely covered by dunes of gypsum grains eroded from evaporite bedrock in a dry climate. In some places, clay and silt particles clump together to form sand-size grains that are blown into dunes.

Dust, the fine-grained particles carried by the wind, is of every conceivable origin. It includes microscopic rock and mineral fragments of all kinds, but most common are the silicate minerals, as might be expected from their abundance as rock-forming minerals. Among these, two types are most important: clays from soils and volcanic dust. Organic materials, such as pollen, hair, bacteria, and a great variety of fragments of plants and animals, are also components of dust. Fragments of charcoal can be seen in dust blown from areas with forest fires. Charcoal fragments in buried oceanic sediments are guides to the extent of forest fires in ancient times. Since the industrial revolution, civilization has been producing a rapidly multiplying dust component consisting of particles of ash from burning coal and many solid chemical compounds from manufacturing processes.

WIND EROSION

Winds need chemical and mechanical weathering coupled with dryness to assist them in eroding and transporting materials. Wet materials are cohesive, the water binding the particles together enough to resist the wind's tendency to pull them apart. By themselves, winds can do little to erode most solid rock exposed at the surface; but once there is some fragmentation of mineral particles, the wind can act.

Deflation

As dry, loose particles of dust and silt are lifted and blown away, the surface of the ground is gradually eroded and lowered. This **deflation** occurs on dry plains and deserts, on temporarily dried-up floodplains of rivers, on tidal flats, and on lake beds. Deflation is characteristic, too, of cold dry periods on the plains of mud, silt, and sand formed by melting glaciers. Where vegetation is firmly established, the effect of wind is small, both because the roots bind soil together and because stems and leaves break up the wind and shelter the ground. But where the vegetation cover is broken naturally by drought, or artificially by construction or deep tracks made by motor vehicles, deflation may attack and scoop out shallow depressions or hollows.* Once started, the hollows grow as wind stress on the surface becomes more efficient, usually helped by rapid water erosion during storms.

Deflation can attack rock surfaces, detaching small fragments, especially from sandstones that weather to form loose grains, but formation of depressions in rock surfaces from this cause alone is slow except in some desert areas underlain by soft, easily disaggregated sandstones. Deflation, by removing the fine particles from a heterogeneous mix of gravel, sand, and silt, can produce a remnant surface of gravel too large for transport. Over thousands of years, the gravel accumulates as a layer of **desert pavement**, which protects the soil or beds below from further erosion (Fig. 8–6). Deflation may be only part of the process that forms such pavements, for there is some evidence that pebbles just below the surface work their way up to the top somehow, leaving behind a thin substrate of sand and silt.

Sandblasting

Armed with blown sand, a wind can effectively wear away and shape solid rock surfaces by the constant impact of grains. Such **sandblasting** accounts for some erosion and rounding of the parts of rock outcrops close to the ground, the frosting of smooth rock surfaces (and glass bottles), and the shaping of **ventifacts**, or wind-faceted pebbles (Fig. 8–7). Such pebbles, which show several curved or almost planar faces that meet at sharp ridges, are almost invariably composed of a fine-grained rock, such as chert or quartzite. The pebbles are faceted by erosion of their windward side. Storms occasionally roll or rotate pebbles, thus exposing a new windward side to be sand-

---

*Modern hiking shoes with rough, knobby soles also break up soil and promote erosion, a result unanticipated by hikers who want to preserve the wild.

Figure 8-6
Desert pavement in the valley of the Little Colorado River, Arizona.
[Photo by Tad Nichols, Tucson, Arizona.]

Figure 8-7
Ventifacts—pebbles and cobbles pitted and faceted by windblown sand. From Sweetwater County, Wyoming. [Photo by M. R. Campbell, U.S. Geological Survey.]

blasted to a plane. Like frosted surfaces, ventifacts buried in deposits give us clues to strong wind action at the time of their formation. The main finds of ventifacts in most parts of the world have been either in relatively recent glacial deposits or in deserts. Ventifacts are rare in older deposits, possibly because they are difficult to recognize in older conglomerates. It may also be that the combination of strong winds, sand for sandblasting, and pebbles to be shaped is a rare association that has been limited to regions affected by the few continental glaciations of Earth's history and to rarely preserved pebbly desert deposits.

DESERTS: MUCH WIND AND RARE WATER

The hot, dry deserts of the world are among the most hostile environments to humans, yet, with their strange forms of animal and plant life, and their barren forms of bare rock and sand dune, they exert a strong fascination for many people (Fig. 8-8). If you go out into the desert, away from automobiles and trailbikes, you may find one of the stillest, quietest places on earth. You may also find yourself in a sudden windstorm, with sand stinging and penetrating everywhere. The desert is where the wind is best able to do its work of eroding and depositing. It does so in partnership with river action that, however infrequent, still does the major part of the work.

Where Deserts Are and How They Evolve

Maps that show average rainfall throughout the world show the great tropical deserts, such as the Sahara and Kalahari of Africa and the Great Australian Desert, all lying near the Tropics of Cancer and Capricorn at latitudes 23°30'N and 23°30'S, respectively (Fig. 8-9). These deserts are under virtually stationary areas of high atmospheric pressure. The stable high-pressure areas are characterized by continuously subsiding air that is warmed up as it sinks, producing low humidity. Tropical deserts normally receive less than 25 millimeters (1 inch) of rainfall each year, in some places less than 5 millimeters (0.2 inch). They are all on the westward sides of continents.

Another belt of deserts and semiarid lands exists in the middle latitudes—between 30° and 50° in each hemisphere—such as the deserts of central Asia and the Great Basin and Mohave des-

erts of the western United States. Here, dryness is a result either of distance from the great reservoir of moisture, the ocean, or of being in the rain shadow of a mountain range. The mountain ranges of the western coastal areas of North America cause moist air coming east from the

Figure 8-8
A Libyan desert landscape. In the foreground, a rocky pavement; in the background, sand dunes are gradually covering a rocky ridge. [Photo by G. H. Goudarzi, U.S. Geological Survey.]

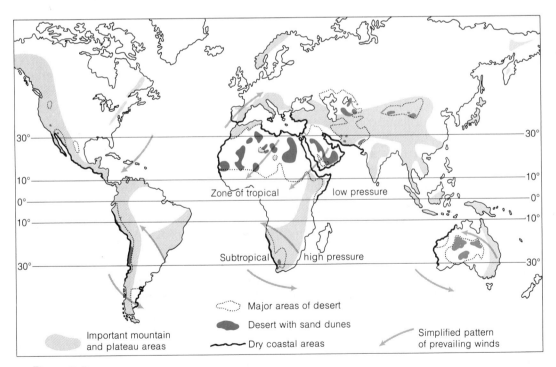

Figure 8-9
Major desert areas of the world in relation to prevailing wind directions and major topographic elevations. Sand dunes are only a small proportion of the total desert areas.
[After K. W. Glennie, *Desert Sedimentary Environments*, Elsevier, 1970.]

Pacific to rise and cool, precipitating rain on their western slopes. When the air descends again on the eastern side, it warms and dries, carrying less total moisture than before. Thus rainfall is limited in the large arid and semiarid regions of the western interior intermontane basins. Another kind of desert forms in polar regions. In these cold, dry areas the frigid air is extremely low in water vapor. An example is the dry valley region of southern Victoria Land in Antarctia, where extreme drought and cold make it the one environment on Earth that most closely resembles a Martian terrain.

Deserts, in a sense, are a matter of tectonics, for the rain-shadow deserts of middle latitudes are there by virtue of mountains produced by plate collisions, and tropical deserts are there because continental drift has placed the continents on which they occur astride the Tropics of Capricorn and Cancer. We can now see another way in which the external and internal heat machines of the Earth are linked. Because of tectonic unrest deserts may evolve in areas that previously have been moist. A rising mountain range on the edge of a continent may interact with wind patterns to cause deserts in its lee, where before there was a more humid coastal plain. A large part of a continent, such as the Sahara, that was once in a moister temperate zone may be moved by plate motions into an equatorial arid belt. Indeed, there is good evidence that long ago, during the Ordovician Period, the northwestern Sahara was covered by glaciers. Several times during the Pleistocene, parts of the Sahara were dry grasslands with springs and lakes, a far more humid regime than at present.

In 1973 millions of people living in the Sahel, the semiarid southern borderland of the Sahara, were devastated by a drought unprecedented in historic times. This resulted in a real encroachment of the Sahara into these borderlands and an expansion of the desert. Natural climatic fluctuations played an important role, but poor grazing and agricultural practices may have had an important additional climatic effect, working to make what was merely a bad situation into a catastrophe.

All told, arid regions amount to one-fifth of all the land area of the Earth. Semiarid steppe and plains regions account for an additional one-seventh.* Because the dynamics of the Earth that produce deserts have been operating through known geologic time, we must infer that deserts have always covered areas as extensive as those covered today. At certain times in the past, perhaps the Earth was even more extensively covered by deserts.

Erosion and Deposition in the Desert

As different as deserts are, there are no geologic processes that are unique to them. Weathering, downslope movements, and transportation operate in basically the same way in deserts elsewhere on Earth, but with a different balance. Soils are thin or absent because of the lack of abundant vegetation and the slowness of chemical decay of rocks, both of which are traceable to the lack of moisture. In desert soils layers rich in organic matter are to be found only directly beneath the widely separated shrubs. What little fine soil is formed is likely to be blown away by the strong winds. Sand, gravel, rock rubble of many different sizes, and bare rock are characteristic of much desert surface.

The coarser, more heterogeneously sized fragments produced by desert weathering form steeper slopes than do the products of weathering in more humid regions covered by soil and vegetation. Desert slopes look, in some way, like mountainous terrain above the timber line (Fig. 8–10). Everywhere the landscape is without the softening effects of vegetation and chemical weathering, which produce the characteristic rounded slopes and rounded rock edges. Instead, one sees steep cliffs and valley walls with masses of angular talus blocks at their bases. Yet chemical weathering is not completely absent. Many iron silicate minerals, such as pyroxene, weather slowly in desert climates, producing the oxidized rusty-iron color of some desert detritus. Clays and soils do form, but slowly. The iron oxides that form from weathering may build up to a hard crust, called **duricrust**. Crusts of calcium carbonate, silica, and other minerals may also form.

Desert varnish is a dark, sometimes shiny patina formed on individual particles and large rock surfaces. A distinct layer on the rock surface, varnish has been shown to consist largely of clay minerals with lesser amounts of manganese and iron oxides dispersed throughout. The result of varying combinations of weathering, impaction of windblown dust, and the action of dew, varnish forms slowly. It seems to take 2000 years to coat some monuments and artifacts.

Even the driest desert gets occasional rain. Rain

---

*The land area of the world is about 145,000,000 km². Desert and near desert areas are about 27,500,000 km² (about 19 percent) and semiarid regions about 21,200,000 km² (about 14.6 percent).

Figure 8-10
Desert mountain slopes, partly covered with alluvium, looking down into Death Valley, California. In the distance is the Panamint Range. [Photo by H. Drewes, U.S. Geological Survey.]

falling on most desert areas may infiltrate permeable rock and surface debris to become groundwater. Some eventually reaches the groundwater table far below—as much as hundreds of meters below, in extreme cases—but most stays in the unsaturated zone, some of it very slowly evaporating. Much of even the shallow groundwater is of marginal quality or too saline to drink because of evaporation in the unsaturated zone. An oasis in the desert generally owes its existence to the groundwater table coming close enough to the surface for roots of palms and other plants to reach it. Some oases are fed by intermittent streams; being dependent on rainfall and runoff, they are seasonal.

Infiltration of rainwater is rapid in sandy areas. It can be slow or virtually absent in areas with much impermeable rock. The rare heavy rain storm produces so much water in so short a time that it cannot all soak in; thus is born the desert intermittent stream. Unhindered by vegetation, the runoff is rapid and may cause flash floods along valley floors that have been dry for years. Because the loose debris on the surface is not held by vegetation, the erosive power of such streams is great, and the streams become so choked with sediment, that they sometimes look more like mudflows than rivers. Because such a high proportion of stream flows in the desert are floods or nearly so, they are efficient eroders of valleys in bedrock. Valleys in the deserts have the same range of shapes, depending on the rock type, as those found elsewhere, but far more of them have steep valley walls caused by the rapid erosion. A

Figure 8-11
A desert dry wash. Such valleys are river channels only briefly during infrequent desert rainfalls. [Photo by F. Press.]

normally dry valley, called a **dry wash** (in the Middle East, a **wadi**), is floored with a flat alluvial fill of coarse debris left from the last water flow; the ordinary features of levees and floodplains are absent (Fig. 8-11). Alluvial fans (see Chapter 7) play a significant role in runoff in mountainous

Figure 8-12
Death Valley, California, a desert playa. Surrounded by mountains and filled with debris from them, playas are intermittently occupied by lakes. The light-colored areas are salts, deposited by a Pleistocene lake. [Photo by Harold E. Malde, U.S. Geological Survey. Courtesy of Charles B. Hunt.]

deserts because of the high sediment load of the streams coupled with the fast infiltration and loss of water in the normally permeable fan itself. Mudflows contribute much to alluvial fans of arid regions.

The drainage systems of deserts are widely spaced because of the relatively infrequent rainfall, but their patterns may be similar to those of other terrains—with one difference: because of water losses by evaporation and infiltration, many of the streams die out, never reaching across the desert to join larger master streams or the ocean. Because many deserts are associated with mountains, this discontinuous or **interior drainage** is compounded by dead-ended intermontane basins called **bolsons.** Such basins—whose drainage exits are blocked by faults in combination with alluvial fans or landslides—may be the sites of temporary lakes, **playa lakes,** formed by storm runoff. Playa lakes may last days, weeks, or longer before complete evaporation leaves a **playa,** a flat lake bed of clay, sometimes encrusted with salts precipitated during evaporation (Fig. 8–12). Playa lakes that are highly alkaline may precipitate so-

dium bicarbonate or sodium carbonate. Others contain a variety of unusual precipitated salt deposits, such as borax, a sodium borate ($Na_2B_4O_7 \cdot 10H_2O$), which is quarried in California from ancient playa-lake deposits. Dessert streams tend to have higher concentrations of dissolved salts than those in humid regions because they redissolve salts deposited by evaporation from earlier runoff. The playas may start out fairly fresh and then become saturated with salts, the particular varieties depending on the kinds of rocks from which the salts were weathered. Because of the unusual nature and great variety of minerals they contain, playa lakes are natural hunting grounds for geochemists interested in the ways in which salts are formed.

Deserts do not necessarily respect shorelines. Along the coast of the Persian Gulf, for example, on the shores of Saudi Arabia, the desert marches right up to the water's edge. There, the interplay of the tides, winds from the water, and sediment produces large flat areas just above the high-water mark that are crusted with a layer of sandy detritus cemented with salt, gypsum, and calcium car-

Figure 8-13
A variety of sabkha flat (salt flat) formed in depressions in the
desert interior. The white is a crust of salt formed by evaporation.
Similar sabkhas form at desert sea margins where sea water laps
onto tidal flats. [From G. H. Goudarzi, U.S. Geological Survey;
photo by Harry F. Thomas.]

bonate. This kind of flat, called by the Arabic
word **sabkha,** is formed by the evaporation of salt
water that has seeped upward by a wicklike
action from a shallow sea-water table below, augmented by the drying of sea spray blown inland.
Sabkhas also form at the edges of inland saline
lakes (Fig. 8-13).

Erosion and sedimentation patterns are responsible for **pediments,** one of the characteristic
landforms of the desert. These are broad, gently
sloping areas of bedrock, spread as aprons around
the bases of mountains. A cross section of a typical pediment and its mountain would show a
fairly steep mountain slope abruptly changing to
the gentle slope of the pediment (Fig. 8-14). A
pediment is an erosional surface covered with
scattered detritus, which is normally more plentiful downslope, the lower surface of the pediment
being covered with compound alluvial-fan deposits merging with the sedimentary fill of the lower
valley or basin. Isolated erosional remnants elevated above pediments (similar to monadnocks)
may be spectacular pinnacles, rounded knobby
masses, or isolated rocky mountains.

Figure 8-14
Stages in the evolution of a typical pediment, an
erosional form produced in arid climates. The surface of the pediment is covered by thin deposits of
alluvial sands and gravels (b), which merge downslope with the depositional surface of the intermontane basin sediments. The mountain slopes retreat
steadily (c and d) but keep the same slope angle,
in contrast to the rounded slopes in humid climates.

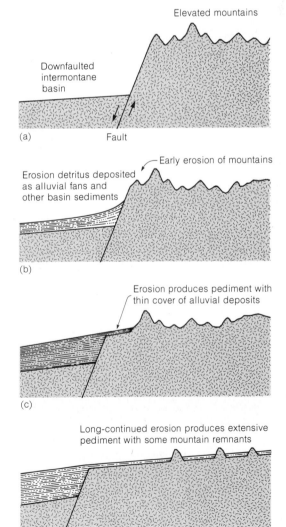

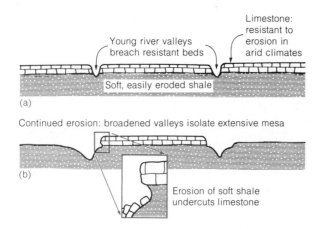

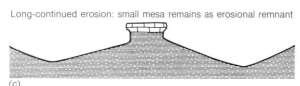

Figure 8–15
The evolution of a mesa is controlled by an erosion-resistant bed overlying weak, easily erodible rocks (a). The border of the mesa retreats as the lower beds are eroded (b), undercutting the resistant beds, which then break off as large talus blocks.

The pediment is formed by running water that builds up an apron of alluvium below as it continues to cut back an erosional platform above. As such, it is an evolutionary product of long-continued headward erosion of rock types (such as sandstones, shales, and volcanic tuffs) that are easily eroded in the desert environment. Just how pediments evolve has been a subject of argument for most of this century, and there is as yet no widely accepted explanation.

For generations, Hollywood and TV westerns made in southern California and Nevada have shown us the kind of arid-land topography that is formed in regions of flat-lying sedimentary rocks: Mesas, tablelike uplands capped by flat erosion-resistant beds and bounded by steep-sided erosional cliffs, rise prominently above pediments and desert flats. The capping bed of a mesa tends to maintain the upland level. If it is breached, however, erosion cuts down rapidly and creates the cliffs. As the cliffs retreat, the upland area shrinks, finally leaving only a few small, isolated mesas above the surrounding lowland (Fig. 8–15).

Large, small, or in between, the valleys and uplands of the dry regions of the world, like those of

humid regions, are forms produced by the eroding action of rivers. Even where water is scarce, it is running water—however infrequent—that does most of the basic work of erosion. The wind helps, but it rarely controls. What the wind does control is the transport and deposition of sand, shaping all the mounds, ridges, and swirling forms of dunes.

DEPOSITIONAL LANDFORMS OF THE WIND: DUNES

Erosion sculpts the rocks of the desert into crags, angles, cracks, and blocks; sand dunes provide the soft forms of undulating slopes and rounded lows, sometimes framed by sinuous, sharp crestlines. The overwhelming impression one gets standing in the middle of a dune field is of an incomprehensible mass of forms with no order, all merging together. As with everything else, there are patterns, but sometimes they can be seen clearly only from the air. The patterns are not accidental: they depend upon the kind and amount of sand available to the wind, upon the underlying rocks, and, most of all, upon the direction, duration, and strength of the wind.

Where Do Dunes Form?

Though we may think of deserts as being unending expanses of sand dunes, a small fraction of most desert land is covered by sand. Only a little more than one-tenth of the Sahara is sand-covered, and sand dunes are far less common in many of the arid lands of the southwestern United States. Most people are more familiar with the dunes that form behind beaches, such as the dunes of Provincetown on Cape Cod, the Indiana dunes on the south shore of Lake Michigan, and the coastal dunes of Oregon. Dunes may also form along the banks of some large rivers.

These places all have in common a ready supply of loose sand. Beach sand deposited by waves is the source of coastal dunes; in river valleys, sand comes from the sandy channels and floodplain deposits of rivers at low water. In the desert, sand is produced by the weathering of sandy bedrock formations. Linked with the supply of sand is wind power. Shorelines are likely places for dunes to form because strong sea or lake winds tend to blow sand inshore away from the beaches. As we have noted, strong winds, sometimes of long duration, are common in deserts.

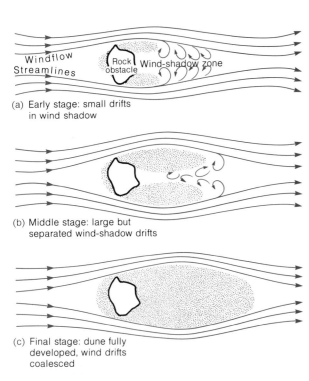

(a) Early stage: small drifts
in wind shadow

(b) Middle stage: large but
separated wind-shadow drifts

(c) Final stage: dune fully
developed, wind drifts
coalesced

Figure 8-16
Formation of sand drift in the lee of an obstacle,
such as a rock. The obstacle, by separating the flow
streamlines, creates a wind shadow in which the
eddies are weaker than the main windflow, thus allowing saltating grains to settle and build up a drift.
The drift is a streamlined body built in response to
the nonstreamlined obstacle. [After R. A. Bagnold,
The Physics of Blown Sand and Desert Dunes, Methuen, 1941.]

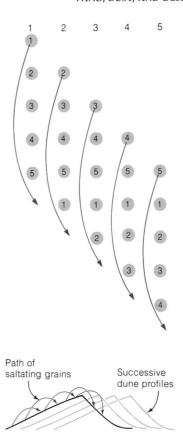

Path of
saltating grains

Successive
dune profiles

Figure 8-17
The movement of a column of people by sequential movements of individuals from rear to
front (above) is roughly analogous to the movement of a ripple or dune by individual grain
movements, though all the sand grains do not
move sequentially in as simple a pattern.

Dynamics of Dunes

Given enough sand and wind, how does a dune
start? Any obstacle to the wind, a large rock, a
clump of vegetation, or even a slightly humped-
up pile of pebbles, can start the formation of a
dune. Wind streams, like water streamlines, separate around obstacles and continue slightly deflected, creating a wind-shadow zone downstream
of the obstacle in which the wind velocity is much
less than in the deflected streams. Blowing sand
grains are deflected with the wind stream. Many
will, as they saltate, come to rest by chance in the
wind shadow, as in Figure 8-16. There they stay,
unable to continue their travels because of the
lower wind velocity, piling up to form a sand
drift. In a strong wind, you can perform this experiment yourself to see exactly how it happens.

If an obstacle is big enough to get a sand drift
started, the drift soon becomes an obstacle itself.
If there is enough sand, and if the wind continues
to blow in the same direction for a sufficient time,
the drift grows into a larger mound of sand—a
dune. Dunes may also form by growth from sand
ripples, much as underwater dunes form in flumes
and streams. Desert and coastal dunes are normally covered by ripples moving over them, just
like the ripples that form on the surface of large
underwater dunes.

As a dune grows, the whole mound starts moving downwind. The sand grains constantly saltate
up the low-angle windward slope to the top, then
fall over into the wind shadow on the lee slope, as
in Figure 8-17. As grains accumulate on the lee
slope, they spontaneously cascade down in little
groups to keep a more or less constant angle of

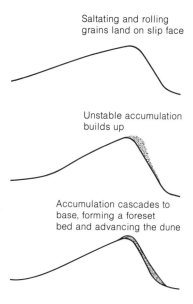

Saltating and rolling
grains land on slip face

Unstable accumulation
builds up

Accumulation cascades to
base, forming a foreset
bed and advancing the dune

Figure 8-18
The formation of the slip face of a dune. The slip
face accumulates an unstable slope as a result of
the deposition of saltating, rolling, and sliding grains.
Intermittently, this accumulation becomes so unstable
that it spontaneously slips—that is, cascades—to the
base, and a new, lower-angle, stable slope is formed.

respose, which is the hallmark of this **slip face** of
the dune (Figs. 8-18 and 8-19). The slip face, deposited at the angle of repose, creates the cross-
bedding of the dunes.

As dunes accumulate, interfere with each other,
and then become buried in a sedimentary sequence, the shapes of the dunes as they appeared
under the wind are lost, but the cross-bedded
structure remains (Fig. 8-20). Cross-bedded ancient sandstone units many meters thick are evidence of high dunes that were most likely windblown. From the direction of the cross-beds, we
can get the average direction of the ancient winds,
or **paleowinds.** From this and other kinds of information, we can start to reconstruct ancient climates—**paleoclimates**—that can be harmonized
with former positions of continental plates relative to the poles.

The height of dunes is determined by a switch-
over from positive to negative feedback. Positive
feedback makes little dunes grow bigger by more
rapid pile-up of saltating grains on the windward
slope than on the slip face, but only to a point.
When the dune reaches a certain height, the
windstream lines become "squeezed" upward
and together by the mound of sand (Fig. 8-21). As
more air has to rush through a smaller space, the
wind velocity increases, ultimately to the point at
which sand grains blow off the top to the advanc-

Figure 8-19
Dunes in the Arabian desert, showing the slip face of a large dune advancing on
a sand-rippled surface. The grooved appearance of the slip face is formed by cascades of sand grains assuming the stable angle of repose. [From Arabian American
Oil Co.]

ing slip face as fast as they are brought up, and a steady state is reached. Exactly what that steady-state height is, of course, depends on the wind speed and the size of the sand grains. Dune heights of 30 meters (100 feet) are not uncommon, and some huge dunes in Saudi Arabia may reach 250 meters (Fig. 8–22).

Dune Types

In an attempt to make some order out of immense diversity, dunes have been lumped into several major types, though there are gradations among them and many irregular shapes that are hard to fit into the scheme. The solitary crescent-shaped dune, the **barchan,** which moves over a flat surface of pebbles or bedrock, is the product of limited sand supply (Fig. 8–23). Where sand is abundant, vegetation is absent, and sand dominates the landscape, dunes form in long, wavy ridges that lie transverse (perpendicular) to the prevailing wind—hence, **transverse dunes** (Fig. 8–24). These dunes have the same easily distinguishable slip faces as barchans, and they may grade into them at the edges of the dune field, where the sand supply is more limited. Typical sandy-beach dunes are transverse dunes; in some places, a series of them may form back from the shore. The sand is

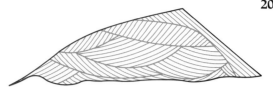

Figure 8–20
Cross section of a sand dune, showing interfering sets of cross-beds, a response to variable wind directions. The appearance can be similar to the interfering trough cross-bedding of Figure 7–11.

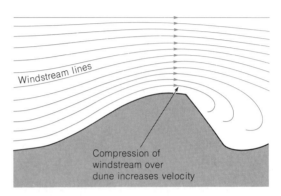

Figure 8–21
Limitation of the height of a sand dune by a compressed windstream. As the dune grows higher, the windstream (shown by streamlines) becomes more compressed and thus travels at a great velocity, which makes it more competent to transport sand grains. Eventually, a height is reached at which the wind speed is so great that all of the sand is transported, and the dune stops growing vertically.

Figure 8–22
Desert sand "mountains" in the Saudi Arabian desert may reach heights of as much as 250 m (820 ft). [From Arabian American Oil Co.]

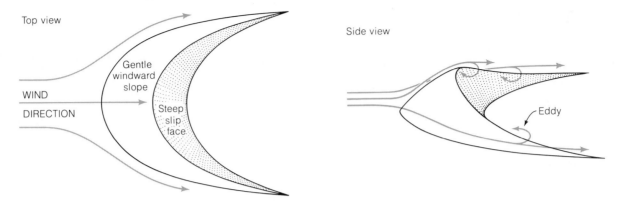

Figure 8-23
Barchans are solitary dunes whose horn shape points downwind. One horn may
elongate as the dune moves downwind; sometimes one barchan may overtake and
coalesce with another.

Figure 8-24
Aerial photograph of transverse dunes on the Saudi Arabian desert, aligned in
a general direction at right angles to the prevailing wind direction. [From Arabian
American Oil Co.]

supplied by the beach. In fairly humid regions, there is a balance between the rate of sand movement and the growth of vegetation, which binds the sand and stabilizes it, preventing further movement. If the vegetation of such stabilized dunes is overwhelmed by sand at some point along the shore, the dune migrates inland, form-

ing a **blowout,** a dune shaped like a parabola. Blowouts may be natural, but these days, when beachfront vegetation is bulldozed away by home or road builders, blowouts are often the result of a crude upsetting of the sometimes fragile balance between vegetation and sand. Those who have tried it can speak with feeling of the difficulty of

(a)

(b)

Figure 8-25
Longitudinal dunes parallel to the prevailing wind direction in the Saudi Arabian desert. Photograph (a) is a view of the dunes as seen from the great height of a satellite (Gemini IV); (b) is a low-altitude airplane photograph. [From Arabian American Oil Co.]

planting vegetation and getting it to grow and stabilize the dune once a blowout is on the move. Over many years, vegetation gets a chance to stabilize the dunes as they move inland, farther from their source of sand on the beach. The hilly wooded or shrub-covered topography of shorelands in many places is one of overgrown, older, stabilized dunes.

Wind can blow sand into long, straight ridges more or less parallel to the prevailing wind—**longitudinal dunes,** or, as Bagnold called them, **seif dunes** (Fig. 8-25). These dunes may reach heights of 100 meters and may extend many kilometers. Moderate sand supply, a rough pavement, and winds of varying direction (but always in the same general quadrant of the compass) combine to form the long ridges.

DUST FALLS

The dust falls in the plains states in the 1930s were a dramatic enactment of processes geologists had years earlier deduced from thick deposits of unstratified fine-grained silt and clay called **loess** (Fig. 8-26). Loess is composed of angular particles of quartz, feldspar, mica, and mafic minerals, mixed with abundant clay minerals. In North America, the origin of Pleistocene loess as wind-

Figure 8-26
Pleistocene loess, showing prismatic jointing, in Colorado. [Photo by H. E. Malde, U.S. Geological Survey.]

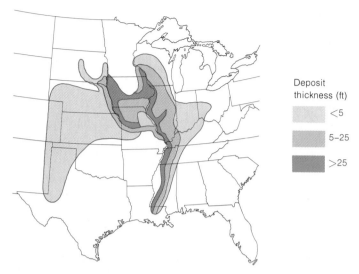

Deposit thickness (ft)

	<5
	5–25
	>25

Figure 8–27
Loess and other eolian deposits in the central United States. The loess thins leeward from its sources along the major valleys that discharged outwash in Pleistocene time. Westward, toward the Rocky Mountains, the eolian deposits on the Great Plains are sandy. [From *Geology of Soils* by C. B. Hunt. W. H. Freeman and Company. Copyright © 1972.]

blown material was originally demonstrated by its pattern of distribution in the upper Mississippi Valley as a blanket of more or less uniform thickness on hills and valleys alike, all in or near formerly glaciated areas (Fig. 8–27). Even more conclusive was the much greater thickness of loess on the eastern sides of major river floodplains than on the western sides and the rapid fall-off in thickness and grain size farther east of the valley. The facts were consistent with deposition from prevailing westerly winds blowing over stream valleys carrying abundant silt and clay from the outwash of melting glaciers. Loess is also found in deserts or downwind of deserts, as in China where loess was deposited to thicknesses of 30 to 100 meters over wide areas by winds blowing over the arid lands of central Asia.

Loess is a distinctive sediment. It tends to form vertical cracks and walls when pieces break off during erosion, and it is frequently filled with thin tubes and nodules of calcite, most of them orien-

ted vertically. The vertical cracking may be caused by a combination of vertical root penetration and uniform downward percolation of groundwater, but the details still await a good explanation. Loess is sometimes picked up and redeposited in river deposits, which, if not too well stratified, may easily be confused with the original windblown material. The correlation between the distribution of thin loess over parts of Illinois, Iowa, and adjacent states and the crop-producing capacity of those areas is not accidental. Soils formed on loess are some of the most fertile. They are also easily eroded into gullies by small streams and deflated by the wind when poorly cultivated. Ultimately, after the tragedy of the dust bowl, the fertility of the land was increased—no consolation to the thousands who had been driven from their farms, like the Joads in John Steinbeck's *The Grapes of Wrath*.

Oceanographic research vessels have measured airborne dust far out to sea, and the mineral composition has been compared with the sediment at the sea bottom. Thus, dust has been discovered to be an important constituent of deep-sea sediments, perhaps accounting for as much as twenty percent. Much of the dust is traceable to volcanic activity. Volcanic dust may be the most important source of particulate matter in the atmosphere. Since particulate matter may increase the reflection of sunlight by the of Earth's atmosphere, it may affect our worldwide climate. There has been some speculation that modern civilization's pollution of the atmosphere by tiny particles of lead, soot, and thousands of other materials might raise the level of dust in the atmosphere enough that the Earth might cool by a small amount, posing at least the possibility of a new ice age. A sober look at the data, however, has convinced most scientists who have studied the question that natural volcanic dust is far more important, and that the amount of it in the atmosphere has probably been fluctuating a good deal for much of geologic time. If doom befalls us in the form of a new ice age, at least it won't be the fault of the dust of modern civilization.

Last but not least, dust in the atmosphere provides us with some marvelous red sunsets.

SUMMARY   1. Winds are important agents of erosion and transportation, particularly in deserts.

2. The distribution of the wind belts of the Earth and the relation of winds to mountain ranges and shorelines are closely related to the locations of such windblown deposits as sand dunes.

3. Wind transports dust by suspension and sand grains by saltation. Air can hold enormous quantities of dust extending high into the atmosphere, but sand is transported in smaller quantities near the ground.

4. Wind erosion deflates arid regions and may produce desert pavements. Sandblasting facets and polishes rocks into ventifacts.

5. Deserts are located in a few tropic regions of stable high pressure and in the rain shadows of mountain ranges.

6. Erosion and deposition operate in much the same way in deserts as they do elsewhere but with different intensity, as a result of dryness, lack of vegetation to bind soil, and rate of river flows. Much of desert erosion is caused by rivers.

7. Drainage of deserts is widely spaced and normally interior, and it may include playas and bolsons. Shoreline deserts form sabkah flats, in which carbonate and evaporite deposits may form.

8. Pediments are characteristic erosional forms of arid lands.

9. Dunes, the constructional products of wind deposition, form by wind action in deserts, behind beaches, and along some sandy river valleys. All dunes are cross-bedded, and they fall into a few general types: barchans, transverse dunes, and longitudinal or seif dunes.

10. Loess is a deposit of windblown dust in deserts and in regions that were glaciated during the Pleistocene.

EXERCISES

1. Do you think a study of the shorelines of continents in the zone of trade winds would show a tendency for coastal sand dune belts to be more abundant and wider on the eastern or on the western coasts? Why?

2. In terms of visibility and sandblast, which vehicle would be more affected by a sandstorm—a low slung sports car or a high freight truck? Do you think the same would hold true in a dust storm?

3. You are to estimate requirements for hauling away windblown sand from a two-lane highway over which a dune is advancing in a region where the wind blows an average of 40 km/hr all day. What tonnage trucks would you provide, and how many trips would you make each week?

4. Describe a set of conditions that might produce a sandstorm in a coastal dune field.

5. Describe the possible changes in the Sahara Desert if the African continent were to move approximately 2000 km toward the North Pole.

6. What changes might occur in the Mojave Desert of the southwestern United States if the California mountain ranges became tectonically inactive and were eroded to lowlands?

7. What evidence would you use to demonstrate that many desert landforms are the product of river action? Would such evidence indicate a formerly humid region?

BIBLIOGRAPHY

Bagnold, R. A., *The Physics of Blown Sand and Desert Dunes*. London: Methuen Publishing Company, 1941.

Glennie, K. W., *Desert Sedimentary Environments*. New York: Elsevier Publishing Company, 1970.

Hadley, R. F., "Pediments and Pediment-Forming Processes," *Journal of Geological Education*, v. 15, pp. 83–89, 1967.

Idso, S. B., "Dust Storms," *Scientific American*, October 1976.

Mabbutt, J. A., *Desert Landforms*. Cambridge, Mass.: MIT Press, 1977.

Starr, V. P., "The General Circulation of the Atmosphere," *Scientific American*, December 1956. (Offprint 841.)

9

The Flow of Ice: Glaciers

The low temperatures and precipitation of snow that are typical of cold polar regions and high mountains contribute to the formation of glaciers and snow fields. As snow accumulates, it becomes compacted, gradually changing from snowflakes to granules to solid, massive ice. As glacial ice moves down valleys, it sculpts the topography by eroding rock and transporting the debris to the terminus, where melting takes place. Glaciers of continental size, like those in Greenland and Antarctica, produce a variety of erosional and sedimentary landforms. The glacial landforms of the recent past are evidence of the Pleistocene glacial epoch, during which huge areas of North America and Eurasia were covered by ice. Advances and retreats of the ice fronts caused large fluctuations in sea level, alternately flooding and exposing shallow ocean margins of the continents. The causes of past ice ages and any predictions of future ones are equally uncertain.

Ice is a rock, a mass of crystalline grains of the mineral ice. That idea should not be too surprising; after all, it is a solid substance that occurs naturally on the Earth. It is hard like most rocks, but its composition makes it much less dense. Like igneous rocks it originates as a frozen fluid; like sediments, it is deposited in layers at the surface of the Earth and can accumulate to great thicknesses; like metamorphic rocks, it is transformed by recrystallization under pressure. Masses of ice may creep, flow, or slide downhill, and just like other masses, they may be folded and faulted. (Fig. 9–1). A large mass of ice that is on land and shows evidence of being in motion or of once having moved is a **glacier.** The motion of glaciers is the clue to the effective work they do in eroding the surface of the Earth into distinctive sculptural forms and in transporting rock debris and depositing it in various forms. The movement of glaciers is now invested with a new and practical interest for humans; early warning of global climatic changes may be indicated by advances or retreats of glaciers. Glaciers are abundant on to-

day's Earth. It is estimated that there are between 70,000 and 200,000 glaciers of all kinds and sizes in the world, covering about ten percent of Earth's land surface.

Much of the richest farmland in America is on glacial deposits. Much of the abundant good water supply of the glaciated terrain is from aquifers of glacial sand and gravel. Sand and gravel deposits are abundant. Lakes, one of our best recreational resources, are another of the gifts of the glacial epoch. But there are liabilities too. Any New England farmer can speak with feeling about all of the rocks in his fields of bouldery till, and geologists are sometimes frustrated by the glacial sediment that covers bedrock and prevents them from mapping geologic formations and mineral resources.

THE FORMATION OF GLACIERS

High altitudes and high latitudes have something in common: both are cold. High altitudes are cold

Figure 9-1
Thrust faulting in glacial ice, showing its rocklike characteristics. The slanting lines
are a series of thrust faults along which the ice on the right was pushed to the left
and upward. Jointing and fracturing are also prominent in this photograph of
Susitna glacier, Alaska. [Photo by B. Washburn.]

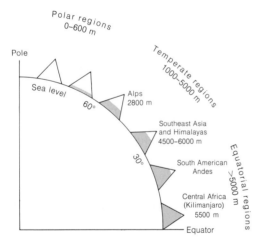

Figure 9-2
The height of the snow line, the altitude above
which snow does not completely melt in summer,
varies with latitude from at or near sea level in
polar regions to heights of more than 6000 m at the
equator. The exact height of the snow line also depends upon local and regional climate.

because the lowest ten kilometers of the atmosphere steadily cools with distance from the Earth,
and high latitudes are cold because the angle of
inclination of the Sun's rays increases toward the
poles. In parts of the Earth that are high enough,
or far enough north or south, not all of the snow
that falls in a year melts. Some of it accumulates
in **snow fields** and is transformed into ice. In the
warm regions of the Earth near the equator, glaciers form only above about 4500 meters (about
15,000 feet). The altitude necessary for glacier formation decreases toward the polar regions, where
ice is stable the year round even at sea level (Fig.
9-2). Because moisture-containing winds tend to
drop most of their snow on the windward side of
a coastal mountain range, the lee side is likely to
be cold but desertlike, lacking in glaciers. For example, in the high Andes Mountains glaciers
occur on the western slopes; but the eastern
slopes are dry, barren of ice. Inland mountains far
from sources of moisture are also unlikely to have
glaciers unless, as in Antarctica, it is so cold all
year round that there is no melting and all snow is
preserved. Thus, even arid climates, if cold
enough, promote the formation of glaciers. Along
the coast of Antarctica, the dry valleys of South
Victoria Land, however, are so devoid of moisture
that any snow sublimes (evaporates without melting), and the ground, though frozen hard, remains
relatively dry and free of ice. These valleys are

isolated from the flow of glaciers out of the Antarctic interior by the Transantarctic Mountains.

Most of us are familiar with ice-making from the common experience of putting water into an ice-cube tray and then freezing it in a refrigerator. Those of us who live in temperate regions see rivers and lakes freeze over in winter. Glaciers, however, do not just freeze; they grow by a different process—the gradual transformation of snow into glacier ice. A fresh snow fall is a fluffy mass of loosely packed snowflakes, small delicate ice crystals grown in the atmosphere. As the snow ages on the ground for weeks or months, the crystals shrink and become more compact, and the whole mass becomes squeezed together into a more dense form, **granular snow** (Fig. 9-3). As new snow falls and buries the older snow, the layers of granular snow further compact to **firn,** a much denser kind of snow, usually a year or more old, with little pore space.* Further burial and slow cementation—a process by which crystals become bound together in a mosaic of intergrown ice crystals—finally produce solid glacial ice. One can think of snow as a sediment that becomes transformed by burial into the metamorphic rock ice by a process of recrystallization, the growth of new crystals at the expense of old ones. The whole process may take as little as a few years but more likely ten or twenty years or longer. The snow is usually many meters deep by the time the lower layers are converted to ice.

In cold glaciers, those formed in the coldest regions, the entire mass of ice is at temperatures below the melting point, and no free water exists. In temperate glaciers, the ice is at the melting point at every pressure within the glacier, and free water is present as small drops or as larger accumulations in tunnels within or beneath the ice. Glaciers may be cold in their upper parts and temperate at their lower ice margin.

Formation of a glacier is complete when ice has accumulated to a thickness (and thus weight) sufficient to make the ice move slowly under pressure, in much the same way that rock deforms plastically deep in the Earth. Once that point is reached, the ice flows downhill, either as a tongue of ice filling a valley or as a thick ice cap that flows out in all directions from the highest central area where most snow accumulates. The

---

*Many of the terms used in glacial geology come from words used by the Swiss and French people of the Alps. The word *glacier* itself comes from the old French of the Alpine province of Savoy. *Firn* is a German-Swiss word used for "last year's snow." The word *névé* means the same as firn, but comes from a French-Swiss dialect word derived from the Latin for "snow."

trip down leads to the eventual melting of ice. In medium and low latitudes, ice melts and evaporates as it flows to lower elevations (Fig. 9-4). In polar climates the ice breaks off in huge chunks at the ocean's edge, and the chunks float out to sea as icebergs, which finally melt (Fig. 9-5).

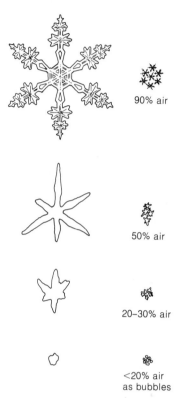

90% air

50% air

20–30% air

<20% air
as bubbles

Figure 9-3
Stages in the transformation of snow crystals to granular ice as snow changes to firn and glacial ice. Accompanying the change of individual crystals is an increase in density by elimination of air. [After H. Bader and others, "Der Schnee und seine Metamorphose," *Beiträge zur Geologie der Schweiz,* 1939.]

The Budget of a Glacier

During winter, a typical glacier grows slightly as snow falls everywhere on the ice surface. In summer, the glacier shrinks, mainly as the snow on the surface of the lower reaches melts and evaporates to uncover solid ice, while the upper reaches stay snow covered. The annual growth budget of a glacier is the amount of solid water added by snow, the **accumulation,** minus the amount lost, called **ablation.** In temperate climates ablation takes place primarily by melting, and in polar climates, mainly by sublimation and by the breaking off of icebergs. The difference between accumulation and ablation is a measure of either growth or

Figure 9–4
Melting of ice at the terminus of a glacier, showing the dark accumulation of debris at the wasting ice front and the formation of meltwater streams (lower left). [Photo by B. Washburn.]

Figure 9–5
As the ice from the Miles glacier, Alaska, meets the sea, it breaks into large irregular masses, many as big as a small island, to become icebergs. [Photo by B. Washburn.]

shrinkage of the glacier. When accumulation minus ablation, the net budget, is zero over a long period, the glacier is in a steady state, accumulating new snow in its upper reaches and constantly moving downslope to the lower reaches, where ablation takes place. Glaciers' budgets fluctuate from year to year, and many show long-term trends of growth or shrinkage in response to climatic variation over periods of many decades. Geologic evidence suggests, however, that many glaciers have remained very roughly in a steady state over the past several thousand years.

Because snow and ice vary so much in internal air space, accumulations are measured by the height of a column of liquid water equivalent to the total amount of precipitation falling over each square centimeter per year. In the cold, dry interior regions of Antarctica, the accumulation may be only 5 centimeters; in the wet maritime climate of Iceland, however, a glacier may accumulate more than 300 centimeters each year.

In the cold of the Antarctic, where melting is minor, ice ablates primarily by the breaking off of icebergs and sublimation. In most temperate climates, however, most ablation occurs by melting under the Sun's rays and less by evaporation. Warm air may blow over the lower reaches of the glacier and speed melting. The air is chilled in the process. If the air is humid, it may precipitate rain over the lower glacial slopes, causing even more ablation by melting. The meltwaters of glaciers in some areas have assumed increasing importance as a possible source of fresh water for irrigation and other purposes. Some geologists think it is feasible to tow icebergs to provide fresh meltwater for water-poor areas. Regardless of the practicality of this idea, the enormous amounts of water stored in glacial ice make it likely that water resource developers will pay more attention to glacial meltwaters in the future.

HOW ICE MOVES

Once the ice on a slope builds to a great enough thickness, it begins to flow downhill. The effect of thickness on flow can be demonstrated with a viscous fluid like honey. A thin layer of honey on a slightly tilted piece of bread flows very slowly, but if more honey is poured on to increase the thickness, it flows much more rapidly. The speed of the flow can also be increased by tilting the bread more. The greater the thickness of glacial ice and the greater its slope, the faster the movement of the glacier.

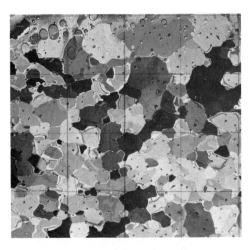

Figure 9-6
Photomicrograph of typical crystalline mosaic of glacier ice. Sample is from the Antarctic ice sheet, depth 193 m, Byrd Station, Antarctica. Each grid measures 1.0 cm. Each small area of uniform white, gray, or black is a single crystal as seen in polarized light. The tiny circular and tubular spots are bubbles of air. [From A. J. Gow, U.S. Army Cold Regions Research and Engineering Laboratory.]

The flow of ice differs from the sliding of a brick down an inclined board in an important way: the brick slides only along its base, whereas the ice moves throughout its bulk by internal sliding or flowing movements, as well as along its base. This internal flow throughout the ice accounts for much of its motion. Under the stress of weight, individual crystals of ice (Fig. 9-6) slip tiny amounts—distances of about a ten-millionth (10^{-7}) of a millimeter in short time intervals. But the sum total of all those small movements in the enormous number of ice crystals over longer time periods amounts to much larger movements of the whole mass. It could be visualized as a large bulk made up of many randomly arranged packs of playing cards, each pack held together by rubber bands, the whole thing moving as the sum of many small slips between cards in the individual packs. This kind of movement is similar to the movement shown by metals, which slowly creep when subjected to a stress—or to the movement of hard rocks at high temperatures and pressures when buried deep in the Earth. This movement is not the whole story. Other processes are at work also. Ice crystals tend to melt and recrystallize a microscopic distance downslope, and other crystal distortions result in movement.

The sliding of a glacier along its base accounts for a significant part of the total movement. Dry-based glaciers are cold glaciers that are thought to

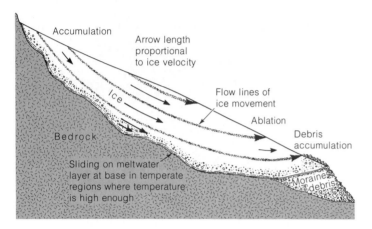

Figure 9-7
How ice flows in a typical valley glacier of temperate regions. The rate of movement decreases toward the base. In temperate climates, where temperatures at the base of the glacier are high enough that the ice pressure causes melting, the entire thickness of the glacier moves by sliding along the liquid layer next to the ground.

be frozen to the ground beneath. In them the ice deforms and moves more easily itself than does the ice-cemented bed contact. The ice at the base of wet-based glaciers is at the melting temperature for the pressure there, so much of the movement takes place along the base (Fig. 9–7). Some of the sliding is caused by the melting and refreezing of ice at the base. The melting results from some combination of the pressure of the overlying ice and the flow of heat upward from the interior of the Earth. If enough melting takes place, a layer of liquid water forms at the bottom; such a layer was found when the bottom of the Antarctic ice sheet was penetrated by a drill at a depth of more than two kilometers below the surface.

In the deeper layers of ice, the solid material that might otherwise tend to be pulled apart by the movement is held together by the compressive force of the overlying mass of ice. But the upper layers of a glacier have little pressure on them, and so the surface ice behaves as a rigid, brittle solid, cracking as it is dragged along by the movement of the plastic ice below. These cracks are the **crevasses** that break up the surface ice into many

Figure 9-8
Crevasses in a valley glacier are deep cracks in the ice. They typically form along the sides of the glacier or, as in right foreground, across the glacier at a bedrock hump in the valley floor. They may be many meters across and tens of meters deep. Upper basin of Muldrow glacier, Alaska. [Photo by B. Washburn.]

small blocks near bedrock walls and at places where the ice surface steepens sharply in the direction of flow (Fig. 9–8). The movement of the surface at these parts of a glacier is a "flow" resulting from slipping movements between these irregular blocks.

The speeds of Alpine glaciers were first measured well over a century ago by placing stakes in the ice and noting the changes in their position over a few years.* The stakes showed clearly that the center of the tongue of ice moved much faster than the edges, where friction of the ice against rock walls hindered the flow (Fig. 9–9). The most rapid movement was about 75 meters (250 feet) in one year. Since then, a wide range of ice speeds have been measured, ranging from a few centimeters to a meter per day.

Today glacier specialists use artificial satellites to measure the change in distance between two transmitters planted in the ice. From such data we know that the ice at the South Pole is moving in the direction of Rio de Janiero, Brazil, at a rate of 8 to 9 meters per year (27 to 30 feet). The fact that the base of a glacier moves more slowly than the upper parts (Fig. 9–7) was originally demonstrated by sinking a straight tube vertically into the ice and measuring its bending as the upper parts moved faster than the lower. Now, airborne radar is used to give a detailed picture of changes in ice surface altitude, thickness, internal layering, and the shape of the ground beneath the ice.

Sudden movements of glaciers, **surges,** sometimes occur after long periods of little movement. Surges last more than two or three years, and during one the ice may travel at rates of more than 6 kilometers (3.7 miles) per year. The movement comes largely from a sudden redistribution of ice from upper to lower parts of the glacier with little change in accumulation rate or in the total size of the ice mass. Surges may be caused by periodic melting of the base of the glacier, allowing it to slide rapidly for a while until freezing slows it, or by intermittent releases of ice that gradually piles up in middle parts of the glacier while the lowest parts are melting. Some glaciologists have hypothesized surges of the Antarctic ice sheet in the past that provoked a cooling of Earth, but such surges have neither been observed nor firmly supported by analyses of older glacial ice or deposits. The large number of surges that have been ob-

*One of the pioneers in the measurement of glacier flow was Louis Agassiz, a Swiss zoologist and geologist. As a young professor, not yet thirty, he built a hut on a glacier and, with the help of his students, mapped its movements. Twelve years later, he emigrated to the United States and became instrumental in founding glacial geology here.

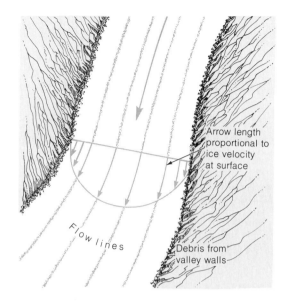

Figure 9–9
A valley glacier moves most rapidly at its center and with much lower velocity along the valley walls.

served in valley glaciers in recent years do not herald an era of glacial advance, a new ice age; these movements are peculiar to a particular variety of glacier that we have come to know better in recent years and whose movements have no global climatic significance.

SHAPES AND SIZES OF GLACIERS

On the basis of size and shape, we can easily divide glaciers into two general types, the **valley glacier** and the much larger **continental glacier.**

Valley Glaciers

The glaciers best known to skiers are valley glaciers, rivers of ice flowing downhill in narrow ribbons (Fig. 9–10). They start as snow fields in the high slopes in amphitheatre-like hollows, **cirques.** As the ice forms from firn in the cirque, it flows down the course of the valley, in some places passing over uneven slopes where the smooth surface of the ice may break up into crevasses. As the glacier moves down its valley, it may be joined by tributary valley glaciers. Unlike the waters of tributary rivers, however, the ice of tributary glaciers does not mix with that of the main glacier. Instead they run along in separate streams, side by side. Another difference between rivers and glaciers is that, although the ice surface is at the same level

Figure 9-10
Valley glaciers showing high snowfields and cirques in the background, tributary glaciers flow-ing down steep valleys, and their junction with the master glacier in the foreground. Harvard glacier, Alaska. [Photo by B. Washburn.]

at the junction of glacier tributaries, the floors of the valleys at that place may be at very different depths below the ice surface. This fact explains the **hanging valleys** that exist where some former glacial tributaries entered a main glacial valley, as in the Yosemite Valley of California, where Bridal Veil Falls plunges over the edge of the steep-walled valley (Figs. 9–11 and 9–12).

Where valley glaciers end on land, streams of meltwater flowing on and under the glacier coa-lesce downstream to form a single river. Some-times melting at the ice margin is uneven, and large blocks of ice may be left isolated as the main part of the glacier melts back in warm weather. Glaciers ending at the water's edge form cliffs of ice, many of them standing 30 to 60 meters or even more above the lake or sea level. Because ice is less dense than water, it floats; the buoyant force exerted upon it, combined with the melting action

Figure 9-11
The hanging valley of Bridal Veil Creek, which emp-ties into the main Yosemite Valley by Bridal Veil Falls. The valley of the creek has not been strongly affected by glacial erosion and retains much of its original V-shape. [Photo by F. E. Matthes, U.S. Geo-logical Survey.]

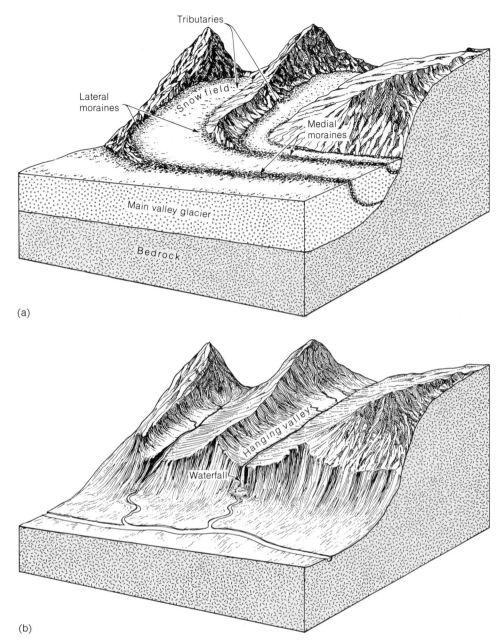

Figure 9–12
The evolution of hanging valleys and their waterfalls by tributary valley glaciers. (a) During glaciation, tributary ice enters a major glacier at different levels. (b) The region after deglaciation.

of sea water and tidal movements, breaks off icebergs from the edge of the ice. The icebergs then drift with ocean currents (a hazard to ships), many ending up in warmer waters where they gradually melt.

The position of the terminus (lower end) of a glacier, which is a response to the balance between accumulation and ablation, can be a sensitive indicator of climate. In the 1930s, for example, many glaciers were observed to shrink fairly steadily by retreat of the ice margin up the valley. This was just a part of a general warming trend

that had gone on since the middle of the nineteenth century. Since 1940 there has been evidence in glacial advances of a slight cooling trend confirmed by world weather records.

Continental Glaciers

Enormous, thick sheets of glacial ice cover much of Greenland and the continent of Antarctica. These glaciers are not confined to valleys, but spread over all the land surface. The 2,800,000

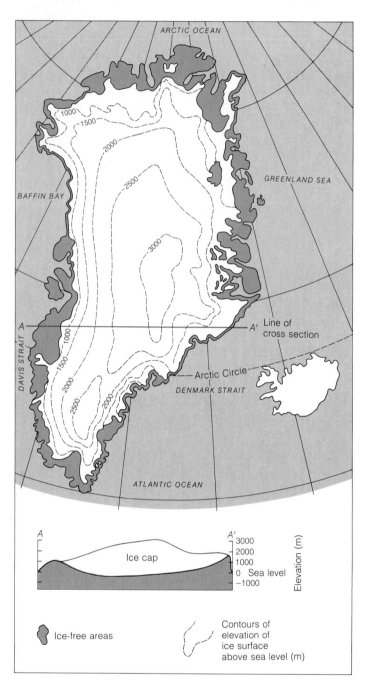

Figure 9-13
The extent of the glacial ice cap and the elevation of the ice surface on Greenland. The generalized cross section of south-central Greenland, *AA'*, shows the central depression of the rock floor and lenslike shape of the ice cap. The ice moves down and out from the thickest section. [Information from J. Haller, *Geology of the East Greenland Caledonides*, Figure 2. Copyright © 1971 by John Wiley & Sons.]

cubic kilometers (672,000 cubic miles) of Greenland ice covers an area of 1,726,400 square kilometers (666,560 square miles), about 80 percent of the total area of the land (Fig. 9-13). In cross section, the ice has the shape of an extremely wide lens, convex on both the smooth upper surface and the rough lower boundary with the ground. At the center of the ice sheet, in the middle of the island, the ice is more than 3200 meters (10,500 feet) thick.

It is in the central area that the snow accumulates in a huge field of firn that gradually transforms to glacial ice. The ice surface is thus built up at the center and slopes to the sea on all sides. The glacier moves down and out in all directions. At the edge of the sea, where much of the coast is rimmed by high mountain ranges, the ice sheet breaks up into the narrow tongues resembling valley glaciers that wind through the mountains to the sea. This glacier is the major source of icebergs in the North Atlantic.

As large as the Greenland glacier is, it is dwarfed by the Antarctic glacier, which covers more than 90 percent of the entire continent (Fig. 9-14). The area covered by ice is about 12,500,000 square kilometers (4,830,000 square miles), and ice thicknesses exceed 3000 meters (10,000 feet) in much of the central area. In some ice-buried valleys, the rock floor of the ice is more than 2500 meters (8200 feet) below sea level. The ground beneath the ice is mountainous over much of the continent; in many places along the margins of the continent, mountains rise through the surrounding ice. In Antarctica, as in Greenland, the ice domes in the center and flows out to the margins, where it ends much as the Greenland glacier. A conspicuous body of ice in the Antarctic, the Ross Ice Shelf, is a thick layer of ice about the size of Texas. The ice shelf floats on the water of the Ross Sea but is attached to the main part of the glacier on land. Scientists working in the Antarctic drilled 420 meters (1378 feet) through the Ross Ice Shelf in 1977 to study the water, life, and sediment beneath it. They discovered a sparse population of fish, crustaceans, and micro-organisms living in sea water that is actively circulating beneath the ice and melting some of its base. The studies of sediment and fossils of the sea floor indicated that the Antarctic continent was at least partially glaciated as early as the middle of the Miocene.

GLACIAL EROSION AND DEPOSITION

As both valley and continental glaciers move over their bedrock floors, they erode large quantities of material and transport it downstream. There it is deposited as the ice melts. An accumulation of rock, sand, and clay riding on the ice or left on the ground after the ice melts is a **moraine** (Fig. 9–15). Glacial erosion and deposition produce a variety of distinctive topographic landforms made up of sedimentary debris left behind when the ice melts. From the study of those forms and materials and study of deposits from modern glaciers came the suggestion that great polar ice caps, much enlarged, may have extended far south into Europe and North America during the Pleistocene.

Abrasion and Sediment Transport by Ice

To anyone who has been in an alpine valley near the end of a glacier, the evidence is plain. As the ice melts, rock material of all sizes is dropped, from the finest clay to huge boulders (Fig. 9–16). The ice itself can be seen to contain all of these materials, with by far the greatest part of the load being near the base and along the valley walls. As the ice melts back from the bare rock pavements over which it moved, it reveals parallel scratches and grooves aligned in the direction of ice movement. In many valleys of the Alps, granitic rocks crop out high in the mountains, and sedimentary rocks make up the lower stretches of the valley.

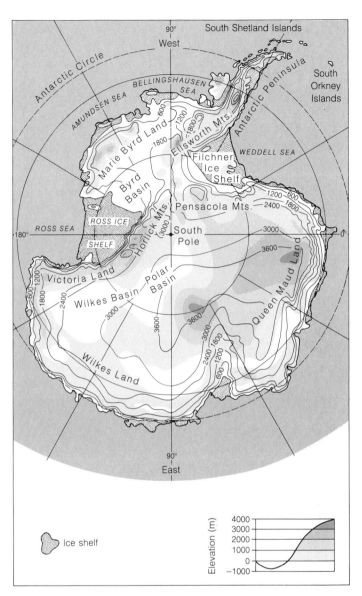

Figure 9–14
Contour map of Antarctica, showing the relief of the continental ice and the land beneath it. The relief of the ice is indicated by the black contour lines, and the height of the land is indicated by various shades of brown. The continental land that lies below sea level is shown in white. The black stippled areas are ice shelves. The black circle around the South Pole is at 80° latitude; the next circle out is at 70° latitude. [From "The Antarctic," by A. P. Crary. Copyright © 1962 by Scientific American, Inc. All rights reserved.]

Figure 9–15
Moraines of the Northern Iliamna glacier, Alaska. In the middle background is the present ice margin, and in the foreground a series of semicircular terminal moraines, called recessional moraines, deposited as the ice melted back from its farthest extent. [Photo by B. Washburn.]

Figure 9-16
A huge erratic boulder carried to Cape Cod during the Pleistocene glaciation. The boulder is a metamorphic rock type typical of areas in southern New Hampshire, 150 km to the northwest. Doan Rock, Eastham, Massachusetts. [Photo by R. Siever.]

Figure 9-17
Glacial polish, striations, and grooves formed on a fine-grained granite surface, Sequoia National Park, California. The direction of ice movement was diagonally toward the right and away from the camera. Since the glaciation, the rock has been disrupted into angular blocks by repeated frost action. [Photo by F. E. Matthes, U.S. Geological Survey.]

Yet in the lower stretches, many of the boulders at the melting ice edge are granitic blocks, some of them as big as a house. All signs point to the conclusion that the ice is the scraper that plucks, grinds, and tears rock particles from the walls and floor of the valley and drags them downslope.

Flowing ice does just as much erosion and transportation work as running water does, and even more efficiently. The capacity of ice to erode is great because it engulfs jointed, cracked blocks in its solid mass and breaks them against the rock pavement below, grinding both pavement and blocks into small grains, much of it into a fine pulverized **rock flour.** The scratches, or **striations,** that are the relic of that action (Fig. 9-17) indicate the direction the ice moved at that point. Small hills of bedrock are smoothed on the upcurrent side and plucked to a rough shape on the downcurrent side. They, too, give us an indication of the direction of ice movement. By mapping striations and other directional patterns over wide areas, we can reconstruct the flow patterns of former glaciers.

As a transporter of debris, ice is most effective because once the material is picked up by the ice, it does not settle out like the load carried by a river. Thus ice can carry huge blocks that no other transporting agent can budge. The carrying capacity of ice is tremendous: some ice is loaded with so much rock material that it looks more like a sediment cemented together with a little ice. Just how far the material is carried depends on the length of the glacier.

Ice erodes and transports material from the sides of its valley with a special efficiency. Not only does it scrape the sides below the ice level, but the ice acts as a conveyor belt for any debris that happens to fall or slide onto the surface of the glacier from higher up on the valley walls. **Lateral moraines,** strips of dirt and rock that flow with the ice along the sides of valley glaciers, are remarkable tracers of the flow of individual tributaries (Fig. 9-18). The lateral moraines of joining ice streams merge and form a single **medial moraine** in the middle of larger flows. The ice arrives at the end of the glacier carrying its load, some of it as clearly marked segregations of lateral and medial moraines, much of it in the basal layers, and the rest dispersed throughout the bulk of the ice.

Sculpture of Topography

Much of the kind of rugged mountainous terrain that has been celebrated for its scenic beauty is the product of glacial erosion (Fig. 9-19). The

Figure 9-18
Forks of Susitna glacier, Alaska, showing how lateral moraines (dark bands), formed at the valley walls of the tributaries, coalesce to become medial moraines in the merged stream of ice. Irregular distortion and folding of bands reflects differential deformation of ice as it flows. [Photo by B. Washburn.]

Figure 9-19
Alpine sculpture of mountainous terrain by glaciers on the approaches to Mt. McKinley, looking up Ruth glacier. This region was completely glaciated in the Pleistocene; since that time the shrinkage of the ice has revealed the sharp, knifelike divides between adjacent valleys and the generally rugged topography. [Photo by B. Washburn.]

Sierra Nevada, the Cascades, the Rockies, and the Alps all have high valleys that were filled with glaciers just a little more than ten thousand years ago, as evidenced by the signs we have learned to recognize: striations, large blocks carried from high in the mountains, moraines, and cirques. In the mountains of northern California, Oregon, and Washington, there are now only small relic glaciers left in the few places that receive enough snow and stay cold enough to preserve ice all year round. The rest of the terrain is exposed, and we can see the effects of Pleistocene glaciation. Farther north, in Canada and Alaska, glaciers become more numerous, but some of the ice-sculptured topography is still exposed.

As glaciers ate away at the mountains, the topography acquired a distinctive appearance. Cirques were formed at the heads of glaciers by ice plucking and undercutting the highest peaks and ridges. Adjacent cirques gradually met to form knife-sharp divides. The result is a jagged, serrated, linear crest. The valleys were excavated to a characteristic **U-shape,** with steep walls and a flat floor, as the glacier scraped and rounded off irregularities (Fig. 9–20). As the ice retreated, hanging valleys of glacial tributaries were exposed. Lakes formed in depressions in cirques. Most spectacular are the **fiords,** arms of the sea that occupy U-shaped valleys that were cut below sea level by valley glaciers descending from coastal mountains.

Depositional Landforms

If glacial erosion is accountable for rugged alpine topography, deposition of glacial debris is responsible for some of the most tried and true country scenes that decorate calendars: rolling hills, small lakes, and fertile fields. Plains in the Dakotas, rolling hills in Iowa and Wisconsin, Finger Lakes in New York, rocky fields in New England are all depositional products of a large continental glacier of a past so recent that its depositional landforms have not been obliterated by erosion. The many different elements that compose such varied landscapes, both in North America and in Europe, led Louis Agassiz and others of the nineteenth century to the hypothesis, fully accepted as theory for the past century, that these parts of the Earth had recently gone through a great ice age (Fig. 9–21).

Glacial deposits differ greatly from river, wind, or ocean deposits in many of their characteristics. In regions formerly covered by ice, the general topography tends to lack organized drainage net-

Figure 9–20
The U-shape of Yosemite Valley is typical of glacially eroded terrain. This view is from Glacier Point Trail above Union Point. [Photo by F. E. Matthes, U.S. Geological Survey.]

works, as though rivers had not had a chance to shape divides and valleys into a systematic, branching tributary pattern. Instead there are hollows with no outlets, some of which are filled with lakes. In some places there are isolated hills, in others long and sinuously winding ridges, none of which seem to have much relation to the underlying bedrock.

The bedrock is rarely seen in some of these regions. It is deeply buried, sometimes 50 meters (160 feet) or more. The ground over the bedrock is mostly a heterogeneous mixture of sand, clay, pebbles, and boulders rather than ordinary soil. Many of the pebbles and boulders are of entirely different composition than the bedrock. Scarce fossil remains of plants, pollen, land snails, and, rarely, the bones of extinct, large, arctic-dwelling elephants, the woolly mammoths, give evidence of life on or near the ice, overridden by the glacier.

There was no way for the geologists of Agassiz's time to relate all of these characteristics to erosion, transportation, and deposition by water or wind. It was this and the striking similarity of some landforms to the moraines and other products of modern valley glaciers that led to Agassiz's conclusion: glacial ice was responsible, but the areal extent of the various landforms is so great that the ice that produced them must have been a

Figure 9-21
Generalized geographic map of North America showing the maximum extent of glaciation
in Pleistocene time. Arrows indicate direction of ice movement. [After U.S. Geological Survey.]

continental glacier like those of Greenland and Antarctica. Once the idea of a former continental glaciation gained acceptance, the many geologists who lived and worked in the northern provinces of Europe and North America began to realize that many of the landforms around them fitted the new concept of continental glaciation.

Glacial moraines proved to be important pieces of the puzzle. The hilly ridges composed of heterogeneous rock fragments, sand, and clay were recognized as **terminal moraines,** similar to those of valley glaciers, marking the farthest advance of the ice. Long before there was a glacial theory, the hetereogeneous materials that make up moraines were thought to have "drifted in" somehow from other areas. The name applied then is still in use: **drift** is the name for *all* material of glacial origin found anywhere on land or at sea. The drift of moraines, heterogeneous and unstratified, was inferred to be the direct deposit of melting ice and was given the name **till,** a Scottish word for much of the rocky soil of Scotland. Mapping an ice front then becomes a matter of finding the limit of till, usually marked by a hilly ridge. Because a terminal moraine consists of till built up at the melting

front of the ice, its height is a rough measure of how long the ice front remained in a certain position, the rate of ablation exactly balancing the rate of advance.

Terminal moraines may be deposited in the relatively shallow waters of continental shelves wherever ice pushes out from land at a level below that of the sea, as does the Antarctic ice sheet today. Cape Cod and Long Island are two of the best known of these moraines. Built up many meters from the sea floor by debris dropped from melting ice, such moraines are left as islands or capes when the ice melts. The areas behind some of these moraines are built up less by glacial deposition than the terminal moraine, so that they become flooded by the sea. Long Island Sound and Cape Cod Bay are two such areas.

In front of terminal moraines, no till is found, but drift of another kind may be—stratified deposits of the same kinds of materials as the till of the moraine, but well sorted by size into distinct layers of gravel, sand, or finer materials. The gravels are rounded and the sands are cross-bedded. This stratified material is glacial **outwash,** the drift caught up and modified, sorted, and dis-

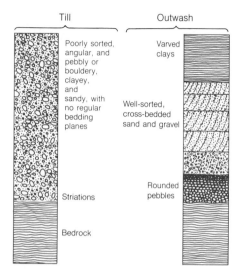

Figure 9-22
Comparison of generalized sections of glacial till and outwash. Till is a mixed, poorly sorted accumulation of material of all sizes left behind by melting ice. Outwash is a sorted, stratified deposit of water-transported drift.

tributed by streams of meltwater running away from the glacier (Fig. 9-22). Some of the outwash forms small hills, **kames,** where sand and gravel accumulated by meltwater streams were dumped near or at the edge of the retreating ice. Other accumulations show the structure and shape of deltas built into lakes. These deltas, called **delta kames,** are left behind as flat-topped hills that are frequently exploited as commercial sand and gravel pits.

The gravelly kames grade into the silts and clays that are characteristic of lake-bottom sediments. The silts and clays typically show the alternating coarse and fine layers that we call **varves** (Fig. 9-23). Putting all of this together, we infer that a stream of glacial meltwater once discharged into a lake near the ice front, building a delta of sand and gravel. The varved sediments of the lake are the finer materials that were carried in suspension beyond the delta. In summer, only the coarser silt was laid down on the lake floor. In winter, when the lake froze over, the water became still, and even the finest clay settled out. Some glacial lakes became very large. When the till dams that created them were breached, they may have drained rapidly, creating huge floods. The great dry channels of the Scablands of east-

Figure 9-23
Varved clay from Ogden, Utah. The varves are layers that form couplets: one part of each couplet is a relatively thick, coarse-grained, light-gray silty clay; it grades upward to the other part, a thin, fine-grained, dark-gray clay. Formed in lakes near the ice edge that freeze over in the winter, the light layers form in the summer and the dark ones in the winter, each pair marking one year. In the early part of this century, Baron G. DeGeer of Sweden correlated the annual banding in late-glacial lakes in northern Europe and determined that about 8700 years had passed since the glacial ice had retreated from all of southern Europe. This estimate was later confirmed by radioactive age methods. [Photo by R. M. Leggett, U.S. Geological Survey.]

ern Washington are relics of a great flood caused by the draining of glacial Lake Missoula. From giant current ripples and other channel characteristics, geologists have estimated flow velocities as high as 30 meters per second and a discharge of 21 million cubic meters per second.*

What of the hollows, the undrained depressions, or **kettles,** found in so many areas of both till and outwash? Many kettles in outwash are steep sided and occupied by lakes or ponds. Again the clue came from observing the ablation of modern glaciers, many of which melt back unevenly, sometimes leaving behind immense, isolated blocks of ice, around which course streams of meltwater carrying glacial sediment. Imagine what would happen if a block of ice about a kilometer in diameter were left behind by a retreating glacier (Fig. 9-24). A mass of ice that big would take a long time to melt—thirty years or more—and it might become partly buried by outwash sand and gravel. By the time it could melt completely, the main ice margin might have retreated so far that little outwash would reach the hole left by the ice block. If the hole were deep enough that its bottom lay below the groundwater table, a lake would form.

The territory behind a terminal moraine has its own kettles and is rich in other forms created by ice movement and by meltwater. **Ground moraine** is a mantle of till deposited underneath the ice. It may be so thick that it completely buries bedrock everywhere, or it may be a relatively thin cover that exposes bedrock knobs and pavements in many places. **Eskers** are long, narrow ridges of sand and gravel in the middle of ground moraine (Fig. 9-25). Varying in height and width from a few meters to tens of meters, they wind for kilometers roughly parallel to the direction the ice moved. The well-sorted materials of eskers look like water-transported outwash, and the eskers themselves have the shape of stream channels; some even show "tributary" eskers. The deduction made from this geologic evidence is that eskers form from meltwater streams flowing in tunnels along the bottom of a melting glacier. The

---

*The theory of the origin of the Channeled Scabland by a giant flood that drained a glacial lake was first enunciated by J Harlen Bretz in the early 1920s. It was considered an outrageous and heretic hypothesis, invoking catastrophe and flood, contrary to the anti-catastrophic and narrowly uniformitarian views that then predominated in geology. Four controversial decades later, a field party of an international congress reassessed the evidence in the field and came to agree completely with Bretz. Bretz lived long enough, into his 90s, to see himself vindicated by the Geological Society of America, which awarded him its greatest honor, the Penrose Medal for excellence in geological research.

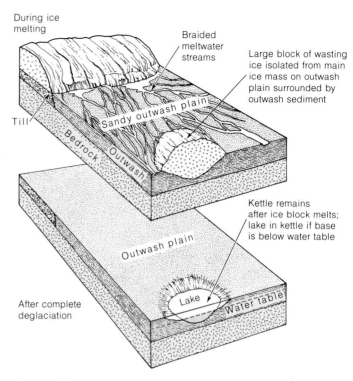

Figure 9-24
Evolution of an outwash kettle. As a glacier retreats, it may leave behind large blocks of wasting ice that are gradually buried by outwash from the receding ice front. After the front has retreated far enough from the region, outwash sedimentation stops, the ice block melts, and a depression remains, filled with water if it is deep enough to intersect the groundwater table.

water seeps through crevasses and cracks in the ice and gradually opens tunnels that run downhill to the end of the glacier. Eskers, like kames, are favorite sources of sand and gravel for road construction and other uses.

Drumlins are streamlined hills of till and bedrock that parallel the direction of ice movement (Fig. 9-26). The shape of a drumlin is like that of a long, inverted spoon with the gentle slope in the downslope direction. They may be 25 to 50 meters (75 to 150 feet) high and a kilometer long. Some drumlins are all till; others are largely bedrock with a cover of till. They may be formed either by ice erosion of an earlier till accumulation or by a subglacial shaping of accumulated till. The nucleus of a drumlin, either a dense till or bedrock knob is thought to have been streamlined by a buildup of till deposited by the ice as a "tail" in the lee of the nucleus. Knobs, drumlins, and irregular elongate hills of ground moraine are some of the different shapes taken by materials of different strengths and resistances to erosion when overridden by an ice sheet.

Figure 9-25
Retreat of the ice margin has revealed an esker formed by a stream that ran under the glacier. The esker is the winding ridge of sand and gravel running from upper right to lower left. The grooves at right are in soft glacial sediment and are not bedrock striations. [Photo by B. Washburn.]

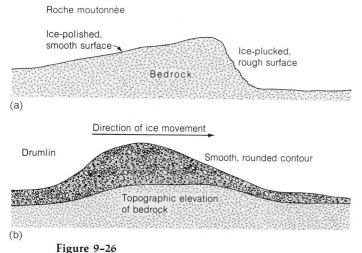

Figure 9-26
Two similarly shaped glacial landforms. (a) Roche moutonnée, a polished rock knob, is formed as the ice rides over a bedrock protuberance, smoothing the side facing the direction from which the ice moved and roughening the downflow side by plucking cracked and jointed rock. (b) Drumlins are composed of till; some have a bedrock base with a streamlined shape formed by ice movement. The steep side commonly faces the direction from which the ice came, but the reverse is also found.

Identification of a glacial landform is a matter of perceiving its shape, which is best done from a hilltop or a good topographic map, and matching the form with the nature of the material making it up. The contrast between the heterogeneous mixture of sizes and shapes of till and the rounded, evenly sorted gravel or sand of waterlaid, stratified drift is a good first guide. Just the presence of many lakes, swamps, and other undrained depressions is sufficient evidence to put one on the watch for other characteristic glacial landforms.

Glacial Sediment Types

The lithology of a glacial sedimentary deposit can be used to infer the particular kind of glacial environment in which it formed. Unstratified, unsorted mixtures of sand, gravel, and clay are correlated with till deposited below actively moving ice. Stratified sand and gravel deposits may be found in association with tills and are evidence of deposition in flowing water on, within, beneath, or in front of the ice. Alternating laminations of silt and clay, varves, are evidence of glacial lake deposition. Bedded marine sediment with occasional ice-rafted pebbles are formed on the sea bottom near glaciers. Detailed studies of these kinds of sediments and their lithified equivalents in ancient rocks have been achieving new importance as many geologists use advances in glacial geology to hunt for clues of former glaciations older than the Pleistocene.

Permafrost

In regions where the summer temperature never gets high enough to melt more than the suface, the ground remains perennially frozen below the surface. **Permafrost** today covers as much as 25 percent of the total land area of the Earth. In Alaska and northern Canada it may be as much as 300 to 500 meters thick. Permafrost includes various proportions of ice and soil and contains aggregates of ice crystals in layers, wedges, and irregular masses. Permafrost is a difficult material to handle in engineering projects, such as roads, foundations, and the Alaska pipeline, because when the surface melts, the water cannot infiltrate the still-frozen ground below and the wet soil may creep, slide, and slump. When large masses of ice grow near the permafrost surface, they form **pingos,** isolated steep-sided hills of ice covered by a thin mantle of silt, sand, or peat. Circular or oval in map view and up to 50 meters in height, they grow in shallow ponds left by rapid drainage of Arctic lakes. Permafrost several hundreds of meters thick occurs as much as 100 to 200 kilometers off the shores of the Arctic coast in places and presents difficult engineering problems for offshore oil exploration.

THE PLEISTOCENE EPOCH AND INDIRECT EFFECTS OF GLACIATION

From an intensive study of landforms and sediment lithologies, geologists deduced that a great continental glaciation had effectively extended both polar ice caps far into what are now temperate climates. They also reasoned that the glaciation was recent, for everywhere the drift is soft, like freshly deposited sediment. It also lies at the surface everywhere and so was one of the last sediments to be deposited, covered only here and there by recent river and other deposits. The matching of varved clays from glacial lakes confirmed a young age: the most recent retreat of ice had taken place between 8000 and 15,000 years

ago. Today the decay of radioactive carbon-14 can be applied to the dating of materials younger than about 60,000 years, and the younger glacial deposits of the Pleistocene Epoch, the ice ages, are open to accurate dating. Deposits of the early Pleistocene, too old for ^{14}C dating, are dated in the same way as the more ancient rocks—that is, on the basis of the fossils they contain. Land deposits have been dated by snails, and marine sediments have been dated by foraminifera, small single-celled organisms that secrete shells of calcite.

In the course of mapping the glacial deposits, geologists found that there were actually several different layers of drift. Furthermore, the older layers contained well-developed soils, including fragments of plants that botanists said were characteristic of warm climates. As the evidence grew, a new conclusion became inescapable: during the ice age, the great continental glaciers alternately advanced and retreated. Between glaciations, there were long interglacial periods during which warm climates returned to a large part of the formerly glaciated lands. At least four major periods of advance and retreat could be deduced on the basis of landforms and stratigraphy of moraines. These were named Wisconsinan, Illinoian, Kansan, and Nebraskan in order of increasing age (Fig. 9-27). Some of the interglacial periods lasted longer than the glacial ones.

More recently, we have learned from the study of deep-sea deposits to measure ancient ocean temperatures. Fossils of Pleistocene foraminifera bear testimony to the coldness of the water by the special ratio of the isotopes of oxygen in the calcium carbonate of their shells. The proportion of ordinary oxygen, ^{16}O, and the heavy oxygen isotope, ^{18}O, depends on the temperature of the water, and the shells from different layers of sediment show a warming and cooling of the sea that correlates with expansion and contraction of glaciers on land. We now know that there have been many shorter, more regular cycles of glaciation and deglaciation. This has been confirmed by ever more accurate radioactive dating of glacial sediments. Moreover, volcanics that are found in some regions between older glacial layers on land have been dated by the potassium-argon method. Even on now-tropical Hawaii, Pleistocene glacial drift, deposited by small ice caps on the tops of volcanoes, has been correlated with worldwide glacial periods in this way.

Some of the indirect effects of glaciation were explored as this picture emerged. The origin of the Pleistocene loess deposits, the thick layers of wind-blown dust, was traced to extensive flood-plain and outwash plain deposits of fine muds

Figure 9-27
Glacial geology of the contiguous United States. [Compiled by C. S. Denny, U.S. Geological Survey.]

that came from rock flour ground up by the glacier (see Chapter 8 and Fig. 8-27). Many river valleys that carried the meltwaters of receding glaciers became choked and filled with that sediment. As the exploration of the seas started, oceanographers found that in the northern and southern parts of the oceans, much of the Pleistocene sediment was like stratified drift, the product of melting icebergs.

Sea-Level Changes

We know from the enormous volume of ice that occupied land during the Pleistocene and from the way the hydrologic cycle works that the sea is the likely source of the ice. Most of the water that precipitated as snow evaporated from the ocean. The more water that was tied up in the ice, the

less the ocean could be replenished by river runoff to keep its steady level. As the ice accumulated, the level of the sea lowered everywhere in the world's oceans.

The total volume of ice on Earth today is a little more than 25 million cubic kilometers. Total ice volume during the height of the ice age can be estimated from the area covered by the ice sheets combined with calculations of the thickness of ice necessary to keep the glaciers moving so many hundreds of miles from their areas of accumulation. Startling as it may seem, the volume must have been about 70 million cubic kilometers. The extra 55 million cubic kilometers that came from the sea to make the additional ice lowered sea level by about 130 meters (425 feet)! Imagine the series of changes caused by that great a reduction in sea level (Fig. 9-28). Much of the shallow sea areas along shorelines were left above the sea

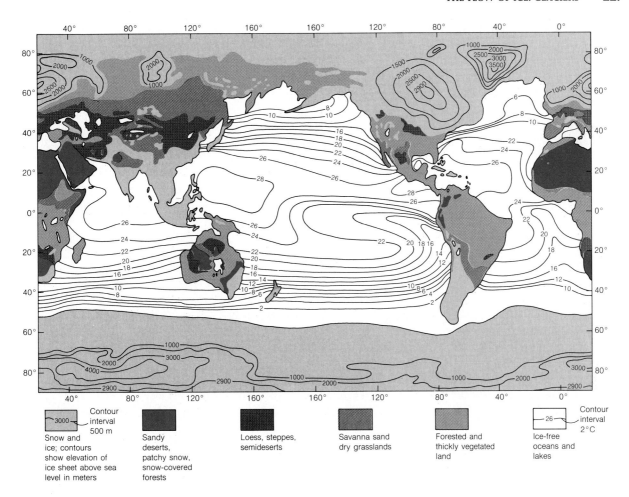

Figure 9–28
Sea-surface temperatures (°C), ice extent, and ice elevation for August, 18,000 years ago. Continental outlines represent sea-level lowering of 85 m. [Modified from CLIMAP, "The Surface of the Ice-Age Earth," *Science*, v. 191, 1976.]

level. In glaciated regions, the ice sheets extended far beyond present shorelines. Some of the newly emerged coastal plains in nonglaciated areas were more than 100 kilometers (62 miles) wide. Rivers extended new channels across them and created new deltas at the water's edge. Wide strips of beaches were formed. Mammoths, huge Pleistocene elephants, roamed those low coastal plains, and no doubt Pleistocene humans did too, for early prehistoric cultures were evolving in the land beyond the ice sheets. The relics of all these happenings have been found by oceanographic research vessels exploring the shallow areas that were reflooded by the sea as the ice melted.

The figures for ice and sea-level changes can be looked at in another way. What if the 25 million cubic kilometers of water now tied up as ice were to melt? The change in sea level would be catastrophic, for the melting of all existing ice would raise the oceans by about 65 meters (210 feet). This would inundate many major cities of the world, such as London, New York, and Tokyo, and the great bulk of the low-lying areas of the continents, where much of the world's population lives. Those who would tinker with the Earth's climate in ways that might inadvertently cause the ice to melt are well advised to remember this simple calculation. We shall see some of its implications in Chapter 12.

But what if it should happen naturally? What if we are really just living in an interglacial period like those during which the Pleistocene ice sheets withdrew? On the other hand, the 25 million cubic kilometers of ice on Earth today is probably much more than existed during most times in the history of the planet so the Earth could be considered to be in a glacial age. It has been only 10,000 years since the last retreat of huge continental glaciers, and we have no reason to suppose that the ice will not advance again. It appears that there are some very real and possibly practical reasons, in addition to our natural scientific curiosity, why we ought to think about the causes of ice ages.

Causes of Ice Ages

For many years geologists have speculated, calculated, and theorized, but no complete agreement yet exists as to the causes of ice ages. The theories themselves have two different parts. One is an explanation of a gradual cooling of the Earth's atmosphere over tens of millions of years that finally led to the glacial epoch. There is evidence that sizable polar ice caps began forming long ago in the Cenozoic and grew throughout the last part of that era. The other part of the theories is an accounting for the oscillations during the Pleistocene—the growth and advance of ice during glaciations, and the shrinkage and retreat during the interglacial periods. The Earth retained its polar ice caps, just as it does now, even during those interglacials. We will cover just a few of the major ideas to give some notion of the kind of thinking that has gone on.

Most theories have concentrated on reasons why the Earth would cool, causing the onset of glaciation. Perhaps most naturally, some have sought an explanation in a possible variation in the heat radiated from the Sun. It has been suggested that there may be periodic changes in the rate at which the Sun burns its nuclear fuel, but no firm evidence for this has been found. It is known that there are short-term fluctuations as measured by sunspot and solar-flare activity. How these fluctuations relate to weather and climatic variation is not at all clear, however. According to another theory, great quantities of volcanic ash in the atmosphere might block a portion of the Sun's rays and so cool the Earth. Though some cooling could be expected from this mechanism, no one has produced strong evidence of a period of sustained volcanic activity in the late Cenozoic that was significantly greater than at other times. Recently an inventory of volcanic ash layers in many cores obtained by the Deep Sea Drilling Program showed no pronounced change in volcanic activity throughout the Cenozoic.

A more attractive theory for general cooling relates to the positions of the continents relative to each other and to the poles. Today there are large land areas in the broad, cold regions surrounding both north and south poles. At certain times in the geologic past, one or the other pole was in the middle of the oceans, where currents and general movement of ocean waters allowed enough exchange of heat with the atmosphere to distribute temperature more uniformly from poles to equator. Once the land areas were near or at the poles, the heat difference between poles and equator was increased, and the polar regions became very cold. If this theory is correct, the events leading to ice ages must have been set in motion at least 20 million years ago, for that was when the continents approached their present positions relative to the poles. Indeed, there is now direct evidence of high-latitude glaciers that started in the Miocene, well before the main, or Pleistocene, ice ages. The marine sediment record indicates that glacial ice began to form in Antarctica as early as the late Eocene. But the main glacial ages had to wait until relative positions of the continents interacted with ocean currents in such a way as to inhibit further the transfer of heat across the globe. How this happened is a problem of joint concern to historical geologists, glaciologists, physical oceanographers, and meteorologists.

The alternation between glacial and interglacial epochs has been explained as the result of astronomical cycles. The shape of the Earth's orbit around the Sun changes periodically, putting us sometimes slightly closer and sometimes slightly farther from the Sun. In addition, the Earth wobbles slightly on its axis of rotation, and this wobble also affects very slightly the heat received by the Earth at different latitudes. These motions, first worked out carefully by Milutin Milankovitch, a Yugoslavian geophysicist, seem to fit with the durations of the glacial and interglacial periods. Milankovitch worked out these ideas in the 1920s and 1930s, but it was not until the past decade that there was a sufficiently long and detailed chronology of the Pleistocene to match with his predictions. That chronology and verification of the theory, announced in late 1976, has been the work of an international project, partly sponsored by the National Science Foundation, called CLIMAP (Climate: Long-range Investigation, Mapping and Prediction). The astronomical theory of oscillations has to be superimposed on a theory of Earth cooling in order to arrive at a reasonable explanation as to why these oscillations took place during the Pleistocene.

The Pleistocene is not the only epoch for which there is a record of glaciation. Ancient glaciations are recorded by rocks that are the lithified equivalents of glacial deposits, especially of till, **tillites.** Such rocks are recognized by their poorly sorted mixture of sand, pebbles, and clay and the association with striated bedrock pavements. The

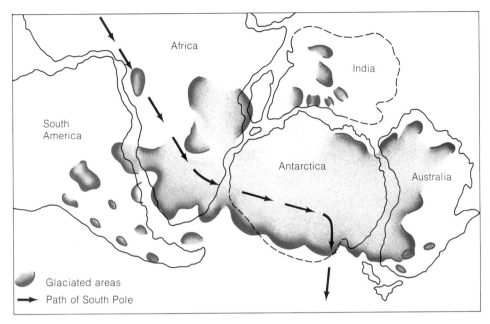

Figure 9–29
Late Paleozoic glaciated areas of Gondwanaland, the southern continent that formed
in the late Paleozoic Era. These continental glaciers, inferred from glacial deposits
and striated bedrock, waxed and waned irregularly for about 90 million years from
Late Mississippian to Middle Permian times, not all areas flourishing at the same
time. The path of the South Pole is reconstructed from paleomagnetic data of
M. W. McElhinny. [After John C. Crowell, "Gondwanan Glaciation, Cyclothems,
Continental Positioning, and Climate Change," *Amer. Jour. Sci.,* v. 278, p. 1346, 1978.]

Dwyka, a late Paleozoic tillite of South Africa,
preserves a record of a glacial epoch affecting the
southern continents, at that time coming together
to form the supercontinent of Pangaea (Fig. 9–29).
Tillites record Ordovician and Precambrian glaciations as well.

Why glacial periods recur every so often in the
history of the Earth remains uncertain. Nor is
there any certainty about the effects that future
glaciations may have on our existence on Earth. If
there is another glacial episode, it will probably
happen on a time scale of thousands of years.
Some think it may happen more rapidly, over a
few hundreds of years but still slowly enough for
all of us to ignore. No matter what the speed, renewed glaciation has implications of yet another

change for the continuous record of human culture. Primitive society was well advanced when
early humans were living in caves in southern
France during the last ice age. Our first sensitive
indicators of an impending change would be the
advance of mountain glaciers in Alaska and elsewhere. But we would have no way of knowing
whether such an advance indicated any more
than a temporary period of slight cooling that
might last only a few years or tens of years and
then be followed by another warm period. This
uncertainty will be countered by increasingly
accurate modeling of climate change, taking into
account all of the effects on atmospheric temperature; from better models we may be better able to
predict a future glaciation.

SUMMARY

1. Glaciers are masses of ice formed by the gradual transformation of snow by recrystallization and pressure.

2. The size of a glacier is determined by the rate at which snow accumulates and the rate
of ablation by melting, evaporation, and iceberg formation.

3. Ice moves downhill partly by internal plastic flow, which is mainly the result of many
small slips along planes of individual ice crystals, and partly by sliding along its base.

4. Surface ice breaks into crevasses as a result of the movement of ice masses over steep
changes in slope and along valley walls. Glaciers move with a wide range of speeds;
some glaciers move in fast surges.

5. Valley glaciers form in alpine mountains and erode U-shaped valleys, hanging valleys, and cirques.

6. Continental glaciers are of great size and exist today only in polar regions, where melting is almost nonexistent, as in Greenland and Antarctica.

7. Glaciers produce many erosional and depositional landforms. Ice abrades material from the walls and floors of valleys and transports it to ice fronts, where it accumulates as moraines.

8. Landforms of Pleistocene continental glaciers include moraines and various forms of till and drift, such as kettles, kames, varves, eskers, and drumlins.

9. There were several glaciations during the Pleistocene, each recorded by a till of a different age and by temperature fluctuations of the oceans. Sea level changed extensively as the ice sheets waxed and waned, at times leaving large areas of the continental shelves of the world exposed as land.

10. Explanations for periods of glaciation have been sought in a general cooling that resulted from movements of continents with respect to the poles. Oscillations during colder periods are explained in terms of astronomical cycles.

EXERCISES

1. Slight differences in color reveal the bedding of glacial ice. What may those differences be due to?

2. Mt. Washington is the highest peak of the White Mountains of New Hampshire, at 1920 meters (6293 feet), and official weather records show winter temperatures there as low as 60°C below zero. Discuss the reasons why it has no glacier and what changes in the current conditions might produce one.

3. One of the dangers of exploring glaciers is the possibility of falling into a crevasse. What topographic features of a valley glacier or its surrounding rocky terrain would you use to indicate that you are on a part of the glacier that might be badly crevassed?

4. What differences in the composition and texture of the glacial till would you expect between two glaciated areas, one on a geologically young series of coastal plain soft shales and loosely cemented sands, the other on an old granitic-metamorphic rock terrain.

5. A number of ancient rock formations have been the subject of controversy concerning their possible glacial origin. Would you expect a glacial till or glacial landforms to be of more use in establishing that a rock 270 million years old is a tillite?

6. You are assigned the job of mapping a terminal moraine of a late Pleistocene glaciation. What characteristics of topography or till composition would you use to distinguish it from ground moraine on the one hand and from outwash on the other? How might you distinguish it from an esker?

7. If you were in charge of a global watch to monitor any indications of a resumption of extensive continental glaciations like those of 15,000 years ago, what kinds of observations would you direct your teams to make? What recommendations would you give to Montreal, Quebec, on the one hand, and to Miami, Florida, on the other, if you detected signs of either a decrease or an increase in the extent of glaciation?

8. It has been estimated that a rise in sea level of 1 meter would result from the melting of 400,000 cubic kilometers of glacial ice. If there were such a melting as a result of slight warming, where would you expect the greatest change in glaciers of the world?

BIBLIOGRAPHY

Field, W. O., "Glaciers," *Scientific American*, September 1955. (Offprint 809.)

Flint, R. F., *Glacial and Quaternary Geology*. New York: John Wiley & Sons, 1971.

Imbrie, J., and K. P. Imbrie, *Ice Ages: Solving the Mystery*. Short Hills, N.J.: Enslow Press, 1979.

Paterson, W. J. B., *The Physics of Glaciers*, 2nd ed. London: Pergamon Press, 1981.

Price, R. J., *Glacial and Fluvioglacial Landforms*. New York: Hafner, 1973.

Robin, G., "Ice of the Antarctic," *Scientific American*, September 1962. (Offprint 861.)

10

Ocean Processes

The sea is the end of the line for the weathered debris of continental erosion. The material brought to the ocean by rivers and worn away from its coasts by wave erosion is distributed along its shores as beaches or transported by currents to the continental shelf and deeper parts of the ocean. Beaches are the expression of a balance between the supply of sand provided by erosion of the shore and the transport of sand by rivers, wave erosion, and longshore currents. The continental slopes, continental rises, and abyssal plains are shaped by turbidity currents that transport much material from the shallower to the deeper parts of the sea. The current patterns of the general circulation of the oceans influence the distribution of fine-grained pelagic sediment that mantles the topography of the deep oceans.

In 1872, H.M.S. *Challenger*, a small wooden warship converted and fitted out for scientific study of the seas, set out from England on a four-year voyage over the world's oceans (Fig. 10–1). From that first systematic study eventually flowed fifty thick volumes on the currents, the underwater topography, the bottom sediments, and the life of the deep sea. The wealth of new information opened the eyes of land-bound geologists. What had been the realm of groundless speculation was suddenly transformed into a matter of serious interest: the observation of ocean phenomena and the building of theories to explain them. Early musings on "featureless plains" beneath the water were abandoned as the discoveries rolled in: the mid-Atlantic ridge and other high ranges of submarine mountains, deep troughs and trenches, vast areas of hilly topography, and, to be sure, some great flat plains. The rocks and sediments of the sea floor were found to be as diverse as the topography: fine-grained brownish clays, blue clays, sands, silts, and, in some places, even pebbles and cobbles. In many parts of the deep sea, the *Challenger* scientists found "oozes," sandy or

H.M.S. Challenger.

Figure 10–1
H.M.S. *Challenger*, as depicted in the official *Challenger* report. [Courtesy Science and Technology Division, The New York Public Library, Astor, Lenox and Tilden Foundations.]

Figure 10-2
A modern oceanographic research vessel, the R/V *Knorr* of the Woods Hole Oceanographic Institution. This ship is equipped for a wide variety of marine geological, geochemical, and geophysical research. [Courtesy Woods Hole Oceanographic Institution.]

muddy sediments dominated by the shells of tiny organisms, some of calcium carbonate, some of silica. Perhaps as surprising was the discovery of igneous rocks and unusual cobble-shaped lumps of heavy rock: manganese nodules.

The broad outlines of the problems of marine geology were laid out by the voyage of the *Challenger*. What was the origin of the varied topography? How could the different arrays of sediment and other rock types on the bottom be explained? What role did ocean currents, tides, and waves play in forming the sea floor? Today, a hundred years after these questions first came to be asked, hundreds of oceanographic research vessels from different countries are exploring the sea, providing some answers to those questions and discovering new facts that raise new ones (Fig. 10-2). More than any other endeavor, it was oceanographic research that has led to our current ideas of the dynamics of the Earth. Plate tectonics, the conceptual framework of this book, could not have been substantiated without oceanographic expeditions. Some 80 percent of the Earth's surface is covered by water, so that the 20 percent available for direct observation is in a real sense atypical. A hundred years ago, the study of the sea was primarily either pure science (science for its own sake) or applied science in the service of the shipping trade. In the decades after World War II, much oceanographic research has been supported financially by nations interested in the military advantage that might accrue to them from a knowledge of the sea, which was soon to become the home of a submarine capable of launching nuclear missiles. Now oceanographic research is made more urgent, both by the possibility that the resources of the sea can be used to support the world's growing population and by the companion danger of pollution of the seas. All the nations of the world, landlocked as well as maritime, are well aware of the importance of the oceans—the United Nations Conference on the Law of the Sea has been in session since 1973. Unfortunately, extension of territorial boundaries to great offshore distances by many nations has hindered research in places.

The study of the oceans—**oceanography**—has united geology, geophysics, and geochemistry with biology, physics, and meteorology to become a field of knowledge much too broad to cover in this book. In this chapter, we shall discuss those aspects of oceanography that continue the story of the erosional, transportational, and sedimentational processes that shape the surface of the Earth's crust beneath the sea as well as on land. We start with the most familiar part, the edge of the sea.

THE EDGE OF THE SEA

The coast, the broad region that is the meeting place of land and sea, can be carved into many kinds of shapes: steep rocky cliffs, broad low beaches, crescents of small beaches in small bays alternating with rocky headlands, or wide sweeping, sandy tidal flats (Fig. 10-3). The forces that shape coasts are essentially the same as those that form other topographies: the destructive processes of erosion operating in conjunction with currents that transport and deposit debris and the tectonic forces that cause uplift or subsidence of the Earth's crust.

Wave Erosion

At the **shoreline,** the line where the water meets the land, the main erosive agent is wave action. The force of high waves breaking against a rocky cliff during a storm can be awesome in its violence; the ground shakes noticeably with the shock (Fig. 10-4). Nearby sensitive seismometers, used to detect far-distant earthquakes, register the small tremors caused by the sudden impact of tons of water on solid rock. Moderately large waves, about 3 meters (10 feet) high at the breaking point, have been measured to produce a force of more than 70 grams per square centimeter (30

Figure 10-3
A rocky headland on Mt. Desert Island off the coast of Maine. To the extreme right is a small sandy pocket beach formed in a recess between headlands. [Photo by R. Siever.]

Figure 10-4
Waves splashing against a rocky shore. Big Sur, Monterey County, California. [Photo by Cole Weston.]

pounds per square inch). So when you see waves break against a concrete seawall and throw water 5 meters (16 feet) into the air, you can estimate an impact force equivalent to dashing about 175 grams (0.4 pounds) against each square centimeter of rock surface. It is not surprising that such sea-walls, built to protect homes just back of the shore, quickly start to crack and constantly have to be kept in repair.

The waves are even more destructive than the force of the water alone, for the waves catapult pebbles, cobbles, and, in intense storms, boulders against the shoreline rocks. The physical destructiveness of the waves is enhanced by the chemical weathering action of sea water that is forced by wave pressure into every tiny crack and crevice. The chemical decay extends and widens the cracks, preparing them for the disintegrating physical beating by the waves.

Where waves break against bluffs of softer unconsolidated material, such as soils or recent glacial deposits, the rate of erosion may be extraordinary (Fig. 10-5). The high sea cliffs of glacial materials facing the open Atlantic along the Cape Cod National Seashore are retreating at a rate of about a meter each year. Since Henry Thoreau first walked the length of those cliffs in the middle of the nineteenth century, about 6 square kilome-

Figure 10-5
Sandy cliffs of glacial deposits at Wellfleet, Cape Cod. Composed of easily eroded unlithified clays and sands, these cliffs retreat about a meter per year under the attack of waves. [Photo by R. Siever.]

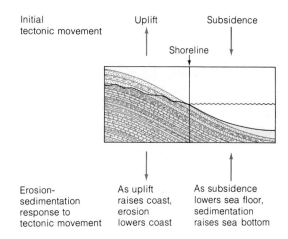

Figure 10-6
Shoreline profiles are a balance between tectonic forces and erosional and sedimentational processes that counter them. The greater the uplift of the land, the greater the erosion; the greater the subsidence of sea floor, the greater the tendency for sedimentation to build up the bottom. Similarly, subsidence of coastal lands decreases erosion, and uplift of sea floor inhibits sedimentation.

ters (2.3 square miles) of land area has been eaten by the ocean.* At that rate, real estate land (or "sea") values on Cape Cod will be worth little in a few thousand years!

The Forces That Shape Shorelines

Just as the surface of the land is the product of the interplay of tectonic forces lifting up the land, erosion wearing it down, and sedimentation filling in the low spots, a shoreline is a result of the same kinds of forces. The shore may be uplifted or it may subside with respect to sea level. If it is uplifted, erosion, principally the work of waves, will attack it by the same kind of negative-feedback process explored in Chapter 5 for landscape evolution: the more the uplift, the more intense the erosion that counters it (Fig. 10-6). If the land subsides with respect to sea level, a different negative-feedback process operates: the low spot tends to be filled in by sedimentation. Uplift and subsidence may operate simultaneously in different parts of the same region. For example, some

*Henry David Thoreau visited Cape Cod four times between 1849 and 1857. In the record of his observations that was published as *Cape Cod*, his description of the beach and other parts of the Cape remains one of the great examples of a nineteenth-century literary form that combined natural history and conversations with the observations of people. One edition is listed in the bibliography for this chapter.

coasts are broken up by tectonic forces into a pattern of large blocks bounded by faults, which allow some blocks to be raised while others are depressed.

Tectonically uplifted coasts, where erosion is active, have a topography of prominent cliffs and headlands that jut out into the sea alternating with narrow inlets and irregular bays, some filled with small rocky beaches. Jagged, irregular shapes dominate in hard, rocky terrain. In softer sediments, the topography is characterized by the gentler slopes of straight bluffs, marked with occasional landslide and slump scars. The waves undercut the cliffs, causing huge blocks to fall into the water, where they are gradually worn away. Along such coasts, the retreat of the sea cliff may leave isolated remnants standing in the sea far from the shore (Fig. 10-4). As the waves erode, they plane the surface just below sea level and create a **wave-cut terrace.** Renewed uplift may expose these terraces (Fig. 10-7).

Once the uplift stops, erosion wears away the land, moving the shoreline back from the sea while the land surface is lowered. Given a very long period of complete stability, the coastline would be reduced to a low coastal plain. This hypothetical state is not reached in nature, for the shoreline, not far from the junction of oceanic and continental crust, is normally involved in upward or downward tectonic movements, in many places in relation to convergence or divergence of lithospheric plates.

The shorelines of virtually the entire Pacific Ocean are notably unstable, as are those of the Mediterranean Sea, a condition characteristic of **active continental margins,** those related to convergent plate boundaries. Much of the eastern shore of North America has been a subsiding coast over the past 100 million years, as has much of the shore of the Gulf of Mexico. The long-term sinking of the crust is characteristic of **passive continental margins,** those that originate at the trailing edges of lithospheric plates at divergent plate boundaries.

Where the land sinks, sediment tends to build up. These coasts have long, wide beaches and wide, low-lying plains of relatively recently deposited sediment behind the shoreline. Typical shoreline landforms along such coasts are sandbars (shallow, narrow ridges of sand parallel to the beach just offshore); low-lying sandy islands; and wide tidal flats that are either sandy or muddy. There is a balance here between the subsidence that creates depressions and the sedimentation of debris eroded from the land that tends to fill up the low places. The relative intensity of the

Figure 10-7
Uplifted marine terraces, Palos Verdes Hills, California. [Photo by John S. Shelton and
R. C. Frampton. From *Geology Illustrated* by John S. Shelton. W. H. Freeman and Company.
Copyright © 1966.]

two processes determines the result. A plentiful supply of sediment from the land combined with only slight subsidence produces a shoreline that is built up and advances seaward. Small amounts of sediment combined with deep subsidence results in deep basins, bays, and other indentations of the coast.

Pleistocene Sea-Level Changes

Thus far, we have covered long-term local or regional movements of the solid crust with respect to sea level, assuming that the general level of the oceans remains constant. Studies of the Pleistocene glacial epoch have revealed that there have been short-term rises and falls of sea level, some more than 130 meters (425 feet), caused by the removal of water from the sea to form huge continental glaciers (see Chapter 9). These worldwide changes in sea level, called **eustatic** changes, affected all coasts. During periods of lowered sea level, erosion dominated in a situation equivalent to tectonic uplift of the land. As shallower parts of offshore areas were exposed, valleys were deepened, formerly submerged terrains were attacked by waves, and sediment was transported to deeper areas of the ocean basins. As sea level rose again when the ice melted, the deepened valleys were flooded far inland, and sediment built up in these shallower areas. Seismic exploration of such sedimentary deposits has allowed the mapping of these changes in sea level. Today, many of the rivers of the central Atlantic coast gradually widen and become mixed with sea water long

before they reach the main shoreline. These long fingers of sea indenting the land are called **drowned valleys** (Fig. 10–8). A drowned valley is one major type of **estuary**—any coastal body of water that is open to the ocean and is diluted by fresh water from the land. To some extent, all of the shores of the world were slightly drowned by the sea-level rise that accompanied the melting of glaciers 10,000 years ago. The shores of northern Europe and North America are now rising tectonically because the Earth's crust is still rebounding from the great weight of earlier glacial ice. This, of course, works counter to the sea-level rise in the formerly glaciated areas. Recent studies show that the sea level from Maine to Virginia has continued rising at a rate of 0.25 to 0.35 centimeters per year; the rate has been much faster than that for some short periods. This is presumably an effect of continued local tectonic subsidence.

With this additional complexity, we now have four major forces shaping shorelines: tectonics, erosion, sedimentation, and eustatic changes in sea level. How erosion and sedimentation interact is most clearly shown by beaches.

BEACHES

Beaches change constantly, and their dynamics can teach us much about the processes that operate at shorelines. The most obvious features of a beach are the materials of which it is made—sand and pebbles—and the action of the waves. Perceptive observation of these two gives a picture of the beach as a dynamic environment, not a fixed,

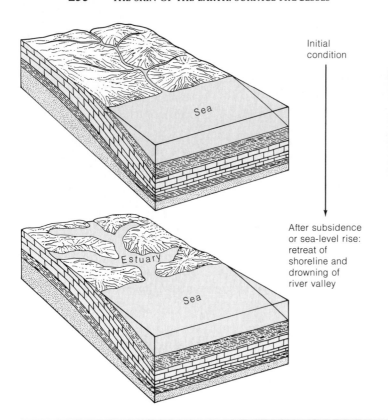

Initial condition

After subsidence or sea-level rise: retreat of shoreline and drowning of river valley

Figure 10-8
Drowned river valleys are greatly elongated estuaries formed either by general subsidence of the coastal region or by a general rise in sea level that floods the river valleys. The photograph shows such a valley, locally called "calanque," in limestone terrain on the Mediterranean coast east of Marseilles, France. [Photo by R. Siever.]

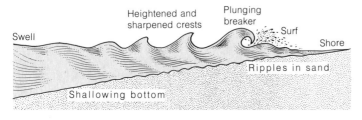

Figure 10-9
Formation of a plunging breaker as swell approaches a gently sloping beach. As low regular swell meets a shallowing bottom, wave crests become higher and steeper until the breaker plunges.

static landform: the waves, in constant motion, move the material of the beach, acting sometimes to destroy the beach and sometimes to broaden and extend it. If we stop to consider the waves, we can begin to see how they work.

Waves

The best place to start looking at waves is from the rise in elevation that is behind every beach. Best is a high bluff, but a low sand dune will do. Once the first sensation of the whole sweep of the shore starts to fade, you can notice some of the elements of regularity about the waves. Waves

appear to form some distance from the shore, build up, and break into splashing **breakers,** or **surf,** near the shore. In tropical areas where coral reefs fringe the shore, the waves break on the outer reef, leaving the main shore protected from the brunt of surf (see Chapter 11). The breakers, separated by troughs, succeed one another in an orderly way. Some distance from the shore, the waves can be distinguished only as low, broad, regular, rounded ridges, called **swell.** The swell becomes higher as it approaches the shore and assumes the familiar sharp-crested wave shape. The crest builds even more as the wave rolls closer to the shore, finally to a steep high wall of water that then breaks forward in a collapse of splashing surf (Fig. 10-9). The time interval between waves, the **period,** may be as short as a few seconds or as long as 15 or 20 seconds. A closer watch on the wave period may reveal more complex regularities, such as the alternation of a few small waves of short period with larger waves of long period. An experienced surfer learns the particular sequence of waves on a beach on any given day and knows that it changes from day to day. In some areas, newspapers include predictions of surf height and period with the daily weather forecast.

Still looking at the shore from some height, you can distinguish a regular horizontal pattern in the

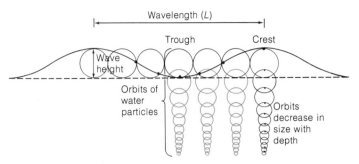

Figure 10-11
Wave forms are produced by orbital motions of water particles in deep water; each water particle continues orbiting about the same position while the wave form travels. The orbits decrease in radius with depth, becoming negligible at a depth greater than half a wavelength.

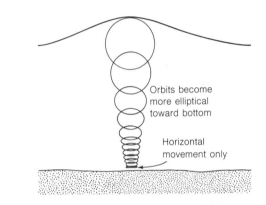

Figure 10-12
Orbits of water particles become elliptical as they approach a shallow bottom. At the bottom, particles move back and forth only.

Figure 10-10
Waves approaching a point of land are bent (refracted) around it. In the lee of the offshore rocks, the waves form an interference pattern. [Photo from U.S. Hydrographic Office Publication 234.]

crests of waves as they approach the shore. Far from shore, the lines of swell are parallel to each other but usually at some angle to the beach. As the waves approach the beach, no matter what the direction or angle, the parallel rows of waves start to bend gradually, so that they approach the shore in a direction more nearly perpendicular to the shoreline; then they finally break into surf and sweep up on shore almost, but rarely exactly, at right angles—that is, the waves themselves lie nearly parallel to the shore when they break (Fig. 10-10).

How do these observations fit together to explain wave patterns? First we must consider the movement of a piece of wood or other light material floating on the water. The piece moves a little forward on a wave crest and then a little backward as a trough between waves passes, but it shows little net movement. The wave form moves steadily toward the beach, but the water itself just moves back and forth. Experiments in large laboratory tanks in which artificial waves are generated show that small floats at different depths in the water all move with a characteristic motion: approximately circular vertical orbits (Fig. 10-11). The orbits have large radii at the top of the water, but they gradually decrease to zero radius at some depth below. Careful experiments of this kind show a relation between the **wavelength** of the

wave pattern, which is the distance from crest to crest, and the orbital motion. At a depth of about one-half the wavelength, orbital motion essentially stops.

Now we can infer what happens as waves approach the shore across a shallowing bottom. At some distance from shore the bottom is only one-half the wavelength from the surface. At that point the orbital motions of the lowest levels of the water start being restricted because the water can no longer move vertically: right next to the bottom, the water can only move back and forth horizontally. In the next higher level, the water can move vertically just a little, this motion combining with the horizontal motion to give a flat elliptical orbit (Fig. 10-12). Higher in the water, the orbits become more circular the farther they are from the bottom.

It is the flattening of the orbits that makes waves "feel" the bottom, for the distortion of the

Figure 10-13
Effect of wind of force 8 (Beaufort scale; see Table 8-1) on the surface of
the sea. Wind speed, 65 km/hr (39 mi/hr); wave period, 6 sec; wave height,
5 m (17 ft). [From Atmospheric Environment Service—Environment Canada.]

orbits into ellipses slows the whole wave. While
the wave slows, its period remains the same—the
waves keep coming in from deeper water at the
same rate. As a consequence, the wavelength
shortens, because all of these waves must follow
the fundamental law of motion

$$\text{wave velocity} = \frac{\text{wavelength}}{\text{wave period}}$$

which is usually abbreviated to $V = L/T$. A further consequence of this change is that the wave
grows higher and steeper and the wave crests become sharper. The wave **steepness,** which is defined as the ratio between its height and the wavelength, increases to the point at which the water
can no longer support itself, at a ratio of about 1 to
7, and the wave breaks with a crash. The depth of
the water at this point is about 1.3 times the wave
height. The distance from shore is highly variable,
depending on how rapidly the bottom shallows.
On a steeply sloping bottom, many small waves
break right at the shore. Sandbars and gently sloping bottoms cause the waves to break farther out.
Thus, the location of the **surf zone,** where the
waves break, is a good guide to the depth of the
water.

Now contrast this picture of regularity with
what you see at the beach during a storm or a
strong wind. Though the waves still break on
shore with some semblance of regularity, the
whole scene is much more wild and confused. In
deeper water, whitecaps are everywhere. These
are not the same as breaking waves. They are
caused by the wind blowing off the top parts of
wave crests. The surface of the water is rough
with ripples, or little waves on big ones, and no

distinct wave pattern can be seen (Fig. 10–13)*. We
can find the key to all by watching a calm sea
surface as still air is freshened by a breeze. Ripples, tiny waves less than a centimeter high, take
shape as the breeze comes up. As the wind increases, the ripples grow into full-sized waves, the
size depending on the speed of the wind (see
Table 8–1), how long it blows in any one direction,
and the distance that the wind blows over the
water, the **fetch** of the wind. The energy of motion
of the wind is thus transferred to the water, much
as a child's energy is transferred to a toy automobile—the longer and harder the child pushes, the
faster it goes.

Storms make irregular choppy patterns of large
waves that radiate outward from the storm area,
just as dropping a pebble into a still pond makes
ripples that move out as ever widening circles
(Fig. 10–14). As the waves leave the storm area,
they become more regular swell of lower height,
which can travel hundreds of kilometers. Some
such swells have been shown to cross the width of
an ocean. Now we can begin to see why waves of
different heights and periods break against the
shore. Let's return to the pebbles dropped in a
pond. Dropping two pebbles in different places in
a pond produces ripples that cross and interfere
with each other in some places but that, in other
areas far from the source, appear to merge into
two different, approximately parallel sets. In a
similar way, two storms produce waves of differ-

---

*Storm waves can build to awesome heights. Sea captains
have estimated wave heights greater than 25 m (80 ft) in hurricanes. One giant wave was estimated, by careful sighting by an
officer of a U.S. Navy ship, to have been about 35 m (115 ft)
high!

ent sizes and periods that interfinger as they approach the shore. Timing the periods and sizes of waves at the shore can give you a rough idea of the many storms near and far that generated the swell. The Pacific shores of North America are far better for surfers because waves from the many storms in various parts of the great expanse of the Pacific Ocean are reinforced by prevailing westerly winds as they roll eastward. Surf along the Atlantic shore is smaller because the westerly winds blow against the advancing waves and decrease their effect. The best surfing on the Atlantic is on its eastern shore—along France's coasts, for example.

Wave Refraction

There is one more piece of the story to be explained: those parallel lines of waves that bend toward the shore as they approach (Fig. 10–10). As a wave approaches the shore at an angle, the part closest to shore feels the bottom first, the circular orbit becoming elliptical, and so that part slows down. Each successive segment along an individual wave crest meets the shallowing bottom in turn and also slows (Fig. 10–15). Meanwhile, the segment closest to shore has moved into even shallower water, and thus slows even more. In this way, not by separate segments but in a continuous transition along the wave crest, the line of the wave bends as it slows, in a process called **wave refraction** because of its similarity to the bending of light rays in optical refraction. The process can be compared with ranks of marching soldiers turning a corner, the ones closest to the corner marching most slowly and the farthest ones most quickly.

Wave refraction produces some special effects on an irregular shoreline with indented bays and projecting headlands (Fig. 10–16). Around the headlands, the water shallows more quickly than the surrounding deeper water on either side. The waves are therefore refracted—bent—toward the projecting part of the shore from both sides. The waves converge around the point of land and expend proportionately more of their energy breaking there than at other places along the shore. Thus, erosion by waves is concentrated at headlands and tends to wear them away more quickly than it does straight sections of shoreline. Conversely, wave refraction operates in a bay to make the waves diverge and expend less energy: the waters there are deeper, so the waves are refracted to either side into shallower water. The minimal wave energy expended along the inner

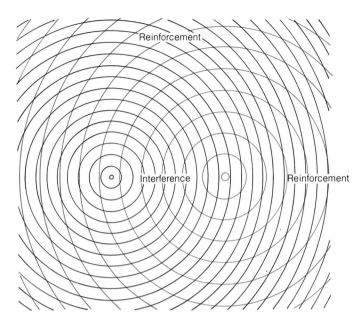

Figure 10-14
Waves from two storm centers interfere and reinforce each other, just like ripples from two pebbles tossed into a pond. At some distance from the storms, the waves will appear to come from the same general direction but with different wavelengths.

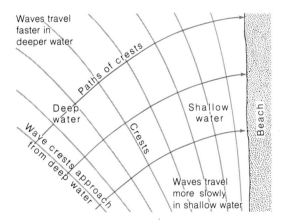

Figure 10-15
Wave refraction, the bending of wave crestlines as they approach the shore from an angle. The part of the wave that first encounters shallow water slows in speed while that part of the wave still in deeper water continues to move with its original speed, making the wave change angle, like a rank of marchers wheeling around a corner.

shores of bays makes the water quieter there, leaving secure places for mooring ships. It is in bays that beaches form on an irregular shoreline of headlands and indentations. To explore more of the interaction of waves and beaches, we should take a closer look at how beaches are affected by the waves.

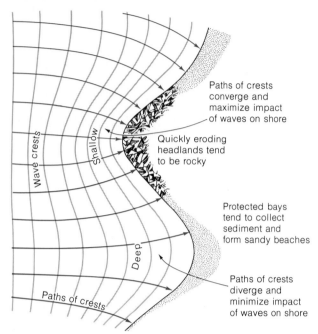

Figure 10–16
Wave refraction around a headland and bay. Because of the configuration of shallow and deep water around projections and indentations of the shoreline, wave energies are concentrated at headlands and dispersed at bays.

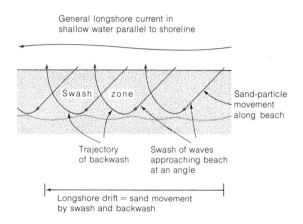

Figure 10–17
Longshore drift is the zig-zag movement of sand grains thrown up on the shore by waves approaching at an angle and washing back out at approximately right angles to the beach, thus causing the sand to progress down the beach in the direction toward which the waves are angled. Longshore current in the shallow water is created in the same way, by the zig-zag movement of water.

Longshore Drift and Longshore Currents

One of the more effortless observations for the student of beaches to make is of how waves roll up onto the sloping front of the beach, the **swash,** and fall back down again, the **backwash.** The swash of a wave has enough current competence to lift sand grains and carry them along, in the same way that river currents do. Strong waves are competent to lift and move large pebbles and cobbles. The backwash carries particles back down again to about their original positions on the slope. Because, despite wave refraction, waves usually break at some small angle to the shoreline, the swash moves up the beach slope at a small angle. But backwash runs down the slope more nearly at right angles, down the steepest slope available to it. Thus the sand grain in one swash-and-backwash cycle is displaced along the shoreline by a small amount. The net result is a zig-zag path in a direction along the shore that is determined by the wave direction (Fig. 10–17). This **longshore drift** of sand on a beach, which is an imperceptibly slow movement to the observer of an hour or a day, has been demonstrated by using a tracer—slightly radioactive sand grains (though not radioactive enough to be dangerous).

Coupled to longshore drift is the **longshore current,** which is induced by waves approaching the shoreline at an angle: the water is transported along the beach, just as the sand grains are, by the combination of swash and backwash. The water transport creates the current, which is strong enough to carry sand grains along in shallow offshore zones. Longshore currents build up with increasing distance down the shore, and the water imperceptibly piles up—the difference in height is too small to see or measure—until a critical point is reached. There, the water breaks out to sea at right angles to the shoreline, through oncoming waves, to form a rapidly moving **rip current** (Fig. 10–18). The combination of angular waves, longshore current, and rip current makes a closed loop along which the water continually moves. A long beach typically embraces many such loops.

The spacing, intensity, and precise location of the rip currents are controlled by the angle, height, and period of the waves and by the topography of the beach, but the relations among these factors are so complex that exact predictions are very difficult to make. Rip currents are dangerous to swimmers and surfers, however, and a knowledge of their characteristics may help to save lives. They can flow at a velocity more than one meter per second, too strong a current for inexperienced swimmers or children to fight against. The best way to get out of them was pointed out by Francis Shepard, an American marine geologist. (Shepard was one of the first to recognize the significance of rip currents and to stop calling them by the colloquial name "rip-tides," for they have nothing to do with tides.) Because they are narrow currents, the best way to avoid being carried out to sea is to swim parallel to the shore and not fight the current at all. In a short distance, the swimmer will be out of the current and can easily swim with the waves to shore.

It is best, of course, to avoid rip currents entirely. They can usually be detected by the absence of breaking waves, for the current both erodes a deeper channel and hinders the advancing waves. A good place to watch for them is at a bend or slight indentation of the beach where the water seems deeper than usual.

The Beach Budget

If sand constantly leaves the beach by longshore drift and current, why doesn't the beach eventually disappear? Though this does not happen on natural beaches, it may very well happen when naïve builders or engineers interfere with nature in certain ways. To understand why, we have to make an input–output budget for the beach, averaged over a significant time, such as a year (Fig. 10-19). Because wind directions change from day to day, the drift may go one way one day and reverse itself the next as the waves approach from the opposite direction. On a few coasts, the net movement for the whole year is zero, but on most there is a prevailing direction, and that is the direction of net movement of sand. In a bay, where waves diverge because of refraction, the sand is transported from both sides to the central beach, where it accumulates. In contrast, any sand at a headland is soon transported away from it, and bare rocky shores are the rule in such places.

What happens at the end of a beach, as where a straight shoreline turns a sharp corner? Any sand in longshore transport will be carried off the end of the beach and into the deeper water beyond, where it gradually builds up a submerged bar. With continued growth, it rises above the surface and so extends the beach as a narrow **spit** of sand. From the shapes of spits and bars, geologists can infer the direction of longshore transport; from that, they can predict the predominant directions of longshore currents and, thus, the average wind and wave directions. Favorite resorts in many

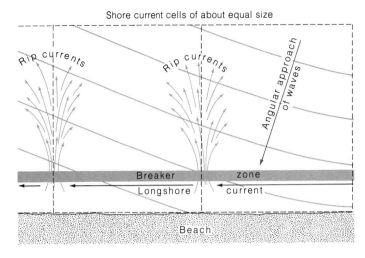

Figure 10-18
Rip currents flow out from the shore at certain points, usually regularly spaced, so that the shore can be divided into cells of water circulation along the beach. As a result of oblique wave attack, which forms the longshore current, the water tends to pile up, forming a rip current that breaks through the surf zone, fans out, and dissipates.

parts of the world are on sand beaches that have been extended for great distances by longshore currents.

There is another output. Any beach will "leak" sand to the deeper water offshore, mainly because of rip currents and intense storms. Once there, the sand is rarely moved, for only the highest waves can stir up sand that lies more than 10 meters deep. This sand, therefore, is permanently lost by the beach. In shallow zones of the beach, sand may move in and out from the shoreline in a pattern related to wave activity. In times of intense storms and high waves, much of the beach above the normal high-water mark may be removed to the submerged nearshore area. In periods of relatively calm weather, that sand is gradually shifted back up the beach, and the beach widens. Typical seasonal fluctuation of this kind is shown by some beaches that narrow during the stormy winter period and broaden during quiet summer weather.

If sand moves down the beach and extends it, there must be an input of sand to balance this output. Erosion of cliffs and headlands supplies some of the material, but much of it is brought to the shore by rivers. A river tends to drop its sediment to form a delta as it meets the ocean. Where long-

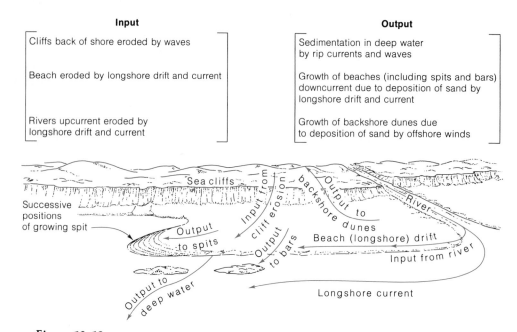

Figure 10-19
The beach budget is a complex balance between erosion and sedimentation in which such forms as spits and bars grow as beach cliffs and other sources supply sand. If input is out of balance with output, the beach tends to grow or to erode.

shore currents are strong, however, deltas never have a chance to form because the sediment is quickly distributed to beaches, to bars, and offshore along the coast. Where sea cliffs are of sandy sediment, much replenishing sand is produced by erosion. Because cliffs of harder rock are broken into sand too slowly to keep pace with longshore transport, beaches do not readily form downdrift from them.

Beaches export sand in another direction— landward, to form sand dunes. Sea breezes and winds blow sand inland to form the dune ridge common to most beaches. That sand may be blown many meters high and accumulate as thin dunes even on top of high bluffs. Neglecting short-term fluctuations from beach to shallow water during storms and calm periods, we can summarize a simplified budget of a beach, or any segment of it, as follows: If input exceeds output for some months or years, the beach will grow; if output consistently dominates over input, the beach will shrink, sometimes to the vanishing point.

In the natural state, beaches tend to maintain a balance between input and output. But most beaches in populous sections of the country are not in the natural state: people build cottages, pave beach parking lots, erect sea walls to "protect" the beach, and construct piers and breakwaters. The usual consequence, unless the develop-

ers or engineers responsible are aware of the dynamics of the beach, is to make the beach shrink in one place and grow in another. The classic example is a narrow pier built out from shore at right angles to it. Much to the surprise of the builders, the sand, during the following months and years, disappears from one side and enlarges the beach greatly on the other (Fig. 10–20). This is an expensive way to discover a longshore current on that beach. The current and drift bring sand toward the pier from the upcurrent direction (the dominant wind direction, usually) and, stopped at the pier, dump the sand there. On the downcurrent side of the pier, the current and drift pick up again and erode the beach. On this side, however, there is no replenishment of sand by the current because it is blocked by the pier, so the beach shrinks and disappears. If the pier is removed, the beach will relax to its original state. It is important to learn that a beach is a dynamic form, and any disturbance of its input–output balance, by altering either the source of material or the transport mechanism, will affect it in predictable ways. Thus construction on (or other alteration of) the dunes or bluffs at the back of a beach may strongly affect the width of the beach by changing its supply of sand.

One of the consequences of oil spills from tankers offshore is the fouling of beaches by heavy oils that gradually alter to tarry lumps and layers. The oil seeps into the pores of the sand and binds it together. Temporarily, the beach may be a disagreeable mess, but sooner or later, because a beach is constantly in motion, it will cleanse itself. The oil lumps are broken up, dispersed, and mixed with the immense volume of sand on the beach. The problem is that the process may take many months or years, which makes the beach's ability to cleanse itself a small consolation to the animals and people that populate it in the meantime.

TIDES

One subject that philosophers and scientists of ancient Greece did not cover was the tide, the daily rise and fall of the sea. Ancient observers of the sea in places as widely separated as China and Iceland, however, devoted much thought to the subject. The Greeks were observant enough—they just happened to live on a large inland sea with almost no tide. The people of civilizations that developed along oceanic coasts were familiar with and affected by the daily alternation of high and low water. Although it had long been known that there was a relation between the position and phases of the Moon, the heights of the tides, and the time of day at which the water would reach its highest level, it was not until the seventeenth century when Isaac Newton formulated the laws of gravitation, that the tides were understood to result from the pull of the Moon and the Sun on the oceans.

The Earth and the Moon attract each other strongly with a gravitational pull that is slightly greater on the sides of the two planets that face each other. The net gravitational attraction of the Moon for any point in the Earth is the vector sum (addition including direction as well as magnitude) of the large constant attractive force between the two bodies that is considered to act as if all of the Earth's mass were concentrated at its center and the small deviation that varies from point to point in the Earth proportional to its distance from the Moon (Fig. 10–21). That deviation is the tide-producing force. This force causes a slight bulge in the solid Earth, the oceans, and the atmosphere (and the Moon, too, of course). The deformation in the solid Earth is too small to be observed, except by sensitive instruments, but the much larger bulge in the water is easily seen as the tide. The net gravitational attraction for the water of the oceans facing the Moon is at a maximum. On the side away from the Moon, the net gravitational attraction for the ocean water at the

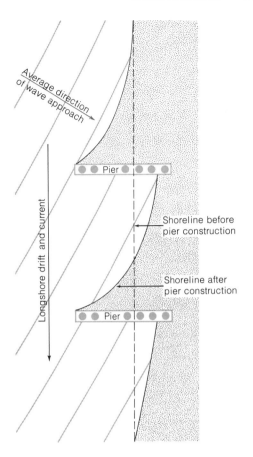

Figure 10–20
Construction of piers (frequently called "groins") along a shore to control erosion of a beach may produce unwanted changes if longshore current and drift are not considered. A typical change induced is erosion downcurrent of the pier.

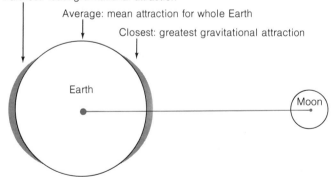

Figure 10–21
Moon's gravitational attraction causes two bulges of water on the Earth's oceans, one on the nearest side and one on the farthest side. As the Earth rotates, the bulges always face the Moon. Thus two high tides (bulge closest to Moon and bulge farthest from Moon) pass each point on Earth's surface each day.

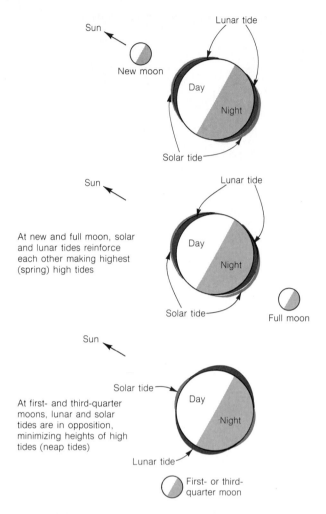

Figure 10-22
The relative positions of Earth, Moon, and Sun determine the heights of high tide during the lunar month. The highest high and lowest low tides (spring tides) come at new and full moons; the lowest high and highest low tides (neap tides) come at first- and third-quarter moon.

surface is at a minimum. As the Earth rotates, the tides move around it, one always facing the Moon, the other directly opposite.

The Sun, though much farther away, has so much more mass that it, too, causes tides. The Sun tides are a little less than half the height of Moon tides. The two sets of tides are not synchronous. Those related to the Sun come every twenty-four hours, once each "solar" day. The time of rotation of the Earth with respect to the Moon is a little longer than the solar day—twenty-four hours and fifty minutes, because the Moon is moving around

the Earth. In that twenty-four hours and fifty minutes, the "lunar" day, there are two high waters, with two low waters in between.

When the Moon, Earth, and Sun line up, the combined gravitational pull of Sun and Moon reinforce each other and produce very high tides, the **spring tides** (Fig. 10-22). Such high tides come every two weeks at full and new Moon. The lowest tides, the **neap tides,** come between, at first- and third-quarter Moons, when the Moon and Sun are at right angles to each other with respect to the Earth.

The above account describes the equilibrium tide, that is, one that is theoretically calculated for a uniform globe. The heights of the actual tides are very different in various parts of the oceans. Because the oceans are of various shapes and sizes, the water of the tide responds in complex ways. It is a bit like connecting a great number of large and small pots and bowls in a complicated pattern and sloshing water back and forth between them. Thus, in Hawaii, the difference between low and high water is only about 0.5 meter; at Seattle, however, it is about 3 meters (10 feet). Extraordinary tides occur in a few places, such as the Bay of Fundy, where the tidal range can be more than 12 meters (40 feet). Inland lakes, such as the Great Lakes, are virtually free of tides. For many of the world's coastlines, especially those with important commercial shipping and harbors, computers have calculated the tides for years ahead with great precision.

Along shallow coasts, tidal movements give up energy through friction of the water with the sea floor—energy that must ultimately come from the rotation of the Earth and Moon. That frictional loss is enough to slow the rotation of the Earth by a very small amount. The average length of the day has been calculated to have lengthened by 0.001 second in the past 100 years. That may seem negligible, but over many millions of years of geologic time, that amount of slowing can mount up. The Earth must once have been rotating much faster, though the time of its revolution around the Sun was unaffected. The Moon's rate of revolution about the Earth would also have been faster, and the Moon would have been closer to the Earth. That means that the tides would generally have been much higher, that there were many more days in a year, and that the days were shorter.

Striking support for this idea came from the world of fossils. Corals, marine invertebrates that secrete calcium carbonate and build modern coral reefs, lay down a microscopically thin layer

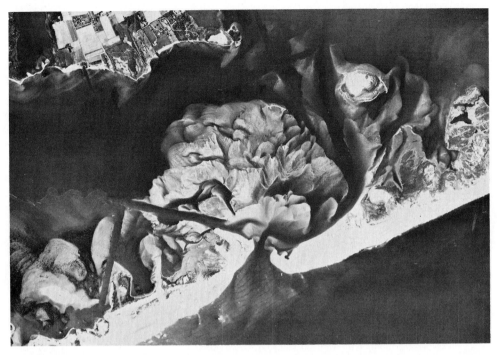

Figure 10-23
A barrier island (white) that was breached by a hurricane in September 1947. After the break, the tidal delta seen at the center formed between the island and the mainland. [Courtesy of M. M. Nichols, U.S. Department of Agriculture.]

of calcium carbonate each day; the layers laid down in summer are thicker than those laid down in winter. Careful counting of these layers in fossil corals has convinced many paleontologists that, 400 million years ago, there were nearly 400 days in a year. Evidence of stronger tides is not so good, but some geologists who study tidal sediments of the past think they can see the effects of stronger **tidal currents,** the rapid movements of water in shallow parts of the oceans.

Tidal Currents

The movement of tidal waters near shorelines causes tidal currents that can reach speeds of a few kilometers per hour. As the tide rises, the water flows in toward the shore as a **flood tide,** moving into shallow coastal marshes and up small streams. As the tide passes the high stage and starts to fall, the **ebb tide** moves out, and low-lying coastal areas are exposed again. Such tidal currents cut channels through **tidal flats,** the muddy or sandy areas that lie above low water but are flooded at high water. Tidal flats may be narrow strips seaward of the beach, or they may be extensive areas covering hundreds of square kilometers. When the tide advances on some wide tidal flats, it may move so rapidly that areas are flooded faster than a person can run. The tidal-flat beachcomber is well advised to consult a local tide table before wandering.

Tidal flats may be separated from the open sea by long sand bars or **barrier islands** parallel to the shoreline (Fig. 10–23). The islands are bars of sand that have been built up above the high-water level and are stabilized by vegetation. The barrier islands or bars are broken by occasional tidal channels through which the water rushes in a strong current. Outside these channels, tidal currents may create tidal deltas of sand transported from the tidal flat during ebb tides. Barrier islands, like reefs, create quiet lagoons between them and the main shore. Like beaches on the main shore, barriers are in states of dynamic equilibrium between inputs and outputs. If their equilibrium is disturbed, either naturally or by human activities, the island may either erode and disappear or migrate toward shore and fill the lagoon. The study of the effects of tidal currents on sediment in the modern world has been applied to sediments of the past. Geologists have thus inferred many deposits to be of tidal-flat, barrier-island, and channel origin.

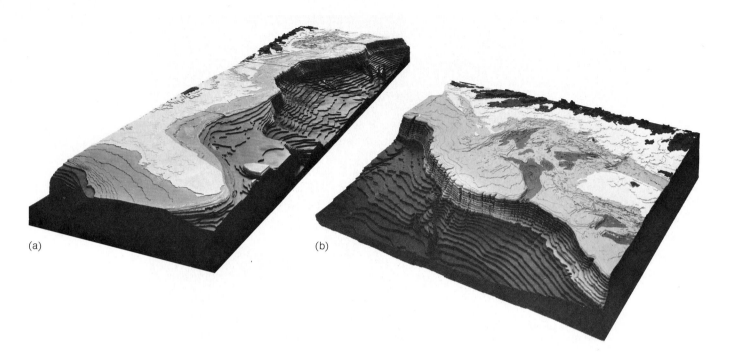

Figure 10-24
Two views of a three-dimensional model of the Atlantic continental shelf, slope, and rise of part of North America. (a) View looking northwest; Florida is in the foreground, Nova Scotia and part of the Grand Banks at the far edge. (b) Closeup of the section from Nova Scotia at right to New York and Long Island at left. Near the left margin is the Hudson submarine canyon, and in the middle foreground is the area of Georges Bank, off Cape Cod. [Courtesy of K. O. Emery and E. Uchupi, Woods Hole Oceanographic Institution.]

CONTINENTAL SHELVES

Bordering the North American coasts of the Atlantic and the Gulf of Mexico are broad, shallow sea-floor platforms that, though submerged, are clearly parts of the continental mass. These platforms, and similar ones around other continents, the **continental shelves,** extend from the edges of the continent to a depth of a few hundred meters, where they give way to steeper slopes that go down several thousands of meters to the main ocean floor. For most of the first century of ocean exploration, marine geologists were satisfied to say that continental shelves extended to depths of about 200 meters (650 feet), where the continental slopes generally started. Then, growing interest in the economic uses of the sea, from fishing to mining, provoked international legal discussions of the territorial rights of nations to the continental shelves. Marine geologists, finding that the old definition was too arbitrary, began to be more precise: in a United Nations report of 1957, geologists recommended a 650-meter depth limit, one that would include all of the world's continental shelves. Legal questions sometimes impose themselves on scientific matters in this way.

Continental shelves vary widely in width, but they average about 65 kilometers (40 miles); the general slope of the surface is slight, the angle of inclination averaging only 0°07′ (Fig. 10–24). Most of the surface is fairly uneven, with small hills and ridges alternating with basinlike depressions, broad valleylike troughs, and occasional narrow, steep-walled valleys called **submarine canyons** (Fig. 10–25). Most areas of the continental shelves were above sea level during the Pleistocene glaciations and, as we noted in Chapter 9, much of erosional and sedimentational character of their surface was formed then. Waves and tidal currents acting on the shelves since the last glaciation have modified the surface, mainly by depositing sand near the shoreline and silts and muds in the deeper water farther from shore. Pleistocene relic sands may be found at shallow depths far out on the shelf, even near the shelf edge, as on Georges Bank southeast of Cape Cod, one of the famous fishing grounds of the Atlantic. These sands are now being dispersed, redistributed, and altered by modern processes. There are about 400 billion tons of this sand in the top 3 meters (10 feet) of sediment off the northeastern coast of the United States, enough to supply the needs of nearby

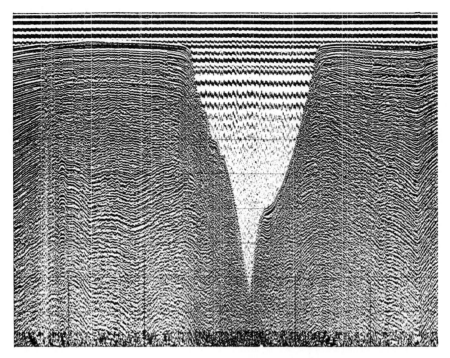

Figure 10-25
An echo-sounding profile of the Congo submarine canyon off the Republic of Zaire on the west coast of Africa. The bottom of the canyon at this point is about 3000 m below the sea floor of the continental shelf; at the top, the canyon is more than 10 km wide. The heavy black horizontal lines in the water at the top of the recording are artifacts of the process. The wavy lines below the sea-floor surface are sound reflections from bedding planes in continental shelf sediments, somewhat deformed because of mild tectonic disturbance. [From K. O. Emery, Woods Hole Oceanographic Institution.]

areas for thousands of years, a valuable sea-bed resource.

Continental shelves may be narrow and broken by isolated deep basins and high ridges or escarpments, as off the coast of California, where water hundreds of meters deep may be within sight of land. This coast bears all of the topographic earmarks of a tectonically disturbed area, a fact that is in keeping with the highly faulted and deformed coastal belt just above the water. Some narrow shelves, however, show little or no evidence of tectonic disturbance. One of the best known of these is off the east coast of Florida, where the powerful Gulf Stream, a rapid ocean current, runs close to shore.

Oceanographers have mapped continental shelves—and the rest of the ocean floor—with a shipboard instrument called an **echo-sounder.** This instrument, sometimes called a **fathometer,** * or **depth-recorder,** works by sending sound waves from the ship to the bottom (Fig. 10–26). The sound waves reflected from the bottom are picked up by sensitive detectors on the ship. From

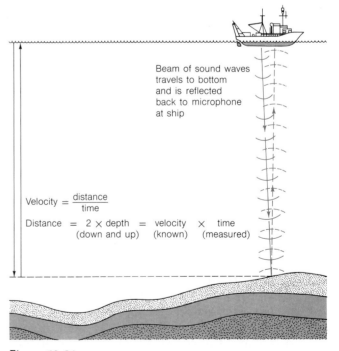

Beam of sound waves travels to bottom and is reflected back to microphone at ship

$$\text{Velocity} = \frac{\text{distance}}{\text{time}}$$

$$\underset{\substack{\text{(down and up)}}}{\text{Distance}} = \underset{}{2 \times \text{depth}} = \underset{\text{(known)}}{\text{velocity}} \times \underset{\text{(measured)}}{\text{time}}$$

Figure 10-26
Echo sounders sense underwater topography by beaming sound waves to the bottom and measuring the time required for the beam to be reflected back to the ship.

---

*From *fathom,* the traditional nautical measure of depth, equal to 6 ft (1.8 m).

Figure 10-27
The deep submersible vessel *Alvin,* designed to conduct geological and biological exploration of the sea floor. [Courtesy of Woods Hole Oceanographic Institution.]

the measured delay in time and the known speed of sound in water, oceanographers can determine the depth. As the ship cruises along a straight course, a two-dimensional profile of the topography is traced out automatically. The topographic map of the ocean floor that emerges from many such profiles is a **bathymetric map.** Submarine geologists also use newer instruments, such as echo-sounders towed along the bottom, to get a much better idea of the topography. These "deep-towed" vehicles, or "fish," are equipped with other geophysical instruments, such as magnetometers, devices for measuring the magnetic field (see Chapter 18). Even better and more detailed information on bottom topography is obtained by the use of side-scanning sonar, a method that uses sound waves sent out so that they graze the bottom at low angles, scanning a wide area on both sides of the ship. Last but not least is direct observation of the bottom by marine geologists in deep-diving submersibles such as that shown in Figure 10-27. In these vessels scientists can see—and photograph—all of the details of the bottom down to depths of thousands of meters. Equipped with mechanical arms, they can

break off pieces of rock or scoop up samples of sediment and bring them back to the laboratory.

The new accurate knowledge of the sea floor discovered by these means is still limited to a few small areas mapped in detail, most of those on mid-ocean ridges. This kind of mapping is extraordinarily difficult and expensive; one marine geologist estimated that in a decade we have looked at only 0.01 percent of divergent plate boundaries at a cost of much more than ten million dollars. Yet without these methods we would know little of the complexity of faults, intrusions and extrusions, hydrothermal vents, and sedimentation patterns that tell so much of the specific dynamics of plate motions.

The internal structure of broad, gently sloping continental shelves was worked out mainly by geophysicists at the Lamont Geological Observatory of Columbia University during the 1950s. They sent very strong sound waves down through the water. These waves penetrated the sediments and were reflected back to their oceanographic research ship in the same way as ordinary echosounders. From the way the sound waves were reflected by layers of sediment and rock the geologists were able to deduce the structure of the shelf: a pile of sediment several kilometers thick built out from the continent (Figs. 10–28 and 10–29). The suspicion that the shelves had such a structure had early been stimulated by speculation on the narrowness of the Florida shelf near the Gulf Stream. There, it was thought, the current in the sea is so fast—up to about 10 kilometers per hour—that the sea floor was kept scoured clean of sediment. Farther north, where the Gulf Stream moves out to sea, the continental shelf widens and consists of sedimentary layers. In 1976, geologists of the United States Geological Survey began a program of drilling at sea on the Atlantic continental shelf. Their findings have corroborated many of the deductions made from geophysical evidence.

In the past few years, the implications of plate-tectonic theory for continental-shelf development have been assuming increasing importance. According to some of these ideas, the continental shelves of the Atlantic Ocean type are thought to have originated as the trailing edges of continental margins on plates that broke apart and have been moving away from each other ever since the Atlantic Ocean started opening about 180 million years ago. The broken, rifted margin was heated by the upwelling of basaltic magma that formed the newly created spreading center, the mid-Atlantic ridge. As the edge of North America moved away from the ridge, it started to cool. As

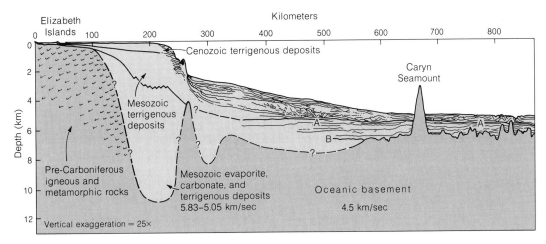

Figure 10-28
A profile of the Atlantic continental margin of North America off southern New England based on seismic data. Stratification, based on continuous seismic profiling, is shown diagrammatically for the younger sediments. [From K. O. Emery and E. Uchupi, *Atlantic Continental Margin of North America*. Copyright 1972 by American Association of Petroleum Geologists.]

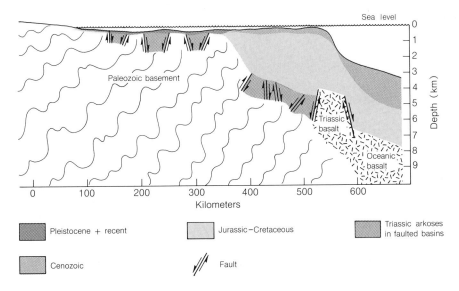

Figure 10-29
Geological cross section, somewhat schematic, of the Atlantic continental shelf from Maine south across Gulf of Maine and Georges Bank to the continental rise. The Jurassic and later sediments were deposited over a multiply faulted, rifted basement of Triassic red beds and basalts that overlie metamorphosed Paleozoic rocks. [After R. D. Ballard and E. Uchupi "Triassic Rift Structures in the Gulf of Maine," *Bull. American Assoc. Petroleum Geologists,* v. 59, pp. 1041–1072, 1975.]

it cooled, the lithosphere contracted, the surface subsided below sea level, and erosional debris from the continent accumulated on its seaward edge. As the Atlantic widened, the shelf, slope, and rise gradually built out from the continent as the edge of the plate continued to contract and subside (see Fig. 1–15).

Narrow, tectonically deformed continental shelves, such as those of the Pacific, are characteristic of continental margins that coincide with the leading edges of lithospheric plates where they end at transform faults or subduction zones. Narrow shelves, many fringed with coral reefs, are found off the shores of volcanic islands of western Pacific subduction zones. On the side of an island facing the subduction zone this shelf gives way to a steep slope into a deep-sea trench (Fig. 10–30). On the opposite side, the shelf gradually slopes to moderate ocean depths without any trench. The sediments that have built up these volcanic island

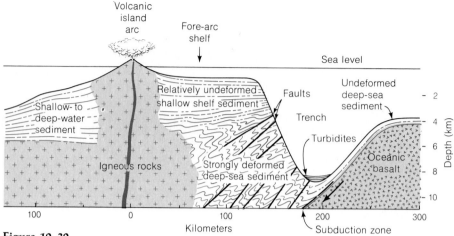

Figure 10-30
Generalized topography of a typical subduction zone in the western Pacific.
The area between the volcanic island and the trench, the fore-arc shelf, is built
up by shallow-water sedimentation; the inner trench wall is built up by the deformed deep-sea sediment scraped off the surface of the descending oceanic lithosphere slab. The horizontal and vertical dimensions of different arcs may vary
considerably.

arc shelves may be largely of limestones formed
from coral reefs or mixed with air- or water-borne
deposits from the nearby volcanoes. In that tectonic environment, continental shelves accumulate sediments, but they are likely to be deformed
as well as affected by volcanism and metamorphism (Fig. 1–16). Much remains to be learned of
these kinds of coasts, where sedimentation in
deeper water plays a relatively more important
role.

CONTINENTAL SLOPE AND RISE

Continental shelves end abruptly at the **continental slopes,** which have an average inclination of
about 4°—a drop of about 70 meters in a horizontal distance of 1 kilometer (Fig. 10–31). That is a
healthy slope, but not exactly a steep cliff, as one
might gather from diagrams that greatly exagger-
ate the vertical scale, such as Figure 10–28. The
sediments of most slopes are muds and silt, derived from continental erosion, that have been carried across the shelf and draped over the edge.
Like the surface shelf sediments, much of the uppermost mud on the slope was deposited during
the Pleistocene glaciations when sea level was
lower. Geophysical exploration of the slopes
below broad continental shelves reveals inclined
layers of sediment that have been deformed a little by slumping and sliding down the incline. In
contrast, continental slopes of tectonic coasts may

be steeper, and older rock may be mantled with
only a thin veneer of sediment. Rock outcrops are
common on such slopes.

The lower parts of many continental slopes
become gentler and merge into a more gradually
sloping apron of sediment extending into the
main ocean basins. These aprons are the **continental rises.** Their average inclination is less than
half a degree, only about one-eighth that of the
continental slope; but they extend for hundreds of
kilometers, whereas continental slopes have an
average width of only 20 kilometers. Rises are
fairly smooth topographically, being broken only
by an occasional seamount, submarine canyon, or
channel. The rises are piles of muddy, silty, and
sandy sediments, kilometers in thickness, that
have been carried from the continental shelf,
down the slope, finally coming to rest on the
deep-sea floor.

The continental slope is typically gullied by
many valleys and interrupted by occasional submarine canyons. The appearance of the slope suggests erosion, as do the canyons, but we cannot
draw again on the Pleistocene lowering of sea
level for an explanation because even the wildest
estimate of sea-level lowering would not envision
the drop of thousands of meters that would be
necessary to expose some of these slopes to land
erosion processes. The erosion must have a marine explanation. The explanation came in a rush
of developments around 1950, when an earlier
hypothesis about a little-known type of sedimen-

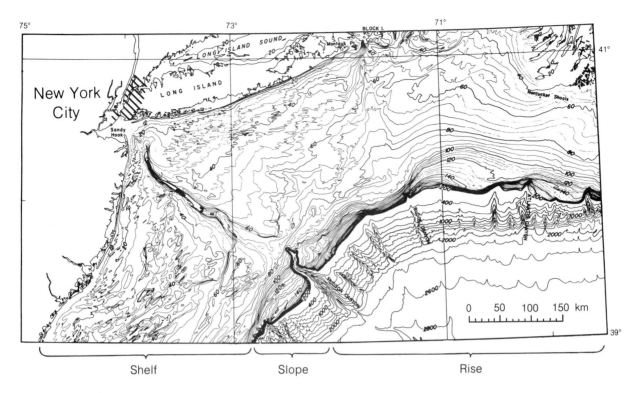

Shelf   Slope   Rise

Figure 10–31
Map of continental shelf, slope, rise, and Hudson Canyon off New York City.
[From K. O. Emery, Woods Hole Oceanographic Institution.]

tation was resurrected to explain erosion and deposition on the deeper parts of continental margins. The hypothesis proposed a special type of current, a turbidity current. Its characteristics ultimately helped to explain submarine canyons, the gullies on continental slopes, the formation of continental rises, and the flat plains that cover large areas of the deep-sea floor, the **abyssal plains.**

TURBIDITY CURRENTS

The idea of turbidity currents was born in the latter part of the nineteenth century in Geneva, Switzerland, where the Rhone River enters Lake Geneva. The river water is cold and full of suspended clay from the meltwaters of alpine glaciers. At that time, when the lake was unpolluted and clear, the river water could be seen through the still lake water above, moving along a channel on the bottom as a distinct muddy current. The river water is denser than the overlying lake water, primarily because of its suspended sediment load and secondarily because it is colder. Being denser, it sinks and runs downhill along the

bottom. The difference in density also acts as a barrier that slows the mixing of the current with the overlying waters, and so the current maintains itself for a long distance until it reaches the flat bottom of the lake. There it fans out in a thin sheet of muddy water and loses momentum, gradually coming to a stop, the suspended sediment slowly settling out on the bottom as a thin layer. This general kind of current, a **density current,** is known also from oceans, where such currents can arise because of differences in temperature or salinity alone; colder and saltier water is denser. The name **turbidity current** is applied to density currents that owe their density to suspended sediment.

The man who sparked the resurgence of interest in turbidity currents in the 1940s was a Dutch geologist, Philip Kuenen, who combined his experience as an oceanographer and field geologist with a series of important laboratory experiments. In the 1930s Kuenen had produced turbidity currents in his laboratory by pouring muddy water into the end of a flume with a sloping bottom. He showed that such currents can move as swiftly as many kilometers per hour, the speed depending on the steepness of the slope and the density of the current. Because of its speed and turbulence, a turbidity current can transport large quantities of sand introduced with the suspension or picked up by the current as it flows over a sandy bottom. The sediment layers produced by settling after the current starts to lose speed formed a **graded bed**—that is, a bed with coarse material at the bottom, grading upward to fine

Figure 10-32
A thick sequence of graded Miocene turbidite beds in the Apennines. The lighter, thicker units are sandstones that are coarsest at an abrupt base and grade upward through finer grain sizes into the darker shale beds. Each pair of sandstone-shale units represents one turbidity flow in a deep marine basin. [Photo by P. E. Potter. From F. J. Pettijohn, P. E. Potter, and R. Siever, *Sand and Sandstone.* Copyright 1972 by Springer-Verlag New York, Inc.]

material at the top. Each turbid flow produces a single graded bed; using a series of flows, Kuenen produced a sequence of graded beds.

The brilliant next step, taken more than a decade later, was the recognition of such graded beds on land, thick marine sandstone and shale sequences produced under water long ago as the product of such currents (Fig. 10-32). The final step was to tie all this together with what was being learned about the sediments and topography of the sea floor and to formulate a hypothesis. Turbidity currents are formed by large slumps of mud and sand draped over the edge of the continental shelf onto the continental slope. The sudden slump or slide, perhaps triggered by an earthquake, throws mud into suspension, creating a dense turbid layer, which starts to flow down the slope, eroding and picking up sediment as it picks up speed (Fig. 10-33). These currents gully the slope and excavate submarine canyons. As the turbidity current reaches the change of inclination where the slope meets the continental rise, the current slows, and some of the sediment starts to settle. Many currents continue across the rise, cutting channels as they go, to reach the level bottom of the ocean basin, the abyssal plain, where they fan out and come to rest. The deposits of the rise and the abyssal plain are graded beds of sand, silt, and clay; **turbidites** is the name coined for them.

Revolutionary as these ideas were, confirmation soon started rolling in. A series of sudden breaks in trans-Atlantic telephone and telegraph cables on the continental slope and rise off the Grand Banks of Newfoundland had taken place in 1929 following an earthquake. The locations of the breaks were plotted on a map and the times of breaking were noted (Fig. 10-34). A pattern quickly became clear. A rapid breaking of cables

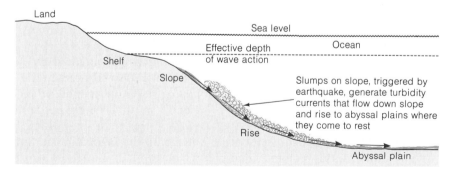

Figure 10-33
Typical formation of turbidity currents in the ocean. Slumps on the continental shelves generate turbidity currents that flow downslope to abyssal plains, where they gradually slow to a stop and deposit turbidite sediments.

high on the slope was followed by a sequence going down the slope and farther from the epicenter of the earthquake. But the breaks downslope occurred much too long after the earthquake to have been caused by earthquake waves. The turbidity-current hypothesis gave the only really plausible explanation: the earthquake provoked a slump that activated a turbidity current so fast and powerful that it snapped cables as it raced down the slope and rise. The current maintained itself for at least 700 kilometers and was calculated to have reached speeds of 40 to 55 kilometers (25 to 34 miles) per hour. Sediment cored from the deep sea near the path of cable breaks showed a thin graded bed at the surface, probably the material settled from the current. Similar patterns of cable breaks on continental slopes in other parts of the world showed up as records of cable companies were searched.

The most compelling evidence of turbidity currents was the distribution of turbidite sediment in the oceans. No other mechanism could explain as simply or satisfactorily how coarse sand particles could be transported to deep waters beyond the reach of waves and tidal currents. How else could graded beds have filled in the abyssal plains, the lowest parts of the ocean basins? The turbidity-current hypothesis was especially attractive because it simultaneously explained the formation of the slope, rise, and plain.

Turbidity currents have not explained everything. Though much is known about the physics of density currents, there remains much to learn about the complex workings of turbidity currents. Turbidity currents may not be the sole agent responsible for excavating submarine canyons in solid, hard rock, and there is also some evidence for the existence of localized deep-ocean currents that can also transport sand. Like other major theories, the turbidity-current idea has become more complicated as it has been worked out in more detail. Now, however, this important simplifying hypothesis, linked to the history of the oceans and continent edges, has become a vital part of plate-tectonic theory and its applications. This has come about as a result of careful consideration of the topographic and geophysical profiles of the oceans.

THE PROFILE OF THE OCEANS

The Atlantic Ocean, which is fairly symmetrical about the mid-Atlantic ridge, is bounded on both sides by continental shelves, slopes, and rises (Fig. 10–35). At the ridge, a high, rough topography of

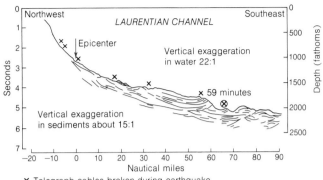

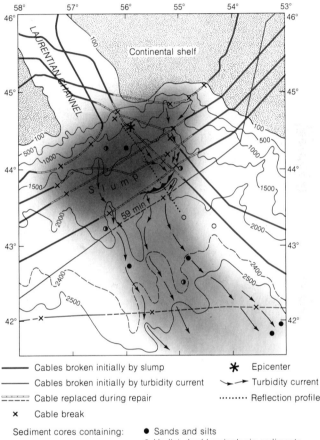

Figure 10–34
Map and seismic reflection profile of Grand Banks area where cable breaks signalled the beginning of a slump and turbidity current. [From B. Heezen and C. L. Drake, Copyright 1964 by American Association of Petroleum Geologists.]

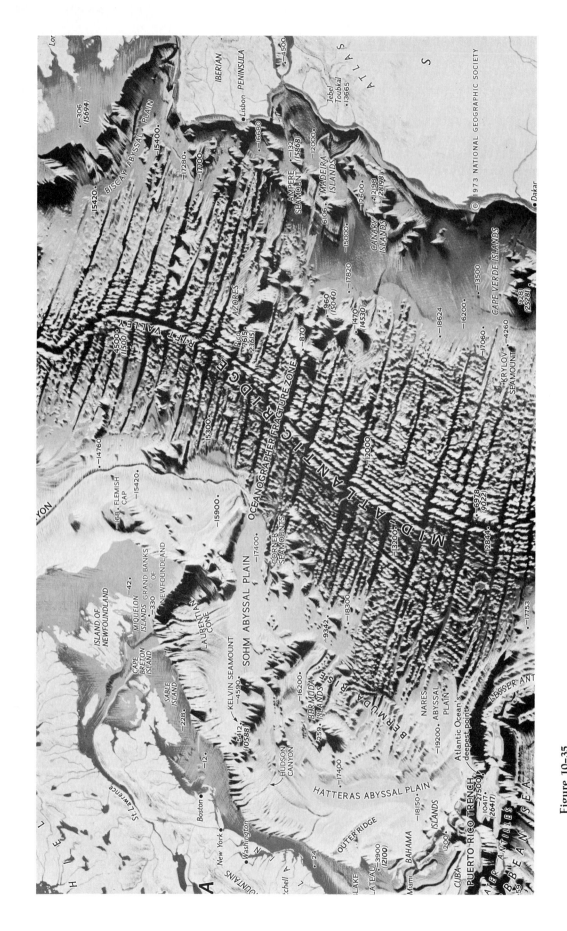

Figure 10-35
An artist's representation of the floor of the North Atlantic Ocean based on bathymetric studies of Bruce C. Heezen and Marie Tharp of the Lamont-Doherty Geological Observatory. Depths shown in feet below sea level. [Painted by Heinrich C. Berann; copyright © 1968 National Geographic Society.]

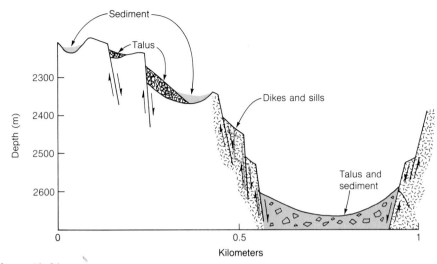

Figure 10-36
Detail of a portion of the central rift valley of the mid-Atlantic ridge in the FAMOUS (French-American Mid-Ocean Undersea Study) area southwest of the Azores. The deep valley, the narrow site where most of the basalt is extruded, is bounded by a series of faults that have created local depressions where basalt talus blocks and pelagic sediment have accumulated. [Modified from "Transform Fault and Rift Valley from Bathyscaph and Diving Saucer," *Science,* v. 190, pp. 108–116.]

basalt is created by the upward movement of material from the partially molten asthenosphere to create new oceanic lithosphere.

Details of the ridge topography and structure in one area were first mapped by direct observation by project FAMOUS (French-American Mid-Ocean Undersea Study) in 1973 and 1974. FAMOUS concentrated on a small area centered around 36° 50′ N, southwest of the Azores. In addition to detailed shipboard geophysical work, three deep-sea submersibles, the French *Archimède* and *Cyana* and the American *Alvin* dove into the rift valley many times to map and collect samples. They discovered that the intense activity of downfaulting, extrusion of lava, and intrusion of dikes and sills was confined to a narrow central portion, little more than a kilometer wide (Fig. 10-36). In 1978 a later expedition explored the next rift to the south, a wide U-shaped valley. They found evidence that the valleys evolve from an initial V-shape to a wider U-shape as spreading continues and new magma rises from the interior to fill the floor. Eventually the valley widens to the point of instability, the center falls, and the edges uplift, returning the valley to V-shape.

Another spreading center, the East Pacific rise, arches upward from the sea floor in the southeastern Pacific Ocean. Though it has the same general character as the mid-Atlantic ridge, it is a broader, lower feature, a reflection of the faster spreading at this zone. As the new crust is created at a more rapid rate, it moves laterally fast enough that it is not uplifted as much as slower-spreading ridges, such as the mid-Atlantic.

The most dramatic evidence of the processes operating on the East Pacific and other fast-spreading rises has come from a series of expeditions using direct observations by submersibles. During dives of *Alvin* in 1977, hot springs and hydrothermally (hot-water) produced mounds of iron-rich clay minerals and manganese dioxide were discovered on the Galápagos spreading ridge near the equator just west of South America. In 1978 the *Cyana* dove to the East Pacific rise at 21° north latitude, south of the mouth of the Gulf of California, and discovered large deposits of iron-zinc-copper sulfides. In 1979, *Alvin,* accompanied by ANGUS, a sophisticated survey camera, and Deep Tow, a powerful array of geophysical sensors, also went to the 21° N area and found the active, ore-carrying hot springs at temperatures of 350°C. These were the "black smokers," mineral chimneys precipitated as waters from hot springs mix with the surrounding sea water (see Fig. 13–10 and Chapters 13 and 22).

In the 21°N area the rise, which is spreading at a rate of about six centimeters per year—about the

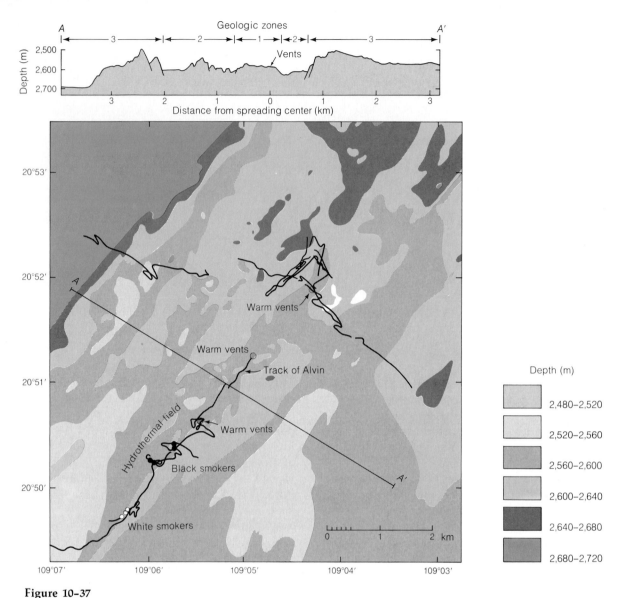

Figure 10-37
Bathymetric chart of the dive site of the *Alvin* in the 21°N section of the East Pacific rise shows the tracks followed by the submersible and the main features of the hydrothermal field. A profile of the sea floor along the line marked *AA'* is at the top. Three geologic zones identified by the dive team are labeled (see text). Warm-water vents are found in the northeast; the hotter smoker vents are clustered in the southwest. Black smoker chimneys are made up of sulfides precipitated from waters having temperatures up to 350°C; white smoker chimneys are built up from burrows of worms and materials precipitated from waters at slightly lower temperatures, up to about 300°C. The warm-water vents are much cooler at 20°C, only about 15° warmer than the sea water. [After "The Crest of the East Pacific Rise" by K. C. Macdonald and B. P. Luyendyk. Copyright © 1981 by Scientific American, Inc. All rights reserved.]

speed human fingernails grow—can be divided into three zones (Fig. 10-37). The first zone, about a kilometer wide on the axis of the spreading center, is a very young volcanic region of basalt pillows and little sediment. Outside this zone is the second zone, where the crust is stretched and cracked with many small fissures lined up parallel to the trend of the rise. Beyond this, in the third zone, major normal faults with almost vertical scarps facing the spreading center dominate the topography.

In 1980, a joint U.S.–Mexican expedition went to the Guaymas Basin on another part of the East Pacific rise in the Gulf of California about 1000 kilometers southeast of San Diego. Here the same kinds of magmatic and hydrothermal activity are modified by a thick blanket of sediment shed from the nearby continental margins. In the Guaymas Basin the lava does not reach the surface but forms mushroom-shaped intrusions within the sediment layer and deforms and domes the overlying sediments and sea floor.

Away from a ridge, the topography becomes lower and less rugged, a region of **abyssal hills,** as the lithosphere subsides as a result of cooling (Fig. 10-38). The slopes become thinly covered with the fine sediment that accumulates slowly over all of the ocean floor. As the lithosphere sinks more, it becomes swamped by turbidity-current deposits that bury the varied hilly topography and create a level abyssal plain (Fig. 10-39). Closer to a continent, the plain becomes a continental rise, with the original basalt topography buried under many kilometers of turbidite.

Isolated volcanic islands dot the ocean floors. Some, like those of the Hawaiian Islands, are active. Others are extinct and may be covered over completely by coral reefs, as are the atolls of the Pacific Ocean. Isolated subsea mountains with the same shape as volcanoes are found rising from the ocean floor. The **seamounts,** like volcanic islands, may occur as clusters or chains; where closely spaced, chains of seamounts merge to become **aseismic ridges.** Many seamounts have flat tops formed by erosion of the volcano top during emergence; such seamounts are called **guyots** (named for the first professor of geology at Princeton University). All of these volcanic mountains are produced by plates riding over hot spots in the mantle, regions where basalt is steadily rising (see Chapter 15).

How this came to be is the subject of much study, the subject of paleo-plate tectonics. In general the picture is as we described earlier in this chapter. As the Atlantic Ocean opened up, the first sediments deposited in the narrow rift-valley ocean, which was perhaps something like the Red Sea is today, were coarse turbidites flowing into deep waters at the steep, newly broken edges of the continental plates. As the plates separated farther, continental shelves were built up over the trailing edges of the subsiding plates, and the sediments being deposited on the rise built up to form a thick pile that merged with the continental slope. As spreading of sea floor continued, the turbidites began to fill in the low spots on the rough, basaltic, oceanic crust, gradually building up to form the abyssal plains. This conception ties together tectonics and turbidite sedimentation in a consistent theory that explains much of what we know about the Atlantic Ocean today.

The Pacific Ocean does not follow such a simple scheme. Continent edges there are narrow, and deep basins may be filled with turbidites very close to shore. Thick accumulations of turbidites are found as submarine fans or subsea deltas where submarine canyons that start near the shore empty onto the deep-sea floor. Turbidite sediments form in trenches, the subduction zones in which oceanic crust is consumed as it turns downward into the asthenosphere. Many of these turbidites are crumpled and folded, scraped off, and plastered against the overriding plate edge (Fig. 1–16). Much of the sediment may not be carried down with the oceanic crust along the subduction zone because the sediment, saturated with water, is much less dense than basalt and tends to "float" at the surface of the lithosphere. Some, of course, may be dragged down, mixed with volcanics and other sediments, and metamorphosed by pressure and heat as it descends (Fig. 10-30).

We have emphasized turbidity currents because they are major sediment eroders and transporters, but they are not the only (or even the major) currents of the oceans, nor are turbidites the only significant sediment type of the deep sea. The general circulation of the oceans, and of the atmosphere above them, is responsible for the patterns of transport of much of the fine-grained sediment of the sea.

OCEANIC CIRCULATION AND TRANSPORT OF TERRIGENOUS SEDIMENT

For centuries, mariners have known that there are surface currents in the sea that can either speed a ship on its course or hinder its progress. At least as early as 800 B.C., the Phoenicians and Greeks knew the currents of much of the Mediterranean Sea. Homer's *Odyssey* mentions the treacherous

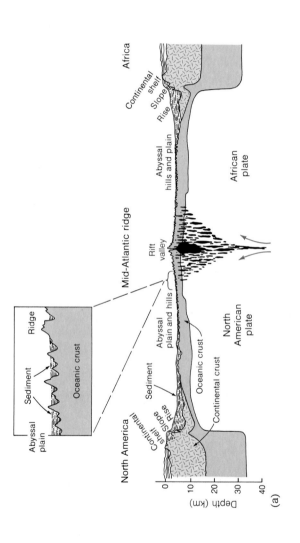

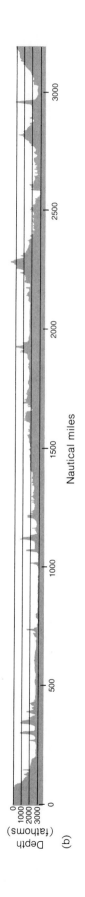

Figure 10-38

Topographic and geophysical profiles of the Atlantic Ocean. (a) Geophysical and topographic cross section deduced from seismic measurements and echo-sounding profiles from New York to the northwest coast of Africa. [After B. C. Heezen, "Physiographic Diagram of the North American Ocean," p. 102, Figure 49. Special Paper 65, Geological Society of America.] (b) Topographic profile of the floor of the Atlantic from Martha's Vineyard, Massachusetts, to Gibraltar. A portion of an abyssal plain extends from 750 miles to about 1000 miles off the east coast of the United States. The island at 2300 miles is one of the Azores. [From B. C. Heezen, "The Origin of Submarine Canyons." Copyright © 1956 by Scientific American, Inc. All rights reserved.]

Figure 10–39

Sedimentation and deformation of sediment in an oceanic trench-subduction zone. Much of the pelagic sediment and turbidity-current deposits in the trench are "scraped off" as the oceanic plate moves down the subduction zone, the sediments being of much lower density than the underlying lithospheric rocks. Some sediment may be carried along with the sub-ducting plate.

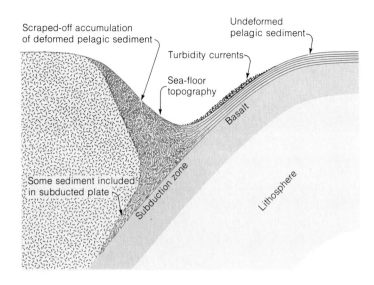

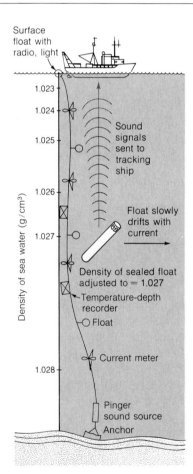

Figure 10–40

Devices such as current meters and floats are designed to measure deep currents in the ocean. Before the float is lowered, its density is adjusted to that of the water at the depth to be tracked. It then sinks to its equal-density level and begins to drift with the current, sending out sound signals as it floats with the current. The density of ocean water varies from place to place depending on the salinity and temperature. Current meters, temperature-depth recorders, and other instruments are moored by cable to an anchor.

waters of the Strait of Messina between Italy and Sicily.* In North America, it was that versatile politician-publisher-scientist Benjamin Franklin who started the serious study of ocean currents by publishing the first chart of the Gulf Stream.

Early investigations of currents made use of drift bottles—floating bottles set free in the ocean with cards inside asking the finder (for a small reward) to return the card with information about the date and place of its recovery from the sea. Modern physical oceanographers have invented instruments to measure the speed and direction of currents directly (Fig 10–40). Current speed is now most frequently measured by rotating vanes attached to a meter that gives the number of rotations per minute and the direction. These current meters can be attached at various depths to a moored cable and the data recorded electronically. Current depth and current speed at depth can also be determined by means of a closed aluminum tube designed to float at various levels in the water and to transmit its movement in a current by a built-in sound source. The sound from the float is picked up by sensitive microphones attached to the research ship. Even satellites have been used to map ocean currents. Using infrared-sensitive instruments, they can make high-quality images of the surface temperatures of the sea and so distinguish warm- and cold-water currents. Bottom currents can be inferred indirectly from

---

*Homer warned that the main danger in navigating the Strait of Messina was the whirlpool Charybdis, (now known as Garofalo). The whirlpool lies opposite Scylla (now Scilla), a rocky ledge near the Italian shore inhabited by a six-headed monster. Hence the expression "between Scylla and Charybdis," descriptive of situations in which the avoidance of one danger means exposure to another.

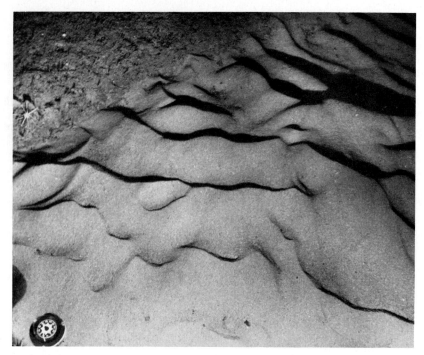

Figure 10-41
Ripples indicative of high flow velocities formed in coarse sand, with
sandy gravel in upper left corner. Ripples indicate a strong westerly
bottom current at a depth of 490 m in the Strait of Gibraltar. [From
G. Kelling and D. J. Stanley, "Sedimentary Evidence of Bottom Current
Activity, Strait of Gibraltar Region." *Marine Geology,* 13:M51-M60.
Copyright © Elsevier Publishing Co., Amsterdam, 1972.]

photographs of rippled and scoured sediment,
though it is difficult to calculate current velocities
exactly from such data (Fig. 10-41).

The surface currents of the oceans can be simplified into a pattern of large closed loops, called
gyres (Fig. 10-42). The main currents are the outside paths of the gyres. Subtropical gyres lie on
either side of the equator, the northern ones rotating clockwise and the southern ones counterclockwise. Between the westward-flowing equatorial currents that are parts of these gyres, are
oppositely directed eastward flows, the Equatorial Countercurrents. North and south of the subtropical gyres are smaller subpolar gyres; each of
these rotates in a direction opposite to that of its
companion subtropical gyre. The close correspondence between the gyres and the directions
of the prevailing winds over the oceans shows
that the winds and currents are linked: the winds
constitute the primary driving mechanism for the
surface currents of the ocean.

The most famous ocean currents are the north-
south sections of the gyres. The north-flowing
Gulf Stream is the strong western part of the
North Atlantic subtropical gyre. The south-flow-
ing California Current is the weak eastern section
of the North Pacific subtropical gyre. The east-
west equatorial currents flow at speeds of 3 to 6
kilometers (2 to 4 miles) per day; thus they move
slowly enough to adjust to the hot latitudes
through which they flow. In contrast, the Gulf
Stream flows more than an order of magnitude
faster, 40 to 120 kilometers (25 to 75 miles) per
day, so that it does not have time to adjust to its
climatic surroundings. Thus the Gulf Stream remains a warm boundary of the subtropical gyre in
the cold North Atlantic, only slowly transferring
heat to the colder air above it. This circulation
contributes significantly to heat transfer from one
part of the globe to another.

The prevailing winds can produce vertical
movements in the sea, sometimes causing deeper
waters to move up to the surface—**upwelling**—or
surface waters to move to lower depths—**sinking.**
Most deeper waters of the ocean are colder than
those at the surface. When they reach the surface,
they cool the air and cause fog to form. This is
what produces the foggy summer weather familiar to those who live along much of the Pacific
coast of North America. Upwelling and sinking
near coasts, in most places, affects the water to a
depth of about 100 to 200 meters (300 to 650 feet).

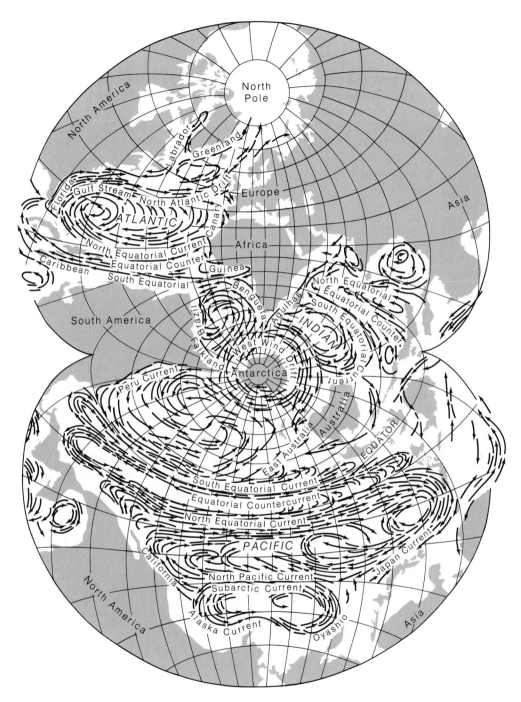

Figure 10-42
Major ocean currents shown on a map centered on the South Pole. [From W. Munk, "The Circulation of the Oceans." Copyright © 1955 by Scientific American, Inc. All rights reserved.]

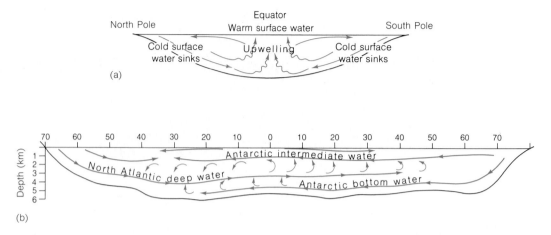

Figure 10-43
Generalized vertical circulation of the Atlantic Ocean. (a) An ideal model of a north-south
ocean that is cooled at the poles and warmed at the equator. Colder waters sink because
they are denser, replacing warmer water, which rises. (b) Simplified model of Atlantic
Ocean circulation, showing how Antarctic bottom water flows along the bottom more
than 20° north of the equator, displacing the Arctic bottom water to an intermediate level.

Upwelling is important for its biological side
effects. Deeper waters tend to be rich in nutrients,
because the sparse marine life at depth is unable
to utilize most of the phosphate, nitrate, and other
dissolved matter in it. When that rich water
reaches the sunlit surface, where organisms can
thrive in abundance, the biological communities
expand in response to the nutrients. Microscopic
plant life blooms, and fish communities that live
on it expand in proportion.

Vertical Ocean Circulation

Imagine the Atlantic or Pacific Ocean as an oval
soup bowl stretched out in the north-south direction with the cold polar waters at opposite shallow edges and the equatorial waters in the middle. The surface waters are warm near the
equator—around 25°C or warmer—while those in
the Arctic and Antarctic are quite cold—about
0°C or a little above. Because cold water is denser
than warm, the polar waters tend to sink and slide
along the bottom toward the equator (Fig. 10–43).
As they do, they push in front of them the deeper
waters, which tend to rise near the equator, being
displaced from both directions. This simplified
model explains how it happens that, just below
the shallow, warm surface waters of the equatorial ocean, the deeper waters are cold. Because the
dense, cold waters move slowly and mix with surrounding waters very slowly, they tend to retain
their original temperature and salinity (saltiness
of the water varies slightly from place to place in

the ocean). Thus the distinctive temperature and
salinity of such waters allows them to be identified as large bodies, or **water masses,** and allows
the place of origin of a water mass to be determined. North Atlantic deep water can be traced
from its polar origins in this way as it sinks and
moves southward. Before it reaches the equator, it
meets and rides over a mass of still colder and
denser water, Antarctic bottom water, which has
worked slowly northward from the cold Antarctic
Sea. The equatorial movement of bottom water
tends to take place mainly on the western sides of
oceans.

The topography of the ocean floor affects deep-
water currents. They may be blocked by ocean
ridges or they may flow through gaps in a ridge.
The water masses are density-driven, so they follow topographic highs and lows just as turbidity
currents do. The relatively new science paleo-
oceanography is based upon the relationship of
bottom topography to deep currents and upon the
ways in which continental masses separate the
oceans and thus affect the general circulation. The
reconstruction of continental drift patterns and
the evolution of ridges and troughs on the sea
floor are the raw materials for hypotheses of the
nature of oceanic circulation patterns in the geologic past. This is particularly so for the Cenozoic
Era, for which we have much information on sediments and continental drift histories. For example, some oceanographers think an important
change came about 50 million years ago, when
Australia broke away from Antarctica and the
Drake passage between Antarctica and South

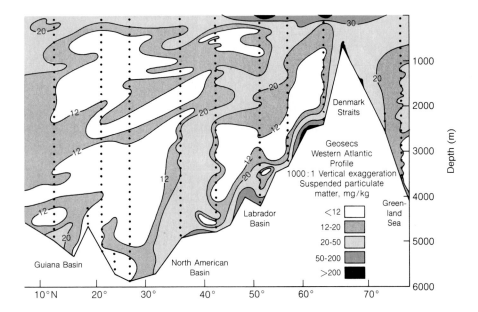

Figure 10-44
A longitudinal section of the North Atlantic Ocean showing the dry weight of particulate matter as mapped by GEOSECS. Vertical lines of dots show position of GEOSECS stations. [After Brewer and others, "The distribution of particulate matter in the Atlantic Ocean," *Earth and Planetary Science Letters*, v. 32, 1976.]

America deepened. The way was cleared for the strong wind that circulates around the Antarctic and for the southern ocean circulation systems that we know today. These circulation patterns had a strong effect on world climate and led to the development of an ice-capped Antarctic continent.

The density-driven vertical circulation is much slower than the wind-driven horizontal circulation at the surface. Deep waters rise at the equator at speeds of only 2 to 5 meters (6 to 15 feet) per year. Polar waters take about 1000 years to circulate, a fact that has become an important practical matter. If pollutants are introduced into high-latitude bottom waters, they will not rise to the surface near the equator for hundreds of years. Thus radioactive waste materials might lose a significant amount of their damaging radiation before they return to the surface. On the other hand, once introduced to such a slow-moving system, toxic materials may remain in circulation for a very long time.

The precise measurement of the chemical composition of sea water in various parts of the ocean is necessary to understanding the details of ocean circulation. The data enable oceanographers to understand how and where natural and synthetic chemical substances are introduced and how they are distributed throughout the world's oceans. GEOSECS (Geochemical Ocean Sections) was designed to do just that. One of the major pro-

grams of the International Decade of Ocean Exploration, GEOSECS was a multinational cooperative study of the oceans from 1970 to 1980. This large-scale geochemical study of the seas taught us much about the movements of water masses, the distribution of radioactive substances in the oceans, and the fate of carbon dioxide as it moves from the atmosphere to the oceans (see Chapter 12). In addition, GEOSECS has given us new understanding of the nature and distribution of particulate matter in sea water (Fig. 10–44); such particles, after all, are part of the raw material of sedimentation in the deep sea.

Deep-Sea Sedimentation

The general circulation of the ocean is too slow to accomplish much erosion, except where such strong currents as the Gulf Stream travel along the shallower edges of the sea. Slow-moving currents do have enough mild turbulence, however, to keep fine-grained sedimentary particles suspended for a long time before they settle through the great depths of water to the bottom. Though sea water may look clear and transparent, careful filtering of large volumes of deep water, such as of 100-liter samples (about 25 gallons), can extract a few thousandths of a gram of suspended mineral solids. In the past few years oceanographers have invented sediment traps that can be attached to

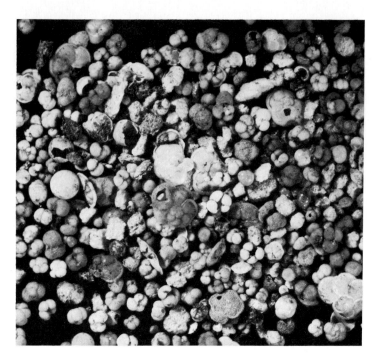

Figure 10-45
Foraminiferal ooze dredged from a depth of 450 m off the coast of Central America. Enlargement about 15×. [Photo by Patsy J. Smith, U.S. Geological Survey.]

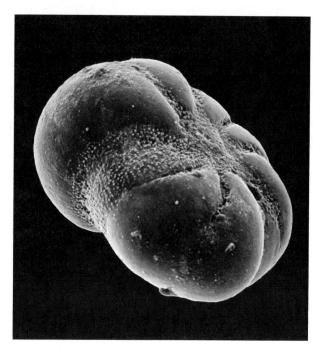

Figure 10-46
Scanning electron micrograph of a Pleistocene foraminifer (*Proteiphidium orbiculare*). Enlargement about 140×. [Photograph by T. Cronin.]

moored lines for many months and collect samples large enough to evaluate how much and what kind of sediment falls to the bottom. The particles are smaller than 0.025 millimeter, and a great many of them are 0.001 millimeter or less. Identification of the particles shows that they include fragments of clay minerals, quartz, feldspar, and other common minerals produced by the weathering of rocks on the continents. Many of the smaller particles are included in larger packages, fecal pellets, produced by organisms grazing the surface waters. These particles in the water are of the same general kind as those that constitute **pelagic sediments,** the gray and brownish muds that cover much of the deep-sea floor. There is little question that the muds on the bottom were formed by the debris through the great depths of water.

Particles so small would take a long time to settle through the water, for their rate of fall is slowed by the frictional resistance of the water. Even in perfectly still water, a particle 0.001 millimeter in diameter would take months to fall several kilometers to the deep-sea bottom. The settling is made even slower by the slow horizontal and vertical movements of the water which may lift the particle up again, extending its lifetime as a suspended particle for many years. Some oceanographers now believe that the abundant fecal

pellets found in sediment traps are the main way in which finer particles are transported to the bottom. Larger and heavier, they fall rapidly through the water column.

The rates at which these particles are laid down as sediment are extremely slow, in most places about 1 millimeter every 1000 years. At that rate, an ocean bottom 4 kilometers deep would take 4 billion years, almost the whole age of the Earth, to fill up! Of course, the ocean bottom does not stay still for that long: as part of a lithospheric plate, it is moving horizontally at the rate of a few centimeters per year, eventually to be subducted into the mantle.

Though geologists have always known that wind transport plays some role in the deposition of pelagic sediments, only in the last decade have they realized how important that role is. Air samples taken at sea by oceanographic research ships and samples of ice from the Greenland and Antarctic glaciers both showed evidence of airborne dust thousands of kilometers from the effects of human industrial pollution. This steady worldwide dust contribution forms about 10 percent of the pelagic sediment. At some times, particularly after violent volcanic eruptions, the proportion may be much greater.

Another important component of pelagic sediment is the calcite (calcium carbonate) shell mate-

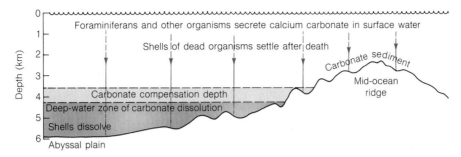

Figure 10-47
The calcium carbonate compensation depth is the level in an ocean below which
the calcium carbonate of foraminifera and other shelled organisms that have settled
from surface water dissolve. The depth, which is a zone rather than a sharp boundary,
varies some from ocean to ocean.

rial of pelagic foraminiferans, tiny single-celled animals that live in the surface waters of the sea (Figs. 10–45 and 10–46). Most are large enough to fall relatively quickly to the bottom when they die. There they accumulate as **foraminiferal oozes.** Very early in the course of marine geologic exploration, it was discovered that oozes are abundant at depths less than about 4000 meters (about 13,000 feet) but rare in the deepest waters of the ocean. This could not be attributed to a lack of shells supplied by foraminifers, which live near the surface and thus are unaffected by the depth of the water. A closer look at many sediments from deeper water showed a few foraminifers, usually the larger and thicker-shelled ones.

At the same time, chemical oceanographers were pointing out that the deeper waters of the oceans are colder, under higher pressure, and contain more dissolved carbon dioxide (from colder polar waters) than shallower waters. Each of these factors contributes to making calcium carbonate more soluble at depth than at the surface. Then came the hypothesis: no matter how abundant the shelled organisms are in the surface water, as their shells fall into the deeper waters that are undersaturated with calcium carbonate, they dissolve, either while settling or soon after they come to rest on the bottom.

A first ingenious test for this hypothesis was made by an oceanographer at Scripps Institution of Oceanography, who carefully machined and accurately weighed perfect spheres of calcite. He lowered them to various ocean depths for various periods of time, and then brought them back up to weigh again. The spheres submerged to shallow depths showed no change, but for those lowered into deeper waters, he observed loss of weight coupled with etched pits and irregularities, which confirmed that calcite was rapidly dissolved

there. The dissolution of calcium carbonate in the deep ocean, now abundantly confirmed by other tests, has become important in consideration of the future impact of the carbon dioxide injected into the atmosphere by our burning of fossil fuels. As more carbon dioxide enters the ocean by absorption from the atmosphere, calcium carbonate shells become soluble—just as the weathering of limestone on land is promoted by carbon dioxide in rain (Chapter 4). The shell dissolution is a way the oceans have of taking up part of the extra carbon dioxide.

The depth of water below which calcium carbonate starts to dissolve is called the calcium carbonate **compensation depth** (Fig. 10–47). It plays an important role in a plate-tectonic interpretation of the evolution of the ocean basins that has resulted from the National Science Foundation's program of deep drilling of sea-bottom sediments. Drilled samples from some parts of the ocean that are now below the compensation depth show that, some millions of years ago, calcium carbonate was abundant in the sediments. The movements of oceanic lithospheric plates offer an explanation for this finding. As part of a plate was formed at a ridge or rise, it would have been topographically high, above the compensation depth and the resting place for much carbonate sediment, just as the mid-Atlantic is today. As that portion of the plate moved away from the ridge, it would have subsided gradually as it cooled, eventually passing below the compensation depth, after which time it would no longer have received much calcium carbonate sediment (Fig. 10–48). Paleo-oceanographers have suggested that, in addition, the compensation depths in the oceans changed with altered circulation patterns of the oceans in response to continental drift and the onset of glaciation in the Cenozoic. Both hy-

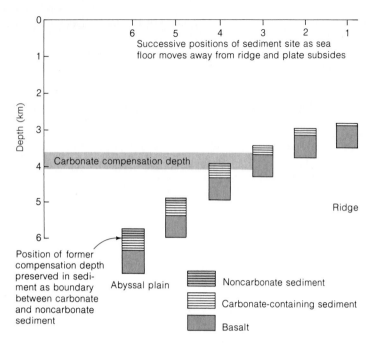

Figure 10-48
As a site of sedimentation is transported away from a mid-ocean ridge by plate movement, plate subsidence moves the site downward with respect to sea level, eventually below the calcium carbonate compensation depth. At that point, carbonate sediment is succeeded by noncarbonate deposits.

potheses are now being actively pursued with the aim of tracing the rates and paths of movement of oceanic plates in many parts of the oceans.

Calcium carbonate oozes are the most abundant of the biologically produced sediments on the sea floor. Oozes of silica—a noncrystalline form of silicon dioxide, like opal—form, in much the same way as calcium carbonate oozes, from the silica shells of diatoms, an abundant class of green, unicellular algae found in the surface waters of the oceans. These **diatom oozes** are found mostly in the Pacific and Antarctic oceans. Their distribution is governed partly by the abundance of nutrients at the surface. Even where the living organisms are abundant, however, diatom oozes are found only if there is little detritus from land erosion. In many places, such detritus forms the bulk of the sediment and so dilutes the slowly accumulating shells of diatoms and other silica-

secreting organisms. **Radiolarian oozes** are formed in areas below surface waters where radiolaria, a microorganism that secretes silica instead of calcite, are abundant.

Some components of deep-sea sediment are formed by chemical reactions of sea water with the surface of the sediment. The most prominent of these are manganese nodules, those black lumpy accumulations first found by the *Challenger* expedition, which form at the surface of the sediment over many large areas of ocean bottom (see Fig. 22–19 and discussion in Chapter 22). Manganese nodules are estimated to cover as much as 20 to 50 percent of the Pacific Ocean floor. They form slowly as dissolved manganese and other metals, such as nickel, precipitate both from the overlying sea water and from the sea water trapped in the pores of recently deposited sediment. The nature of the chemical reactions that form the nodules has long been a subject of scientific research. Much of the manganese may be contributed by hydrothermal waters of mid-ocean ridges.

As much as we have now learned about the oceans, we must learn a great deal more about the deep sea than we now know to adequately treat two concerns of practical importance. The utilization of the ocean for its mineral and food resources and the pollution of the ocean by using it as the garbage can of the world are intertwined. It is impossible to use the ocean's resources without interfering with it in some way. The imperative is to use it in such a way that the alteration will not be damaging or irreversible. The production of oil on continental shelves or of manganese from the deep sea need not permanently disrupt the biological population of the oceans nor destroy the adjacent beaches, *if* there is sufficient geological knowledge and rigorous enforcement of fail-safe engineering procedures for the prevention of accidents. The oceans can be self-cleansing if ocean dumping and oil spills are few and far between. It is heedless dumping of wastes on a global scale that will overload the capacity of the world ocean to maintain its equilibrium.

SUMMARY

1. Wave action is the major erosional agent operating along coasts. Rates of erosion by waves are fastest on shorelines of unconsolidated materials.

2. The shape of a shoreline is determined by wave erosion, modified by tectonic uplift or subsidence, sedimentation, and eustatic changes in sea level.

3. Waves are generated by wind and are described by their wavelength, height, period, and velocity. Surf is the product of waves entering shallow water. Waves radiate outward from storms at sea, their places of formation.

4. Wave refraction, the bending of waves as they approach shorelines, is controlled by the shape of the shoreline and the wind direction. Wave refraction affects erosion and sedimentation on beaches and is responsible for longshore currents and rip currents.

5. A beach is a dynamic form that is the result of a balance between material eroded from it and material transported to it along the shoreline. It can be described by a budget relating input and output. Human interference can drastically alter the budget of a beach and, thus, its shape and size.

6. Tides originate from the gravitational pull of the Moon and the Sun. The currents that develop are important distributors of sediments in tidal channels and tidal flats.

7. Continental shelves, slopes, and rises characterize the edges of continents. They are formed by sedimentation of erosional debris from the continents accumulating at the deep-water margin.

8. Formed by the suspension of mud in waters, turbidity currents are density currents that erode submarine canyons, gully the continental slopes, and deposit the distinctive graded sediments called turbidites on the continental rise and abyssal plain. Such currents are the agents of transport of coarser materials to the deep sea.

9. Plate tectonics explains the development of the profile of the oceans in terms of crust forming at the ridges, moving laterally away, and being covered with deep-sea sediment. Passive continental margins develop on the trailing edges of diverging continental plates. Active continental margins develop along subduction zones at plate convergences.

10. The horizontal circulation of the oceans is wind-driven into major and minor gyres. The Gulf Stream is the northward-flowing boundary of one of these, the North Atlantic subtropical gyre.

11. The vertical circulation of the oceans is a response to the greater density of masses of colder water formed in the polar oceans.

12. Pelagic sediments are combinations of fine-grained terrigenous clays, some waterborne and some transported by wind, combined with the calcareous or siliceous shells of small organisms, foraminifers and diatoms that live in the uppermost layers of the oceans.

EXERCISES

1. How would shoreline processes act to modify a seaward-facing scarp formed by a fault along the coastline?

2. After a period of calm winds and no breakers along the northeastern shore of the United States, an intense storm with high winds passes over the area, followed by another period of calm. Describe the state of the surf along the beach during the storm and several days afterward.

3. What time(s) during one month that you have to spend at a rocky seashore would you pick to observe the maximum exposure of a wave-cut terrace? What time(s) to observe maximum erosion of a headland?

4. Describe the changes that would be expected on the continental shelf off the eastern coast of North America if there were a eustatic drop in sea level of 100 meters. How could such a sea-level drop affect the shoreline of a volcanic island arc in the western Pacific?

5. How might you account for the contrast between the broad continental shelf off the eastern coast of North America and the narrow, almost nonexistent one off the western coast?

6. Hoover Dam was built across the Colorado River in southern Nevada. When the river valley was flooded for many kilometers upstream, forming Lake Mead, engineers noted turbid muddy waters at the foot of the dam beneath clear waters above. How might this mud suspension have been produced?

7. Given strong waves obliquely approaching a coast, would you expect longshore currents to develop in a lagoon behind a long barrier island or on the open ocean side of the barrier? Why?

8. An international consortium has given you a grant of ten million dollars to explore new areas of the ocean for hydrothermal sulfide deposits by manned submersible. Other than areas of the East Pacific rise, where would you recommend exploration?

9. Why would you expect thicknesses of pelagic sediments in depressions near the mid-Atlantic ridge to be much less than in depressions at the flanks, far distant from the ridge?

10. Where would you be more likely to find foraminiferal oozes on the sea floor, on a mid-ocean rise or on an abyssal plain? In which of these two provinces would you be more likely to find turbidites? In what particular situation might you possibly find both kinds of sediment interlayered?

BIBLIOGRAPHY

Davis, R. A., Jr., *Coastal Sedimentary Environments.* New York/Heidelberg: Springer-Verlag, 1978.

Heezen, B. C., and C. D. Hollister, *The Face of the Deep.* New York: Oxford University Press, 1971.

Heirtzler, J. R., and W. B. Bryan, "The Floor of the Mid-Atlantic Rift," *Scientific American,* August 1975.

Inman, D. L., and B. M. Brush, "The Coastal Challenge," *Science,* v. 181, pp. 20–32, 1973.

MacDonald, K. C., and B. P. Luyendyk, "The Crest of the East Pacific Rise," *Scientific American,* May 1981.

Moore, J. R., ed., *Oceanography: Readings from Scientific American.* San Francisco: W. H. Freeman and Company, 1971.

Ross, D., *Introduction to Oceanography,* 2nd ed. Englewood Cliffs, N. J.: Prentice-Hall, 1977.

Shepard, F. P., *Submarine Geology,* 3rd ed. New York: Harper & Row, 1973.

Thoreau, H. D., *Cape Cod.* New York: T. Y. Crowell (n.d.).

Turekian, K. K., *Oceans,* 2nd ed. Englewood Cliffs, N. J.: Prentice Hall, 1976.

11

Sedimentation and
Sedimentary Rocks

Sedimentation is the final stage of a process that begins with erosion and transportation of eroded materials to sites of deposition. Physical sedimentation is the deposition of such materials in the lowest places to which air and water currents can transport them. Chemical sedimentation is mainly the process by which sea water keeps a constant composition by depositing precipitates to balance the dissolved weathering products brought in by rivers. Calcium carbonate makes up the largest volume of chemical sediment, much of it extracted from sea water by invertebrates and secreted as shells. Silica-rich sediments are also largely produced by organisms, mainly the diatoms. Bacteria play an important role in sulfide sedimentation in environments lacking oxygen. Gypsum and salt form from the evaporation of sea water in isolated basins. Depositional patterns of sediments are strongly affected by the geomorphic environments in which they were deposited. Tectonism controls both subsidence in the depositional area and weathering at the source of the erosional debris. Chemical and physical changes after deposition convert soft sediment to rock and cause many other alterations of composition and texture.

Sedimentation, the settling out of suspension or deposition in a layer, is a word with many associations. The engineer designing a dam on a river for flood control worries about sedimentation filling the reservoir in back of the dam. The ship's captain navigating through sandy areas near shore is concerned that the shifting patterns of sedimentation may have built up shoals on which the vessel may run aground. Geologists are concerned with all of the kinds of sedimentation on the surface of the Earth and how they play a role in the overall dynamics of the planet. Sediments include both solid materials physically deposited by wind, water, or ice and dissolved substances chemically precipitated from oceans, lakes, or rivers. Physically deposited sediments have been discussed in relation to erosion and transportation in Chapters 5 through 10; here we shall con-

sider them in relation to chemically deposited sediments and in the context of the accumulation of sediments in general.

SEDIMENTATION AS
A DOWNHILL PROCESS

Physical sedimentation starts where transportation stops. When the wind dies down, dust settles; when water currents slow, sand settles. On Earth, physical transportation and sedimentation follow a general downhill trend in response to gravity, from rock falls and mass movements downslope to river systems, and then down to the sea (Fig. 11-1). In running water, sedimentation is a one-way street, each temporary stage of transport and sedimentation carrying the sediment farther

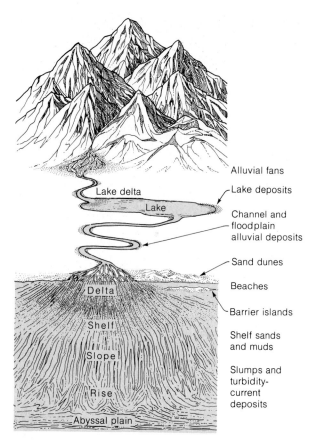

Figure 11–1
The downhill path of transportation and deposition
takes solid particles from the heights of mountains
to the depths of the oceans, many different sedimentary environments of physical deposition being encountered along the way.

toward the bottom of the deep sea. Much of it is
dropped permanently along the way and never
reaches the end of the line. Eolian sedimentation
is different, for winds may blow material from
low to high places and back again. But in the long
run, eolian sedimentation is effectively a one-way
street, too: once windblown material drops to the
ocean surface, it is trapped. It settles through the
water and cannot be picked up again.

Chemical sedimentation is also a downhill
process, but the driving force is chemical rather
than gravitational. A major aspect of weathering
and erosion is the chemical decay of rocks exposed to the water and carbon dioxide of the atmosphere. In the course of decay, ions from the
rocks are dissolved, and rivers carry them to the
sea (see Fig. 6–19). The ocean may be thought of as
a huge chemical reservoir: water continually
evaporates from the surface, and fresh river water
runs in to replenish it. Although that keeps the
amount of water fairly constant, it works also to
enrich the sea in the dissolved ions: evaporation

takes away only the water; the ions do not evaporate.* Yet the sea maintains the same salinity. It
does so because of sedimentation of the dissolved
material as chemical precipitates. Totalled over
all of the oceans of the world, those precipitates
must balance the total inflow of ions released by
weathering and brought in by rivers.

Understanding the nature of both physical and
chemical sedimentation processes, geologists can
work backward from the properties of a sedimentary rock to deduce its history. They can read that
history in terms of the distribution of land and
sea, mountains and plains, deserts and swamps—
in other words, the **paleogeography** of the ancient
time when the sediment was formed.

Many sedimentary geologists start by analyzing
the physically deposited sediments, for they are
the most useful in deciphering the major tectonic
elements of paleogeography—the uplands that are
the source of detritus, the low basins that receive
the sediment. They also are a key to the nature
and geography of the transportation network by
which the solid products of erosion are carried to
sedimentary environments.

CLASTIC SEDIMENTS

How are we to discover what we want to know
about transportation and deposition of solid particles, particularly in relation to the tectonic and
erosional forces that produce detritus? **Clastic
sediments,** those made up of particles broken and
weathered from preexisting rock, tell that story,
much of which has been covered in earlier chapters on erosion and transportation. These sediments are also frequently referred to as **detrital,**
for they are made up of the detritus of erosion.
Clastic sediments—shales, sandstones, and conglomerates—are far more abundant than the
chemical precipitates. They constitute well over
three-quarters of the total mass of sediments in
the Earth's crust. Of the clastic sediments, shale is
by far the most abundant rock type, being about
three times as common as the coarser clastics.
Clastics dominate because mechanical erosion is
constant and widespread over the Earth and because much of the crust is composed of silicates
that are relatively insoluble at Earth's surface
temperatures and pressures.

---

*Though the dissolved salt in sea water does not evaporate,
the air above the ocean always contains some small crystals of
sea salt that are left floating after droplets of sea spray from
breaking waves have evaporated. This salty air causes automobiles near the ocean to rust much more than those inland, because the salt is corrosive to metal.

Table 11-1
Classification of sand by grain diameter

Classification	Grain size (mm)
Very coarse sand	1.0–2.0
Coarse sand	0.5–1.0
Medium sand	0.25–0.5
Fine sand	0.125–0.25
Very fine sand	0.0625–0.125

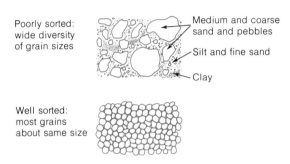

Figure 11-2
Sorting of sedimentary particles produces sediments with a small size range of particles. See also Figures 4-15 and 11-3.

Sedimentation of particles is largely controlled by the strength of the currents that carry them. The larger the particle, the stronger the current must be in order to carry it. It was natural, then, for geologists to distinguish the various clastic sediments on the basis of particle size. Such a division reflects the conditions of sedimentation. A gravel, or its lithified equivalent conglomerate, calls to mind strong currents of rapidly flowing rivers in mountains, or high waves on a rocky beach. Mud, or its rock equivalent shale, suggests quiet waters that allow the finest particles to settle. Sand and sandstone imply moderate currents, such as the water currents of rivers and shores, or the strong winds that blow sand into dunes.

Sand and Sandstone

Sandstone, more than any other clastic sediment, contains easily read information about its origin. Sand particles are large enough to be seen with the naked eye. Under a microscope, the minerals can be identified by their optical characteristics, and their textures can be studied. In contrast, most of the particles in shale are too fine for optical study, even with high-power microscopes. Sandstones are more abundant and widespread than conglomerates, and they tend to occur as erosion-resistant formations that make easily observed outcrops.

The range of sizes of sand and sandstone grains has been studied extensively as an aid in deciphering their origin. Table 11-1 shows how sands are divided on the basis of the average diameter of grains. The grain-size groups are divided on a geometric basis rather than an arithmetic one; each group contains particles ranging from the minimum size up to twice that minimum size. This kind of system takes into account the differences in relation to the sizes, thus distinguishing, for example, between the big difference 0.1 millimeter makes in a 0.5 millimeter sand grain and the negligible difference 0.1 millimeter makes in a 5 millimeter pebble.

Average size does not tell the whole story. If all of the grains in a sample are close to the average size, the sand is considered *well sorted*; if there are many grains much larger and smaller than the average, the sand is poorly sorted (Fig. 11-2). Because **sorting** is related to the kind of depositing current, we can distinguish between well-sorted beach or dune sands and poorly sorted turbidite sands. The degree of sorting can be measured quantitatively; this allows fine distinctions to be made as to the kind of current that transported the sand.

The shapes of sand grains can be important guides to their origin. From observation of modern sands and from experiments with laboratory-simulated transportation, we know that sand grains are abraded as they are knocked against each other by currents. Grains lose their sharp edges and corners, their angular aspect, and become rounded. At the same time, they tend to assume more spherical shapes. From a look at the rounded, spherical grains of some ancient sandstones, such as that in Figure 11-3, a sedimentologist can infer that they had a long, arduous journey. Angular grains of various shapes imply a short transportation distance.

Surface textures such as frosting of desert or dune sands can be clues to origin. The scanning electron microscope is used to study shapes and patterns of fractures, pits, and smoother areas as guides to desert, glacial, beach, and other environments.

The bedding of sand and sandstone gives even more useful information, for we can recognize cross-bedding, ripple marks, and those characteristic bedding-plane structures that are associated with turbidites (see Chapter 10). These help identify the sedimentary environment in which the deposit was laid down. Cross-bedding and other

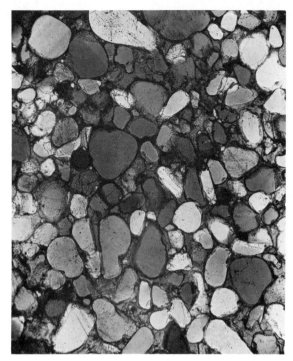

Figure 11-3
A microscopic view of a thin section of St. Peter Sandstone (Ordovician) of Illinois. All of the grains are well rounded and well sorted by the action of transporting currents. Almost all of the grains are quartz. (Magnification 44×.) [From *Sand and Sandstone* by F. J. Pettijohn, P. E. Potter, and R. Siever. Copyright 1972. Springer-Verlag, New York, Inc.]

current structures also are guides to the current direction. They can be measured in the field to make a **paleocurrent** map, showing the directions of sediment transport (Fig. 11-4). If the sandstones were deposited by a river, the paleocurrent map can be viewed as a map of the general land slope down which the rivers ran. If the bed structures of a turbidity current are measured, we can construct a paleocurrent map of the submarine slopes down which the muddy suspension flowed. Thus we can map part of the route of sediment migration from erosional source to depositional basin.

The mineralogy of sandstones is important in reconstructing the nature of the source areas that produced the detritus. In Chapters 4 and 5 we saw that chemical weathering can remove much of the feldspar from an eroding granite, so that the resulting detritus is mainly quartz, eventually producing a quartz-rich sandstone. That same granite in a rapidly eroding terrain on a mountainside can produce a feldspar–quartz mixture that will make an arkosic sandstone. But mineralogy also tells us something more of the source—the kind of rock

being eroded. Certain minerals typical of metamorphic rocks, such as the silicates kyanite and staurolite (see Appendix IV), are diagnostic of metamorphic terrains formed at various temperatures and pressures. Thus we can glean something of the paleogeology of the source area by estimating the proportion and kinds of igneous, metamorphic, and sedimentary rocks being eroded there.

On the basis of mineralogy and such textural attributes as sorting, geologists have classified sandstones into several major groups. The **quartz arenites** (arenite = sand-size material; drawn from *arena,* the Latin word for sand) are sandstones made up almost entirely of quartz grains, usually well sorted and rounded. **Arkoses** contain much feldspar; the grains tend to be poorly rounded and less well sorted than quartz arenites. **Lithic arenites** have a high content of fine-grained rock fragments, mostly from shales, slates, schists, or volcanics. **Graywackes** are made up of sand grains of quartz, feldspar, and rock fragments surrounded by a fine-grained clay matrix; much of this matrix is thought to be formed by the squashing and alteration of soft rock fragments, such as those of shale, after deep burial of the sandstone formation. Many graywackes are turbidites, and some of the matrix may have been formed from the clay deposited by the muddy turbidity current.

As discussed in the next section sands are deposited in many different environments: rivers, deltas, estuaries and tidal flats, beaches and offshore bars, continental shelves, slopes, and rises, and deep-water turbidite basins. The properties of sand accumulations are particularly helpful for deducing the environment of deposition. Paleocurrents, grain size and sorting, mineralogy, and many other characteristics of sandstones are studied intensively by petroleum geologists, for much of the world's oil is found in the pore spaces of sandstones. Knowledge of how the sand was deposited guides geologists in their exploration for new oil fields and assists them in the reservoir engineering required to extract the oil from the ground. Such knowledge is also a necessity for the geologist seeking groundwater supplies in porous sandstone beds.

Gravel and Conglomerate

If sand grains are easy to study, the pebbles of a conglomerate are that much more so (Fig. 11-5). The pebbles are as varied as the different kinds of

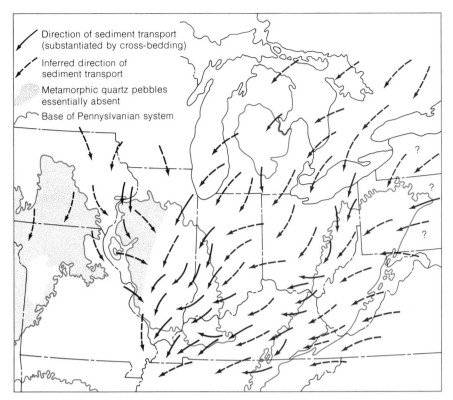

Figure 11–4

Map of paleocurrent directions based on cross-bedding in the basal Pennsylvania sandstones of part of the northeastern United States. General directions correspond to areas distinguished by presence or absence of metamorphic quartz pebbles. [After P. E. Potter and R. Siever, "Sources of Basal Pennsylvanian Sediments in the Eastern Interior Basin," *Jour. Geology*, v. 64, p. 242, 1956. Copyright © 1956 by University of Chicago Press.]

In the map legend:
- Direction of sediment transport (substantiated by cross-bedding)
- Inferred direction of sediment transport
- Metamorphic quartz pebbles essentially absent
- Base of Pennyslvanian system

outcrops from which they were eroded, and the information they provide is explicit. A granite pebble is sure evidence of the existence of a mass of exposed granite in the drainage basin of the rivers that brought the pebbles to the site of deposition. The number and size of the pebbles is directly related to the strength of the current that transported them. Because there are relatively few kinds of places in which currents strong enough to transport pebbles can flow, we can often infer the origin of the pebbles with some confidence. Mountain streams carry gravels, and alluvial fans are among the important places where they are dropped. Many river channels carry pebbles in their upper and middle reaches. Beach gravels are common along rapidly eroding coasts of hard rocks. Glacial outwash typically contains much gravel.

Because pebbles and cobbles abrade and become rounded very quickly in the course of transport, their shapes and the roundness of edges and corners are good guides to the distance they have traveled from their source area. Maps of average

Figure 11–5

Conglomerate and associated sandstone, Puente Hills, California. [From *Geology Illustrated* by J. S. Shelton. W. H. Freeman and Company. Coyright © 1966.]

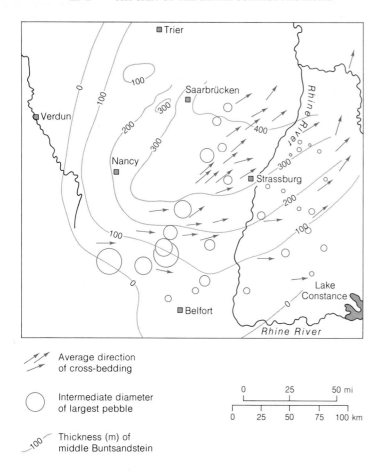

Average direction
of cross-bedding

Intermediate diameter
of largest pebble

Thickness (m) of
middle Buntsandstein

0   25   50 mi

0   25   50   75   100 km

Figure 11-6
Change in pebble size associated with cross-bedding
directions in the Triassic Bunter Sandstone of France
and Germany. Note the increase in the thickness of
sandstone in the down-dip direction of cross-bedding;
this increase is associated with deposition of an alluvial plain. [From P. E. Potter and F. J. Pettijohn,
Paleocurrents and Basin Analysis. Copyright 1963
by Springer-Verlag, New York, Inc.]

or maximum pebble size show a steady decrease
in pebble diameter in a certain direction (Fig. 11–
6). In some studies, cross-bedding in the formation shows the same direction. Because there is a
limit to the size of pebble any river can carry, generally about 25 centimeters in diameter, one can
project upstream from measured sizes, knowing
the rate of size increase from the map, to the position where 25-centimeter pebbles would be
found. That would be the source area. In this way,
ancient mountain fronts have been deduced from
sediment characteristics.

Conglomerates can be of local origin within the
sedimentary environment. Strong storm waves
can tear up previously deposited, partly compacted sediment and redeposit the irregular pebbles in a new matrix of sand or mud when the

storm dies down. This happens most commonly
with such fine-grained deposits as shales. The
boulders and cobbles of a coral reef that accumulate along its margin as a talus slope may also
form conglomerates. Gravel talus slopes resting
against continental mountains are continually
stripped away by erosion and so are poorly preserved in the rock record.

Mud and Shale

The fine-grained sediments of the Earth, which are
the most abundant and widespread, reveal the
least about their formation. All too often, their outcrops are completely covered with soil or rubble.
The material itself must be studied mainly by
x-ray diffraction or chemical methods. Much of
what we can determine from shales about their
formation is a general deduction from their fine
grain size: they are the result of slow settling from
very gentle transporting currents.

On a river floodplain, muds and silts may be
deposited from a waning flood or in an oxbow
lake formed from a cut-off meander. Muds are left
behind by ebbing tides along many tidal flats
where wave action is not great. Below the depth
of effective wave transport, mud settles over
much of the deeper parts of continental shelves,
slopes, and rises. This slow, steady deposition of
fine particles also makes its contribution to the
extensive deep-sea sediments that blanket ridges,
trenches, abyssal hills, and abyssal plains. Mud
deposited by waning turbidity currents accounts
for much of the sediment of abyssal plains.

Because muds are more likely than other grain
sizes to contain small bits and pieces of organic
matter remaining from the decay of organisms,
they are attractive to organisms that live in or
upon them to gain their food. Worms, burrowing
clams, crustaceans, and other kinds of organisms
feed by eating sediment and digesting the minute
quantities of organic matter found in it. They ex-
crete the unused inorganic bulk of the sediment,
leaving a record of burrows, tracks, and trails,
which the geologist recognizes as the hallmarks of
bioturbation, the reworking and modification of
sediment by organisms. The nature of the bioturbation serves as a guide to the types of organism and, to the extent that organisms are found
in specific environments, tracks and trails are
used to determine the sedimentary environment.

Many muds and shales are mixed with chemical sediments; calcareous shale and shaly limestone are the most prominent types. Cherts, evaporites, and other chemically precipitated rocks

may be mixed with mud, and some sediments grade from one pure rock type to the other with all the intermediate transitions. Frequently, it is this chemical component that gives the shale a distinctive character and, thus, a name. Black or organic shales are those that contain abundant organic matter, evidence of formation in a poorly oxygenated environment that allowed the organic compounds to escape oxidation and decay. Ferruginous (iron-rich) shales are those rich in such iron minerals as hematite.

Perhaps the major significance of shales lies in their mineral composition. Most shales are at least one-third clay minerals; thus, they represent what we might call the dregs of weathering. Each shale bed is evidence of the weathering of a great volume of feldspar and other silicate minerals to produce the major groups of clay minerals. The clay mineral type may provide some basis for interpreting the weathering regime that produced it and hence some idea of climate. But clays may be altered by chemical changes after long or deep burial in the crust, and so the record of origin may be obliterated.

Though shales may not provide the best indication of the geological conditions at their time of formation, they are economically very useful as the raw material for brick, tile, china, and pottery of all kinds. Mixed with limestone, shale becomes the raw material for Portland cement, the common cement used in making mortar and concrete. At this point in history, one kind of shale stands out as most valuable—oil shale. One of the world's major oil-shale reserves is in the western United States. The Green River Shale owes its high oil content to the geological transformation of abundant algae and other organisms that grew in an enormous lake covering parts of Wyoming, Colorado, and adjacent states. The Green River Shale has, in addition to its high organic content, a most unusual group of chemically precipitated minerals, the product of alkaline conditions in the lake.

CLASTIC SEDIMENTARY ENVIRONMENTS

How does one make order out of the almost bewildering array of different kinds of sediments of all ages and take advantage of all the information they contain about what the earth was like when they were deposited? One way is to use the concept of the **sedimentary environment**—that is, all of the important physical and chemical factors or variables that control the formation of a particular sedimentary type.

We might describe a sedimentary environment in a limited way by making a statement about its average kinetic energy—that is, the energy of motion of its water. We might thereby lump together deposits of point bars in rapidly flowing rivers, sandbars off a beach, and some turbidity-current deposits on the continental rise. Or we might characterize a chemical environment by saying that the water is relatively pure—that is, fresh. That would put together soils, sediments formed in rivers and lakes, and permeable rocks through which fresh groundwater percolates. These kinds of environmental definitions limited to one variable generally fail to be useful because they lump rather diverse sediment types and are not easily used for reconstructing the past.

A more useful general definition of a sedimentary environment is one that emphasizes the geomorphologic approach as a meaningful way to group the variables. Thus we might describe a lagoonal environment as one of limited area, separated by some barrier from the open sea, in which the water is relatively calm and the depths shallow. Some varieties may be further characterized by restricted circulation and lack of oxygen in bottom waters, and reef lagoons may be distinguished as a special type. This approach, which unites all the variables by the geometric shape of the place and its relation to adjacent environments and larger geographic and geologic elements of the Earth's surface, has proved most useful. For convenience we shall divide environments into two broad categories corresponding to the division of sediments into clastic and chemical.

Clastic sedimentary environments, those dominated by deposition of gravels, sands, silts, and muds, include the nonmarine alluvial, desert, and glacial environments; the mixed marine and nonmarine deltaic and shoreline environments; and the wholly marine shelf, turbidite, and pelagic environments. These could be subdivided into smaller units, down to small areas that have been called "microenvironments." Environments may not be entirely sedimentary. For example, environments around volcanic islands may be characterized by mixtures of shallow- and deep-water sediments (derived by the erosion of the volcanic rocks on land) with submarine volcanic flows and pyroclastic deposits from violent eruption. What is important in any such use of a particular environmental designation is that a set of rock characters—the only evidence a geologist sees—should be associated with a geomorphologic environment in which certain well-defined processes operate. In the past decade sedimentologists have refined and sharpened definitions of environments and which of the many sediment properties are particularly diagnostic of them. Those studies have compared observations of modern environments and the sediments found in them with ancient sedimentary rocks and their stratigraphy. In the sections below we describe the most important environments from this perspective, in a different way than we have looked at processes and landforms in Chapters 7 through 10.

Alluvial Environments

Because of the range of processes associated with rivers (Chapter 7), the alluvial environment includes river channels, meander belts on flood plains, alluvial fans, and alluvial plains formed by rivers migrating widely over lowlands. In Figure 7–26 the relation of channel, point bar, and floodplain deposits to migration of the river channel is shown. Therein lies the basis for deducing an environment from its sediment pattern. As the channel migrates during the buildup of sediment on a valley floor, it leaves behind a distinctive sedimentary sequence. The channel-floor deposits, coarse sand and gravel with large-scale cross-bedding, are overlain by point-bar deposits, finer-grained sands with smaller-scale cross-bedding, and finally by the sand, silt, and mud deposits of the floodplain at the top. The successively finer sediments give the name to this sequence, the **fining-upward alluvial cycle,** that is characteristic of this environment. It has been recognized in many stratigraphic sections, from early Precambrian to Recent.

Alluvial fans show slightly different sequences deposited by migrating distributaries on the fan. More gravelly, many tend to show poorly sorted mixtures of clay, sand, and gravel that result from debris flows at the sides of the fan after heavy rains. Alluvial plains show sequences of interfingering fining-upward cycles over broad areas. Braided channels, thinner and more variable than meandering ones, show sequences of coarse, cross-bedded channel deposits overlain by finer deposits formed on longitudinal and transverse bars. Alluvial systems are parts of other environments too; they grade into deltaic, glacial, and desert environments.

Desert Environments

The dryness of deserts allows sand to be blown by the wind (Chapter 8). The characteristic dune forms have their own patterns of cross-bedding, much of it on a very large scale. The sand grains are fine to very fine, well sorted, and many are frosted. Dune deposits grade into alluvial deposits of desert rivers and into playa lakes or sabkhas in interior basins or along arid coasts. The sabkha deposits are typically saline sediments such as salt or gypsum. Extensive dune deposits with little else are formed in the large expanses of sandy desert found where there is an abundant source of sand.

Glacial Environments

In addition to the special settings on, in, and under the ice, glacial environments include alluvial environments in front of the ice, the glacio-marine environment where glaciers calve icebergs in the sea, and the eolian environment where glacial rock flour of alluvial outwash flats is transported by strong off-glacier winds and deposited as loess (Chapter 9). The deposits under the ice, the tills, are recognized by their heterogeneous, unsorted character. Identification of a glacial environment is confirmed by the presence of striated bedrock pavements. Though the landforms of eskers, drumlins, and kames are rarely recognizable in pre-Pleistocene glacial deposits, their sediments are identifiable as alluvial, modified by the glacial associations.

Deltaic Environments

Deltas are a major dropping point for river sediment (Chapter 7). The delta environment is complex, including the alluvial delta-plain, the distributary channels, overbank, and splay deposits formed during floods. These interfinger with salt-marsh and shallow-marine interdistributary areas, the delta-front area just offshore of the distributary mouths, and the sea floor farther out that is overridden by the advancing toe of the delta. The key to identifying ancient deltas is the stratigraphic pattern of alluvial fresh-water and fossiliferous marine deposits. Bar-finger sand sequences typically show coarsening-upward cycles developed as the river mouth advances, depositing coarser sands of the channel over finer silts and muds offshore. In the marine zones carbonate sediments may be abundant, and in interdistributary areas brackish- or fresh-water swamp deposits may be found. Some ancient coal beds are found with deltaic sands as a product of this association.

Beach and Bar Environments

Sediments of the coastal zone are dominated by sands on the beaches and barrier islands, with finer-grained silts and muds in lagoons (Chapter 10). The sands are well sorted, rounded, and perhaps frosted if found on backshore dunes. Cross-bedding, bedding gently inclined toward the sea, and in the surf zone, oscillation ripples are the rule. Gravels can also be found on beaches. A typical sequence formed from a shoreline that is building out into the sea would be a fine-grained subtidal sediment, one formed below low tide, overlain by tidal zone deposits, typically medium- to coarse-grained sand. The sequence then continues relatively coarse-grained upward to surf zone bars and beach sands and is topped by salt-marsh organic-rich muds or dune sands. If a lagoon is present behind a bar, then fine-grained muds with abundant organic matter may be found between the bar and shoreline deposits.

In these as in other sandy environments, the size range of the sand depends on the particular sizes of sand brought to the area from the source.

Shallow Marine Environments

The dynamics of sedimentation on continental shelves are determined by the action on the bottom of waves produced under prevailing winds, storms, and hurricanes, and the tidal currents of different tidal ranges. Sands, sandy muds, and muds are found in relation to these currents and waves. Medium- and fine-grained sands are distributed on the shallower parts of the shelves as thin ribbons (centimeters thick by a meter wide and several hundred meters long) and as cross-bedded dunes up to ten meters high. Rippled finer-grained sands and silts may be interspersed with patches of shelly material in areas of weaker currents. Muds are typically found in depressions too deep for wave action and sheltered from other currents. All sediments may be bioturbated by the abundant marine life on and in the sediment.

Turbidite Environments

The name of turbidite environments is given by the kind of current that transports and deposits the sediments on the continental slopes, rises, abyssal plains, and trenches of the oceans (Chapter 10). All turbidite sequences show the same order, but some units may be absent depending on the particular subenvironment. Sequences begin with abrupt bases of coarse, structureless sand, in some places containing pebbles or ripped-up pieces of the previously deposited sediment. This grades upward to medium-grained, cross-bedded sands, then to finer-grained rippled sands, and is topped by silts and muds, the latter typically bioturbated. The upward grading of grain size and sedimentary structures in a continuous sequence, the products of waning turbidity current strength, are the criteria for identifying these environments. The coarseness of the units and completeness of the sequence are guides to the nearness of the slump that initiated the flow. The muddy top is missing from the sequence in areas nearest the slump, and the coarse base from the sequence farther away. Patterns of sequences identify upper and lower parts of submarine fan deposits at the foot of a continental slope and abyssal-plain sediments. Turbidites may be interbedded with pelagic sediments.

Pelagic Environments

The nonturbidite clastic deposits of the deep sea are fine-grained "red" (actually a range of reddish-brown) clays. The color is the result of slow oxidation of iron in the clay, which is exposed to sea water for long times as the rate of sedimentation is so slow (a few millimeters per thousand years). Clays are finely laminated, and manganese crusts and nodules are common. Volcanic ash beds are common in association with plate convergences. Pelagic clays are interbedded with carbonate oozes above the carbonate compensation depth and with siliceous oozes at any depth.

Sedimentary Facies

We know that a variety of different sedimentary environments exist at the same time in a region. For instance, we can go to a continental margin and see well-sorted sand being deposited at the shore and in very shallow near-shore waters; farther out, in deeper waters of the continental shelf, silts and muds are being laid down. Still farther out, on the continental slopes and rises, turbidity currents are depositing sand, silts, and mud. To characterize such sets of simultaneously deposited sediments, we use the word **facies** to emphasize the lithologic aspects of sediments deposited in different environments at the same time in history. Geologists have concentrated on this idea of sedimentary environment and facies because it has proven immensely useful in reconstructing

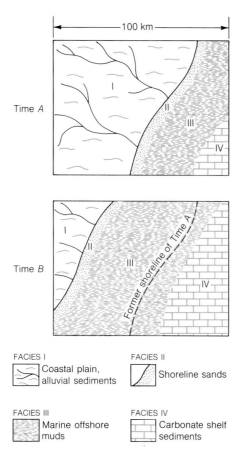

FACIES I
☐ Coastal plain,
   alluvial sediments

FACIES II
☐ Shoreline sands

FACIES III
☐ Marine offshore
   muds

FACIES IV
☐ Carbonate shelf
   sediments

Figure 11–7
Paleogeographic maps based on environments deduced from sedimentary facies for
two time periods, A and B, where B may be
a million years later than A.

the history of the Earth in past epochs. The construction of a paleogeographic map is based on
the analysis of the facies that formed at the same
time in the past in different sedimentary environments (Fig. 11–7).

The historical geologist who deduces an offshore shale facies, a nearshore and shoreline
sandstone facies, and an alluvial sandstone-silt-
stone facies and can plot them on a map has
thereby reconstructed the position of the ancient
shoreline and the land and sea environments associated with it. If at the same time one can infer
the composition of the source rocks of the clastic
sediments, paleogeology can be added to the
paleogeographic map.

Drawing maps of the same area for two successive time periods in Earth history shows changes,
commonly a shift in the position of the shoreline.
A shift showing that the sea later covered more of
the continent's edges, or even its interior, than
previously indicates a **transgression** of the seas on

the continents. A map constructed for a still later
period might show that the transgression was followed by a **regression,** during which the seas
withdrew (see Fig. 11–7). Transgressions and regressions may be the result of local tectonic
movements that elevate or depress that portion of
the continent relative to sea level. **Seismic stratigraphy,** the use of high-resolution seismic exploration methods to construct profiles of sedimentary
sequences of continental margins, has revealed as
never before the patterns of transgressive and re-
gressive sedimentation associated with sea-level
changes. Paleontologic correlation showing simultaneous sea-level changes in widely separated
regions of the world has indicated that some are
eustatic—caused by worldwide sea-level changes
(see Chapter 10). One well-known cause of
eustatic changes is a glacial age, when the conversion of huge amounts of sea water to ice decreases
the volume of sea water (see Chapter 9). Another
is a change in the oceans' capacity, which can be
related to periods of faster and slower sea-floor
spreading (see Chapter 19). The Cretaceous Period, for example, has long been known as a time
of extensive transgression of the seas on continental platforms. Evidence from the oceans indicates
that it may have been a time of rapid spreading,
when the volume of rapidly moving oceanic lithosphere would have been puffed up due to heat
expansion, thereby lowering the general volume
of the ocean basins and raising sea level. By linking local changes in environment to larger causes
of sea-level change, the historical geologist can
define the complex way in which the surface of
the planet responds to its interior motions.

CHEMICAL SEDIMENTS: CARBONATES AND THEIR DEPOSITIONAL ENVIRONMENTS

The most abundant chemically precipitated sediment, limestone, best illustrates how and where
chemical sediments form. Limestone, $CaCO_3$, and
the related rock dolomite, $CaMg(CO_3)_2$, are often
called **carbonate rocks,** or, more simply, **carbonates.** The ocean is the scene of immensely varied
kinds of carbonate sedimentation, from the formation of pelagic foraminiferal oozes in the deep
sea (Chapter 10) to shallow-water accumulations
of sand and mud made up of calcium carbonate.
The chemical basis for carbonate sedimentation
is the relative abundance of calcium and bicarbonate ions in sea water. The equation for carbonate sedimentation is:

calcium ion + bicarbonate ion →

$$Ca^{2+} \qquad\qquad HCO_3^-$$

calcium carbonate + hydrogen ion

$$CaCO_3 \qquad\qquad H^-$$

At equilibrium (or saturation) there are, together with the solid calcium carbonate, concentrations of calcium and bicarbonate ions in solution whose multiplication product is constant. The precipitate may form whether there is more calcium than bicarbonate, or vice versa, or even if the concentrations are equal as long as their product exceeds the saturation value.

Two forms of calcium carbonate can precipitate, calcite or aragonite. If the precipitation proceeds undisturbed and very slowly from slightly supersaturated solutions, calcite forms and remains stable indefinitely—a criterion for chemical equilibrium. In most precipitations, however, aragonite forms first, and then very slowly—sometimes over a period of many years—transforms to calcite. Most modern carbonate sediments are mixtures of the two forms, partly because the shells of organisms that contribute to carbonate sediments may be either or both. Almost all ancient limestones are calcite, because enough time has elapsed for all of the unstable aragonite to have changed to calcite.

In most of the oceans, sea water is fairly close to being saturated with calcium carbonate—that is, a chemist would expect that calcium carbonate would precipitate if very many calcium or bicarbonate ions were added to it. The warm surface waters of many tropical areas are slightly supersaturated, whereas colder waters in high latitudes or at great depths are slightly undersaturated. This makes reasonable the notion that the ocean as a whole stays near the saturation point by a steady-state process (Fig. 11–8). What would happen, then, if the inflow of river water stayed the same but its content of calcium and bicarbonate ions were to drop for some reason? The oceans would become undersaturated with calcium carbonate, and carbonate sedimentation might be expected to slow or stop. Similarly, if the influx of those ions were greater, we would expect the ocean's composition to move toward supersaturation, increasing calcium carbonate sedimentation.

Thus, at first sight, the ocean seems to be a reasonable chemical system whose behavior is predictable. A closer look, however, shows that an important part of the system has not been taken into account—the biological world.

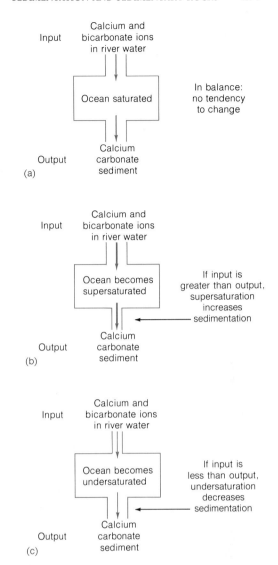

Figure 11-8
The feedback system that keeps the oceans approximately saturated with calcium carbonate. (a) If input and output are balanced, the ocean is saturated and there is no tendency to change. (b) If output decreases, the ocean becomes supersaturated, and output increases to bring it back to saturation. (c) If input decreases, the ocean becomes undersaturated, and output decreases to bring it back to saturation.

Biological Precipitation

Almost everywhere in the oceans, carbonate sediment is made up of the shells of organisms rather than of inorganic precipitates. Apparently, the conditions for precipitation are right, but it is accomplished by the organisms extracting calcium carbonate from the water for their shells rather than by inorganic precipitation directly from the water, as in the laboratory. Extensive carbonate

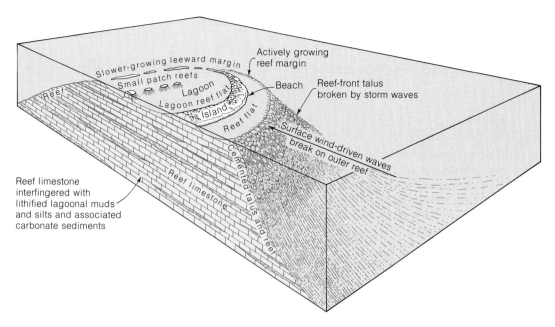

Figure 11-9
A reef. The reef tends to grow more rapidly on the windward side,
where waves bring fresh nutrients from deeper, open waters to the organisms.

sedimentation occurs only in warm tropical seas, a finding that shows the general correspondence in such places between the chemical condition of slight supersaturation and the high populations of organisms whose shells make the sediments. Shelled organisms live in colder undersaturated waters, but the shells tend to dissolve after the organisms die; the shells of pelagic foraminiferans, for example, disappear after they settle into deeper waters below the calcium carbonate compensation depth, as we pointed out in Chapter 10. Carbonate-secreting organisms and carbonate sediments are also found in fresh-water lakes, some of which are saturated and some of which are not.

An enormous array of organisms, from primitive one-celled animals to the common oysters, clams, and other highly evolved invertebrates, secrete some calcium carbonate. Carbonate-secreting plants range from primitive algae to some higher plants, such as some aquatic grasses. Vertebrates (and some invertebrates) do not secrete calcium carbonate but, rather, harder calcium phosphate minerals to form bone and teeth. The physiology of carbonate and phosphate secretion varies greatly among the many groups of organisms, and much remains to be learned about it.

Carbonate-shelled organisms live in dependence on the sea or lake water around them, for it provides nutrients and the dissolved materials from which they make their shells. They also live in a complex interrelation with the other organisms in their habitat. The entire web of interdependence of organisms and their environment is the subject of the science of ecology. Because the ecological study of carbonate-secreting organisms and carbonate sediments in various environments has been carried on vigorously in the past two decades by geologists and paleontologists, we have learned how limestones are made. One sort of habitat, in particular, has always captured the attention of scientists—the coral reefs of tropical seas.

Coral Reefs

From 1831 to 1836, Charles Darwin sailed on the *Beagle* as naturalist–scientist for a British surveying expedition, a voyage famous because it allowed Darwin to observe and collect the great variety of organisms that laid the foundation for his theory of organic evolution. One of the products of the voyage was his analysis of Pacific coral reefs. Darwin was one of the first to explore in detail the relation of organisms to their geologic environment. Many of his ideas on coral reefs are still accepted today.

The major physical characteristics of a coral island or **atoll** are: (1) an outer wave-resistant reef front, a steep slope facing the open ocean; (2) a flat reef platform in back of the reef and extending toward the island; (3) a shallow lagoon behind the reef platform, protected from the waves by the reef; and (4) the island itself (Fig. 11–9). Some atolls have no central island, just a more or less circular reef, parts of which may be above water and forested, surrounding a central lagoon.

The main part of the reef tract is made up of actively growing coral, an organism that grows as colonies of great numbers of individuals joined to each other (Fig. 11–10). The coral secretes its calcium carbonate as it grows, cementing itself to the carbonate remains of dead coral below, and the reef grows outward. The living stalks of coral get their food from tiny bits of organic matter brought to them by waves coming in from the open ocean (Fig. 11–11). Reef-building corals need light to grow because they live in *symbiosis* (that is, in a mutually life-supporting relationship) with small green algae (zooxanthellae) that live in the translucent coral tissue and need light to live. The coral provides protection for the algae and the algae provides photosynthetic oxygen for the coral. The light requirement limits reef-building coral growth to shallow waters, less than about 20 meters (65 feet) because, as transparent as it may seem, sea water gradually filters out more and more light the deeper it gets.

Living in association with the coral are various kinds of coralline algae, which also secrete carbonate and thus help cement the whole structure into a massive rock terrace. On the reef platform live many species of shelled and soft invertebrates. In the lagoon live many other kinds of organisms. The organisms feed upon each other, and, in the process, the shells or other hard parts of the prey organisms are broken up into sand-, silt-, or clay-size particles. These, together with the physical erosion of the reef front, which produces fragments of coral and coralline algae, provide the carbonate sand of the island beaches and some of the fine carbonate mud that floors parts of the lagoons. Some of the mud comes from microscopic particles secreted by algae; some of the rest may be produced by nonbiological precipitation from warm lagoon waters that become supersaturated with calcium carbonate as a result of evaporation.

A seasoned observer can deduce where samples of reef sediment came from just by looking at the sediment and the shells in it. Hard, cemented rock made up of coral and coralline algae come from the reef front. Sediments from the main reef platform may contain many kinds of clams and other shells, sometimes with some carbonate sand from the erosion of the reef. The lagoon is dominated by muds, sometimes built up into small mounds, occasionally spotted with little coral "heads"

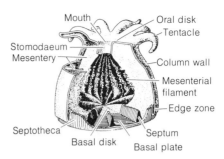

Figure 11–10
A typical coral, showing the relation of the soft body parts (mesentery) to the skeleton parts of calcium carbonate; basal plate and septum. [After *Treatise on Invertebrate Paleontology*, Geol. Soc. Am., and University of Kansas Press, 1956.]

Figure 11–11
An underwater photograph of a coral reef knoll in a few meters of water near North Rock, Bermuda. The scale is 50 cm long. Below the scale is a sheet-like encrusting coral colony. At left center and lower right are large heads of brain coral. The branching, treelike forms are sea whips and sea fans. [Photo supplied by Peter Garrett.]

Figure 11–12
The Guadalupe Mountains of western Texas from the south. The drawing below is a reconstruction of the sedimentary environment associated with a great reef that existed here in Permian times. [From *Geology Illustrated* by John S. Shelton. W. H. Freeman and Company. Copyright © 1966.]

growing here and there. The beaches are where the sand accumulates. The geologist uses this information to infer the nature of ancient reefs formed in the seas of the past—such as the reefs of the Silurian Period, the remains of which lie in a belt across what is now Indiana, Illinois, and Wisconsin, or the great reefs of Permian age in western Texas (Fig. 11–12).

Origin of Coral Reefs

Ever since Darwin first studied reefs, geologists have wondered about how they form. At first sight, there appears to be a paradox. How can corals, which need light to live, build islands that reach to the surface from the floor of the deep sea thousands of meters below? There is no conceivable way that the level of the sea could once have been low enough for coral to have formed so near the bottom, nor is there any evidence that much of the floor of the ocean was ever structurally elevated to sea level to allow the corals to begin growing. The clues Darwin saw were **fringing reefs,** reefs similar to atolls growing around the edges of volcanic islands. The theory he proposed over a century ago remains the most probable explanation.

The process starts with a volcano building up from the sea floor (Fig. 11–13). As the volcano stops its eruptive activity—temporarily or permanently—corals and coralline algae colonize the shore and build a fringing reef around the volcanic island. Erosion may then lower the central volcano almost to sea level. If such a volcanic island slowly sinks below the waves as a result of some tectonic movement, actively growing coral and coralline algae may keep pace with the sinking and build up the reef. In this way, the volcanic center may disappear, to be replaced by an atoll with a central lagoon. Deep drilling on several atolls gave striking confirmation of this theory: volcanic rock was penetrated by the drills deep below the coral. Recently, as we will see in Chapter 19, plate-tectonic theory has proposed an explanation of the subsidence of volcanic islands on oceanic lithospheric plates as part of the process of sea-floor spreading. Of course, some coral islands have a much more complicated history, and sea-level changes during the Pleistocene affected all of the reefs then in existence.

Fascinating as coral-reef islands of the open ocean are, they account for a relatively small proportion of the carbonate sediment produced in the ocean. By far the largest fraction of carbonate sediments similar to the ancient limestones

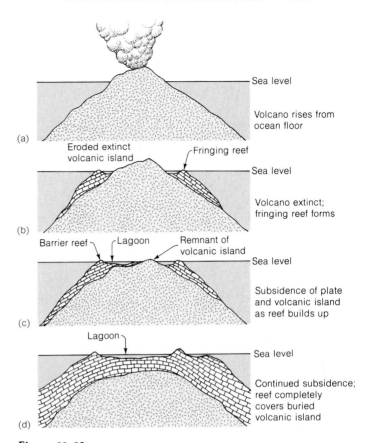

Figure 11-13
Evolution of a coral reef from a subsiding volcanic island. In most respects this chain of events was first proposed by Charles Darwin in the nineteenth century. The framework of plate tectonics provides an explanation for the subsidence of the islands.

formed on the continents are formed on shallow-water platforms near or attached to the continents.

Carbonate Platforms

One of the best examples of a carbonate platform is the area of the Bahama Islands and the shallow banks surrounding them (Fig. 11–14). This region of about 155,000 square kilometers (about 60,000 square miles) is a flat-topped shallow platform, on the west separated from the mainland of Florida by the narrow, deep Straits of Florida and on the east dropping abruptly more than 3600 meters (about 12,000 feet) to the deep ocean floor. Most of the platform is only a few meters below sea level, and it is dotted with many small islands. Coral and algal reefs are found on the eastern edges of the platform where the trade winds blowing from the southwest bring open-ocean water rich in nutrients to the organisms of the actively growing

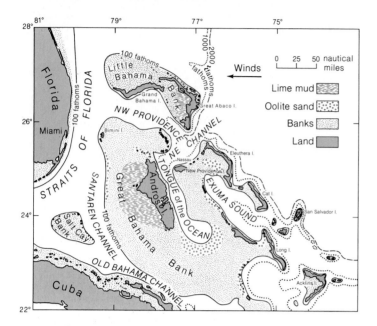

Figure 11-14
Map of carbonate sediments in the Bahama Islands.
The reef fronts and islands are on the eastern margins of the banks, facing the easterly trade winds.
[After N. D. Newell and J. K. Rigby, "Geological Studies on the Great Bahama Bank," Soc. of Econ. Paleontologists and Mineralogists Spec. Paper No. 5, 1957.]

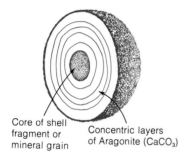

Core of shell fragment or mineral grain

Concentric layers of Aragonite (CaCO₃)

Figure 11-15
Cross section of an oolite, made up of concentric shells of calcium carbonate (aragonite) precipitated as the grain was rolled and suspended in current-agitated waters. Typical oolites range in size from 0.25 to 2.0 mm in diameter.

reef tracts. Much of the area of the shallow banks to the west is covered with carbonate sand and mud particles with essentially no land-derived detrital material, for the supply of sediments from the continent is cut off by the Straits of Florida.

A distinctive kind of carbonate sand, **oolite,** is formed where strong tidal currents flood the banks each day. Oolites are more or less spherical grains of aragonite that are made up of concentric layers surrounding a nucleus, like the structure of an onion (Fig. 11-15). Most of the nuclei are frag-

ments of shells. These nuclei are coated with layers of aragonite as the tides bring in cold ocean waters that become supersaturated with calcium carbonate as they warm and evaporate on the shallow banks. The currents are strong enough to keep the grains rolling or suspended in the water, and, as a result, they become coated with precipitate on all sides. The limit to growth of the oolites is determined by the maximum grain size that can be moved by the current.

The variety of types of carbonate sediments found on the Bahamas is exemplified by a cross section of one of the islands, Andros, and the banks west of it (Fig. 11-16). The reefs on the east shore protect a narrow muddy lagoon, which lies in front of a sand beach on the island, like the lagoon of an atoll. The western side of the island leads down to a tidal flat. The tidal flat is covered with muds, sometimes finely rippled, interrupted by tidal channels, and in many places overgrown with **algal mats,** interwoven gluey surfaces of blue-green algae. This sticky surface traps fine carbonate sediment and, as the algae grow around and upward through the carbonate, the sediment is incorporated in a layered mixture (Fig. 11-17). In this way, the algae produce the layered structures called **stromatolites** that are found in many tidal-flat limestones. These structures have achieved great importance as evidence of algal life far back in the Precambrian. Also found on tidal flats are areas in which the calcite and aragonite of the original sediment have been transformed into the mineral dolomite, CaMg(CO₃)₂. In these areas, the sea water in the pores between the grains just below the surface has become enriched in magnesium ions owing to evaporation. That increase in magnesium promotes the exchange of some magnesium for calcium ions in the carbonate mineral, converting it to dolomite. Many ancient dolomites show evidence of this kind of tidal-flat origin.

Below the tidal flats west of Andros are extensive shallow banks, only a meter or two deep in many places, where the water is quiet and the bottom is carbonate mud. Some of the mud is in the form of soft pellets, produced as fecal pellets by the many kinds of organisms living on and ingesting sediment on the bottom. In the mud area lives a variety of algae that secretes microscopic needles of aragonite, about 0.003 millimeters long. Aragonite needles of similar size and shape occur abundantly in the mud. A small part of the carbonate mud may also be the result of nonbiological precipitation from the warm, supersaturated water of the banks, which can also produce aragonite needles. Most of the fine sediment is from

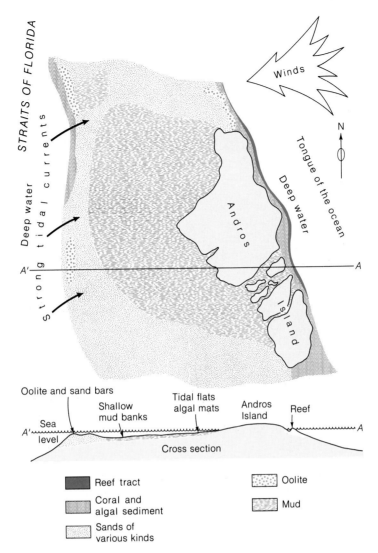

Figure 11-16
Map and cross section of Grand Bahama Bank west of Andros Island.

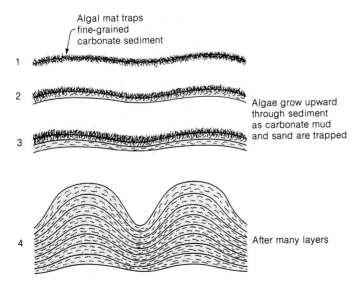

Figure 11-17
Stages in the growth of a stromatolite formed by the sediment-trapping action of algae. As the sediment accumulates, the algae continue to grow upward while the previously deposited layers become cemented and lithified. The convex upward structures vary greatly in all dimensions and in their regularity, ranging in size from a few centimeters to more than a meter in diameter. The layers are typically a few millimeters thick. A good example of one kind is shown in Figure 12-5.

algal precipitation and the wearing down of coarser shell materials by abrasion and animal ingestion.

On the far western side of the banks west of Andros, the movement of the water increases, and the currents winnow out the fine material. Waves and tidal currents move sand consisting of broken pieces of shells into submarine dunes and bars much like those formed by currents in large rivers or those found on shallow continental shelves in noncarbonate terrains.

Armed with the information about Pacific atolls and carbonate platforms like the Bahamas, geologists have learned to recognize many different kinds of ancient limestones that originated long ago in similar situations. Thus, they have been able to make maps of ancient reef tracts, extensive tidal-flat areas, sand banks, and the other kinds of shallow-water areas where animals, plants, and the environment interact to produce carbonate deposits.

Shipboard geologists have studied the carbonate oozes of the deep oceans formed from the remains of the small organisms floating in the surface waters of the sea. Most abundant are the foraminiferal oozes (Chapter 10). **Nannofossil oozes** are made up of organisms of the smallest size, from 0.06 to less than 0.005 millimeters diameter. These are primarily **coccoliths,** microscopic plates of calcite produced by coccolithophores, single-celled algae. Samples of Cenozoic and Mesozoic deep-sea sediments cored by the Deep Sea Drilling Program show the past distributions of these oozes, the older and more deeply buried of them having been transformed into their lithified equivalent, **chalk.** The White Cliffs of Dover are made up of a Cretaceous chalk that underlies much of southern England and parts of northern France.

Not all carbonate deposits form in the sea. Calcium carbonate is deposited as crusts and layers around hot springs, both by algae that grow in the hot water and by nonbiological precipitation. The deposits are **tufa,** which is porous and easily broken, and **travertine,** a denser variety that is familiar because of its wide use as a decorative stone for building facings and table tops. Carbonate is also found in **stalagmites** and **stalactites** in caves formed by groundwater in limestone formations (Fig. 11–18). Groundwater dripping from fractures in the cave roof forms the needlelike stalactites. The stalagmites build up from the floor as the drips fall on that spot. The groundwaters in these limestone formations, saturated with respect to calcium carbonate, contain more carbon dioxide

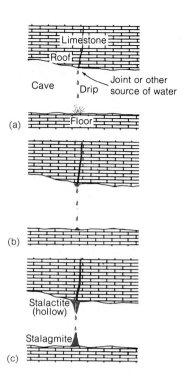

Figure 11–18
The formation of a stalactite and a stalagmite follows a drip of calcium-carbonate-saturated water from a cave ceiling.
(a) Groundwater saturated with calcium carbonate, high in carbon dioxide, leaks into the cave. (b) Drops lose some carbon dioxide to air, become supersaturated, and precipitate small amounts on ceiling and floor.
(c) Continued precipitation builds up the stalactite and stalagmite.

than is in equilibrium with the cave atmosphere. When the drops enter the cave they lose some of the excess carbon dioxide to the air in the cave, thereby becoming supersaturated with carbonate, and precipitate small amounts to form the stalactites and stalagmites.

Carbonate sediments can form in a wide variety of lakes as a result of algal activity and nonbiological precipitation. They range from such salty lakes as the Great Salt Lake and the Dead Sea to the fresh-water lakes of northern temperate regions, such as those of Minnesota and Wisconsin.

OTHER CHEMICAL SEDIMENTS

Although calcium carbonate is the most abundant biological sediment, there is a great variety of other chemical sediments whose formation is wholly or partially controlled by the biological world.

 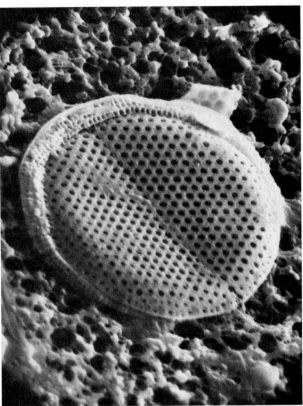

Figure 11-19
Two species of modern diatoms filtered from sea water. These photographs were taken with a scanning electron microscope at very high magnifications. The long dimension of each species is about 10 micrometers. [Photos by C. Stein and A. Eaton.]

Silica

The deposition of silica, SiO_2, has some similarities to the deposition of carbonates. Most of the silica of biological origin is secreted by single-celled organisms. The most important are small algae called **diatoms** that populate much of the surface layers of the ocean and fresh-water lakes (Fig. 11-19). These little organisms are extraordinarily efficient in removing silica from sea water to make their shells, which are an amorphous form of silica similar to opal in crystal structure. In contrast to the marine carbonate-secreting organisms, which live in waters that are nearly saturated with calcium carbonate, diatoms can extract silica from very undersaturated solutions such as surface sea water. How they do it is still a mystery. In addition to diatoms, there are small single-celled animals called **radiolarians** that are related to calcareous foraminifera and live much like them. Some kinds of sponges secrete silica,

too. When diatoms and radiolarians die, they sink to the bottom of the sea and accumulate in the sediment. Under areas of high biological productivity, where silica-secreting organisms are abundant because of a high supply of silica and other nutrients in the water, the silica shells of dead organisms rain down and form silica-rich **diatom ooze** and **radiolarian ooze.** When these oozes become cemented and hardened into rock, they are called **diatomite** and **radiolarite.** Diatomite is a useful rock: because of its high porosity, it has been used in filters and as fillers, and it is sometimes used as a mild abrasive. Some of the best known diatomites are in the Monterey Formation, which is exposed in many places along the coastal regions of central and southern California. Many ancient **cherts,** hard and dense silica-rich rocks, originated as diatomites or radiolarites and later became strongly lithified as they recrystallized to a very fine-grained rock composed of microcrystalline quartz.

Sulfide

Organisms may control chemical sedimentation indirectly by changing the chemical conditions in the environment. One important example is the formation of the mineral pyrite, FeS_2, by the indirect action of bacteria. The bacteria are in the same group as those that cause the noxious smell of rotten eggs. The smell is the gas hydrogen sulfide, H_2S. This gas is produced in sediments or sea water when these bacteria chemically change sulfur from its oxidized state, as the sulfate, SO_4^{2-}, dissolved in sea water, to its reduced state, sulfide S^{2-}. These primitive bacteria, which can grow only in the absence of oxygen and so cannot respire as most organisms do, get their energy by converting the oxidized sulfur ion from its $+6$ state in sulfate, where the atom has lost 6 electrons, to the -2 state in sulfide, where the atom has gained 2 electrons. The hydrogen sulfide produced, itself a powerful reducing agent, changes ferric iron Fe^{3+} to ferrous iron Fe^{2+} and precipitates the highly insoluble pyrite (Fig. 11–20).

The presence of decaying organic matter is necessary for the H_2S bacteria to thrive; this decay keeps the immediate environment free of oxygen. In most normal oxygenated or aerated environments, organic matter decays by using up oxygen, essentially as a slow combustion process similar to the way we respire:

$$\text{organic carbon} + \text{oxygen} \rightarrow \text{carbon dioxide}$$
$$\text{C} \qquad\qquad \text{O}_2 \qquad\qquad \text{CO}_2$$

As the oxygen is used, it is replenished by free mixing with the atmosphere, which is one-fifth oxygen. In most shallow parts of the ocean, the mixing of the water by waves and currents keeps the water well aerated. In the deep ocean, the rate of replenishment is slower, taking as much as a thousand years; but, because there is relatively little organic matter in the deep sea, the oxygen does not completely disappear there.

The oxygen *can* totally disappear in two situations. If there is a great amount of decaying organic matter in the sediment, such as in some biologically productive areas of tidal flats, the oxygen slowly permeates into the sediment by diffusion and gradually is depleted by biological activity. The sediment becomes reducing—that is, lacking in oxygen. When that happens, the bacteria that produce hydrogen sulfide become active. This is the explanation for the black sands and muds just below the surface of many tidal-flat sediments. A quick scoop of the sediment will let you smell a whiff of hydrogen sulfide. The blackness comes from the first form of ferrous sulfide precipitated, a finely dispersed black substance that slowly transforms to crystalline pyrite with time and burial.

The second situation in which a deoxygenated environment can form is in some areas of the sea where deep basins are cut off from replenishment by aerated waters by a **sill,** a topographic ridge or barrier that prevents free circulation and mixing (Fig. 11–21). In such a place, the waters of the basin do not mix well with oxygenated surface

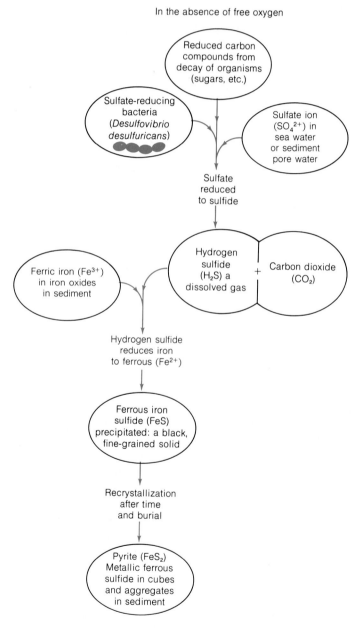

Figure 11-20
Pyrite, FeS_2, is formed by a chain of reactions that starts with the reduction of sulfate (SO_4^{2-}) from sea water by sulfate-reducing bacteria living in an environment free of oxygen.

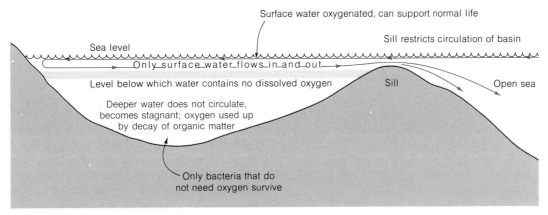

Figure 11-21
Sea basins separated from the main body of open sea water by sill may become deoxygenated as the water below the sill level stagnates from lack of circulation.

waters because the depth is too great for the deep waters to be affected by waves, and the greater density of those waters (caused by low temperatures) prevents general mixing by convection. Here again, as organic matter decays, oxygen is used up and replenishment by mixing with the aerated waters above is too slow; thus, bottom waters and sediment become reducing. Many fiords—drowned valleys scoured well below sea level by glaciers—have sills at the seaward end formed by terminal moraines or by the limit of the scouring action of the glacier. This causes them to be reducing below the level of the sill.

The deoxygenation of surface waters can also be caused by some forms of pollution. When abundant phosphate, one of the important nutrients of organisms, pollutes a lake or river, the algae and many different aquatic plants will flourish, sometimes to the point of glutting the surface waters, a process called **eutrophication.** When these organisms die, their abundance is transferred to the bottom, where the extensive decay uses up oxygen faster than it can be replenished. The result is a great shift in animal communities, as fish and other organisms that need oxygen are cut off from much of the lake. The lake is not "dead," a term that many people have applied to Lake Erie; but its community has been drastically altered, and many species have disappeared, especially popular food and sport fishes, while "scummy" algae and weeds thrive.

Coal

Lack of enough oxygen to decay vegetation completely on land is responsible for the formation of one of our most valuable sediments, **peat,** and its lithified equivalent, **coal.** The abundant plant fossils in coal beds indicate that they originated in swamps, much as peat bogs form now. As the luxuriant plant growth of such a swamp dies, it falls to the waterlogged soil. The water and rapid burial by falling leaves protect the twigs, branches, leaves, fruit, and seeds from oxidizing, just as thick leaf mats in some forests protect the lower layers from complete decay. The vegetation accumulates and gradually turns into peat, a porous brown mass of organic matter in which twigs, roots, and other plant parts can still be recognized (Fig. 11-22). After more burial and chemical transformations of the aging organic matter, the peat becomes **lignite,** a very soft coallike material. Longer times and the higher temperatures that accompany greater depths of burial may ultimately metamorphose the lignite into **subbituminous** and **bituminous coal,** or **soft coal,** and, in extreme cases, to **anthracite,** or **hard coal.** The greater the metamorphism, the harder and brighter the coal and the higher its heat value. In the chemical reactions responsible for these physical transformations, gaseous compounds containing carbon, hydrogen, and oxygen are liberated, and the coal becomes richer in carbon, well over 90 percent in anthracite. The amount of this volatile matter remaining in the coal is a measure of the **rank,** the degree of metamorphism of the coal. Rank is a geological guide to temperature and hence to depth of burial and the geothermal gradient. Reflectivity, the degree to which coal reflects light, has been found to be a reliable measure of rank also. This method has been applied to the brightest component in coal, vitrinite. Vitrinite also occurs in small amounts as detritus in sandstones and can be used to measure the temperatures of burial reached by the sandstone.

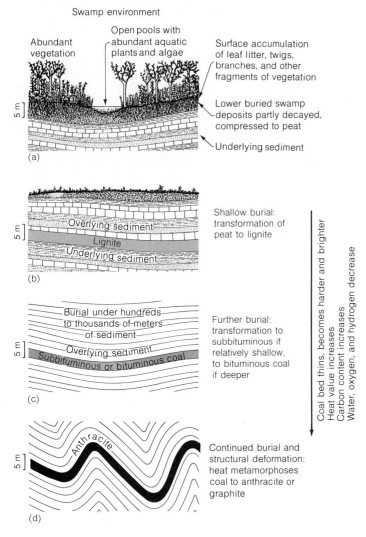

Figure 11-22
The process by which coal beds form begins with the deposition of vegetation. Protected from complete decay and oxidation in a swamp environment, the deposit is later buried and subjected to mild metamorphism, which transforms it successively into lignite, subbituminous coal, bituminous coal, or anthracite, depending on depth of burial, temperature, and amount of structural deformation.

There is little doubt that, as our energy needs continue to grow and as oil and gas are less able to satisfy the demand for fuel, coal is becoming once again, as it was up to about forty years ago, an extremely important chemical sedimentary rock (Chapter 22).

Evaporites

In an arid climate, the same general kind of barred basin that restricts circulation and causes deoxygenation can be the scene of a different kind of sedimentation—salt deposition. Sediments composed of the mineral halite or rock salt, NaCl, the calcium sulfate minerals gypsum, $CaSO_4 \cdot 2H_2O$ and anhydrite, $CaSO_4$, and many other less abundant salts formed by the evaporation of sea water are called **evaporites.** The thin crust of salt formed when a pond of sea water dries up is easy to observe along many shorelines, but not much salt forms in such a situation. The ratio of salt to water is such that, for every liter of sea water evaporated, about 35 grams (a little more than 1 ounce) of solid salt crystallizes. How much sea water had to evaporate, then, to produce an evaporite formation of reasonable dimensions, say 100 meters thick, 50 kilometers wide, and 100 kilometers long? The volume of the evaporite, about 5×10^{11} cubic meters, at an average density of about 2.3 grams per centimeter, weighs about 1.15×10^{18} grams. This represents the salt evaporated from about 3.3×10^{16} liters of sea water (approximately 8.7×10^{15} gallons)! A single pond containing that much water would have a volume of well over 30,000 cubic kilometers. In other words, the water would have had to be more than 6 kilometers deep over the area covered by the evaporite formation, much more than the average depth of the oceans. This absurd conclusion alone led geologists to suspect that there must be some natural mechanism that constantly replenishes the sea water as it evaporates.

The study of the minerals formed by sea-water evaporation provided another lead. The first experimental evaporation of sea water was performed in 1849 by Usiglio, who showed how a sequence of minerals would be precipitated as evaporation proceeded, starting with calcium carbonate and calcium sulfate, progressing to sodium chloride, and then, in the final stages, to magnesium and potassium minerals.* The important point, one that was amplified and carefully worked out in detail much later, was that a distinctive sequence involving a complex order of precipitation of many minerals would characterize complete evaporation. Some evaporites show those sequences and are interpreted as simple products of complete drying. Yet a great many

---

*Usiglio simply filled a bucket with water from the Mediterranean Sea, took it to his laboratory, and let it evaporate. Though he did crystallize many minerals found in natural salt deposits, his results were erratic since his methods were primitive by modern standards. Half a century later, a Dutch chemist, van't Hoff, tackled the same problem with more advanced laboratory methods but, more importantly, with a deep theoretical understanding of the newly developing science of physical chemistry. His results, a detailed sequence of minerals precipitated as evaporation proceeded, still stand today as a triumph of experimental geochemistry.

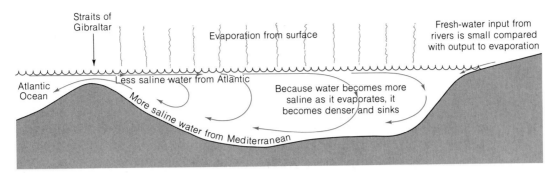

Figure 11-23
Simplified diagram of circulation of the Mediterranean Sea through the Straits of Gibraltar. Because influx of fresh water from rivers into the Mediterranean does not balance evaporation from the surface, there is an inflow of water from the Atlantic Ocean to replace sinking surface water that has become more saline and thus denser as a result of evaporation.

thick evaporite deposits consist only of gypsum, or anhydrite, with halite or of a combination of the three. Did incomplete evaporation go only so far as to produce halite? If so, an even greater volume of water would have been required. Why are there so few evaporites that contain the last salts to precipitate just before complete drying?

A resolution of these difficulties came with the theory that typical evaporite sequences are deposited in relatively shallow arms of the sea or in basins connected to the sea by narrow openings. During times of normal salinity, limestone deposition is the rule. As surface water evaporates from the basin, the dissolved materials become more concentrated until an evaporite precipitate, typically gypsum, starts to form. The concentrated solution formed at the surface, called a **brine,** is denser than the water below it because of its greater content of dissolved salt, and so it tends to sink to the bottom. As it does, normal sea water flows in at the top, through the seaway from the open ocean. A continuous circulation pattern is set up by the denser brine flowing out through the bottom of the seaway. Thus, a steady-state system is created in which gypsum is continually removed as a precipitate while sea water circulates through the evaporite basin. Circulation of this sort occurs in the Mediterranean Sea and the Straits of Gibraltar (Fig. 11-23). Atlantic Ocean water flows in at the top of the straits to the saltier Mediterranean, which has a high evaporation rate; saltier Mediterranean bottom water flows out to the Atlantic through the bottom of the straits. The Mediterranean does not now evaporate sufficiently to form gypsum. However, a recent hypothesis, based on the abundant evaporite found in Deep Sea Drilling Program holes in the Mediterranean, proposes that about 6 million

years ago, the Mediterranean was cut off from the Atlantic at Gibraltar and dried up completely to form a temporary salt-covered desert.

Alternative circulation patterns have been proposed, including ones in which the dense brines move down into and through permeable sediment below the basin and reappear along lower slopes of an adjacent ocean basin. This mechanism also works for evaporites formed on evaporating tidal flats, where the brine formed by evaporation sinks into the upper unsaturated groundwater zone, slowly precipitating gypsum as it moves down.

Important evaporite deposits of the past are the Permian salt beds of North Germany and the North Sea, the Permian salt formations of western Texas and New Mexico, and the Silurian salt beds of Michigan, Ontario, and western New York State. Salt deposits in the Precambrian testify to the long-continued salinity of the oceans. Evaporites and salinity are closely tied, for it is evaporite sedimentation that removes most of the chloride from the ocean to keep it in a steady state.

BURIAL AND ACCUMULATION OF SEDIMENT

A bar in a river may exist for only a few months, from the high-water stage that formed it to the next one, which destroys it. Of all the sediments that are made day by day, only a small fraction is preserved to record geologic time. Sedimentary environments differ in the probability of preservation, and the differences are related, in part, to the rapidity of sedimentation and burial. In the deep sea, deposits accumulate at very low rates, only a few millimeters every thousand years.

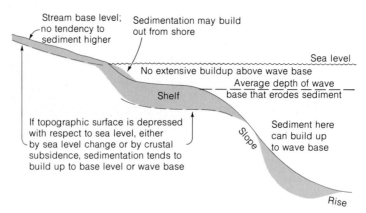

Figure 11-24
Sedimentation is closely dependent upon subsidence in those environments where the site of deposition is close either to base level of erosion on land or to wave base in the sea.

Once deposited, many of these deep-water sediments are unlikely to be eroded and redeposited because bottom currents strong enough for the task are unusual. That is why deep-sea pelagic sediments offer the most complete historical records of organic evolution, of temperature changes during ice ages, and of other geologic patterns.

In contrast, the shallower waters of the continental shelf deposit sediments at much higher rates—many centimeters every thousand years, on the average. These sediments are buried faster, but waves and currents are much more likely to scour the bottom and rework it mechanically. Furthermore, the higher the sedimentary accumulation, the more likely the waves are to rework and redistribute it; so that there is a level of sedimentation above which accumulation stops (Fig. 11-24). The importance of **subsidence** of the crust is that it allows sedimentation to continue and the accumulation to build up. This process is one of slight positive feedback, in the sense that tectonic subsidence is reinforced by the sedimentary load. The further sedimentation proceeds, the greater the weight on the crust, because each layer of sediment is much denser than the sea water it replaces. The effect of the weight is to push down the crust (see the discussion of isostasy in Chapter 18), allowing still more sedimentation to proceed. This process does not continue endlessly, nor can it account for all of the subsidence shown by most continental shelves. The reason for this is that, as subsidence continues and sediments are laid down, more and more of the crust at that point consists of sediments; because these are much less dense than average crust, the effect of additional sediment tends to become negligible. Subsidence initiated by tectonic mechanisms and enhanced by sediment weight is evidenced by great thickness of sediments.

Geosynclines

The first to theorize about the origin of sediments of great thickness was the State Geologist of New York in the middle of the nineteenth century, James Hall. As a result of extensive field mapping of sediments of the Appalachian Mountains, correlating them stratigraphically, and measuring their thicknesses, Hall concluded that there were many tens of thousands of feet of sediment that had accumulated in a long, relatively narrow trough bordering the continent. He envisioned it as having been something like a large downfold in the surface that steadily accumulated sediment as it subsided. Later, the name **geosyncline** was applied to the sediment-filled trough.

For the past century the term has undergone a long and complex series of redefinitions tied to changing notions of the origins of the supposed troughs, mechanisms of subsidence in relation to sedimentation, the evolution of igneous and metamorphic rocks in the geosynclines, and of the nature of the structural deformation of these thick sedimentary sections. In the 1930s geosynclines were subdivided into an inner belt closer to the continental platform, the **miogeosyncline,** and an outer belt closer to an ocean basin, the **eugeosyn-**

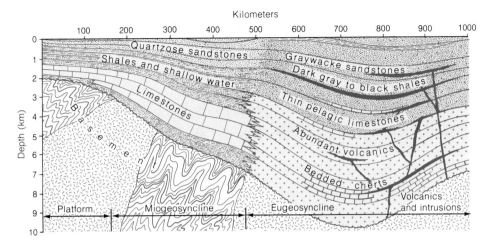

Figure 11–25
Diagrammatic cross section of thicknesses and rock types of a geosyncline and adjoining platform area. The platform and miogeosyncline differ largely in thickness of sediment; miogeosyncline and eugeosyncline differ also in kinds of sedimentary rocks. Deformation patterns, not shown here, also differ: the miogeosyncline is moderately folded; the eugeosyncline more intensely folded and faulted.

cline. The miogeosyncline was characterized by lesser thicknesses of limestones, shales, and alluvial sandstones, though other kinds of sediment could also be found (Fig. 11–25). The greater thicknesses of the sediments characteristic of the eugeosyncline included turbidite sandstones and shales, pelagic limestones and shales mixed with volcanic rocks, submarine basalts, and ash falls and flows.

An explosion of information on continental margins has accompanied the rise of plate-tectonic theory and its amplification in the last decade. This knowledge has led to new definitions, solidly based on geophysical and geological theory and observation of both oceans and continents. We now see geosynclines—there is still much of value in keeping the name—as thick, linear piles of sediments deposited along continental margins at present or former plate boundaries. The sandstones, shales, and carbonates of the continental shelves of passive continental margins accumulate to thicknesses of several kilometers as the edge of a rifted continent moves away from a mid-ocean ridge, cools, and subsides. These sediments correspond roughly to the miogeosyncline. The volcanics, turbidites, and fore-arc sediments of active continental margins are mixed with scraped-off pelagic sediments at subduction zones. This is the thicker sequence of the eugeosyncline. The two are brought together by a change in plate motions as a formerly passive continental margin converges with an active continental margin. When the two continents collide, their margins crumple first. Continental-rise

turbidites of the one continent are mashed and faulted against the turbidites, volcanics, and pelagic sediments of the other (Fig. 11–26). A collisional mountain chain is created of the thick sedimentary accumulations of the geosynclines.

Structurally deformed, metamorphosed, and injected with igneous intrusions, geosynclines were the forerunners of the major mountain chains of the world: the Alps, the Himalayas, the Appalachians, and the Western Cordilleran belt of the Americas. Only in the youngest mountain belts are the youngest geosynclinal beds preserved. As the mountains wear away, older and older sedimentary rocks are stripped off by erosion, in many places uncovering igneous and metamorphic rocks below. As geologic time goes on, the less likely it is that sediments uplifted into mountain ranges will survive erosion. In general, less and less sediment is found preserved as one moves back through the history of the Earth.

Continental Platforms

Wide areas of the North American continent between the Appalachians and the Rocky Mountains have been relatively stable at least since the Cambrian Period began. Sedimentary accumulations are much thinner than in the geosynclines along the continental margins to the east and west, though there are a number of sedimentary basins, such as the Illinois Basin, that have received as much as 3000 meters (10,000 feet) of sediment. Deep-water formations are conspicuously

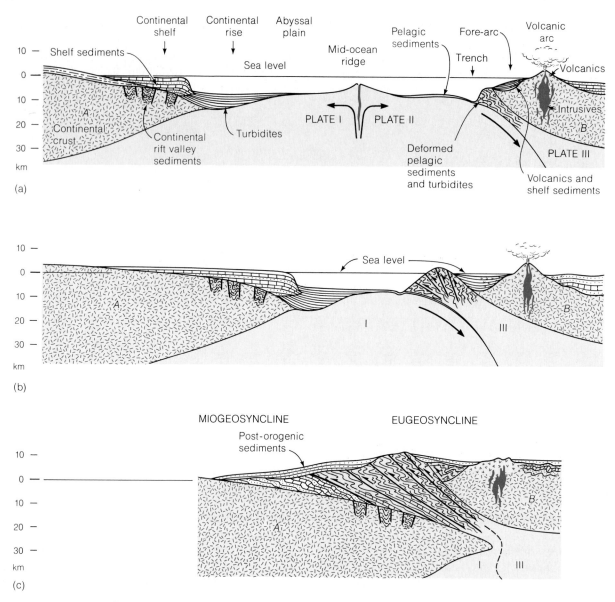

Figure 11-26
Three stages in the evolution of a geosynclinal mountain belt. (a) A spreading ocean, plates I and II, separate continent *A* with a passive margin from continent *B* with an active margin. (b) Spreading stops, the ocean closes, plate II disappearing by subduction. Sediments on both margins have grown in thickness. (c) The two continents collide, margin sediments are deformed, the active margin rocks thrust over the passive margin sediments, and a high mountain range is formed. The former subduction zone between plates I and II has disappeared and the former plate boundary is represented by a junction of the two continents.

absent, and many of the sediments are shallow carbonate-platform sediments and alluvial (or near-shore marine) sandstones and shales. Individual formations more or less continuous with those in geosynclines are much thinner on the platform. Subsidence has been much less on the platform, and no great mountain-making episodes have affected it. Conversely, erosion has not eaten so deeply into the surface sedimentary rocks, so

that Precambrian igneous and metamorphic rocks, the basement, are exposed in only a few places, such as in the Ozark Mountains or the Black Hills. The key feature of platform sedimentation is not so much in sediment types and the environments in which they formed, for there is almost complete overlap with geosynclines in that regard, but in the thinness of the accumulation, a response to only mild and intermittent subsidence. The origin of this mild subsidence in areas far from plate boundaries remains in doubt.

Subsidence, which influences environment and thickness of accumulation, is only one aspect of the tectonic control of sedimentation. Just as important is the indirect effect that tectonics has on the mineral composition of detrital sediments.

TECTONICS AND SEDIMENT COMPOSITION

Contrasts in the mineral composition of clastic sediments can be startling. For example, compare two alluvial sandstones, an arkose and a quartz arenite. The arkose may contain as much as 30 to 40 percent feldspar, the remainder being quartz and other mineral and rock fragments. The grains are coarse and angular, showing little sign of abrasion. The quartz sandstone may be more than 95 percent quartz with no feldspar. The grains are well sorted and well rounded. The contrast is not in the environment, for both are alluvial, but in the source of the erosional debris. The source of the arkose is likely a granitic igneous rock; the rocks eroded to make the quartz sand might well have been another sandstone.

As shown in Chapter 5, feldspar decomposes chemically to clay minerals, given enough time for reaction with rain and soil water. In comparison, quartz is essentially unaffected by most chemical weathering. The survival of feldspar (and thus the quartz-feldspar ratio) in a weathering terrain depends upon the rate of chemical decomposition in relation to the rate of mechanical erosion. That ratio is related to the topography of the terrain—the more elevated and rugged the mountains, the more rigorous the mechanical erosion. In relating the detrital quartz-feldspar ratio to mountainous topography, we have tied it to the product of tectonic activity. Tectonically quiet low-lying areas not subject to severe mechanical erosion produce detritus low in feldspar; much of the feldspar decomposes to clay before it can be transported.

This tectonic background to sediment composition is a major basis for the reconstruction of ancient tectonic events, of high mountain chains that once shed abundant detritus and are now gone. For example, the Devonian sandstones of the Catskill Mountains of New York are a very thick accumulation of alluvial and deltaic sands that were transported from a great mountain range to the east, roughly parallel to the present low hills of the Taconic Mountains east of the Hudson River. The mineralogy of the sandstones, including the types of feldspar, mica, other silicates and rock fragments, indicates a mixture of granitic igneous and high-grade metamorphic source rocks that are typical products of the mountain-building that results from continent-continent plate collision. Continental drift reconstruction shows this mountain range to have been formed by the junction of eastern and western continents at the beginning of the assembly of Pangaea. A similar analysis has been applied to the Tertiary formations of the San Joaquin Valley of California. Their mineralogy indicates volcanic island arc source rocks. The stratigraphic and structural relationships, coupled to plate tectonic reconstructions of the sea floor off California, show this sequence to have been formed in the fore-arc region of a subduction zone formed by convergence of a former oceanic plate, now entirely gone, and the North American plate. In Canada geologists have reconstructed fragmentary patterns of vanished mountain ranges, some more than three billion years old, from Precambrian sediments. In these ways, geologists have learned to use sediments as the hieroglyphics of geologic history.

SEDIMENT INTO ROCK: DIAGENESIS

Once a sediment is deposited and buried by other sediments, it is not immune to change. One has only to compare freshly deposited muds and sands with ancient shales and sandstones to see the obvious differences in hardness, cohesion, and porosity. The many processes that produce the changes in a rock's composition and texture after deposition are lumped together in the term **diagenesis.** Generally, they operate to harden the soft sediment into rock—that is, to **lithify** it. Diagenesis may also alter the mineral composition by dissolving some of the original minerals and precipitating new ones. The nature of oil, gas, and coal is almost completely the result of diagenesis of original sedimentary organic matter.

The major physical diagenetic change is **compaction,** a decrease in porosity caused by mineral grains being squeezed closer to each other by the weight of overlying sediment (Fig. 11-27). Sands are fairly well packed as they are deposited, so they compact little, even when buried deeply. Newly deposited muds, however, are highly porous. They have a high water content when deposited, often well over 60 percent, and their layers thin drastically by compaction. The porosity of a sediment may also decrease greatly from the precipitation of diagenetic minerals in its pore spaces.

Chemical diagenetic changes are the result of two general tendencies. The first is a gradual approach toward chemical equilibrium of the non-

equilibrium mixture of diverse minerals that have been brought together as detritus in the sediment (and, perhaps, mixed with some chemical precipitate from the environment, such as calcite). Thus, sedimentation may mix minerals from two very different kinds of igneous rocks, minerals that would have been incompatible under the original conditions of formation of the igneous rocks, such as a sodium-rich plagioclase feldspar from a granite and a calcium-rich one from a basalt. Diagenetic chemical changes tend to dissolve the calcium-rich feldspars and precipitate sodium-rich feldspars, thus moving the rock toward chemical equilibrium—namely, a more homogeneous plagioclase composition. A different example of lack of equilibrium is a grain of aragonite in a carbonate sediment, which, given enough time, tends to change to calcite, the form of calcium carbonate stable at low pressures and temperatures. This tendency to equilibrium results in many other chemical reactions between incompatible or unstable minerals, which result in the formation of new minerals and, thus, bring the rock to a new composition in equilibrium with its surroundings.

The second tendency is for a sediment to be buried more or less deeply in the crust. As a sediment is buried, it is subjected to increasingly high temperatures—on the average 1°C for each 30 meters (100 feet) of depth on the continents (see Chapter 13)—and high pressures—on the average about 1 atmosphere for each 4.4 meters of depth (1 pound per square inch for each foot of depth). As minerals and the surrounding groundwater in pore spaces are heated and put under greater pressure, they tend to react chemically to form new minerals. This process, when carried far enough, becomes metamorphism, in which the entire character of the rock alters. The boundary between diagenesis and metamorphism is somewhat arbitrary, usually drawn at a temperature of about 300°C.

The list of specific diagenetic changes is great. Few of these are universal; which ones occur depends on the geologic situation. A few important examples include the change of unstable opaline silica shells of diatoms to quartz, the stable form; transformation of the unstable swelling clay, smectite, to a stable mica type of clay, illite; the precipitation of calcite and quartz as pore-filling cement in sandstones; and the alteration of volcanic glass to clay minerals and zeolite minerals. All of these changes must be understood before we can deduce the nature of the original sediment that was laid down, as we must do in order to interpret geologic history.

Diagenesis is of practical value, because recov-

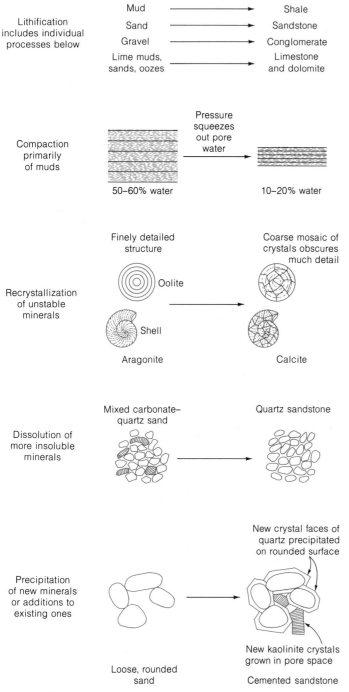

Figure 11-27
Some changes in composition and texture that are produced by diagenetic processes. Most of the changes tend to transform a loose, soft sediment into a hard, lithified sedimentary rock.

ery of oil, gas, or groundwater from formations depends on the porosity of the rock, which may have been largely determined by diagenetic cementation. The formation of oil itself is a diagenetic process by which organic remains of many kinds of organisms are gradually altered to liquid petroleum or natural gas, or both. The transformation of peat through bituminous coal to anthracite is another diagenetic process. Understanding all of these processes better enables us to search more intelligently and successfully for new mineral and energy resources.

SUMMARY

1. Physical sedimentation is controlled by gravity; and chemical sedimentation, which keeps the ocean constant in composition, is dominated by precipitation from sea water.

2. Clastic sediments are divided into conglomerate, sandstone, and shale on the basis of grain size, which reflects current strength.

3. A study of the grain size and bedding characteristics of sand and sandstone gives information on current type and direction. Paleocurrents are inferred from cross-bedding and other directional structures.

4. Gravel and conglomerate are formed by strong currents characteristic of rapidly flowing rivers and strong waves. Pebble size and roundness indicate the distance of transport from the sediment source area.

5. Mud and shale are deposited by settling of fine material from slow or waning currents on river floodplains and in deeper marine environments. Many shales are mixed with chemical components. The clay minerals of shales come from the weathering of silicate minerals.

6. Sedimentary environments are described in terms of geomorphology, which links important sedimentary characteristics to particular geographic locations. Clastic sedimentary environments are alluvial, desert, glacial, deltaic, beach and bar, shallow shelf, turbidite, and pelagic.

7. Carbonate sediments form from supersaturated sea water in warm parts of the ocean's surface, primarily through the activities of organisms that secrete carbonate.

8. Coral reefs and atolls are built by large communities of carbonate-secreting corals and coralline algae. Their structure of reef front, lagoon, and island is made by the complex interaction of many kinds of organisms and the sea. Reefs form around originally volcanic islands, which may later erode and subside to be covered over by reef.

9. Carbonate platforms near or attached to continents in tropical seas account for most of the shallow-water carbonate sediments of the world. Oolite, carbonate sands, muds, and tidal-flat deposits, such as stromatolites, are characteristic of the various environments of the shallow bank areas of the Bahama Islands.

10. Silica is secreted by diatoms, radiolarians, and sponges. Siliceous oozes and cherts are formed by the sedimentation of the skeletons of these organisms on the sea floor.

11. Precipitation of sedimentary sulfides is accomplished by the action of sulfate-reducing bacteria on organic matter in such deoxygenated environments as silled basins or tidal flats.

12. Coal is formed by the partial decay and transformation of organic matter laid down in swampy environments. Vegetation is first changed into peat and then, in an ascending series of metamorphic stages, to lignite, bituminous coal, and anthracite.

13. Evaporites are deposited in restricted basins of the sea in which evaporation continues to produce precipitates of gypsum and salt as sea water is replenished by a connection to the open sea. Some evaporites form from complete desiccation. Others form by evaporation on tidal flats in arid environments.

14. Sediment accumulates as a result of subsidence in the depositional area. Geosynclines, which are the sites of great accumulations of sediments, are formed by deposition of

shelf, slope, and rise sediments at passive continental margins and subduction zone sediments at active continental margins.

15. A continental platform is a large area of a continent over which subsidence has been slight for a long time. As a result, sedimentary accumulations on such platforms are thin.

16. Tectonic control of sediment composition is evidenced by the ratio of quartz, a stable mineral, to feldspar, an unstable mineral in weathering environments. Mountain-building results in more rapid mechanical erosion of feldspar; thus, abundant feldspar in sediments indicates tectonism.

17. Diagenesis, the process by which sediments are changed physically and chemically and are lithified, is partly a result of burial and increases in temperature and pressure.

EXERCISES

1. Assume a large oil spill off an atoll led to the death of all of the corals on the reef. What events might characterize the further evolution of the reef? Could you recognize these events from the sediment produced?

2. What reasons can you think of why a passive margin continental shelf would have few coral atolls of the Pacific Ocean type?

3. Early in the Precambrian era there were no carbonate-secreting organisms. Assuming the oceans, with that exception, were much like those of today, where and what kind of carbonate sediment, if any, would you expect to have been deposited?

4. The Black Sea is known to lack oxygen in its deep waters. What does this suggest about the nature of the Bosporous and Dardanelles, the straits that connect the sea with the Mediterranean? What kinds of sediment might you expect to find at the bottom of the Black Sea?

5. The general oceanographic circulation conditions of some ancient evaporite basins seem to have been similar to those of some present-day deoxygenated basins, yet only some of the evaporite basins contain much organic matter and sulfides. Why might this be so?

6. A geologist has mapped a formation of a certain age as changing from a coarse conglomerate inter-

fingering with lake deposits on the western edge of her map area to 10- to 30-meter thick, cross-bedded, fine- to medium-grained sandstone beds and little else on the eastern edge. What deductions might be made about the depositional environment and paleogeography of the formation?

7. In the Mississippi delta country of Louisiana, oil wells have been drilled 7000 meters or more to penetrate Miocene formations of nearshore shallow-water sandstones and shales. How do you think this much sediment has accumulated in such a time interval?

8. How do you think the processes that have made mud into shale might generally differ from those that have made sand into sandstone?

9. A number of geologists estimate that 10 percent of the ocean's salt may have been removed as evaporite deposits during the Permian Period. If the ocean's volume were then the same as now, 1.37×10^{21} liters, how many kilograms of salt would have been deposited?

10. Some sedimentary environments are adjacent, others always separated by intervening ones. Make a list of pairs of environments that are not adjacent and those that are.

BIBLIOGRAPHY

Blatt, H., G. Middleton, and R. Murray, *Origin of Sedimentary Rocks*, 2nd ed. Englewood Cliffs, N.J.: Prentice-Hall, 1980.

Garrels, R. M., and F. T. Mackenzie, *Evolution of Sedimentary Rocks*. New York: W. W. Norton, 1971.

Goreau, T. F., N. I. Goreau, and T. J. Goreau, "Corals and Coral Reefs," *Scientific American*, August 1979. (Offprint 1434.)

Kuenen, P. H., "Sand," *Scientific American*, April 1960. (Offprint 803.)

Laporte, L. F., *Ancient Environments*. Englewood Cliffs, N.J.: Prentice-Hall, 1968.

Reading, H. G., ed., *Sedimentary Environments and Facies*. New York: Elsevier North-Holland; Oxford: Blackwell Scientific Publications, 1978.

Selley, R. C., *Ancient Sedimentary Environments*, 2nd ed. Ithaca, N.Y.: Cornell University Press, 1978.

12

Earth and Life

Earth is probably only one of many billions of planets in the universe on which the conditions necessary for the development of life were met. The compositions of the crust, the atmosphere, and the oceans have not only determined the course of biological evolution but have themselves been affected by life processes. The primitive Earth had an oxygen-free atmosphere conducive to the origin of life. Some time after primitive life began, photosynthesis evolved, and oxygen began to build up in the atmosphere, eventually to near present levels. Life depends on the environments in which it evolved and to which it has adapted. Human activities may affect those environments seriously by increasing the levels of carbon dioxide, which might in turn alter the global climate, and by introducing large quantities of toxic metals into the biosphere.

The Earth gives us permission to live. It provides the basics: the air we breathe, the water we drink, and the food we eat—the plants and animals that share the planet with us. For modern civilizations, it provides the mineral and energy resources necessary for innumerable systems that we take for granted, from the fertilizers required by modern high-productivity agriculture to the jet fuel required by airliners. The surface environment of Earth plays a profound role in our lives, as it does for all organisms. We can survive only in a narrow range of temperatures and pressures, and we are dependent on the abundance of many chemical compounds—water in particular. It has become abundantly clear, particularly in the past decade, that the relation between life and the environment is not one-way. Not only does the physical environment control life; the biological world also has a profound effect on the environment and has had since life first appeared billions of years ago. Rich, black soil—an environmental necessity for productive agriculture—is formed largely by the interaction of organisms with elements of the physical environment. The oxygen in

the atmosphere—an element of the environment required for respiration of all complex forms of life—has been manufactured by green plants in the process of converting sunlight to usable chemical energy by means of photosynthesis. The human population of the Earth is increasingly altering the global environment. We add more and more carbon dioxide to the atmosphere by burning coal and oil, for example, and we have dispersed chemical pollutants throughout the waters of the continents and oceans.

In this chapter, we shall explore the interaction between Earth and life. Why is there life here and not on the Moon? How did life start, and how did it evolve? Above all, how precarious is our existence on this planet, and what are the possible effects of our growing population and its technology on the environment?

LIFE BEGINS

Throughout most of human history, the subject of the origin of life was exclusively a mythological or a theological concern. It centered on discussion of the revealed truth in the sacred texts or oral traditions of the Judeo-Christian and other religions. All explanations were framed in terms of the actions of a creator. The rise of geology in the eighteenth and nineteenth centuries (as part of the

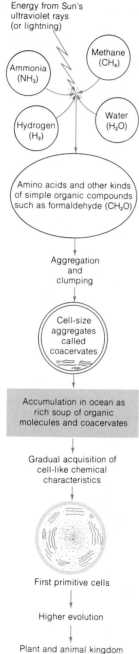

Figure 12-1
Origin of primitive organic compounds, followed by precellular evolution leading to the first cells as first envisioned by A. I. Oparin. The requirement for a reducing atmosphere is no longer considered necessary, and carbon dioxide may be substituted for methane and hydrogen.

Age of Enlightenment) led to theories about the age of the Earth in scientific terms rather than in terms of a literal interpretation of the Bible. Yet it was not until the twentieth century that scientists seriously opened the subject of the origin of life to nontheological discussion. Louis Pasteur waged a decisive battle against the nineteenth century idea of spontaneous generation of life, challenging the then commonly held view that, by some such process, living maggots, flies, and bacteria appeared spontaneously on rotting garbage. Pasteur showed that neither bacteria nor any other kind of life can arise from any kind of dead organic matter if it is first sterilized by heat. Life originates only from other life, he argued. Right as he was about life as we now know it, he and others simply ignored the question of whether there was ever a beginning to the process.

In 1924 a young Russian biochemist, A. I. Oparin, opened up the whole matter by asserting that there must have been a beginning of life sometime in the early history of the Earth and that it is possible to make some intelligent guesses about how that beginning might have come about. Oparin theorized that, in its early evolution, the Earth's atmosphere lacked oxygen but contained many reduced gases, such as ammonia NH_3, hydrogen, H_2, and methane, CH_4. In that kind of atmosphere, energy in the form of ultraviolet rays from the Sun or lightning might have created organic molecules similar to the basic building blocks of life, such as the amino acids, the molecules that form proteins when strung together chemically in long chains (Fig. 12–1). Once the building blocks were formed, they might have tended to clump together in larger and larger units, which would then have begun to attain some of the characteristics of a primitive cell. All of this gradual synthesis would have taken place in the early ocean, which he described as a "soup" of organic molecules. In that soup, the primitive cell-like aggregates would gradually have acquired more lifelike attributes, a process culminating in the appearance of the genetic substance that governs reproduction and inheritance. This stage would have marked the beginning of life. Oparin thus showed how a series of small, statistically probable chemical events could lead to the formation of life.

About thirty years later, a young American graduate student in chemistry, Stanley L. Miller, devised a crucial experiment that showed how it really might have happened (Fig. 12–2). He built an apparatus in which he could simulate lightning bolts in a reducing atmosphere by setting off powerful electric sparks in a glass chamber filled with a mixture of water, ammonia, and methane. After sparking such a mixture for many hours, Miller was able to identify a number of amino acids. Since Miller's experiment the study of the origin of life has become an active science linking astronomy, biology, biochemistry, and geology.

Today the general outlines of Oparin's theory remain, but many components have changed. Newer experiments and calculations have shown that synthesis of organic compounds can take place in other chemical surroundings than the reducing atmosphere Oparin postulated. Most interest has been shown in an early atmosphere consisting mainly of water and carbon dioxide with little hydrogen or oxygen. Such a composition is favored by many Earth scientists as expectable from the outgassing of the primitive Earth (Chapter 6). Others have performed experiments suggesting that organic compounds could have formed at very low temperatures in comets and come to Earth from this source early in Earth's history.

The major energy source required to link small molecules together to make larger ones was most probably ultraviolet radiation. In an early atmosphere with little oxygen, ultraviolet would have been far more intense than it is today. On today's Earth, most ultraviolet light is filtered out by a layer of ozone, a three-atom oxygen molecule, O_3, formed in the stratosphere from the ordinary two-atom oxygen molecule (Fig. 12-3). In the

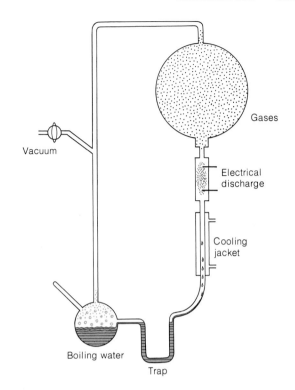

Figure 12-2

S. L. Miller's experiment in which amino acids were made by circulating methane (CH_4), ammonia (NH_3), water vapor (H_2O), and hydrogen (H_2) past an electrical discharge. The amino acids, collected at the bottom of the apparatus, were detected by paper chromatography. [From "The Origin of Life" by G. Wald. Copyright © 1954 by Scientific American, Inc. All rights reserved.]

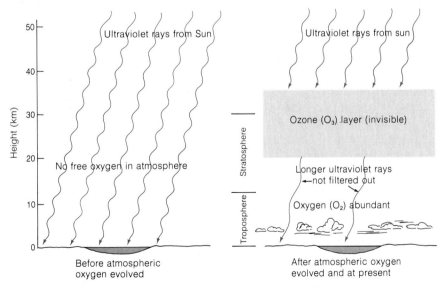

Figure 12-3

The evolution of oxygen in the atmosphere led to formation of an ozone layer in the stratosphere as oxygen atoms and molecules combined to form the three-atom molecule O_3. This layer absorbs much of the shortwave-length ultraviolet rays from the sun and protects surface life from this cell-damaging radiation.

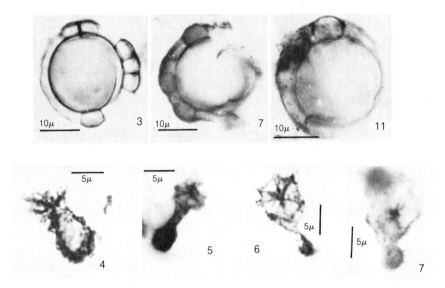

Figure 12-4
Microorganisms from the Precambrian Gunflint Chert of Ontario, about two billion years old. The top row shows sections through three different individuals of the same algalike organism *Eosphaera tyleri,* thought to be a free-floating colonial photosynthetic organism. The bottom row shows four different individuals of a peculiar primitive organism, *Kakabekia umbellata,* first discovered in the Gunflint Chert. [Photos by E. S. Barghoorn. From E. S. Barghoorn and S. A. Tyler, "Microorganisms from the Gunflint Chert," *Science,* v. 147, pp. 563–577. Copyright 1965 by the American Association for the Advancement of Science.]

ozone layer, some oxygen molecules are split by absorption of the energy of ultraviolet rays to form single oxygen atoms. These then combine with oxygen molecules to form the ozone, which itself absorbs ultraviolet rays very effectively. Were it not for the ozone layer, we would all be irradiated with powerful ultraviolet light, which would induce such intense sunburn that life on the surface of the land would be impossible.* Strange that the rays that now tend to destroy life worked then to build it.

The early ocean "soup" was fairly dilute, perhaps with a thin scum of oily organic matter on the ocean surface. Much current research is concerned with the next step: the making of large aggregate molecules of proteinlike or DNA like materials of the sort that led to the evolution of the first cell. Once DNA was formed, cells could divide, each keeping a strand of DNA, which would cause it to resemble its parent and, at the same time, would offer the opportunity for change and evolution through random alteration of the

strand. The changes in the organism produced by alterations of its DNA were important, for they allowed the cell to interact differently with the physical environment. Some changes are beneficial to the organism in its interaction with the environment and others are not; thus some individual organisms survive and others die. This phenomenon, called *natural selection* by Darwin, was instrumental in directing the course of evolution.

Whatever the steps in building these lifelike molecules, the information from the Viking expedition to Mars in 1976 suggests that they either did not take place on that planet or were aborted by the long-term lack of liquid water. No form of life familiar to us has been found on the Martian surface. Earth's evolution as a planet may have been unique in this solar system in providing the conditions for life to begin and flourish.

THE OLDEST FOSSILS

When did all of this take place? Paleontology gives us an important date. In stromatolitic rocks of a sedimentary sequence in western Australia that is about 3.5 billion years old, there are microscopic forms of carbonaceous organic matter—fossils of primitive cells. They indicate that the whole process of life began relatively early in the

---

*Some scientists believe that the existence of the ozone layer may be threatened by introduction of Freon (chlorofluoromethanes) used as a refrigerant and as aerosol spray propellant and by the nitrogen oxide gases originating from the heavy use of nitrogen fertilizers. Both result in ozone-destructive chemical reactions that might be difficult or impossible to counteract. The ever-increasing use of nitrogen fertilizer may ultimately be limited because of this unforeseen side effect.

Earth's history, perhaps as early as 4 billion years ago. Although there are even older microscopic carbonaceous objects in the oldest known rocks, the 3.8 billion year old metamorphosed sediments of the Isua complex in southwest Greenland, most experts in Precambrian life doubt their biological origins. Just as the search continues for the oldest rock, there is the possibility that very old sediments will be found that will allow geologists, with certainty, to push the origin of life even farther back than 3.5 billion years.

We are sure that by about 2 billion years ago, well-organized algal life was thriving. A formation that old exposed on the north shore of Lake Superior, the Gunflint Chert, contains forms that are undoubtedly biological (Fig. 12–4). Formations of stromatolitic limestones of about the same age have been mapped in Canada and South Africa; the stromatolites are of the same sort as those formed by algal mats today (Fig. 12–5). By one billion years ago, a higher form of life may have been established: the cell with a nucleus. (Cells of this sort are the ones with which most of us are familiar, but they represent relatively advanced bacteria and higher organisms; the earliest life forms, such as primitive bacteria and algae, had only nonnucleated cells.) We know of organisms from the Bitter Springs Formation of central Aus-

tralia and elsewhere that contain what look like remnants of nuclei. The rocks of that formation have the same lithology as other Precambrian sedimentary rocks in which the oldest fossils are preserved—unmetamorphosed cherts with excellent preservation of organic structures (Fig. 12–6).

Figure 12–5
Algal stromatolites from the Bulawayan Limestone of Rhodesia, which is three billion years old. Typical crescent-shaped stromatolitic laminations are visible on the weathered rock surface. [Photo by Keith Kvenvolden.]

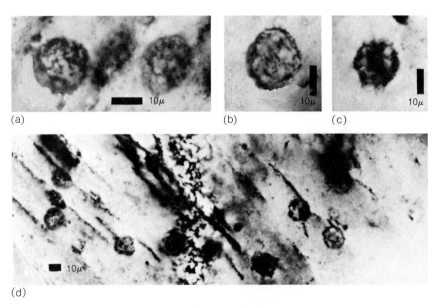

(a) (b) (c)

(d)

Figure 12–6
Organic, algalike microscopic fossils in black chert of the early Precambrian Fig Tree Series near Barberton, South Africa. (a) Spheroidal and distorted fossils showing typical irregular surface texture; (b) section through one spheroid showing cell wall of variable thickness; (c) spheroid containing "coal"-like organic material interpreted as the altered remnants of the original cell material; (d) spheroidal fossils showing varying degrees of preservation. Diagonal layers are laminae of organic matter parallel to bedding. [Photos by E. S. Barghoorn. From J. W. Schopf and E. S. Barghoorn, "Alga-like Fossils from the Early Precambrian of South Africa," *Science*, v. 156, pp. 508–512. Copyright 1967 by the American Association for the Advancement of Science.]

Figure 12-7
A late Precambrian fossil from rocks of the Ediacara Hills, South Australia. This annelid worm is one of many soft-bodied organisms whose development preceded the evolution of the shelled invertebrates of the Cambrian Period. [Photo by M. F. Glaessner.]

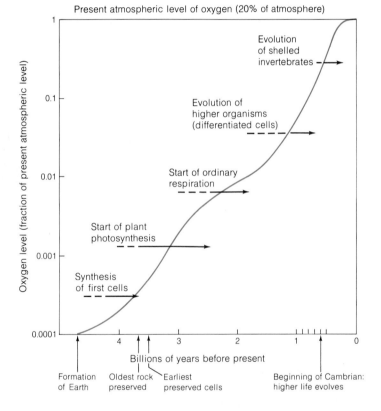

Figure 12-8
One hypothesis of the evolution of oxygen in the atmosphere in relation to the origin of life and evolution of higher organisms. There is as yet no general agreement on exactly when and to what levels oxygen accumulated in the Precambrian and how close the levels of oxygen in Phanerozoic time came to the present-day level.

The first shelled organisms appeared at the beginning of the Cambrian Period. Cambrian fossils include an incredible diversity of species of invertebrates. Just before that time, in very late Precambrian sediments transitional to the Cambrian, higher forms of multicellular (or **metazoan)** life without shells had developed (Fig. 12-7). Sometime between 600 million and one billion years ago, higher forms of life evolved so rapidly that compared to the rates of earlier times, the rate of evolutionary development, could be described as explosive. From a world populated by primitive algae and bacteria, there emerged many thousands of species of animals, from sponges and mollusks to worms and trilobites. Then began the march of evolution that paleontologists have mapped so well: the appearance of the first vertebrates, land plants, flowering plants, the dinosaurs, the mammals, and, finally, a few million years ago, the genus *Homo*.

Evolution of Atmospheric Oxygen

There was a radical transformation from the first life that developed in an atmosphere with little or no oxygen to the kinds of organisms of the Cambrian and later periods that needed oxygen for respiration. The first organisms must have lived without oxygen. Some such primitive bacteria exist today—the sulfate-reducing species, for example that are instrumental in precipitating pyrite (Chapter 11). At some time around three billion years ago, some primitive species must have developed one of the great biological inventions: photosynthesis.

Photosynthesis is a complex chemical process by which plants that contain chlorophyll convert carbon dioxide and water into carbohydrates, the sugars needed for biological energy. In the equation below, we show the production of CH_2O, a simple formula representing all of the many carbohydrates that actually form.

$$\text{carbon dioxide} + \text{water} \xrightarrow[\text{sunlight and chlorophyll}]{\text{(in the presence of}}$$
$$CO_2 \qquad\qquad H_2O$$

$$\text{carbohydrates} + \text{oxygen}$$
$$CH_2O \qquad\qquad O_2$$

For each 30 grams of carbohydrate produced in this reaction, approximately 112,000 calories from sunlight is converted to chemical energy tied up in sugars, the energy that is the basis of life processes. At the same time, one molecule of oxygen is produced for every atom of carbon incorporated into organic matter.

At the beginning of the evolution of photosynthesizing plants, there were no animals to breathe oxygen, so much of the oxygen produced by photosynthesis must have started to accumulate in the atmosphere.* This free oxygen made possible a new method of utilizing energy: respiration. Respiration is oxidation of carbohydrate, which releases the energy stored by photosynthesis.

carbohydrate + oxygen →
$$CH_2O \qquad O_2$$

$$\text{carbon dioxide} + \text{water}$$
$$CO_2 \qquad H_2O$$

In this reaction, one molecule of oxygen is used up in the conversion of each atom of organic carbon to carbon dioxide. The first respirers were undoubtedly primitive bacteria that could live on small amounts of oxygen, perhaps as little as a hundredth the amount now present in the atmosphere. This development may have taken place as early as 2.7 billion years ago (Fig. 12–8). For a long time oxygen must have continued to accumulate slowly, as photosynthetic production exceeded its depletion by respiration. As a result, we can deduce that an equivalent amount of organic carbon was buried as organic matter in sediments (Fig. 12–9). That organic matter can be looked upon as the buried excess of photosynthesis over respiration. In its simplest form the total budget since oxygen first started accumulating is given by this subtraction:

$$\begin{pmatrix} \text{number of carbon atoms} \\ \text{in organic matter} \\ \text{produced by photosynthesis} \end{pmatrix} = \begin{pmatrix} \text{number of} \\ \text{molecules} \\ \text{of oxygen} \\ \text{produced} \end{pmatrix}$$

minus   *minus*

$$\begin{pmatrix} \text{number of carbon atoms} \\ \text{in organic matter} \\ \text{oxidized by respiration} \end{pmatrix} = \begin{pmatrix} \text{number of} \\ \text{molecules} \\ \text{of oxygen} \\ \text{used up} \end{pmatrix}$$

equals   *equals*

$$\begin{pmatrix} \text{number of carbon atoms} \\ \text{in organic matter} \\ \text{preserved in sediments} \end{pmatrix} = \begin{pmatrix} \text{number of} \\ \text{molecules} \\ \text{of oxygen} \\ \text{accumulated} \\ \text{in atmosphere} \end{pmatrix}$$

The total amount of carbon in organisms living today is so small compared to the total for all the

---

*Chemical reactions in the upper atmosphere split a small number of water molecules into hydrogen and oxygen gas. A small amount of atmospheric oxygen may have originated in this way, but the rate of production by these reactions is so slow as to be, for practical purposes, negligible.

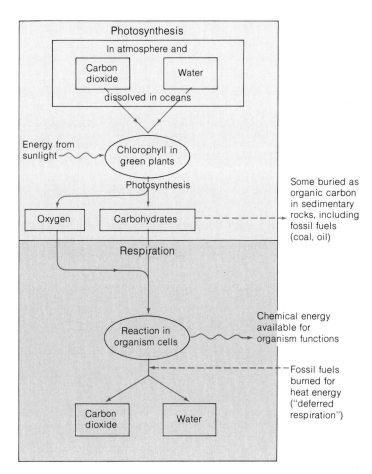

Figure 12–9
The photosynthesis-respiration couple that links organisms to the oxygen and carbon dioxide in the atmosphere and oceans.

organisms that have lived in the past that it can be neglected.

Sometime just before the Cambrian, atmospheric oxygen reached levels close enough to today's to allow for the rapid evolution of the higher invertebrates. For the rest of geologic time, the oxygen in the atmosphere has been maintained by the photosynthesis of the green plants of the world, much of it by green algae in the surface waters of the ocean. This story is one of the most dramatic examples of how far-reaching the consequences of the origin and evolution of life were for the surface of the Earth.

Periodically geochemists are called upon to demonstrate that our oxygen supply is vast and that human activities will not deplete it even in the distant future. Some biologists are concerned that chemical pollutants interfere with photosynthesis of some green algae. Others are worried that the forests of the world are so cut that the total amount of photosynthesis will drop. The erroneous reasoning follows that, if photosynthesis

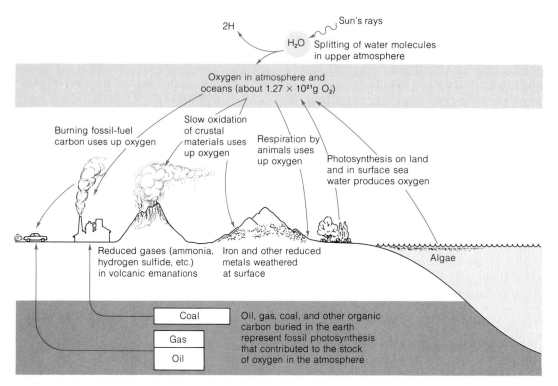

Figure 12–10
Input and output of oxygen in the atmosphere and oceans. Photosynthesis and respiration are in approximate balance and far outweigh other inputs and outputs. The amount of oxygen in the atmosphere is so great in proportion to the output rate that even if all photosynthesis were to stop, our oxygen supply would last well over 2000 years.

stopped all over the world, we would "soon" use up all the atmosphere's oxygen by respiration. The concern is groundless, for there is such an immense amount of oxygen in our atmosphere and oceans (about 10^{21} grams) that, even if all photosynthesis stopped tomorrow and all other respiring life went on as before, it would be several thousand years at least before oxygen would be significantly depleted by respiration and all of the other reactions in which oxygen is used (Fig. 12–10). In addition, the assumptions behind the feared decrease in photosynthesis have been shown to be wrong. Cutting forests does not seriously affect total photosynthesis, and there is no evidence that the Earth's major photosynthesizers, oceanic algae, are detectably affected by chemical pollution. Some atmospheric oxygen *is* used up by the oxidation of metals that occurs in the weathering of rocks, but the amount per year is very small compared with the huge quantities used by respiration and provided by photosynthesis. Accelerated burning of fossil fuels also uses up oxygen, but this amount is also too small to affect the overall picture seriously. So far, we

have used up only 7 of every 10,000 oxygen molecules by burning coal, oil, and gas; if we were to keep burning fuel at an increasing rate until the year 2000, we would still have used up only 20 of every 10,000 oxygen molecules. Some concern about the environment is justified, but there need be no fear of running out of oxygen.

EXTRATERRESTRIAL LIFE

The evolution of life on Earth was an expectable consequence of the evolution of the planet. Earth is large enough for its gravitational attraction to hold an atmosphere. The surface temperature of the primitive Earth was low enough that water was liquid. All of the chemical elements needed for our varied biological processes were present. Given this conjunction of circumstances, was it not inevitable that life forms based on the organic chemistry of carbon and water would arise? If so, then is it not also inevitable that life is not a strange accident on a unique planet but an event that we would expect to be repeated in various places and at various times in the universe? This is how Harlow Shapley, one of the great American astronomers of the past generation, started to think about the subject. He then made a simple calculation. How many stars had about the same composition and quality of luminosity as our Sun, and thus could heat and light a planet such as ours? From his knowledge of our galaxy and the rest of the universe, Shapley estimated that there were about 10^{20} such stars. How many of those stars might have a series of planets like the ones in our solar system? One in a thousand is a conservative estimate. How many of those planets might be placed, as the Earth is, just the right distance from their stars for them to have the right temperature? One in a thousand, to continue being conservative. How many of those planets might be big enough to hold atmospheres but not so big as to hold all of their hydrogen, like Jupiter? Again, one in a thousand. After all these and other estimates, Shapley ended with a most conservative estimate of the number of planets in the universe on which life might be expected to develop: 100,000,000. With such a large number, he thought it exceedingly probable that life does indeed exist out there somewhere. Since Shapley's estimates, others have modified the estimates of the likelihood of life existing elsewhere in the universe. Some have modified the estimates upward. However one calculates, the conclusion is inescapable: there is a reasonable chance that life exists somewhere else in the universe.

Part of the excitement about the first trips to the Moon was the prospect of confirming our ideas about the origins of life. The Moon lacks an atmosphere; it is too hot on the day side and too cold on the night side; and there is no detectable free water. We did not expect to find life there, and we didn't. The most careful analysis of the first Moon rocks failed to reveal any sign of life, neither cells of organisms nor telltale organic molecules that could have been formed only by biological processes earlier in the Moon's history. So we have important negative evidence that gives us some confidence in our reasoning.

Mars is another story. Because Mars has a thin atmosphere (mostly carbon dioxide) and water (as ice), the chances that some form of life exists—or once existed—are better than on the Moon. But no form of even very primitive life, at least as we know it on Earth, could survive for long on Mars today. One of the instruments that was landed on the Martian surface was designed to detect small quantities of amino acids in soil. The results were negative. Results of other instrumental analysis

shed further doubt on the existence of life, though some believe that we may yet turn up evidence for it. The verdict for the present: no life on Mars now, and no evidence that conditions at the surface were ever favorable for long enough to generate life.

Life in outer space? It's doubtful. But astronomers have detected small quantities of about forty simple organic compounds in the nearly empty space between stars. One of them is formaldehyde, CH_2O, the simplest building block of carbohydrates. Some of these organic compounds are also found in meteorites. Thus it is apparent that the basic units of the compounds of life can be formed in space and transported by meteorites. Now we must indeed recognize the probability that life exists throughout the universe. Whether we shall ever detect it, by radio or other means, remains only guesswork.

Back here on Earth, the importance of all of these ideas is that they make us focus on why we are here and what the conditions are for our continued existence on this planet. How stable is our environment? Can we, the members of one species, so alter the Earth that we make life more difficult for ourselves?

INTERACTIONS OF LIFE WITH THE ENVIRONMENT

Once metazoans evolved, the well-known fossil record began—a record of a host of evolutionary events, as new organisms arose by the interaction of the genetic material of each species with its environment, for the physical and chemical environment of the Earth's surface is the arena of natural selection that determines which species survive and which cease to exist. But the environment is itself acted upon by the organisms. The surface of the Earth, its atmosphere, and its oceans, have been profoundly affected by developments in organic evolution.

The shelled organisms did more than mark history with their fossil remains. In the oceans these organisms secreted enormous quantities of calcium carbonate, calcium phosphate, and silica to make their shells. As they died, new types of sediment were thus created, as were constructional features, such as reefs. In the mid-Paleozoic, the higher plants evolved and the land surface became vegetated with coniferous trees, ferns, and other early plant species. Lushly vegetated swamps became possible for the first time, producing another kind of biological sediment, coal.

Vertebrate life had its origin in the early Paleozoic, beginning with the fishes, followed by the amphibians, and then the climactic evolution of the reptiles, the age of dinosaurs, in the Mesozoic. Mammals appeared in the Mesozoic too, but they reached dominance only after the dinosaurs disappeared. None of these vertebrates had any significant effect on Earth's physical or chemical environment; that was to come with the next development: the evolution of humankind. The genus *Homo* appeared on the scene only a few million years ago, and evolved to *Homo sapiens,* our species, only within the last million.

Paleontologists have long sought to tie major events of organic evolution to major geologic events. Hundreds of species of invertebrates, vertebrates, and plants became extinct at various times in geologic history, the most massive extinctions coming at the ends of the Paleozoic and Mesozoic eras. What caused such wholesale disappearance of species? The assembly of the supercontinent of Pangaea in the late Paleozoic has been linked to the Permo-Triassic extinctions. As continents collided to form Pangaea, most of the expanse of shallow continental shelf surrounding each continent disappeared, leaving only one narrow perimeter around the supercontinent. During the Paleozoic the shelves had harbored the most productive biological communities. Climatic extremes, including glaciation of parts of what are now Africa, Australia, and South America, worked with the geographic constriction of shelves to create environmental stresses great enough to decimate many species. When Pangaea rifted apart, forming wide new expanses of hospitable shelves, the survivors founded the new stocks of the Mesozoic world.

A visitor from space is credited with the extinctions at the end of the Cretaceous in one of the newest geological mechanisms hypothesized. In 1980 a group of researchers from Berkeley, California led by Luis Alvarez, a physicist, and his son Walter, a geologist, announced the finding of extraordinarily high concentrations of the element iridium—thirty times normal—in marine clays deposited in several places exactly at the end of the Cretaceous. They ascribe the iridium and some other anomalous element concentrations to the impact of an asteroid of about ten kilometers in diameter. They hypothesize that it hit the Earth traveling at about 90,000 kilometers per hour, throwing tens of quadrillions of tons of pulverized rock up into the stratosphere. Because of this dust, they speculate, sunlight would have decreased to about 10 percent of full moonlight for several years, killing off plants on land and in the sea and leading to extinction of many species from dinosaurs to foraminifera. Though the chemical evidence for such an impact may be strong, paleontologists point out that different groups of organisms became extinct at different times in different places near the end of the Cretaceous, not instantaneously. Much research is being done to evaluate the implications of this provocative hypothesis and how well it explains the geological and paleontological record.

The major geologic event associated with human evolution and early history was, of course, the Pleistocene glaciation. Early humans learned first to take advantage of natural protected habitats, such as caves, for refuge from the snow and ice and cold. The next step was learning to build shelter. And that step ultimately led to our huge buildings, highways, dams, canals, and all of the other new "landforms" of the modern Earth. *Homo sapiens* is, of course, a natural species, and in that sense anything we do is "natural," but the landscape of our civilization goes so far beyond any other organism's work in modifying "nature" that our influence is clearly of a different order of magnitude. Our influence has had its negative aspects too, for inadvertently our modification of the environment to suit ourselves may have produced hazards as well as benefits.

HAZARDS TO THE ENVIRONMENT

"Environment and ecology" sections of public libraries and bookstores are disturbing places to visit. Titles are made up of words like "crisis," "survival," "threat," and "danger." Perhaps more revealing is the use of such words as "frail" and "fragile" to describe the Earth. How do we reconcile the doomsday view with the geological evidence that the Earth is a dynamic global system that is remarkably stable in its steady-state equilibrium? To what extent have people suddenly wakened to the real possibility that we are so efficient at controlling our own environment that we may foolishly be tipping "the balance of nature" on a global scale, and to what extent have they simply become aware of what the Earth has been like all the time?

Some environmental hazards are local and regional; others are global. Landslides and earthquakes are natural and local phenomena. We describe, in Chapter 5, how poor engineering and construction practices can precipitate landslides and, in Chapter 17, how minor earthquakes can be provoked by pumping fluids at high pressures into subsurface formations. At the opposite extreme are the global hazards, such as the worldwide increase of carbon dioxide in the atmosphere, that might influence the Earth's climate.

Many concerns for specific regions or types of environment are purely esthetic—the preservation of natural beauty, wilderness areas, and, in general, "unspoiled" landscape. To the extent that such concerns are based on uniqueness of the landscape, a knowledge of geology may inform our decision to preserve or modify any specific element of that landscape. We know that river gorges may be common in a variety of terrains, but there is little question that there is only one that has the magnificence of the Grand Canyon of the Colorado.

Environmental Effects on Health

The environmental hazards of greater importance are those to public health. Are the levels of lead in our drinking water or food much above the natural level, and, if so, are they potentially a medical danger? How widespread is mercury pollution, and how toxic are a few parts per billion of mercury in drinking water? In such matters the job of the scientist is dual. One is to measure the spread of a pollutant through the Earth-surface system of atmosphere, surface waters, biological communities, and sediments, and to determine how much human activities are altering that spread. The other is to answer a series of medical questions. What bodily damage may result from long exposure to low levels of toxic metals or other pollutants? At what levels do obvious disease symptoms appear? At the extreme, what are the lethal doses? And at the opposite extreme, we must consider the ill effects of deficiencies of some metals in our diet.

Most perplexing are the maps that show the incidence of certain diseases, from cancer to kidney stones, for each county or region of a country. These maps show that people living in certain areas are seemingly more prone to certain diseases than those living elsewhere. Does the incidence of a disease have to do with the kinds of people who settled in a region, or with their descendants, or does it have to do with the local geological environment that affects the water, food, dust, and anything else that may affect health?

Toxic Metals

The toxic metals are some of the most serious potential offenders. Mercury, for example, which is known to be toxic, first hit the headlines some years ago when a Canadian graduate student discovered high levels of the metal in the tissues of lake fish. At about the same time we became aware that acute mercury poisoning was occurring in people living around several bays in Japan where shellfish, severely contaminated by mercury-laden industrial wastes, constituted a large portion of the diet. Mercury occurs in small amounts in many rocks—about 0.2 part per million in granite and less than half that amount in the average crustal rock. The mercury in rocks is steadily released in small amounts to natural waters by ordinary chemical weathering processes. Most natural waters contain only a few parts per billion and thus are harmless. Some part of the mercury in water is naturally converted to an organic form, methyl mercury, which is the form most harmful to organisms. Medical data indicate that chronic mercury poisoning may arise from high levels (many parts per million) of the metal dissolved in water, much of it as methyl mercury. At these levels, mercury affects the nervous system in hidden ways, with few well-defined symptoms shown for long periods of time after the exposure. For these reasons, the World Health Organization has recommended the maximum permissible mercury level for human food, including fish, at 0.5 part per million.

Enough work has been done on the circulation of mercury at the Earth's surface to show how the biological world and the mercury in the physical environment interact (Fig. 12–11). In the past fifty years, many thousands of tons of mercury have been mined for use in electrical equipment, chemical processing plants, and pesticides. We can look on this as an accelerated weathering process by which much more mercury than normal is being released from rocks. Though some of the mercury in some chemical processes is reused, a great deal of it escapes into natural waters or is vaporized into the atmosphere. From there, it is distributed to lakes, the ocean, and various sedimentary environments. A fraction of the mercury, converted to methyl mercury by bacteria, is ingested by organisms and accumulates in their tissues. As larger animals eat smaller ones, more of the metal accumulates in the larger ones, so that very large fish, such as tuna or swordfish, may contain relatively large concentrations, perhaps a few hundredths of a part per million. In waters that are polluted by industrial waste, the levels may be higher. Eventually, this material is absorbed by sedimentary particles, particularly the clays, and is buried out of reach of the biological system.

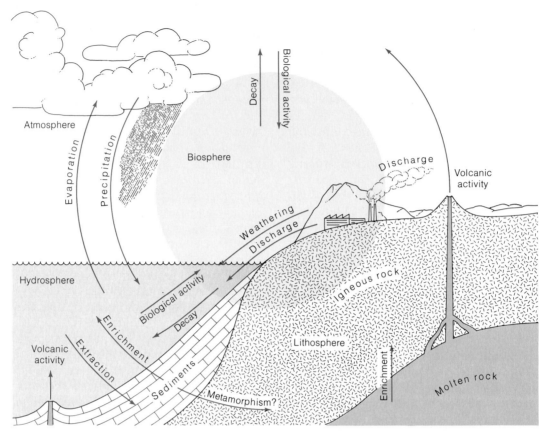

Figure 12-11
The mercury cycle disperses the metal throughout the lithosphere, hydrosphere, and atmosphere and through the biosphere, which interpenetrates all three. Mercury is present in all spheres in trace amounts, but it tends to be concentrated by biological processes. Human activities—in particular, certain industrial processes—may now present a threat by significantly redistributing the metal. [From "Mercury in the Environment" by L. J. Goldwater. Copyright © 1971 by Scientific American, Inc. All rights reserved.]

The serious question about mercury is, by just how much does the mercury in our waters exceed natural levels, and how widespread are such high levels? It is by no means clear that, in most places in the world, natural levels have been exceeded greatly. We know of a few places where industrial wastes have been uncontrolled and the pollution has reached dangerous levels; but we have only recently begun to pay much attention to these elements. We need to monitor unpolluted and polluted natural waters and their biological communities much more carefully. Like other toxic metals, mercury has always been present in the human diet at very low levels. Exactly what the range of those levels is—and by how much they can be exceeded without danger—remains to be determined. Minimum levels of some metals are required for good health, if, for fear of absorbing "poisonous" substances, we were to use only distilled water for drinking and food preparation, we might produce more harmful consequences than those caused by most impurities in water. We are, after all, adapted to a natural habitat in which

small amounts of almost every element are found.

Other metals have the same general cycles as mercury. The most important, from a medical point of view, are lead, cadmium, arsenic, chromium, and nickel. Lead is particularly important, both because it is very toxic and because it has been dispersed throughout the atmosphere by automobile and industrial emissions (Fig. 12–12). Two aspects of the problem are apparent: locally, in inner cities or near major highways, the lead levels from emissions may be very high compared to those in suburban or rural regions; and globally, there is evidence that the entire atmosphere is being charged with more and more lead. The United States has for some years been converting to lead-free gasoline for automobiles. If and when other countries follow and this source of atmospheric lead is cut down, the problem will disappear, for lead is rained out of the atmosphere very quickly.

Some geological materials are harmful to health as inhaled dust, especially by miners and industrial workers. *Silicosis* results from breathing

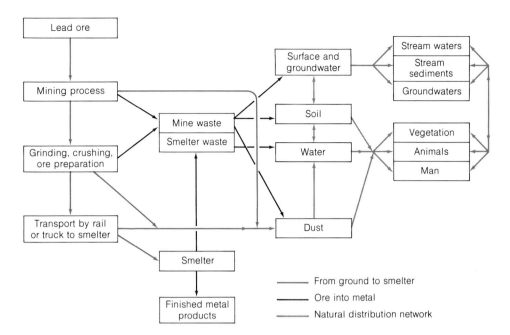

Figure 12–12
Movement of lead through surface environments, beginning with the mining of lead ore.
At each stage of processing or transport, lead-rich material may get into the air or into
surface waters and soil, some of it eventually to be ingested by plants and animals.
[Modified from B. G. Wixson, 1973.]

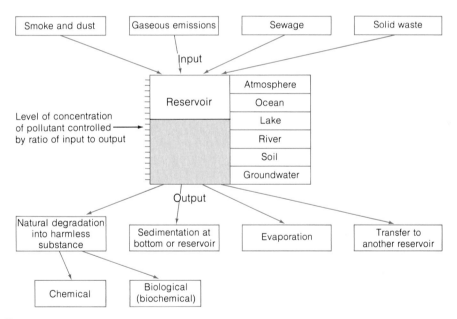

Figure 12–13
Inputs and outputs of pollutants in relation to concentration levels in reservoirs.
The level is controlled by the input-output ratio. The rate of change of level
depends on the rates of the processes involved.

quartz dust, *black lung* from coal dust, and *asbestosis* from asbestos minerals. A recent discovery of an extremely high rate of an unusual lung disease in a village in Turkey was linked to the abundance of a zeolite mineral, erionite, in the volcanic tuffs used locally to build houses. As we find out more about these kinds of diseases, we learn to be more prudent in the use of potentially hazardous materials.

All environments are self-cleaning by sedimentation. Sooner or later, the contaminants we worry about will settle out of the atmosphere, lakes, and oceans (Fig. 12–13). That is small consolation to most of us because rates of sedimenta-

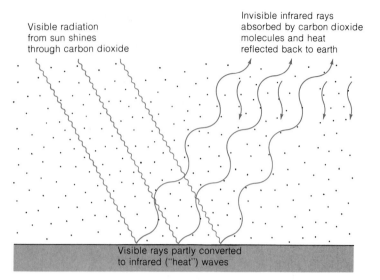

Visible radiation from sun shines through carbon dioxide

Invisible infrared rays absorbed by carbon dioxide molecules and heat reflected back to earth

Visible rays partly converted to infrared ("heat") waves

Figure 12–14
The greenhouse effect. Just as the glass of a greenhouse transmits light rays but holds in heat, the carbon dioxide of the atmosphere transmits visible radiation from the sun but absorbs and reflects back to Earth the infrared rays from the surface.

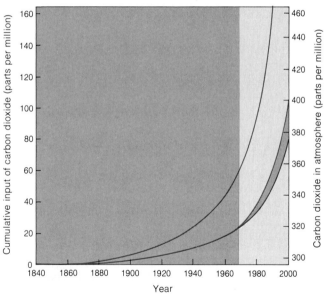

Figure 12–15
The lower curve shows the increase in atmospheric carbon dioxide since 1860, with a projection to the year 2000. The upper curve shows the cumulative input of carbon dioxide. The difference between the two curves represents the amount of carbon dioxide removed by the ocean or by additions to the total biomass of vegetation on land. [From "The Carbon Cycle" by B. Bolin. Copyright © 1970 by Scientific American, Inc. All rights reserved.]

tion are so slow for most dissolved pollutants that, even if sources of excessive pollution are cut off, dangerously high levels may remain for a hundred years. In a particular environment, any pollutant tends toward a steady-state level that is determined by the balance between input rate from natural and human activity, dispersal through the environment, and sedimentation rates in that environment. Historical records show whether the system is out of balance, whether the levels are increasing because total input is greater than output. In the future monitoring may also show how levels decrease as artificial inputs are lowered and the natural output continues. Input-output relations are shown well by the carbon dioxide system.

CARBON DIOXIDE AND CLIMATE

The small amount of carbon dioxide in the atmosphere (a little over 320 parts per million) has a profound effect on our climate. The atmosphere is relatively transparent to the incoming visible rays of the Sun. Much of that radiation is absorbed at the Earth's surface and then reemitted as infrared, invisible long-wave rays that radiate back away from the surface (Fig. 12–14). The atmosphere, however, is relatively opaque and impermeable to infrared rays because of the combined effect of clouds and carbon dioxide, which strongly absorbs the radiation instead of allowing it to escape into space. This absorbed radiation heats the atmosphere, which radiates heat back to the Earth's surface. This is called the "greenhouse effect," by analogy to the warming of greenhouses, whose glass is the barrier to heat loss. Any process that alters the amount of carbon dioxide in the atmosphere may conceivably affect Earth's climate.

Since the start of the industrial revolution, about the beginning of the nineteenth century, we have been pumping carbon dioxide into the atmosphere at an accelerating rate by our burning of coal, oil, and gas (Fig. 12–15). The carbon dioxide level of the atmosphere has been increasing, as shown by systematic measurements in various places in the world. The amounts of carbon dioxide added to the atmosphere have also been calculated from the figures for fuel consumption. There is a pronounced discrepancy: the level in the atmosphere has not risen as much as would have been predicted by the additional supply. This suggests some loss or output from the atmosphere, something absorbing the extra carbon dioxide, moderating the effect of the increased input.

Much of the carbon dioxide that is "missing" from the atmosphere has been mixed with the oceans. Gas molecules of carbon dioxide in the air are in equilibrium with dissolved gas molecules in the water. As the concentration of gas in the air increases, there is a tendency toward reestablishing equilibrium: the water dissolves more gas, taking some of the excess from the air. In this way, the oceans are absorbing some of the carbon dioxide produced by the industrial revolution (about 50 percent) and keeping the atmosphere from departing much farther from its natural levels. Nevertheless, in spite of the ocean's moderating effect, carbon dioxide levels are expected to reach about 375 parts per million by the year 2000, a significant increase over 320 parts per million in 1970 and 295 parts per million in the middle of the nineteenth century. The possible increase in average global temperature as atmospheric carbon dioxide builds up is not great, because of many mediating effects such as the greater cloudiness produced by a rise in temperature, which tends to lower the radiation coming in from the Sun. But even small changes in temperature may have large climatic effects.

At the present rates of fossil fuel burning we may expect a doubling of the carbon dioxide level sometime in the next century. Calculations based on models of atmospheric and oceanic dynamics indicate such a carbon dioxide level may increase the global average surface temperature by 1.5 to 4°C depending on the many uncertainties in the calculations. Such an increase in temperature might have drastic effects on the world's climate and weather patterns, shifting arid and temperate zones, possibly changing the frequency of droughts, and changing the distribution of water supplies. Agriculture might be profoundly affected. Even more important is the potential warming of glaciers, releasing meltwaters to the oceans. One mass of ice, the West Antarctic ice sheet, is particularly vulnerable to melting and breakup because its rock bed lies far below sea level. We have evidence of episodes of shrinkage and expansion of this sheet that affected sea level in the recent geologic past. If warmer climates and consequent warmer oceans in the next century should cause the West Antarctic ice sheet to disintegrate over a period of decades to centuries, sea levels would rise by three to six meters. Large parts of the world's coastal cities would be flooded, a disaster hard to imagine.

But calculations of this kind are subject to many uncertainties. The increase in carbon dioxide supplied to the atmosphere depends on the size of the world's population and its use of fossil fuel for energy. The rise in carbon dioxide concentration in the atmosphere depends on the degree of oceanic absorption and the balance between photosynthesis and respiration on land and in the sea. The climatic response to a rise in atmospheric carbon dioxide concentration is calculated using most imperfect models of the atmosphere-ocean dynamic system, and many of the quantities used in the models are poorly known. Complicating matters further, geologic processes unrelated to carbon dioxide in the atmosphere may affect climate over short times scales. One of the most uncertain factors is the role of dust. These tiny particles high in the stratosphere may reflect much sunlight away from the Earth and so decrease the available radiant energy that warms the surface. The amount of dust put into the atmosphere by human activity has become measurable, but at present it is still small compared to that of volcanic dust from major eruptions (Fig. 12–16). The Earth's atmosphere, at least in historic times, has quickly recovered from a few relatively large eruptions as rainfall washed the dust out of the air. It is possible, however, that an exceptionally long string of big eruptions, such as those geologists know of in the past, could have an important long-term cooling effect. The injection of much additional dust by industrial and agricultural activities over the next decades could also have an effect.

The issue of climate change has deep implications for the social, political, and economic life of the next century. It is too important to be left to calculations that depend upon so many uncertainties. Levels of carbon dioxide and dust continue to be monitored by research laboratories and work on models of the atmosphere and ocean is being stepped up so that we may better predict what will happen in the future.

USING LAND WITH GEOLOGICAL FORESIGHT

Life is not all hazards. Part of our concern for the environment stems from feelings that we may not be making the most sensible decisions about our use of land. A geologist can foresee the unhappy consequences of scalping a hillslope of soft, potentially waterlogged formations and building many small homes on the resulting unstable tract. The siting of highways, dams and reservoirs, strip and underground mines, large-sale tract housing, and many other enterprises is best not left to chance development, for the costs of unwise choices are too often borne by the community as a

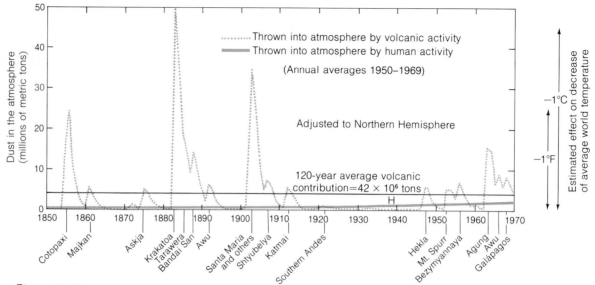

Figure 12-16

Worldwide load of atmospheric dust thrown into the stratosphere by volcanic eruption since 1850 compared to that contributed by human activity. Peaks of volcanic activity may have an effect on world average temperature, as shown in the estimates on right side of graph. [After J. M. Mitchell, "Pollution as a Cause of the Global Temperature Fluctuation," in *Global Aspects of Pollution.* Copyright 1970 by Springer-Verlag New York, Inc.]

whole. Though zoning regulations on building are traditional in many urban areas, they rarely take geological realities into account.

Any construction has to be firm on its foundations, but that is only one aspect of the problem. Builders have to take into account what will happen to the foundation if water from many septic tanks waterlogs the soil. Underground piping of gas, water, and sewage may be affected by slumping, landslides, or earthquakes. Engineering in areas of permafrost requires special precautions, as builders of the Alaska pipeline found. Sewage discharged from septic tanks may find its way into subsurface water supplies, contaminating them with dissolved chemical substances. Garbage disposal is an increasing problem as communities move against open dumps and uncontained burning of trash. One response, spreading the garbage and covering it with soil as sanitary landfill, may also result in groundwater contamination if the site is not well chosen. The formations below the landfill should be impermeable enough to prevent downward seepage into the water table. These problems are more severe in the disposal of toxic chemical wastes. Love Canal in Niagara Falls, New York, a shallow trench filled with such waste, came under scrutiny in the late 1970s as the possible cause of unusually high incidences of miscarriages and birth defects among those people living in the neighborhood of the dump.

The underground disposal of radioactive waste has become one of the most important issues in waste disposal. The main concern is the disposal of high-level waste from nuclear weapons manufacture and nuclear power reactors, but disposal of low-level radioactive waste from hospitals and laboratories also constitutes a growing problem. Most plans for the disposal of high-level waste call for incorporating the waste in a glass or other relatively insoluble canister encased in corrosion-resistant metal and buried a few hundred meters. The burial repository would be chosen for its ability to serve as a natural barrier to migration of any of the radioactive materials should they leak from the canister. Repositories in pure, dry salt beds or salt domes, granites, shales, or tuffs might serve to isolate the waste from possible leakage for at least several thousands of years and, it is hoped, for up to a million years.

The radioactive waste disposal issue is crucial to any large expansion of nuclear power plants. It is particularly critical for the planned development of fast breeder reactors, which produce the element plutonium. A major component of atomic bombs, plutonium is also one of the most toxic substances known. As little as a few thousandths of a gram (a few hundred-thousandths of an ounce) is enough to kill a person.

Other land-use planning must take larger regions of the land into account. Is it wise, for example, to plan for the urbanization of an area known to be the recharge of an underground aquifer that supplies drinking water to the population? Another question is one of surface-water storage. Builders are often tempted to drain marshes and swamps and build on the reclaimed land, but such places are important storage depots for water (see Chapter 6). Once the wetlands are drained and paved, the area is much more subject to flash flooding: because there is no opportunity for the water to soak into the ground, it runs off immediately.

There are many other choices to be made that involve evaluations of the suitability of land for agriculture, recreation, or transportation. Once the suitability of a given piece of land is deter-mined, we must still face the priority problem: we must determine whether one use of the land is more important than another. Those priorities, such as designating land for recreation rather than building development, are socially and politically determined, but the choices among the alternatives have to be illuminated by geological knowledge. If we are to live with our decisions on development and utilization of land and resources for different purposes, they ought to be based on planning that is as intelligent as possible. The geological contribution is an understanding of how all contemporary surface processes are related, and how the crust of the Earth beneath has been patterned by past geological events.

SUMMARY

1. Life appeared on Earth early in the planet's history as a response to the composition of the primitive atmosphere. Energy, probably as ultraviolet radiation, produced small- to medium-size organic molecules, such as amino acids, which then grew to larger sizes. Ultimately, the molecules became extremely complex and evolved into true cells.

2. The oldest fossils are primitive cells of Precambrian bacteria and algae found in chert beds. The oldest known are 3.5 billion years. By 2 billion years ago, algal life was well established. By the beginning of the Cambrian period, higher forms of life had evolved, and shelled organisms started to provide a good geologic record of evolution.

3. Atmospheric oxygen started to evolve as a result of the development of photosynthesis by early single-celled plants; during the Precambrian, it gradually accumulated to levels close to that of today. Our oxygen supply is now so great that there is no cause for fear that it will be depleted.

4. It is likely that life exists elsewhere in the universe. No life has been found on the Moon, nor do we have any evidence of it on Mars.

5. Environmental hazards—local or global—may be primarily esthetic, or they may be related to problems of health or safety. Environmental geology relates levels of potentially harmful substances, such as lead or mercury, to sources of pollution, to dispersal in the environment, and to rates of removal of pollutants by sedimentation.

6. Global environmental problems are illustrated by carbon dioxide, which has been increasing in the atmosphere as a result of burning fossil fuels. Some of the increase has been absorbed by the oceans. The small increase in temperature expected to be produced by the carbon dioxide increase may seriously affect the world's climate and cause melting of glacial ice and a rise in sea level.

7. Intelligent decisions on the use of land in urban and rural areas for various purposes depend on a knowledge of geological processes and the local geological situation.

EXERCISES

1. Devise a scenario for a geological effect that would lead to a lowering of global temperature. If the oceans were to freeze over completely, how would life be affected?

2. Predict the probable changes in the composition of the Earth's atmosphere over the next several thousand years if all photosynthesis were to stop suddenly today.

3. If you were part of a team of scientists planning a hypothetical manned landing on Mars, what sort of things would you ask the astronauts to look for as evidence of life of any kind on the surface? What kinds of organic molecules would you presume to be evidence of life?

4. An unsubstantiated report suggested that very high amounts of aluminum in the diet might be harmful to health. From what you know of geochemical and environmental cycles, as well as your knowledge of common aluminum products, suggest possible routes by which people might be subject to unusually high aluminum intake.

5. If you were the mayor of a coastal city, how might you react to a report that some engineers wanted to melt the polar ice caps to produce large quantities of fresh water?

6. A land developer would like to build many houses on a shoreline strip of low-lying coastal marsh below bluffs of relatively unconsolidated shales and sandstones 100 meters high. What geological arguments can you think of against such a development?

7. A waste disposal contractor wants to dump chemical waste in a 10-meter deep trench in the surface rock, a 30-meter thick sandstone. Can you testify on the wisdom of this choice?

BIBLIOGRAPHY

Bolt, B. A., W. L. Horn, G. A. Macdonald, and R. F. Scott, *Geological Hazards*. New York/Heidelberg: Springer-Verlag, 1975.

Laporte, L. F., *The Earth and Human Affairs*. San Francisco: Canfield Press, 1972.

Goldwater, L. J., "Mercury in the Environment," *Scientific American*, May 1971. (Offprint 1221.)

McHarg, I. L., *Design with Nature*. New York: Natural History Press, 1969. (Also issued in paperback by Doubleday/Natural History Press.)

Oparin, A. I., *The Origin of Life*. New York: Macmillan, 1938. (Reprinted in paperback by Dover Publications, New York, 1953.)

Schopf, J. W., "The Evolution of the Earliest Cells," *Scientific American*, September 1978. (Offprint 1402.)

Shklovski, I. S., and C. Sagan, *Intelligent Life in the Universe*. New York: Holden-Day, 1966. (Reprinted in paperback by Dell, New York.)

Wald, G., "The Origin of Life," *Scientific American*, August 1954. (Offprint 47.)

PART III

THE BODY OF THE EARTH: INTERNAL PROCESSES

The materials that make up the atmosphere, the ocean, and the rocky skin of the Earth, and the forces that deform its outermost layers, have their origin in the deeper interior. The internal heat of the Earth provided the energy that led to the formation of plates and their embedded continents, and it provides the energy that keeps the plates moving. These deep-seated forces are so powerful as to cause earthquakes and volcanic eruptions and to form deep-sea trenches, geosynclines, and mountain belts, mainly along plate boundaries. The geologist can only infer the properties of the deep interior, using as clues the nature of plate motions, the materials brought up by volcanoes, and physical measurements of such phenomena as heat flow, variations in the magnetic field and gravity, and the travel times of seismic waves. Our first explorations of other planets will enable geologists to test theories of Earth's evolution in the perspective of other bodies that have evolved differently.

13

The Internal Heat of the Earth

Earth's internal engine is driven by radioactively generated heat and heat left over from the differentiation that took place early in the planet's history. The work done by the engine registers on the surface in the form of plate motions, earthquakes, mountain-making, and volcanic eruptions. In fact, almost all of geology is ultimately related to Earth's internal heat.

The evidence of Earth's internal thermal energy is everywhere: volcanoes, hot springs, and the elevated temperatures in mines and boreholes document the flow of heat from the interior to the surface (Fig. 13–1). We could add such examples as global plate motions, earthquake activity, and the uplift of mountains, all of which represent mechanical work whose ultimate source of energy is the heat in the interior.

Despite these clear manifestations of internal heat, we cannot positively reconstruct Earth's thermal history by working backward in time from present-day observations. To do so would be like writing a history of civilization on the basis of today's newspaper. An infinite number of thermal histories could have brought our planet to its present state. Nevertheless, we can make plausible models by applying our general knowledge of the Earth and the other planets and using data gained from studies of heat flow, radioactivity, the transmission of heat by rocks, rheological (flow) properties of heated rock, and travel times of seismic waves. We have no guarantee that these models correspond to anything that happened, but they do provide insight into the present thermal state of Earth and its possible history.

Before we consider the source of Earth's internal heat and how this heat is transferred to the surface, we give a brief overview of the thermal history of the Earth, summarized from Chapter 1.

Figure 13–1
Geothermal power plant at The Geysers, Sonoma County, California. Heat from the interior converts water to steam, which is piped through turbines to generate 600 megawatts of electrical power. [From U.S. Geological Survey.]

Table 13-1
Radioactive heat production in common igneous rocks

Kind of rock	Amount of radioactive element in rock (parts per million)			Amount of heat produced (ergs/gram/year)
	Uranium	Thorium	Potassium	
Granite	4	13	4	300
Basalt	0.5	2	1.5	50
Peridotite	0.02	0.06	0.02	1

*One calorie (cal) = 4.18×10^7 ergs.

HEAT PRODUCTION AND TRANSFER IN THE EARTH

Heat Sources

Radioactivity and the conversion of gravitational to thermal energy are thought to be the major sources of internal heat. The processes of planetary accretion and compression warmed the interior, and Earth started on its evolutionary course about 4.7 billion years ago with an initial temperature that may have been somewhere near 1000°C. Radioactivity then took over, and the internal temperature began to rise. Core-mantle separation was triggered perhaps 4 or 4.5 billion years ago, when the temperature rose to the melting point of iron. The sinking of vast drops of iron to the core would have liberated some 2×10^{37} ergs of gravitational energy in the form of heat, a huge amount—the energy equivalent of 10^{15} one-megaton nuclear explosions. This heat was enough to produce extensive melting and reorganization, yielding a differentiated Earth, zoned into core, mantle, and crust.

The principal types of rocks that make up the Earth's crust are granites and basalts. Seismologists and petrologists tell us (Chapters 14, 15, and 17) that peridotite is an excellent candidate for the major constituent of the mantle. The heat produced in these rocks by the radioactive disintegration of uranium, thorium, and radioactive potassium can be determined in the laboratory. Table 13-1 summarizes the results. Because the radioactive elements like uranium are concentrated in granite as a by-product of Earth's early differentiation, this common rock leads in radioactive heat production.*

---

*Uranium is one of the heaviest elements, yet it became concentrated in the outermost layers of the Earth instead of sinking to the core. This is an example of how chemical affinity as well as gravity can be important in determining where elements end up. Uranium has a strong attraction for oxygen, which is most abundant in the crust. Oxygen migrates upward because it forms lightweight and easily meltable compounds with calcium, sodium, potassium, aluminum, and silicon.

Within each gram of granite about 300 ergs of thermal energy is produced per year. It is easy to see by multiplying this number by 2.7×10^{25} grams—the amount of granite in a hypothetical spherical shell 20 kilometers (12 miles) thick and of the same diameter as Earth—that some 10^{28} ergs of thermal energy can be produced by an outer layer of granite only 20 kilometers thick. This is equal to the total heat reaching the Earth's surface from its interior each year as determined by measurements. It is about 1000 times the energy released each year in earthquakes and about 250,000 times the energy of a 1-megaton nuclear explosion! This simple calculation not only demonstrates the importance of radioactivity as a present-day heat-producing agent but also gives some measure of the total volume of granite in the Earth's crust, for a layer much more than 10 to 20 kilometers thick would generate more heat than is observed. We will see that much of the heat flowing out of the continents originates in the radioactivity of near-surface granitic rocks, whereas heat flow from the sea floor, where there is no granite, has a deeper source.

Heat Flow by Conduction

Heat is energy in transit. It may seem obvious that heat flows from regions of high temperature to regions of low temperature—or in the example of our planet, from the interior to the surface. Not so obvious is the mechanism of heat flow or the rate at which it occurs. Heat energy in a solid exists as the vibration of atoms. The intensity of the vibrations determines **temperature.** Heat is conducted when the thermally agitated atoms and molecules jostle one another, thus mechanically transferring the vibrational motion from the hot region to the cool one (Fig. 13-2).

The quantity of heat transferred per unit time between two points is proportional to the temperature difference per unit distance and to the property called **thermal conductivity,** which differs for

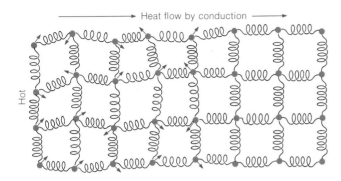

← Heat flow by conduction →

Hot

Figure 13-2
Schematic representation of heat flow by conduction through a solid. Interatomic forces that bind atoms to each other are indicated by springs. Heat applied at the left induces thermal agitation of the atoms. Heat is conducted as the vibrations gradually spread to the right.

Figure 13-3
A familar example of convection is seen when water is heated in a coffee pot.

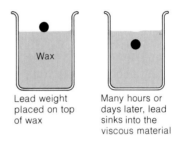

Wax

Lead weight placed on top of wax

Many hours or days later, lead sinks into the viscous material

Figure 13-4
Materials can behave as solids over short times and as viscous fluids over long intervals. Cold wax is a solid, but a lead ball placed on top of the wax slowly sinks into the interior in a few hours or days, as if the wax were a viscous fluid. Earth's mantle is supposed to show this dual behavior, premitting slow convection to occur.

each substance and is a measure of the ability of a substance to conduct heat. Rock is a very poor heat conductor: it has a very small thermal conductivity, which is why underground pipes don't freeze and why underground vaults have nearly constant temperature despite large seasonal temperature changes at the surface. A lava flow 100 meters (300 feet) thick would take about 300 years to cool. Heat entering one side of a plate of rock 400 kilometers (250 miles) thick would take about 5 billion years to flow out the other side. In other words, if the Earth cooled by conduction only, heat from depths greater than about 400 kilometers would not yet have reached the surface!

When a substance is so hot that it begins to glow, like a red-hot poker, heat can be transferred by radiation. Most radiated heat is emitted as electromagnetic waves in the near infrared and visible region of the spectrum. This **radiative transfer** can be a more efficient process than conduction in some materials, but because the minerals in Earth's interior are relatively opaque, radiative heat loss is probably less important than loss by conduction. There is no question that heat conduction must be an important agent of heat transfer in the interior, but it may be overshadowed by convection.

Heat Flow by Convection

The phenomenon called **convection** (Fig. 13-3) is rather common. One can see it happening in a rapidly heating kettle of water. Because liquids conduct heat poorly, a kettle of water would take a long time to heat to the boiling point if convection did not distribute the heat rapidly. Convec-

tion is at work when a chimney draws, or when warm tobacco smoke rises, or when cumulus clouds form on a hot day. All of those examples of convection are governed by the fact that a heated fluid, either liquid or gas, expands and rises because it becomes less dense than the surrounding cooler and heavier material. Thus instead of heat being transferred slowly by conduction, it is transported more rapidly upward by the moving heated material. Colder material flows in to take the place of the rising fluid, is in turn heated, and rises to continue the cycle. The regular flow circuit of rising warm fluid and sinking cold fluid shown in Figure 13-3 is called a **convection cell.**

Under certain conditions solids can also "flow" (Fig. 13-4). Over short terms, like seconds to years, the Earth's mantle behaves as a rigid solid, transmitting seismic waves efficiently and responding elastically to the tidal pull of the Moon. But when stresses are applied over millions of years, the mantle is weak; we will see direct evidence for this when we discuss "floating" mountains in

Box 13-1   CONVECTION

Lord Rayleigh, one of the great English physicists of the last century, found that a fluid placed between a hot lower surface and a cool upper surface, as in Figure 13-3, would convect when certain conditions were met. Convection is fostered by a large temperature difference and a high **coefficient of thermal expansion**—a measure of how much a material expands when its temperature is raised. Because the expanded material is lighter, it tends to rise, displacing the colder, heavier material, which sinks. Increasing the distance between the hot and cold boundaries also encourages convection. Convection is inhibited by **viscosity**—a fluid's resistance to flow—and by a high thermal conductivity, which would make heat transfer by conduction more efficient, and convection, in a sense, less "necessary."

Chapters 17 and 19. Thus under long-enduring conditions of high pressure and temperature, the mantle "creeps" and behaves as an extremely viscous substance, so that convection is indeed a possibility.*

Now for the key questions—ones that have been debated by geophysicists for years: Is convection an important process of heat transfer within the Earth? Is convection now occurring or has it occurred at any time in the past? The discovery of sea-floor spreading and plate tectonics, offers, in a sense, direct evidence of convection although the flow pattern in the mantle is still unknown. The rising hot matter under mid-ocean ridges builds new lithosphere, which cools as it spreads away, eventually to sink back into the mantle where it is resorbed (Fig. 13–5). This *is* convective transport, in that heat is carried from the interior to the surface by the motion of matter. The process is enormously more complicated than is indicated by the simple convection cells shown in Figure 13–3. It involves melting, upwelling, solidification, horizontal movement, sinking, and resorption. And instead of the smooth flow of water, the motion may be jerky as a result of sticking and slipping at plate boundaries.

For most geophysicists the debate over convection no longer concerns its existence as much as its scale and periodicity. Does convection involve the entire mantle or only the upper few hundred kilometers? Is it continuous—that is, repetitive, cycle after cycle—or transient, occurring once every few hundred million years when the internal temperature builds up to some critical value? We don't know the answers, but some possible models will be discussed later in this chapter. In any case, convection in connection with the creation and spreading of the sea floor is an important mechanism for the transfer of heat from the mantle under the ocean.

The British geologist Arthur Holmes was among the first to propose convection as the driving mechanism of continental drift. When he did so in the 1930s, Holmes was 30 years ahead of his time; corroboration had to wait for the extensive exploration of the sea floor that began after World War II and the massive gathering of data that led to the concept of sea-floor spreading.

THERMAL STATE OF THE INTERIOR

Heat-Flow Observations

Except for the heat received from the Sun, the heat flow from the interior is the most important terrestrial energy source. About 2×10^{20} calories, or 10^{28} ergs, of energy per year reaches the surface from the interior. This averages out to 1.5 microcalories (μcal) per square centimeter of surface area per second. This is about 100 times the annual energy release of earthquakes and about 3 times the total amount of energy we now use. It is more than a thousand times the energy required to lift the Rocky Mountains by 1 centimeter (0.4 inch). Although internal heat energy is more than enough to raise mountains and make earthquakes, it is a puny amount compared to the energy received from the Sun. So far as controlling climate is concerned, it is the Sun, delivering five thousand times the energy from the interior, that is the dominant factor. Ultimately both solar and internal heat are radiated into space. Solar heat has its geological consequences, however, in driving the atmosphere and hydrosphere, the chief agents of erosion. In a real sense, the Earth's internal heat engine builds mountains, and its external heat engine, the Sun, destroys them.

---

*The silicon compound Silly Putty illustrates how a solid can flow over long periods. It can be bounced like a ball or broken by a sudden blow, but overnight a ball of Silly Putty will flow into a pancake shape under its own weight.

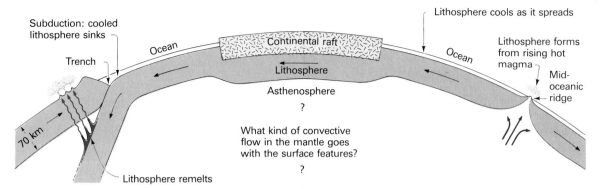

Figure 13–5
The motion of plates, spreading from mid-ocean ridges and sinking in subduction zones, is the surface manifestation of convection currents in the interior. The nature of the flow in the interior is uncertain.

Several thousand measurements of heat flow have been made on land and on the sea floor. One obtains a heat flow value by first measuring the rate of temperature increase with depth in the Earth (the temperature gradient). This value multiplied by the thermal conductivity of the rock gives the outward heat flow per unit area and per unit time. For example, at one place in the Rocky Mountains, the temperature increases with depth by 25°C in 1 kilometer (10^5 centimeters). The conductivity of the rock, measured from a sample brought to the laboratory, was 0.008 cal/cm · sec · deg. The heat flow is given by the product

$$\frac{25°C}{10^5 \text{ cm}} \times \frac{0.008 \text{ cal}}{\text{cm} \cdot \text{sec} \cdot \text{deg}} = 2 \times 10^{-6} \text{ cal/cm}^2 \cdot \text{sec}$$

On the land the temperature gradient is measured by lowering thermometers into boreholes in the Earth. Sea-floor measurements are more plentiful, not only because the oceans cover more area, but also because measurements there are easier to perform. Because temperatures at the land surface undergo both diurnal (daily) and seasonal variations, deep holes are needed to remove these effects and the effect of flowing groundwater, which may cool or heat the rocks abnormally. Because the ocean shields the sea floor from these effects, it is possible to obtain accurate temperature gradients in holes only a few meters deep. Drilling is unnecessary, since the deep-sea mud can be easily penetrated by the temperature probe as it falls into the sea bottom under its own weight (see Fig. 13–6).

Continental Heat Flow

The continental crust is mostly granite in the uppermost portion, and granite is radioactively the "hottest" rock, as Table 13–1 shows. We should

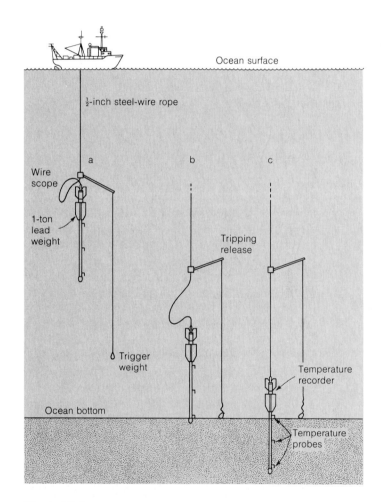

Figure 13–6
Heat flowing out of the sea is measured by plunging a core tube about 10 meters long into the sediments. Thermometers on the side of the tube record the temperature increase with depth, and the thermal conductivity of the sediments is measured when the core is retrieved. The product gives the heat flow.

Box 13-2 UNSCRAMBLING THE SOURCES OF HEAT FLOW FROM THE INTERIOR

When total heat flow is plotted against radioactive heat production in surface rock for each geological province, the plots are all linear like the one shown. Plots for the different regions are very nearly parallel (that is, of the same slope), but they have different intercepts. The lines are given by the linear equation

surface heat flow
(μcal/cm^2/sec)

$$= \text{deep heat} + \frac{\text{depth of}}{\text{granite}} \times \frac{\text{granitic heat}}{\text{production}}$$

$$(\mu\text{cal/cm}^2/\text{sec}) \qquad (\text{cm}) \qquad (\mu\text{cal/cm}^3/\text{sec})$$

or

$$q = q_0 + d \times A$$

Granitic heat production A is obtained by laboratory measurements of rock samples. The intercept q_0 corresponds to the point where there is no contribution from near-surface radioactivity (that is, $A = 0$) and so equals the heat originating in the deeper crust and mantle. The slope d is the thickness of the radioactively hot surficial granitic layer.

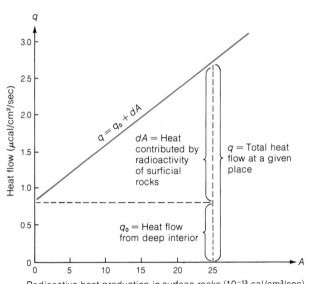

expect, therefore, that some of the heat flowing from the continents originates in the granitic layer, but not all of it does. The main problem is to determine the heat budget—that is, to assess how much heat originates in the granite, how much comes from the deeper mantle, and how this all relates to the geologic history of a region.

Geophysicists have recently developed a remarkably simple way of unscrambling the diverse factors. They sorted all available heat-flow data for North America according to geological region. For each region heat flow was plotted against the radioactive heat production from actual measurements of surface rock. By fairly simple mathematical analysis of the plots that resulted (Box 13-2), it became possible both to estimate the thickness of the radioactively "hot" surficial granite layer and to distinguish the heat originating below the crust (q_0) from that due to radioactivity in surficial rocks. The depth of the surficial layer turns out to be nearly constant throughout the continent at about 10 kilometers (6 miles), showing the remarkable extent to which differentiation has concentrated the radioactive elements at the top of the crust. The value of q_0, the deep heat, is not constant across the continent. For the Canadian shield q_0 is about 0.7 μcal/cm^2/sec compared with the average heat flow reaching the surface in this

region of 0.9 μcal/cm^2/sec. We see that 25 percent of the heat flux originates in the granitic layer of this geologically old (2 to 3 billion years), inactive province and 75 percent comes from below. Because of the low radioactivity of the basalts and peridotites underlying the granitic layer in the lithosphere, most of this deep heat must be from below the lithosphere—that is, from regions deeper than about 100 kilometers. The quantity q_0 could include heat delivered by convection in the mantle.

The average surface heat flow q for the Basin and Range province (Arizona, New Mexico, Utah, and Nevada) is about 2 μcal/cm^2/sec, of which 1.4 μcal/cm/sec is contributed by deep heat q_0. Here too, in this geologically young (0 to 65 million years) and active region, some 70 percent of the heat has a deep source. But twice as much deep heat is being delivered here as in the geologically inactive Canadian shield. Some geologists believe that the rising plume of a convection current lies under the Basin and Range province and accounts for such observations as the excess heat flux, the thin crust with evidence of recent fracturing, the recent volcanism, and the earthquake activity in that region. Does this mean that California will one day split from the rest of the country, with an ocean basin opening somewhere in Utah and New

Mexico? At the rate of divergence—a few centimeters per year—we need not worry about the political consequences.

The Sierra Nevada, near the California-Nevada boundary, shows a surprisingly small deep-heat flux, $q_0 = 0.4$ μcal/cm^2/sec, about half the total heat flow for this mountain region. Some geophysicists have speculated that because of an ancient plate collision, the Sierra Nevada block is underthrust by an old and still cool plate, reducing the flow of deep heat (Fig. 13–7).

The first two examples seem typical of the rest of this planet: low heat flow ($\sim$1 μcal/cm^2/sec) in geologically old and inactive areas, such as the large area of exposed Precambrian rocks in central and eastern Canada; high heat flow ($\sim$2 μcal/cm^2/sec) in regions of more recent mountain-building or volcanism, such as the Alps or the western United States. Special cases like the Sierra Nevada, however, require unique explanations.* In each region the highly radioactive rocks at the top of the continental crust account for less than 50 percent of the total outward flow of heat. The mantle contributes the remainder, the amount being about twice as much in regions of recent orogeny (mountain-building) and volcanism than in geologically old and stable provinces. Combining all of the regions, the average heat flow for the continents is about 1.4 μcal/cm^2/sec.

The separation of continental heat flow into contributions from near-surface granites and deeper mantle sources and the identification of different geological provinces as "hot" or "cold" are new approaches to the understanding of intracontinental geology that are just beginning to be used.

Heat Flow from the Sea Floor

The continents may be likened to rafts embedded in large plates. The rafts have grown through geological time. The oldest rocks found on Earth have been preserved on continents for nearly 4 billion years. Continents are difficult to destroy; they may be deformed, but they survive plate convergence because they are light enough to keep afloat. In marked contrast, the sea floor is created

---

*Heat flow is also low in regions of rapid sedimentation, such as along the Gulf Coast of the United States. The sediments blanket the heat from the interior, temporarily insulating the surface from the heat sources below. The flow of heat from the interior into thick accumulations of sediments is essential for the transformation of organic matter in the sediment into oil and gas.

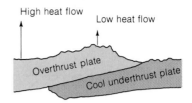

Figure 13-7
To account for the small heat flow found in the Sierra Nevada Mountains, K. E. Torrance and D. L. Turcotte propose that because of an ancient plate collision the Sierras rest on an old and still cool plate, reducing the heat flow from the deep interior.

at mid-ocean ridges and destroyed in subduction zones on a time scale of 100 to 200 million years. Furthermore, sea-floor crust and lithosphere are made of rocks like basalts and peridotites, which show far less radioactivity than the continental granites. From these considerations we should expect the sea floor and continents to show completely different heat-flow characteristics.

Like the heat-flow observations made on land, the oceanic observations show a pattern, and from them a correlation with the geology of the sea floor has been found. Oceanographers tell us that although there is some variability in the observations, the average heat flow for the major oceanic provinces is as follows. Newly created ocean ridges, younger than 5 million years >3 μcal/cm^2/sec; ocean basins 50 to 100 million years old, $\sim$1.4 μcal/cm^2/sec; sea floor older than 125 million years and farthest from the ocean ridges, $<$1.1 μcal/cm^2/sec. The reader should be able to anticipate the explanation of the decrease in heat flow from ridge crest to ocean basin to deep-sea trench. Mid-ocean ridges sit atop the rising plume of hot rock and magma that carries heat from the deeper mantle, This "mush" cools, solidifies, and becomes attached to the oceanic lithosphere. As the lithospheric plate spreads away from the ridge, it loses its heat by conduction to the sea floor and gradually cools (Fig. 13–8). For this reason, heat flow should decrease with the age of the sea floor, hence with distance from a mid-ocean ridge. The oldest parts of the sea floor should have the lowest heat flow, and these parts should be found farthest from ridges and closest to deep-sea trenches, where the cold slabs sink back into the mantle.

Oceanic heat flow is thus dominated by the process of cooling of the recently created oceanic lithosphere. Geophysicists believe that this form of convection may account for as much as 60 per-

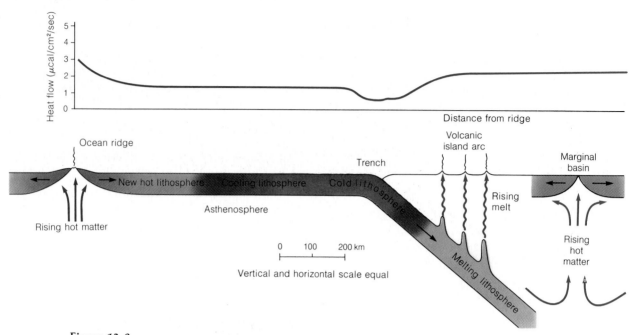

Figure 13-8
The pattern of heat flow out of the sea floor. High values are observed over mid-ocean ridges. As the sea floor spreads and cools, less heat reaches the surface. On the side of deep-sea trenches where the lithosphere plunges into the asthenosphere, rising magma from the melting lithosphere produces higher heat-flow values. A small region of secondary spreading may develop in a marginal basin behind the subduction zone. An example is the Sea of Japan, which separates the Japanese Island Arc from the Asian continent, is such a region.

cent of the total heat flow from the Earth and that this may represent a major mode by which the Earth has cooled.

Hot Springs and the Sea Floor

Much of the heat in the ocean lithosphere is thought to be dissipated when cold sea water percolates into the many fissures associated with sea-floor spreading along mid-ocean ridges. The cold water sinks several kilometers, encounters hot basalt, surges upward, and emerges on the sea floor as hot springs enriched in dissolved minerals and gases leached from the magma. This hypothesis was verified by recent spectacular discoveries of such hydrothermal (hot-water) vents at several places along mid-ocean ridges. The hot springs seem to occur in two forms. In the Galapagos Islands rift zone the springs flow gently from cracks at maximum temperatures of about 16°C (60°F). On the East Pacific Rise near Baja California, superheated water (380°C or 716°F) spouts forcefully from mineralized chimneys. The chimneys are built up from the dissolved minerals that precipitate around the hot jet as it mixes with the near freezing ocean bottom waters (Fig. 13-9). The deep-diving submarine *Alvin* was nearly de-

stroyed when it first approached a superheated vent, not expecting such high temperatures.

Hydrothermal vents have been called the discovery of the decade for several reasons. They represent a major ore-forming process, possibly an important new source of minerals, on the sea floor. The entire ocean cycles through such hydrothermal systems once every 8 million years, profoundly affecting ocean chemistry. The ecology of these vents is completely different from that of the dark, near-freezing, barren, deep sea floor. Dense colonies of exotic life forms populate the warm water surrounding the vents. Among them are new species of giant worms, clams, and crabs. The vent animals live on unusual bacteria that draw energy from the hydrogen sulfide, carbon dioxide, and oxygen in the vent water. A new chapter of ocean science was opened by the deep-sea explorers who discovered the hot springs of the sea floor.

Temperatures in the Earth

The average increase of temperature with depth, as measured in boreholes or mines, is about 2 or 3°C per 100 meters (300 feet). How can we estimate temperatures in the Earth at depths greater

Figure 13-9
Plume of hot, mineral-laden water spouts from mineralized chimney on East Pacific Rise.
The chimney and adjacent formations build up from dissolved minerals that precipitate
around the hot jet as it mixes with the near-freezing ocean bottom waters. [Photographed
from the deep diving submarine *Alvin* by Dudley B. Foster, Woods Hole Oceanographic
Institution.]

than those we can reach with a thermometer—
that is, below about 8 kilometers (5 miles)? The
problem is not an easy one, and the experts need
all the help they can get, especially for the lower
mantle and core.

Temperatures in the continental crust are fairly
well known from measurements of surface heat
flow, which can be separated into heat contributions from the radioactive surficial rocks and
from the underlying region by means of graphs
like the one in Box 13-2. A knowledge of these
two sources of heat is sufficient to fix the temperatures with some assurance for the continental
crust. **Geotherms,** or temperature–depth profiles,
typical of such geologically old and stable regions
as the eastern United States and of such young
and active regions as the Basin and Range province, are shown in Figure 13-10. The higher heat
flow in the tectonically active region shows up in
the more rapid increase of temperature with
depth. Forty kilometers below the surface the
temperature rises to almost 1000°C in the active
region. This is close to the point of initial melting
of such deep crustal and mantle rocks as basalt
and peridotite. In contrast, the temperature at the
same depth under the stable region reaches only
the relatively low value of about 500°C. The association of tectonic activity and volcanism with

subcrustal temperatures near the melting point,
and of long-term geological stability with relatively cool regions of the crust, is an important
discovery. For example, it supports the speculation that the Basin and Range province may overlie the rising plume of a convection current.

Temperatures in the suboceanic lithosphere are

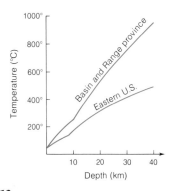

Figure 13-10
Temperature variation with depth in different
regions, according to the ideas of R. F. Roy, D. D.
Blackwell, and F. Birch. The higher heat flow in the
Basin and Range province implies higher temperatures at depth in this young and active region. The
temperature rises to almost 1000°C at 40 km. In the
geologically older and more stable eastern United
States, the temperature reaches only 500°C at the
same depth.

Box 13-3 ANCHORING THE GEOTHERM

In order to "anchor" the geotherm at a few points to keep it from drifting too far from reality, the following facts are used:

1. Seismology tells us that the asthenosphere is partially molten. We therefore "fix" the temperature at a depth of 100 km (60 miles) at the **solidus,** or point of incipient melting, about 1100 to 1200°C. This is just about the temperature of lavas issuing from oceanic volcanoes, which provides a nice check.

2. At depths greater than 300 kilometers (200 miles) in the mantle, the temperatures must be below the solidus. We know this because seismic shear waves cannot pass through a partially molten region without being severely attenuated, and no such deadening is observed below the asthenosphere. We therefore use an estimate of the solidus recently inferred from laboratory studies as an upper limit for mantle temperatures below the asthenosphere.

3. Rapid increases in seismic wave velocities are observed at depths of about 400 and 700 kilometers (250 and 450 miles). These important clues will be discussed in Chapter 17 on seismology. They are caused by a more dense packing of the atoms of a major mineral constituent of the mantle—olivine. The sudden increase in density, or **phase change,** occurs when critical pressures and temperatures are reached. A few years ago the phase change at 400 kilometers was actually duplicated in the laboratory by squeezing and heating rocks to the pressure and temperature at that depth. Therefore, we know the pressure and temperature at which the phase change takes place. This anchors the geotherm at 400 kilometers at about 1500°C.

The conditions for the phase change at 700 km can be estimated theoretically. Using this estimate and seismological data, geophysicists estimate the temperature to be 1900°C at this depth.

4. We will see in Chapter 17 that the core is primarily iron, molten in the outer part and solid in the inner part. Recently a new melting-point curve for iron has been used by geochemists to fix temperatures at the high pressures in the core (see figure). The temperature at the mantle-core boundary must exceed the melting point of iron to account for the liquid core, and it must be below the melting point of the mantle to account for the mantle's solidity. It is an interesting and useful coincidence that these upper and lower bounds on temperature are close together at the mantle-core bound-

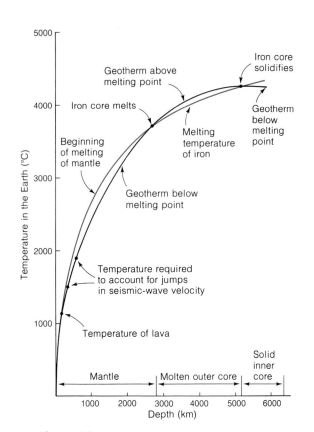

A geotherm (black curve) showing how temperature may increase with depth from the surface to the center of the Earth. The probable melting point temperature in the mantle and core are depicted by the brown curve. Where the geotherm is below the melting point, as in most of the mantle or the inner core, the material is solid. Where the geotherm is at the melting point, at depths of 100–300 km, the upper mantle is partially molten. The outer iron core is hot enough to be fully molten. Temperatures at Earth's center reach about 4300°C.

ary, fixing the temperature at about 3700°C. Moreover, the geotherm must cross over and fall below the iron melting-point curve at 5100 kilometers (3200 miles) in order to provide for the solid inner core insisted upon by the seismologists. This gives a temperature of about 4300°C. It seems unlikely that the temperature at Earth's center could be much above this value.

---

dominated by the process of cooling down from the high temperature of the magma that builds the slab in the first place. Referring again to Figure 13-8, we could make a good guess that the temperature under the ocean would be close to 0°C (the temperature of the bottom water) at the top of the slab, and close to 1200°C (the melting point) at the bottom of the slab, where it is in contact with the partially molten asthenosphere. Where the slab plunges deep into the mantle, we would expect it to warm gradually to the high temperatures of the material surrounding it.

One of the important tasks of the next generation of geophysicists is to find a method for di-

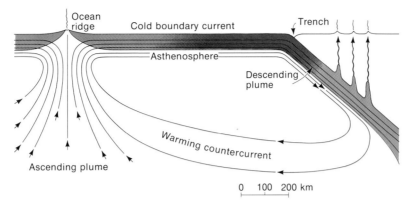

Figure 13-11
Possible convection flow in the upper mantle, according to D. H. Turcotte and E. R. Oxburgh. An ascending hot plume spreads laterally under the ocean ridges; cooling by conduction to the ocean, it solidifies to form a boundary layer, the cold, rigid lithosphere. The descending cold plume coincides with the sinking lithosphere of the subduction zone.

rectly determining the temperature in the deep interior. Until this is done we can use only indirect evidence to speculate on the nature of convection within the Earth. We know that we can't simply extrapolate the temperature curves for the crust (Fig. 13-10) by straight-line extension all the way to the depth of the Earth's core, at 2900 kilometers (1800 miles); that would give a temperature of at least 25,000°C at the Earth's center, which would require that most of the Earth be molten. This impossible situation shows once again that the important radioactive elements are concentrated in the outermost layers and that the rate of temperature increase with depth (temperature gradient) must lessen below this radioactively hot zone. More efficient heat transfer in the deeper layers will also decrease the temperature gradient so as not to give impossibly high values.

At present all we can do is sketch a possible geotherm, showing temperatures as they might occur all the way down to the center of the Earth where the temperature reaches 4300°C (Box 13-3).

The Tectonic Engine

It is widely accepted that tectonic phenomena—that is, plate motions, mountain-building, and earthquakes—are accounted for by internal convection currents. Many schemes for convective flow have been advanced. They vary somewhat in scale, some involving the entire mantle and others limiting the flow to the upper few hundred kilometers.

Some fascinating calculations have led to a proposed scheme of convection in which the flow of matter takes place in the outer several hundred kilometers of the Earth. The model explains many of the features summarized earlier in Figure 13-8; these features are repeated in Figure 13-11, with the convective flow lines superimposed. An ascending hot plume spreads laterally, cools by conduction to the ocean, and solidifies to form a cold, brittle boundary layer—the lithosphere. A descending cold plume, the subducted lithosphere, begins the return flow. The cycle of convection is completed by a counter-current several hundred kilometers below the surface, which heats up and rises again. Buoyancy forces of the light, hot ascending plume and of the heavy, cold descending plume drive the flow. The energy source is radioactivity plus perhaps some of the original heat left over from the early years of the planet. The velocity of the flow, which is calculated for the model, agrees with the sea-floor spreading rates, and the heat flow at the surface matches the observational data quite well. The results encourage us to believe that the geophysicists who suggested these ideas may be on the right track in their search for the mechanism that makes the tectonic engine run.

The question of the driving force behind plate tectonics has become so important that it was the basis of an international scientific program called The Geodynamics Project. Many of the questions posed in this chapter are being attacked in a concerted, multinational research effort that is now underway.

SUMMARY

1. The internal heat of the Earth is derived from the radioactivity of uranium, thorium, and potassium. Heat left over from the early years of the planet may be an additional source.

2. Much of the internal heat reaches the surface by convective flow in the mantle; the large-scale surface manifestation of this process is the spreading sea floor. New lithosphere created from rising hot rock and magma cools as it spreads, eventually to sink back into the mantle, where it remelts. Measurements of high heat flow over mid-ocean ridges and low heat flow over deep-sea trenches are consistent with this picture. The creation of new sea floor and percolation of water through mid-ocean ridges is an important mechanism by which the Earth's internal heat reaches the surface.

3. Heat flow on the continents correlates with regional geology—low values in old stable regions and high values in regions of recent rifting, orogeny, and volcanism.

4. Temperatures increase rapidly with depth in the outermost 100 kilometers (60 miles) of the Earth, reaching the melting point in the asthenosphere (100 to 300 kilometers or 60 to 190 miles). Below 300 kilometers, the geotherm must rise less rapidly or the Earth would be extensively molten. It is everywhere below the melting point until the core is reached. Here the temperature is about 3700°C—above the melting point of iron, as required for a liquid iron outer core. The temperature at the center is about 4300°C, according to current views.

EXERCISES

1. Give some "kitchen physics" or common-experience examples of the following concepts: (a) uses of low-conductivity and high-conductivity materials; (b) convection; (c) materials of different viscosity; (d) work done by heat. *Example:* Hot tea is easier to handle in a porcelain cup than in a metal cup because porcelain is a poorer conductor of heat and less likely burn the hand or the lip.

2. Enlarge upon the statement "in a real sense Earth's internal heat builds mountains, and external heat from the Sun destroys them."

3. Almost twice as much heat flows out of the ground in Utah as in Massachusetts. Why?

4. If you were prospecting for geothermal energy, where would you look? If 0.1 percent of the heat reaching the surface from the interior could be used for energy, would it represent a significant contribution to the world's needs? How about 0.1 percent of the solar energy reaching the surface?

5. Although geologists have direct access only to the upper few kilometers of the Earth, they can infer temperatures at greater depths. How?

6. It is widely believed that convection currents in the mantle provide the driving forces for plate motions. What would be the connection between rising, warm convection plumes, sinking cold plumes, and sea-floor spreading and subduction?

7. If the heat flow from the ocean basin floor came solely from the radioactivity of basalt, how thick a layer of basalt would be needed to account for it? (Assume the density of basalt is 3 grams per cubic centimeter, and see Table 13-1 for basaltic heat production.)

BIBLIOGRAPHY

MacKenzie, D., and F. Richter, "Convection Currents in the Earth's Mantle," *Scientific American,* pp. 72–89, November 1976.

Sclater, J. G., C. Jaupart, and D. Galson, "The Heat Flow Through Oceanic and Continental Crust and the Heat Loss of the Earth," *Reviews of Geophysics and Space Physics,* v. 18, pp. 269–311, 1980.

Spiess, F. N., et al., "East Pacific Rise: Hot Springs and Geophysical Experiments," *Science,* v. 207, pp. 1421–1432, 1980.

Turcotte, D. L., and E. R. Oxburgh, "Continental Drift," *Physics Today,* v. 22, pp. 30–39, 1969.

14

Formation of Igneous Rocks

Localized heat and pressure deep in the Earth cause materials of the deep crust and mantle to melt, forming magmas. As they intrude higher portions of the crust or erupt at the surface, magmas crystallize to form igneous rocks of different compositions. Experimental studies of the crystallization behavior of melts indicate that this diversity of igneous rocks is a result of partial melting of the original material and chemical differentiation. In this way, silicic magmas can be produced from originally mafic melts. Pressure and the presence of water in the melt greatly affect melting temperatures and the formation of partial melts.

The connections among the interior heat of the Earth, volcanoes, and the formation of igneous rocks (those crystallized from a molten state) seem to be straightforward. Simple observation of many volcanoes reveals how cooling and congealing of molten lava produce the rock we call basalt. More than 200 years ago, geologists drew the conclusion that ancient basalts were also of volcanic origin, even though, in many cases, all traces of the volcanoes that produced them had long since vanished. Other volcanoes produce lavas that crystallize to form rocks of different composition, ranging from andesite to rhyolite. Somehow, the composition of the melt varies from place to place, and this determines the composition of the rock (see Fig. 3–17). The explanation for diversity of compositions lies in the ways rocks melt, how and where molten masses accumulate and move, and how they crystallize to igneous rocks. First we look at the geology of granite as a guide to its origin.

GRANITE

It is not easy to demonstrate that granite, one of the commonest of coarsely crystalline materials,

is a rock formed by crystallization from a hot, molten mass. Granite is found in high mountain chains, such as the Sierra Nevada and the Alps, and in the cores of deeply eroded ancient mountains, such as those of the Scottish highlands, but this in itself is no great clue to its origin. The details of its relations to other rocks, however, were important clues to James Hutton as he worked in the field in Scotland. He saw cross-cutting relations, in which sedimentary rocks were somehow fractured and invaded by the granite, as if it had been forced into them as a liquid (Fig. 14–1). Even more significant were the mineralogical and textural changes in the sediments bordering the granite, which he inferred were the result of great heat.

Hutton guessed that the granite was formed by molten material rising from great depths. As it rose, it intruded the rocks of the upper part of the crust and cooled slowly to form crystals of quartz, feldspar, mica, and other minerals. The texture, a mosaic of interlocking crystals, could not be confused with the texture of a sandstone consisting of grains of the same minerals (Fig. 14–2). Many of the minerals in granite show sharp crystal faces, and the crystals are interlocked and held together like the pieces of a jig-saw puzzle. Sandstone, in

Figure 14-1
Cross-cutting contact of granite (below) with shales and tuffs (above). Bedding dips gently to the right. A small, nearly vertical, light-colored dike is at the right (arrow). Shasta County, California. [Photo by J. P. Albers, U.S. Geological Survey.]

Figure 14-2
Photomicrograph of a thin section of granite in polarized light. The large light crystal at lower left and the dark one above it are microcline, a variety of potassium feldspar. The more homogeneous light- and medium-gray crystals at upper right are quartz. This interlocking mosaic of crystals results from crystallization of a melt. [Photo by A. H. Koschman, U.S. Geological Survey.]

Table 14-1
Chemical analysis of a granite from Cumberland, England

Element	Formula*	Relative proportion (%)
Silicon	SiO_2	68.55
Aluminum	Al_2O_3	16.21
Ferric iron	Fe_2O_3	2.26
Magnesium	MgO	1.04
Calcium	CaO	2.40
Sodium	Na_2O	4.08
Potassium	K_2O	4.14

*Formulas are conventionally shown as oxides, though the elements are not actually present in this form.

Table 14-2
Mineralogical analysis of a granite from Cumberland, England

Mineral	Formula	Relative proportion (%)
Quartz	SiO_2	24.4
Feldspar		
Orthoclase	$KAlSi_3O_8$	36.2
Plagioclase	$\begin{Bmatrix} NaAlSi_3O_8 \\ CaAl_2Si_2O_8 \end{Bmatrix}$	33.6
Mica		
Biotite	$K(Mg,Fe)AlSi_3O_{10}(OH)_2$	5.8

contrast, consists of grains rounded by abrasion and bound together by a mineral cement between the grains. The minerals of granite are so insoluble in water that it would be hard to conceive of the rock having been precipitated from sea water, like an evaporite, which may have the same kind of crystalline texture. Thus granite, like basalt, came to be recognized as the product of crystallization of a **magma,** a large body of very hot molten rock.

In the second half of the nineteenth century, after the polarizing microscope had been invented and used for mineral identification of small crystals in rocks, geologists described the multitude of igneous rocks of the granite family, discovering in them as much variation as they saw in the extrusive rocks formed by volcanism (Tables 14-1 and 14-2). Techniques of chemical analysis were joined with microscopic examination and field study to produce a classification scheme of the sort we still use (see Fig. 3-17). The major division of igneous rocks into fine-grained (as typified by basalt) and coarse-grained (as typified by granite) was based on their differing origins. Volcanism gave rise to extrusives, which cooled quickly at the relatively low temperature of the Earth's surface to form glasses or fine crystals. Plutonism—the crystallization of a magma at depth where temperatures and pressures are higher—gave rise to intrusives, which cooled slowly to form coarse crystals.

The explanations for the variety of compositions of the granite series, from granite to gabbro, were more elusive. Complicating the picture further were the ultramafic intrusives composed of olivine and pyroxene, which have few corresponding extrusives. What accounts for the formation of magmas of such a variety of chemical compositions? Did they arise from melting of different kinds of rocks? Or are there processes producing variety from an originally uniform starting product? What temperatures and pressures are required for the formation of melts of different kinds? If these temperatures and pressures could be estimated, then we might have some idea of how deep in the Earth the magmas formed and crystallized.

MELTING AND CRYSTALLIZATION

By the early twentieth century, as geological data on igneous rocks accumulated, answers to questions about where and how magmas were formed started coming from the young science of physical chemistry. Thermodynamics, the science of heat and energy, was applied to crystallization from solutions and melts of geological interest. The new way to approach the problem was by experimentation. Scientists mixed chemicals in the proportions in which they were known to occur in igneous rocks, melted them in high-temperature furnaces, and allowed them to crystallize, observing carefully the temperatures of melting and crystallization and the compositions of the minerals that were formed.

Plagioclase Feldspar

Igneous rocks contain many elements. In addition to the abundant elements—oxygen, silicon, aluminum, iron, sodium, potassium, magnesium, and calcium—there are a host of minor elements. The experimentalists wisely chose not to tackle the terribly complicated natural mixtures immediately but to take a few very simple compositions and to work them out first. One of those is the plagioclase feldspar system, a group of feldspars ranging from albite, a sodium aluminosilicate, $NaAlSi_3O_8$, to anorthite, a calcium aluminosilicate, $CaAl_2Si_2O_8$. Plagioclase is a major mineral

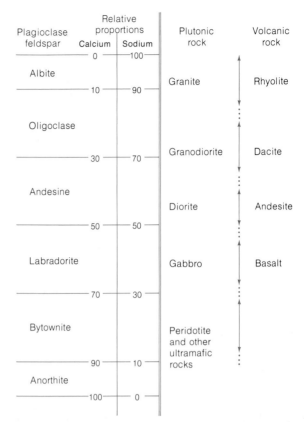

Figure 14-3
Igneous rock types are classified according to the relative proportions of the feldspars albite (sodium plagioclase) and anorthite (calcium plagioclase). The boundaries between rock types are gradual; the divisions between the feldspars are arbitrarily defined.

component of igneous rocks, and its systematic change of composition is basic to classifying the range of the intrusives, from granite to gabbro, and the range of extrusives, from rhyolite to basalt (Fig. 14-3). The results of the experiments with plagioclase were most instructive.

The experiments showed that melts of various plagioclase compositions, if allowed to crystallize very slowly, would form crystals through a complicated path of crystallization (see Box 14-1). The first crystals to form were rich in the calcium component, anorthite. Their formation slightly depleted the calcium present, so that the melt became relatively richer in sodium. As crystallization proceeded, the earlier crystals were gradually transformed to crystals containing relatively more and more sodium, until, when crystallization was complete, the final mass was homogeneous and had the same composition as the original melt. The key to the process was a continuous reaction series, a sequence of transformations of the earlier crystals, by reactions with the remaining

Box 14–1 CRYSTALLIZATION OF PLAGIOCLASE FELDSPARS

The figure summarizes the results of many crystallization experiments with melts of compositions corresponding to those of various members of the plagioclase series. The chemical composition is shown on the horizontal axis of the graph as the percentage of the two components, albite and anorthite, in a simple mixture of the two. Natural plagioclases are a solid-solution series; that is, there is a continuous range of compositions from one side to the other, the basic crystal structure remaining the same throughout (see Table 3–2). The melts of albite and anorthite are completely miscible in all proportions as are the solid minerals.

Because we are interested in temperatures of crystallization, we have plotted temperature on the vertical axis. The graph tells us what to expect if a melt of composition X (about 30 percent anorthite), heated to about 1500°C, is allowed to cool slowly. There is no change as the liquid cools until a temerature of about 1370° is reached. At that point the first crystals of plagioclase form. These crystals are very different in composition from the liquid; they turn out to be a little over 70 percent anorthite. As the first crystals form, because the proportion of anorthite to albite being taken out of the melt is much higher than the original composition at X, the liquid becomes a little depleted in anorthite and starts to move down the upper curve.

As more crystals grow, they form from a liquid of slightly changed composition and so are a little less rich in anorthite. In the meantime, the first-formed crystals are no longer in equilibrium with the liquid because the melt is now more albitic. If crystallization is very slow, these first crystals will continuously react with the changing liquid; and, as new crystals form, the composition of all of the crystals will travel down the lower curve. This process continues—the liquid moving down the upper curve and the solid moving down the lower curve—until, at a temperature of about 1190°C, the last drop of liquid (by now containing only about 5 percent anorthite) forms crystals of the original composition, X. After that, the entire crystalline mass continues to cool without further change. The melt has crystallized into a plagioclase of the same composition as the original melt, but by a complicated route. When a solid plagioclase is melted, it follows exactly the same route in reverse. The first liquid formed is of the same composition as the last drop of liquid in the crystallization process, and liquid and crystals move up their respective curves until the last crystal is melted, when the liquid attains the composition X.

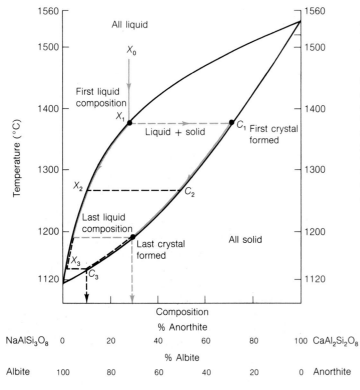

Crystallization diagram for the cooling of a plagioclase feldspar melt. The original melt at composition X_0 (about 30 percent anorthite) cools to a temperature of about 1370°C (X_1), at which point the first crystal of composition C_1 is formed. As successive crystals are formed, the liquid changes in composition to that indicated by point X_2, at which time all crystals have reacted to form crystals of composition C_2, the last liquid forming a crystal with the composition of the original melt. If crystals of composition C_2 are withdrawn at that point, the liquid will change to composition X_3, and the final crystal will be of composition C_3.

melt, to new compositions, so that, at any point in the course of crystallization, all of crystals present had the same composition.

Mafic Minerals

Magmas do not just crystallize to form plagioclase feldspars; they also produce such mafic minerals as olivine, pyroxene, amphibole, and black mica (biotite). Melting-crystallization relations among those minerals are a little more complex, but their order of crystallization can produce differentiation too. In a process in which slow cooling allows all crystals to react completely with the liquid, olivines crystallize first; then, when the liquid reaches a certain point, pyroxene starts to form, and all the olivine is converted to pyroxene (Fig. 14-4). In such a discontinuous reaction series, reactions take place between minerals of two definite compositions at a particular temperature (a little over 1550°C for magnesium olivine and pyroxene), rather than over a continuous range of compositions and temperatures, as with the plagioclase series.

Partial Melting

In the laboratory experimenters can heat samples in a furnace through many hundreds of degrees to melt them completely. In nature large rock volumes are rarely subjected to such a range of temperatures. If a mixture of minerals, for example those in a sedimentary rock, is heated, the first minerals to melt might be a sodium-rich plagioclase. Other silicates of low melting points, such as micas, might also melt at the same time. A **partial melt** is formed by either continuous or discontinuous reaction series (or both) in which the proportion of liquid to remaining solid depends upon the composition and melting temperatures of the source rocks and upon the temperature and pressure. As one can see from the plagioclase and pyroxene-olivine series, the composition of a partial melt may be significantly different from both the source rock composition, which would be achieved by complete melting, and from the compositions of partial melts produced at different temperatures and pressures.

MAGMATIC DIFFERENTIATION

Different patterns of crystallization and melting provide the mechanisms for producing different

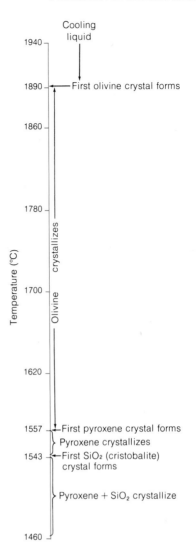

Figure 14-4
Sequence of events in the crystallization of a cooling liquid of magnesium and silica. This sequence forms one part of the discontinuous reaction series by which different mafic minerals are successively crystallized from a melt.

compositions from a uniform starting material, the process of **differentiation,** or *fractionation.* When N. L. Bowen published his pathfinding experimental work on feldspar crystallization in 1913, he had natural geologic processes and the possibility of differentiation very much in mind. He focused on the course of crystallization in situations in which the feldspars did *not* continuously react with the liquid to change compositions. This would occur, for example, if a magma cooled so rapidly that there was time for only the outer surface of the earlier-formed crystals to react with the changing liquid. Then only the outer rim of each crystal would change composition during the crystallization, each successive layer being

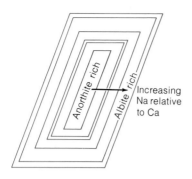

Figure 14-5
A zoned plagioclase crystal is formed during
the fractional crystallization of a silicate
melt in which crystallization was too rapid
to permit continuous reaction of already
formed crystals with the changing melt composition. As the melt congeals, the layers of
the crystal become richer in albite.

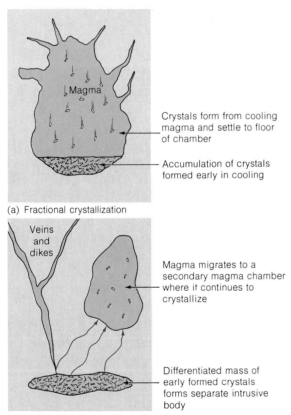

(a) Fractional crystallization

(b) Deformational squeezing out
of remaining fluid from crystal mush

Figure 14-6
Two stages in the evolution of a differentiated
magma. In the first stage (a), crystals formed early
settle to the floor of the magma chamber. During
cooling, structural deformation may squeeze remaining liquid from the chamber and separate the early-
formed crystals as a distinct intrusive body (b),
while the liquid migrates elsewhere to form veins,
dikes, and other magma chambers, where it continues to crystallize.

covered by a layer richer in albite. The result of
such a process is a mass of **zoned crystals,** which
have anorthite-rich interiors and gradually grade
to albite-rich exteriors (Fig. 14–5). The zoning of
plagioclase has another implication. If the anorthite-rich cores of the growing crystals are unable
to react with the liquid, the liquid is richer in albite than it would be in a slow, continuous reaction. If, for example, at any stage of crystallization, the crystals already formed were to be
removed somehow, the solution at that point, already albite enriched, would behave as though it
had just started to crystallize, and the next crystals to form from it would be more albitic. Then, if
the reaction thereafter were slow and continuous,
the remaining liquid would ultimately form crystals with an anorthite content much lower than
those that would have formed from the original
liquid, had it cooled slowly in a continuous reaction from the outset.

Bowen proposed a theory of chemical differentiation based on this idea. Though the mechanisms he suggested are no longer accepted for the
differentiation of most igneous rocks, his ideas
served as a foundation for most later work and
are instructive today. Bowen suggested that if the
early crystals formed deep in the crust in a large
opening filled with melt, a **magma chamber,** the
crystals might settle to the bottom and thus be
withdrawn from further reaction (Fig. 14–6). Another possibility is that structural deformation
midway in the crystallization process might
squeeze the remaining liquid away from the crystal mush into other places, where it would continue to crystallize. Whatever the exact details of
the geologic process, **fractional crystallization** of
the melt—that is, crystallization in which crystals
do not continuously react with the melt—could
account for albite-rich plagioclases forming from
an originally anorthite-rich melt. Basaltic magmas
are anorthite-rich melts, and granites contain albite-rich plagioclases. Magmatic differentiation
relates the two in the sense that the basaltic
magma is the starting material that could gradually differentiate to a more silicic melt by a fractional crystallization process that, if carried far
enough, would form granite. Just as in the continuous reactions of plagioclase formation, the first-
formed crystals of olivine in a discontinuous reaction series may settle out or become coated by
later-formed pyroxene crystals, and so be withdrawn from further reaction. Then the melt becomes enriched in silica and finally crystallizes to
form the high-temperature silica mineral cristo-
balite (or, at lower temperatures, tridymite or

quartz). All this is possible, but where could a geologist check it with real rocks?

Facing the city of New York on the west bank of the Hudson River is the Palisades, a massive cliff some 50 miles long and in places more than 1000 feet from top to bottom. The Palisades is a formation of basalt that was intruded as a melt almost horizontally into nonmarine sedimentary rocks. The variation in mineral composition from top to bottom of this formation makes it a classic example of how laboratory experiments serve to explain field observations (Fig. 14-7). After the intrusion was emplaced at a temperature near 1200°C, which is the melting temperature of a rock of this composition, parts within a few feet of its upper and lower contacts cooled rapidly to become a fine-grained rock, preserving or "quenching in" the chemical composition of the original magma. The molten interior cooled more slowly, so that large crystals could form. The first mineral to crystallize was olivine, which is heavy and sank through the melt to the bottom of the intrusion, where it can be found today in the olivine-rich layer just above the chilled, or fine-grained, basalt zone. The reaction proceeded to the stage of pyroxene crystallization in some parts of the intrusion, and the pyroxene accumulated in the lower third, probably by gravity-settling. After the olivine crystals settled out, changing the composition of the remaining magma, the reaction proceeded to the stage of plagioclase feldspar in the cooling of the melt. This explains the enrichment of plagioclase feldspar found in the upper third of the intrusion. This explanation of the Palisades was one of the successes of the early versions of magmatic differentiation theory, and it firmly established the ties of field observations to laboratory experiments and to thermodynamic theory.

The Reaction Series

Bowen combined the continuous and discontinuous fractional crystallizations of the major minerals of igneous rocks into a general scheme for magmatic differentiation, the Bowen reaction series. A high-temperature magma of basaltic composition starting to cool and crystallize gradually differentiates by fractional crystallization along two paths—the continuous plagioclase-feldspar series and the discontinuous mafic-mineral series (Fig. 14-8). The melt converges toward a final low-temperature granitic composition of orthoclase and albite feldspar, muscovite mica, and

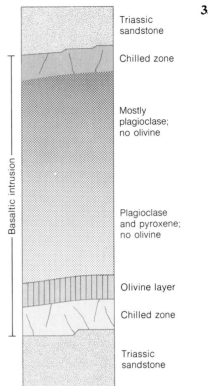

Figure 14-7
Diagrammatic section of the Palisades on the west bank of the Hudson River. The formation is a basalt that intruded sediments as a melt some 200 million years ago. The interpretation of the vertical variation in texture and mineral composition in this formation is a classic example of fractional crystallization of the discontinuous reaction series. [After F. Walker, 1940.]

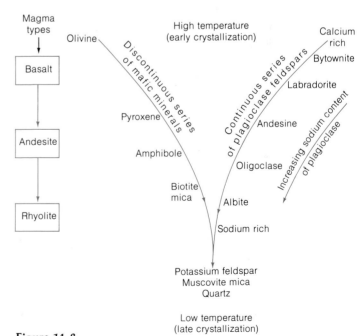

Figure 14-8
Bowen's reaction series, showing how the sequence of fractional crystallization of a melt could lead to the formation of differentiated magmas.

quartz. After the early crystallization and separation of olivine crystals, the magma corresponds to a diorite intrusive in composition—or to its fine-grained relatives, andesite volcanics, if the magma reaches the surface at that stage. At later stages, the magma becomes more silicic, ending with granite and its extrusive equivalent, rhyolite. By the time this has happened, so many crystals have formed that only about 10 percent of the original liquid is left. Associated with the reaction series is a tendency for the structures of the discontinuous series to become increasingly complex as temperature drops. Olivines are isolated silica tetrahedra; pyroxenes, single chains of tetrahedra; amphiboles, double chains; and micas, continuous sheets (see Chapter 4). Fractionation of minor chemical elements also accompanies the series.

This differentiation theory accounted better than any other theory at the time for many field relations and for the range of compositions of igneous rocks. It was consistent with the known temperatures of magmas of different compositions. A number of layered intrusive bodies, such as the Palisades sill, are known in which the bottom layers are olivine and successively higher layers contain lower-temperature minerals arranged in order of the reaction series. Sometimes rhyolite volcanics erupt from large basaltic volcanoes toward the end of an eruption episode. Zoned plagioclase crystals are common in granites and granodiorites, as one would expect. In addition, there are very few extrusive equivalents of the most mafic intrusives of olivine and pyroxene. Their absence suggests that most mafic plutonic bodies are not formed from a magmatic liquid—which might be expected to erupt somewhere—but are accumulations of crystals in a magma chamber. As will be discussed in Chapter 15, the temperatures of basaltic lavas are much higher than those of more silicic compositions and are approximately in harmony with the experimental results.

This general theory of differentiation, elaborated and extended by much work in the field and laboratory, proved to be workable. It seemed to explain the occurrences of many igneous rocks of the continents, and it integrated field observation, experimental evidence, and sound physical-chemical theory.

Many of the facts, however, were difficult to reconcile with simple fractional crystallization as the major mechanism for silicic magma generation. One objection is the almost infinite time required for small crystals to settle through a highly viscous magma to form a segregated layer. Lay-

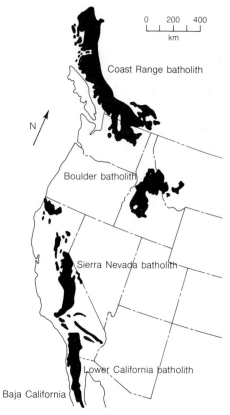

Figure 14-9
Major Mesozoic (mostly mid-Cretaceous) granitic batholiths (solid black) of western North America. Enormous volumes of granite are included in these large plutons. [From *Petrologic Phase Equilibria* by W. G. Ernst. W. H. Freeman and Company. Copyright © 1976.]

ered intrusives are more complex than would be predicted. Another objection is that huge batholiths—enormous bodies of granite, such as those found in California or Idaho—seem to require the existence of vastly larger bodies of basaltic magma from which they would have differentiated (Fig. 14-9). Hence, many believe that these granites did not originate by differentiation from basalt. Oceanic basalts, typically those formed by upward movement of the asthenosphere at mid-ocean ridges, do not show differentiation. Great areas of granite in regionally metamorphosed areas merge into gneissic metamorphosed bodies of the same composition; there, much of the granite seems to be of metamorphic origin (see Chapter 16).

Igneous petrologists now recognize that differentiation does exist but that the mechanisms are more complex than we have described. Partial, rather than complete, melting is the rule and a dominant process in differentiation. Magma does not cool uniformly but may have a wide range of temperatures within a chamber. Those differ-

Table 14-3
Compositions of average oceanic crust and average continental crust
(Averages include compositions of sedimentary, granitic, and basaltic layers.)

Crust	Composition (%)							
	SiO_2	Al_2O_3	Fe_2O_3	FeO	MgO	CaO	Na_2O	K_2O
Continental	60.2	15.2	2.5	3.8	3.1	5.5	3.0	2.9
Oceanic	48.7	16.5	2.3	6.2	6.8	12.3	2.6	0.4

Source: Data from A. B. Ronov and A. A. Yaroshevsky, *Chemical Composition of the Earth's Crust,* American Geophysical Union Monograph 13, 1969.

ences in temperature induce thermal diffusion of ions by which some elements migrate toward the cooler regions and others toward the hotter. Thus chemical zones are created within a single magma. Certain melt compositions are now known to be immiscible—that is, insoluble in each other—so that magma chambers can be filled with two melts of different composition, each giving rise to its own crystallization products. The oxygen concentration in the magma also affects the course of crystallization by determining the oxidation state of iron. Ferric and ferrous ions enter into different crystal structures and thus control the amounts of iron in the crystallized rock. Convective motion within the magma chamber may distribute crystals as layers around the walls and ceiling rather than as a uniform layer at the bottom. Two magmas of somewhat different compositions may mix as they rise, causing a crystallization path different from that of either magma alone. Finally, not all granitic rocks stem from the differentiation of a single kind of magma; melting of varied source rocks is responsible for much variation. Large granitic bodies associated with plate-convergence zones are now thought by many to have been formed by melting of some combination of sediments, igneous rocks, and metamorphic rocks that would give a melt of granitic composition. We shall consider further the origin of the many kinds of granite after we turn to the origin of the magmas themselves.

ORIGIN OF MAGMA

Up to this point, we have taken for granted that magma occurs deep in the Earth, but we have not asked where, when, and how. We must reject the older notion that all of the interior of the Earth is molten, because the data we have used to infer its nature tell us that Earth is solid for thousands of kilometers down to the boundary of the core (see Chapter 17). Magma chambers under active volcanoes have been surveyed by geophysical methods,

and it is clear that magma occurs as discrete liquid regions in the otherwise solid interior.

Mid-ocean ridges and rises are major zones of basaltic magma generation: **flood basalts** flow intermittently from their rifts and fissures. The same kind of flood basalts are spread extensively over continental tectonic belts and over some stable shield areas as well. Granite, granodiorites, and other silicic intrusives occur almost exclusively on the continents, where they are accompanied in places by silicic extrusives. The differences in the distribution of rock types are reflected in the average compositions of the continental and oceanic crusts: the continental crust (including sedimentary, granitic, and basaltic layers) is more silicic as contrasted with the more mafic composition of average oceanic crust (Table 14-3).

The differences in composition can be related in part to melting temperatures. All magmas must form deep in the Earth where temperatures are high. Basaltic magmas form at temperatures well over 1000°C; granitic melts may form at temperatures several hundred degrees lower. Some idea of the depths of formation of these magmas can be obtained from the geothermal gradient shown in Box 13-3. To get a better picture, we must consider other factors that influence melting and crystallization, including pressure and the presence of water in the melt.

Effects of Water and Pressure

At atmospheric pressure, pure albite in the presence of small amounts of water remains solid up to temperatures a little over 1000°C. Above that temperature, the albite melts to form a liquid containing water, an aqueous melt. If more water is present, the melting temperature of the albite is lower (see Fig. 14-10, in which the relative amount of water present is indicated by the pressure of water vapor). In the same way, the melting temperatures of all of the feldspars and other sili-

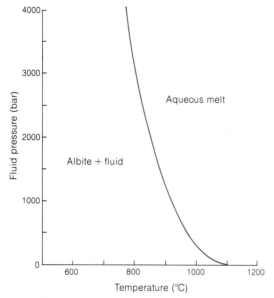

Figure 14-10
The melting curve for albite in the presence of
water, as measured by fluid pressure in bars (1 bar
equals 0.987 atmosphere). As fluid pressure increases,
the melting temperature decreases. These lowered
melting temperatures are characteristic of silicate
materials in the presence of water. [After *Earth
Materials* by W. G. Ernst. Copyright © 1969 by
Prentice-Hall, Inc.]

cate minerals drop considerably in the presence
of large amounts of water. Thus, melting can
occur in some places in the lower parts of the
crust, both because temperature and pressure are
higher there than in the upper parts of the crust
and because there is more water present than in
the relatively dry mantle below (Fig. 14-11). Pressure alone also has an important effect. Figure
14-12 shows the melting of dry basalt as a function of pressure and temperature, plotted together
with geothermal gradients. The effect of increased
pressure is to raise the dry melting points of basalt
and the silicate minerals. At a depth of several
hundred kilometers, we can expect some melting
of relatively dry mantle material. This is consistent with the inference of a partially molten zone
found at these depths in the asthenosphere from
seismological methods. This partial melting
would account for the widespread occurrence of
basaltic magma.

Geologic field evidence long ago provided evidence that the role of water in melts is important.
Numerous hydrothermal veins branch off from
the sides and tops of many intrusive bodies. Such
veins are filled with minerals that contain much
chemically bound water, which, together with a

variety of other evidence, shows that abundant
water was associated with the magma. Some of
the water may be from the magma, but much may
be groundwater circulating around the intrusive
body. Hydrothermal veins are of great importance
in the geology of many metal ores, which are
found as rich deposits in such veins. Some of the
most valuable deposits of mercury, gold, silver,
lead, and copper are mined from them (see Chapter 22). The hot-water solutions percolating upward from the intrusives carried these metals
with them. The metals were precipitated along
the walls of joints and cracks as the solutions rose
and cooled. Hydrothermal veins are commonly
associated with silicic intrusives on the continents. Hydrothermal activity recently discovered
on the East Pacific rise is related to basaltic
magmas. Here the water is sea water circulating to
depths of several kilometers (Chapters 10 and 22).

Water-rich silicic magmas, such as those that
produce hydrothermal veins, cannot be produced
by simple differentiation of deep mantle material
that has been melted into basalt. They are likely
produced by partial melting of a mixture of
quartz-rich, water-rich portions of the crust—sedimentary rocks, in particular. Sediments are
richer in silica and aluminum than the parent
rocks from which they were eroded.

Basaltic and Silicic Melts

In the 1960s, experimental studies of the melting
behavior of wet and dry mantle materials at extraordinarily high pressures (well over 30,000
atmospheres) were made possible by newly developed high-pressure equipment. In the 1970s, a
new generation of diamond anvil high-pressure
cells enabled experimenters to reach pressures of
a megabar—approximately one million atmospheres. The results of these studies have led to our
present ideas on basaltic magma generation at
mid-ocean ridges and elsewhere. One likely candidate for upper-mantle composition is peridotite,
an olivine-pyroxene rock. A small degree of partial melting of this material—up to only about 2 or
3 percent liquid—produces a liquid of basaltic
composition. The remaining solid is depleted in
the components of the basalt. On the way up
through upper mantle and crust, the basaltic liquids may undergo some fractional crystallization
at various temperatures and pressures, leading to
the specific compositions found in different basalts. The ascent of the basaltic liquid through different regions of the mantle under mid-ocean
ridges, on oceanic islands, and in lava plateaus on

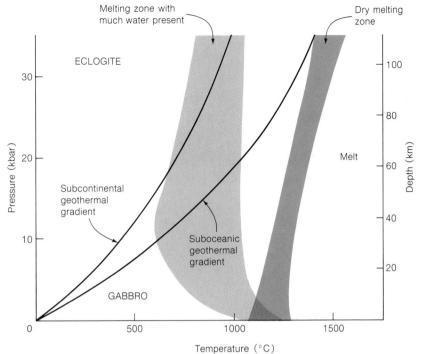

Figure 14-11
The melting temperatures of rocks of the same chemical composition as basalt vary greatly with pressure, temperature, and the amount of water present. The melting region for water-containing rocks of these compositions is at much lower temperatures than for dry melts. These are shown in relation to subcontinental and suboceanic geothermal gradients. It can be seen that basalt melting can take place in water-rich melts at much lower pressures—and hence at shallower depths—under the oceans than under the continents. The solid rocks shown are gabbro, the coarse-grained equivalent of basalt (at low pressures), and eclogite, one of the presumed materials of the mantle (at high pressures). [Simplified and modified from *Petrologic Phase Equilibria* by W. G. Ernst. W. H. Freeman and Company, Copyright © 1976.]

the continents accounts for some of the differences in these kinds of basalts. Magma mixing, immiscibility, and thermal segregation may account for other changes in the compositions of the crystalline rocks produced. All these processes combine to give slight but distinctive differences in composition that allow petrologists to recognize the components of their origin and, in ancient rocks, to recognize basalts of mid-ocean ridge and other origins.

Trace elements, those present in amounts less than 0.1 percent in general, play an important role in defining the nature of the mantle rocks from which the basaltic partial melts were derived. They are valuable indicators because the extent to which different minerals incorporate these elements varies sharply as a function of temperature, pressure, or composition of the fluid from which the minerals crystallized. Of special interest are barium, rubidium, strontium, zirconium, and the rare earth elements, those with atomic numbers from 57, lanthanum, to 71, lutetium. The relative abundance of the isotopes of strontium provides more information: the ratio of strontium-87 (derived from the radioactive decay of rubidium-87) to strontium-86, a nonradioactive isotope, is an

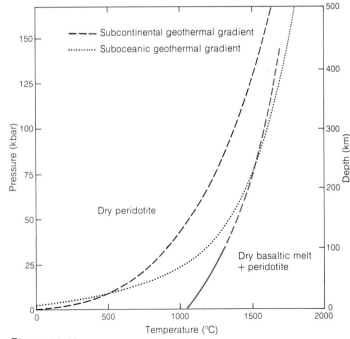

Figure 14-12
Calculated geothermal gradients under continents and ocean basins. These curves are somewhat speculative, for the melting of basalt in the absence of water has been investigated experimentally only in the pressure-temperature region indicated by the solid line. [After *Earth Materials* by W. G. Ernst. Copyright © 1969 by Prentice-Hall, Inc.]

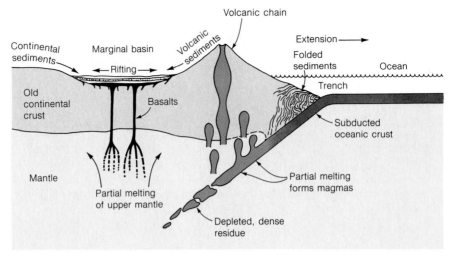

Figure 14-13

A marginal basin often opens behind volcanic island arcs where oceanic crust is being subducted under a continental margin. Some of the subducted crust melts, forming a magma. Most of the magma solidifies near the base of the continental crust, but some of it overflows as lava from volcanoes. As subduction ceases, a basin opens up behind the volcanic arc and fills with lavas and sediments. [From "The Oldest Rocks and the Growth of Continents" by Stephen Moorbath. Copyright © 1977 by Scientific American, Inc. All rights reserved.]

index of the rock's age and the original rubidium content (Chapter 2). In the past ten years trace elements and isotopic compositions have been used to show that the mantle is not homogeneous. It may be a mixture of an original, primordial composition from the first evolution of the mantle, some of it depleted by partial melting, and other components derived by mixing with continental and oceanic crustal materials brought down by subduction. These chemically different reservoirs are the sources of isotopically distinct basalts. This active area of research promises to answer many questions about the origin of the mantle, its age, and variations in its composition. It has also focused attention on subduction zones as regions where the crust is mixed with the mantle. In these areas generation of magmas leads to differentiation of basaltic fluids to give other magma types.

Returning to our questions about the origins of granite and other igneous rocks, we can conclude that basaltic magmas are only the first step in production of other magmas. We have long known of the voluminous andesite volcanics in active tectonic belts, like those that ring the Pacific Ocean. In the western North American Cordillera the average composition of the many batholiths comes close to andesite. These rocks must have come from a magma of andesitic composition. Such a composition could have been derived from partial melting of pyroxene-olivine-garnet rocks of the mantle if it occurred in the presence

of water. Alternatively, if partial melting took place after mantle rocks were contaminated by mixing with some water-rich sedimentary rock materials, andesite and other differentiates could be produced. At depths of 35 to 40 kilometers, temperatures of about 600°C, and water pressures of about 10,000 atmospheres, silicic melts could form from mixtures of sedimentary and metamorphic rocks that are compositionally equivalent to granodiorite.

Subduction zones at plate convergences are the obvious sites for partial melting and other differentiation processes to alter mixtures of mantle partial melts and sedimentary and metamorphic materials. As a lithospheric plate moves down, carrying with it the oceanic crust, some sediment, and water, partial melting occurs as the plate sinks into the asthenosphere (Fig. 14-13). This melting produces magmas of various compositions in or above the descending plate at various depths, depending on the amount of water, the temperature, and the pressure. A marginal basin may open behind the island arc and be partly filled with basaltic lavas. These magmas are then intruded or extruded in the mountain belts associated with crustal deformation near the subduction zone. This theme will be further explored in the next chapter and integrated with a general view of magmatism and volcanism in relation to plate tectonics.

SUMMARY
1. Intrusive igneous rocks of many compositions are formed by emplacement of different kinds of magmas in the crust.

2. Slow crystallization of plagioclase feldspars involves continuous reaction of already formed crystals with the surrounding melt until complete crystallization results in all of the crystals having the same composition as the original melt. In fractional crystallization earlier-formed crystals do not continuously react with the melt.

3. Magmatic differentiation may occur as a result of fractional crystallization, partial melting, differential cooling, liquid immiscibility, variations in oxygen content, and mixing of different magmas. This differentiation can produce silicic magmas from originally mafic ones.

4. Differences in the composition of basaltic melts and the places where they originate in the crust or mantle are accounted for by the effects of pressure and water content on the melting temperatures of basalt and the silicate minerals; different source rocks give rise to partial melts and to various differentiation processes as liquids ascend through crust and mantle.

5. Magmatic differentiation explains the occurrence of granite and silicic volcanics. Many granites were probably formed by metamorphism or by partial melting of a mixture of sediments and earlier granitic crust in a subduction zone, in either the crust or the upper mantle.

EXERCISES

1. Compare those aspects of the mineralogical composition of granites that suggest an origin from a melt with the evidence from field relations.

2. What evidence could you use to decide which of two intrusions formed at different temperatures crystallized from the hotter magma and which from the cooler?

3. What kinds of intrusive igneous rocks would you expect on a hypothetical planet of the same crustal composition as Earth, on which all magmas cooled extremely slowly and always maintained equilibrium between crystals an remaining melt?

4. Why would you not generally expect basaltic magmas to be generated in the upper ten kilometers of continental crust?

5. What is the evidence for appreciable quantities of water being included in some magamas?

6. The compositions of many basalts extruded in continental rift valleys differ from those produced in rift valleys at oceanic plate divergences. How can you account for these differences?

7. Would you expect a partial melt formed at higher temperatures to contain more sodium-rich or calcium-rich plagioclase feldspar than one formed at lower temperatures?

BIBLIOGRAPHY

Cox, K. G., Bell, J. D., and Pankhurst, R. J., *The Interpretation of Igneous Rocks*. London: George Allen and Unwin, 1979.

Ernst, W. G., *Earth Materials*. Englewood Cliffs, N.J.: Prentice-Hall, 1969. (Chapters 2 and 5.)

Ernst, W. G., *Petrologic Phase Equilibria*. San Francisco: W. H. Freeman and Company, 1976.

O'Nions, R. K., P. J. Hamilton, and N. M. Evenson, "The Chemical Evolution of the Earth's Mantle," *Scientific American*, May 1980. (Offprint 947.)

Wyllie, P. J., "Experimental Petrology: An Indoor Approach to an Outdoor Subject," *Journal of Geological Education*, v. 14, no. 3, June 1966.

Wyllie, P. J., *The Dynamic Earth: Textbook in Geosciences*. New York: John Wiley & Sons, 1971. (Chapters 7 and 8.)

15

Volcanism

*Volcanoes can be beautiful, informative, and both beneficial and danger-
ous. They serve as windows through which we can dimly perceive the
interior. Volcanism occurs in many different styles, principally near plate
boundaries.*

For decades geologists have debated the source of
magma, the hot, mobile material that is generated
within the Earth and that solidifies into igneous
rock. There is now little doubt, so the seismologists tell us, that the asthenosphere, which extends from depths of about 75 to 250 km, is the
one large region in the mantle that is partially
molten. It is reasonable, therefore, to identify this
layer as a main source of magma. Remelting of
sections of the lithosphere may provide another
source of magma. At certain places, perhaps
where the lithosphere is fractured or otherwise
weakened, the magma rises, squeezed up by the
weight of the overlying crust, some of it eventually reaching the surface where it erupts as **lava**
(Fig. 15–1). Lava differs from the parent magma in
that lava has lost some volatile constituents
(gases) to the atmosphere or ocean and has gained
or lost other chemical components en route to the
surface. Despite these differences, lava provides
important clues to the chemical composition and
physical state of the upper mantle.

The 500 to 600 active volcanoes of the world are
not randomly distributed, but show a definite pattern that correlates strongly, though not exclusively, with the plate boundaries, where lithosphere is created or destroyed (Fig. 15–2). Not long
ago we saw in volcanism mainly the grandeur and
the spectacular nature of visible eruptions. We
now think that the creation of the sea floor is, in
terms of volume, the most important volcanic
process, one that is hidden almost entirely from

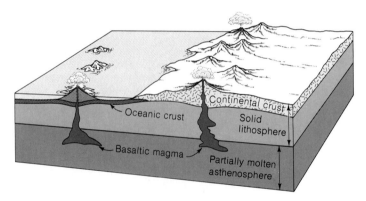

Figure 15–1
Basaltic magma, which originates in the partially
molten asthenosphere, rises through the lithosphere
to erupt as basaltic lava on the surface.

direct observation by the overlying ocean. At the
end of this chapter, we will return to this global
pattern of volcanism: the different compositions
and eruptive styles of lavas and their association
with plate boundaries and other geologic features.

VOLCANIC DEPOSITS

Volcanic rocks are important not only for what
they reveal about the interior but also because the
chemical-mineralogical composition of lavas affects eruptive styles and the kinds of landforms
they build upon freezing. To understand the vari-

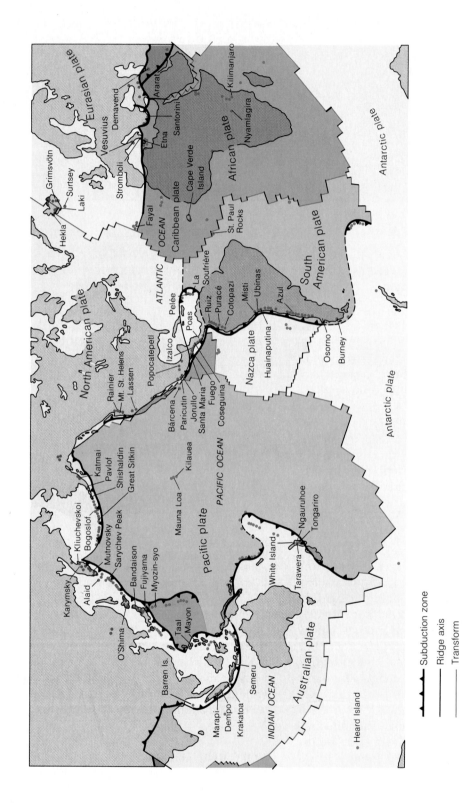

Figure 15-2
The active volcanoes of the world are not distributed randomly on Earth's surface; they tend to be associated with plate boundaries.

Subduction zone
Ridge axis
Transform
· Volcanoes

· Heard Island

Eurasian plate

Grimsvötn
Surtsey
Hekla
Laki

Ararat
Demavend
Vesuvius
Etna
Santorini
Cape Verde
Island
Fayal
ATLANTIC
OCEAN
Caribbean plate
St. Paul
Rocks

African plate
Kilimanjaro
Nyamlagira

North American plate

Rainier
Mt. St. Helens
Lassen
Popocatepetl
Izalco
Poas
Pelée
La
Soufrière
Ruiz
Puracé
Cotopazi
Misti
Ubinas
Azul
Huainaputina
Osorno
Burney

South
American plate

Nazca plate

Bárcena
Paricutin
Jorulo
Santa Maria
Fuego
Coseguina

Katmai
Pavlof
Shishaldin
Great Sitkin

Mauna Loa
Kilauea

PACIFIC OCEAN

Pacific plate

Kliuchevskoi
Bogoslof
Mutnovsky
Sarychev Peak
Bandaison
Fujiyama
Myozin-syo

Karymsky
Alaid
O'Shima

Taal
Mayon

Barren Is.

White Island
Tarawera
Ngauruhoe
Tongariro

Australian plate

INDIAN OCEAN

Marapi
Dempo
Krakatoa
Semeru

Antarctic plate

Antarctic plate

ety of shapes of volcanoes—why some eruptions are explosive and others gentle—we must begin with the rocks themselves.

Figure 15–3 reviews the volcanic rock types. Igneous rocks are classified into four major groups on the basis of silica content and mineralogy (see Chapter 3): felsic, intermediate, mafic, and ultramafic. The major minerals associated with these rocks are listed in Figure 3–17. Texture is also important in rock classification. Very slow cooling gives rise to the large mineral grains found in intrusive rocks; faster cooling results in a fine-grained rock; and very rapid cooling in glass, since there is no time for crystal growth. The coarse-grained, intrusive, analogs of lava, which are important in discussing the origin of magma, are: rhyolite–granite, andesite–diorite, and basalt–gabbro. Ultramafic magmas almost never reach the surface through volcanism. The principal lavas are basalt and andesite, the mafic and intermediate types, respectively.

During eruptions, magma moves in response to pressures within the volcano and flows onto the surface as lava or is violently ejected as ash. Rapid cooling follows, and indeed we find that most volcanic rocks are fine grained, glassy, or pyroclastic (fragmented deposits of material ejected or exploded into the air from a volcanic vent).

Lava Flows

Like all liquids, lavas flow downhill. Basaltic melts are highly fluid and remain fluid longer than the more felsic lavas; basalt can flow fast and far (Figs, 15–4 and 15–16). Flow velocities as high as 100 kilometers per hour (60 miles per hour) have actually been observed, but velocities of a few kilometers per hour are more common. Lava streams extending more than 50 kilometers have been witnessed in historic times. On flat terrain, basaltic lava spreads out in thin sheets, successive flows often piling into immense lava plateaus. The great Columbia Plateau of Oregon and Washington is one example; the lunar maria* also formed in this way. Such sheets can sometimes be confused with sedimentary strata. In one sense they are like sediments, for they obey the laws of original horizontality and superposition. Basaltic magma erupts with temperatures of 1000 to 1200°C.

Rhyolite, the most felsic magma, has a lower melting point and erupts with temperatures of 800 to 1000°C. It is much more viscous than basalt and

*Plural of *mare*, from the Latin for "sea." These are the large dark areas of the Moon.

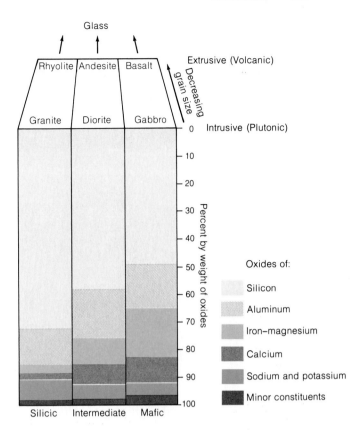

Figure 15–3
Average composition of the major types of lava (basalt, andesite, and rhyolite) and their intrusive counterparts (gabbro, diorite, and granite). Note the systematic increase in silica and the decrease in iron-magnesium oxides in going from basalt to andesite to rhyolite.

flows very slowly, forming thick, bulbous deposits. Andesite, with intermediate silica content, shows intermediate properties.

Lava flows can be distinguished by a variety of interesting and sometimes bizarre surface and internal features. The basaltic flows on Hawaii fall into two categories according to their surface forms: **pahoehoe** (pronounced pahoyhoy) and **aa** (ah ah). These euphonic terms are among the few Hawaiian words in the scientific vocabulary. *Pahoehoe* in Hawaiian means "ropy"; *aa* is what it says—or what a barefooted Hawaiian says while walking on it: "ah . . . ah . . . oh . . . ah," Pahoehoe is a highly fluid lava that spreads in sheets. A thin, glassy elastic skin that forms on the surface is dragged into ropy, filamented folds as flow continues below the surface (Fig. 15–5). Aa is a slower-moving lava flow whose thick skin is broken into an indescribably rough, jagged, clinkery surface. The blocks ride on the viscous, massive interior, advancing as a steep front of angular boulders like a tractor tread. Many a mainland

Figure 15-4
Basaltic lava flowing from a fissure on the east flank of Kilauea volcano and cascading across fault scarps. [Photo by Agatin T. Abbott, Hawaii Institute of Geophysics.]

Figure 15-5
Lava flows of aa (left) and pahoehoe from Kilauea volcano in Hawaii. The dark, jagged aa, about 3–4 m thick, lies on top of the glistening, ropy pahoehoe surface. [Courtesy of R. Decker and B. Decker from *Volcanoes*, San Francisco: W. H. Freeman and Company, 1981.]

Figure 15–6
Pillow lava, characteristic of underwater volcanic eruptions, photographed at an ocean depth of 3300 m off Hawaii. Area of photograph approximately 1.25 × 1.25 m. [Photo by J. G. Moore, U.S. Geological Survey.]

haole (nonnative) who, "site" unseen, has paid dearly for an acre of Hawaiian aa. His shoes would be cut to ribbons if he tried to walk on his own land. Aa differs from pahoehoe because it has lost its volatiles and consequently its fluidity. Single flows commonly grade from pahoehoe near their source into blocky aa "downstream." The scale of the blockiness varies from small, football-sized blocks (as at Mt. Etna, Italy) to house-sized blocks (at Hekla, Iceland). The blocks, regardless of scale, are loose and unstable, and their surfaces are covered with gas cavities, making them jagged. The topography of blocky aa flows is commonly extremely rough with ups and downs of 10 feet or more. They are truly treacherous to cross. A good pair of boots has a mean lifetime of about a week on aa, and the traveler or geologist can count on cut knees and elbows. The instability of house-sized blocks makes some flows downright hostile and sufficiently dangerous to be treated with respect.

Pillow lavas, piles of ellipsoidal, sacklike blocks about a meter in dimension, are characteristic of underwater eruptions of basalt and andesite (Fig. 15–6). Apparently, tongues of lava, chilled by contact with water, develop a tough, plastic skin that is convex upward and filled with liquid. On cooling, the interior becomes crystalline, and the skin glassy and often radially cracked. Identification of pillow structure in a geological section is a sure indication of underwater deposition. It may also identify top and bottom in deformed sequences.

In 1971 Dr. James Moore of the U.S. Geological Survey and some colleagues performed the remarkable feat of swimming up to the advancing underwater lava flow of the Kilauea Volcano and observing the formation of pillows. Their report describes what they saw there:

The flow is advancing under water as a wall of rubble . . . , the base of [which] is covering over the old ocean floor to a depth of about 100 feet. Tongues of lava, circular in cross section, extend down the front of this rubble slope. Some of these are as long as 200 feet and they are 3 to 4 feet in diameter, generally. Budding off of them are typical pillows. We could approach one of these elongated lava tongues, in which lava was actually flowing inside and whose surface was too hot to touch on the outside. Generally, the surface was dark, but periodically it would crack and bright incandescence could be seen inside for a few seconds until the crack solidified and then it would crack in some other place. There was a continuous loud mixture of noises: hissing, cracking, small explosions, and rumbling noise as the lava went down through these tubes. General water temperature was slightly elevated around the flowing lava tongues from a normal of 76°F to about 80°F, but in cracks and around behind these lava tubes the water was boiling, and many bubbles were coming up out of cracks in the tubes. One surprising feature was the concentration of marine life including fish, lobsters, and eels which were crowded ahead of the advancing rubble flow at the base of the rubble, apparently having been displaced from their normal habitats along the coast.

Figure 15-7
A chunk of vesicular lava, Newberry Caldera, Oregon. [From *Volcanic Land Forms and Surface Features* by J. Green and N. M. Short, Springer-Verlag, New York. Photo by J. Green.]

Figure 15-8
Giants Causeway shows columnar jointing in lava. [Courtesy of Northern Ireland Tourist Board.]

Lavas show many other features. They can be glassy, fine grained, or even coarsely crystalline, depending on the rate at which they cooled. When water vapor or other gases are released, ovoid gas cavities, or **vesicles,** can form (Fig. 15-7). **Pumice,** an extremely vesicular, generally rhyolitic lava, occasionally has so much void space that it is light enough to float. **Xenoliths,** foreign inclusions picked up by a magma, provide important clues about the source region and the rock formations through which the lava passed en route to the surface. When lava cools and contracts, shrinkage cracks (**joints**) often form. Some flows develop impressive **columnar jointing** perpendicular to the cooling surface (Fig. 15-8). Where flows are thin and cool rapidly, the joints may be highly irregular. On Hawaii one can walk into **lava caves,** or **tubes.** These form when the source of an enclosed stream of lava is cut off and the remaining lava drained from the channel. Some geologists think the deep valley on the moon where the Apollo 15 landed, Hadley Rille, was formed when the roof of a lunar lava tube collapsed. **Spatter cones** are steep-sided, conical hills built from the spatter of **lava fountains** spewed out of vents (Fig. 15-9).

Pyroclastic Deposits

Water and dissolved gases are important constituents of magma. Before eruption, the confining pressure of the overlying rock keeps these volatiles from escaping. When the magma rises close to the surface and the pressure drops, the volatiles may be released with explosive violence, shattering the lava into fragments of various sizes, shapes, and textures. This is more common in the felsic, volatile-rich, viscous rhyolitic and andesitic lavas than in basalts, which are more fluid and release volatiles more quickly.

Pyroclasts are the fragmentary volcanic rock materials that are ejected into the air. The particles, whether rocks, minerals, or glass, are classified according to size. The finest are called **dust;** fragments up to about 2 millimeters (generally the same as sand) are called **ash;** and chunks larger than about 6 centimeters are called **bombs** (Fig. 15-10). Bombs have several characteristic shapes: ellipsoidal, discoidal, and irregularly rounded; they range from baseball to basketball size. Some volcanic bombs, however, can be blockbusters. Blocks as heavy as 100 tons are known to have been thrown more than 10 kilometers in violent eruptions. Volcanic ash, and especially dust, can be carried great distances. Within two weeks the dust from the 1980 eruption of Mt. St. Helens could be traced around the entire world by orbiting Earth satellites.

Sooner or later pyroclasts must fall, mostly near the source, building up deposits. When the fragments become cemented together (lithified), the rocks formed from the smaller fragments are called **volcanic tuffs;** those formed from the larger ones are called **volcanic breccias** (Fig. 15–11). There is as much variety among volcanic deposits as among sedimentary ones, and there are many similarities between the two. Some volcanic deposits show graded bedding not unlike that of certain sedimentary rocks, the coarse fragments settling first, followed by the finer debris. Volcanic beds often show lateral sorting, a gradation from coarser material near the vent to finer material farther away. Even dune structures and cross-bedding are found, formed by winds blowing at the time of fallout and also by the vigorous radial (outward) expansion of the volcanic eruption cloud. Similar bedded and dunelike deposits were observed around some of the sites of atomic bomb explosions. From details of the forms of those deposits, it is possible to estimate the wind velocities produced by the blasts.

A spectacular and often devastating form of eruption occurs when hot ash, dust fragments, and gases are ejected in a glowing cloud that rolls downhill with amazing speed. The solid particles are actually buoyed up by the hot gases, so that

Figure 15–9
Lava fountain building a cinder and spatter cone during the 1960 eruption of Kilauea, Hawaii. [Photo by T. McGetchin, M.I.T.]

Figure 15–10
A field of volcanic bombs and other ejecta. Haleakala volcano, Maui, Hawaii. Source cone is in the background. [From *Volcanic Land Forms and Surface Features* by J. Green and N. M. Short, Springer-Verlag, New York. Photo by J. Green.]

Figure 15–11
Volcanic breccia, Augustine Volcano, Alaska. [Photo by H. -U. Schmincke, Ruhr-University, Bochum, Germany.]

Figure 15–12
Nuées ardentes, "glowing clouds" of hot volcanic ash, dust, and gas, plummeting down the slopes of Mount Mayon in the Philippines during the eruption of 1968. [Photo by W. Melson, Smithsonian Institution.]

Figure 15–13
Hand specimen of welded tuff from an ash-flow sheet. The dark lenses are collapsed pumice fragments enclosed in a light-colored matrix composed mostly of welded glassy shards. Laboratory experiments indicate that welded tuffs form at temperatures in the range 700–900°C. [Photo from R. G. Schmidt and R. L. Smith, U.S. Geological Survey.]

there is little frictional resistance to this incandescent pyroclastic flow, or **nuée ardente** (Fig. 15–12). In 1902 a nuée ardente with an internal temperature of 800°C flowed down from Mont Pelée on Martinique at a speed of 160 kilometers per hour (100 miles per hour). In one minute and with hardly a sound, it enveloped the town of St. Pierre and killed 28,000 people. Deposits left behind by a nuée ardente are poorly sorted, without much sign of bedding, as would be expected. Typically, the fragments are still hot, hence soft, when deposited and become compacted and stuck together to form **welded tuffs** or **ignimbrites** (Fig. 15–13). The welding agent that binds the fragments together is glass. Vast ignimbrite sheets, also called ash-flow deposits, more than 100 meters thick, covering tens of thousands of square kilometers, are known on the continents. These pyroclastic deposits are the felsic counterpart of plateau basalts.

ERUPTIVE STYLES

Knowing something of the several types of volcanic deposits that have poured out or been blown out from the interior, we can go on to consider the different ways that eruptions occur and the different results they produce. Eruptions do not always form the majestically symmetrical cone of a Fujiyama, which has become a symbol to a nation. The hundreds of thousands of square kilometers of monotonous **flood basalt** that make up the Columbia Plateau represents another variant. Eruptions like those of Hawaii's Kilauea have been heralded by warning events that preceded relatively quiet outpourings of lava, which move slowly and predictably, destroying property but not lives. Pyroclastic eruptions, like that of Mont Pelée, or **phreatic explosions** (violent steam blasts), like that of Krakatoa in 1883, can wipe out entire populations (Fig. 15–14). Teams of archeologists and marine geologists have pieced together the story of the volcano Thera (formerly called Santorin), situated in the Aegean Sea. It appears that the eruption of Thera in the fourteenth century B.C. was an event of super-Krakatoan proportions. The volcanic debris and sea waves that were produced caused destruction in coastal settlements over a large part of the eastern Mediterranean. The mysterious disappearance of the Minoan civilization has been attributed by some scientists to this cataclysm, and it is thought that the legend of the lost continent of Atlantis may have its origin in this event.

Figure 15-14
A phreatic explosion on the volcano Surtsey in the Atlantic Ocean off the coast of Iceland. [Copyright by Solarfilma, Reykjavik, Iceland.]

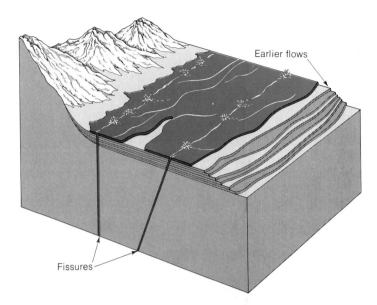

Figure 15-15
In a fissure eruption of highly fluid basalt, lava rapidly flows away from fissures and forms widespread layers, rather than building up volcanic mountains. [After R. S. Fiske, U.S. Geological Survey.]

Figure 15-16
The area covered by the Columbia River flood basalts. [After R. S. Fiske, U.S. Geological Survey.]

Fissure Eruptions

Either lava or pyroclastic materials may emanate from long, narrow fissures or groups of such fissures (Fig. 15-15). The lavas are usually basaltic; the pyroclastics originate from more felsic magmas. The rather fluid basaltic lavas flow away from the fissures rather than forming volcanoes; they often flood extensive areas, building lava plains or piling up into lava plateaus—hence the names "flood basalts" and "plateau basalts" (Figs. 15-16 and 15-17). The basalt flows that made the Columbia Plateau buried 130,000 square kilometers (50,000 square miles) of the preexisting topography, which had a relief of about 1.5 kilometers (1 mile). Some individual flows were more than 100 meters thick, and some were so fluid that they spread more than 60 kilometers from their source. The total volume of basalt poured out in this Miocene volcanic episode has been estimated at more than 100,000 cubic kilometers (25,000 cubic miles)! An entirely new landscape with new river valleys has since evolved on top of the old surface. Other famous continental plateaus or flood basalts are the Deccan of India, which has an area of more than 500,000 square kilometers (200,000 square miles), and the Parana of Brazil and Paraguay, covering more than 750,000 square kilometers.

When rocks are dredged from a mid-ocean ridge or samples are collected from such occasional oceanic ridge peaks as the Azores, which stick out above water, they almost always turn out to be basalt. The basalt originates in fissure erup-

Figure 15-17
View across Imnaha River Canyon and part of the Columbia Plateau in Oregon.
Flood basalts form the layered rocks. [Photo by H. -U. Schmincke, Ruhr-University,
Bochum, Germany.]

Figure 15-18
The volcano Surtsey was born in the sea south of Iceland in 1963. Some 300×10^6
m^3 of lava and $600 \times 10^6 \, m^3$ of ash poured out of the sea floor to build this new
island. [Copyright by Solarfilma, Reykjavik, Iceland.]

tions in connection with sea-floor spreading. The
process is unspectacular because it is seldom
seen, but volumetrically it dwarfs all of the continental outpourings combined. The fissures occur
along the ridge crests and often are marked by
median **rift valleys** and seismic belts (see Fig.
1–13). The global oceanic ridge fissure system can
be traced over distances totaling some 50,000

kilometers (see inside back cover). Enormous
amounts of basalt have poured out of this world-
encircling sequence of cracks, enough to build the
crust of all of the present sea floor in the past 200
million years.

Iceland, an exposed segment of the mid-Atlantic ridge, provides an unexcelled opportunity to
view the process of fissure eruption and sea-floor

spreading directly. The island is composed mostly of basalt with small amounts of the more felsic lavas. Repeated geodetic surveys show that Iceland is in a state of tension, literally being pulled apart, one half moving eastward with the Eurasian plate, the other westward with the American plate. Tensile cracks develop, magma flows in from below, and overflows onto the surface.* There is ample evidence of basalt flooding from great fissures in the geologic record, but the only historically recorded flooding occurred on Iceland. In a single eruption in 1783, a crack 32 kilometers (20 miles) long, the Laki fissure, opened and spewed out some 12 cubic kilometers (3 cubic miles) of basalt. At the conclusion of such an episode the lava solidifies to form a nearly vertical **dike** in the fissure and nearly horizontal beds on the adjacent surface. With the next episode of lateral spreading, a new crack opens, and another flow pours out over the old one. This is the way Iceland grows, primarily by repeated eruptions from long fissures, but also by eruptions from localized vents (Figs. 15–18 and 15–19). Although the details may differ under water, it is reasonable to expect the sea-floor crust to grow in a similar fashion. The upwelling basalt frequently produces ridges centered on the deep valleys of the fissures rather than plateaus like those in Iceland. The ridges gradually subside into ocean basins as they move away from the zone of spreading, to be replaced by fresh ridges. This is discussed further in Chapter 19, which summarizes what we know about plate tectonics.

Fissure eruptions of pyroclastic materials rather than lavas are more likely when the parent magma is more felsic than basalt. Such eruptions have produced extensive ignimbrite sheets, although we have never witnessed one of these spectacular events. The early Tertiary ignimbrites of the Great Basin in Nevada and adjacent states cover an area of about 200,000 square kilometers (80,000 square miles), and are as much as 2500 meters (1.5 miles) thick in some places. In Yellowstone National Park, successions of forests were buried by broad ignimbrite sheets (Fig. 15–20).

Central Eruptions

Unlike fissure eruptions, which emanate from linear sources, central eruptions are much like point sources of magma. The extrusive materials issue

---

*In 1977 magma entered a borehole drilled in a geothermal field in northern Iceland. Over a period of 20 minutes, 3 tons of vesicular, glassy cinders spurted from the hole in this tiny induced volcanic eruption.

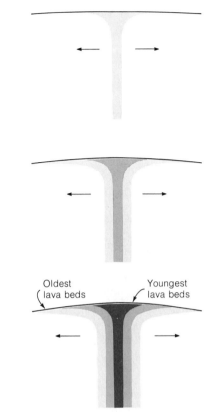

Oldest
lava beds

Youngest
lava beds

Figure 15–19
Iceland is an exposed part of the mid-Atlantic ridge. This schematic diagram shows the mechanism of its growth, repeated fissure eruptions and lateral spreading.

Figure 15–20
Ignimbrites, or ash-flow sheets, are deposited when intensely hot, gas-charged volcanic dust, ash, and pumice spread swiftly over the surface to settle and cool as flat sheets. The example shown is on the Pajarito Plateau, near Los Alamos, New Mexico. Source of the ash lies beyond the mountains in the background. [Photo from R. G. Schmidt and R. L. Smith, U.S. Geological Survey.]

from a **central vent,** or **pipe,** and give rise to the most familiar of all volcanic features—the **cone.**

Lava cones are built by successive lava flows. Basaltic lava is relatively free flowing and spreads widely. If there is a large supply, a broad shield-shaped volcano can be built up (Fig. 15–21), many tens of kilometers in circumference and more than one or two kilometers in height. The slopes are relatively gentle, 6 to 12°. Mauna Loa, on Hawaii (Fig. 15–22), is the classic example of a **shield volcano.** Although it rises 4 kilometers (2.5 miles) above sea level, its actual size is hidden by the ocean. Measured from the sea floor, Mauna Loa rises 10 kilometers (6 miles) and has a basal diameter of 100 kilometers (60 miles)! It grew to this enormous size by the accumulation of individual lava flows, each only a few meters thick. The island of Hawaii is actually formed by the overlap of a group of shield volcanoes. When the lava supply is less plentiful, a smaller lava cone is the result. Most basaltic volcanoes probably begin as fissures and develop central vents later since it is easier for the magma to ascend as a column than as a sheet.

Felsic lavas are so viscous that they can just barely flow, as is shown by the shapes of the **volcanic domes** they produce (Fig. 15–23). The domes give every appearance that the lava had been

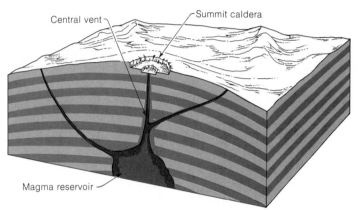

Figure 15-21
Schematic diagram showing how a shield volcano is built up from the accumulation of thousands of thin basaltic lava flows that spread widely and cool as gently sloping sheets. Each layer shown represents the accumulation of many hundreds of thin lava flows. [After R. G. Schmidt, U.S. Geological Survey.]

Figure 15-22
The shield volcano Mauna Loa on Hawaii, the giant among the active volcanoes of the world. Smaller craters lead upward to Mokuaweoweo caldera, the large depression at the summit. [U.S. Air Force photo.]

squeezed out of a vent like toothpaste, with very little lateral spreading, to form a bulbous steep-sided mass. Domes often collapse soon after formation or are destroyed by explosions as occurred in the 1980 eruptions of Mt. St. Helens.

Cinder cones are built of pyroclastic material explosively ejected from a vent (Fig. 15–24). The profile of the cone is determined by the angle of repose (Fig. 5–2)—that is, the maximum angle at which the debris remains stable instead of sliding downhill (Figs. 15–25 and 15–26). The larger fragments, which fall near the summit, can form stable slopes exceeding 30°. Finer particles are carried farther from the vent and form gentle slopes of about 10° at the base of the cone. The beautiful classic, concave-shaped, volcanic cone reflects this variation in the angle of repose.

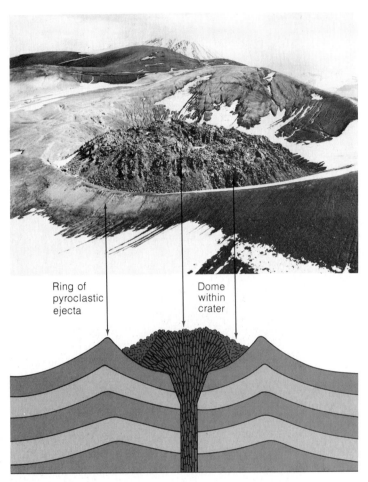

Figure 15-23
Volcanic domes are bulbous masses of lava. The lava masses are so viscous that instead of flowing they pile up over the vent. Shown here is Novarupta Dome in the Katmai region of Alaska. [After A. Chidester and R. G. Schmidt, U.S. Geological Survey.]

Figure 15-24
Ejecta building a cinder cone on the flank of Mt. Etna, Sicily, during the 1969 eruption. Fragments with dimensions of 0.1 to 1 m³ were thrown out of the crater with a velocity of about 50 m/sec. [Bernard Chouet, MIT.]

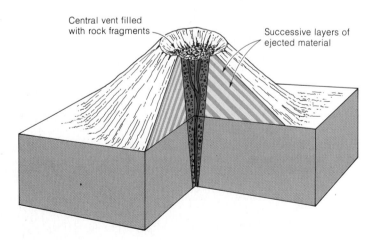

Figure 15-25
In a cinder cone ejected material is deposited as layers that dip away from the crater at the summit. The vent beneath the crater is filled with fragmental debris.

Figure 15-26
Cerro Negro in 1968. This volcano, near Managua, Nicaragua, is a cinder cone
built on an older terrain of lava flows. [Photo by Mark Hurd Aerial Surveys.]

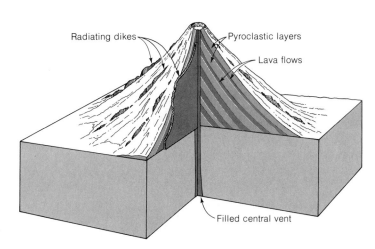

Figure 15-27
Structure of a composite volcano built up of layers
of pyroclastic material and lava flows. Lava, solidified in fissures, forms riblike dikes that strengthen
the cone. [After R. G. Schmidt, U.S. Geological
Survey.]

When a volcano emits lava as well as pyro-
clasts, a **composite cone** or **strato-volcano** is built
of alternating lava flows and beds of pyroclasts
(Fig. 15-27). This is the most common form of the
large continental volcanoes such as Fujiyama, Vesuvius, Etna, and Mt. St. Helens (Fig. 15-28).

If the central vent of a volcano grows too high

or becomes plugged, the lava may find an easier
route by opening a subsidiary vent and building a
small cone on the flank of the main volcano. Hundreds of subsidiary cones pimple the slopes of
Etna, Mauna Loa, and Mt. Newberry (central Oregon), to cite a few examples. Flank eruptions can
also originate from fissures.

Craters form the summits of most volcanoes,
usually centered over the vent. During the eruption of a lava volcano, the upwelling lava overflows the crater walls; when eruption ceases, the
lava that remains in the crater sinks back into the
vent. When pyroclastic cones erupt, the material
is literally blasted out of the crater, which later
becomes partially filled by debris falling back
into it. Because crater walls are steep, they may
cave in or retreat by erosion after an eruption,
becoming enlarged to several times the vent diameter. Craters may be up to a kilometer in diameter
and hundreds of meters deep. Etna, for example,
has a central vent 300 meters (1000 feet) in diameter and at least 850 meters (2800 feet) deep.

Calderas are large, basin-shaped depressions of
volcanic origin. Most calderas are collapse features that form after a magma chamber empties so
that it no longer supports part of the overlying
cone. The subsidence can occur catastrophically
or gradually, leaving a steep-walled, flat-floored
depression that can be as much as several kilometers in diameter. Later activity can lead to the
growth of new cones in the caldera floor. Figure
15-29 diagrams the evolution of a caldera. Some

Figure 15-28
Mt. St. Helens, a composite volcano in southwestern Washington, photographed
before the catastrophic eruptions of 1980. Spirit Lake, in the foreground, is
dammed by a volcanic mudflow deposit (lahar) formed in the eruptions
between 800 and 1200 B.C. [From U.S. Geological Survey.]

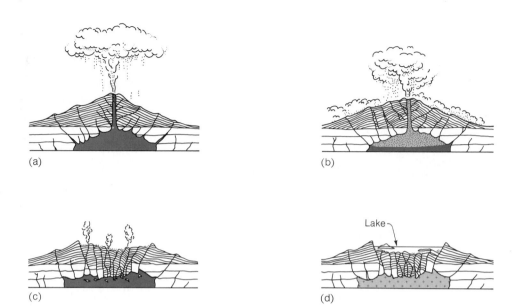

(a)

(b)

(c)

(d)

Lake

Figure 15-29
The evolution of the caldera at Crater Lake, Oregon. (a), (b) Eruptions from the summit of the vanished former Mount Mazama deplete the magma chamber below. (c) A caldera results from collapse of the mountain summit into the empty magma chamber. (d) The lake forms in the caldera; minor eruptions into the lake continue until the magma below the surface finally solidifies.

Figure 15-30
Crater Lake, Oregon, fills the caldera in the stump of the vanished Mt. Mazama. [From U.S. Geological Survey.]

famous examples are the still-active Kilauea, the dormant Crater Lake (Fig. 15–30), and the newly active Krakatoa (Indonesia). Perhaps the largest and most spectacular caldera known is the one on Olympus Mons, a huge volcano photographed on Mars from the Mariner 9 spacecraft (Fig. 21–15).

At one time it was thought that a caldera was formed by a huge explosion, as if the volcano had "blown its top." Geologic mapping of the debris around volcanoes, however, has shown that more volcanic rock subsided than was blasted out. The pattern of down-faulting is more consistent with the roof caving in than being blasted out.

This is not to say that volcanic explosions have not left their mark on the Earth's surface. Hot, gas-charged magma encountering groundwater or sea water forms a highly explosive mixture. The vast quantities of superheated steam generated cause phreatic explosions. The destructive eruptions of Krakatoa and Taal (Philippines) are two of the many examples of such cataclysmic events. Another is Bandai-san (Japan), which blew out in 1888, leaving an amphitheater 2.5 kilometers (1.5 miles) across with walls 350 meters (1200 feet) high.

Diatreme is the name given a volcanic vent, or pipe, that is filled with breccia as a result of the explosive escape of gases from the deep interior. Shiprock, which towers 515 meters (1700 feet) over the surrounding, flat-lying sedimentary rocks of New Mexico, is a diatreme exposed by the erosion of its enclosing sedimentary rocks. As viewed by transcontinental air travelers, Shiprock looks like a gigantic black skyscraper in the red desert (Fig. 15–31). The eruptive mechanism that produces diatremes has been pieced together in much detail from the geologic record to be found in the ejecta and the diatremes. Rock and mineral data tell us that some diatremes are produced by the upward movement of materials from great depths—well within the upper mantle. The evidence indicates that they are formed by gas-charged magmas that melt their way to the surface, finally ejecting gases, lava fragments from the vent walls, and fragments from the deep crust and mantle—all with explosive energy and sometimes with supersonic velocity. One can imagine such an eruption to look much like the exhaust jet of a Saturn 5 rocket upside down in the ground, blowing rocks and gas into the air. This was observed by a Chilean Air Force pilot when the volcano Nilahue erupted in 1955. The jets lasted for about 30 minutes, each blast followed by a pause of about the same duration. Activity stopped after about a month, presumably after the volcano literally ran out of gas.

Figure 15–31
Shiprock is a diatreme, or volcanic pipe, exposed by erosion of its enclosing rock. It towers 515 m above the surrounding, flat-lying sediments of New Mexico. [From U.S. Geological Survey.]

The underground workings of the fabled Kimberley diamond mines of South Africa penetrate diatremes, which contain mantle materials (including diamonds*) and fragments of rock the magma encountered en route to the surface. Geologists view this pipe much as they would a 300-kilometer (180-mile) long drill core in which the rocks had been scrambled by eruption. Careful field and laboratory work on such scrambles of rock has enabled geologists to unravel their mysteries and deduce something of the layering of the Earth's crust and mantle well beyond the reach of the deepest drill hole.

Other Volcanic Phenomena

When a nuée ardente meets a river, it can be transformed into a landslide or mudflow of volcanic debris called a **lahar**. Lahars can also be produced when the wall of a crater lake breaks, suddenly releasing water, or when glacial ice is melted by lava flows or nuées ardentes. One formation in the Sierra Nevada contains 8000 cubic kilometers (2000 cubic miles) of material of lahar

---

*Diamond, the hardest substance known, is formed when carbon is subjected to high pressures, corresponding to depths of about 125 km in the Earth's mantle. Gem-quality diamonds only occur naturally, whereas small industrial diamonds are now produced synthetically.

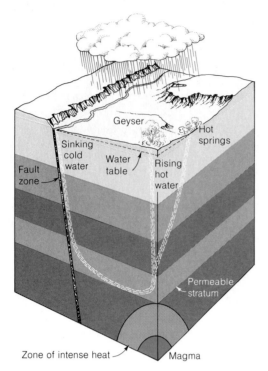

Figure 15-32
Surface water percolating down along fractures occasionally reaches the vicinity of a magma chamber.
The heated fluid rises through other channels to
erupt from geysers and hot springs as steam and hot
water. [After D. E. White, U.S. Geological Survey.]

Figure 15-33
Old Faithful Geyser, Yellowstone National Park. Hot
water and steam are violently ejected about once
every 65 minutes. Between eruptions hot water presumably fills underground cavities. Further heating
converts some of the water into steam, generating
the pressure that forces the discharge. [From U.S.
Geological Survey.]

origin. Lahars have been known to carry huge
boulders for tens of kilometers. When Kelut (Java)
erupted in 1919, the crater lake containing 380,000
cubic meters (14 million cubic feet) of water was
displaced, and the resultant lahar rushed down
the flanks killing some 5500 people.

Volcanic gases have been studied by coura-
geous volcanologists.* Water vapor is their main
constituent (70 to 95 percent), followed by carbon
dioxide, sulfur dioxide, and traces of nitrogen,
hydrogen, carbon monoxide, sulfur, and chlorine.
Enormous amounts of volatiles are released.
Parícutin (Mexico) emitted some 18,000 tons of
water in a single day.

The nature and origin of volanic gases are of
considerable interest and importance because
these gases formed the oceans and the atmos-

phere. Some volcanic gas may be derived from
deep within the Earth and may be making its way
to the surface for the first time (**juvenile gas**);
some may be recycled groundwater and ocean
water, atmospheric gas, or gas that had been
trapped in rocks.

Emissions of gas and vapor unaccompanied by
the eruption of lava or pyroclastic matter often
mark the late stages of volcanic activity. Circulating groundwater reaches and is heated by buried
magma (which retains heat for a long time), producing **hot springs** and **geysers** (Fig. 15-32). A geyser is a hot-water fountain that spouts intermittently with great force, frequently accompanied
by a thunderous roar. The best-known example is
Old Faithful in Yellowstone Park (Fig. 15-33),

*In 1938 two Russian volcanologists, V. F. Popkov and I. Z.
Ivanov, made measurements of lava temperatures and collected gas while floating down a river of lava on a raft of colder
lava for a distance of 2 kilometers. The surface temperature of
their raft was 300°C, and the lava river showed a temperature
of 870°C! The occasion was the eruption of Biliukai on Kam-
chatka. Pliny the Elder may have been the first scholar to lose
his life studying a volcano. He was killed in the great eruption
of Vesuvius, which destroyed Pompeii in A.D. 79.

which erupts irregularly, about every 65 minutes, sending a jet of hot water as high as 60 meters. **Fumaroles** are vents that emit gas fumes or steam. All of these volcanic emanations contain dissolved materials that precipitate as the water evaporates and cools, forming various sorts of encrusting deposits (for example, travertine), some of which contain valuable minerals.

CASE HISTORIES OF VOLCANOES

The stories of individual volcanoes make fascinating reading, not only for their descriptions of eruptions but also for their human interest. Some volcanoes have changed the course of ancient history—for example, Thera and Vesuvius.

Kilauea—An Instrumented Volcano

The giant shield volcano Mauna Loa and the smaller Kilauea on its eastern flank make up the southern half of the island of Hawaii. Because the U.S. Geological Survey operates a volcano observatory on the rim of Kilauea caldera, this volcano is perhaps the best studied in the world, and what has been learned from it has influenced our notions of volcanic processes profoundly. A modern network of seismographs (see Chapter 17), has been installed to provide information on the internal structure and activity of Kilauea. Tiltmeters (instruments that measure tilting of the ground) are used to observe the swelling and deflation of the volcano due to the accumulation and underground movement of magma. A geochemical laboratory follows the systematic variations in the chemistry and petrology of the lavas and the chemistry of the emitted gases.

The U.S. Geological Survey used these facilities to monitor the Kilauea eruption of 1959–1960. Between August 14 and 19, 1959, the seismographs detected a swarm of small earthquakes at a depth of 55 kilometers beneath Kilauea caldera. This marked the entrance of magma into the channels leading from the asthenosphere, through the lithosphere, to the surface. The upward migration of the magma could be traced by weak seismic disturbances originating at depths of 5 to 15 kilometers. Between August and October, the tiltmeters showed that the volcano was beginning to swell as the magma rose and filled a shallow reservoir below the caldera.

The first sign of an outbreak occurred in September, when seismographs picked up a series of small earthquakes. By November this swarm of

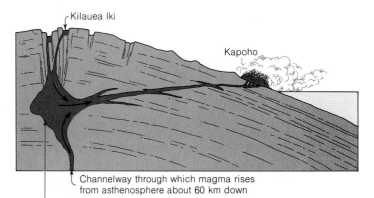

Figure 15-34
Schematic diagram of the 1960 eruption on the flank of Kilauea, which destroyed the village of Kapoho on Hawaii. In an earlier stage, lava filled Kilauea Iki crater on the summit, forming a lake 125 m deep, but more than half drained back into the vent. Following the cessation of activity at Kilauea Iki, the flank eruption at Kapoho, 45 km away, covered about 10 km² with lava. [After J. P. Eaton and R. G. Schmidt, U.S. Geological Survey.]

tremors exceeded a thousand per day, and the summit of Kilauea was swelling three times faster than in prior months. Then on November 14 the quakes increased tenfold in number and intensity, signifying that the eruption fissure was splitting its way to the surface. That evening the lava broke out of a kilometer-long fissure in the wall of Kilauea Iki, a crater just east of the caldera. The earthquakes stopped, but the seismographs recorded an almost continuous **harmonic tremor** (release of seismic energy) that typically accompanies the movement of lava. In the next seven days about 30 million cubic meters (1 billion cubic feet) of lava poured into Kilauea Iki, accompanied by spectacular lava fountains. The tiltmeters showed the volcano to be deflating as the reservoir emptied. The cycles of swelling and eruption repeated until December 21, when Kilauea Iki became dormant. The Geological Survey scientists expected more trouble, however, because the volcano began to inflate even more than it had in November. More lava was in the shallow reservoir below the caldera than had been there when the eruption began. The pressure would have to force open new rifts, and dikes of lava would split their way to the surface, signalled by earthquakes. In early January of 1960, a new swarm of earthquakes was detected, this time not far from the village of Kapoho, about 40 kilometers (25 miles) east of the caldera. Then on January 13, as expected, a flank eruption broke out (Fig. 15-34). It came through a crack about a

Figure 15-35
View of St. Pierre toward steaming Mont Pelée, after it was destroyed by a nuée ardente in the eruption of May 8, 1902. [Copyright 1902 by Underwood and Underwood. Library of Congress.]

kilometer long, a few hundred meters north of the village. Within the next four weeks, more than 100 million cubic meters (3.5 billion cubic feet) of lava poured out of the fissure, destroying Kapoho but causing no casualties. A new landscape was created as the lava flowed to the sea. Twenty-foot walls were built in a futile attempt to divert the lava and save a seashore community. When it was all over, the tiltmeters showed that the reservoir under the caldera had been drained in the Kapoho eruption. Confirmation was given by the fact that the amount by which the summit subsided was equal in volume to the lava erupted.

Geologists believe that the basaltic magma in the partially molten asthenosphere is lighter than the rock in the overlying lithosphere. The magma could therefore be squeezed to the surface by the weight of the lithosphere, if an escape route were available. A fracture under Hawaii taps this source of magma, and the lava rises to fill the shallow chamber under the caldera. The rest of the story is as given above.

Mont Pelée

The adjective **peléan,** taken from the name of this volcano, is used to describe violent eruptions and explosions of gas-charged and very viscous

magmas. It is aptly chosen, for on May 8, 1902, at 8:02 A.M., the Caribbean coastal town of St. Pierre below Mont Pelée on Martinique was destroyed by a nuée ardente, and all but two of its 28,000 inhabitants perished within minutes. Once the nuée ardente exploded laterally out of the mountain with very little warning, nothing could be done to save the panicked victims. A hot-gas hurricane avalanche plunged down the slopes at a speed of about 100 kilometers per hour and engulfed the town with a searing 800°C emulsion of gas (mostly CO_2), glass, and dust (Fig. 15-35). It is sobering to scientists who render advice to others to recall the statement of Professor Landes of St. Pierre's College, issued the day before the cataclysm: "The Montagne Pelée presents no more danger to the inhabitants of Saint Pierre than does Vesuvius to those of Naples." Professor Landes perished with the others.

The following is the account of an officer aboard a ship in the harbor, quoted by K. Wilcoxon:

> As we approached St. Pierre we could distinguish the rolling and leaping of red flames that belched from the mountain in huge volumes and gushed into the sky. Enormous clouds of black smoke hung over the volcano. . . . There was a constant muffled roar. It was like the biggest oil refinery in the world burning up on the mountain top. There was a tremendous explosion about 7:45, soon after we got in. The mountain was blown to pieces. There was no warning. The side of the volcano was ripped out and there was hurled straight toward us a solid wall of flame. It sounded like a thousand cannon.
>
> The wave of fire was on us and over us like a flash of lightning. It was like a hurricane of fire. . . . The fire rolled in mass straight down upon St. Pierre and the shipping. The town vanished before our eyes.
>
> The air grew stifling hot and we were in the thick of it. Wherever the mass of fire struck the sea, the water boiled and sent up vast columns of steam. The sea was torn into huge whirlpools that careened toward the open sea. . . . The blast of fire from the volcano lasted only a few minutes. It shrivelled and set fire to everything it touched. . . .
>
> Before the volcano burst, the landings of St. Pierre were covered with people. After the explosion, not one living soul was seen on the land.

In 1976 the volcano La Soufrière on the Caribbean island of Guadeloupe erupted with a series of worrisome phreatic explosions. Fearful of another Mont Pelée catastrophe, the French government ordered the evacuation of some 70,000 people at a cost of many tens of millions of dollars. A peléan eruption did not occur, however, and the inhabitants were permitted to return after several weeks. La Soufrière is being watched closely for signs of a dangerous turn—though tourists now come perilously close to the smoking summit.

Krakatoa

The 1883 explosion of Krakatoa, in the strait between Java and Sumatra, was one of the greatest ever witnessed. Krakatoa was an island formed from nested volcanic cones in an ancient caldera. The caldera, 6 kilometers across, was a remnant of a prehistoric andesitic stratovolcano. On August 27, after many smaller explosions, Krakatoa blew its top in a phreatic explosion with a TNT equivalent of about 100 megatons. The explosion was heard in Australia, nearly 2000 kilometers away, and atmospheric pressure waves were recorded on weather barographs on the other side of the Earth. Nearly 20 cubic kilometers (5 cubic miles) of debris was discharged into the air, much of it falling as ash over an area of some 700,000 square kilometers (300,00 square miles). Almost total darkness settled on Jakarta, 150 kilometers (100 miles) away when the dust blotted out the Sun. Fine dust rose to the stratosphere and drifted around the Earth, producing brilliant red sunsets for several years and lowering Earth's mean annual temperature a few degrees by partially blocking solar radiation. The explosion generated a **tsunami,** or giant sea wave that reached a height of almost 40 meters (130 feet) and drowned 36,000 people in nearby coastal towns. The sea waves were recorded on tide gauges as far away as the English Channel. After the eruption, most of Krakatoa had disappeared, leaving in its place a water-covered basin 300 meters (1000 feet) deep. It is believed that much of the energy was provided by the violent expansion of hot steam after the walls of the volcano first ruptured, letting sea water into the magma chamber. The result can be viewed as one of the biggest steam-boiler explosions in recorded history.

After forty-four quiet years, another cycle of volcanic activity began in 1927. A new island, Anak Krakatoa (Child of Krakatoa), is now building up by repeated eruptions and has reached a height of more than 100 meters (330 feet).

Mt. St. Helens

Long before the 1980 eruptions, geologists knew Mt. St. Helens to be the most active and explosive volcano in the contiguous United States. They could piece together a 4500-year long history of destructive lava flows, hot pyroclastic flows, lahars, and distant ash falls by examining the geologic record. Beginning March 20, 1980, a series of earthquakes under the volcano signalled the start of a new eruptive phase after 123 years of dormancy. The earthquakes moved the United States Geological Survey (USGS) to issue a formal "hazard alert." The first eruptive outburst of ash and steam from a newly opened crater on the summit occurred one week later. The following weeks saw dozens of new fumaroles and many phreatic explosions caused when groundwater and melt water from glaciers came into contact with hot rocks and gases inside the volcano. Up to this point no fresh magma was found in the ejecta, a reassuring sign that the volcano was not in its most dangerous phase. However, beginning in April, harmonic tremor showed up on the seismographs, indicating that magma was moving beneath Mt. St. Helens, and an ominous swelling of the northeastern flank was noticed. By April 23, the bulge had grown to 100 meters (328 feet) and was expanding at a rate of 1.5 meters (6 feet) each day. The USGS issued a more serious warning, and people were ordered out of the vicinity.

On May 18 the climactic eruption abruptly began (Fig. 15-36). A magnitude 5 earthquake apparently triggered a massive landslide on the large unstable bulge, forming a huge debris flow that plummeted down the mountain. At the same time, as if a plug had been removed, high-pressure gas and steam was released in a tremendous lateral blast that blew out the northern flank of the mountain. From his observation post 8 kilometers to the north, USGS geologist David A. Johnston must have seen the advancing blast wave when he radioed a last message: "Vancouver, Vancouver, this is it!" A northerly directed jet of superheated (500°C) ash, gas, and steam roared out of the breach with hurricane force, devastating a zone 20 kilometers (12 miles) outward from the volcano and 30 kilometers wide. A vertical eruption sent an ash plume to a height of 25 kilometers. The ash cloud drifted to the east and northeast with the prevailing winds, bringing darkness at noon to an area 250 kilometers (150 miles) to the east and depositing ash up to 10 centimeters (4 inches) deep on much of Washington, northern Idaho and western Montana. The force of the volcanic explosion was equivalent to a hydrogen bomb of about 30 megatons, 1500 times as powerful as the bomb that destroyed Hiroshima. The volcano's summit was destroyed, reducing its elevation by 350 meters (1150 feet).

The local devastation was spectacular: Within an inner 10-kilometer blast zone the thick forest

was denuded and buried under several meters of pyroclastic debris. Beyond this zone, out to 20 kilometers, trees were stripped of their branches and blown over like broken match sticks aligned radially away from the volcano. Twenty-six kilometers away the hot blast was so intense that it overturned a truck and melted its plastic parts; it seriously burned fishermen who survived only by jumping into a river. More than 60 others were killed by the blast and its effects.

The lahar, formed when the landslide and pyroclastic debris were fluidized by groundwater, melted snow, and glacial ice, flowed 28 kilometers (17 miles) down the valley of the Toutle River, filling the valley bottom to a depth of 60 meters (200 feet). Beyond this debris pile, muddy water flowed into the Columbia River, where sediments clogged the ship channel and stranded many vessels in Portland.

A lower level of activity continued for months after the cataclysmic eruption. Mt. St. Helens may go on erupting for months or years in its present episode of activity.

USGS scientists have enjoyed a much better record in predicting the behavior of Mt. St. Helens than did Professor Landes of Mont Pelée. In a report written two years before Mt. St. Helens erupted, USGS scientists D. R. Crandell and D. R. Mullineaux warned of the likelihood of an eruption this century and accurately predicted the extent and nature of the damage. The first edition of this book also cited Mt. St. Helens as a potential hazard. Since the major eruption in 1980, the USGS has continued to monitor the volcano and has successfully predicted several small eruptions using harmonic tremor and gas composition as indicators.

THE GLOBAL PATTERN OF VOLCANISM

As we observed at the outset of this chapter, most volcanoes occur in connection with plate boundaries (Fig. 15–2), 15 percent in belts of tension, where plates diverge, and 80 percent in belts of

Figure 15–36
The cataclysmic eruptions of Mt. St. Helens on May 18, 1980. (a) 8:27:00 A.M., view looking southwest. (b) 8:32:37 A.M. The north slope of the volcano collapses as an earthquake triggers a massive landslide and debris flow. (c) 8:32:41 A.M. High-pressure gas and steam explode horizontally out of the north face of the breach with hurricane force. (d) 8:32:51 A.M. Gas-steam jet extends outward leveling forests in its path. It was followed by a surge of pyroclastic flows and debris. [© Gary L. Rosenquist 1980.]

compression, where plates converge. To a lesser extent, volcanoes are also found within plates. Examples are Hawaii, which lies within the Pacific plate, and the eastern rift zone, which lies within the African plate. The types of lavas that erupt and consequently the styles of eruptions also correlate with the locations of volcanoes. Most lavas that issue from vents in oceanic divergence zones and from intraoceanic volcanoes are derived from basaltic magma. Where ocean plate collides with ocean plate, basalts and andesites predominate (see Fig. 15-3 for a review of the classification of lavas). Near the zone where an ocean plate converges against a continental margin, rhyolitic ignimbrites are found as well as basalt and andesite. If we are allowed to present heuristic arguments (reasonable ones, but without proof), make some broad generalizations, and not try to account for all details, this worldwide pattern can be interpreted in a consistent manner. But first we must say something about mafic, intermediate, and felsic magmas and their possible modes of origin.

Magmas—Where Do They Come From?

Our current views of the principal zones of the upper few hundred kilometers of the Earth are summarized in Fig. 1-12. The continents float like rafts on the rigid plates of the lithosphere. The continental crust is made up mainly of granite and other felsic rocks. Such rocks are almost entirely absent from the sea floor, the oceanic crust consisting mostly of the mafics basalt and gabbro. The crust is separated from the rest of the lithosphere by an abrupt change in chemical composition called **M discontinuity.** Beneath this discontinuity the lithosphere consists of ultramafic rock (even less silica and more ferromagnesian minerals than mafic), underlain by the weak, partially molten asthenosphere, which is also ultramafic. We will give more details about these zones in later chapters.

Basalt amounts to about 40 percent of all the rock in the Earth's crust and is the principal rock produced by volcanism. We surmise that basalt must be the molten component of the asthenosphere for several reasons. Recalling the case history of Kilauea on Hawaii, we know that the magma taps the asthenosphere and that basalt is the eruptive result. Peridotite is our best guess for the ultramafic rock that makes up most of the upper mantle. When peridotite is heated in the laboratory to the temperature at which melting begins, basalt is the first liquid to form.

The felsic lavas at the siliceous end of the scale of volcanic rocks, such as rhyolite, must have another origin. Their composition is so different from that of basalt that the location and nature of the parent rocks or the evolution of the magma must be different. That rhyolites are found on continents and on some islands, such as Iceland and the Azores, but not on the sea floor (where a felsic crust is absent) is an important clue. One way to make rhyolite is obvious—by completely remelting granitic continental crustal rocks to produce felsic magma. Alternatively, sediments derived from these rocks and deposited on the sea floor adjacent to continents could be reheated to yield felsic magma. One favorable environment for the formation of felsic magma would therefore be a convergence zone where the leading edge of one of the colliding plates is a continental margin. All the ingredients for "cooking up" felsic magmas are present there: heat, deformation, and an abundance of parent materials of the proper composition (see Fig. 1-16).

Felsic lavas could also be made by "refining" basaltic magma. Suppose a basaltic magma cools in a magma chamber until crystals just begin to form. The first mineral to crystallize is olivine, which sinks to the bottom, leaving a more felsic magma. This process of fractional crystallization and gravitational settling of heavy crystals can continue as long as new minerals crystallize. At each stage the remaining liquid, or **differentiate,** would have a different composition, depending on which minerals had formed and separated from it. In principle, the differentiation series represented by the transformation from basalt to andesite to rhyolite is possible, and it has been observed on some oceanic islands. It is doubtful whether rhyolite, the ultimate, most siliceous, differentiate, commonly originates from a parent basaltic magma, but an intermediate differentiate, andesite, is found on oceanic islands, where it undoubtedly formed as a differentiate from the much more abundant basalt. In fact, Icelandic geologists have shown that the longer the period between eruptions, the more felsic is the eruptive, since there is more time for fractional crystallization and settling to occur (see Chapter 14 for a discussion of fractional crystallization).

As was mentioned in the preceding chapter, andesite might also be the product of partial melting of mantle materials in the presence of water, perhaps derived from the uppermost layers of a subducting plate. Because andesites are roughly intermediate in composition between basalt and rhyolite, they can also result from a mixture of mafic and felsic magmas or from assimilation of surrounding granitic rock by a mafic magma.

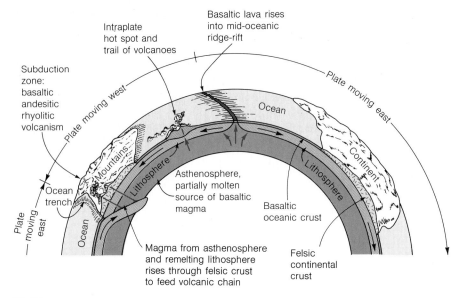

Figure 15–37
Schematic representation of volcanism on a global scale, showing the association with plate boundaries, particularly the regions of formation and destruction of the lithosphere.

The occurrence of large volumes of andesite in conjunction with subduction zones argues for still another source, perhaps the most important one. In recent years it has been suggested that a downgoing, cold oceanic plate begins to remelt when it plunges diagonally into the asthenosphere. Heat from the hotter asthenosphere and heat generated by friction along the contact between the upper surface of the descending plate and the overriding plate would supply the thermal energy for melting. Thus the basaltic crust of the old sea floor, enriched with silica by its clinging veneer of wet ocean-bottom sediments, could melt into andesitic magma, which would rise buoyantly to erupt from volcanoes in the overlying volcanic arc.

Ocean-Ridge Volcanism

A worldwide system of fissures, along which basalt erupts, coincides with the ridge-rifts depicted on the inside of the back cover and in the schematic section of Figure 15-37. The fracture between the separating plates extends to the asthenosphere. A mush of peridotite and basaltic magma rises buoyantly into the fracture to form new lithosphere. As the hot peridotite from the asthenosphere rises in the gap between the separating plates, the reduced pressure induces further melting, adding to the supply of basalt. The overflow forms ocean ridges, volcanoes, the basaltic seafloor crust and basaltic plateaus such as Iceland.

Intraplate Volcanism

Although most volcanic activity is concentrated at plate boundaries, volcanoes within plates can also produce enormous volcanoes of lava. An example is the island chain that begins with the active volcanoes on Hawaii and continues as a string of progressively older, extinct, eroded, and submerged volcanic ridges and mountains (Figure 15–38). Except for Hawaii itself, where earthquakes are triggered by volcanic activity, the chain is essentially aseismic (without earthquakes) and is called an **aseismic ridge,** different from the seismic mid-ocean ridges where seafloor spreading takes place.

Aseismic ridges, which also occur elsewhere in the Pacific and in the other large oceans, were difficult to incorporate into the framework of plate-tectonic theory. Jason Morgan of Princeton and J. Tuzo Wilson of Toronto introduced the concept of **hot spots** to explain aseismic ridges and other volcanic centers within continents. They propose that hot spots are the surface manifestations of jets or plumes of hot material that rise from deep within the mantle, drill through the lithosphere, and emerge as volcanic centers. These columnar currents are supposedly fixed in the mantle and do not move with the plates. As a result, the hot spot leaves a trail of extinct, progressively older volcanoes as the plate moves over it. If hot spots are indeed fixed in the mantle, the trail of volcanoes carried away from the hot spot provides a

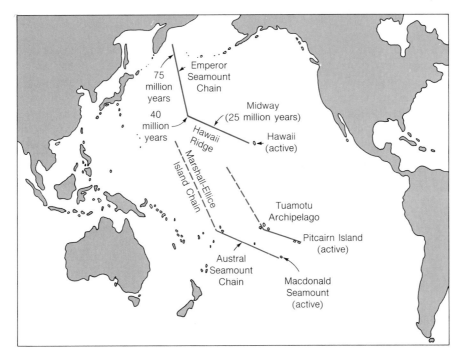

Figure 15–38
If hot spots are stationary, the "tracks" of the Hawaiian Island chain and the parallel Tuomotu and Austral chains would trace movement of the Pacific plate over the hot spots marked by the active volcanoes on Hawaii, Pitcairn Island, and MacDonald Seamount. All three chains would indicate the same motion of the Pacific plate; the bends in the chains formed 40 million years ago record a change in direction of plate motion at that time. Broken segments indicate inconclusive evidence of progressively older age sequence. [After "Hot Spots on the Earth's Surface" by K. C. Burke and J. T. Wilson. Copyright © 1976 by Scientific American, Inc. All rights reserved.]

powerful method of measuring the velocity and direction of plate motion (Figs. 15–37 and 15–38). The assumption that hot spots are immobile is now being examined very carefully by geophysicists. In a general way the hot-spot origin of the Hawaiian and Emperor Seamount Chains has been confirmed by deep-sea drilling which has established the progressively older periods of volcanism along the chains.

A second type of ocean-bottom formation, extinct undersea basaltic volcanoes rising kilometers or more above the sea floor, are common intraplate volcanic features. More than 10,000 are known in the Pacific Ocean alone. They are thought to have originated as active volcanoes near spreading centers and to have become extinct as the plate carried them away. Some initially projected above sea level but have been eroded to flat-topped cones by wave action. As the oceanic lithosphere spread, cooled, and contracted, the sea floor subsided, dropping the truncated volcanoes below sea level. These are the guyots mentioned in Chapter 10 and depicted in Figure 15–39.

Volcanic centers within continental plates take different forms. Fissure eruptions of basalt, like

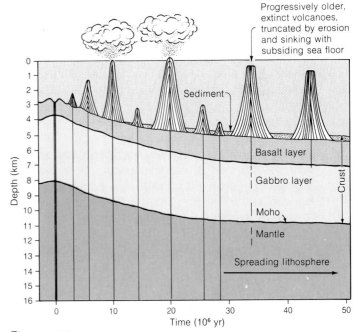

Figure 15–39
Guyots probably began as undersea volcanoes that grew near spreading centers. Many projected above sea level, became extinct, and were eroded down to flat-topped cones by wave action. The volcanoes rode along on the moving plate as they grew, were then truncated, and finally sank beneath the sea surface as the plate that carried them subsided. [After "The Deep Ocean Floor" by H. W. Menard. Copyright © 1969 by Scientific American, Inc. All rights reserved.]

those in the Pacific Northwest of the United States, suggest that a fracture penetrated the continental lithosphere and that the basalt spurted rapidly to the surface without much contamination from the felsic crust. Perhaps fractures developed in the early stages of an abortive spreading episode. Eruptions of basalt that mark the intial stages of continental rifting, or breakup, can be documented in several parts of the world. For example, basalt flows are found today in troughs in eastern North America bounded by ancient faults. These flows formed at the beginning of the breakup of Pangaea some 200 millions years ago. Basalt is also found in association with the rift valleys of East Africa—a feature that some geologists interpret as representing an abortive stage in the breakup of that continent.

The vast ignimbrite sheets, which are the rhyolitic counterpart of flood basalt, have no easy explanation. Perhaps magma rising slowly from the upper mantle assimilates large quantities of the siliceous granitic or sedimentary crust and in this way is transformed into a rhyolitic magma.

Convergence-Zone Volcanism

The many phenomena that occur where plates converge are now being sorted out by scientists. One of the main features of a region of plate collision is the chain of volcanoes parallel to the adjacent deep-sea trench that marks the edge of the overriding plate. Where two oceanic plates converge, an island arc builds up from the sea floor, typically by the extrusion of basalts and andesites. The basalts probably derive from the asthenosphere above the descending plate, and the andesites might come from the partial melting of the basaltic crust and the ocean-bottom sediments attached to the descending plate, as described earlier. The Philippine and the Marianas arcs are examples of this type of convegence zone. The growing island arc contributes erosion products and volcanic debris to the offshore region and beyond into the trench. In a later stage this pile can be compressed, heated, intruded, and uplifted to form new continental crust and mountains. The arc of islands that makes up Japan is a prototype of this process; Fujiyama is an andesitic cone rising within a growing arc. Another example is Mt. Pelée on Martinique, which squeezes out viscous felsic lavas and explosive nuées ardentes, which are deposited as ignimbrites.

When an ocean plate is overridden by a plate carrying a continent on its leading edge, as in Figure 15–37, an arcuate mountain chain is thrown up in the zone of compression near the continental margin. The Chilean trench and the adjacent Andes Mountains are examples of such a feature. Magmatic activity typically involves the ejection of huge quantities of ash and the eruption of andesitic and basaltic lavas. Basalt would occur if magma from the asthenosphere could reach the surface without being contaminated by sediments or the intervening felsic crust. The more felsic eruptives seem to indicate remelting of the subducted plate, contamination by sediments, or actual melting of continental crust. Mt. St. Helens, Parícutin (Mexico), Cosequina (Nicaragua), Irazu (Costa Rica), Cotopaxi (Equador), and Calbuco (Chile) are examples (see Fig. 15–2). The subduction of the small Juan de Fuca plate under the North American plate gives rise to the volcanoes of the Cascade Range, which includes Mt. St. Helens (Fig. 15–40).

The Growth of Continents

It is basic to the theory of sea-floor spreading that the sea floor is ephemeral: it is created by spreading at ridge-rifts and destroyed by subduction. Furthermore, continents and island arcs can be eroded, but the materials stay in the lithosphere because they are too light to sink back into the mantle. Continents grow on their leading edges by sedimentation and magmatism of the kind just described. Island arcs build up from the sea floor. According to the theory, ocean basins eventually close up; continent runs into continent, trapping arcs and sediment-filled troughs between them. Perhaps this mechanism also contributes to continental growth. Even the erosion products may eventually be recycled back to the continents after spending some time on the sea floor (see Fig. 1–16).

VOLCANISM AND HUMAN AFFAIRS

Can volcanic eruptions be predicted? This question is of real concern. Some 200,000 people have been killed by volcanoes in the past 500 years. The eruption of Mt. St. Helens in a relatively thinly populated area of the world killed over 60 people and did $1 billion worth of damage. With the increased concentration of population, the toll could go higher (Fig. 15–41). The geologic record shows that ignimbrite sheets have been deposited on a vast scale—something that humans have not yet recorded. If the Katmai (Alaska) eruption of 1912 had been centered in

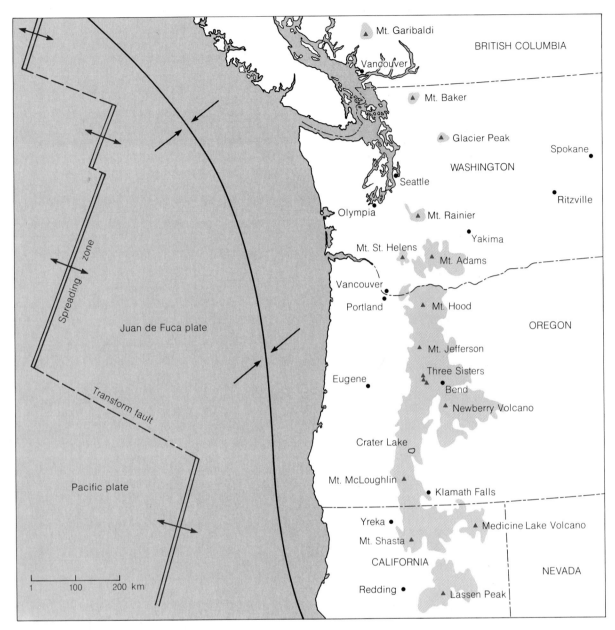

Figure 15-40
The small Juan de Fuca plate is sandwiched between the Pacific and the North American plates. Subduction of the Juan de Fuca plate under North America gives rise to the volcanoes (small triangles) of the Cascade Range, including Mt. St. Helens. [After "The Eruptions of Mt. St. Helens" by R. Decker and B. Decker. Copyright © 1981 by Scientific American, Inc. All rights reserved. Data from U.S. Geological Survey.]

New York (chosen only to show the scale of the event), the city would have been destroyed, one foot of ash would have covered Philadelphia, and fumes would have reached Denver. Imagine a major eruption of Mt. Ranier, not far from the Seattle-Tacoma metropolitan district!

Volcanoes kill and destroy in several ways: lava flows, nuées ardentes, and phreatic explosions and their consequences, such as tsunamis, ash falls, and lahars. A shield volcano like Kilauea,

whose eruptive style is the release of basaltic lava, has the best chance of successful prediction. The warning phenomena are many: earthquakes, ground tilt, seismic tremor. There may be no way to prevent Kilauea from damaging property, but it need not be a killer. Erupting basaltic lava usually flows slowly enough for people to get out of the way, and in some situations the lava streams might even be diverted from property. Perhaps the most successful attempt to control volcanic

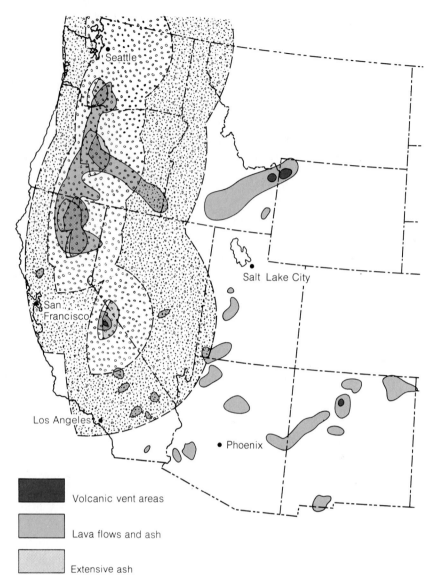

Volcanic vent areas

Lava flows and ash

Extensive ash

Figure 15-41
Volcanic hazards of the United States occur in Alaska, Hawaii, and the western states shown in the map. Dark brown zones are volcanic vent areas that have had extremely explosive and voluminous eruptions within the last 2 million years. Medium brown zones are subject to lava flows and small volumes of ash from groups of volcanic vents. Zones in light brown would get most of the ashfall from nearby relatively active and explosive volcanoes. The inner dashed line encloses areas subject to 2 in. or more of ash from a large eruption, and the outer dashed line encloses area subject to 2 in. or more of ash from a very large eruption. [D. R. Mullineaux, U.S. Geological Survey.]

activity was made on the Icelandic island of Heimaey during the eruption of January 1973. By spraying the advancing lava with sea water, Icelanders cooled and slowed the flow, preventing the lava from destroying homes and blocking the port entrance.

Volcanoes situated at plate boundaries, erupting gas-charged, felsic (hence viscous) magma and pyroclasts, are particularly dangerous. They explode, and the instant of the explosion is almost impossible to predict. The beginning of a period of dangerous activity might be foretold by seismographs, tiltmeters, temperature monitors, gas detectors, or changes in gravity and magnetic fields. Harmonic tremor and gas analysis have been good predictive indicators for Mt. St. Helens. The eruption of pyroclasts containing fresh magmatic glass rather than altered, reworked old material is a danger signal. At the first indication of activity, a local population, especially downwind of the volcano, might at least be alerted for possible evacuation. This, however, would be more like long-range or climatological forecasting than a weather report for the next day. Phreatic explosions must be treated similarly, for the danger zone can extend over hundreds of kilometers, especially if the volcano is one that, like Krakatoa, has a history of generating tsunamis.

Lahars occur in conjunction with eruptions near ice fields or bodies of water. The paths of lahars should be predictable, both from topography and recent history. Even though some minutes or hours of warning might be possible after an eruption has occurred, it seems better to prevent habitation of possible lahar paths below volcanoes.

Although it will be difficult, it may soon be possible to predict future eruptions of active volcanoes. But what of dormant or long-extinct volcanoes? Will they come to life suddenly—as Vesuvius and Mt. St. Helens did? Some inactive but potentially dangerous volanoes are Mt. Baker (Washington), Mt. Hood (Oregon), Mt. Shasta (California), and several other peaks in the Cascade Range of the Pacific Northwest. An even more difficult problem of prediction is posed by eruptions like that of Parícutin, which rose up from a flat farm field. Learning how to sense the movements of deep lava in relation to possible new outlets to the surface will be a real challenge.

Beneficial Aspects of Volcanism

We have seen something of the beauty of volcanoes and also something of their destructiveness. Volcanoes contribute to our well-being in many ways. The atmospheres and the oceans originated in volcanic episodes of the distant past. Soils derived from volcanic materials are exceptionally fertile. Emissions of volcanic rock, gas, and steam are also sources of important industrial materials and chemicals, such as pumice, boric acid, ammonia, carbon dioxide. The thermal energy of volcanism is being harnessed more and more. Most of the houses in Reykjavik, Iceland, are heated by hot water tapped from volcanic springs. Geothermal steam is exploited as a source of energy for the production of electricity in Italy, New Zealand, the United States, Mexico, Japan, and the U.S.S.R. Exploration for new geothermal energy concentrations is actively underway in the United States and Mexico.

SUMMARY
1. Volcanism occurs because molten rock inside the Earth rises to the surface, squeezed up by the weight of the overlying layers.

2. Lavas are classified as felsic, intermediate, or mafic based on decreasing silica content and increasing iron-magnesium mineral content. The chemical composition and gas content of lava are important in determining the form an eruption takes.

3. Basalt can be highly fluid and erupt in sheets from fissures, often building lava plateaus. Shield volcanoes grow from repeated eruption of basalt from vents. Silicic magma is more viscous and, when charged with gas, tends to erupt explosively. The pyroclastic debris piles up into cinder cones or covers extensive areas with ignimbrites or ash-flow sheets. Stratovolcanoes are built of alternating layers of lava flows and pyroclastic deposits.

4. Oceanic volcanoes are almost entirely basaltic. The formation of the sea floor from basaltic magma rising from the asthenosphere into fissures of the ocean ridge-rift system is, in terms of volume, the most significant form of volcanism. Intermediate and felsic eruptives tend to occur on islands and continental margins, in regions where plates collide. They may originate by remelting of the sinking lithosphere or by assimilation of felsic sea-bottom sediments or silicic continental crust in rising magma.

EXERCISES

1. The asthenosphere has been identified as a major source of magma. Why? What forces magma to rise to the surface?

2. What is the difference between magma and lava? Give some examples of types of lava and their coarse-grained, intrusive counterparts.

3. Describe the principal styles of volcanic eruptions and the deposits and landforms each style produces.

4. What is the association between plate boundaries and volcanism? Can the eruptive style and composition of volcanic deposits be correlated with plate boundaries?

5. About 6 cubic kilometers of basalt pours out on the sea floor each year (3 times the amount of lava that spews out on continents). Assuming constant outflow and a uniform spread over all the present oceans, how thick a layer would be laid down on the sea floor over 200 millions years? Obtain the area of oceans from Appendix III.

6. From the experience of the Mt. St. Helens eruptions, what new public policy initiatives would you recommend be undertaken (for example, in zoning, land use, insurance, warning systems, and public education)?

BIBLIOGRAPHY

Burke, K. C., and J. T. Wilson, "Hot Spots on the Earth's Surface," *Scientific American,* August 1976.

Crandell, D. R., and D. R. Mullineaux, "Potential Hazards from Future Eruptions of Mt. St. Helens," *U.S. Geological Survey Bulletin* 1383–C, 1978.

Decker, R., and B. Decker, "The Eruptions of Mt. St. Helens," *Scientific American,* March 1981.

Decker, R., and B. Decker, *Volcanoes.* San Francisco: W. H. Freeman and Company, 1981.

Green, J., and N. M. Short, *Volcanic Landforms and Surface Features.* New York: Springer-Verlag, 1972.

Heiken, G., "Pyroclastic Flow Deposits," *American Scientist,* September 1979.

Herbert, D., and F. Bardossi, *Kilauea: Case History of a Volcano.* New York: Harper & Row, 1968.

Peck, D. L., T. L. Wright, and R. W. Decker, "The Lava Lakes of Kilauea," *Scientific American,* October 1979.

Rosenfeld, C. L., "Observations on the Mt. St. Helens Eruption," *American Scientist,* v. 68, pp. 494–508, 1980.

U.S. Geological Survey, *Atlas of Volcanic Phenomena.* U.S. Geological Survey, Washington, D.C. 20242.

U.S. Geological Survey, *Volcanoes of the United States* (a pamphlet). U.S. Geological Survey, Washington, D.C. 20242.

Wilcoxon, K. H., *Chains of Fire—The Story of Volcanoes.* Philadelphia: Chilton Books, 1966.

16

Plutonism and Metamorphism

The structure and form of plutonic bodies and the metamorphic rocks that are associated with them give evidence of the processes of igneous intrusion and metamorphism by heat and pressure. Plutons take a variety of shapes and sizes, depending on the kinds of rocks intruded, the volume of magma, and the deformation accompanying emplacement: dikes, sills, laccoliths, lopoliths, and batholiths are all geological expressions of intrusion. Metamorphic rocks, associated with mountain belts and igneous activity, are of three main kinds: cataclastic rocks are mechanically broken and ground up rocks produced by high deformational pressures; contact metamorphic rocks are produced in the baked zones of high temperature that border intrusions; and regionally metamorphosed rocks are made by the combination of pressure and temperature operating over wide areas. As is true for igneous rocks, texture and mineralogy are the pressure gauges and thermometers of metamorphism.

Much is known of the mechanisms by which extrusive rocks are formed because the eruption of volcanoes and the ejection of lava have been open to direct study in many places at the Earth's surface. The mechanisms by which igneous and metamorphic rocks are formed at depth, however, can be studied only indirectly from evidence gathered long after the rocks have cooled. The forms that molten igneous masses take when they intrude the crust are deduced from direct evidence gained in geological field work done millions or billions of years after the rocks were emplaced. Similarly, metamorphism, the geological effects of regional increase of heat and pressure, has to be studied in the same way, long after the fact. The biologist dissects his specimen to see what goes on inside, but the geologist must, in a sense, wait for Earth to dissect itself. Field study of igneous and metamorphic rocks can proceed only where the crust has been uplifted and subjected to deep erosion. Thus the study of these

kinds of rocks is indissolubly linked with structural geology and the history of mountain belts.

Holes have been drilled into the Earth's crust in various places where heat flow and seismic activity indicate active igneous and metamorphic processes are now operating. One of these is under the Salton Sea, in southern California, where the North American lithospheric plate abuts the Pacific plate along the San Andreas transform fault. The temperatures in the drill hole are so high that superheated steam and water are produced, along with a great quantity of unusual chemical substances—including many metallic elements—all dissolved in water. From the chemical composition of the waters, geochemists have inferred that a magmatic body at depth is metamorphosing the sediments around it. Drill holes and mine workings are of chief importance, however, not in regions of current igneous activity, but in regions where they extend the reach of geologists to older rocks that are now cooled but

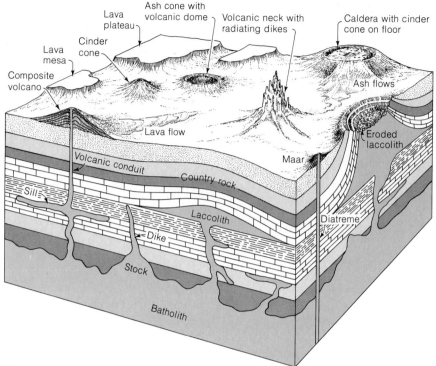

Figure 16-1
Plutonic and volcanic structures. [After R. G. Schmidt and H. R. Shaw, U.S. Geological Survey.]

still buried. There the details of structure, form, and composition are mapped as a guide to the igneous, metamorphic, and structural events that shaped the geologic history of the region. Such a history is necessary for further exploration for the many great mineral resources of the Earth that are formed by igneous and metamorphic activity.

We first turn to the many shapes and sizes of igneous intrusions. Their forms, dimensions, and distribution are the physical counterparts to the chemical processes of magmatic crystallization discussed in Chapter 14. Intrusions are linked to metamorphism as well as to structural deformation. The heat of intrusions causes some kinds of thermal metamorphism. The mechanical force of a magma invading rocks leads to structural deformation, which alone can metamorphose rocks. The three related processes, intrusion, metamorphism, and deformation, are ultimately tied to the origin of mountain belts at plate convergences. These relationships plus geologists' field observations of rocks and their structures can lead to reconstructions of past plate motions and their consequences.

PLUTONS

Large igneous bodies that have congealed from magma underground are called **plutons.** When

uplift and erosion expose them to study, or when they are examined in mines and boreholes, they are found to be variable in size and shape and in their relationship to the **country rock**—that is, the invaded rock surrounding them. Plutons include laccoliths, lopoliths, stocks, and batholiths. Figure 16-1 depicts some of these different kinds of plutons as well as the smaller igneous bodies, dikes and sills.

Magma rising through the crust makes space for itself in several ways. It can push its way upward by breaking off blocks of the invaded rock, which sink and are assimilated, that is, melted into the magma, at depth. **Stoping,** as this process is called (the name is borrowed from the mining term for the removal of ore by working upward), can heat the invaded rock enough to make it flow plastically out of the way, and it can forcibly split rocks apart or bow them up. The structure of overlying rocks often gives the clue to the process of intrusion.

Sills, Laccoliths, and Dikes

There are probably few open voids or cavities at depths much greater than eight or ten kilometers. The pressure due to the weight of the overlying rock would tend to close them. Since there are no vacant spaces for sills, laccoliths, and dikes to fill,

Figure 16-2
Diabase sills in Victoria Land, Antarctica. The concordant sill in the center of the mountain face sent discordant tongues into its floor and turned abruptly into a thick dike toward the left side. [Photo by W. B. Hamilton, U.S. Geological Survey.]

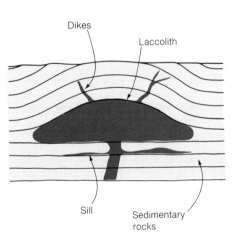

Figure 16-3
A laccolith intruded between sedimentary rock layers, bowing up overlying rocks in a domelike structure.

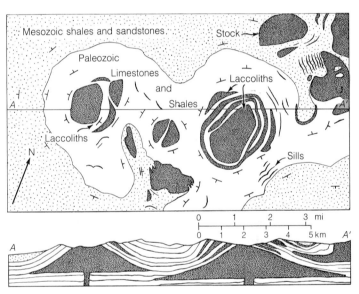

Figure 16-4
Map and cross section of laccoliths and stocks in the Judith Mountains, Montana. [After W. H. Weed and L. U. Pirsson, U.S. Geological Survey.]

these intrusions must make their own space by wedging open and penetrating cracks, joints, or planes of weakness, the force coming from the pressure that drives the magma. A **sill** is a tabular pluton that has been formed by injection of magma between beds of layered rock in a **concordant** fashion; that is, its boundaries are parallel to the layering whether or not that layering is horizontal. Sills range in thickness from a mere centimeter to hundreds of meters and can extend over considerable areas (Fig. 16-2). The Whin Sill in northern England is the type example. It averages about 25 meters (80 feet) in thickness and has an area of some 4000 square kilometers (1550 square miles). The 300-meter (1000-foot) thick Palisades

Sill, which overlooks the Hudson River in New York and New Jersey, is another example

Sills can be distinguished from sheets of extrusive rock because they lack the ropy, blocky, vesicular, or pillow structure of lava flows. Moreover, sills are typically coarser grained because of their slower rate of cooling, and the rocks above and below them show thermal effects, such as baking or bleaching (see Contact Metamorphism in this chapter).

Like sills, **laccoliths** are formed by the injection of magma along bedding planes of flat, layered rocks, but instead of being tabular they typically have a mushroom shape (Figs. 16-3 and 16-4). Unlike sills, laccoliths dome the overlying layers upward. Sills and laccoliths are both injected at high enough pressures to overcome the weight of the overlying rocks, which are lifted to accommodate the magma.

Dikes are tabular plutons that are discordant; that is, they cut across the layering of the country rock. Dikes can force open preexisting fractures, but more often their channels follow cracks opened by the pressure of magmatic injection. Some individual dikes can be traced aross country for tens of kilometers. Widths vary from centimeters to many meters (Fig. 16–5). When country

rock is deformed by intrusion, many cracks are formed, all of them possible channels for the invading magma. For this reason, dikes rarely occur alone; more typically, large numbers, or swarms, of hundreds or thousands of dikes are found in a region.

Some dikes crop out in roughly the form of circles or ellipses. Called **ring dikes,** these features are thought to be erosional remnants of intrusions that filled cylindrical fractures. Supposedly, a fracture of this sort forms when a crustal block with roughly circular section sinks into a depleted magma chamber. The resulting ring dike bounds a circular area of subsidence. Ring dikes up to 25 kilometers in diameter have been found.

Lopoliths and Batholiths

These are more massive intrusions, generally formed deeper in the crust than the plutons discussed in the preceding section. A **lopolith** is a large, floored intrusive whose center has sagged downward, both roof and floor, to form a bowl-shaped body (Fig. 16–6). The Duluth lopolith, the type example, is a huge intrusion of gabbro. It crops out on both sides of Lake Superior's western end and is inferred to continue beneath the lake. It is estimated to be 250 kilometers across, 15 kilometers thick, and to have a volume of about 200,000 cubic kilometers. Geophysical data suggest that, buried beneath younger rocks, it extends all the way to Kansas. If so, it is many times larger than estimated above.

Batholiths, the largest of plutons, are discordant intrusives that are, by definition, at least 100 square kilometers in area (Fig. 16–7). Similar, but smaller, discordant plutons are called **stocks.** The bottoms of most batholiths are unexposed and

Figure 16-5
Dikes of dark igneous rock cutting across biotite gneiss. Gunnison County, Colorado. [Photo by W. R. Hansen, U.S. Geological Survey.]

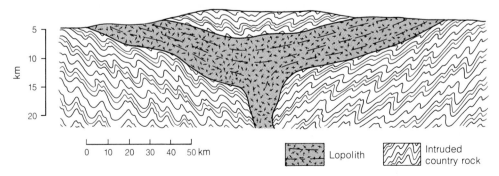

Figure 16-6
Cross section of a lopolith, a large saucer-shaped intrusion with a central feeder channel. Lopoliths are composed of mafic rocks, sometimes layered intrusions. Borders may be concordant or discordant.

hence unplumbed by the field geologist. Whether their cross sections remain constant or increase with depth is known only by geophysical inference, but many are estimated to extend perhaps 10 to 30 kilometers into the crust. Some batholiths seem to be less thick; the Boulder batholith, for example, is no more than 15 kilometers deep and may be much less.

Lopoliths and batholiths are generally coarse-grained—a consequence of the slow cooling that deeply buried bodies experience. Most lopoliths are differentiated, showing conspicuous zones or bands of contrasting minerals. During the slow crystallization that takes place in these igneous masses, crystals become segregated; the heaviest and first to form sink through the melt under gravity to a level governed by their density (Fig. 16-8; see also Chapter 14). Convection currents in the cooling magma might also influence the mineral stratification by sorting out groups of minerals.

Some batholiths have sharp contact with country rock and show flow structures in which elongated, needle-shaped crystals are arranged in parallel alignment, like logs aligned parallel to the flow of a river. These features imply that the batholiths were formed by injection of an igneous magma. Other batholiths grade into country rock without sharp contacts and show structures that vaguely resemble those of sedimentary rocks. Such features suggest that preexisting sediments were **granitized**—converted in place to granite by partial melting and invasion by hot solutions and gases percolating up from great depths. The origin of batholiths, linked with the origin of granite (Chapter 14), is one of the most durable of controversies; it is still hotly debated. Geologists today believe that some plutons are the result of igneous intrusions and that some are produced by granitization of preexisting rocks in place.

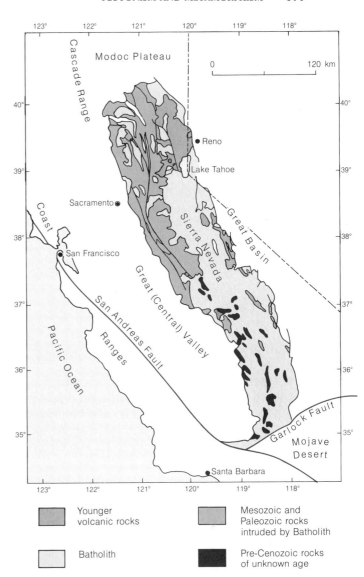

Figure 16-7
Generalized geological map of the Sierra Nevada and adjacent areas, showing the extent of this large batholith, now exposed at the surface by subsequent erosion. [After P. C. Bateman and J. P. Eaton, *Science*, v. 158, p. 1407, 1967. Copyright 1967 by the American Association for the Advancement of Science.]

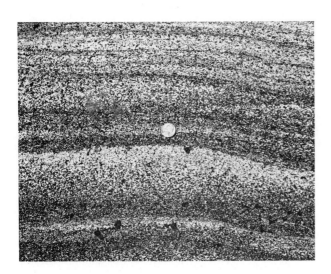

Figure 16-8
Banded structure, or layering, in granitic intrusion, possibly due to gravity settling. Sierra Nevada batholith. [From P. C. Bateman, L. D. Clark, N. King Huber, J. G. Moore, and D. Rinehart, U.S. Geological Survey.]

Batholiths are found in the cores of many mountain ranges (see Fig. 14–9). Moreover, immense regions of exposed Precambrian rocks in continental interiors are made up of successions of overlapping batholiths, many of which show the characteristics of roots of ancient mountains long since removed by erosion. All of this implies a connection between large-scale plutonism and the mountain-making process—a connection that will be further explored in Chapter 20.

The structures and compositions of all plutons are related to the pressures and temperatures where they were intruded. From the combination of pressure and temperature we can estimate the depth in and thermal regime of the crust. The mineral assemblages and textures of metamorphic rocks give us more clues to the pressures and temperatures in the crust.

METAMORPHISM

What happens to a rock when it is baked and squeezed in the Earth? The change in mineral composition and texture may be slight or so great that the rock is completely altered in appearance. The degree of change depends mainly on the pressures and temperatures to which the rock was subjected. One good place to see the effect of heat alone is in an outcrop where a shale has been intruded by a dike or sill. At some distance from the intrusion, the shale begins to show a loss of bedded structure, and its clay minerals have recrystallized to micas. At the contact with the igneous rock, the shale has lost all its original texture. It is now a coarsely grained, crystalline rock containing pyroxenes and anhydrous (lacking water) aluminosilicate minerals like andalusite, Al_2SiO_5, the product of **contact metamorphism.** The conclusion is inescapable that the high temperature of the intrusion caused the disappearance of clay minerals, which were originally formed by low-temperature weathering processes, and the appearance within the rock of new minerals, such as pyroxenes, that are characteristic of high temperatures.

Along fault planes, where rock is sheared and ground up to a paste, the transforming effect of pressure and tectonic movement alone on rocks is obvious. The mineralogy may be unaffected even though the rock has been pulverized (Fig. 16–9). These **cataclastic** rocks, the mechanically fragmented metamorphic products of intense folding and faulting associated with a variety of tectonic structures, are the products of **dynamic metamorphism.**

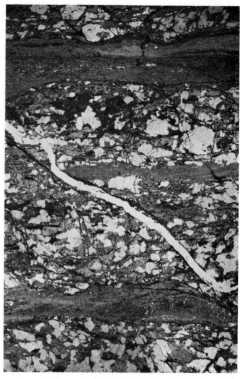

Figure 16–9
Photomicrograph in polarized light of a cataclastically deformed rock, a gneiss taken from a fault zone. Bands of coarse, granulated gneiss with large, light crystals alternate with darker bands of finely granulated mylonite. Gilpin County, Colorado. [Photo by W. B. Braddock, U.S. Geological Survey.]

The great bulk of metamorphic rocks have been subjected to both high temperature and high pressure over large regions, with the consequent destruction of original igneous or sedimentary structures and the growth of new minerals. These rocks are the products of **regional metamorphism** rather than of localized reactions along fault planes or the borders of igneous intrusions, with which cataclastic and contact metamorphic rocks are associated (Fig. 16–10). Regionally metamorphosed rocks are found in the cores of deeply eroded, folded, and intruded mountain belts in company with granitic batholiths. The formation of new minerals goes hand in hand with structural deformation, and both produce characteristic textures.

METAMORPHIC TEXTURES

How the crystals in a metamorphic rock have been rearranged is an indicator of the processes that produced the new rock from its sedimentary or igneous parent. The grain and bedding textures

Figure 16-10
Structurally deformed and regionally metamor-
phosed Paleozoic rocks in eastern Massachusetts.
[Photo by R. Siever.]

Figure 16-11
Slaty cleavage in a schist. [Photo by W. B. Hamilton,
U.S. Geological Survey.]

in detrital sediments are mainly the products of
fluid flow regimes, and the crystalline textures of
igneous rocks and some chemical sediments are
the result of primary crystallization from a melt
or solution. The textures of metamorphic rocks, in
contrast, are the result of recrystallization or conversion of one mineral to another in the solid
state.

Foliation and Fracture Cleavage

In a large variety of deformed metamorphic rocks
the most conspicuous textural feature is the set of
parallel planes that generally cut the rocks at an
angle to the bedding of the original sediment,
though these planes may coincide with bedding in
some places. This **foliation** may be so weak as to
be barely seen, or it may so dominate that it is the
only visible structure. Many shales, compacted
and hardened by diagenesis, part fairly easily
along bedding planes, a property called **fissility**
that has some resemblance to foliation. As many
such shales are traced into structurally deformed

mountain belts, they become transformed into
slates, harder rocks that typically show a **fracture
cleavage.** This variety of foliation is named from a
tendency to cleave or break into moderately thin
sheets at more or less regular intervals. The cleavage is not only at an angle to the bedding but
sometimes can be seen to offset it, as a microscopic fault, by a small amount at each fracture.
The cross-cutting relationship is evidence of the
deformational origin of the foliation. The almost
perfect cleavage of some slates makes them ideal
for use as roofing tile. Others that cleave into
thicker sheets are more suitable for flagstones.
The cleavage is sound textural evidence of metamorphism.

If the slate can be traced into an even more deformed region, the foliation is found to become
finer and more all pervasive, the rock showing
flow or **slaty cleavage** (Fig. 16-11), characterized
by very thin sheets along which platy minerals,
such as mica, have grown parallel to the foliation
planes. Slaty cleavage is **penetrative;** that is, it
occurs throughout the body of a rock, as opposed
to fracture cleavage, which is **nonpenetrative,** de-

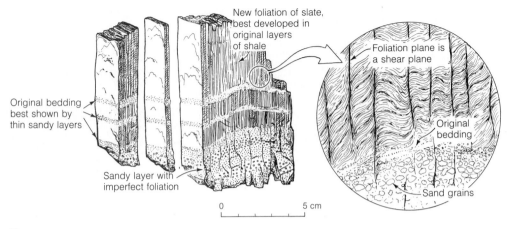

Figure 16-12
Fragments of slate (left) showing foliation (vertical lines) and relics of the original bedding. The enlargement shows small faultlike offsets of the bedding along the surfaces of foliation of the cleavage. [From *Principles of Geology* (4th ed.) by J. Gilluly, A. C. Waters, and A. O. Woodford. W H. Freeman and Company. Copyright © 1975.]

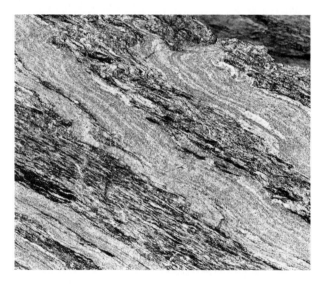

Figure 16-14
A banded gneiss of metamorphosed sandstone and schist. The darker beds are a mixture of flattened lenses of dark biotite schist and lighter gray quartz-feldspar rock similar to the lighter metamorphosed sandstone layers. Great Smoky Mountains, South Carolina. [Photo by J. B. Hadley, U.S. Geological Survey.]

Figure 16-13
Intricately folded veinlet of quartz cutting across planes of schistosity in a mica schist. [Photo by J. C. Reed, U.S. Geological Survey.]

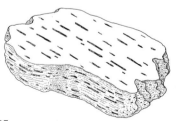

Figure 16-15
Lineation is a metamorphic texture formed by parallel orientation of prismatic or needle-shaped crystals of such minerals as pyroxenes or amphiboles. Viewed end on, the needles appear as dots.

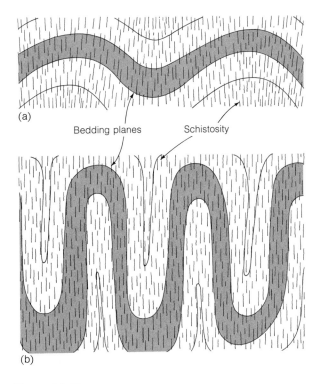

(a)

Bedding planes   Schistosity

(b)

Figure 16-16
(a) In moderately folded bedded rocks, the angle between schistosity and bedding planes varies from the limbs to the apex of a fold. (b) In tightly folded rocks, the schistosity is essentially parallel to bedding over most of the fold. In both cases shown, the planes are perpendicular to the page.

Figure 16-17
A hornfels breccia. Darker fragments of hornfels were broken during the intrusion and recemented as a friction breccia with a lighter hornfels. Both kinds show a granular texture and absence of foliation. [Photo by H. C. Granger, U.S. Geological Survey.]

veloped only in parts of the rock, as along certain layers (Fig. 16-12). Growth of crystals to larger size, visible to the naked eye, accompanied by some segregation of minerals into lighter and darker bands, produces **schistosity,** the coarse foliation that defines **schists** (Fig. 16-13).

The coarsest foliation is shown by **gneisses,** in which coarse bands of segregated light and dark minerals are prominent throughout the rock (Fig. 16-14). There is little or no tendency to part or split along these bands, and there are relatively few platy minerals parallel to the bands.

Amphiboles form needlelike crystals in many metamorphic rocks. In some the needles show a marked **lineation,** or parallel alignment (Fig. 16-15). Lineation can also be formed by the intersection of cleavage and bedding planes. In schists this texture is oriented parallel to foliation. Where foliation is not well developed, the lineation may be the most prominent textural feature. Good foliation is most common in the mica-bearing slates and in schists formed from shales. Good lineation is best developed in rocks of mafic composition, which metamorphose to form large numbers of elongate amphibole crystals.

Foliation, lineation, and other metamorphic textures are the product of **preferred orientation** of crystals, the pronounced tendency toward parallel alignment of crystals. The orientation is related to the directions of the compressional forces of deformation responsible for crystallization or recrystallization. In simplest relation to folded structures, mica flakes grow with their shortest dimension perpendicular to the major compressional force. The plane of the flake or sheet is parallel to (and is the main cause of) the cleavage or foliation, which is itself parallel to the plane of the folding, as shown in Figure 16-16 (see also the section Folds in Chapter 20). Preferred orientation probably results from a combination of crystal growth in the preferred direction, some plastic deformation or flowage of an existing crystal, and the physical rotation of a crystal during deformation.

Mineral Textures

Because most contact metamorphic rocks have undergone little or no deformation, they lack foliated structures. Their texture is nondirectional; platy or elongate crystals are randomly oriented. Contact rocks of this type are called **hornfels,** and the texture is **hornfelsic** (Fig. 16-17).

Figure 16-18
Feldspar porphyroblasts and thin quartz veins in biotite schist. The light-colored feldspar crystals have grown large at the expense of the surrounding fine-grained matrix. [Photo by J. E. G. W. Greenwood, U.S. Geological Survey.]

Figure 16-19
Augen gneiss. The augen (eyes) are large white or light-gray potassium feldspar crystals that grew as porphyroblasts and were then sheared and abraded as the rock was structurally deformed. [Photo by H. L. Foster, U.S. Geological Survey.]

In both contact and regionally metamorphosed rocks, new minerals may grow into large crystals surrounded by a much finer-grained matrix of other minerals. These **porphyroblasts** form in the solid state from the chemical ingredients of the rock, and they grow at the expense of the matrix surrounding it (Fig. 16–18). Porphyroblasts look something like **phenocrysts,** the large crystals in mainly fine-grained igneous rocks, but the textural relations are the opposite. Porphyroblasts grew after the main rock matrix was formed, whereas phenocrysts formed first, and the matrix crystallized around them.

Some metamorphic rocks have developed an interlocking mosaic of crystals with no outstanding preferred orientation or foliation. **Marbles,** metamorphosed limestones, may develop this texture with some dark mineral banding. Similarly altered silicate rocks are called **granulites** to define their textures, though some geologists use the term to imply specific mineral compositions.

Cataclastic Textures

The shearing and grinding associated with faulting and intense folding produce a broken, granulated texture in which preexisting crystals are pulverized and strung out in bands or streaks. Coarse-grained rocks of this type, **friction**

breccias, consist of particles ranging from a millimeter to a meter in diameter (Fig. 16–17). The finer-grained, more pulverized variety, **mylonite,** is composed of grains 0.01 to 0.1 millimeter in diameter. A distinctive texture, **augen** (from the German for "eye"), is formed by the shearing and abrasion of large porphyroblasts. They give the name to a strongly banded, sheared **augen gneiss** (Fig. 16–19).

All of these foliation and mineral textural details serve as evidence of the pressure changes that took place during or after heating and of the rapidity of heating during metamorphism. Though our knowledge of the mechanisms by which all of the many textures are formed is still far from complete, the textures themselves nevertheless constitute some of the best information that the field geologist has for interpreting the extent to which pressure and temperature increased and in what sequence. They are also indispensable tools of the trade for structural geologists as they unravel the patterns of folding and faulting.

REGIONAL METAMORPHISM

Geologists have learned to use minerals as pressure gauges and thermometers. This practice has been well developed in the study of metamorphism. By mapping characteristic **index minerals**

| | Greenschists | Amphibolites | Pyroxene granulites |

Chlorite

White mica (mainly muscovite)

Biotite

Garnet

Staurolite

Kyanite

Sillimanite

Albite (sodium plagioclase)

Quartz

(a)

Increasing metamorphic grade →

Chlorite isograd — Chlorite rocks — Biotite isograd — Biotite rocks — Garnet isograd — Garnet rocks — Staurolite isograd — Staurolite rocks — Kyanite isograd — Kyanite rocks — Sillimanite isograd — Sillimanite rocks

(b)

Figure 16-20
(a) Changes in mineral composition of shales metamorphosed under conditions of intermediate pressure and temperature. (b) Idealized map view of a regionally metamorphosed terrain in which shales have been metamorphosed under the same conditions. The isograd lines mark the first appearance of the index mineral and correspond to the diagram in (a).

in the field, geologists were able to define broad zones or belts of metamorphism, ranging from least to most intense. On a typical traverse from weakly to strongly metamorphosed shales, slates, and schists, one may go from detrital micas and other unchanged sedimentary minerals in the shale toward the metamorphic rocks and encounter the first new mineral, chlorite, in the slates. Moving in the direction of increasingly intense metamorphism, one may meet first biotite and then garnet in the schists, followed by staurolite, kyanite, and finally sillimanite, the schists becoming progressively more coarsely foliated. Lines on a map connecting points where an index mineral first appears are **isograds,** signifying a change in metamorphic grade (Fig. 16-20). The pattern of isograds on the map tends to follow the structural grain of the region (Fig. 16-21). The isograds are rough measures of pressure and temperature. More precise guides are given by mineral assemblages, two, three, or many minerals found in a rock, whose textures indicate that they were formed at the same time in chemical equilibrium. These assemblages are calibrated by reference to laboratory experiments. Metamorphic minerals do not everywhere appear in exactly the same

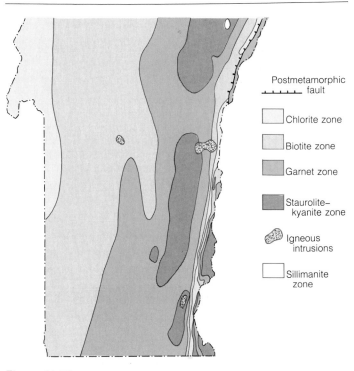

Postmetamorphic fault

Chlorite zone

Biotite zone

Garnet zone

Staurolite-kyanite zone

Igneous intrusions

Sillimanite zone

Figure 16-21
Metamorphic zones in the southern part of Vermont. The trend of those zones generally parallels the structural lineaments of the terrain. [Simplified from "Centennial Geologic Map of Vermont." Copyright © 1961, State of Vermont.]

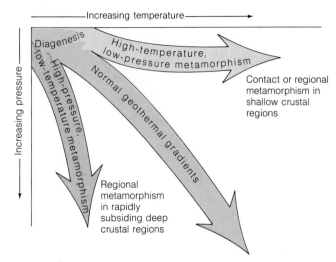

Figure 16-22
Different metamorphic routes of pressure and temperature increase may be induced by combinations of tectonic and igneous activity. The different routes produce different groups of metamorphic rock types.

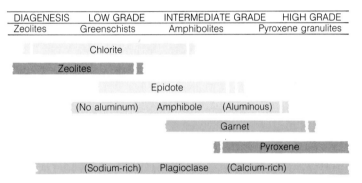

Figure 16-23
Changes in mineral composition of basalts and the other mafic rocks metamorphosed under conditions of intermediate pressure and temperature. Compare these with the mineral assemblages of shales metamorphosed under the same conditions (Fig. 16-20) to see the effect of original composition on mineralogy.

sequence as metamorphic grade increases because pressure and temperature may not increase at the same rate as metamorphism becomes more intense. Thus, in some areas, high pressure may have built up with relatively little increase in temperature; in other places, temperature may have increased greatly with relatively little pressure increase (Fig. 16-22). Most common are the regionally metamorphosed rocks that have formed by a combination of moderately high temperature and pressure. This kind of metamorphism is a more severe version of the mild changes caused in sediments by diagenesis during moderate and deep burial. The dividing line between diagenesis

and metamorphism is thus more or less arbitrarily drawn at low temperatures and pressures (see Chapter 11). Metamorphic rocks formed at high temperature and pressure may later be subjected to another set of metamorphic conditions, perhaps as a part of a second orogeny. If the second episode is at a lower temperature and pressure, the rocks will be lowered in metamorphic grade, a process called **retrograde** metamorphism.

The course of metamorphism is partly determined by the chemical (and thus mineralogical) composition of the original rock. The examples given above were drawn from shales. Volcanic rocks rich in mafic minerals show a different progression of associated minerals with increasing metamorphic grade (Fig. 16-23). The lowest-grade rocks contain a variety of **zeolite** minerals, complex hydrous aluminosilicates formed by the alteration of mafic volcanic glasses. The next higher grade, **greenschist,** contains chlorite (a sheet silicate), epidote (an aluminosilicate), and actinolite (an amphibole)—all minerals containing much iron, magnesium, and calcium—plus albite (sodium plagioclase feldspar). Following the greenschists in grade are the **amphibolites,** characterized by hornblende (an amphibole), sodium-calcium plagioclase feldspar, and garnet. The highest metamorphic grade of mafic volcanics includes the **pyroxene granulites,** containing pyroxenes and anorthitic plagioclase. Although the textures are different, this metamorphic mineral assemblage does not differ greatly in mineral composition from gabbro, the intrusive equivalent of basalt.

If a mafic volcanic is subjected to high temperatures but relatively low pressures, pyroxene granulites are the common product. In metamorphic belts where pressures are high and temperatures low, the typical metamorphic rock is a **blueschist,** characterized by glaucophane, a blue amphibole. At extremely high pressures and variable moderate to high temperatures **eclogites** form. These rocks, which many geologists think are characteristic of many parts of the earth's mantle, are pyroxene-garnet rocks.

Many of the same mineral groups are represented in all of these varied metamorphic rocks. The amphiboles, pyroxenes, and garnets exist as many species of variable chemical composition. Each is more or less characteristic of a different rock type. It is this range of compositions that makes these mineral groups so useful in applying the results of experimental synthesis of metamorphic rocks at controlled temperatures and pressures, for the mineral compositions are strongly affected by pressure and temperature as well as by the total chemical composition of the rock. The

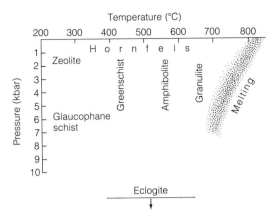

Temperature (°C)

Figure 16-24
Generalized and simplified metamorphic facies diagram, showing the distribution of metamorphic rock types in relation to the temperature and pressure fields in which they are formed. There are no sharp boundaries between any of these facies.

experimental work gives the basic pressure-temperature data. Metamorphic geologists first obtain samples of rocks in the field, noting carefully where they were collected, and then take them to the laboratory, where they analyze the exact composition of each mineral by a number of methods. The most powerful techniques makes use of the electron microprobe to give the chemical analysis of a small area of a single crystal only a few thousandths of a millimeter in diameter. From this field and laboratory information, a geologist can plot pressure-temperature lines on a map to show the direction in which the intensity of metamorphism increased or decreased.

All of this work on pressures and temperatures of metamorphic rock formation resulted from a concept enunciated in 1920 by a great Finnish metamorphic geologist, Pentti Eskola. He drew a chart on which groups of different metamorphic rocks—which he called metamorphic facies—could be related to pressure and temperature. Figure 16-24 is a modified version of his chart. There are no exact boundaries because the rock types depend partly on the total, or bulk, chemical composition of the rock. This idea has been one of the most durable in metamorphic geology and remains the cornerstone of modern interpretation.

Current usage tends to define metamorphic grades in somewhat the same way as metamorphic facies, on the basis of temperatures and pressures as indicated by changes from one mineral assemblage to another. In general an increase in pressure also increases the temperature necessary for mineral transformations. Assemblages below about 300°C are referred to as very low grade; those between about 300 and 500°C, as low grade;

Figure 16-25
A veined gneiss of migmatite origin. The darker layers are strongly metamorphosed sedimentary rocks. The lighter layers are quartz-feldspar rock introduced by invasion of partly molten material. [Photo by J. B. Hadley, U.S. Geological Survey.]

and those between about 500 and 600°C as medium grade. High grades, those above 600°C, merge into rocks that have been partially melted. Rocks of this high grade were formed at extreme depths in the crust, and their intermediate character between extreme metamorphic and igneous rocks has been difficult to interpret.

Extreme Metamorphism

The most far-reaching effects of extreme metamorphic transformation can be seen where erosion has cut deeply into the crust in such ancient mountains as the Adirondacks of New York. The rocks are badly deformed, stretched, and contorted and show much evidence of having flowed as plastic solids. In addition, the rocks are shot through with veins and intrusions of igneous rock. In such country it is difficult to say which rocks are metamorphic and which are igneous, particularly for a kind of veined gneiss frequently called **migmatite** (Fig. 16-25). This term is used for rocks of mixed origin. Such a rock is partly metamorphic in that it has been converted to its present condition largely as a solid, but it is also partly igneous in that parts of it have at times been molten. Arguments have raged over the origin of these rock types for 50 years, for many of them look remarkably like intrusive granites.

Some veined gneisses are truly intrusive, formed by the injection of magma in veins that show contact metamorphic margins or evidence of having forced the surrounding rock apart. Others seem to have been transformed at high temperatures in the solid state by the diffusion of elements, an exchange process in which individual atoms or ions migrate slowly through the crystalline material. In this process elements like sodium and potassium replace others like calcium and magnesium and convert the rock to a granitic composition. The rocks may retain their original textures, and there may be no evidence that they have disrupted the surrounding rock, a combination of features that suggests they are products of recrystallization rather than melting. More problematic gneisses seem to have formed by partial melting, and others seem to be extreme examples of the kind of mineral segregation into bands that is typical of lower-grade schists and gneisses. Many geologists believe that large volumes of batholithic granites are migmatites rather than true intrusives. Careful, detailed mapping coupled with analysis of the pressures and temperatures of mineral formation is needed to evaluate the origin of these rocks in any particular area.

CONTACT METAMORPHISM

Igneous intrusions are surrounded by margins of altered rock called **aureoles.** The width and nature of an aureole depends on the temperature of the intrusive magma and on the depth in the crust where the intrusion took place, which determines the pressure and temperature of the rocks being intruded. Aureoles are most prominent around plutons injected at temperatures of about 1000°C in shallow crustal rocks only a few kilometers from the surface. In such a situation, the temperature falls off rapidly away from the magma contact. Magmas intruded deep in the crust are not that much hotter than the surrounding rocks, and aureoles are obscured by general regional metamorphism.

Contact aureoles have sequential zones of index minerals somewhat like regional metamorphic isograds. The particular minerals formed by recrystallization are strongly influenced by both temperature and the bulk chemical composition of the rock. Thus limestones show patterns much different from those of shales, because their compositions are so different. Chemical compositions may change locally as ions in the country rock are exchanged with those of the intrusion in the aureole. Pressures are generally low to moderate. The

zones may be as narrow as a few centimeters adjacent to small dikes and sills or as much as several kilometers wide bordering a large granite pluton.

The metamorphism of limestone to marble in aureoles leads to a number of changes in the mineralogy of the rocks, many of them involving clay or silica impurities in the calcite or dolomite of the sediment. One of the minerals formed by such metamorphism is wollastonite (calcium silicate), a product of the reaction between silica impurities (in the form of quartz silt or chert) and the calcium carbonate of the limestone:

calcite + silica →

$CaCO_3$ SiO_2

wollastonite + carbon dioxide

$CaSiO_3$ CO_2

The carbon dioxide gas this reaction produces generally escapes. Though silica and calcium carbonate have no tendency to react at low temperatures, we know from experiments at near-surface pressures that at around 500°C they do combine. Like many other reactions, the temperature at which these substances begin to react rises as the pressure increases.

If the rock contains dolomite, the mineral diopside (calcium-magnesium pyroxene) forms in the same way as wollastonite (Fig. 16–26). If diopside forms near the contact, it gives way farther out to tremolite (calcium-magnesium amphibole), a mineral that contains the same elements but that crystallizes at lower temperatures. Different patterns of zones form from the interplay of compositional and temperature factors in these banded rocks, called **skarns,** formed by contact metamorphism of carbonate rocks. A common pattern shows gradation from the unaltered dolomite to a tremolite zone a few hundred meters wide, then to a somewhat narrower diopside zone, and finally to an inner contact zone of wollastonite that may be only a few meters wide.

Contact metamorphism of silicate rocks produces similar patterns but of different minerals (Fig. 16–27). Outer zones contain micas, amphiboles, and calcite. Inner zones are characterized by pyroxenes, including wollastonite and the aluminosilicate andalusite, Al_2SiO_5. The general nature of contact aureole patterns illustrates the tendency for the minerals that contain **volatiles,** the easily escaping elements or compounds (for example, carbon dioxide bound chemically in calcite), to be found in outer zones and for dry, gas-free minerals to be found in inner, hotter zones.

METAMORPHIC CHEMICAL CHANGES

The mineral transformations of regional and contact metamorphism illustrate the general tendency for systematic chemical changes that accompany increased temperature and pressure. One of the major processes affecting sediments is the loss of volatiles, mainly water and carbon dioxide. The metamorphism of some continental mafic volcanics involves the opposite change: many of them erupt as relatively dry, volatile-free lavas and become hydrated and carbonated (combination with carbon dioxide to form carbonate minerals). Organic materials containing carbon and hydrogen (such as the hydrocarbons in petroleum) lose their volatile gas, hydrogen.

Because silica and two alkali metal elements, sodium and potassium, are highly soluble in the hot aqueous fluids associated with metamorphism, they may be dissolved and transported from deeper to shallower zones in the crust. These solutions combine with partial melting to form migmatites.

Even though these chemical transformation processes are known to take place, a great deal of regional and contact metamorphism seems to have taken place with little or no bulk chemical change other than loss of water and carbon dioxide. It is this fact that enables metamorphic geologists to "look through" the metamorphic mineral assemblage to reconstruct the original mineralogy of the sediment or igneous rock that was transformed. Knowing the total mineral composition of the rock, they can recast it, using experiments on mineral reactions, such as the conversion of calcite and silica to wollastonite, to infer the earlier mineral assemblage that would have been stable at lower temperatures and pressures. In this way, former rock patterns, including sedimentary facies, can be reconstructed even though almost all direct evidence has been obliterated. The study of metamorphism illustrates again how knowledge of process—how things happen—is inextricably tied to the history of the Earth, how things *did* happen.

The link between rock formation and the history of the Earth leads to questions about the nature of the forces causing regional metamorphism and their association with mountain-building. That conjunction has led metamorphic geologists to use plate-tectonic theory as a conceptual framework for their ideas. Greenstones, metamorphosed basalts, have been dredged from such zones of plate divergence as the mid-Atlantic ridge. At these active zones of basalt upwelling, localized heating transforms newly extruded ba-

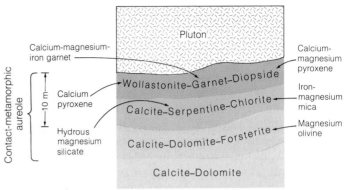

Figure 16-26
Skarns are banded rocks produced by contact metamorphism of limestones and dolomites. Grading from unaltered rock toward the contact, they change progressively from pure carbonate marble to bands composed of different calcium-magnesium silicate minerals, finally to a carbonate-free silicate rock.

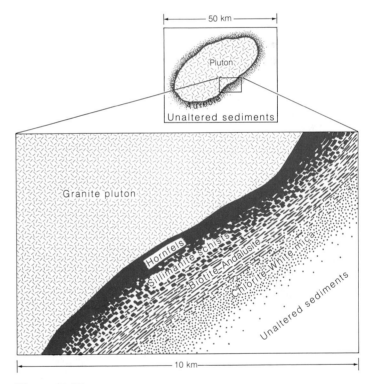

Figure 16-27
A contact aureole in sandstones and shales intruded by a granite pluton is a series of zones characterized by mineral assemblages formed at progressively lower temperatures with greater distances from the intrusive.

salts into low-grade metamorphic rocks. Hydro-thermal solutions derived from sea water circulating through the basalt pile play a role in this basalt metamorphism. Blueschists, metamorphosed sediments whose minerals indicate that they were formed under very high pressures but relatively low temperatures, are recognized as the result of rapid and deep subsidence of sediments along the surface of a cool lithospheric slab at a subduction zone (Fig. 19–12). The subducted plate, carrying deepwater marine sediments, such as turbidites and cherts, moves down so rapidly that its great mass heats up very slowly while the pressure increases rapidly. Just the opposite pair of conditions, high temperature and low pressure, characterizes the metamorphic rocks of volcanic island arcs at plate convergences. Here the igneous rocks that are generated by melting at plate margins come close to the surface and transform shallowly buried volcanics and sediments.

The two kinds of metamorphism, attributable to high- and low-pressure zones, occur as paired metamorphic belts in ancient rocks. Two belts of contrasting metamorphic type were first distinguished in Japan about 20 years ago by A. Miyashiro, who pointed out their length, several hundred kilometers, and their orientation. The oceanic side was the belt of high-grade, high-pressure rocks of mafic igneous composition. We now recognize this as the metamorphosed fore-arc subduction zone, where the heat flow is low and therefore pressures are relatively high for a given temperature. The landward side was the belt of lower-pressure high-grade rocks of shale-like composition, now interpreted as the remnants of the volcanic arc itself. It would have had high heat flow, and therefore relatively low pressures for a given temperature. These broad metamorphic zones along ancient continental margins can thus be used to reconstruct something of the pattern of ancient plate motions. We discuss the full story of the relationships between the theory of plate tectonics and igneous and metamorphic processes in Chapter 19, after the stage has been set by exploring how seismological and other geophysical investigations led to the theory.

SUMMARY

1. The forms of igneous intrusions, such as plutons, dikes, and sills, are a guide to their method of emplacement at depth. Such bodies may intrude by stoping, by heating surrounding rock so that it gives way by plastic flow, or by fracturing or bowing up.

2. Sills and laccoliths are concordant; dikes are discordant.

3. Lopoliths and batholiths are very large plutons. Some batholiths appear to be granitized sediments.

4. Metamorphic rocks are of three kinds: contact, dynamic, and regional. Contact metamorphism is produced by heat at the borders of an igneous intrusion. Dynamic metamorphosed, or cataclastic, rocks are mechanically fragmented by folding and faulting. Regionally metamorphosed rocks are the product of generalized heat and pressure associated with orogeny.

5. Metamorphic textures include foliation, the general term for parallel planes, fracture cleavage, flow cleavage, schistosity, and lineation. These textures are formed by the preferred orientation of platy and needle-shaped crystals of new minerals grown during metamorphism. Hornfelsic textures form in contact metamorphic rocks. Porphyroblasts are large crystals formed at the expense of a fine-grained matrix. Granulites have little or no preferred orientation.

6. Cataclastic rocks include friction breccias, mylonites, and augen gneiss.

7. Regional metamorphism produces a series of mineral assemblages that can be used to map isograds that reflect pressure and temperature conditions. Metamorphic facies are a grouping of rock types by their mineral assemblages in relation to pressure and temperature of formation.

8. Extreme metamorphism can produce migmatites, which may be a major component of some batholiths.

9. Contact metamorphic aureoles are altered margins around plutons in which zones of index minerals appear in definite sequences.

10. Systematic chemical changes normally accompany regional metamorphism, mainly the loss of volatiles, carbon dioxide, and water.

11. Metamorphic belts are interpreted in the framework of plate tectonics, primarily as the result of alteration at subduction zones. Paired belts of high- and low-pressure metamorphic rocks are used to identify fore-arc and arc portions of former volcanic arc systems.

EXERCISES

1. What differences in the texture of sills and lava flows can be used to distinguish them?

2. What properties of some granitic batholiths are difficult to reconcile with formation by the simple emplacement of a large body of differentiated magma?

3. What kinds of metamorphism might you be most likely to find along a continental rift valley?

4. Which kind of pluton would you expect to have produced the most intense contact metamorphism of surrounding rocks—a granite intrusion very deep in the crust or a gabbro intrusion at moderate depths? Why?

5. Would mineral assemblages or chemical composition be better to use to infer the original nature of a group of metamorphic rocks of medium and high grade?

6. Would you choose foliation or chemical composition as the better property to use in determining metamorphic grade? Why?

7. Would you expect to find cataclastic rock in the contact aureole of a mafic pluton intruding a limestone at a depth of 8 kilometers? Why?

BIBLIOGRAPHY

Billings, M. P., *Structural Geology*, 3rd ed. Englewood Cliffs, N.J.: Prentice-Hall, 1972. Chapters 16, 17, 18, and 19.

Ernst, G., *Earth Materials*. Englewood Cliffs, N.J.: Prentice-Hall, 1969.

Mason, R., *Petrology of the Metamorphic Rocks*. London: George Allen and Unwin, 1978.

Miyashiro, A., *Metamorphism and Metamorphic Belts*. London: George Allen and Unwin, 1973.

Turner, F. J., *Metamorphic Petrology*, 2nd ed. New York: McGraw-Hill, 1981.

Winkler, H. G. F., *Petrogenesis of Metamorphic Rocks*, 5th ed. New York: Springer-Verlag, 1979.

17

Seismology and the Earth's Interior

The basic causes of earthquakes are strains induced by plate motions. By their locations and the nature of the ruptures they produce, earthquakes define the plate boundaries. Analysis of seismic waves, together with laboratory studies of rocks, helps us to infer the composition and state of the Earth's interior.

Seismology is the study of earthquakes and seismic waves. Because earthquakes are among the scourges of humankind, seismologists are concerned with minimizing their destructiveness. They do this by assessing seismic risks in different geographic regions, so that sensible building and zoning codes can be written, and by researching the problems of tsunamis, earthquake prediction, and even earthquake control. But reducing the hazards of earthquakes is not the only job of seismologists.

By studying the pattern of earthquakes, seismologists have provided one of the essential clues to the development of the concept of plate tectonics: earthquake belts outline plate boundaries, the zones along which plates collide, diverge, or slide past one another. The modern seismograph, which records the waves generated by earthquakes and explosions, is the most important means of probing the deep interior.

SEISMOGRAPHS

The seismograph is to the Earth scientist what the telescope is to the astronomer—a tool for peering into inaccessible regions. The ideal seismograph would be a "skyhook," a device fixed to a frame outside of the Earth, so that when the ground shakes, the seismic vibrations could be measured by the changing distance between the fixed device and the ground. Because this fixed base is impossible to achieve, a compromise is struck by coupling a mass so loosely to the Earth that the ground can vibrate without causing much motion of the mass. The mass is coupled to the Earth by means of a pendulum (Fig. 17–1) or by suspending it from a spring (Fig. 17–2). When the ground moves, the mass tends to remain stationary because of its inertia, and the displacement of the Earth relative to the stationary mass is used to sense ground movement. The most advanced electronic technology* is used to magnify this motion, so that modern instruments can detect ground displacements as small as 10^{-8} centimeter—an astounding feat, considering that such small displacements are of atomic size. This is far more sensitivity than can actually be used on Earth because the ground moves in a continual state of unrest, shaken by the action of the winds, ocean waves, and machinery. In most places such sensitive seismographs would be driven off scale by Earth's "noise." The Moon is the place for this most advanced seismograph; there, no winds, no ocean sounds, no mechanical vibrations can over-

---

*For the electronics hobbyist, we might mention that the magnification of a seismograph is achieved by means of moving-coil and variable-reluctance transducers, like those on phonograph pickups, and low-noise, ultra-high-gain, solid-state amplifiers. The output is recorded photogalvanometrically on film, on paper with direct-recording pen motors, on AM, FM, or digital magnetic tape.

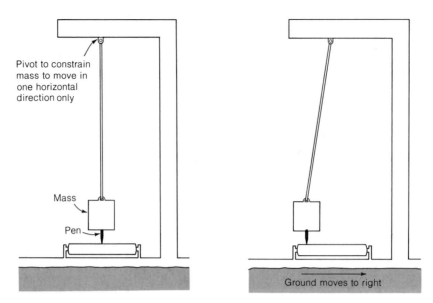

Figure 17-1
The principle of the pendulum-mounted seismograph. Because of its inertia, the mass does not keep up with the motion of the ground. The pen traces the difference in motion between the mass and the ground, in this way recording vibrations that accompany seismic waves. The instrument schematized here records horizontal ground motion.

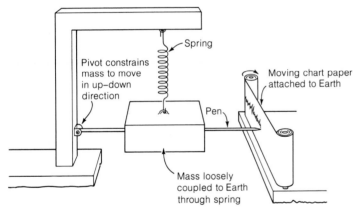

Figure 17-2
Spring-mounted seismograph to record vertical ground motion.

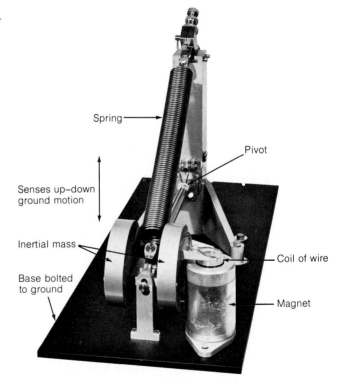

Figure 17-3
The Press-Ewing seismograph, an example of a modern instrument. Airtight cover, electronics, and recording system not shown.

load it. The lunar seismographs left behind by the astronauts can detect the seismic waves generated by a 1-kilogram (2.2-pound) meteor striking anywhere on the Moon's surface. A modern seismograph is shown in Figure 17–3.

The strain seismograph shown schematically in Figure 17–4 was developed by H. Benioff, a leader in the design of electronic musical instruments and seismographs. By electronic measurement of the change in distance between two concrete piers about 30 meters (100 feet) apart, this instrument can detect stretching and compression of the ground caused by seismic vibrations or by the pull of the Moon on the solid Earth (body tides, just like those to which the ocean responds but much smaller). A record made by a strain seismograph is shown in Figure 17–5, and an actual installation is shown in Figure 17–6.

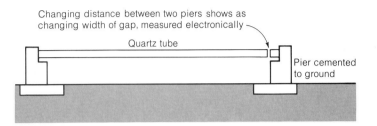

Figure 17–4
The Benioff strain seismograph. The distance between two piers attached to the ground changes due to tides in the solid Earth and the passage of seismic waves. These variations are measured by placing an electronic motion detector in the gap.

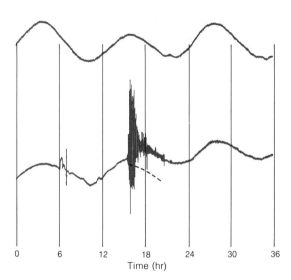

Figure 17–5
Record made by a strain seismograph. The slow periodic movements are the Earth tides; the more rapid vibrations are the seismic waves from an earthquake. [From "Resonant Vibrations of the Earth" by Frank Press. Copyright © 1965 by Scientific American, Inc. All rights reserved.]

Figure 17–6
A strain seismograph installation in an underground tunnel. This system is so sensitive that it could detect a change of 1 mm in the distance between New York and California.

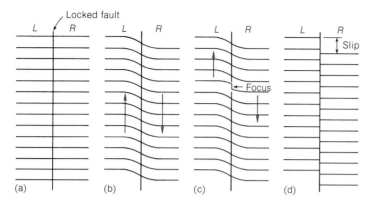

Figure 17-7
The elastic rebound theory of an earthquake. The two simulated crustal blocks L and R are being forced to slide past each other (a). Friction along the fault prevents slip (b), but the deformation builds up until the "frictional lock" is broken (c) and an earthquake slip occurs (d).

Figure 17-8
The earthquake of 1906 was caused by slip along the San Andreas fault. The offset fence shown here shows a slip of nearly 3 m. The scene is near Bolinas, California. [Photo by G. K. Gilbert; courtesy of R. E. Wallace, U.S. Geological Survey.]

EARTHQUAKES

What Is an Earthquake?

The rupture of the San Andreas fault that devastated San Francisco in 1906 also provided the evidence that led H. F. Reid, one of the official investigators of that catastrophe, to advance his **elastic rebound theory** of earthquakes. Earthquakes are associated with large fractures, or faults, in the Earth's crust and upper mantle. Consider the fault between the two hypothetical crustal blocks in Figure 17-7. Suppose that surveyors had located lines running perpendicular to the fault from block L to block R, as shown in part a of the diagram. Blocks L and R are moving in opposite directions, but because they are pressed together by the weight of the overlying rock, friction locks them together, just as a brake can lock the wheel of a car if enough force is applied. Instead of slipping along the fault, the blocks are deformed near the fault, and the surveyors' lines are bent as shown in Figure 17-7b. As the rock is strained, elastic energy is stored in it in the same way that it is stored in a wound-up watch spring. The movement continues, the strain builds up until the frictional bond that locks the fault can no longer hold at some point on the fault, and it breaks (Figs. 17-7c, 17-7d, and 17-8). The blocks suddenly slip at this point, which is the **focus** of the earthquake (Fig. 17-9). Once the rupture begins, it travels at a speed of about 3.5 kilometers per second (7200 miles per hour), continuing for as much as 1000 kilometers. In great earthquakes, the **slip,** or offset, of the two blocks can be as large as 15 meters (50 feet). Figure 17-7d shows the two blocks after the earthquake, displaced by the amount of slip. Once the frictional bond is broken, the elastic strain energy, which had been slowly stored over tens or hundreds of years, is suddenly released in the form of intense seismic vibrations, which con-

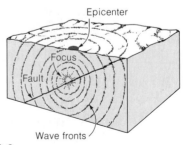

Figure 17-9
The focus of an earthquake is the site of initial slip on the fault. The epicenter is the point on the surface above the focus. Seismic waves radiate from the focus.

stitute the earthquake. The vibrational waves are propagated large distances in all directions from the fault. Near the focus the waves can have large, destructive amplitudes. The process can be likened to storing elastic energy by slowly drawing out the rubber band of a sling shot and then releasing it suddenly to propel a pebble.

Strictly speaking, the elastic rebound theory is an incomplete one. The reason for this is that the pressure holding the blocks together is so great that the frictional bond is actually stronger than the rock itself. In other words, the block should more easily break elsewhere than slip along the fault. Yet faults do exist, and movement occurs along them periodically. To complete the theory, we need a means of "lubricating" the fault or reducing the locking pressure. Geologists working in rock mechanics are currently trying to remove some of the mystery of the mechanics of faulting.

Earthquakes—How Big and How Many?

The time between great earthquakes is about 50 to 100 years in California and somewhat less in more active seismic regions, such as Japan or the Aleutians. Thus the time required to build up the elastic strain energy in the rocks along a fault is enormous compared with the time that elapses during the release of stored energy, for earthquakes last only a few minutes. The amount of stored energy can be gauged in several ways. The two most common methods are to measure the distortion of surveyed lines, as in the example just discussed, or to measure the energy of the released seismic waves. About 10^9 ergs of elastic strain energy is released from each cubic meter (1.3 cubic yards) of rock at the time of an earthquake—the equivalent of a fire cracker per cubic meter. This may seem unimpressive until one adds up the cubic meters affected by a great earthquake. Suppose the fault is 1000 kilometers (600 miles) long, extends 100 kilometers downward, and distorts surveyed lines as far as 50 kilometers on either side of the fault. This amounts to a strained volume of 10^{16} cubic meters, each cubic meter contributing 10^9 ergs, which gives a total of 10^{25} ergs. This is one big fire cracker indeed—about the equivalent of 10,000 nuclear explosions of the strength of the bomb exploded at Hiroshima.

Energy release gives the most precise measure of the size of an earthquake, but it is a long, complicated process to determine the fault dimensions, the slip, and the other factors needed to compute the total energy involved. Seismologists have therefore adopted the **Richter magnitude**

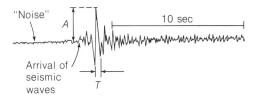

Figure 17–10
Determination of earthquake magnitude from a seismograph recording. Dividing A, the maximum trace motion, by the magnification of the seismograph gives the maximum ground motion a, measured in micrometers (10^{-4} cm). T is the duration of one oscillation, or the period of the seismic wave in seconds. Magnitude $m = \log(a/T) + B$, where B is a factor that allows for the weakening of seismic waves with increasing distance from the earthquake. *Example:* An earthquake 10,000 km away ($B = 6.8$ from the table of data) produced a ground motion $a = 10 \ \mu m$ (micrometers) with period $T = 1$ sec. Thus $m = \log 10 + 6.8 = 7.8$. The correction factor B is found empirically, so that a seismograph located any place in the world would give the proper magnitude of an earthquake, regardless of distance to it.

scale, which is based on the amplitude of seismic waves recorded by seismographs. Actually, magnitude (m) is based on the logarithm of the maximum amplitude adjusted by a factor that takes into account the weakening of seismic waves as they spread away from the focus. Thus seismologists all over the world can study their records and in a few minutes come up with nearly the same value for the magnitude of an earthquake. Seismographs are sensitive enough to detect earthquakes of magnitude less than 1 quite easily. The largest earthquakes yet recorded have Richter magnitudes of about 8.5. Because these magnitudes are based on a logarithmic scale, an increase in magnitude of one unit corresponds to a tenfold increase in the size of an earthquake as measured by the amplitude of seismic waves. An earthquake of magnitude 8 would be 10,000 ($10^8/10^4$) times one of magnitude 4. Figure 17–10 shows how the magnitude of an earthquake is determined in practice.

Table 17–1 relates magnitude and energy to earthquake effects and indicates the number of earthquakes of given magnitudes each year. The table demonstrates the fortunate fact that most earthquakes are small. Each year, 800,000 little tremors are recorded by instruments but not felt by humans. Great earthquakes, those with magnitudes exceeding 8, occur about once every 5 to 10 years. Damage begins at magnitude 5 and increases to nearly total destruction in nearby settlements for earthquakes with $m > 8$ ($>$ means "greater than"). The San Fernando (Los Angeles)

Table 17-1
Earthquake magnitudes, energies, effects, and statistics

Characteristic effects of shallow shocks in populated areas	Approximate magnitude	Number of earthquakes per year	Energy (ergs)
Damage nearly total	≥ 8.0	0.1–0.2	$> 10^{25}$
Great damage	≥ 7.4	4	$\geq .4 \times 10^{24}$
Serious damage, rails bent	7.0–7.3	15	$0.04–0.2 \times 10^{24}$
Considerable damage to buildings	6.2–6.9	100	$0.5–23 \times 10^{21}$
Slight damage to buildings	5.5–6.1	500	$1–27 \times 10^{19}$
Felt by all	4.9–5.4	1,400	$3.6–57 \times 10^{17}$
Felt by many	4.3–4.8	4,800	$1.3–27 \times 10^{16}$
Felt by some	3.5–4.2	30,000	$1.6–76 \times 10^{15}$
Not felt but recorded	2.0–3.4	800,000	$4 \times 10^{10}–9 \times 10^{13}$

Source: Data from B. Gutenberg.

earthquake of 1971 was only of magnitude 6.6, yet the damage bill amounted to a billion dollars. As damaging as the earthquake was, the seismic energy release was only about one-thousandth that of some truly great earthquakes, such as those in San Francisco (1906), Tokyo (1923), Chile (1960), Alaska (1964), and China (1976). It is no wonder that Californians worry about the great shock that should visit them about every 50 to 100 years!

Table 17-1 also points up the interesting fact that the few large earthquakes each year release more seismic energy than the hundreds of thousands of small shocks combined. This should put to rest the notion that small earthquakes act as a safety valve, gradually releasing strain in harmlessly small amounts and thus forestalling a big shock. About 10^{26} ergs of seismic energy are released each year. This is about 1 percent of the heat energy reaching the Earth's surface from the interior, and about 3 percent of the energy used by humankind.

Earthquakes—Where Do They Occur?

A **seismicity** chart showing the map locations, or **epicenters,** of almost 30,000 earthquakes that occurred between 1961 and 1967 is reproduced in Figure 17-11. Seismologists have known for decades that earthquakes tend to occur in belts—for example, the "ring of fire" surrounding the Pacific Ocean. In recent years, however, it has become possible to detect the more numerous, smaller earthquakes and to improve methods of locating epicenters, so that seismic belts can now be defined more accurately. Interestingly, the increase in the number of seismic observatories and the use of computers to store and analyze seismic data that made this possible were stimulated by

research done in the 1960s, during negotiations for a nuclear test-ban treaty; the purpose was to determine whether small underground nuclear explosions could be detected and distinguished seismically from earthquakes.

The new high-quality seismicity maps showed that narrow belts of epicenters coincide almost exactly with the crest of the mid-Atlantic, the east Pacific, and other oceanic ridges, where plates separate. (Compare Fig. 17-11 with the map inside the back cover.) Earthquake epicenters are also aligned along transform faults, where plates slide past each other. But earthquakes that originate at depths greater than about 100 kilometers (60 miles) typically occur near margins where plates collide (Fig. 17-12). The foci of these earthquakes are distributed on well-defined planes that dip into the mantle, and these occur in close association with deep-sea trenches, island arcs, young mountains, and volcanoes. It is a basic tenet of the theory of plate tectonics that these deep earthquakes actually define the positions of subducted plates, which are plunging back into the mantle beneath an overriding plate (Fig. 17-13). This global correlation between topography, geology, and seismicity provided essential data for defining the boundaries of the lithospheric plates. It may seem like a simple matter now to draw a line through the seismic belts and so define the plates depicted on the inside of the back cover, but this important advance could not have been made without the "knowledge explosion" in seismic data and the imaginative, uninhibited, and synthesizing minds of about ten workers who sifted through those data in the years following 1967.

Although most earthquakes are recorded at plate boundaries, the seismicity map shows that a small percentage originate within plates. Some of these have been quite destructive, as is indicated

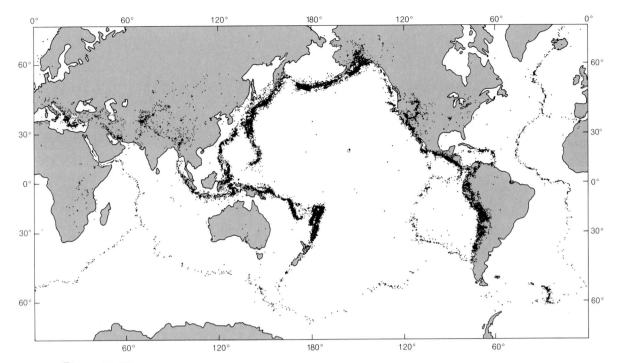

Figure 17–11
Epicenters of some 30,000 earthquakes recorded in the years 1961–1967, with focal depths between 0 and 700 km. [Epicenters by the U.S. Coast and Geodetic Survey. Computer plot by M. Barazangi and J. Dorman, Columbia University.]

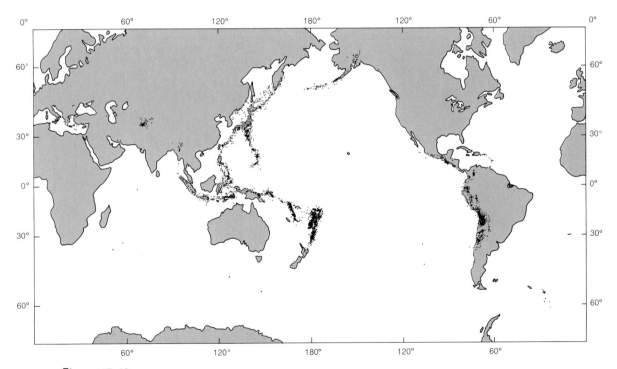

Figure 17–12
Subset of earthquakes from preceding figure with focal depths greater than 100 km. These deep earthquakes typically originate near margins where plates collide and thus serve to identify such plates. [Computer plot by M. Barazangi and J. Dorman, Columbia University.]

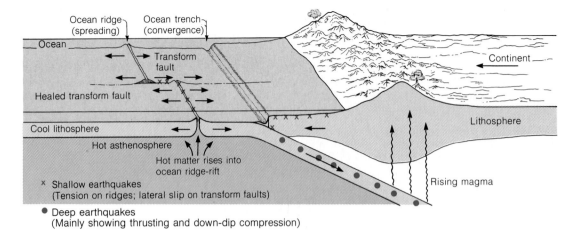

Figure 17–13
The association of earthquakes with three types of plate boundaries: ocean ridges, transform faults, and trenches.

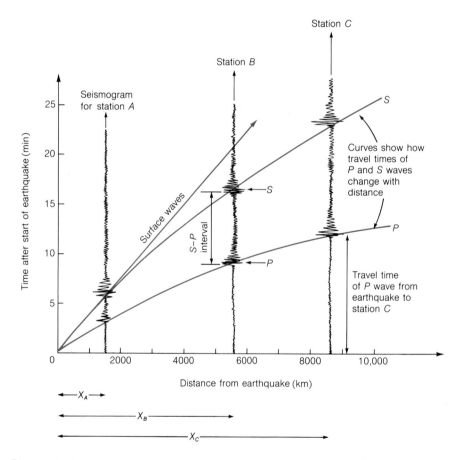

Figure 17–14
The time required for P, S, and surface waves to travel a given distance can be represented by curves on a graph of travel time against distance over the surface. To locate an earthquake epicenter, the time interval observed at a given station is matched against the travel-time curves for P and S waves until the distance is found at which the separation between the curves agrees with the observed S–P time difference. Knowing the distance from the three stations A, B, and C, one can locate the epicenter as in Figure 17–15.

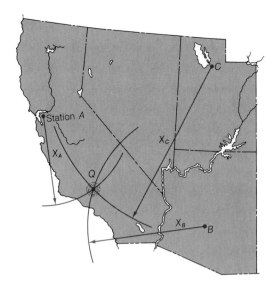

Figure 17-15
Knowing the distance, say X_A, of an earthquake from a given station, as by the method of the preceding figure, one can say only that the earthquake lies on a circle of radius A, centered on the station. If, however, one also knows the distances from two additional stations B and C, the three circles centered on the three stations, with radii X_A, X_B, X_C, intersect uniquely at the point Q, the epicenter.

by these examples: New Madrid, Missouri (1812), Charleston, South Carolina (1886), Boston, Massachusetts (1755), T'ang-shan, China (1976). Apparently, stresses within the lithospheric plates occasionally build up until they exceed the rock's strength, producing one of these rare intraplate earthquakes.

Locating the Epicenter

The principle involved in identifying a quake's epicenter is quite similar to deducing the distance to a lightning bolt from the time interval between the flash and the sound. Later in this chapter we will describe the two types of seismic waves that travel through the body of the Earth—the first-arriving P and the slower-traveling S waves, which take almost twice as long to reach the seismograph. Surface waves, a third type, skirt along the surface. The lightning flash may be likened to the P waves of earthquakes and the thunder to the S waves. The time interval between the arrival of P and S waves therefore increases with the distance traveled by the waves, and with each S–P time interval there is associated a definite distance to the epicenter. This is indicated on the travel-time chart for P and S waves in Figure 17-14, which shows diagrammatically how the travel times of P and S waves depend on distance and how the S–P interval increases with distance.

To get an approximation of the epicenter, the seismologists simply read off the S–P interval on the seismogram from a given station and use a graph like that in Figure 17-14 or a table to get the distance to the epicenter. Knowing the distances from three or more stations enables them to pinpoint the epicenter (Fig. 17-15). They can also deduce the time of the shock at the epicenter because the arrival time of the P waves at each station is known, and from a graph or a table it is possible to read about how long the waves took to reach the station. Once an approximation has been made, the exact location can be found by making refinements.

Obtaining Stress Patterns

When an earthquake occurs, one block slips relative to an adjacent one along a **fault plane** (Fig. 17-16a). The orientation of the fault plane and the slip direction are of great interest because they provide information about what is happening at plate boundaries.

If the concept of plate tectonics is correct and seismicity is primarily associated with boundaries along which plates separate, collide, or slide past each other, then the fault orientations and slip directions should differ for each type of plate junction. Earthquakes in divergence zones should result from tension, as if the plates were being

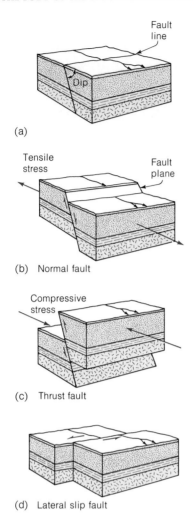

(a)

(b) Normal fault

(c) Thrust fault

(d) Lateral slip fault

Figure 17–16
The types of fault movement and the
stresses that cause them: (a) situation
before movement takes place; (b) normal
fault due to tensile stress; (c) thrust (or
reverse) fault due to compressive stress;
(d) lateral slip (or strike slip) fault due
to shearing stress.

pulled apart, and **normal faults,** in which the
overlying block moves down the dip of the fault
plane, should characterize the earthquake mechanism (Fig. 17–16b). Many earthquakes in convergence zones, where plates collide, should show a
compressive mechanism—for example, **thrust
faulting,** in which the overlying block moves up
the dip of the fault plane (Fig. 17–16c). Where
plates slide past each other along transform faults,
the earthquake mechanism should be simple lateral (sideways) slip along nearly vertical planes
(Fig. 17–16d). Faults will be discussed further
when we take up structural geology in Chapter 20.
 Seismologists have learned to deduce which of
these earthquake mechanisms is involved from
seismograms. This ability is especially conven-

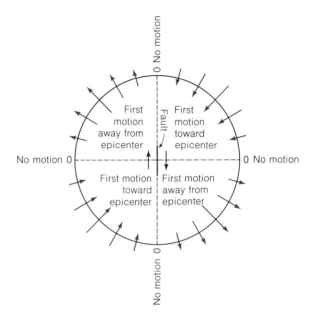

Figure 17–17
The initial motion of seismic waves shows a characteristic movement, a push away from the source or
a pull toward it, depending on the orientation of the
fault, the direction of slip along the fault, and the
direction of the seismograph from the earthquake.
The fault orientation and slip direction can be obtained from the dashed lines that separate the stations into quadrants of initial motion toward the
epicenter or away from it. Because most earthquakes
originate below the surface, seismologists must rely
on this principle in analyzing their records to find
the orientation of the fault and the direction of slip.

ient, since very few earthquake faults break
through to the surface, where the slip direction
and fault orientation can be observed directly,
and many small faults are not always so clearly
and unambiguously displayed in the field as they
are in our beautiful illustrations. Seismographs
located in different directions from an earthquake
epicenter characteristically record seismic waves
with an initial motion of the ground either to-
wards the epicenter or away from it, as shown in
Figure 17–17. This "radiation pattern" is shown in
idealized form as a distribution of outward and
inward motions into four quadrants defined by an
extension of the fault plane. From such a distribution seismologists can deduce the orientation of
the fault plane and the direction of slip, as indicated in the figure.* Although Figure 17–17 simpli-

---

*Actually, a fault perpendicular to the one shown in the fig-
ure would produce the same pattern of up and down motions if
the slip directions were ⇌ rather than ↑↓. Fortunately, however, the correct orientation of the fault plane can be deduced
either from field evidence, if there is any, or from the after-
shocks that follow the main event, for they originate in the fault
plane, and their positions define the true orientation of the
fault.

Figure 17-18
Destruction caused by the San Fernando, California, earthquake of 1971.
[Photo by R. E. Wallace, U.S. Geological Survey.]

fies nature somewhat and real seismograph stations are not situated in simple circles around faults, seismologists know how to allow for natural complexities and can uncover the true source mechanism of an earthquake from the radiation pattern of *P* waves. What they find is the following: When the topography of mid-ocean ridges is examined in detail, the ridges are often found to be segmented, the segments being offset by transform faults (Fig. 17–13). Earthquake epicenters coincide with the ridge crests and with the transform faults between the offset ridge segments, as shown in the figure. In 1967 seismologists at Columbia University found that the pattern of *P* wave motions radiated from earthquakes on ridge crests indicated that they originate in normal faults that run parallel to the ridge crest. This means that the axis of tension is perpendicular to the trend of the fault, as we have seen in Figure 17–16. This finding is just as predicted by the plate-tectonics concept, which specifies that the ridge crests mark the boundary between separating plates. Furthermore, earthquakes in transform faults between ridge crests were found to show lateral slips, just as would be expected for a region where plates slide past each other in opposite directions. What elegant support seismology gives to the notion that plates spread from ridge crests! Outside the region between the ridge

crests, the transform fault becomes **aseismic—** that is, it produces no earthquakes. This, too, is to be expected since in this region the plates move in the same direction on both sides of the fault. In a sense the fault has healed and is evidenced only by topography, usually by a scarp (cliff or steep slope), as shown in Figure 17–13.

What about the leading edges on the other side of the moving plates, where collisions occur? Seismologists have found that many of the deep earthquakes that originate within the subducted plate show the predicted compressive mechanism in the direction of the dip of the downgoing plate (Fig. 17–13). They have also devised methods of mapping the downgoing plate by tracking the seismic waves, which are guided more efficiently up the cold plate from the earthquake focus to the surface than through the adjacent mantle, which is hot and muffles the waves.

**Earthquake Destructiveness—
Can It Be Controlled?**

Earthquakes cause destruction in several ways. Ground vibrations can shake structures and stress them to the point of failure and collapse (Fig. 17–18). The ground accelerations caused by great earthquakes can approach and even exceed that

Figure 17-19
Foundation failure as a result of soil liquefaction caused these buildings in Niigata, Japan, to topple during the earthquake of 1964. The structures themselves were built to withstand earthquakes; they toppled intact. [Photo by G. Housner, California Institute of Technology; National Science Foundation.]

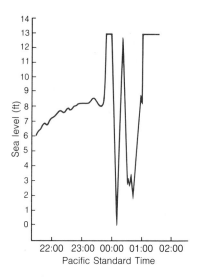

Figure 17-21
The tsunami from the great Alaskan earthquake of 1964 reached California, where several people perished and some coastal damage resulted from the waves. A tide gauge at Crescent City, California, which usually serves to record ocean tides, made this record of the tsunami, showing sea-level changes of almost 2 m more than normal.

Figure 17-20
Destruction at Seward, Alaska, caused by the tsunami generated during the earthquake of 1964. Although a warning system exists to alert people on distant coasts to the danger of tsunamis, it cannot yet function rapidly enough to help residents in the epicenter region. [Photo by F. Press.]

Figure 17-22
Monument standing atop the debris of an avalanche that buried the town of Khait in the Tadzhik Republic of the USSR following the earthquake of 1949; 12,000 people were killed. The slide, more than 30 m thick, moved over the town with a velocity of 100 m/sec. No sprinter could have outrun it. [Photo by F. Press.]

Table 17-2
Some of the world's worst earthquakes (in lives lost)

Year	Place	Deaths (est.)	Year	Place	Deaths (est.)
856	Corinth, Greece	45,000	1915	Avezzano, Italy	30,000
1038	Shansi, China	23,000	1920	Kansu, China	180,000
1057	Chihli, China	25,000	1923	Tokyo, Japan	99,000
1170	Sicily	15,000	1930	Apennine Mountains, Italy	1,500
1268	Silicia, Asia Minor	60,000	1932	Kansu, China	70,000
1290	Chihli, China	100,000	1935	Quetta, Baluchistan	60,000
1293	Kamakura, Japan	30,000	1939	Chile	30,000
1456	Naples, Italy	60,000	1939	Erzincan, Turkey	40,000
1531	Lisbon, Portugal	30,000	1948	Fukui, Japan	5,000
1556	Shen-shu, China	830,000	1949	Ecuador	6,000
1667	Shemaka, Caucasia	80,000	1949	Khait, U.S.S.R.	12,000
1693	Catania, Italy	60,000	1950	Assam, India	1,500
1693	Naples, Italy	93,000	1954	Northern Algeria	1,600
1731	Peking, China	100,000	1956	Kabul, Afghanistan	2,000
1737	Calcutta, India	300,000	1957	Northern Iran	2,500
1755	Northern Persia	40,000	1960	Southern Chile	5,700
1755	Lisbon, Portugal	30,000–60,000	1960	Agadir, Morocco	12,000
1783	Calabria, Italy	50,000	1962	Northwestern Iran	12,000
1797	Quito, Ecuador	41,000	1963	Skopje, Yugoslavia	1,000
1822	Aleppo, Asia Minor	22,000	1968	Dasht-e Bayaz, Iran	11,600
1828	Echigo (Honshu), Japan	30,000	1970	Peru	20,000
1847	Zenkoji, Japan	34,000	1972	Managua, Nicaragua	10,000
1868	Peru and Ecuador	25,000	1976	Guatemala	23,000
1875	Venezuela and Columbia	16,000	1976	T'ang-shan, China	500,000(?)
1896	Sanriku, Japan	27,000	1976	Philippines	3,100
1897	Assam, India	1,500	1976	New Guinea	9,000
1898	Japan	22,000	1976	Iran	5,000
1906	Valparaiso, Chile	1,500	1977	Rumania	1,500
1906	San Francisco	500	1978	Iran	15,000
1907	Kingston, Jamaica	1,400	1980	Algeria	3,500
1908	Messina, Italy	160,000	1980	Italy	4,000

of gravity near the epicenter, and very few man-made structures can survive without severe damage. Certain kinds of soil lose their rigidity and "liquefy" when subjected to repeated seismic shocks. The ground simply slides away, taking buildings, bridges, and everything else with it (Fig. 17-19). As was mentioned earlier, coastal earthquakes occasionally generate the awesome waves called tsunamis, which travel across the ocean at speeds of up to 800 kilometers per hour (500 miles per hour) and form walls of water as much as 20 meters high as they sweep over low-lying coastal areas (Figs. 17-20 and 17-21). Avalanches, mudflows, and fire may accompany earthquakes and take their toll (Fig. 17-22). Of the 99,000 fatalities in the Tokyo earthquake of 1923, 38,000 were due to fire. Table 17-2 lists the human losses of historical earthquakes. A seismic-risk map of the United States, based on the earthquake history of the country, is shown in Figure 17-23. Many readers may be surprised to find that they live in a zone where there is risk of earthquake damage.

Can anything be done about reducing earthquake hazards? Seismologists in the United States, Japan, the Soviet Union, and China are working hard to find answers to this question. Even though no major scientific breakthroughs have been made so far, damage and loss of life can be mitigated by encouraging sound building practices. Figure 17-24 shows something that should not be allowed—a residential area built in an active fault zone, the San Andreas, the most dangerous in the United States. Construction on unstable soils or in avalanche-prone areas should be prohibited. Engineers can design structures

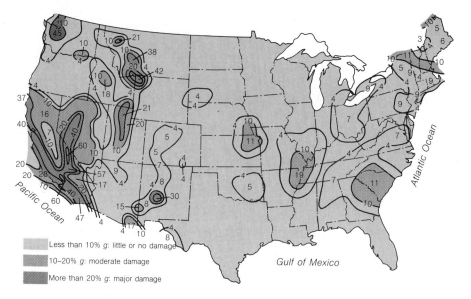

Less than 10% g: little or no damage

10-20% g: moderate damage

More than 20% g: major damage

Figure 17-23
Expected level of earthquake-shaking hazards. The levels of ground shaking for different regions are shown by contour lines that express the maximum amount of shaking likely to occur at least once in a 50-year period as a percentage of the force of gravity. Damage begins to occur at about 10 percent g. An acceleration of 0.1 percent g or more is perceptible to people. [Modified from U.S. Geological Survey Chart.]

Figure 17-24
Housing tracts constructed within the San Andreas fault zone, San Francisco peninsula. The white line indicates the approximate fault trace, along which ground ruptured and slipped about 2 m during the earthquake of 1906. [Photo by R. E. Wallace, U.S. Geological Survey.]

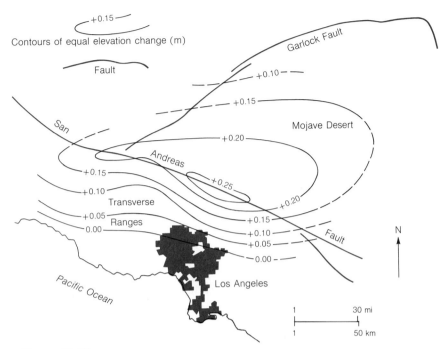

Figure 17-25
The uplift that has taken place in the vicinity of Los Angeles since 1960. Some seismologists are concerned that this extraordinarily large and rapid uplift signals preparation for a major earthquake. [After U.S. Geological Survey.]

that will withstand most earthquakes, and building codes should require that only such building be done in high-risk areas.

Ten years ago only astrologers, mystics, and religious zealots were concerned with earthquake prediction. Today seismologists in many countries are actively working on this problem and can claim a few successes. In February 1975 an earthquake was predicted five hours before it occurred near Haicheng in northeast China. Several million people, prepared in advance by a public education campaign, evacuated their homes and factories in the hours before the shock. Although many towns and villages were totally destroyed, only a few hundred lives were lost. Western scientists who have since visited the region estimate that tens of thousands of lives were saved. The Chinese have successfully predicted other earthquakes, but unfortunately they were able to provide only a long-term warning (within 5 years) of the great T'ang-shan earthquake of August 1976—not enough accuracy to save the 700,000 people estimated to have lost their lives. Smaller earthquakes have been predicted in the United States and the Soviet Union, in connection with the research programs in those countries.

The scientific basis of earthquake prediction is in the process of being established in laboratory and field experiments. A promising model proposes that with the accumulation of strain, a grad-ual weakening of the fault zone takes place before the sudden rupture that results in an earthquake. One manifestation of this would be a slow deformation or slip in the fault zone. Observations of ground tilt, strain, or creep might pick up these premonitory changes and provide a warning days or hours before the earthquake. Other promising methods make use of unique systematic patterns in the occurrence of small earthquakes prior to the impending main shock. When the "dangerous" pattern of foreshocks appears, a warning would be issued. A method that has had some recent success in Mexico and Alaska predicts the location of a future earthquake, but predicts the time only to within a few years or so. Called the gap method, it identifies places where strain has built up because of plate motions, yet the historic record of earthquakes indicates that a major shock has not occurred recently and is overdue.

An ancient method of prediction used for centuries is based on unusual animal behavior in the few days or hours before an earthquake. Because of recent reports by trained observers in China and Italy, scientists no longer dismiss this "peasant wisdom" but are trying to check if, and why, it works.

A matter of some concern in California is a bulge that has appeared along the San Andreas fault near Los Angeles (Fig. 17-25). Also worrisome are more recent surveying results, which

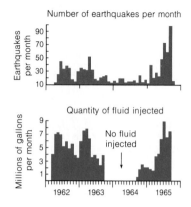

Figure 17-26
Correlation found by D. M. Evans between the quantity of waste water injected into a deep well and the number of earthquakes per month in the vicinity of Denver, Colorado. This unplanned "experiment" opens the distant possibility of earthquake control by fluid injection.

imply a reduction of the forces that squeeze the North American and Pacific Plates together, as if the San Andreas fault were becoming unlocked in this region. Increased emissions of radon from wells were observed at the same time. Could these phenomena signal a coming disastrous earthquake? Because of these concerns a major earthquake prediction and hazard mitigation program was initiated by President Carter and Governor Brown of California in 1980. Similar activities are underway in other countries with high seismic risk.

If earthquake prediction research has just become respectable, earthquake control continues to boggle the imagination. Yet some chance discoveries have opened this intriguing possibility. In 1966 a dramatic correlation was found between the rate of high-pressure injection of waste fluids into a deep well and the frequency of earthquakes in the vicinity of Denver, Colorado (Fig. 17-26). Apparently the earthquakes were triggered by reduction of frictional resistance to faulting. The pressure exerted by the injected fluids "unlocked" a preexisting fault, and strains that had built up earlier were released. Perhaps someday earthquake-control wells will be spotted every few miles along the San Andreas fault, and fluid injected so as to cause the fault to creep continuously and slip frequently. These controlled earthquakes would prevent strain from building up over periods of 50 to 100 years and being released in a large, damaging shock. Much research will be needed to achieve this important (and perhaps impossible) goal; the first steps have already been taken.

EXPLORING THE INTERIOR WITH SEISMIC WAVES

To appreciate the importance of seismic waves in revealing the properties of the interior, one need only reflect on what the state of knowledge would be in the absence of this key tool. We might surmise the existence of a crust from the observation that most surface rocks are light and felsic or mafic compared to the more dense ultramafic intrusions that seem to invade the surface layers from below, but we could only speculate on its thickness. The sea-floor crust would of course be terra incognita. We probably would have guessed the mantle to be composed of ultramafic rocks, but we could only wonder about its physical state, structure, and thickness. From the clues provided by iron meteorites, the large relative abundance of iron in the cosmos, and our efforts to explain the Earth's density and magnetic field, we might have been led to hypothesize the existence of a molten iron core, but this would have been argued extensively. Its depth would be uncertain, and the inner solid core would be unknown. Continental drift and sea-floor spreading would be debated, but the overall concept of plate tectonics—especially the fate of plates in subduction zones—would probably have escaped us.

Types of Seismic Waves

As early as the 1800s, mathematicians proved with pencil and paper the existence of compressional and shear waves in elastic bodies. Not until the close of the nineteenth century, however, did seismologists devise instruments sensitive enough to detect such waves in the Earth—the P waves and S waves generated by sudden slip along a fault. Figure 17-27 shows the faster-traveling P wave as the propagation of a volume change—a squeezing and unsqueezing of the medium; the individual particles vibrate to and fro in the direction of wave propagation. In Figure 17-28, the S wave is shown as a traveling shearing disturbance, the material distorting in shape rather than changing in volume; the particles vibrate back and forth at right angles to the direction of propagation.

Figure 17-29 depicts the paths of P waves as they travel from the source of an earthquake or explosion into the interior, emerging again at distant points. These wave paths and their travel times have been determined empirically from the seismographic records of earthquakes all over the world. The Earth's core deflects the waves and in

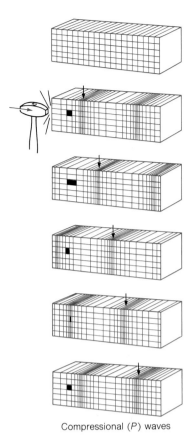

Compressional (P) waves

Figure 17-27
Stages in the deformation of a block of material with the passage of compressional P waves through it. The undeformed block is shown at the top. In the sequence from top to the bottom, a crest of compression, marked by an arrow, moves through the block with the P-wave velocity. It is followed by an expansion, and any small piece of matter, like the marked square, shakes back and forth in response to alternating compressions and expansions as the wave train moves through. A sudden push (or pull) in the direction of wave propagation, indicated by the hammer blow, would set up P waves. [After O. M. Phillips, *The Heart of the Earth,* Freeman, Cooper & Co.]

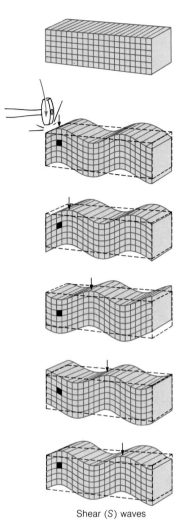

Shear (S) waves

Figure 17-28
Stages in the deformation of a block of material with the passage of shear waves, or S waves, through it. A wave crest, marked by an arrow, moves through the block with the S-wave velocity as vertical planes shake up and down. Any small piece of matter, like the marked one, shakes up and down and experiences a shearing deformation (from a square to a parallelogram in the figure) as the shear wave passes through. A sudden shear displacement, indicated by the hammer blow at right angles to the direction of wave propagation, would set up S waves. [After O. M. Phillips, *The Heart of the Earth,* Freeman, Cooper & Co.]

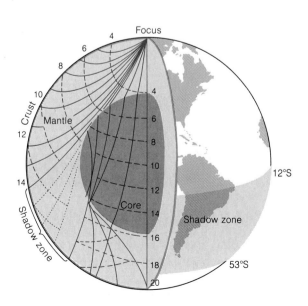

Figure 17-29
Cutout showing the pattern of P-wave paths through the Earth's interior. The numbers show the travel time in minutes for the waves to reach the associated broken line. Note the shadow zone, a region not reached by P waves (for this hypothetical earthquake at the North Pole) because they are deflected by the Earth's core. [After *Internal Constitution of the Earth* by B. Gutenberg (ed.), Dover Publications, 1951.]

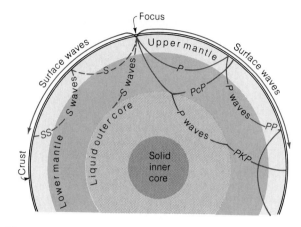

Figure 17-30
P and S waves radiate from an earthquake focus in many different directions. Waves reflected from the Earth's surface are called PP or SS. PcP is a wave that bounces off the core, and PKP is a P wave transmitted through the liquid core. S waves cannot travel in a liquid.

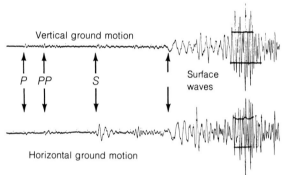

Figure 17-31
Seismograph recording of P, S, and surface waves from a distant earthquake.

effect casts a shadow where very little P-wave energy reaches the surface. The existence of such **shadow zones** suggested that the core is molten because compressional waves decrease sharply in velocity when they pass from a solid into a liquid of the same composition. The suggestion became a firm pronouncement when seismologists found that shear waves do not penetrate this region. Liquids transmit P waves but not S waves because the fluids elastically resist and recover from squeezing but do not resist shearing.

When P and S waves encounter a boundary like that between the core and the mantle, they are in general both reflected back and transmitted across it, just as light may be partly reflected and partly transmitted at a water surface. If in the new medium the wave velocity is different, the waves are bent, or **refracted.** Because of all of these possibilities of reflection, transmission, and refraction, P and S waves break up into several types as

they travel through the Earth, as shown in Figure 17-30. Follow the wave PcP in the figure as it bounces like a radar beam—from the Earth's core and yields the depth of the core from the round-trip time. The wave PKP, which penetrates the core, is useful for exploring that region. Many of the P-wave trajectories and travel times are sketched in Figure 17-29.

In addition to P and S waves, there is a third category of seismic wave, the surface wave, that travels along the Earth's surface. Surface waves are guided by the surface and the outer layers of the Earth, just as the motion of ocean waves is mostly surficial. Interestingly the chief difference seismologists found between earthquakes and underground nuclear explosions is in their ability to generate surface waves. Explosions excite surface waves with less efficiency than do earthquakes—a discovery that may stimulate statesmen to agree to an underground nuclear test-ban treaty. Figure 17-31 shows seismograms in which many seismic waves are labeled.

Pluck a violin string and a tone is emitted; strike a bell with a hammer and it rings. The Earth also rings when it is disturbed by a great earthquake that causes the entire globe to vibrate like a bell for as long as several weeks. The tones of Earth's vibrations are pitched too low for the human ear to hear, but modern seismographs are sensitive enough to detect these low-frequency oscillations. The Earth can vibrate in different ways, or modes, actually an infinite number of them. Some are shown in Figure 17-32. The mode with the lowest pitch is the "football," or *spheroidal,* mode, which takes 53 minutes to execute one vibration. For our musician-readers, we add that in terms of what one might call music of the spheres, that vibration corresponds to E flat in the twentieth octave below middle C. The "balloon," or *radial,* mode has a frequency of one vibration in 20 minutes; and the twisting, or *torsional,* mode, a frequency of one in 44 minutes.

Finding Earth Models from Travel Times and Vibration Frequencies

Using thousands of sensitive seismographs and highly accurate clocks, seismologists around the world measure precisely the travel times of P, S, and surface waves and the frequencies of the vibrations of the Earth. From these measurements they can plot travel-time curves of the kind shown in Figure 17-14 for the different kinds of seismic waves. The travel times depend on how the velocities of compressional and shear waves

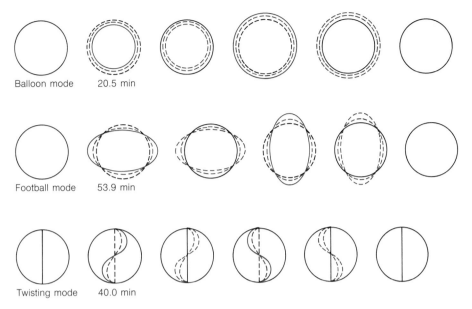

Balloon mode   20.5 min

Football mode   53.9 min

Twisting mode   40.0 min

Figure 17-32
Three of Earth's vibrational modes. The schematic illustration shows how the planet changes shape and gives the time it takes to complete the sequence of vibrations shown. The actual movements are only a small fraction of a millimeter.

Box 17-1 FINDING OIL WITH SEISMIC WAVES

Exploration for oil is an important application of seismology. In offshore prospecting, a ship tows a sound source and underwater phones. Sound waves are generated by a pneumatic device that works like a balloon burst. The sound waves bounce off rock layers below the sea floor and are picked up by the phones. In this way subsurface sedimentary structures that trap oil, such as faults, folds, and domes, are "mapped" by the reflected waves. This technique is used extensively to explore the submerged continental shelves and shallow seas for oil and gas deposits. Oceanographers use this method to study the sedimentary layers on the continental slope and rise and on the floor of the deep sea.

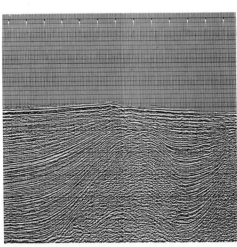

A section of the Gulf of Mexico, 30 km long and 10 km deep, in which folded sedimentary layers are revealed by reflected seismic waves. [From Petty Geophysical Engineering Co.]

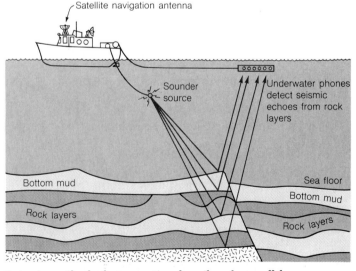

Seismic method of prospecting for oil and gas offshore. [After U.S. Department of the Interior.]

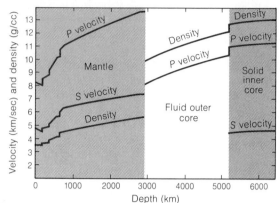

Figure 17-33
Estimate of the variation in density and in P- and S-wave velocities in the Earth's mantle and core. Uncertainty is probably only a few percent of the actual value. [Prepared by A. M. Dziewonski and D. L. Anderson for the Standard Earth Committee of the International Union of Geodesy and Geophysics.]

change as they pass through materials of different elastic properties. The vibration frequencies of the Earth depend on the velocities of these waves as well as on the density of different parts of the interior, just as the tone of a bell depends on its elasticity and density. Once all the seismological data are accumulated, the next step is to find Earth models whose P-wave and S-wave velocities and internal densities are consistent with the data.

Solving this "inverse problem," as it is called by mathematicians, is something like being told that a driver made a trip from Los Angeles to San Francisco in seven hours, in bad weather, on a Monday, and having to make a best guess of the route he took. The mathematics of this process cannot be explained here, but the techniques are powerful enough to allow us to make a best estimate of an Earth model.

A plot of internal densities and P- and S-wave velocities for the Earth is given in Figure 17-33. It is quite likely that these curves represent the real

---

Box 17-2 HIGH PRESSURE AND SHOCK EXPERIMENTS

Even if we knew in detail how density and seismic velocities change with depth in the Earth, we would still want to identify the materials and describe their physical conditions. To do this we also need information on the densities of different materials and the velocities with which seismic waves travel through them under the high pressures and temperatures that exist in the interior of a planet. The pressure at the center of the Earth is nearly 4 million times atmospheric pressure,* and temperatures there range to several thousand degrees. Using a hydraulic press, geophysicists can squeeze rocks in the laboratory to pressures of about 100 kilobars, heat them to temperatures of about 1000°C, and at the same time measure many of their properties. This procedure duplicates conditions at depths of about 300 kilometers. A recent technical breakthrough now makes it possible to increase laboratory pressures to 1.7 megabars and temperatures to 3000°C, conditions similar to those in the Earth's core. The experiment is similar to that depicted in the figure except that a diamond anvil is used. The sample is squeezed between two cut diamonds and heated by a laser beam. Another technique for compressing rocks to very high pressures happens to be the very same method used to compress uranium when triggering an atomic explosion. An ordinary chemical explosive, such as dynamite, is wrapped around the rock. When the dynamite is detonated, the shock wave squeezes

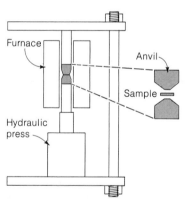

The Bridgman "squeezer," a device for subjecting minerals to pressures of a few tens of thousands of atmospheres and temperatures of several hundred degrees. The low pressure of the hydraulic press is amplified by concentrating the total force on the small area of the anvil. A furnace surrounding the anvil supplies heat. This apparatus simulates environments deep in the Earth's crust.

the rock, raising the pressure and temperatures to the high values needed to duplicate conditions at great depths. The rock is destroyed in the process, but in the few millionths of a second before it falls apart, data needed to calculate the density, pressure, and shock velocity (which is simply related to the seismic wave velocities) are obtained electronically from sensors on the rock. Other types of experiments are used to determine such things as strength, thermal, electrical, and elastic properties at high pressures and temperatures.

---

*Pressure is measured in atmospheres (atm), bars, or pounds per square inch (p.s.i.). 1 atm = 1.01 bars = 14.7 p.s.i. Geologists tend to use bar, kilobars (10^3 bars) and megabars (10^6 bars).

Earth within a percent or so. As more complete and more precise data become available, these curves should become more accurate, perhaps someday converging to give us a perfect model of the real Earth. But what can we say *now* about the interior?

Composition, Structure, and State of the Interior

Velocity and density models are important mainly as a means to an end; the ultimate goal is to understand the composition, structure, and state of the Earth's interior. Laboratory experiments make the connection between seismology on the one hand and petrology and geochemistry on the other. High-pressure equipment and shock waves generated by explosives are used, as described in Box 17–2, to learn how velocity and density would vary in different rocks, either in the solid or in a partially molten state, and either near the surface or in the deep interior. With this information the Earth model shown in Figure 17–33 can be interpreted to give information about the materials and their state.

The major divisions, crust, mantle, and core (Fig. 17–30), were discovered from the analysis of reflected and refracted *P* and *S* waves and have been known for more than 60 years. The boundary between the crust and the mantle is called the **Mohorovičić discontinuity** (**M,** or **Moho,** for short) after the Yugoslavian seismologist who discovered it in 1909. It separates rocks in which *P* waves have velocities of about 6 to 7 kilometers per second (3.8 to 4.4 miles per second) from underlying mantle rocks, in which *P* waves have a velocity of about 8 kilometers per second (5 miles per second). The field method of measuring these velocities is described in Box 17–3. From geological sampling to find all possible crustal and mantle materials and from laboratory measurements of the properties of these materials, we have learned to associate *P*-wave velocities with composition, as indicated in Table 17–3. We conclude from these measurements that the continental

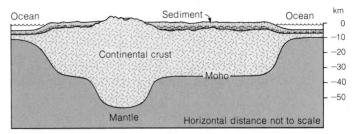

Figure 17–34
The lithosphere is topped by a relatively lightweight crust. Seismology reveals that the crust varies in thickness; it is thin under oceans, thicker under continents, and thickest under high mountains.

crust consists mostly of granitic rocks, with gabbro appearing near the bottom and that no granite occurs on the floor of the deep ocean, the crust there being entirely basalt and gabbro. The mantle below the M discontinuity is almost certain to be primarily the dense ultramafic rock peridotite. The crust is a distillate of the mantle and therefore differs chemically from its parent. In this sense, the Moho is a chemical boundary located by seismic waves.

Nowadays seismologists are excited by the finer details, the variations within the crust, mantle, and core. The variation in crustal thickness in a section like the one shown in Figure 17–34 is one of the most important recent seismological results. The thickness of the crust varies from about 35 kilometers to 10 kilometers in a section extending from continent to ocean. Under a high mountain the crust thickens to as much as 65 kilometers. If Figure 17–34 suggests to you that the continental crust floats on the denser mantle like an iceberg on the ocean, you have made a good observation. Icebergs float because they are less dense than sea water; flotation comes from the large volume of ice below the sea surface. When Archimedes' principle of buoyancy is applied to the flotation of continents and mountains, it becomes the **principle of isostasy,** which holds that the relatively light continents float on a more dense mantle; most of a continent's volume lies below sea level for the same reason that most of an iceberg lies below the ocean surface. Nature has contrived that large topographic loads such as mountains and continents are *compensated*—that is, supported primarily by buoyancy rather than by the strength of the crust. Rocks, which we know to be solid and strong over the short term (seconds or years), are, over the long term (thousands to millions of years), weak and flow like a viscous fluid when loaded. When continents grow or mountains are pushed up, a supporting root

Table 17–3
Correspondence between composition and *P*-wave velocity in igneous rocks

Composition	*P*-wave velocities (km/sec)
Felsic (granitic)	6
Mafic (gabbro)	7
Ultramafic (peridotite)	8

Box 17-3   SEISMOLOGICAL SOUNDING OF THE EARTH'S CRUST AND UPPER MANTLE

Seismologists have developed a field procedure for measuring the thickness of the crust and the velocity of P waves in the crust and at the top of the mantle. Small seismometers are placed on the surface in a line extending away from a "shot point," where an explosion is set off to generate P waves. The waves leave the explosion in all directions—some traveling along the surface, others along the top of the mantle, as shown in the upper figure. A travel-time curve can be plotted on which each point represents the travel time required for the waves to reach a seismometer. The plot of waves that travel along the surface is a straight line through the origin of the graph with a slope of $1/V_1$. The slope of the line is measured to obtain the speed V_1 of P waves in the crust. The waves traversing the man-

tle give a line with slope $1/V_2$ and intercept T. Values of V_1, V_2, and T obtained from the graph are used to calculate the thickness from the formula

$$D = \frac{T}{2} \frac{V_2 V}{\sqrt{V_2{}^2 - V_1{}^2}}$$

A student familiar with trigonometry and Snell's law might try deriving this simple but important equation.

The lower figure shows seismic waves recorded at a distance of 163 kilometers from an explosion of TNT. The traces are recorded from six seismometers placed 100 meters apart. The waves that have traveled through the crust and mantle are indicated by P_1 and P_2, respectively.

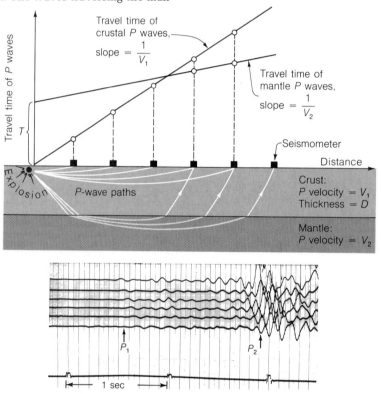

must develop as part of the process to provide buoyancy and keep the new load from sinking.

There is one variant to this general mechanism. If for some reason—for example, regional heating—part of the upper mantle becomes less dense than the adjacent mantle, it will also exert a buoyant force that can support elevated topography above it without the need for a crustal root. In a sense, the lower-density mantle serves as a root. This mode of isostatic compensation seems to be operating in the Basin and Range province (Utah, Arizona, Nevada) of the United States. The high heat flow in this region is consistent with the ex-

planation. Seismological studies of crustal thickness have provided quantitative corroboration for the mechanism of isostasy.

In the years 1965 to 1970, geologists and geophysicists the world over concentrated research efforts on the upper thousand kilometers of the Earth as part of the International Upper Mantle project. This concerted attack led to many exciting discoveries about a region that had previously been poorly known. We can illustrate the more important of these by discussing the shear-wave velocity model. According to this model the mantle is divided on the basis of shifts in velocity into

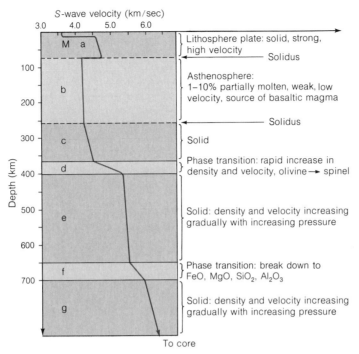

Figure 17-35
A modern view of the structure of the outermost 700 km of the Earth is illustrated by a plot of S-wave velocity against depth. Note how changes in velocity mark the important zones: lithosphere, partially molten asthenosphere, transitions to more dense molecular structures.

zones a to g (Fig. 17-35). Zone a is the **lithosphere,** a slab about 70 kilometers (45 miles) thick in which the continents are embedded. Crust forms the uppermost part of this outer shell of the Earth. Its lower boundary is marked by an abrupt decrease in shear-wave velocity. The lithosphere is characterized by high velocity and efficient propagation of seismic waves, both of which imply solidity and strength.

Zone b is the **asthenosphere,** or zone of weakness. It is also called the **low-velocity zone** for the obvious reason that the shear-wave velocity there is low. Seismic waves are attenuated more strongly in the asthenosphere than anywhere else in the Earth. Because laboratory experiments show that seismic waves are slowed and absorbed in a crystalline-liquid mixture, a "slush," most geophysicists and petrologists think that the asthenosphere is partially melted, perhaps 1 to 10 percent. We have already noted that the Earth's lithosphere is made up of about ten distinct plates, created along mid-ocean ridges and destroyed in subduction zones. A solid slab underlain by a weak layer might be more easily movable, and perhaps this accounts for the mobility of the lithospheric plates.

Velocity and density in both the lithosphere and asthenosphere fit a peridotitic composition.

These two zones, therefore, do not differ so much chemically as they do in physical state; the boundary at 70 kilometers marks the solidus (where melting begins), separating the solid lithosphere from the not-so-solid asthenosphere. In Chapter 15 it was proposed that the melt of the asthenosphere is the primary source of basaltic magma, which fits the picture nicely.

The asthenosphere ends at a depth of about 250 kilometers (155 miles), and the rocks become solid again in zone c. The velocity increases slightly with depth in this region because of the effect of increasing pressure.

Zone d, about 400 kilometers (250 miles) below the surface, is thin but very important. The rapid increase in velocity there correlates with the rapid increase in density in Figure 17-33. This transition is too abrupt to be accounted for by a composition change. A change of phase—that is, a closer repacking on the atomic level—is required. The theoretical explanation was beautifully verified in 1969 when E. A. Ringwood and S. Akimoto squeezed olivine in their laboratories and found that at critical pressures and temperatures its atoms take up a more compact arrangement, changing into the spinel structure (Fig. 17-36). Olivine is the principal mineral in peridotite, and at a depth of about 400 kilometers in the Earth, condi-

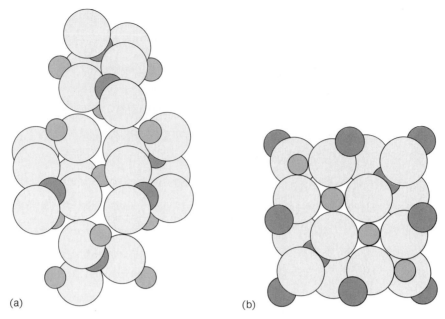

(a)   (b)

Figure 17-36
Olivine (Mg_2SiO_4 with Fe_2SiO_4) is a major mineral in the Earth's mantle. Part (a) shows olivine in its low-pressure form. The large pale-brown atoms are oxygen, the medium-brown atoms are silicon, and the dark-brown ones are magnesium (or iron). When the pressure reaches a critical value, corresponding to a depth of about 400 km in the Earth, the molecule collapses into a denser form (b). Note the decrease in void space in the high-pressure form, in which the oxygen atoms are more closely packed. Seismologists have found where this transition occurs in the Earth, and petrologists have observed the transition in high-pressure laboratory experiments.

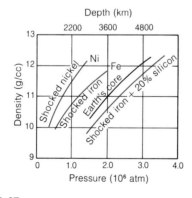

Figure 17-37
Density in the Earth's fluid core plotted against depth below the surface and against pressure (black curve). Comparison with the densities for iron, nickel, and iron-silicon mixtures measured in laboratory studies enables seismologists to conclude that the core is mostly iron but slightly less dense than pure iron, as if a small amount of a "lightening" element like silicon were present.

tions are just right for it to change phase. This is an excellent example of how the collaboration of specialists with different backgrounds (seismologists and petrologists) can, little by little, remove some of the mystery of Earth's interior.

Zone e is one of small change with depth, but zone f, near 700 kilometers, shows another rapid transition. In 1974 experimenters in Japan and the U.S. reached the required pressures in the laboratory and found that olivine breaks down into dense, simple oxides like FeO, SiO_2, and MgO at this depth. The entire region from 400 to 700 kilometers, containing zones d, e, and f, is sometimes called the **transition zone.**

The lower mantle, zone g, extending from 700 kilometers (435 miles) to the core at a depth of 2898 kilometers (1800 miles), is a region that changes little in composition and phase with depth. Density and velocity increase gradually, again due to increasing pressure.

The Earth's core is far away but not out of the reach of seismic waves. We know that its outer region is fluid and its inner one solid (Fig. 17-30). To obtain its composition, we use the same approach that has already proved so useful—comparison of laboratory experiments and seismological data. Look at Figure 17-37 to see how this is done.

The density in the fluid core is plotted. Also shown are the densities of nickel, iron, and a mixture of iron and silicon, determined by "shocking" these materials in the laboratory with explosives, as was described in Box 17–2. We see that nickel is too dense. Iron is better but needs to be lightened by adding perhaps 15 percent silicon. Would other elements fit the data? Perhaps, but our choice is limited by the relative abundance of elements. Because the core accounts for one-third of the mass of the Earth, it must contain relatively abundant elements. Iron is the only abundant element that approaches the required density under the pressure of millions of atmospheres at these great depths. It is a little too dense, as Figure 17–37 shows, so a plentiful "lightening" element like silicon must be added. Oxygen or sulfur might also be possible lightening elements.

In this way seismological observations and laboratory measurements of the properties of materials combine to give an incomplete but nevertheless good approximation of the Earth's interior. A zoned, differentiated Earth is found in which the major components are a metallic iron core and a rocky mantle consisting primarily of iron-magnesium silicates. The mantle includes a transition zone in which atoms are forced into closer packing, a partially molten asthenosphere, and most of the outer lithosphere. A thin, lightweight crust—the end-product of the differentiation process—caps the mantle.

SUMMARY

1. Most earthquakes originate in the vicinity of plate boundaries. The mechanism of earthquakes is governed by the kind of plate boundary: fracture under tensile stress occurs at boundaries of divergence, fracture under compressive stress at boundaries of convergence, and lateral slip along transform faults.

2. Great earthquakes release in a few minutes huge amounts of elastic strain energy that had been slowly stored in the rocks of the fault zone over tens or hundreds of years. The source of this strain is plate motions.

3. Richter magnitudes are determined from the size of the ground motions, as measured when seismic waves are recorded on seismographs. The three types of seismic waves are P waves, S waves, and surface waves. The entire Earth can be set into global vibration by great earthquakes.

4. From a study of the travel times of seismic waves and the frequency, or pitch, of the global oscillations, seismologists have found that the Earth is divided into shells—that is, it is a zoned, differentiated planet, with
   a. A slablike, mostly ultramafic lithosphere, broken into large, mobile plates.
   b. A partially molten asthenosphere, the primary source of basaltic magma, as evidenced by reduced velocity and high absorption of seismic waves.
   c. A transition zone, where atoms are forced into a closer packing by the high pressures.
   d. A lower mantle, mainly iron-magnesium silicate.
   e. A fluid outer core, mostly iron but with one or more "lightening" elements.
   f. A solid iron central core.

5. The continents with their lightweight felsic crusts—the end-products of the differentiation process—are embedded in the lithosphere.

EXERCISES

1. What is an earthquake? How is its magnitude measured? How many earthquakes cause serious damage each year?

2. How does the distribution of earthquake foci correlate with the three types of plate boundaries?

3. Seismograph stations report the following S–P time differences for an earthquake: Dallas, S–P = 3 minutes; Los Angeles, S–P = 2 minutes; San Francisco, S–P = 2 minutes. Use a map of the U.S. and travel time curves (Fig. 17–14) to obtain a rough epicenter.

4. Taking into account the possibility of false alarms, reduction of casualties, mass hysteria, economic depression, and other possible consequences following an earthquake prediction, do you think the objective of predicting earthquakes is a worthwhile goal?

5. You wish to determine the depth to the water table before drilling a well. Using small explosions and seismographs you find that the P-wave velocity in the surface sediment is 600 meters/second and the velocity in a subsurface layer, presumably the water table, is 1500 meters/second. The intercept time, T is 0.8 second. How deep is the water table? (See Box 17–3.)

6. Draw a cross section of the Earth, showing to scale the crust, lithosphere, asthenosphere, transition zone, core-mantle boundary, fluid outer core, solid inner core. What are the major characteristics of each region?

BIBLIOGRAPHY

Anderson, D. L., "The San Andreas Fault," *Scientific American,* November 1971. (Offprint 896.)

Bolt, B. A., "The Fine Structure of the Earth's Interior," *Scientific American,* March 1973.

Isacks, B., J. Oliver, and L. R. Sykes, "Seismology and the New Global Tectonics," *Journal of Geophysical Research,* v. 73, pp. 5855–5899, 1968.

Office of Earth Sciences, National Research Council, *Trends and Opportunities in Seismology.* Washington, D.C.: National Academy of Sciences, 1976.

Press, F., "Earthquake Prediction," *Scientific American,* May 1975.

Richter, C. F., *Elementary Seismology.* San Francisco: W. H. Freeman and Company, 1958.

Scientific American, *Earthquakes and Volcanoes.* San Francisco: W. H. Freeman and Company, 1980.

Tectonophysics, "Forerunners of Strong Earthquakes," *Tectonophysics,* v. 14, pp. 177–344, 1972.

18

The Earth's Magnetism and Gravity

Earth's magnetic field originates in the fluid iron core. Many rocks "remember" the direction of the magnetic field that existed at the time of their formation, and this remanent magnetization provides not only a history of the magnetic field but also a history of the movements of the lithosphere. Variations in the gravity field over the surface correlate with subsurface geological changes. Mountain roots, an example of buoyant support of mountains, were discovered because of the gravity anomalies they produced.

Earth is influenced by two fields of internal origin—magnetic and gravitational. The Earth's gravitational field, like that of any body, attracts other bodies through interaction with their gravitational fields. Although just why this occurs is unknown, since Newton's time we have been able to calculate the effects of gravity—such as the acceleration of an apple falling from a tree, the orbits of planets, the trajectories of spacecraft, and so on (Fig. 18-1). Similarly, without knowing its basis, we observe and use the fact that magnetism is associated with moving electric charges, such as an electric current in a wire or electrons spinning around the nucleus of an atom (Figs. 18-2 and 18-3).

In this chapter we shall see how a planet's magnetic and gravitational fields yield information on conditions and processes in its interior. For instance, just a few years ago the Earth's magnetic field provided the key that unlocked the secret of sea-floor spreading. Gravity observations led to the discovery that continents and mountains are buoyantly supported—the principle of isostasy.

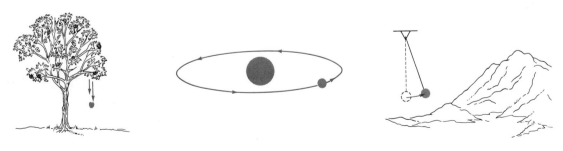

Figure 18-1
Three examples of gravitational attraction: an object falling in Earth's gravity field, a planet held in orbit by the Sun's gravity, and a pendulum deflected by the gravitational attraction of the mass of a mountain.

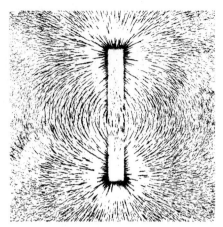
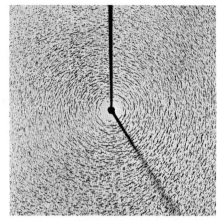

Figure 18-2
Magnetic fields of a bar magnet and of a wire carrying an electric current, made evident by the alignment of iron filings on paper. The filings are aligned parallel to the magnetic field lines. [From *PSSC Physics,* 3rd edition, D. C. Heath and Co. Reprinted by permission of the publisher.]

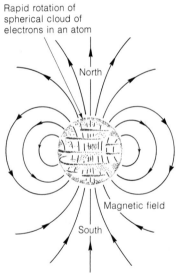

Figure 18-3
Moving electrons are equivalent to electric currents and create magnetic fields as they whirl around the nucleus of an atom. [After *Magnets* by F. Bitter, Doubleday Anchor Books. Copyright © 1959 by Educational Services, Inc.]

EARTH'S MAGNETIC FIELD

The Earth as One Big Magnet

Twenty-five hundred years ago, the Greeks discovered the magnetic properties of lodestone, the mineral called magnetite, one of the ores of iron. It took about another thousand years for the Chinese to invent the first crude magnetic compass, which was simply a piece of lodestone suspended on a string. Travelers of the fourteenth century brought news of this discovery back to Europe, where it was adopted as a navigational aid and made possible the great feats of such men as Columbus and Magellan during the Age of Exploration. It remained for William Gilbert, physician to Queen Elizabeth I, to explain in his book *De Magnete,* published in 1600, how the magnetic compass works. He offered the proposition that "the whole Earth is a big magnet" whose field acts on the small magnet of the compass needle to align it in the north–south direction.

The Earth's magnetic field can be fairly well described by the model of a small but powerful permanent bar magnet located near the center of the Earth and inclined about 11° from the geographic axis, as shown in Figure 18-4. The lines of force of the magnetic field, shown in the figure, indicate the presence of a magnetic force at each point in space. A magnetic needle that is free to swing under the influence of this magnetic force rotates into a position parallel to the local line of force, approximately in the north–south direction.

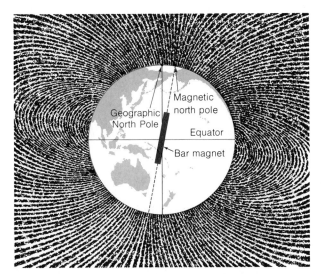

Figure 18-4
Earth's magnetic field is much like the field that
would be produced if a giant bar magnet were
placed at the Earth's center and slightly inclined
(11°) from the axis of rotation.

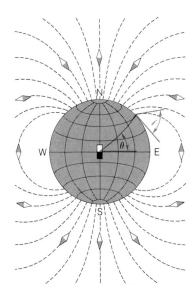

Figure 18-5
A suspended magnet needle aligns itself in the direction of the Earth's magnetic field. This diagram
shows that the inclination *i* of the needle would
vary with latitude from vertical over the pole to horizontal at the equator. [After *Continental Drift* by D.
and M. Tarling, Doubleday Anchor Books. Copyright
© 1971 by G. Bell and Sons, Ltd.]

The north-seeking end of the needle points to the
magnetic north pole. A compass needle that is
free to swing in a horizontal plane does not point
to true geographic north but deviates slightly, east
or west, depending on where the observer is, as
Figure 18-4 implies; the angle of east or west deviation from true north is called the **declination.** In
New York, the compass points about 10° west of
geographic north; in California, the declination is
about 20° east. The declination would be zero
everywhere, making a navigator's life simple, if
the magnetic north pole coincided with the geographic North Pole in Figure 18-4; that is, if the
hypothetical bar magnet at the center of the Earth
lined up perfectly with the axis of rotation.

A magnetized needle free to pivot in a vertical
plane (called a **dip needle**) measures the **inclination** of the magnetic field—that is, the angle the
field makes with the surface of the Earth. At the
magnetic poles, the dip needle swings into a vertical position (Fig. 18-5), which is how early explorers located these two points on the Earth's surface. Near the equator, where the inclination is
zero, the dip needle takes a horizontal position.
For the configuration shown in Figure 18-5, where
the magnetic pole coincides with the geographic
pole, the angle of inclination *i* is related to the
latitude θ by the simple formula $\theta = \frac{1}{2} \tan i$.

The compass and dip needle together give the
direction of the geomagnetic field. To characterize
the field fully, one must also measure its intensity

or strength with a device called a **magnetometer.**
Some versions of this instrument simply measure
the force exerted by the Earth's field on a small,
standard magnet. The unit of magnetic field
strength is the gauss (named after the great mathematician Karl Friedrich Gauss). Ordinary horseshoe magnets have fields of about 10 gauss. The
Earth's field is about 0.5 gauss near the surface.

The Puzzle of the Earth's Magnetism

Unfortunately, though a good description of the
field can be given assuming a permanent magnet
at the center of the Earth, this model has a fatal
defect: heat destroys magnetism, and magnetic
materials lose their permanent magnetism when
the temperature exceeds a certain value called the
Curie point (after Pierre Curie). For most magnetic materials, the Curie point is about 500°C, a
temperature exceeded below depths of about 20
or 30 kilometers in the Earth. In other words, the
Earth cannot be permanently magnetized below
this depth, so the notion that there is something
like a bar magnet near the center of the Earth is
eliminated, even though it represents the field
nicely.

Another way to make magnetic fields is with
electric currents, as Figure 18-2 shows. About a
billion amperes of current would be needed to
produce the Earth's magnetic field, and this is

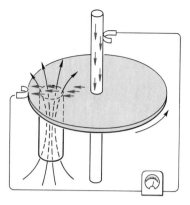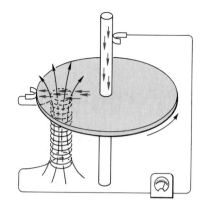

Figure 18–6
Simple dynamo generates an electric current (brown arrows) when a copper disk
is turned through the magnetic field of a bar magnet. If the bar magnet is replaced
by a coil of wire, the same electric current creates a magnetic field in the coil,
which keeps the system going. As long as the disk keeps rotating, the current flows,
and the magnetic field remains in this self-exciting dynamo. The Earth's magnetic
field is thought to originate in a self-exciting dynamo, but one that is enormously
more complicated. [From "The Earth as a Dynamo" by W. M. Elsasser. Copyright
© 1958 by Scientific American, Inc. All rights reserved.]

1580	1660	1820	1970
11° east of north	Due north	24° west of north	7° west of north

Figure 18–7
A compass needle does not remain fixed in direction
for all time, but drifts gradually in response to slow
changes in the Earth's magnetic field. The example
above shows how a compass in London varied in
direction between 1580 and 1970.

nearly as much electric current as the total
amount generated today. Dynamos in power
plants make electricity using an electrical conductor rotated by steam or falling water through a
magnetic field. Where inside the Earth is there a
dynamo with the capacity to generate this much
current?

The obvious place to look is the Earth's fluid
iron core because fluids can move readily and
iron is a good conductor. Walter Elsasser and
Edward Bullard proposed that a dynamo actually
exists in the Earth's core—a **self-exciting dynamo,**
which generates its own magnetic field. Figure
18–6 shows the principle of such a dynamo. No
one believes that a rotating disk like the one in the
figure actually exists in the core; what is proposed
is that the fluid iron is stirred into convective motion by heat generated from residual radioactivity
in the core. A small, stray magnetic field would

interact with the moving fluid iron to produce
electric currents, which would then create their
own magnetic field, starting up a self-exciting
dynamo. This description sounds vague because
it is: the dynamo theory for the core is exceedingly difficult to spell out in detail—in fact, the
mathematical equations have never been completely solved. We do know enough, however, to
show (1) that the mechanism is feasible; (2) that
the magnetic pole and the geographic pole (about
which the Earth rotates) must nearly coincide, as
they do today; and (3) that the polarity of the magnetic field is a matter of chance—that is, that the
north magnetic pole could just as easily be a south
magnetic pole. Large-scale fluid motions in the
core seem to offer the best possibility for explaining the Earth's magnetic field.

Changes in the Earth's Magnetic Field

Because the magnetic compass was a key factor in
navigation, governments have long subsidized the
science of geomagnetism. As a result, magnetic
observations have been made since the sixteenth
century. This historic documentation of the geomagnetic field over the last 400 years shows that
declination, inclination, and field strength vary
gradually the world over—a phenomenon called
secular variation. For example, in London the
compass needle has swung from 11° east of true
north in 1580, to due north in 1660, to 24° west of
north in 1820, and back to 7° west of north in 1970,
as Figure 18–7 shows. This amounts to a drift of
about 0.1° per year, a rapid rate of change compared to most geological processes. The solid

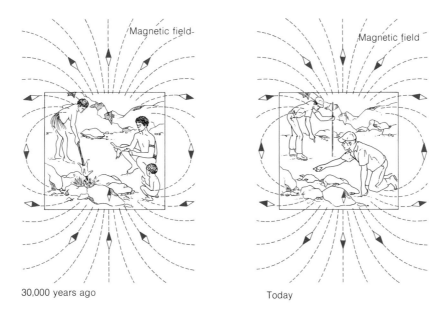

30,000 years ago   Today

Figure 18–8
Earth's magnetic field 30,000 years ago was the reverse of today's. We know this
from the discovery of reversely magnetized rocks found in the fireplace of an
ancient campsite. The rocks, cooling after the last fire, became magnetized in
the direction of the ancient magnetic field, leaving a permanent record of it,
just as a fossil leaves a record of ancient life.

parts of the Earth, even when they creep plastically, change much more slowly. Because fluid motions can occur this fast, again we resort to the fluid core as the likely source of the secular variation. Thus fluid motions in the core not only account for the origin of the main field, but also for its fluctuations, which originate perhaps in small eddies within the large-scale convective motion.

Fossil Magnetism

Not too long ago, an Australian graduate student found a fireplace in an ancient campsite where the aborigines had cooked their meals. He carefully removed several stones that had been baked by the fires, first noting their orientation. Then he measured the direction of magnetization of the stones, and he found that they were magnetized exactly opposite to the present geomagnetic field. He proposed to his disbelieving professor that, as recently as 30,000 years ago, when the campsite was occupied, the geomagnetic field was reversed from the present one—that is, that magnetic north was south.

Surprisingly enough, Earth scientists have discovered how to find out what the geomagnetic field was like in the past, not just thousands but millions of years before there were instruments for recording it. Recall that the Curie point is the

temperature at which permanent magnetism is lost when a magnet is heated. An important property of many very hot magnetizable materials is that, as they cool below the Curie point, they become magnetized in the direction of the surrounding magnetic field. This is **thermoremanent magnetization,** magnetization "remembered" by the rock long after the magnetizing field has disappeared. Thus the Australian student was able to determine the direction of the field when, after the last meal was cooked in that ancient fireplace some 30,000 years ago, the stones cooled and took on the magnetization of the surrounding geomagnetic field (Fig. 18-8). One might also think of an ancient volcano in eruption, say 100 million years ago. When the lava solidified and cooled below the Curie point (Fig. 18-9), it became magnetized, leaving us with a permanent record of the geomagnetic field in mid-Cretaceous time, just as a fossil leaves a record of ancient life.

Some sedimentary rocks can also take on a remanent magnetization. Recall that marine sedimentary rocks are formed when sedimentary particles that have settled through the ocean to the sea floor become lithified. Magnetic grains among the particles—chips of the mineral magnetite, for example—would become aligned in the direction of the Earth's field while falling through the water, and this orientation would be incorporated into the rock with lithification. The **depositional remanent**

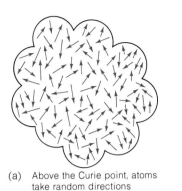
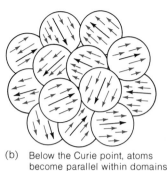
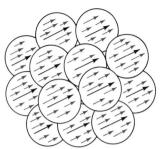

(a)   Above the Curie point, atoms take random directions

(b)   Below the Curie point, atoms become parallel within domains

(c)   Below the Curie point, in the presence of an external magnetic field, domains line up

Figure 18-9
Magnetic properties of magnetizable material. (a) At temperatures above the Curie point, heat agitates the atoms; they take random directions, and their randomly oriented magnetic fields cancel. (b) Below the Curie point, domains form in which the atoms line up, giving each domain a definite magnetization represented by the larger arrows. (c) If the material cools below the Curie point in the presence of an external magnetic field, the domains line up parallel to the external field, and the material becomes permanently magnetized.

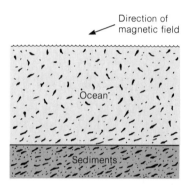

Direction of magnetic field

Ocean

Sediments

Figure 18-10
Sedimentary rock can take on a magnetic field. Magnetic mineral grains, transported to the ocean with other erosion products, become aligned with the Earth's magnetic field while settling through the water. The initial orientation is preserved in the lithified rock, which thus "remembers" the field that existed at the time of deposition.

magnetization of a sedimentary rock would then be due to the parallel alignment of all of these tiny magnets, as if they were compasses pointing in the direction of the field prevailing at the time of deposition (Fig. 18-10).

With this important tool of **fossil magnetism** or **paleomagnetism** for exploring the ancient magnetism of the Earth, scientists roamed the surface of the planet, in the 1950s and 1960s, collecting old rocks on every continent and determining their magnetism and ages in order to reconstruct the history of the geomagnetic field. This worldwide effort led to the discovery of three startling phenomena: (1) polar wandering, (2) reversals of the geomagnetic field, and (3) the sea floor as a magnetic "tape recorder."

Polar Wandering

From the direction of remanent magnetism in a rock, it is possible to determine the position of the magnetic pole at the time the rock was magnetized. For example, suppose that in a formation 600 million years old remanent magnetic inclination is zero; that is, $i = 0°$. This means that regardless of the present location of the land mass in which the formation is situated, it was close to the equator 600 million years ago; that is, the latitude $\theta = 0$, as Figure 18-5 shows. Knowing the remanent magnetic inclination of a formation, one can use the formula given earlier in this chapter to deduce the latitude the formation occupied at the time it became magnetized, which gives the distance to the ancient magnetic pole. The remanent magnetic declination tells us the direction to the ancient pole.* It is assumed, of course, (1) that the formation hasn't been rotated, folded, or magnetically disturbed and (2) that the geomagnetic field was always of the same simple configuration shown in Figure 18-5, with coincident magnetic and geographic poles. The first assumption can be checked for each sample. The second assumption is generally accepted on theoretical grounds, but also because, even though the magnetic pole has deviated slightly from the geographic pole over the past 5 million years (as today), the average position seems to center on the geographic pole.

---

*Unfortunately we cannot recover the original longitude of the sample. The magnetic inclination for a given latitude is the same for all longitudes, and the remanent declination points to the pole and thus defines the direction of a line but doesn't specify which longitude.

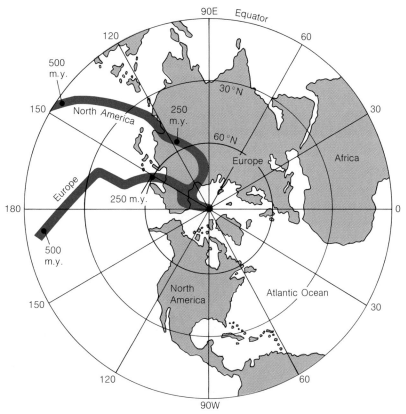

Figure 18-11
Apparent polar-wandering curves based on North American and European paleomagnetic data showing apparent positions of the magnetic pole from the present to 500 million years ago. Most geophysicists believe that the pole didn't actually wander, as the figure implies; it remained fixed, close to the geographic pole, while the "map"—that is, the outer layers of the Earth—drifted over the interior. If the Atlantic had not opened and continental drift had not occurred, the European and North American polar-wandering curves would coincide prior to Triassic time. Thus the polar-wandering curves really indicate the simultaneous northward drift of these continents and the east–west opening of the Atlantic Ocean. The polar-wandering curve, though fictitious, is a convenient way of summarizing paleomagnetic data.

Thus paleomagnetism tells us the apparent positions of ancient geographic poles.

Using these two assumptions, ancient pole positions can be plotted for remanent inclinations and declinations found in rock formations of different ages, and the apparent **polar wandering** for hundreds of millions of years can be charted as in Figure 18-11. The magnetic pole (and presumably the geographic pole) seems to have moved extensively in the past. Some 600 million years ago, it was in the mid-Pacific equatorial region. It moved to its present position gradually, following the path shown in the figure. In the absence of any external influences, however, the direction of the Earth's axis of rotation (which defines the geographic pole) cannot shift but remains fixed in space for all time. This is a requirement of the law of conservation of angular momentum, which was mentioned in Chapter 1. The only way out of this

dilemma is for the "geography" to have moved relative to the axis of rotation. To an observer in space, the geographic pole would remain fixed, and polar wandering would show as a movement of the surface features of the Earth over the globe. The continents would appear to "wander" about 90° to the north in some 600 million years, and not the magnetic pole as Figure 18-11 implies. The sites of Denver and Paris would have been south of the equator 600 million years ago. Actually, it is the plates of the lithosphere that "wander" with respect to the axis of rotation, carrying the geography along.

The apparent polar-wandering curve of Figure 18-11, though fictitious, is a convenient way of summarizing paleomagnetic data. Paleoclimatic data—that is, the rock record of ancient climates—such as the occurrence of coal beds (which indicate humid climates) in polar regions

Figure 18-12
Coal outcrop below sandstone, Victoria Land, Antarctica. The presence of coal indicates tropical or subtropical conditions, suggesting that the Antarctic continent was once closer to the equator. [Photo by W. E. Long, Ohio State University.]

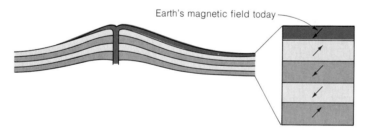

Earth's magnetic field today

Figure 18-13
Lava beds preserve the record of reversals of the Earth's magnetic field. The modern flow at the top shows the direction of the field today. Earlier flows record the directions of ancient fields that existed when the lavas cooled.

(Fig. 18–12) and glacial deposits near the equator, may be taken as additional evidence for the movement of continents, as implied by the polar-wandering curve.*

Polar wandering is really a manifestation of the mobility of the plates of the lithosphere. An observer in space would not simply see the map slip over the globe; he would also see it distorted, with the continents colliding and drifting apart and the oceans closing and opening. In a sense, paleomagnetism enables the geologist to view continental movements as would the fictitious observer in space.

Reversals of the Geomagnetic Field

Imagine the state of shock and consternation if navigators and hikers the world over were to see their compass needles flip around with the north-seeking pole pointing south. Such reversal of the Earth's magnetic field will take place in the next few hundred thousand years, but not as rapidly as the preceding sentence implies. Erratically, but roughly every half-million years, the Earth's magnetic field changes polarity, taking perhaps a few thousand years to reverse its direction.

The reversals are clearly indicated in the fossil magnetic record of layered lava flows, as shown in Figure 18–13. In the sequence of rocks, each layer represents a progressively earlier period of geological time whose age can be determined by radiometric methods. The direction of remanent magnetism can be obtained for each layer and, in this way, the time sequence of flip-flops of the field—that is, the **magnetic stratigraphy**—can be deduced. The history of reversals going back almost 7 million years has been worked out in this way (Fig. 18–14). It has been found that about half of all rocks studied are magnetized opposite to the Earth's present magnetic field, which implies that the field has flipped frequently over geologic time and that normal and reverse fields are equally likely. Although normal and reverse **magnetic epochs** (each of which is named after a famous

*The polar-wandering curve based on European paleomagnetic data is roughly parallel to the North American curve in Figure 18–11 but separated by about 30° in longitude for much of Paleozoic and Triassic time. This is because the Atlantic Ocean opened by this amount after the Triassic, that is, after the Paleozoic and Triassic rocks were magnetized. The two polar-wandering curves become approximately coincident if the paleomagnetic pole directions are rotated so as to close the Atlantic and bring the continents together. This observation, as well as similar results for other continents buttressed the case for continental drift in the middle 1950s before the sea-floor data became available.

magnetician) seem to last on the order of one-half million years, superimposed on the major epochs are transient, short-lived flips of the field, known as **magnetic events,** which may last anywhere from several thousand to 200,000 years. The Australian graduate student apparently found a new reversal event within the present Brunhes normal epoch.

Just why the field reverses is unknown. We are even uncertain whether the field dies down and builds up again in the opposite direction or simply tilts over. The simple self-exciting dynamo can have either polarity, and perhaps this indicates that the Earth's internal dynamo is easily changeable.

Some specialists on the microfossils of the sea floor have reported a curious correlation between extinctions of certain faunal types and reversals of the field. It is a matter of some concern if the disappearance of certain organisms is related to polarity changes. One explanation that has been advanced is that penetration to the Earth of lethal cosmic rays may increase when the geomagnetic field is near zero. Most cosmic rays, however, are absorbed by the atmosphere rather than deflected by the magnetic field, so that effect on life is likely to be small. Perhaps the constituents of the microfauna somehow depend on the magnetic field in their internal biochemical reactions. Perhaps an external common cause, like a large meteorite impact, affects both the fauna and the field. Perhaps the correlation is either nonexistent or pure coincidence.

Although the cause of reversals remains for future scientists to explain, their occurence has made possible an important discovery about the sea floor.

The Sea Floor as a Magnetic Tape Recorder

During World War II, extremely sensitive airborne magnetometers were developed to detect submarines by their magnetic fields. With slight modification these same instruments were adopted by oceanographers. Towed behind research ships, the magnetometers measure two things: the main planetary or geomagnetic field (see Fig. 18–5) and the local magnetic disturbance, or **magnetic anomaly,** due to magnetized rocks on the sea floor. The general, broad geomagnetic field could easily be subtracted, leaving a record of a the magnetized rocks beneath the sea. Rocks strongly magnetized in the normal direction would show as a positive anomaly; reversely magnetized rocks would show as a negative anomaly.

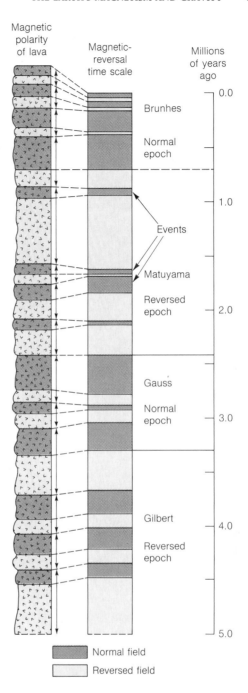

Figure 18–14
Magnetic polarities of lava flows are used to construct the time scale of magnetic reversals over the past 5 million years. In no one place is the entire sequence found; the sequence is worked out by patching together the ages and polarities from lava beds all over the world. The radiometric-magnetic time scale has recently been extended to 7 million years.

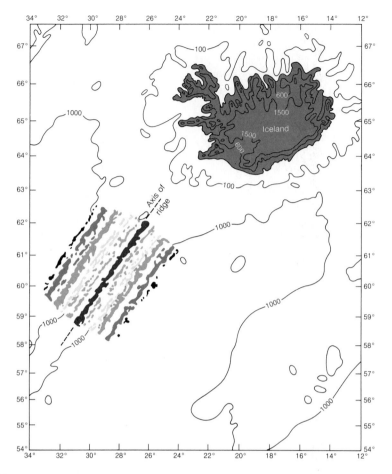

Figure 18–15

Magnetic anomaly pattern associated with the Reykjanes Ridge, a part of the mid-Atlantic ridge southwest of Iceland. The white spaces correspond to rock formations on the sea floor that are reversely magnetized. The strips shown in tones of brown or gray correspond to normal magnetization—that is, similar to the present-day direction. The almost perfect symmetry with respect to the ridge axis is emphasized by the matching tones on opposite sides; the youngest rock is brown and the oldest black, in accordance with the hypothesis of sea-floor spreading. [After "The Origin of the Oceanic Ridges" by E. Orowan, Copyright © 1969, by Scientific American, Inc. All rights reserved.]

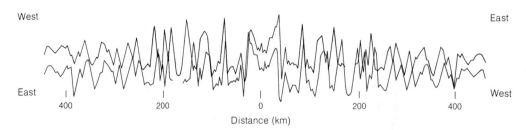

West   East

East   West

| 400 | 200 | 0 | 200 | 400 |

Distance (km)

Figure 18–16

Symmetry of the magnetic anomaly of sea-floor rocks on both sides of the ridge axis is demonstrated by reversing a record covering about 900 km of both flanks (brown) and superposing it on the record as ordinarily shown (black). [After "Sea-Floor Spreading" by J. R. Heirtzler. Copyright © 1968 by Scientific American, Inc. All rights reserved.]

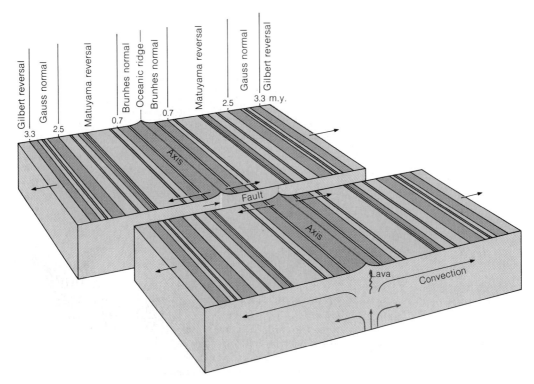

Figure 18-17
The succession of normal and reversed magnetic anomalies on the sea floor provides strong support for the concept of sea-floor spreading. Rocks of normal, or present-day, polarity are shown in shades of brown; rocks of reversed polarity are in shades of gray. The symmetry of the magnetic stripes and the correlation with the time scale of reversals worked out from lava flows on land suggest that molten rock upwelling along the ridge axis became magnetized as it cooled, was pushed out on both sides, and gradually moved outward with the separating plates. The separation of the two blocks represents a transform fault. The diagram is based on the studies of many scientists. [After "The Confirmation of Continental Drift" by P. M. Hurley. Copyright © 1968 by Scientific American, Inc. All rights reserved.]

Steaming back and forth across the ocean, sea-going scientists discovered amazing magnetic anomaly patterns like the one shown in Figure 18-15. In many areas, the bands of positive and negative magnetic anomalies are linear over hundreds of miles and show an almost perfect symmetry with respect to the crest of the mid-ocean ridge. If a map of such anomalies is folded along the ridge axis, the magnetic bands on one side fall nearly on top of those on the other side, almost like the ink-blot pattern of a Rorschach test. Figure 18-16 shows a magnetic profile across a ridge, which illustrates these features in greater detail. Note the amazing correspondence of the sequence of peaks and troughs of the magnetic-anomaly curve on both sides of the ridge.

This peculiar magnetic pattern puzzled scientists for several years until 1963, when two Englishmen, F. J. Vine and D. H. Mathews, and, independently, two Canadians, L. Morley and A. Larochelle, made a startling proposal. They reasoned that the positive and negative magnetic zones correspond to bands of rock on the sea floor that were magnetized during ancient episodes of normal and reversed magnetism of the Earth's field and that the magnetic stripes could be used as evidence in support of the theory of sea-floor spreading (Fig. 18-17). They argued that the ocean progressively widens as new sea floor is created along a crack that follows the crest of mid-ocean ridges. Lava flowing up from the interior solidifies in the crack and becomes magnetized with either normal or reversed magnetization, depending on the direction of the Earth's field at the time. As the sea floor spreads away from the ridge, approximately half the newly magnetized material moves to one side and half to the other, forming two symmetrical, magnetized bands. Newer material fills the crack, continuing the process. In this way, the sea floor acts like a tape recorder that encodes, by magnetic imprinting, the history of the opening of the ocean in terms of the history of reversals of the geomagnetic field. Using magnetic stratigraphy (Fig. 18-14) worked out from lavas on land to

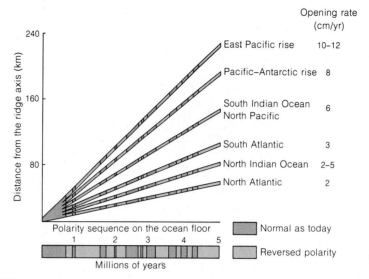

Figure 18-18
The rate of ocean spreading from magnetic anomaly patterns. The same magnetic polarity zone occurs at various distances from different ridges in different oceans, depending on the rate of spreading. Since the ages of the polarity zones are known from the magnetic-reversal time scale, the spreading rates can be calculated. The spreading rate of the Pacific Ocean along the East Pacific rise is the most rapid, amounting to 10–12 cm/yr. The North Atlantic shows the slowest rate, 2 cm/yr. [After *Continental Drift* by D. and M. Tarling, Doubleday Anchor Books. Copyright © 1971 by G. Bell and Sons, Ltd.]

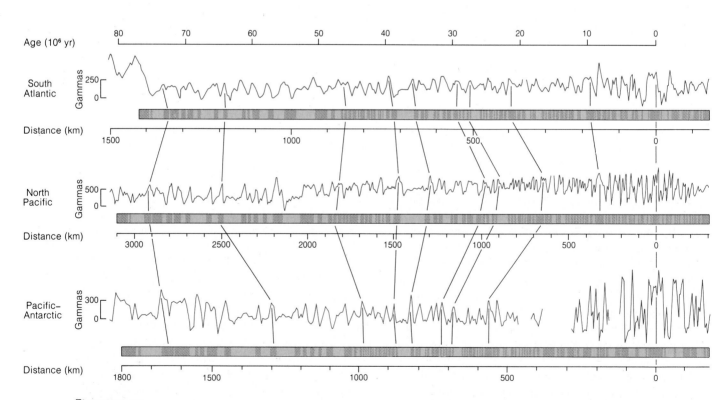

Figure 18-19
Magnetic anomalies recorded over the oceans reveal the same succession of magnetized lava formations on the sea floor, depicted by the dark and light stripes. The spacing may differ because the spreading rates vary, but each ocean shows the same sequence of some 171 reversals over the past 76 million years. The illustration shows the correlation of the anomalies from ocean to ocean. [After "Sea-Floor Spreading" by J. R. Heirtzler. Copyright © 1968 by Scientific American, Inc. All rights reserved.]

date magnetic bands on the sea floor, geophysicists learned to "replay the tape" and showed that the magnetic data recorded how fast the ocean opened up. We would expect the rocks on the crest of the ridge to be normally magnetized, because they were extruded during the present normal magnetic epoch. Reversely magnetized rocks corresponding to a magnetic epoch of about one million years ago have been displaced some distance from the ridge—say, about 15 kilometers on each side of the ridge crest if the spreading rate is 1.5 centimeters per year. The reversal time scale can actually be followed through many oscillations of the Earth's field, and the corresponding magnetic bands extend out to many hundreds of kilometers from the ridge crest (Fig. 18–18).

The magnetized strips of sea floor provided a powerful tool to extend the history of magnetic-field reversals back in time almost 200 million years into the Jurassic Period. Precise dating of reversals (and therefore of spreading rates) goes back only about 7 million years because the radiometric method of dating lava flows on land loses accuracy beyond this time. By assuming the spreading rate determined for the past 7 million years to be representative of a much longer period, the older normal and reversed magnetic epochs could be correlated over all the oceans and assigned dates. Figure 18–19 shows the sequence of 171 reversals, extending back 76 million years to the Cretaceous Period, as they were worked out by this method. The magnetized strips on the sea floor are shown schematically below the observed magnetic anomalies. Note that the spacing of the strips differs in the several oceans because the sequence of reversals is compressed or stretched out according to slower or faster spreading.

After geologists study a region, they often present the sequence of rock layers extending back in geological time as a stratigraphic section. Magnetic anomalies on the sea floor enable geophysicists to plot a "magnetic–stratigraphic" section, as in Figure 18–20, showing the historical sequence of worldwide magnetic reversals back in geological time to the Jurassic Period.

The power and convenience of the new magnetic stratigraphy in working out the history of ocean basins cannot be overemphasized. Simply by steaming back and forth, measuring the magnetic field of the magnetized rocks of the sea floor, and correlating the pattern of reversals with the sequence shown in the preceding two figures, ages can be assigned to different regions of the sea floor without even examining rock samples! All one needs is a good magnetic record, and these

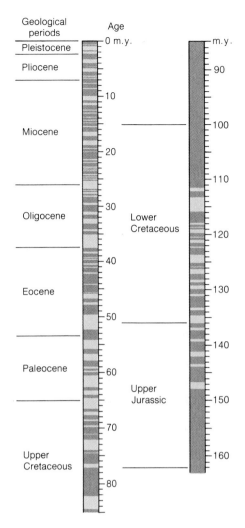

Figure 18–20
Geomagnetic reversal time scale from the present to the base of the Upper Jurassic 162 million years ago. Shaded intervals are normal; white intervals indicate reversed fields. [After J. R. Heirtzler, R. L. Larson, and W. C. Pitman III, *Bulletin of the Geological Society of America*, v. 83, 1972.]

have already been obtained and interpreted for large sections of the world's oceans.

Isochrons (contours of the age of the sea floor) obtained in this way show the time that has elapsed and the amount of spreading that has occurred since the magnetized rocks were injected as lava into a mid-ocean rift. Note how the isochrons in Figure 18–21 show progressively older sea floor on both sides of the ridge-rifts. The more widely spaced isochrons of the east Pacific signify a faster spreading rate than those of the Atlantic. We will see in the next chapter that these "magnetic ages" were verified and extended when the deep-sea drilling project brought back rock samples from the sea floor that could be dated in the laboratory using fossils and radiometric methods. What a coup for the scientists who discovered this tool!

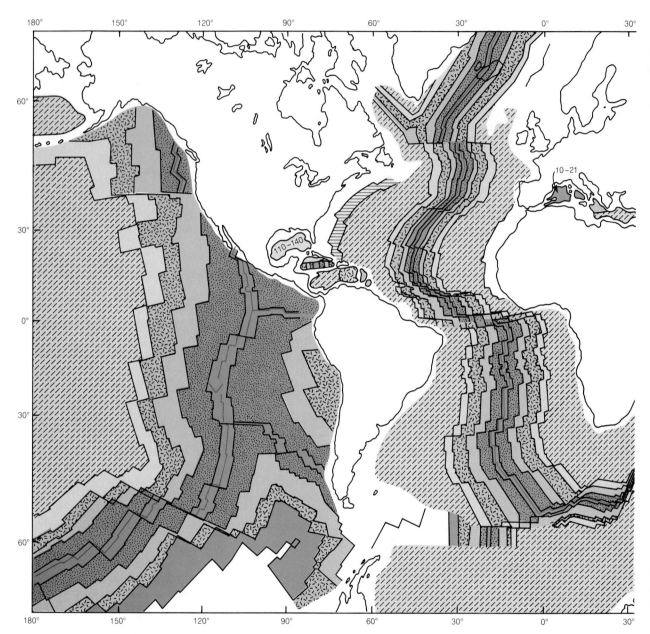

Figure 18-21
A worldwide pattern of sea-floor spreading is revealed by the age of the sea floor as determined from magnetic data. Mid-ocean ridges, along which new sea floor is extruded, coincide with the youngest sea floor. The dating of magnetic anomaly stripes makes it possible to establish isochrons (shown by different colors and textures) that give the age of the sea floor in millions of years since creation at the

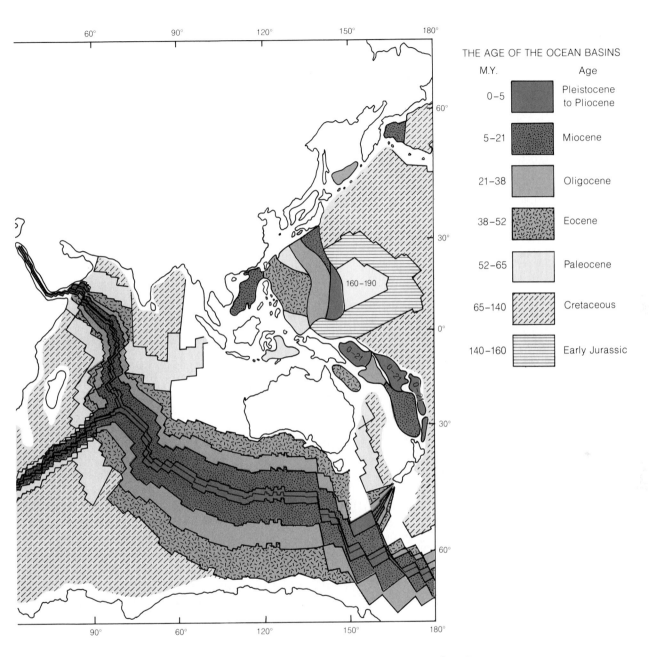

THE AGE OF THE OCEAN BASINS

M.Y.		Age
0–5		Pleistocene to Pliocene
5–21		Miocene
21–38		Oligocene
38–52		Eocene
52–65		Paleocene
65–140		Cretaceous
140–160		Early Jurassic

160–190

0–21

ridges. The Atlantic Ocean is symmetrical about the mid-Atlantic ridge. Asymmetry of the pattern in the Pacific is partly caused by subduction in the Aleutian trench south of Alaska, in the Peru-Chile trench along the west coast of South America, and in many trenches in the Western Pacific. [After map prepared by John Sclater and Linda Meinke of MIT.]

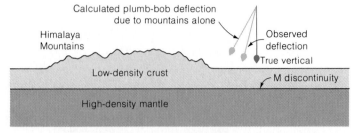

Figure 18-22
A plumb line ordinarily hangs in a vertical position. Near a mountain system we would expect the plumb bob to be deflected toward the mountains because of the gravitational attraction of their mass. The observed deflection is typically less than expected, a discrepancy whose explanation led to an important discovery. The diagram exaggerates the amount of deflection, which is small but readily measurable.

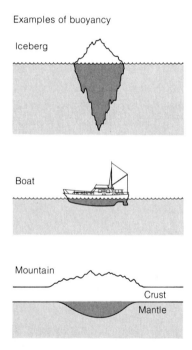

Figure 18-23
Examples of buoyancy. Icebergs and ship hulls float because the volume submerged is lighter than the volume of water displaced. Similarly, the volume of relatively light crustal rock projecting into the denser mantle provides a buoyant force that supports the mountain mass above.

Magnetized Moon Rocks—A Puzzle

Unlike the Earth, the Moon has no planet-wide magnetic field. There is no question about this. Soviet and American spacecraft have been unable to detect such a field after several efforts. Yet magnetized rocks have been found lying on the lunar surface. Discordant data are the stuff of great discoveries, and planetary scientists are vying to explain these seemingly contradictory results. The leading hypothesis at this time proposes that the Moon rocks in their remanent magnetism "remember" an earlier period of lunar history some 3 to 4 billion years ago (the age range of the rocks), when the Moon did have a planetary magnetic field. This magnetic field implies in turn the existence at this early time of a small liquid iron core that has since cooled and solidified. Is there a better way to manifest the power of modern geological and geophysical methods than to fetch a rock from the lunar surface, date it, measure its magnetic field, and then describe the physical state at the center of the Moon billions of years ago?

EXPLORING THE EARTH WITH GRAVITY

The Indian Puzzle

Some 150 years ago during the great land survey of India, a curious discrepancy was uncovered by British surveyors. The distance between Kaliana, some 100 kilometers (60 miles) south of the Himalaya range, and Kalianpur, 600 kilometers (375 miles) farther south, was determined in two precise ways—by measurement over the surface and by reference to astronomical observations—and the results disagreed by some 150 meters (500 feet) in 600 kilometers. This may seem a small amount, but it was an intolerable surveying error even by nineteenth-century standards. The astronomical method of measuring distance uses the angles of stars with respect to the vertical, which is defined by a plumb line (a weight suspended on a string). To account for the difference, it was proposed that the plumb line was tilted toward the Himalayas because of the gravitational attraction of the mountains on the plumb bob, causing an error in the distance measurement. When the effect was actually calculated, it was found that the mountains should have introduced an even larger error—one of about 450 meters (1500 feet)—thus compounding the puzzle (Fig. 18-22).

In 1865 no less a figure than the Astronomer Royal, Sir George Airy, came forward with an explanation for this discrepancy that contained the basis of the principle of isostasy discussed in Chapter 17. Airy proposed that the enormously heavy mountains are not supported by a strong, rigid crust below, but that they "float" in a "sea" of denser rock. Stated otherwise, the excess mass of the mountains above sea level is **compensated** by a deficiency of mass in an underlying root. This root provides the buoyant support, in the manner of all floating bodies, just as a ship with a deep hull is buoyed up (Fig. 18-23). The plumb bob "feels" both the excess mass on top and the deficiency of mass below, hence the reduced deflection (Fig. 18-24). The resolution of the Indian puzzle not only led to the concept of isostasy but also introduced **gravity surveying** as a method for detecting mass variations in the interior by their corresponding gravity variations.

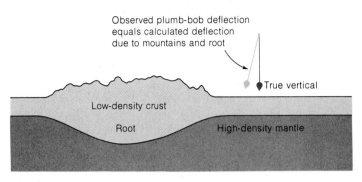

Figure 18-24
The discrepancy between the observed and expected deflection of the plumb bob in Figure 18-22 can be reconciled if the excess mass of the mountain is compensated by a deficiency of mass in a "light" crustal root below. The root provides buoyant support for the mountain, which otherwise would sink into the mantle.

Gravimeters

The local value of gravity, g, can be obtained from the period* of a swinging pendulum or from the acceleration of a falling weight—an experiment performed in every elementary physics course. Pendulums have been used in gravity surveys, but they have mostly been supplanted by the modern gravimeter. This is a device no more complex than a weight on a spring that stretches or contracts as gravity increases or decreases from place to place (Fig. 18-25). Although the principle is simple, the engineering of the gravimeter is most elegant: This device, not much larger than a thermos bottle, can detect gravity variations as small as 10^{-8} Earth's gravity (g). The standard unit of acceleration used in gravity surveys is the milligal, which is 0.001 centimeter per second per second. A modern gravimeter can easily measure the difference in gravity between a table top and the floor, even though the table top is only one meter farther from the center of the Earth!

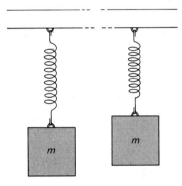

Figure 18-25
The gravimeter is simple in concept but elegant in implementation. A mass attached to a spring experiences a larger or smaller pull as gravity varies. The corresponding extensions or compressions of the spring are measured very precisely, so that small changes in gravity can be observed.

Gravity Surveying

The millions of dollars required to develop the sensitivity achieved by the modern gravimeter were underwritten by the oil industry. This was the direct result of the recognition some forty years ago that the buried geologic structures in

which oil is trapped (such as folds, faults, and salt domes) often produce variations in the normal gravitational field that are detectable by sensitive instruments. Such a **gravity anomaly** is caused by a change in subsurface mass due to a mountain root, a salt dome, or any other lateral geological change (see Fig. 18-26). The idea is to use its gravity effect to find and describe any anomalous mass—anything more or less dense than the average rock—and thus to explore the subsurface geology by making gravity surveys. Before a profile like that of Figure 18-26 can be produced, how-

---

*Time required for one complete oscillation.

ever, three important corrections must be applied to the value of g from each station at which the gravimeter is read; otherwise, the most interesting anomalies would be obscured.

1. If the Earth were spherical and nonrotating, the gravitational attraction of the planetary body would be the same everywhere on the surface. Because of centrifugal force, however, things tend to fly outward from a rotating body. Hence, gravity is less at the equator; that is, things weigh less there than at the poles. The same centrifugal force makes the Earth bulge outward at the equator and

Gravity anomaly

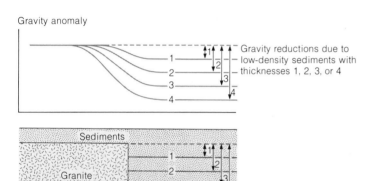

Figure 18-26
Schematic illustration of a gravity anomaly. The value of gravity changes across the structure shown because the less-dense sediments contain less mass than an equal volume of granite. The thicker the sedimentary deposit, the greater the decrease in gravity, as the curves show.

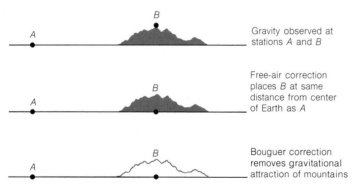

Figure 18-27
Gravity is measured at A and B to see if there is a difference in subsurface mass. To emphasize subsurface effects, corrections are made to the value of gravity at B, as if to bring B to the same elevation as A and also to remove the obvious gravitational attraction of the mountain. Any remaining gravity difference between A and B is ascribed to a change in subsurface geology.

flatten at the poles. Remember that g is proportional to $1/R^2$, in which R is the distance to the center of the planet. Because R decreases by 21 kilometers going from the equator to the pole of our flattened or spheroidal planet, g increases with latitude. An international formula has been adopted that best describes g everywhere on the Earth's surface, taking into account the Earth's shape and rotation.* This formula value is subtracted from the gravimeter reading in the search for anomalies, which could otherwise be masked by the big effects of rotation and flattening.

2. Because of topographic variations, gravity stations generally differ in elevation—that is, in distance R from the center of the Earth. In our search for anomalous subsurface masses, we must remove the obvious effect of elevation on the local gravity value (using the fact that g varies as $1/R^2$ to get the correction), and all readings are corrected as if they were made at sea level. This is called a **free-air correction,** and it amounts to adding 0.31 milligals for each meter of elevation above sea level.

3. Finally, in order to highlight subsurface effects, it is important to account for all obvious near-surface masses that affect gravity. The free-air correction reduces the reading to sea level only partially, because it allows for the distance to the center of the Earth but not for the attraction of the mass of rock between the station and sea level. The correction that completes the subtraction of the local topography is called the **Bouguer correction.** It amounts to subtracting about 0.1 milligals for each meter of rock between sea level and the point of the reading. For gravity surveys on the ocean, where g is measured at sea level, the Bouguer correction corrects for the low density and, therefore, the low gravitational attraction of water by increasing the gravimeter reading by the amount necessary to "convert" the ocean to rock. In this way, the obvious gravity deficiency of the ocean is removed in order to emphasize anomalous suboceanic masses. The free-air and Bouguer corrections are depicted diagrammatically in Figure 18-27.

If there were no local variations in mass in the interior, the sum of all these corrections and the gravimeter reading would be close to zero—there would be no gravity anomaly, because everything

---

*By international agreement the value of $g = 978, 049(1 + 0.0052884 \sin^2 \phi - 0.0000059 \sin^2 2\phi)$ milligals, in which ϕ is the latitude. Thus, gravity is about 0.5 percent stronger at the North Pole where $\phi = 90°$ and $\sin \phi = 1$ than at the equator where $\phi = 0°$ and $\sin \phi = 0$.

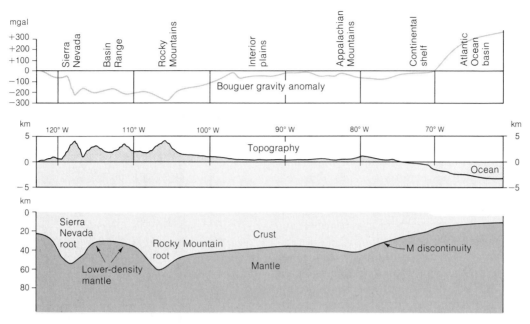

Figure 18-28
Transcontinental gravity survey from the Pacific Ocean to the Atlantic Ocean. The negative gravity anomalies over the mountainous regions, the near-zero values at low elevations, and the positive values over deep oceans, mirroring the topography, demonstrate the role of isostatic compensation in shaping Earth's surface features.

would have been taken into account. However, where the sum differs from zero, we have found an anomalous mass—an ore body, an intrusion, a sedimentary basin, or a mountain root. The shape and magnitude of the gravity anomaly helps define the dimensions and density of the anomalous rock mass.

Imagine driving across the United States, stopping every mile to measure gravity. You would thereby produce a gravimetric profile of the country crossing many geological provinces, and it might be interesting to see how the gravimeter can aid us in interpreting the geological features. This has actually been done, and the resultant gravity-anomaly profile is shown in Figure 18-28, which includes an extension into the Atlantic Ocean made with a shipboard gravimeter. The top and middle sections of the figure show the gravity profile and the topography, respectively. All the corrections have been applied to the gravity curve—that is, all the land mass above sea level and all of the *lack* of land mass in the oceans have been accounted for—so this wiggly curve should reflect subsurface mass variations. Note the negative values—that is, the deficiency in gravity—across the Sierra Nevada, the Rocky Mountains, and the Basin and Range province. The anomaly is close to zero in the low plains, becomes slightly negative under the Appalachian

Mountains, approaches zero again in the coastal plains and continental shelf, and zooms to large positive values over the deep Atlantic.

The striking feature revealed by the gravity anomaly and shown in the bottom section of Figure 18-28 is the way the Mohorovičić discontinuity mirrors the topography (except in the Basin and Range province). The gravity anomaly is strongly negative where the crust thickens to provide buoyant support for mountains. What causes the negative anomaly in these topographically high places is the mass deficiency due to the displacement of denser mantle by the less dense crustal root. The high positive gravity values over the ocean basin signify the presence of excess mass; dense mantle rock is much closer to the surface here. This feature has been called an **anti-root,** and it mirrors the "negative" topography (water instead of rock) of the ocean basin. The Appalachians show a modest negative anomaly, which indicates that they have a shallow root. This is appropriate for an old mountain system; its root (and g anomaly) is disappearing as its topography erodes away. One might have expected the structurally high Basin and Range province, with its average elevation of about one kilometer, to have a slightly thickened crust to go along with its negative anomaly. Actually, many geophysicists thought this to be the case until experiments with seismic waves revealed the thin crust. Because the seismic information reveals that there is no crustal root for the Basin and Range province, we deduce that the mass deficiency there is a result of the area being underlain by mantle material of relatively low density, as shown in the figure.

Many of these results were anticipated in Chapter 17, where we discussed isostasy in the context of the structure of the crust and upper mantle as determined by seismic methods. Actually, the concept of isostatic compensation with its notions of "floating" continents and still higher floating mountains was discovered from gravity observations such as these. But, seismological data contributed much to our understanding by clearing up such questions as where the mass deficiencies are located and whether compensation involving crustal roots or compensation via low-density mantle is the isostatic mechanism responsible for them. Low-density mantle seems to go with a tectonic setting that includes recent volcanism, high heat flow, and low seismic velocities—which implies, perhaps, a partially molten mantle directly below the Moho. Some geologists suggest that these features, as they occur in the Basin and Range province, indicate that tension-producing forces, perhaps due to a spreading or divergence zone, are active within a continent. Compensation involving crustal roots is the predominant isostatic mechanism for continents as a whole, as well as for high mountains.

In this way, gravimetry and seismology combine to reveal the importance of isostasy to the study of geology on a regional and continental scale. The following examples show what kinds of insight the combination can offer. Eroding mountains will be pushed up by the excess buoyancy of the root until both root and mountain range nearly disappear. Such loads as ice caps or sediments filling a basin can depress the crust. Where gravity studies indicate that a positive or negative load (a depression, for example) is not compensated, some force must be present that helps support it and keeps it from subsiding or rising. Continents are unlikely to be destroyed by subduction; their ability to float allows them to "ride out" repeated episodes of splitting and collision.

The Fennoscandian Uplift— Nature's Experiment with Isostasy

If you depress a cork floating in water with your finger and then release it, the cork pops up almost

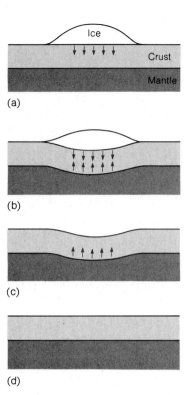

Figure 18-29
The mechanism of postglacial uplift. (a) A continental glacier grows, loading the crust. (b) The crust sags, and a root develops to support ice load isostatically. (c) The glacier disappears, but the root remains because of viscosity of mantle. The root is evidenced by negative gravity anomaly. (d) Buoyancy of the root leads to slow uplift. The root disappears, the surface assumes its original level, and the gravity anomaly disappears. Arrows depict the direction of forces due to ice load and root. Fennoscandia today is between stages c and d. Figure not to scale: the crust is about 40 km thick; a 3-km thick glacier would produce a root about 1 km thick—less if the strength of the crust supported part of the weight.

instantly. A cork floating in molasses would rise more slowly; the drag of the viscous fluid would slow down the process. If we could perform a similar experiment on the Earth, we could learn much about the viscosity of the mantle and how it affects rates of uplift and subsidence. How convenient it would be if we could push the crust down somewhere, remove the force, and then sit back and watch it rise.

Nature has been good enough to do this experiment for us. The load is a continental glacier—an ice sheet 2 to 3 kilometers thick that can appear with the onset of an ice age in the geologically short period of a few thousand years. The crust is depressed by the ice load, and a downward bulge develops on its underside to provide buoyant support. At the onset of a warming trend, the glacier melts rapidly. With the removal of the load, uplift of the depressed crust begins (Fig. 18-29). The rate of uplift can be documented by dating ancient beaches that are now well above sea level (see Fig. 10-7). Such raised beaches can tell us how long ago a particular stretch of land was at sea level. A negative gravity anomaly tells how much of a relic root remains and how much more uplift will occur before the root disappears.

Such depression and uplift has occurred in Norway, Sweden, and Finland, as well as elsewhere in glaciated regions. The ice cap retreated from these regions some 10,000 years ago, and the land has been rising since. Figure 18-30 shows how much upwarping took place in the past 5000 years. The most intense upwarping has occurred near west-central Sweden, which is believed to have been overlain by the thickest ice. Some 200 meters of uplift has occurred in 10,000 years, an average rate of 2 centimeters per year. The remaining negative gravity anomaly of −50 milligals implies that part of the root still remains and that about 200 meters more uplift must occur before isostatic compensation is complete. A few minor earthquakes occur in the region, additional evidence, perhaps, that stresses due to too much buoyancy are still present. Geophysicists have used these data to show that the weak zone (which "flows" so that the crustal root could develop) coincides with the partially molten asthenosphere. This region is important not only as a factor in the mobility of the lithospheric plates and as the source of basaltic magma; its ability to yield makes it the key factor of isostatic compensation.

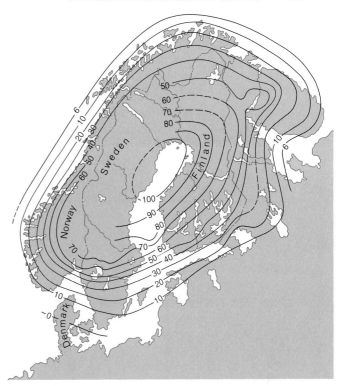

Figure 18-30
The uplift (in meters) in Fennoscandia in the past 5000 years, according to M. Sauramo. The crust, depressed by the weight of the ice cap of the last ice age, is still rebounding some 10,000 years after the ice disappeared.

No human could have designed a better experiment to demonstrate the isostatic mechanism than nature's demonstration with the Fennoscandian postglacial uplift.

SUMMARY

1. Motions in the fluid iron core somehow set up a dynamo action that generates the Earth's magnetic field. The field can be fairly well described by a hypothetical bar magnet located near the center of the Earth and approximately aligned with its axis of rotation.

2. Many rocks became magnetized in the direction of the geomagnetic field that prevailed when they were formed. If the rocks are dated radiometrically, the history of the magnetic field can be recovered from this remanent magnetization.

3. The remanent magnetization of old rocks suggests that the Earth's magnetic pole occupied different positions in the past. Actually, the magnetic pole probably did not wander but stayed fairly close to the geographic pole. The apparent polar wandering is probably an indication that the lithospheric plates have been moving, changing the geography of the surface. Evidence such as coal in Antarctica and glacial deposits near the equator support this idea.

4. Remanent magnetization has also led to the discovery of reversals in the magnetic field. The history of reversals since Cretaceous time has been worked out. Although unexplained, these reversals have become a very important tool in dating the sea floor. When the sea-floor crust is formed at mid-ocean rifts, it becomes magnetized. This

magnetic imprinting stays with the crust as it spreads away from the rift. The sequence of reversals shows as positive and negative magnetic anomalies, which a surveying ship can readily detect. Using the history of reversals, we can determine the age of the underlying sea floor and the rate of sea-floor spreading.

5. Gravitational changes over the surface of the Earth are due to the planet's oblateness, its rotation, its topography, and differences in its subsurface mass. The first three factors can be allowed for, so that the remaining gravitational anomalies indicate subsurface geological differences.

6. Gravity anomalies associated with continents, oceans, and mountains show that the Earth's crust is not strong enough to support topographic loads over long periods. Roots, or downward bulges of the crust, develop and provide buoyant support. This is an example of isostatic adjustment. Another example is the depression of the Fennoscandian crust by the weight of a continental glacier. Although the ice cap disappeared some 10,000 years ago, uplift is still continuing in that area and will continue until the relic root disappears.

EXERCISES

1. What evidence supports the hypothesis that the Earth's magnetic field originates in a fluid iron core? The remanent magnetization in some meteorites and Moon rocks has yet to be explained. Would you hazard a guess as to its origin?

2. In a region where the geothermal gradient—that is, the temperature increase with depth—is 3°C per hundred meters, at what depth would you expect the rocks to lose their magnetism?

3. A sedimentary formation is found to have remanent magnetism, with an inclination of 45° measured from the bedding plane. What was the latitude of the formation when it became magnetized? What was its longitude? Would your answer change if the bed had been tilted after it was magnetized?

4. What is the connection between the sequences of magnetic reversals worked out on land and the bands of positive and negative magnetic anomalies found on the sea floor? There are regions of the sea floor known as "magnetic quiet zones" where no reversals in magnetic anomalies occur. Can you guess the age of the crust in these regions from the magnetic reversal time scale in Figure 18–20?

5. Would the gravity anomaly (after making free-air and Bouguer corrections) show large negative values, values near zero, or large positive values at each of the following places? (a) Rocky Mountains. (b) East coast of the U.S. (c) Middle of an ocean basin. Explain your answers.

6. You observe a near-zero gravity anomaly atop a mountain after making the free-air and Bouguer corrections. What would you conclude?

BIBLIOGRAPHY

Carrigan, C. R., and D. Gubbins, "The Source of the Earth's Magnetic Field," *Scientific American,* February 1979.

Cox, A., ed., *Plate Tectonics and Geomagnetic Reversals.* San Francisco: W. H. Freeman and Company, 1973.

Garland, G. D., *The Earth's Shape and Gravity.* New York: Pergamon Press, 1965.

Heirtzler, J. R., "Sea-Floor Spreading," *Scientific American,* December 1968. (Offprint 875.)

LePichon, X., J. Francheteau, and J. Bonnin, *Plate Tectonics.* New York: Elsevier Publishing Company, 1973.

Strangway, D. W., *History of the Earth's Magnetic Field.* New York: McGraw-Hill, 1970.

Takeuchi, S., S. Uyeda, and H. Kanamori, *Debate about the Earth,* rev. ed., San Francisco: Freeman, Cooper & Company, 1970.

19

Global Plate Tectonics: The Unifying Model

Geologists are gradually rejecting the picture of a rigid Earth with fixed continents and ocean basins. Most now believe that the Earth's lithosphere is broken into about a dozen plates, which for reasons not fully understood move over the interior. Plates are created along the crests of mid-ocean ridges and are pushed down into the mantle near deep-sea trenches. Continents, embedded in the lithosphere, drift along with the moving plates. Plate tectonics gives new life to the old ideas of continental drift and explains the distribution of many large-scale geological features and zones of activity—narrow belts of mountains, volcanic and seismic activity—in terms of their association with plate margins.

Just a few years after the proposal of plate tectonics theory, on the occasion of an international geophysical meeting in Moscow, an interesting exchange took place between two western participants—a younger man who had achieved prominence because of his work on plate tectonics and a well-known older scientist. The setting was a party in the apartment of a Soviet geophysicist, and the conversation was well lubricated by vodka. The din of cocktail party chatter stopped suddenly when the younger man called out to his older colleague, "Dr. ———, everyone tells me how brilliant you were in your younger days. If that's the case, why didn't you discover sea-floor spreading and plate tectonics twenty years ago?" The explosive response of the older man needn't be recorded, but the question, properly generalized, is indeed thought-provoking. Why did this concept, which unifies so much of geological thought, "arrive" so late in the history of the subject?

Actually, a key element of the concept—large-scale displacement of continents—had been around for a long time. The jigsaw-puzzle fit of the coasts on both sides of the Atlantic had not escaped the notice of early natural philosophers.

Francis Bacon remarked on the parallelism of the facing shores of the Atlantic in 1620. As early as 1858 Antonio Snider published maps in France depicting continental drift. At the close of the nineteenth century, the Austrian geologist Eduard Suess put some of the pieces of the puzzle together and postulated the former existence of a single giant continent—Gondwanaland, made up of the combined present-day southern continents. Early in this century, Alfred Wegener, a German meteorologist, cited as further evidence of continental drift the remarkable similarity of rocks, geological structures, and fossils on opposite sides of the Atlantic. In the years following, Wegener continued to build the case for continental drift; he postulated that a supercontinent called Pangaea, once made up of all the present continents, began to break up some 200 million years ago, with ocean filling the widening gaps (Fig. 1–14).

Although the theory received serious attention for about a decade, "continental drift" never caught on except among some geologists in Europe and South Africa. Although they could point not only to geographic matching but also to geological similarities in rock ages and structural trends, the proponents could not come up with a plausible driving force. Drift advocates buttressed their speculation with special pleading, selecting evidence patently favorable to their views, evidence that was far from incontrovertible. But there were significant arguments—accepted now as good evidence of drift—based on fossil and climatological data. The evolution of vertebrates and land plants showed similarities in development on different continents up to the supposed breakup time; thereafter they showed divergent evolutionary paths. The distribution of Permian glacial deposits in South America, Africa, India, and Australia was difficult to explain in terms of separate glaciers, some close to the equator. Drift advocates noted that if the southern continents are reassembled into Gondwanaland in the south polar region, a single continental glacier could account for all the glacial deposits. "It has always happened that after several distinguished palaeontologists have presented evidence favourable to continental drift, some other equally distinguished ones have proceeded to point out other facts that are made more difficult to explain"—so argued Sir Harold Jeffreys in his influential book *The Earth* (1929). Geology and paleontology were not enough. Independent, diverse, corroborative evidence from geophysics would be needed to persuade the scientific establishment to abandon prevailing ideas and elevate an unorthodox speculation to the level of a generally accepted theory.

In 1928, Arthur Holmes invoked the mechanism of thermal convection in the mantle as the driving force. Holmes proposed that subcrustal convection currents "dragged the two halves of the original continent apart, with consequent mountain building in the front where the currents are descending, and the ocean floor development on the site of the gap, where the currents are ascending." Holmes came close to expressing the modern notions of plates, divergence, and subduction when he speculated that a subcrustal basaltic layer serves as a conveyor belt that carries a continent along to the place where the belt turns downward into the mantle, leaving the continent resting on top. Nevertheless, Holmes himself recognized the tenuous nature of his views. He wrote that "purely speculative ideas of this kind, specially invented to match the requirements, can have no scientific value until they acquire support from independent evidence."

Convincing evidence began to emerge as a result of extensive exploration of the sea floor during the years following World War II. In particular, the mapping of the mid-Atlantic ridge and the discovery of the deep, cracklike valley, or rift,

running down its center line sparked much speculation. In the early 1960s Harry Hess of Princeton University suggested that sea floors separate along the rifts in mid-ocean ridges and that new sea floor forms by upwelling of mantle materials in these cracks, followed by lateral spreading (see Fig. 15–37). Vine and Mathews' work showed how the oceanic magnetic patterns (Chapter 18) could be explained by Hess' concept. Thus was born the theory of sea-floor spreading. Within a few years abundant confirmation was available from the study of such diverse evidence as that provided by worldwide magnetic-anomaly surveys, the observation of earthquake mechanisms, the measurement of heat flow, and the determination of the thickness and age of the sedimentary layers of the sea floor.

It remained for the next generation of geophysicists to broaden the concepts of continental drift and sea-floor spreading into the more general theory of plate tectonics. Beginning about 1967, they extended the idea of Hess and Canadian geophysicist J. T. Wilson about the mobility of the lithosphere by identifying the separate lithospheric plates and discussing their relative motions and the phenomena that occur at their boundaries. By the end of the 1960s the evidence became so persuasive that most Earth scientists, except for a few prominent holdouts, embraced these concepts. Textbooks were revised, and specialists began to think of the implications that the new discoveries held for their own fields.

Let us return to the question of why these new concepts became generally accepted so late in the history of geology. There are different styles among scientists. Some scientists—those with particularly inquiring, uninhibited, and synthesizing minds—perceive great truths before others. Although their perceptions may turn out to be false, these individuals are often the first to see the great generalizations of science. Most scientists, however, proceed more cautiously and wait out the slow process of gathering supporting evidence. The concepts of continental drift and sea-floor spreading were slow to be accepted simply because the audacious ideas came so far ahead of the firm evidence. The oceans had to be explored, a new worldwide network of seismographs had to be installed and used, the magnetic stratigraphy had to be painstakingly worked out, and the deep sea had to be drilled before the majority could be convinced. In a well-known European laboratory, a list was compiled (in good humor) of the names of Earth scientists in the order of the date of their acceptance of sea-floor spreading as a confirmed phenomenon. The names of scientists of distinction appear at both the top and the bottom of the list.

PLATE TECTONICS: A REVIEW AND SUMMARY

Plate tectonics is the conceptual framework of this book, and we have already introduced the basic ideas in earlier chapters. In this chapter we draw together and review the diverse lines of evidence that support the theory of plate tectonics, using primarily illustrations you have already seen in earlier chapters. We will begin a discussion of rock associations and orogeny within the framework of plate tectonics. The fragmentation of Pangaea since the Jurassic will be reviewed together with some speculation about continental drift and extinct plates in the pre-Jurassic. The chapter will close with some brief remarks on the driving mechanism of plate tectonics, but the reader shouldn't expect more than vague speculations, for the subject is just beginning to receive serious study.

The Mosaic of Plates

According to the theory of plate tectonics, the rigid plates whose outlines are shown in the illustration inside the back cover slide over a partially molten, plastic asthenosphere in the general directions shown. According to the relative motions of adjacent plates, we can define three kinds of plate boundaries: (1) zones of divergence or spreading, typically ocean ridges; (2) fracture zones, or transform faults; and (3) zones of convergence (Fig. 17–13).

Zones of divergence are boundaries along which plates separate. In the process of plate separation, partially molten mantle material upwells along linear ocean ridges, and new lithosphere is created along the trailing edges of the diverging plates. Such boundaries are characterized by active basaltic volcanism, shallow-focus earthquakes caused by tensile (stretching) stresses, and high rates of heat flow. The outpouring of magma along ocean ridges and the building of the oceanic lithosphere are volumetrically the most significant form of volcanism. Figures 1–13, 1–17, 13–8, 13–11, 15–19, and 15–37 emphasize the different aspects of divergence zones.

Typically transform faults are boundaries along which plates slide past one another, with neither creation nor destruction of lithosphere. Sometimes marked by scarps, transform faults are characterized by shallow-focus earthquakes with horizontal slips. Occasionally there occur "leaky" transforms, in which some volcanism and slight plate separation accompanies the transform. Examples are in Figures 1–17, 17–13, and 18–17.

Zones of convergence are boundaries along which the leading edge of one plate overrides another, the overridden plate being subducted, or thrust into the mantle, where lithosphere is resorbed. The thrusting mechanisms that operate along these collision boundaries tend to produce deep-sea trenches, shallow- and deep-focus earthquakes, adjacent mountain ranges of folded rocks, and both basaltic and andesitic volcanism. Convergence zones are illustrated in Figures 1–16, 13–8, 13–11, and 15–37.

Each plate is bounded by some combination of these three kinds of zones, as can be seen on the inside of the back cover. For example, the Nazca Plate in the Pacific is bounded on three sides by zones of divergence, along which new lithosphere forms, and on one side by the Peru–Chili trench, where lithosphere is consumed. Continental margins may or may not coincide with plate boundaries. If they do, the continents tend to remain "afloat" because continental plates are thicker and too buoyant to be readily subducted. Where two plates with continents at their leading edges converge, the crust thickens to form great mountain ranges like the Himalayas.

The global sum of plate creation and consumption is approximately zero. The Earth would otherwise change size in order to accommodate the new sea floor, and this doesn't seem to be happening. Instead, the plates form and disappear and change in size and shape as they evolve.

The Structure and Evolution of Plates

Figure 19–1 depicts some of the structural details of a rigid lithospheric slab, a plate, from its region of generation at a ridge axis to its region of subduction, where it is resorbed. Both oceanic and continental crust cap the plate; the continent, embedded in the moving plate, is carried along passively by it. Thus, in a real sense, continental drift is simply a consequence of plate movements. Underneath is the plastic, partially molten asthenosphere—source of the raw materials that build new lithosphere. Once heated and partially melted, subducted lithosphere becomes a source of magma, which rises to feed the overlying vol-

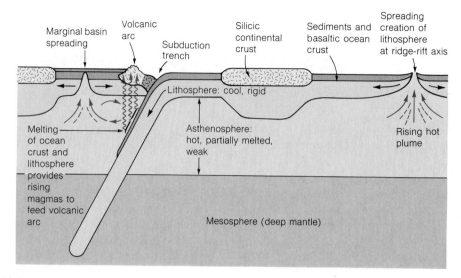

Figure 19–1
Cross section of the upper mantle. The lithosphere is a plate of solidified rock that rides on the partially molten asthenosphere. It is approximately 70 km (40 mi) thick under oceans and perhaps 100 to 150 km thick under continents. The continent is embedded in the plate and moves along with it. The lithosphere forms at mid-ocean ridges from a rising plume of partially molten rock; it sinks back into the mantle in subduction zones, where it remelts. Arrows in the plastic asthenosphere indicate directions of possible convective motions. Secondary convection currents may form small spreading centers in a marginal basin. The Sea of Japan, for example, formed in this way. Located behind an island-arc volcanic chain, it separates the chain from a continent.

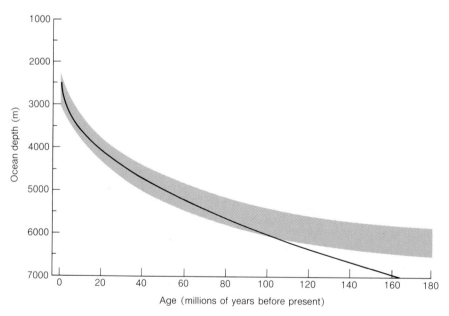

Figure 19–2
Mean ocean depths for the Atlantic and Pacific oceans plotted as a function of the age of the sea floor are shown by the brown band. The theoretical curve (black) assumes the depth is proportional to the square root of age. [After B. Parsons and J. G. Sclater.]

canic chain. A generalized heat-flow profile (see Fig. 13-8) shows a large amount of heat emerging along the ridge axis, less from the older, cooled slab, and more from the volcanic chain of the subduction zone and the marginal basin behind it, where a small region of secondary spreading occurs.

Geophysicists have made theoretical studies and computer models of the evolution of a plate, from its creation out of hot rising matter at ocean ridges, through its spreading and cooling phase to its subduction, with reheating, melting and final resorption in the underlying mantle. The models help explain some important geological and geophysical observations: the major features of the ocean floors, the variation of heat flow from the sea floor, the occurrence of volcanism at plate margins, and the location and mechanism of earthquakes in the subducted slab.

Ocean depths increase with age, t, of the sea floor in a remarkably simple manner (Fig. 19-2). For the first 80 million years the data fit a curve in which ocean depth increases as $\sqrt{t}$. This is precisely the relationship predicted if a plate cools and contracts as it spreads. Beyond 80 million years ocean depths tend to flatten out, compared to the theoretical cooling curve, as would be expected if a small amount of heat is flowing into the plate from the underlying hot asthenosphere.

When a cold plate is subducted, it remains cooler than the surrounding hot mantle for about 12 million years, only gradually warming as it penetrates more deeply. Slow-moving plates heat up and are assimilated at shallower depths, perhaps 400 kilometers (250 miles), than fast-moving plates, which can penetrate to about 700 kilometers before heating to the point of assimilation. The process of subduction involves very large forces, and in a general way these forces must be responsible for the deep-focus earthquakes that occur only in downgoing plates (see Chapter 17). The sudden failures associated with earthquakes take place until the plates become so warm that stress is relieved by slow plastic deformation rather than by faulting. This seems to be the likely explanation for the fact that no earthquakes occur below 700 kilometers (Fig. 19-3).

Rates of Plate Motion

The velocities of moving plates are measured by dating ocean-floor magnetic anomalies (using the time scale of magnetic stratigraphy) and dividing the age of each anomaly into the distance be-

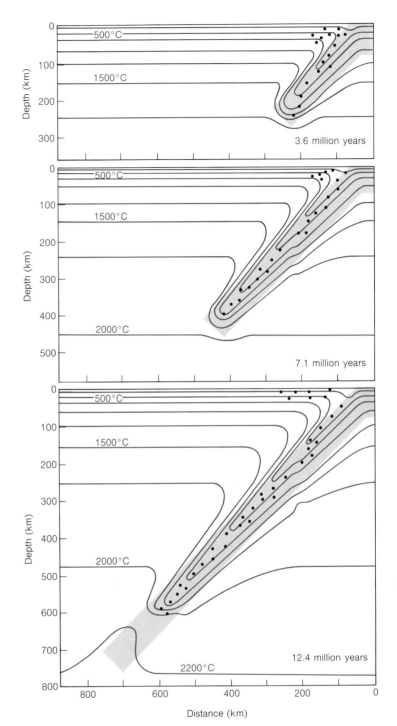

Figure 19-3
The evolution of descending slabs is described by computer models made by M. N. Toksöz. These diagrams depict the fate of a plate subducted at a rate of 8 cm/yr. Contours show computed temperatures in plate and adjacent mantle. Note that temperatures in the plate are several hundred degrees cooler than those in the adjacent mantle. After 12 million years, the plate reaches the temperature of the surrounding mantle at a depth of 600–700 km and loses its original identity. At shallow depths, earthquakes (dots) occur in the shear zone between the colliding plates and in the overriding plate. The deeper earthquakes occur in the cooler, brittle center of the slab, but none occur deeper than 700 km where the plate is assimilated.

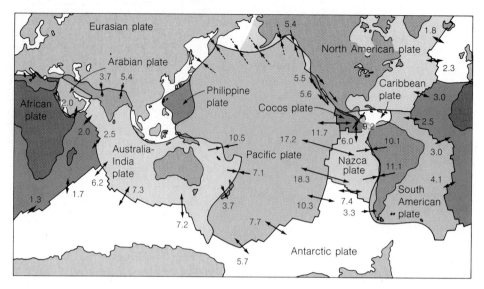

Figure 19-4
Relative velocities and directions of plate separation and convergence in centimeters per year. Opposed arrowheads indicate convergence at trenches, except for Himalayas. Diverging arrowheads indicate plate separation at ocean ridges. Parallel and opposed arrowheads, as along the San Andreas fault in California, indicate transform faults, where plates slide past each other. [From "Convection Currents in the Earth's Mantle" by D. P. McKenzie and Frank Richter. Copyright © 1976 by Scientific American, Inc. All rights reserved.]

tween it and the ridge axis. The procedure was outlined graphically in the preceding chapter (Figs. 18–17, 18–18, and 18–21).

The worldwide pattern of sea-floor spreading is being worked out by using a combination of magnetic, seismic, and bathymetric data. The charts used earlier (Fig. 18–21 and inside back cover) map the world's zones of spreading, subduction, and fracture; their geographic locations were obtained from the positions of ocean ridges, deep-sea trenches, earthquake epicenters, and other indications of activity. On the basis of spreading rates determined from magnetic data, isochrons (contours that connect points of the same age) were drawn to show the age of the sea floor in millions of years. The distance from a ridge axis to a 50-million-year isochron, for example, indicates the extent of new ocean floor created in that period. In Figure 18–21, note the closer spacing of the isochrons in the Atlantic than in the Pacific, where the spreading rate is higher. Because the fracture zones offset the isochrons, the age of the sea floor changes abruptly as one crosses a fault. A summary of the rates and directions of plate motions, measured in centimeters per year relative velocity, is given in Figure 19–4. The largest spreading velocity, 18.3 centimeters per year, occurs between the Pacific and Nazca plates.

The first results announced by the Deep Sea Drilling Project represented a great triumph for the magneticians who worked out spreading rates.

The goal of this joint project of major oceanographic institutions and the National Science Foundation was to drill through the sediments of the sea floor at many places in the world's oceans. Studying the sedimentary cores makes it possible to work out the history of the ocean basin directly, in contrast to the indirect methods of magnetic anomalies. Since sedimentation begins as soon as an ocean forms, the age of the oldest sediments in the core, those closest to the basaltic bedrock, dates the ocean floor at that spot. The age is obtained from the fossils found in the cores. No sediments older than about 150 million years have been found, attesting to the "youth" of the sea floor. The sediments become older with increasing distance from mid-ocean ridges, confirming the prediction of the sea-floor-spreading hypothesis. Figure 19–5 is a plot of the ages determined from drill cores from the Atlantic and Pacific Oceans against ages predicted from magnetic data. It is remarkable how closely the experimental points approach the straight line with slope of one, which represents perfect agreement. This agreement clinches the concept of magnetic stratigraphy and the hypothesis of sea-floor spreading.

As an interesting aside, we have included a photograph of the drilling vessel *Glomar Challenger* (Fig. 19–6). It is 400 feet long, and amidships it carries a drilling derrick 140 feet high. The only ship of its kind in the world, it can lower drill pipe

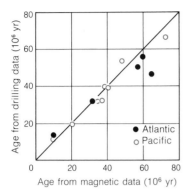

Figure 19-5
A comparison of ages of igneous rocks at different distances from mid-ocean ridges with ages obtained from fossils in the sediments immediately above the igneous rock. The igneous rocks were dated from their magnetic anomaly pattern. The sediments were recovered by deep-sea drilling operations. The 45° line is a theoretical one, implying perfect agreement between these two methods of dating the sea floor. The confirmation of the magnetic ages by deep-sea drilling, shown by the close fit of the experimental points to the theoretical line, lends strong support to the concept of sea-floor spreading. [After C. L. Drake.]

Figure 19-6
The deep-sea drilling vessel *Glomar Challenger*, a unique facility capable of recovering cores of sediment and underlying igneous rock from the floor of the deepest oceans. The Deep-Sea Drilling Project was originally an American one; it is now jointly supported and operated by the United States, the U.S.S.R., Germany, England, Canada, and other countries. [From National Science Foundation.]

several kilometers to the sea floor and drill thousands of meters into the sediments and underlying volcanic rock. For the ship to accomplish such a feat required a technological breakthrough. A means had to be devised to hold the ship stationary, regardless of current, wind, or waves, during drilling. Otherwise, the drill pipe would break off. The problem was solved by developing a positioning device that uses sound waves from acoustic beacons planted on the sea floor. Any change in the ship's position is sensed by a computer that monitors the time of arrival of the sound pulses. The same computer controls bow and stern side thrusters and the ship's main propulsion to keep the vessel on station. The *Glomar Challenger* was the answer to those who said when lunar exploration started, "Better to explore the ocean's bottom than the backside of the Moon." We ended up doing both.

Geometry of Plate Motion

If the individual plates behave as rigid bodies, as seems a reasonable first assumption, several interesting and useful geometric consequences follow. By "rigid" we simply mean that the distances among three points on the same plate—say, New York, Miami, and Bermuda—do not change, no matter how the plate moves. But the distance between New York and Lisbon, of course, increases

because the two cities are on different plates that are being separated along a narrow zone of spreading on the mid-Atlantic ridge. Listed here are some geometric principles, mostly self-evident, that govern the sliding of plates on a planet.

1. Along transform-fault boundaries, no overlap, buckling, or separation occurs; the two plates merely slide past each other without changing the surface area. Look for a transform fault if you want to deduce the direction of plate motions because the orientation of the fault *is* the direction of relative sliding of two plates, as Figures 1–17 and 17–13 show. Surface area obviously changes at zones of convergence or divergence where plates are subducted or created. The plates can move perpendicularly or obliquely to the trend of convergent boundaries, which are therefore not as reliable indicators of directions of movement as transform faults or divergence zones.

2. Magnetic anomaly stripes and isochrons are roughly parallel and are symmetrical with respect to the ridge axis along which they were created. Look at Figure 18–17 to see why this must be so. Since each magnetic strip or isochron marks the

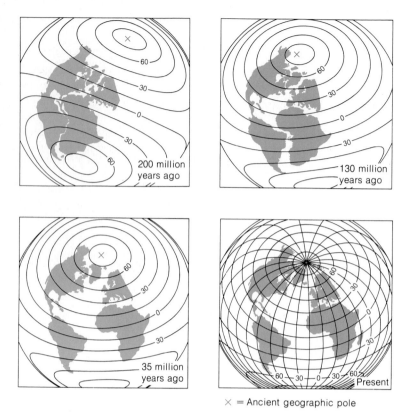

Figure 19-7
Magnetic and deep-sea drilling data are used to chart the northward drift of the continents and the opening of the Atlantic Ocean over the past 200 million years. The Central Atlantic, the Caribbean, and the Gulf of Mexico began to form about 200 million years ago in Triassic time, when Africa and South America drifted away from North America. The South Atlantic opened about 150 million years ago with the separation of South America from Africa. As the continents drifted apart, they also migrated in a northerly direction to their present positions. Note that the equator passed through the southern parts of the United States and Europe in Triassic time. [From J. D. Phillips and D. Forsyth, *Bull. Geol. Soc. America*, v. 83, p. 1579, 1972.]

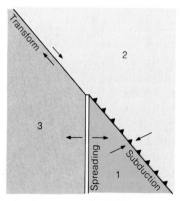

Figure 19-8
A triple junction. Plates 1, 2, and 3 meet at the intersection of a spreading zone, a subduction zone, and a transform fault. The arrows depict relative motion between adjacent plates.

edge of an earlier plate margin, isochrons that are of the same age but on opposite sides of an ocean ridge can be brought together to show the positions of the plates and the configuration of the continents as they were in that earlier time. By this means we can reconstruct, for example, the opening of the Atlantic Ocean, as shown in Figure 19-7.*

3. The point at which three plates meet is called a **triple junction.** Figure 19-8 shows an example of a point at which a spreading zone, a subduction zone, and a transform fault meet. If the relative motion between two pairs of plates is known, we can solve for the third by using a simple equation (see Box 19-1).

---

*The Great Pyramid of Egypt is aimed slightly east of true north. Did the ancient Egyptian astronomers make a mistake in orienting the pyramid 40 centuries ago? Probably not. Over this period of time Africa drifted enough to rotate the pyramid out of alignment with true north.

Box 19-1 SOLVING FOR THE RELATIVE MOTIONS OF PLATES

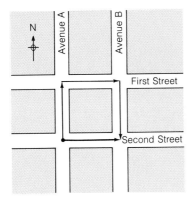

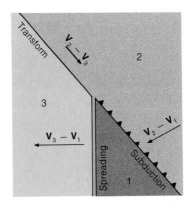

$$(V_2 - V_1) = (V_3 - V_1) + (V_2 - V_3)$$

Velocity is a **vector** quantity, one that has both direction and magnitude. A simple example of how vectors are added is shown in the figure above. If a man walks single blocks north on Avenue A, east on First Street, south on Avenue B, he ends up at a place he could have reached directly by walking one block east on Second Street. In vector addition this direct route is equal to the sum of the segments that make up the long route.

In the figure at the right (part a), let the velocities of the three plates meeting at a triple junction be V_1, V_2, V_3. The velocity of plate 2 relative to plate 1 is $V_2 - V_1$; $V_2 - V_3$ and $V_3 - V_1$ are the other possible relative plate velocities. It is clear that the sum of the relative velocities of the plates, taken in order around the triple junction, must be zero, since

$$(V_2 - V_1) + (V_3 - V_2) + (V_1 - V_3) = 0$$

or, rearranging this,

$$(V_2 - V_1) = (V_3 - V_1) + (V_2 - V_3).$$

This enables us to solve for $V_2 - V_1$, the direction and amount of relative motion across the subduction zone because the directions of the vectors $V_2 - V_3$ and $V_3 - V_1$ are parallel to the transform and perpendicular to the spreading axis respectively, and the magnitudes of the relative motions can be obtained from the magnetic stripes. Part b of the figure below shows the solution for $V_2 - V_1$ in terms of vector addition.

The point where the Pacific, Cocos, and Nazca plates meet (see inside back cover) is an actual triple junction. Three spreading zones meet at this junction, as shown in the enlarged view in Figure 19-9. The unknown motion, found by vector addition, was that between the Nazca and Pacific plates, the motions between the Pacific–Cocos and Cocos–Nazca plates having been worked out from transform faults and magnetic-anomaly stripes. The arrows show the resultant plate movements. Note also how the isochrons bend to become parallel to the spreading centers, where they originated, and how they are offset by the transform faults. The spacing of the isochrons reflects the spreading rates, which are largest for the Pacific–Nazca plates and least for the Cocos–Nazca plates.

Up to this point we have considered plates sliding on a plane. Although much can be learned about plate motions by making this simplification, plates actually move on the Earth's spherical sur-

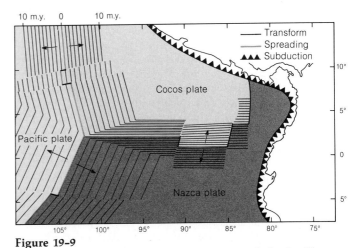

Figure 19-9
Triple junction formed by the intersection of the Pacific, Nazca, and Cocos plates, three spreading zones in the southeast Pacific Ocean. The schematic isochrons parallel the ridge axes from which they migrate as the sea floor ages and spreads. The spacing between isochrons is a measure of the spreading velocities. [After R. N. Hey, K. S. Deffeyes, G. L. Johnson, and A. Lowrie, "The Galapagos Triple Junction," *Nature*, v. 237, p. 20, 1972.]

Box 19-2 PLATE MOTIONS ON A SPHERICAL EARTH

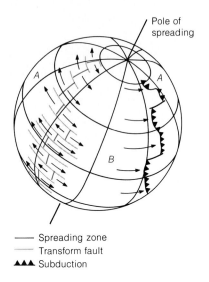

—— Spreading zone
—— Transform fault
▲▲▲ Subduction

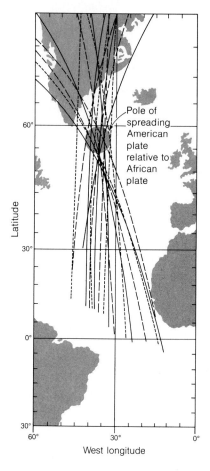

After W. J. Morgan, "Rises, Trenches, Great Faults, and Crustal Blocks," *J. Geophysical Research,* v. 73, pp. 1959–1982, 1968.

Geometry allows us to describe the separation of two plates on a sphere—for example, plate *A* and plate *B* in the figure—as a rotation of *B* with respect to *A* about some pole of rotation, called a pole of spreading. Note on the diagram of plates (inside back cover) that along mid-ocean ridges where plates separate, the axis of spreading is not continuous but is offset by transform faults, approximately at right angles to the axis. Why this occurs is not fully understood, but it appears to be easier for plates to break apart this way with the plates typically sliding by each other at the transform fault, rather than pulling apart or overlapping there. Because of this geometry, if one imagines latitudes and longitudes drawn with respect to the pole of spreading, the transform faults lie on lines of latitude, and lines perpendicular to them are longitudes that converge at the pole of spreading. To understand why this must be so, consider the following analogy: If a tennis ball were sliced in two parts and put back together, one could rotate the two parts along the cut (as on a transform fault). The cut would also describe a latitude centered on a pole of rotation, which can be located by drawing longitudes perpendicular to the cut. The intersection of two or more such longitudes is the pole of rotation. On a model of the Earth, if great circles are drawn perpendicular to transform faults between a pair of plates, their intersection locates the pole of spreading, which together with the spreading rate completely describes the relative motion of the two plates. The spreading rate is zero at the pole of spreading and increases to a maximum 90° away at the equator of spreading, as the

figure indicates. This maximum equatorial value is frequently cited as the spreading rate between plates.

To see how a pole of spreading is located in practice, refer again to the inside of the back cover, which shows the zone of spreading and the transform faults that separate the African and American plates. Great circles perpendicular to the transform faults are drawn in the figure below. They intersect near the point 58°N, 36°W, off the southeast coast of Greenland. This is the pole of spreading of these two great plates. Don't bother going there, for there is nothing to be seen. The pole of spreading has no physical significance. It serves only as a construction point, a convenience for describing the relative motion of plates merely by giving the latitude and longitude of this point.

face. Box 19-2 explains how plate movements on a sphere can be described. With the application of these geometric principles to find spreading directions and magnetic anomalies to deduce spreading rates, the relative motions of the lithospheric plates are being worked out worldwide. Some results have already been pictured in Figures 18-21 and 19-4. However, geophysicists are searching for ways to measure the absolute motions of individual plates rather than their motions relative to each other. If the hot spots discussed in Chapter 15 turn out to be fixed in the mantle below plates, then the string of extinct volcanoes trailing from the hot spot would record the movement of individual plates as they glide over the mantle (Figure 15-38). This is currently a subject of active research.

SEA-FLOOR SPREADING AND CONTINENTAL DRIFT: RETHINKING EARTH HISTORY

One of us (F. P.) once helped write a paper dealing with the permanence of ocean basins. If he were allowed to expunge from the scientific record the one contribution he regrets the most, this would be it. The notion of the stability of global geographic features was not only a main tenet of the old geology but seems to be firmly rooted in the human psyche. We now know that on the geological time scale the sea floor is far from permanent. The present ocean basins are being created by spreading and recycled by subduction on a time scale of about 200 million years, which is about 4 percent of the age of the Earth. The likelihood of finding extensive older remnants of sea floor is slight. Continents, on the other hand, are mobile but permanent features. They are too buoyant to be subducted. They may be fragmented, moved, reassembled, deformed, and eroded at their surfaces, but their bulk does not seem to be much diminished. Old terrains with ages of around 3.5 to 3.7 billion years can still be found. Continents grow with time by the gradual accumulation of materials along their margins. New continental strips can therefore be added on in different places at different times, depending on the history of fragmentation, movement, and reassembly.

With the emergence of these revolutionary ideas, geologists are rethinking Earth history. Most of the evidence for plate tectonics comes from the sea floor, a relatively simple place compared to the enormously complicated continents. Just how plate tectonics explains continental ge-

ology is now receiving much attention. New developments reported in nearly every issue of the geological journals show that the subject has definitely been revitalized. Rock associations, volcanism, metamorphism, the evolution of mountain chains—all are being reexamined in the framework of plate tectonics. Some of the new interpretations that we describe in this chapter may not stand the test of time. In this connection, future editions of this book may show some changes, not so much in the big picture of plate tectonics as in the details of fitting regional geology into the overall framework. The student (as well as the authors of this book) should be cautioned against calling on plate tectonics for easy explanations of everything geological. It is not clear, for example, how or whether the origin of such structures as the Ozarks, the Black Hills, the Colorado Plateau, or such intracontinental, sediment-filled depressions as the Michigan Basin are related to plate movement.

Rock Assemblages and Plate Tectonics

The only record we have of past geologic events is the incomplete one found in the rocks that have survived erosion or subduction. Since only sea floor younger than 200 million years (the last 4 percent of Earth history) has survived subduction, we must focus on the continents to find the evidence for most of Earth history. Some of the methods of reading the rock record have been described in earlier chapters. Here we explore the nature of the rock assemblages that characterize different plate-tectonic regimes as a first step in unraveling the history of past plate motions. Our aim is to reconstruct the process of continent fragmentation and ocean development, to locate the sites of vanished oceans, and to recognize the sutures that mark ancient plate collisions.

Of the three kinds of plate boundaries, we might expect distinct suites (assemblages) of rocks to be associated with plate divergence and convergence. At transform faults no distinct or characteristic rock assemblages are to be expected. Discontinuities across the fault are found, however, since rock formations formed and altered elsewhere have slipped past one another, and once-continuous formations or structural features are displaced.

Think of all that happens at a zone of divergence, where plate accretion and spreading occur, and you can predict the kinds of rock that would characterize the place and the process. Because

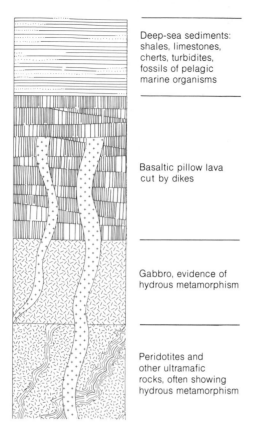

Figure 19-10
Idealized section of an ophiolite suite. The combination of deep-sea sediments, submarine lavas, and mafic igneous intrusions indicates a deep-sea origin. Many geologists now believe ophiolites to be fragments of oceanic lithosphere emplaced on a continent as a result of plate collisions.

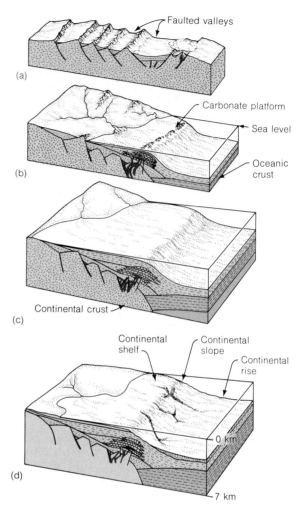

Figure 19-11
The development of a geosyncline on a rifted continental margin off the Atlantic coast of the United States. A rift develops in Pangaea as the ancient continent splits. Volcanics and Triassic nonmarine sediments are deposited in the faulted valleys (a). Seafloor spreading begins, the lithosphere cools and contracts, and the receding continental margins subside below sea level. Evaporites, deltaic deposits, and carbonates (b) are deposited and then covered by Jurassic and Cretaceous sediments derived from continental erosion (c and d).

there is extensive undersea volcanism, one would expect to find submarine basaltic lava, perhaps pillow lavas, the volcanic rock formed when hot lava is quenched by cold sea water (Chapter 15). Suboceanic crust and mantle are created here; dredge hauls and geophysical data show these layers to consist of mafic rocks, such as gabbro and peridotite, often showing evidence of alteration in a water environment (hydrous metamorphism). A carpet of deep-sea sediments would cover all of this. From Chapters 10 and 11 we remember that these deposits are recognized by thin layers of shale, limestone, and the siliceous rock chert, often with thin, discontinuous turbidites between them. Some or all of these layers may contain fossil remains of open-ocean marine organisms. A combination of deep-sea sediments, submarine basaltic lavas, and mafic igneous instrusions like that shown in idealized section in Figure 19-10, is called an **ophiolite suite.** The presence of narrow ophiolite zones in convergence features like the Alpine-Himalayan belt

and the Ural and Appalachian belts may indicate that slices of oceanic crust and mantle originally produced at accreting plate margins were thrust onto land when an ancient ocean finally disappeared as two continents converged. It is generally believed that the Appalachians, for example, mark the site at which the ancestral Atlantic Ocean (called *Iapetus* for one of the Greek gods) closed when North America and Africa converged about 375 million years ago. The Atlantic reopened a few hundred kilometers east of this old suture, about 200 million years ago, in a spreading episode that is still underway.

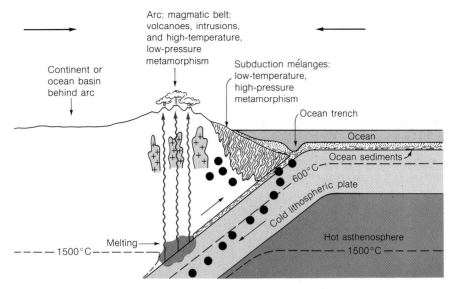

Figure 19-12
Geologic features and activities associated with plate collisions and subduction: ocean trenches, mélange deposits, magmatic belts, metamorphism, volcanism, earthquakes (dots). The drawing is not to scale; the thickness of lithosphere is about 70 km, depth of the ocean trench 10 km, and the distance from trench to arc is 300–400 km.

Continental-shelf deposits are sedimentary rock assemblages that are laid down in an orderly sequence under tectonically quiet conditions in a geosyncline at a receding continental margin. Figures 19–11 and 10–29 show the orderly sequence of deposits in the geosyncline that is still forming off the Atlantic coast of the United States. The continental margin there was formed when the American plate separated from the European plate about 200 million years ago. Resting on the offshore shelf is a wedge-shaped deposit of sediments eroded from the continent and carried into shallow water. Because the trailing edge of the continent slowly subsides as the spreading lithosphere cools and contracts, the geosyncline continues to receive sediments for a long time. The load of the growing mass of sediment further depresses the crust isostatically, so that the geosyncline can receive still more material from land. For every three meters of sediments received, the crust sinks two meters. The result of these two effects is that the geosynclinal deposits can accumulate in an orderly fashion to thicknesses of 10 kilometers or more. At the same time, the supply of sediments is sufficient to maintain the shallow-water environment of the geosyncline, or miogeosyncline, as we called it in Chapter 11.

The deposits show all of the characteristics of shallow-water conditions (Chapter 11). At the bottom of the entire sequence are rift valleys containing basaltic lavas and nonmarine deposits formed during the early stages of continental fissuring. In the early stages of shelf deposition, sandy materials started to fill the depression.

Much was dropped on the continental slope, only to be moved later to the continental rise by turbidity currents. In deep water, very thick deposits can be built up in this way. As the shelf miogeosyncline builds up, deposition may become dominated by shales and carbonate platform deposits—indicators of a decrease in the supply of detritus from the continent.

Think what might happen to these geosynclines if the orderly, sequentially layered, gently dipping sediments were to become the leading edge of a plate in collision. In the following sections we describe some of the many possibilities.

Just as the events that take place in a convergence zone are different from divergence-zone phenomena, so do the rock assemblages have different characteristics. The main features of ocean-ocean or ocean-continent collision are shown in transverse section in Figure 19–12. Thick marine sediments, mostly turbidites, eroded from the continent or the island arc, rapidly fill the long marginal depressions. In descending, the cold oceanic slab stuffs the region below the inner wall of the trench with these sediments and with deep-sea materials brought with the incoming plate. Regions of this sort are enormously complex and highly variable, as they include turbidities and ophiolitic shreds scraped off the downgoing slab by the edge of the overriding plate—all highly folded, intricately sliced and metamorphosed. They are difficult to map in detail but recognizable by their distinctive mix of materials and structural features. Such a chaotic mess has been called a **mélange.** The metamorphism is the kind characteristic of high pressure and low temperature because the material may be carried relatively rapidly to depths as great as 30 kilometers, where recrystallization occurs in the environment of the cold slab. Somehow, perhaps by buoyancy and mountain-building, the material rises back to the surface much later. Find a mélange and you can't be too far from the place of downturn of an ancient plate, long since consumed, but leaving this relic of its existence.

Refer again to Figure 19–12. Parallel to the mélange is a magmatic belt that makes up the arcuate system of volcanoes, intrusions, and metamorphic rocks formed on the edge of the overriding plate. Here the conditions are dominated by the rise of magma from the descending plate. At the interface, where the descending plate slides past the overriding one, perhaps friction is great enough to melt the upper part of the downturned slab, including the subducted wet sediments and ocean crust. The liquids rise buoyantly from depths of 100 to 200 kilometers to erupt and build the volcanic chains on the leading edges of plates. The characteristic igneous rocks produced are andesitic lavas and granitic intrusives. Island arcs, built up from the sea floor, may contain larger amounts of basalt; continental margins typically erupt rhyolitic ignimbrite and are intruded by granitic batholiths below (see Chapter 15). In contrast to that in a mélange, the metamorphism in the magmatic belts is typically the result of recrystallization under conditions of high temperatures and low pressures. This is because the hot fluids rise close to the surface, delivering much heat to a low-pressure environment.

Paired belts of mélange and magmatism (Fig. 19–12) are the signature of subduction. The essential elements of these features of collision have been found in many places in the geologic record. One can see mélange in the Franciscan Formation of the California Coast Ranges and magmatism in the parallel belt of the Sierra Nevada to the east (Fig. 19–13). This paired belt marks the Mesozoic boundary between the colliding Pacific and American plates. It even shows the polarity of the convergence by the location of mélange on the west and magmatism on the east; the Pacific plate was the subducted one. Other paired belts—for example, in Japan—can be found along the continental margins framing the Pacific basin. The central Alps, a European example, were produced by the convergence of a Mediterranean plate with the European continent.

Seismic reflection profiles (see Box 17–1) are beginning to provide "x-ray" views of layers deep

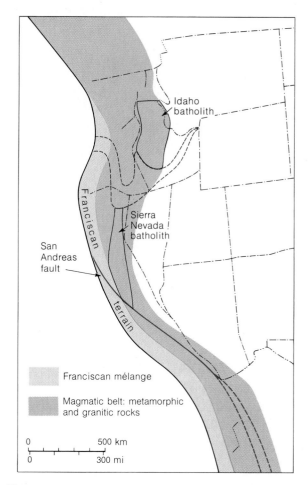

Figure 19–13
This paleogeologic map of the western United States shows the geology of the region as it was at the beginning of Tertiary time. The paired mélange and magmatic belts indicate a collision of the Pacific and American plates in Mesozoic time, the Pacific plate being the subducted one. [After W. Hamilton and W. B. Myers, "Cenozoic Tectonics," *Reviews of Geophysics,* v. 4, p. 541, 1966.]

within the crust. Figure 19–14, a remarkable example of this new technique, shows the Australian plate being subducted under the Eurasian plate at the Java trench.

Orogeny and Plate Tectonics

Orogeny means mountain-making, particularly by folding and thrusting of rock layers. In the framework of plate tectonics, orogeny occurs primarily at the boundaries of colliding plates, where marginal sedimentary deposits are crumpled and magmatism and volcanism are initiated.

Consider first some scenarios of plate convergence. In Figure 19–15a, a plate with a continent at

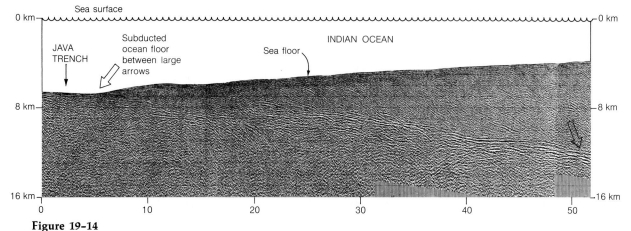

South   North

Figure 19-14
Seismic reflection profile across the Java Trench subduction zone south of Bali, along longitude 112°E.
Subducted ocean floor (between large arrows) dips about 6° under overthrust wedge of highly deformed
sediments. The ocean floor can be followed from the beginning of subduction at the north wall of the
trench to a depth of 12 km below sea level. [Courtesy of R. H. Beck and P. Lehner, Shell Internationale
Petroleum.]

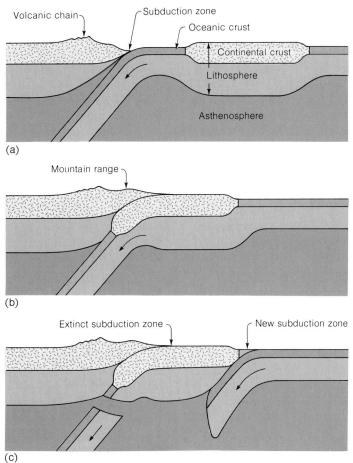

Figure 19-15
Possible stages in plate collisions. (a) Convergence between plates with continental and
oceanic lithosphere at leading edges. Magmatic belt, folded mountains, and mélange deposits
are features of the overriding continental boundary. (b) Collision of continents, producing
a mountain range, magmatic belt, and thickened continental crust. Since the continent is
too buoyant to be carried down into the mantle, plate motions may be brought to a halt.
(c) Alternatively, the plate may break off and a new subduction zone be started elsewhere.
An extinct subduction zone may show as a scar in the form of a mountain belt within a
continent. Examples are the Ural Mountains and the Himalayas. [After "Plate Tectonics"
by J. F. Dewey. Copyright © 1972 by Scientific American, Inc. All rights reserved.]

the leading edge collides with another plate carrying a continent. In the early stage, during which the convergence is between continent and subducted oceanic lithosphere, a magmatic belt, folded mountains, and mélange deposits may be features of the overriding continental boundary. An example exists today along the Pacific coast of South America, where the American and Nazca plates are colliding. Look at the illustration inside the back cover to see the setting of the plates. The Andes, from which the name of the volcanic rock andesite is derived, lie in the magmatic belt; subduction is taking place under the Peru–Chile trench.

In a later stage, continent may meet continent, as shown in Figure 19-15b. Since continental crust is too light for much of it to be carried down, the plate motions could be slowed or halted. Another possibility, the one depicted in the figure, is that the plate motions continue, with subduction ceasing at the continent-continent suture but starting up anew elsewhere. Cold and dense as the descending slab is, chunks of it may break off, fall freely into the mantle, and be resorbed. As Figure 19-15c shows, the suture is marked by a mountain range made up of either folded or thrust rocks, or both, coincident with or adjacent to the magmatic belt, and by a much-thickened continental crust. A prime example of continent-continent collision is the Himalayas, which began forming some 25 million years ago when a plate carrying India ran into the Asiatic plate (the collision and uplift are still going on). This may be how the root underlying the Himalayas originated (see Chapter 18). The plate-tectonic cycle of the closing of an ocean basin, a continent-continent collision, and the formation of an intracontinental mountain belt has been called the **Wilson cycle,** after the Canadian geologist J. Tuzo Wilson, who first suggested the idea that an ancient ocean closed to form the Appalachian mountain belt and then reopened to form the present day Atlantic Ocean.

Displaced Terrains

Geologists have come across blocks within continents whose rock sequences, fossils, and paleomagnetism are alien to their surroundings. The rock assemblages and the fossils indicate different environments and ages than the surrounding terrain, and the paleomagnetic poles imply that the block originated in a different latitude. These are now believed to be fragments of other continents or of ocean crust that were swept up and plastered onto a continent in the process of plate

collisions and separations. Coastal New England and Newfoundland may be slices of Europe; parts of Alaska, British Columbia, and Nevada may have been scraped off Asia; and central Florida may be a fragment of Africa. Displaced terrains have also been found in Japan, Southeast Asia, China, and Siberia, but their original locations have yet to be worked out.

The Grand Reconstruction

At the close of the Paleozoic, some 250 million years ago, there was a single supercontinent Pangaea, stretching from pole to pole (Fig. 1-14). The fragmentation of Pangaea as a result of plate tectonics and continental drift over Mesozoic and Cenozoic time to form the modern continents and oceans is documented in the well-preserved record of magnetic reversal stripes on the ocean floor. But what of the pre-Pangaean distribution of continents? What were their shapes and where were they located? There is growing evidence that Pangaea was formed by the collision of continental blocks—not the same continents we know today but continents that existed earlier in the Paleozoic. The ocean-floor record for this period has been destroyed by subduction, so we must rely on the older evidence preserved on continents to identify and chart the movements of these paleocontinents. Old mountain belts like the Appalachians and the Urals mark the collision boundaries of the paleocontinents. Rock assemblages there reveal ancient episodes of rifting and subduction. Rock types and fossils also indicate the distribution of shallow seas, glaciers, lowlands, mountains, and climatic conditions. Paleomagnetic data can be used to find the latitude and the north-south orientation of the paleocontinents. Latitudes can also be checked by paleoclimatic data. Although it is not possible to assign longitudinal position to the paleocontinents, the relative sequence of continents around the globe can be pieced together from the fossil record. One of the first efforts to depict the pre-Pangaean configuration of continents using these methods is shown in Figure 19-16. The ability of modern science to recover the geography of this strange world of hundreds of millions of years ago is truly impressive. Geologists may be able to continue to sort out more details of this complex jigsaw puzzle, whose individual pieces change shape over geologic time.

Figure 19-17 reconstructs the most recent breakup of Pangaea as we now understand it. Fig-

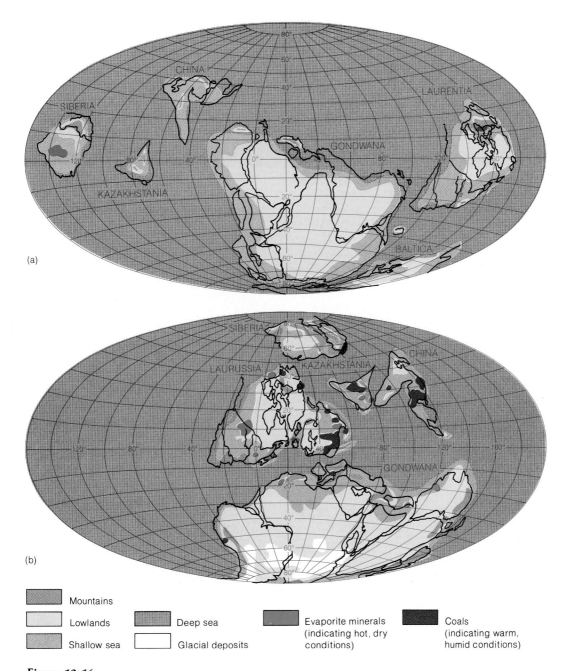

Figure 19-16
(a) Paleocontinents in the Middle Ordovician, about 475–490 million years ago. At that time the continents consisted of Gondwana (made up of South America, Southern Europe, Africa, the Near East, India, Australia, New Zealand and Antarctica), Laurentia (North America and Greenland), Baltica (most of Northern Europe and European Russia), Kazakhstania (Central Asia), China (China, Malaysia), and Siberia. (b) Paleocontinents in Early Carboniferous, about 340–360 million years ago. Gondwana has moved across the South Pole entering the opposite hemisphere; Baltica has collided with Laurentia to form a larger continent Laurussia. The continents are assembling for the collisions that formed the supercontinent Pangea at the end of the Paleozoic. [After R. K. Bambach, C. R. Scotese, and A. M. Ziegler, *American Scientist,* January 1980.]

ure 19-17a shows the world as it looked in Permian times, a little more than 200 million years ago. Pangaea was an irregularly shaped land mass surrounded by a universal ocean called Panthalassa, the ancestral Pacific. The Tethys Sea, between Africa and Eurasia, was the ancestor of part of the Mediterranean. The fit of North and South America with Europe and Africa is very good in detail when taken at the outer edge of the continental shelves, instead of at the present shorelines, which are some distance from the original rift. It is the fit for which we have the

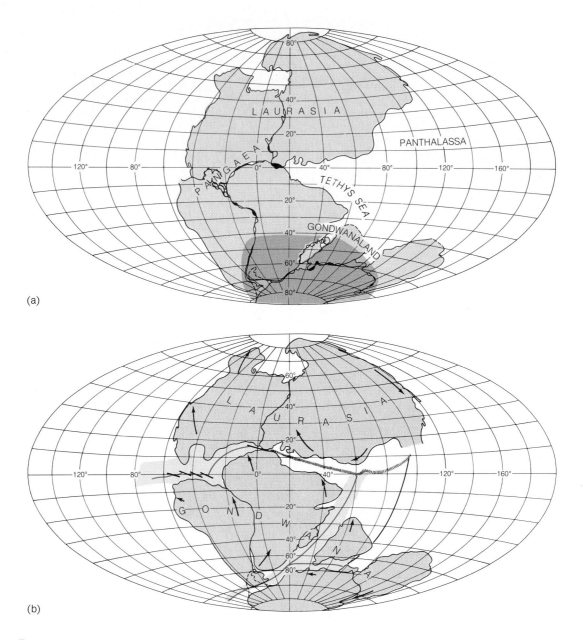

(a)

(b)

Figure 19–17

The breakup of Pangaea. (a) The ancient landmass Pangaea, meaning "all lands," may have looked like this some 200 million years ago at the close of the Paleozoic and beginning of the Mesozoic. Panthalassa ("all seas") evolved into the present Pacific Ocean, and the present Mediterranean Sea is a remnant of the Tethys. Permian glacial deposits are found in widely separated areas, such as South America, Africa, India, and Australia. This distribution is simply explained by postulating a single continental glacier flowing over the south polar regions of Gondwanaland in Permian time, before the breakup of the continents. The probable extent of the glacier is shown by shading.
(b) One view of world geography at the end of the Triassic Period, 180 million years ago, after some 20 million years of drift. New ocean floor is shown in color. Spreading zones are represented by dark brown lines, transform faults by black lines, and subduction zones by hatched lines. Arrows depict motions of continents since drift began. (c) World geography at the end of the Jurassic Period, 135 million years ago, after some 65 million years of drift. Ocean floor created in the preceding 45 million years is shown in color. (d) World geography at the end of the Cretaceous Period, 65 million years ago. Color indicates new ocean floor created after some 135 million years of drift. (e) World geography today. Color shows sea floor produced during the past 65 million years, in the Cenozoic Period. [After "The Breakup of Pangaea" by R. S. Dietz and J. C. Holden. Copyright © 1970 by Scientific American, Inc. All rights reserved.]

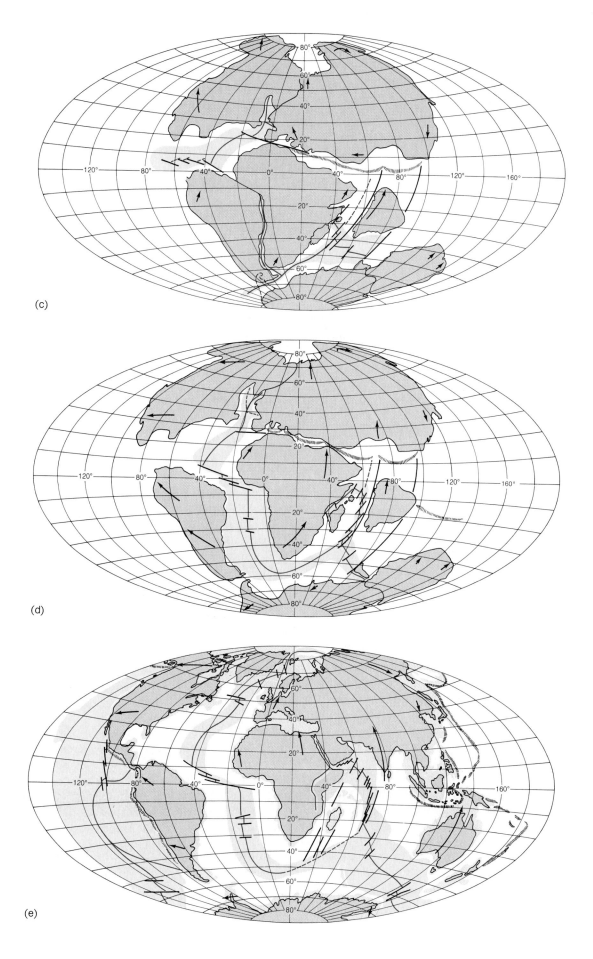

(c)

(d)

(e)

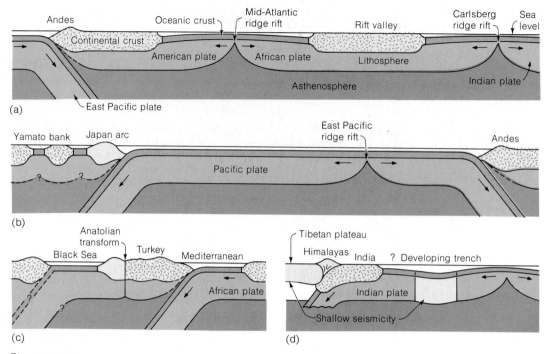

Figure 19-18
Schematic sections showing modern plate, ocean, continent, and island-arc relationships.
[After J. F. Dewey and J. M. Bird, "Mountain Belts and New Global Tectonics,"
J. Geophysical Research, v. 75, pp. 2625–2647, 1970.]

firmest evidence. The positions of Central America, India, Australia, and Antarctica are less certain.

The breakup of Pangaea was signaled by the opening of rifts from which basalt poured. Relics of this great event can be found today in the Triassic basalt flows all over New England. Radioactive dating of these flows provides the estimate of about 200 million years for the beginning of drift.

The geography of the world after 20 million years of drift—at the end of the Triassic some 180 million years ago—is sketched in Figure 19-17b. The Atlantic has opened, the Tethys has contracted, and the northen continents (Laurasia) have all but split away from the southern continents (Gondwana). New ocean floor has also separated Antarctica–Australia from Africa–South America. India is off on a trip to the north.

By the end of the Jurassic period, 135 million years ago, drift had been underway for 65 million years. The big event at this time is the splitting of South America from Africa, which signals the birth of the South Atlantic (Fig. 19-17c). The North Atlantic and Indian oceans are enlarged, but the Tethys Sea continues to close. India continues its northward journey.

The close of the Cretaceous period 65 million years ago sees a widened South Atlantic, the split-

ting of Madagascar from Africa, and the close of the Tethys to form an inland sea, the Mediterranean (Fig. 19-17d). After 135 million years of drift, the modern configuration of continents becomes discernible.

The modern world, produced over the past 65 million years, is shown in Figure 19-17e. India has collided with Asia, bringing its trip to an end. Australia has separated from Antarctica. Nearly half of the present-day ocean floor was created in this period. Figure 19-18 shows several schematic sections that summarize modern plate, ocean, continent, and island-arc relationships for the American, African, Eurasian, and Indian plates.

Most of the modern Pacific Ocean basin consists of the Pacific plate side of the East Pacific rise spreading zone, as can be seen inside the back cover and in Figure 19-18b. This implies that an area equal to most of the Pacific Ocean has disappeared by subduction under the Americas in the past 130 million years. As much as 7000 kilometers (4300 miles) of Pacific sea floor may have been thrust under North America!

Not one branch of geology, except perhaps crystallography, remains untouched by this grand reconstruction of the continents. Economic geologists are using the fit of the continents to find mineral and oil deposits by correlating the formations

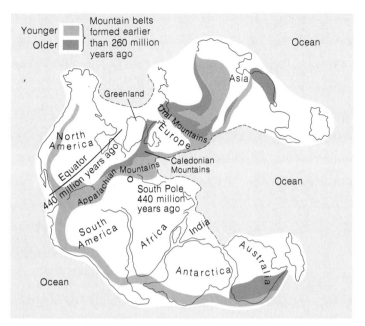

Figure 19-19
When the ancient continent of Pangaea is reconstructed, the Caledonian Mountains of Europe and the Appalachian Mountains form a continuous belt believed to have been formed in a plate collision between the pre-Pangaean continents. The continuity of the mountain belts that extends from the Americas across Antarctica and western Australia supports the reconstruction. The Urals and other old mountains contain ophiolite zones, marking these sutures as the sites of vanished oceans.
[After "Plate Tectonics" by J. F. Dewey. Copyright © 1972 by Scientific American, Inc. All rights reserved.]

in which they occur on one continent with their predrift continuations on another continent. Paleontologists are rethinking some aspects of evolution in the light of continental drift. For example, during most of the age of reptiles, the continents we know today were grouped together in two supercontinents, Laurasia and Gondwanaland. These continents were fragmented during most of the age of mammals, with faunas developing on the daughter continents isolated from one another. Is this why mammals diversified into so many more orders than the reptiles did, and in a much shorter period of time? Structural geologists and petrologists are extending their sights from regional mapping to the world picture, for the concept of plate tectonics provides the means of interpreting such geological processes as sedimentation and orogeny in global terms. For example, the Caledonian mountain belt that runs along the northwest margin of Europe is the predrift continuation of the Appalachian belt, and the trend of the Andes may be followed into Antarctica and Australia, as Figure 19-19 shows.

Oceanographers are reconstructing currents as they might have existed in the ancestral oceans, to understand better the modern circulation and to account for the variations in deep-sea sediments. Paleoclimatologists are "forecasting" backward in time to describe temperature, winds, the extent of continental glaciers, and the level of the sea as they were in predrift times. What better testimony to the triumph of this once-outrageous hypothesis than its ability to revitalize and shed lignt on so many diverse topics!

THE DRIVING MECHANISM OF PLATE TECTONICS

Up to this point everything we have discussed might be categorized as descriptive plate tectonics. The geometry and rates of plate motions, the consequences of plate separation and collision have been described. But what drives it all? We will not fully understand plate tectonics until we can answer this question. The International Geodynamics Project enlisted the efforts of thousands of scientists in seeking the underlying cause of plate motions.

It is generally accepted that most of the mantle is a hot solid, capable of flowing like a liquid at a speed of about a centimeter per year, about the

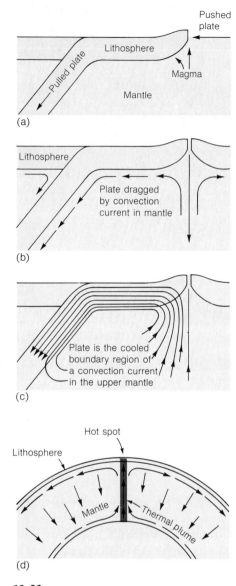

Figure 19-20
Possible driving mechanisms of plate tectonics.
(a) The plate is pushed by the weight of the ridges
at centers of spreading or is pulled by cool, heavy
downgoing slab or both. (b) The plate is dragged
by convection current in mantle. (c) The plate is the
cooled, brittle, boundary region of a convection current in the hot, plastic upper mantle. (d) Jetlike thermal plume rises from great depth, causes hot spots
at mid-ocean ridges, and spreads laterally dragging
the plates. Downward return flow occurs throughout
the rest of the mantle.

rate your fingernails grow. The lithosphere is broken into rigid plates, somehow responsive in their
motions to the flow in the underlying mantle.

As is generally the case when there is an abundance of data in search of a theory, many hypotheses have been advanced. Some would have
plates pushed by the weight of the ridges at the
zones of spreading or pulled by the heavy downgoing slab at subduction zones. Others hold that
the plates are dragged along by currents in the
underlying asthenosphere. Figure 19-20 shows
some of these ideas. In line with the discussion in
Chapter 13, we agree with those who view the
process not in piecemeal but as a highly complex
convective flow, involving rising, hot, partially
molten materials and sinking, cool, solid materials, under a variety of conditions ranging from
melting to solidification and remelting. A significant part of the mantle must be involved, for slabs
are known to penetrate to depths of some 700 kilometers before being completely resorbed. Figure
19-20c, shows one of the first computer models of
the process—one that neglects some of the effects
just mentioned, but that nevertheless accounts for
many observations. A rising plume of hot material, heated from below, reaches the surface at a
center of spreading. It moves away from the center, cools near the surface, and the cooled boundary becomes solid, strong lithosphere. Finally
becoming heavier after it has cooled, the lithospheric slab sinks back into the mantle in a subduction zone, where it is reassimilated, to be
heated and to rise again in the future. Another
theory (Fig. 19-20d) proposes that hot, narrow, jetlike plumes rise from the bottom of the mantle,
feed the growing plate, and drive it laterally away
from spreading centers where the plumes mostly
occur. These same plumes are evidenced at the
surface by hot spots. Among the problems left to
the next generation of Earth scientists is the incor-
poration of such important details as the shapes
of plates, the history of their movements, and the
formation and growth of continents into an explanation of the distribution of convective currents
in time and space.

SUMMARY

1. According to the theory of plate tectonics, the lithosphere is broken into about a dozen
rigid, moving plates. Three types of plate boundaries are defined by the relative motion
between plates: boundaries of divergence, boundaries of convergence, and transform
faults.

2. In addition to earthquake belts, many large-scale geological features are associated
with plate boundaries, such as narrow mountain belts and chains of volcanoes. Boundaries of convergence are recognized by deep-sea trenches, inclined earthquake belts,

mountains and volcanoes, and paired belts of mélange and magmatism. The Andes Mountains and the trenches of the west coast of South America are modern examples. Divergent boundaries (for example, the mid-Atlantic ridge) typically show as seismic, volcanic, mid-ocean ridges. A characteristic deposit of this environment is the ophiolite suite. Transform faults, along which plates slide past one another, can be recognized by their topography, seismicity, and offsets in magnetic anomaly bands. Ancient convergences may show as old mountain belts, such as the Appalachians.

3. The age of the sea floor can be measured by means of magnetic-anomaly bands and the stratigraphy of magnetic reversals worked out on land. The procedure has been verified and extended by deep-sea drilling. Isochrons can now be drawn for most of the Atlantic and for large sections of the Pacific, enabling geologists to reconstruct the history of opening and closing of these oceans. Based on this method and on geological and paleomagnetic data, the fragmentation of Pangaea over the last 200 million years can be sketched.

4. Although plate motions can now be described in some detail, the driving mechanism is still a puzzle. An attractive hypothesis proposes that the upper mantle is in a state of convection with hot material rising under divergence zones and cool material sinking in subduction zones. The plates, according to this model, would be the cooled, upper boundary region of the convection cell.

EXERCISES

1. Summarize the principal geologic features of subduction zones and divergence zones.

2. Explain the following in the context of plate tectonics: (a) Iceland. (b) San Andreas fault of California. (c) Ural Mountains. (d) Aleutian trench. (e) Earthquakes in Italy and Turkey. (f) Andes Mountains.

3. How do we know that spreading along the East Pacific rise is faster than along the mid-Atlantic ridge?

4. What would an astronaut look for on Mars to find out if plate tectonics is an active process on the planet?

5. How would one recognize the boundaries between ancient plates no longer in existence?

6. Can you think of a way not mentioned in the text, by which to measure absolute motions of individual plates rather than relative motions between plates?

BIBLIOGRAPHY

Bambach, R. K., C. R. Scotese, and A. M. Ziegler, "Before Pangea: The Geographies of the Paleozoic World," *American Scientist*, v. 68, pp. 26–38, 1980.

Beloussou, V. V., "Why Do I Not Accept Plate Tectonics?" *EOS*, v. 60, pp. 207–210, 1979. (See also comments on this paper by A. M. S. Sengör and K. Burke on same pages.)

Cox, A., ed., *Plate Tectonics and Geomagnetic Reversals*. San Francisco: W. H. Freeman and Company, 1973.

Dewey, J. F., "Plate Tectonics," *Scientific American*, May 1972. (Offprint 900.)

Hallam, A., *A Revolution in the Earth Sciences: From Continental Drift to Plate Tectonics*. Oxford: Clarendon Press, 1973.

Le Pichon, S., J. Francheteau, and J. Bonnin, *Plate Tectonics*. New York: Elsevier Publishing Company, 1973.

Marvin, U. B., *Continental Drift*. Washington, D.C.: Smithsonian Institution Press, 1973.

Meyerhoff, A. A., and H. A. Meyerhoff, "The New Global Tectonics: Major Inconsistencies," *American Association of Petroleum Geologists Bulletin*, v. 56, pp. 269–336, 1972.

Smith, A. G., and J. C. Briden, *Mesozoic and Cenozoic Paleocontinental Maps*. Cambridge: Cambridge University Press, 1977.

Tarling, D., and M. Tarling, *Continental Drift*. Garden City, N.Y.: Doubleday, 1971.

Vine, F. J., "The Continental Drift Debate," *Nature*, v. 266, pp. 19–22, 1972.

Wilson, J. T., ed., *Continents Adrift: Readings from Scientific American*. San Francisco: W. H. Freeman and Company, 1970.

Wylie, P. J., *The Way the Earth Works*. New York: John Wiley & Sons, 1976.

20

Deformation of the Earth's Crust

Forces in the crust have been active throughout geologic time. Rocks respond to these forces by folding and fracturing. In a typical mountain-making episode, a thick section of sediments, deposited in a geosyncline along the margin of a stable block, is crumpled. Mountains are raised as the belt is deformed and the sediments are thrown into a series of folds and faults. Intrusion and metamorphism are typical. After the highlands are eroded down, a final stage of rejuvenation by uplifting often occurs. The cycle of geosyncline formation and orogeny can be integrated with the plate-tectonic concept of opening and closing oceans in a simple and natural way.

What a spectacular movie could be made by a camera set to film the Earth's surface by snapping one frame every thousand years! Replayed at standard speed, a movie made this way would condense almost 50 million years of geologic time to an hour-long feature. Geologic movements that normally are invisibly slow could be viewed and really appreciated in such a fast-motion replay. One could see the drift of continents accompanying the opening of some oceans and the closing of others. In some regions, uplift, tilting, faulting, or folding would give rise to plateaus, mountains, rift valleys, and other geologic structures, and the continuous interplay between tectonic activity and erosion would be evident. In other places subsidence would lower portions of a continent beneath the sea. The idea is not as far-fetched as it sounds, because movies of this sort have been made in the manner of animated cartoons. Each frame of such a film is a sketch based on interpretation of the geologic record.

Geologists examine the geologic record and make a geologic cross section, showing beds in various attitudes—flat-lying, tilted, folded, faulted, perhaps intruded by igneous rocks, metamorphosed, or covered by lava flows. They do so to unravel the geologic structure of the region and then to deduce the sequence of events that we call geologic history (Fig. 20-1). They recognize those features (such as lithology, bedding, subaqueous slump, ripple marks, or mud cracks) that are connected with the original formation of the sedimentary beds. Deformation structures, such as folds and faults, which are imposed by later events, can then be interpreted in terms of the sequence in which they occurred. This chapter takes up the subject of geologic structures—particularly the changes ranging in extent from meters to hundreds of kilometers that take place in rocks due to deformation of the Earth's crust.

HOW ROCKS DEFORM

Folding and faulting are the most common forms of deformation of the rocks that make up the outer layers of the Earth's crust. For years, geologists were baffled by the problem of how rocks, which seem strong and rigid, could be distorted by crustal forces into folds or broken along faults (Fig. 20-2). The question still does not have a complete answer, but laboratory experiments in

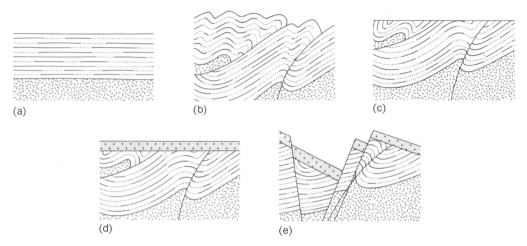

Figure 20-1
Stages in the development of the Basin and Range province, as postulated by W. M. Davis.
(a) Deposition of stratified sediments. (b) Folding and thrusting. (c) Development of an erosional surface. (d) Sheets of lava cover the eroded surface. (e) Block faults develop, breaking up the earlier features. A geologist sees only the last stage and attempts to reconstruct from the structural features all of the earlier stages in the history of a region. [After *The Evolution of North America* by P. B. King. Copyright © 1977 by Princeton University Press. Redrawn with permission of Princeton University Press.]

Figure 20-2
Small-scale folds in interbedded shales and cherts of the Franciscan Formation in central California. [From A. M. Johnson, *Physical Processes in Geology*, Freeman, Cooper & Company. Copyright © 1972.]

which rocks were squeezed under the conditions of high pressure and temperature known to occur in the crust have contributed much to our understanding of the processes involved.

Figure 20-3, for example, shows what happens to a small cylinder of marble when it is squeezed by applying **stress** at the ends. Stress is a force divided by the area over which it is applied.* The experiment consists of applying stress by pushing down on one end of a sample of rock with a piston while at the same time applying pressure to all sides of the sample. This simulates the confining pressure to which materials deep in the crust are subjected because of the weight of the overlying rock. Temperature can also be raised to realistic, deep-Earth values, usually to several hundred degrees centigrade. The figure shows a normal cylindrical sample and two samples that have been strained by 20 percent—that is, changed in length (in this case shortened) by 20 percent by the applied stress. Under low confining pressures, which simulate shallow depths in the crust, the middle sample fractured; under high confining pressures equivalent to conditions at greater depths in the crust, the third sample deformed smoothly and continuously into its shortened, bulging shape. The investigators concluded that,

---

*For example, a weight of 4 kilograms applied over an area of 2 square centimeters exerts a stress of 2 kilograms per square centimeter. Pressure is also measured as force per unit area. One kilogram per square centimeter is approximately equivalent to atmospheric pressure.

(a)   (b)   (c)

Figure 20-3
Marble cylinder deformed in a laboratory by pushing down at one end. From left to right: (a) undeformed; (b) 20 percent strain (shortened by 20 percent of the original length), 270 atmospheres confining pressure; (c) 20 percent strain, 445 atmospheres confining pressure. Note how the sample is ductile at higher confining pressure and brittle at lower confining pressure. [Photo by M. S. Paterson, Australian National University.]

if a bed of this particular marble were subjected to crustal forces near the surface, it would tend to deform by faulting; but, if it were deeper than a few kilometers, it would fold. Experiments also show that if rocks are hot when stresses are applied, smooth, continuous deformation occurs more readily.

Rocks, like most solids, can be classed as *brittle* or *ductile*. A brittle material breaks suddenly, rather than deforming smoothly, when forced beyond a critical value. Ductile substances are capable of considerable smooth, continuous deformation before they break. Glass, near room temperature, is a familiar brittle material, and modeling clay is obviously ductile. The marble discussed in the previous paragraph is brittle at shallow depths, but it would respond to stress as a ductile substance deeper in the crust.

Nature is more complex, of course, than the conditions under which such simple experiments are performed. For example, stresses produced by crustal deformation may continue for thousands or millions of years, whereas laboratory geologists have never squeezed individual rock samples for much longer than a few years. Rocks would be ductile under stresses much lower than those required in the laboratory if the stresses were applied for long periods, and would behave this way more readily if wet or hot. Nevertheless, the experiments shed some light on how rocks respond to stresses, and they give us more confi-

dence in our interpretations of field evidence. When we see folds and faults in the field, we can remember that some rocks are brittle, that others are ductile, that the same rock can be brittle at shallow depths and ductile deep in the crust, and that every rock behaves one way or the other, depending on the stresses and the conditions under which they are applied. Other things being equal, laboratory experiments teach us that we should expect to find that granite is less deformable than sandstone, that crystalline basement rocks are more brittle than the ductile young sediments that may cover them, and that sedimentary rocks increase in ductility from limestone and sandstone through clay and shale to salt and gypsum (Fig. 20-4).

Folds and faults are the details of the patterns of deformation, strong and weak, that geologists map in the field as clues to the larger panorama of tectonics. We now go on to describe the forms of these folds and faults before discussing larger-scale regional deformation.

FOLDS

The term "fold" implies that a structure that originally was planar, like a sedimentary bed, has been bent. The deformation may be produced by either horizontal or vertical forces in the crust. Folding is the most common form of deformation of layered rocks, and its most typical manifestation is in mountain belts. In young mountain systems, where erosion has not yet erased them, majestic, sweeping folds can be traced, some of them with dimensions of many kilometers (Fig. 20-5). On a much smaller scale, very thin beds can be crumpled into folds a few centimeters long (Fig. 20-6). Folds can be gentle or the bending can be severe, depending on the magnitude of the applied forces and the ability of the beds to resist deformation. Ancient mountains, long since eroded away, have left a partial record that enables geologists to reconstruct their former grandeur— the remnants of intense folding found today in the stable Precambrian terrains of continental interiors (Fig. 20-6).

In order to characterize and interpret a deformed rock layer, a geologist must first describe it. Ideally, it would be easiest to visualize a deformed stratum if the elevation of its surface could be mapped and contoured, just as a topographic map charts the Earth's surface in terms of contours (Appendix V). Sometimes, particularly when a search for oil or minerals is involved, there is enough information to do just this, using such evidence about the depth of the layer as data obtained from mines, boreholes, outcrops in valleys, or perhaps the echoes of seismic waves (see Box 17-1).

More often, the geologist has to reconstruct the geometry of folds and faults from sparse surface information only (Fig. 20-7). The process begins with a topographic map showing the contours of

Figure 20-4
Rocks respond differently to forces in the Earth's crust. In this tilted section, sandstone beds (light colored) were able to withstand deformation, but shales (dark colored) show evidence of flowage and changes in original thickness. Zumaja, Spain. [Photo by R. Siever.]

Figure 20-5
Large-scale folds in Lower Paleozoic graywacke, northeast Victoria Land, Antarctica. [Photo by W. B. Hamilton, U.S. Geological Survey.]

the Earth's surface. From field observations, which involve much walking and surveying, the geologist superimposes on the topographic map the pattern of outcrops of the different formations. The geologist is ever hopeful that the soil cover doesn't hide all of the outcrops. If there are hills and valleys and the beds are horizontal, the outcrop of a given stratum always falls at the same topographic level—that is, on the same contour line (Figure 20-8). More often, the beds are tilted, folded, or faulted, and the same formation

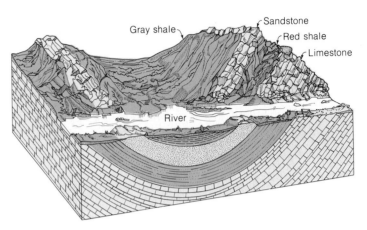

Figure 20-7
Geologists typically work from surface outcrops of rock formations to reconstruct subsurface structures. This block diagram shows schematically one example of the surface expression of a fold in a sequence of sedimentary rocks.

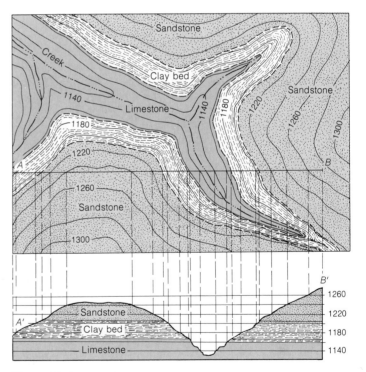

Figure 20-8
Construction of a geological section from a geological map. In this example, the formations are horizontal and therefore crop out at the same topographic level—that is, between the same elevation contours. [After *Principles of Geology* (4th ed.) by J. Gilluly, A. C. Waters, A. O. Woodford. W. H. Freeman and Company. Copyright © 1975.]

Figure 20-6
Folds in Precambrian iron formation, Beresford Lake area, Manitoba. This is a relic of an ancient mountain-making episode in the continental interior, a region that has remained relatively stable for approximately the past billion years. [From Geological Survey of Canada, Ottawa.]

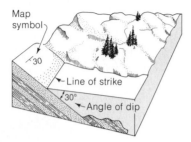

Figure 20-9
The strike and dip of an inclined bed.

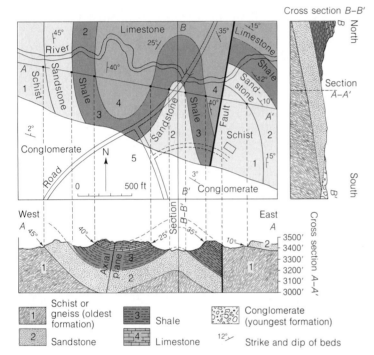

Figure 20-11
Geological map with formations, strikes, and dips indicated. Geological sections A-A' and B-B' are inferred from the map. Note how the surface features indicated on the map along the lines A-A' and B-B' are transferred to the corresponding cross sections and how the subsurface reconstruction is consistent with the surface observations. [After U.S. Geological Survey.]

Figure 20-10
Geologist measuring the dip of inclined beds using a clinometer. The dip is given by the angle between the horizontal and the edge of the instrument, which is aligned parallel to the bedding planes. The horizontal is found by rotating a level bubble until the bubble is centered. [From U.S. Geological Survey.]

the plane of the bed with the horizontal plane. Figures 20-9 and 20-10 show how dip and strike are measured in the field and plotted on maps; topographic and geological maps are described in more detail in Appendix V.

Once the formations are mapped and the dips and strikes recorded from many places, the geologist can attempt to reconstruct the deformed shape of the bed, even when erosion has removed sections of it. It is like putting together a three-dimensional jigsaw puzzle with missing pieces. Figure 20-11 shows a geological map with formations, strikes, and dips indicated. Below it is a cross section inferred from the map, with eroded portions indicated by dashed lines. The upfolds and downfolds of the beds show in the section as they would appear if a vertical cut could be made through the Earth in the direction indicated on the map.

Upfolds or arches of layered rocks are called **anticlines,** and downfolds or troughs are **synclines** (Fig. 20-12). A steplike bend in otherwise gently dipping or horizontal beds is a **monocline** (Fig. 20-13).

In order to discuss the different types of folding with an economy of language, we need to define the parts of a fold. The two sides of a fold are its **limbs.** The **axial plane** is a surface that divides a fold as symmetrically as possible. The line made by the intersections of the axial plane with the

will crop out at many different elevations. Using such criteria as relative age (indicated by fossils or radiometric dating) and lithology, the geologist keeps track of the sequence of layers. Figuring out the structure of the deformed beds requires information about the geometric attitude or orientation of the beds at outcrops. The geologist records this by marking on a map the **dip** and **strike** of the beds wherever they outcrop. The dip is the angle of inclination of the bed from the horizontal in the direction of steepest descent: water flows in the down-dip direction. The strike is at right angles to the dip direction. It is the intersection of

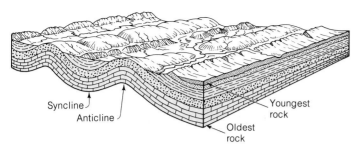

Figure 20-12
Upfolds, or anticlines, and downfolds, or synclines. Erosion of the folds produces parallel ridges where resistant rocks crop out and parallel valleys on the easily eroded formations. [After U.S. Geological Survey.]

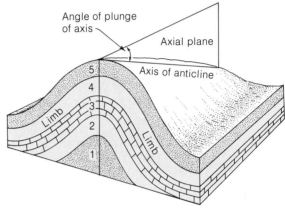

Figure 20-14
Parts of a fold.

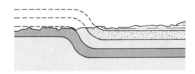

Figure 20-13
Schematic illustration of a monocline.

beds is the **axis** of the fold. If the axis of a fold is not horizontal, the fold is said to **plunge** by an amount given by the angle the axis makes with the horizontal. Parts of a fold are illustrated in Figure 20-14.

It is easy to see (Fig. 20-11) that, for a folded sequence of layers, the oldest beds would be found at depth in the core (or central axis) of the anticline and the youngest rocks on the surface over the axis of the syncline. On a geological map or cross section, eroded anticlines would be recognized by an axial core of older rocks, bordered on both sides by younger rocks dipping away from the axis. In the core of a syncline, the map would show younger rocks on the axis with older rocks dipping inward on both sides, as in Figures 20-11 and 20-12.

Types of Folds

We should not expect every fold to have a vertical axial plane with limbs dipping symmetrically away from the axis. When horizontal force is applied to layers that have an initial dip or that change laterally, say, in thickness, strength, or ductility, or when the force itself is uneven, the folds can be thrown into **asymmetrical** shapes, with one limb dipping more steeply than the other (Fig. 20-15). This is a common situation. When the

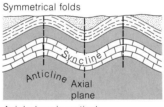

Symmetrical folds

Axial plane is vertical

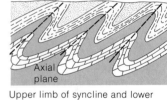

Overturned folds

Upper limb of syncline and lower limb of anticline, tilted beyond vertical, dip in same direction

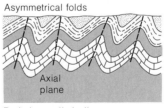

Asymmetrical folds

Beds in one limb dip more steeply than those in the other

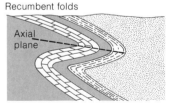

Recumbent folds

Beds in lower limb of anticline and upper limb of syncline are upside down; axial plane is nearly horizontal

Figure 20-15
Symmetrical, asymmetrical, overturned, and recumbent folds.

Figure 20-16
Overturned syncline in Cretaceous limestone, Lake Lucerne, Switzerland.
[Photo by W. B. Hamilton, U.S. Geological Survey.]

Figure 20-17
A recumbent fold, north of Grandjeans Fiord, Greenland. From top of cliff to valley
bottom, the difference in altitude is about 800 m. [Lauge Koch Expedition; courtesy
of J. Haller, Harvard University.]

deformation is intense, the fold may be **over-turned,** with the lower limb of an anticline or the upper limb of a syncline tilted more than 90° from its original attitude. Both limbs of an overturned fold dip in the same direction, as in Figures 20-15 and 20-16, and the overturned limb shows an inverted sequence in an outcrop, with older beds on top of younger ones. In **recumbent folds,** the axial plane is horizontal, or nearly so, and one limb has been rotated into a completely upside-down sequence (Fig. 20-17). Figure 20-18 shows how an overturned fold might evolve. The spectacular recumbent folds of the Alps, some of the first ones to be mapped in detail, are particularly famous among geologists.

Domes and basins are folds in which beds dip radially away from, or toward, a point, respectively. The outcrops of such formations on the surface would tend to be circular or elliptical (Fig. 20-19). Domes are very important in oil geology

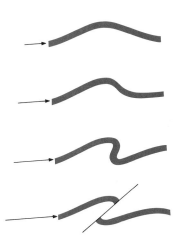

Figure 20-18
How an overturned fold might evolve as a result of an increasing horizontal force. In the bottom sketch the formation has actually ruptured.

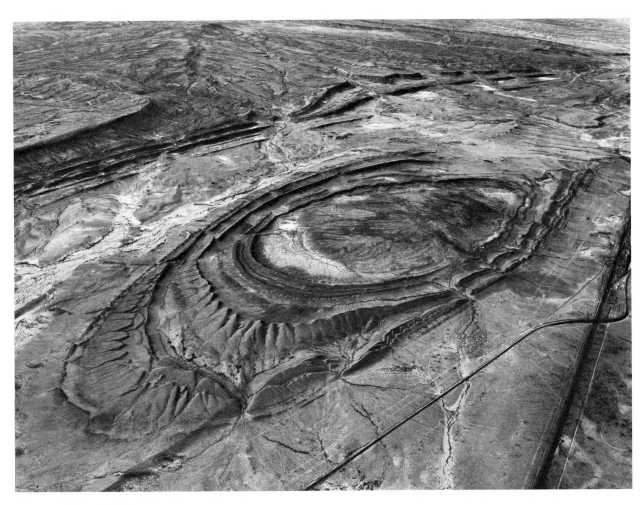

Figure 20-19
Dome in strata 6 miles east of Rawlins, Wyoming. Highway and railroad at right give scale.
[From *Geology Illustrated* by John S. Shelton. W. H. Freeman and Company. Copyright © 1966.]

because oil tends to migrate upward and become trapped against impervious rocks at the high point of a dome.

Follow the axis of any fold, and sooner or later the fold will die out. When an anticline disappears, for example, its axis descends and the folds become smaller and smaller, as in Figure 20-20. The axis of a syncline ascends as the fold disappears. The outcrops of eroded remnants of a plunging fold show a characteristic pattern like the bow of a canoe because the limbs converge where the fold disappears (Fig. 20-20b). Successions of (eroded) plunging anticlines and synclines show on geological maps or air photographs as a zig-zag pattern of outcrops (Figs. 20-20 and 20-21).

Folds tend to occur not in isolation but in elongated groups. Finding a **folded belt** suggests to a geologist that the region at one time was compressed by horizontal crustal forces. The Appalachians are an excellent example of a folded mountain belt (Fig. 20-22), and one current hypothesis, discussed later in this chapter, invokes a plate collision to account for the wrinkling of the once flat sheets of Paleozoic sediments.

FRACTURES

Rocks yield to deforming forces in several ways. Some layers crumple into folds, some break, some fold first and then break if the applied forces build up beyond a critical point. Fractures may be divided into two categories, joints and faults. A **joint** is a crack along which no appreciable movement has occurred. If there is displacement of the rocks on both sides of a fracture and parallel to it, the fracture is a **fault.**

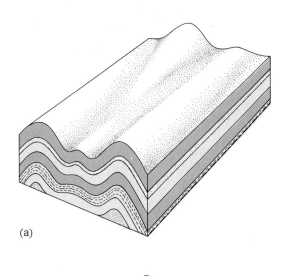

(a)

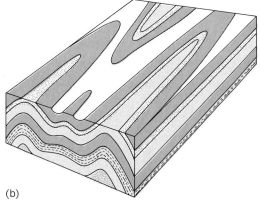

(b)

Figure 20-20
(a) Schematic block diagram of plunging folds.
(b) The eroded remnants of plunging folds show a characteristic pattern in which the limbs converge like the bow of a canoe.

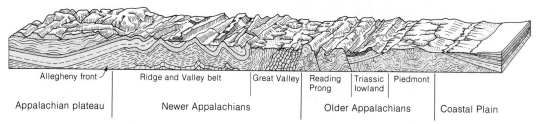

Figure 20-22
Block diagram of a section of the newer Appalachian Mountains, an example of the remnants of a folded mountain belt. [After *Stream Sculpture on the Atlantic Slope* by D. Johnson. Columbia University Press. Copyright © 1931.]

Figure 20-21
The erosional remnants of plunging folds show a zig-zag pattern in this view taken in the Valley and Ridge belt of the Appalachian Mountains, 30 miles northwest of Harrisburg, Pennsylvania. In the drawing below, the imaginary trench reveals the subsurface structure. [From *Geology Illustrated* by J. S. Shelton. W. H. Freeman and Company. Copyright © 1966.]

Figure 20-23
Intersecting joints form rectangular blocks in volcanic rocks near Braintree, Mass. [Photo by J. Haller, Harvard University.]

Joints

These structural features are found in almost every outcrop. In some places, joints are randomly irregular, but they are particularly interesting to the geologist if they show a regional pattern—for example, if all the cracks in a particular area are approximately parallel. Often two or more sets of joints intersect, breaking the rocks into large rectangular blocks or parallelepipeds. Regularity in a joint system often implies uniformity in the stress system that produced it. For example, the compression of sedimentary layers produced by the weight of overlying beds can produce one set of joints and the decompression caused by erosional unloading can establish another set. Regional stresses—compressional, tensional, or shear—that have long since vanished may leave their imprint in the form of distinct sets of joints. The contraction of layers of cooling lava may result in columnar jointing (see Fig. 15–8).

Joints provide channels for the flow of water through rock. Because of their greater surface

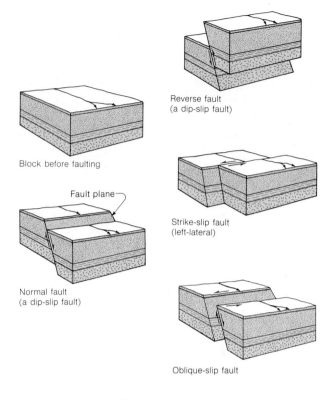

Figure 20-24
Types of faults.

area, the irregular blocks formed by intersecting joint systems are more readily attacked by air and water than are massive, unjointed formations; thus, weathering is speeded by jointing (Fig. 20-23). Joints may also provide underground channels for the flow of magma, often leaving as evidence a swarm of parallel dikes, or cracks filled with igneous rock. Magma injected under high pressure can enlarge existing joints or open new ones.

Faults

Because faults have already been discussed in the chapters on seismology and plate tectonics (Chapters 17 and 19), the review here will be brief. Like folds, faults are a common feature of mountain belts, particularly where the deformation is intense. As discussed in Chapter 19, margins where plates collide, pull apart, or drift past each other are the sites of subduction zones, rift valleys, or transforms, all of which involve faults. Some transform faults—the San Andreas fault of California, for example—show such large displacements that the offset of the two plates may amount to hundreds of kilometers. The different categories of faulting are distinguished by the di-

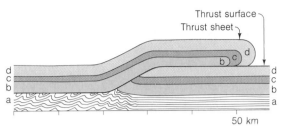

Figure 20-26
Schematic diagram of large-scale overthrust sheet, modeled by C. K. Longwell after examples in California and southern Nevada.

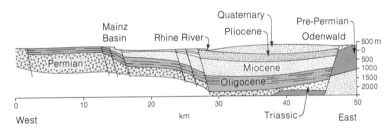

Figure 20-27
Block diagrams of a graben and a horst.

Figure 20-25
Thrust fault in limestone, Gusta, Norrland, Sweden. Offset of once-continuous bed, indicated by the distance between the brown layers, is about 0.8 m. [Photo by J. Haller, Harvard University.]

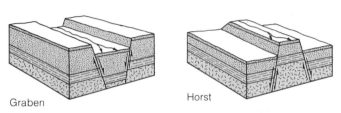

Figure 20-28
Geological cross section of the Rhine graben. [After P. Dorn.]

rection of motion along the fracture plane (Fig. 20-24). A **dip-slip fault** involves displacements up or down the dip of the plane. A **strike-slip fault** is one in which the movement is horizontal, parallel to the strike of the fault plane. A combination of dip-slip and strike-slip movements would describe an **oblique-slip fault.**

Faults need a further characterization since the movement can be up or down, or right or left. A **normal fault** is one in which the rocks above the fault plane move down relative to the rocks below. A **reverse fault,** then, is one in which the rocks above the fault plane move up relative to the rocks below. A reverse fault in which the dip is small so that the overlying block is pushed mainly horizontally is a **thrust fault** (Fig. 20-25). Finally, if, as one faces a strike-slip fault, the block on the other side is displaced to the right, then the fault is a **right-lateral fault; left-lateral faults** are displaced in the opposite direction (see Figs. 1-18, 17-8, and 20-25 for some photographs of faults).

Low-angle thrusts or overthrusts with displacements of many kilometers are often found in intensely deformed mountain belts. These thrusts are an expression of compressive forces, which shorten the crust. The surficial layers accommodate the crustal shortening, in this case, by break-

ing, with one sheet overriding another. A diagram of the thrusting in California and southern Nevada is shown in Figure 20-26.

Grabens—a term sometimes used synonymously with rift valleys—are long, narrow troughs bounded by one or more parallel normal faults. Tensional crustal forces, literally pulling the crust apart, produce these down-dropped fault blocks. The East African rift valleys, the mid-ocean ridge-rifts, and the Rhine River Valley are famous examples. A **horst** is the opposite, a ridge formed by parallel reverse or, more commonly, normal faults (Figs. 20-27, 20-28, and 20-29).

Because a fault involves dislocation along a fracture, it can be recognized in the field by the disruption of formations on either side of the frac-

Figure 20-29
The Red Sea divides to form the Gulf of Suez on the left and the Gulf of Aquaba on the right. The Arabian Peninsula on the right, splitting away from Africa on the left, has opened these great rifts, now flooded by the sea. [Photo by Apollo 7 astronauts; from National Aeronautics and Space Administration.]

Figure 20-30
Slickensides on rhyolite, Braintree, Mass. [Photo by J. Haller, Harvard University.]

ture. In some places, the two segments of an offset formation can be found, and the amount of relative displacement can be measured directly (Fig. 20-25), but this becomes difficult with large offsets. In establishing the time of faulting, geologists use the simple rule that a fault must be younger than the youngest rocks it cuts and older than the oldest undisrupted formation that covers it (see Fig. 20-1).

Crushed or ground up rock is often found in a fault zone. In some places, friction along a fault plane has produced polished and striated surfaces. These surfaces are called **slickensides** (Fig. 20-30). Because they align in the direction of relative movement of the fault blocks, they tell what this direction was, though not with complete reliability. If movement has occurred recently, the effect on topography, drainage patterns, vegetation, and human artifacts can often be recognized (see Figs. 1-18 and 17-8).

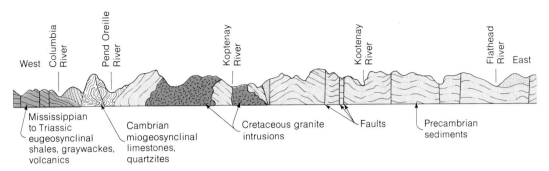

Figure 20-31
Geological cross section from the Cascade Mountain front of northeast Washington to the Rocky Mountains of northwest Montana. Deformed eugeosynclinal and miogeosynclinal deposits and granitic intrusions characterize much of this region. [After R. G. Yates, G. E. Becraft, A. B. Campbell, and R. C. Pearson, Canadian Institute of Mining and Metallurgy.]

Several cross sections of folded and faulted belts are shown in Figures 20-22, 20-31, 20-32, and 20-33 to illustrate some of the ways rocks deform in response to crustal forces. Figure 20-34 shows the surface appearance of such folded mountains.

Topographic Expression of Deformation

Deformation—in the form of mountain belts with their folds and faults, plateaus, grabens, and strike-slip faults—leaves its unmistakable mark on the configuration of the Earth's surface. These topographic expressions are often a guide to the structures that control them. Even for such relatively small-scale features as the shapes of hills and valleys and the courses of streams, the controlling factors are the structural elements in a complex interaction with erosion.

It should also be apparent that, the older the structure, the more likely it is that erosion has erased it and the less evident is its physiographic expression. Thus, present-day surface relief is largely due to movements that occurred during and since the Tertiary Period. These movements and earlier erosion have tended to obscure Mesozoic and Paleozoic structures. Precambrian deformation no longer shows as mountains but only as relic remnants of folds and faults in the basement rocks of the continental interior. Tertiary deformation still shows marked physiographic expression in the Alpine and Himalayan belts, the island arcs and deep-sea trenches of the Pacific, the great rift valleys of Africa, the rejuvenated Late Paleozoic Appalachians, the Rocky Mountains, and the Pacific Coast Ranges (see the figure in Box 20-2).

Although most features of topography can be related to some particular structural element, one

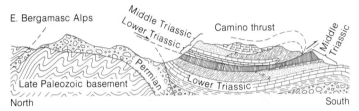

Figure 20-32
Unconformities, folds, and thrust sheets in a section of the southern Alps. [After *Structural Geology* by L. U. De Sitter. McGraw-Hill Book Company. Copyright © 1964.]

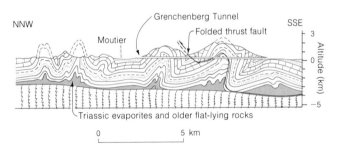

Figure 20-33
Folds and faults in a section of the Juras, a folded mountain belt. According to one hypothesis, the beds overlying the Triassic were sheared off and deformed independently of the older rocks below.

should not expect the crests of anticlines always to form ridges and the troughs of synclines always to become valleys. When stratified rocks are deformed, the resistance of the individual beds to weathering and erosion are important factors in controlling landforms, as well as whether the layers are flat-lying, tilted, folded, or faulted. In the block diagram of the Appalachians (Fig. 20-22), one can see every combination of anticlinal and synclinal hills and valleys.

Figure 20-34
The Jura Mountains, in a view looking northeast over the town of Moutier, Switzerland. The near and distant ridges are anticlines of upper Jurassic limestone. The crest of the anticline can be seen in the water gap in the distant ridge. The town is in a synclinal valley. See Figure 20–33. [Photo by J. Haller, Harvard University.]

REGIONAL MOVEMENTS OF THE EARTH'S CRUST

Orogenic Movements

A striking feature of continents is that they are marked by long, relatively narrow mountain chains of strongly folded and faulted rocks. The movements that produce these deformed belts are termed **orogenic,** and they are generally thought to be the result of plate collisions. The Appalachian Mountains, the Alps, the Urals, and the Cordillera of the western Americas are examples. A typical case involves compressive deformation (folding, thrust faulting) accompanied by episodes of intrusion and metamorphism. We will further discuss orogenic belts later in this chapter.

Epeirogenic Movements

Sedimentary rock sequences all over the world record the same kind of history of downward and upward movements of the crust that we can see in the Grand Canyon (Chapter 2). **Epeirogeny** is the term used to describe these movements. Great thicknesses of marine deposits of shallow-water

origin, once buried hundreds or thousands of meters below the sea floor, are well above sea level. These deposits give evidence of continued slow subsidence during sedimentation. In many cases, later elevation to their present position above the sea is due to simple uplift. Gaps in the rock sequence represented by unconformities (Fig. 2–15) tell of uplifts of large regions that stopped sedimentation and started stripping away previously deposited rocks. Fossil trees and other plants embedded in coal deposits now mined deep in the Earth tell of earlier times when they grew in the sunlight of the surface. Beaches and marine shells now found at elevations of 250 meters above sea level substantiate the postglacial uplift of Fennoscandia. Other evidence of vertical movements are incised meanders, coral reefs raised above sea level, and drowned river valleys (see Figs. 10–7, 20–35 and Chapters 5 and 7).

Along seacoasts, geologists are careful to distinguish between movements in which one portion of the crust went up or down relative to another part of the crust and the worldwide sea-level changes that accompanied Pleistocene glacial and interglacial periods. Whether the land goes up or the sea goes down (or vice versa) makes no difference in shaping such landforms of seacoasts as terraces or eroded headlands.

Figure 20-35
Regional uplift, evidenced by a delta raised about 30 m above sea level. An old delta is being dissected by streams as a new delta is being deposited. Antarctic Sund, E. Greenland. [Lauge Koch Expedition; courtesy of John Haller, Harvard University.]

Though many of the vertical movements are connected with orogeny, epeirogenic movements commonly affect large regions without extensive folding or faulting. A typical product of this kind of downward movement, which is usually slow and intermittent, is a **basin,** a bowl- or spoon-shaped depression that fills with sediment. Such gradual and intermittent downward movement of the crust occurs in one part of the relatively unde-formed or stable broad region between the Appalachians and the Rockies—the Michigan Basin. Covering much of the lower peninsula of Michigan, this circular area of about 500,000 square kilometers has subsided throughout much of Paleozoic time and received sediments more than 3 kilometers thick in its central, deepest part. The structure of the formations in this basin has been likened to a pile of saucers (Fig. 20-36).

Upward movement with little or only moderate faulting or folding can produce broad uplands and plateaus. The Colorado Plateau, the Adirondack Mountains, and the Black Hills of South Dakota were all produced by general upward movements. The Black Hills structure is an oval **dome**—an area of uplift sloping off more or less uniformly in all directions from the highest point—that rises more than 2 kilometers above the Great Plains surrounding it. The larger area in

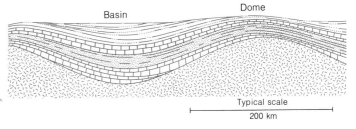

Figure 20-36
Idealized section of a dome and basin, evidence of vertical movements in the relatively undeformed interior of the United States.

which the Black Hills are located was once lower; it subsided to receive a blanket of Paleozoic and Mesozoic sediment more than 2 kilometers thick. Sometime between late Cretaceous and Oligocene, the whole area of about 50,000 square kilometers was uplifted, as if pushed up from below by a piston, without being extensively crumpled or broken. Subsequent erosion has stripped away the sediments overlying the central part of the uplift, exposing the Precambrian igneous and metamorphic rocks below. Here is where the famous Homestake Gold Mine was discovered in the Precambrian rocks near Lead, South Dakota

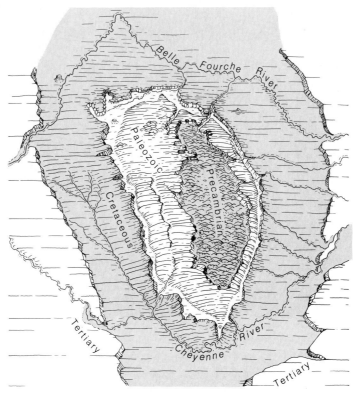

Figure 20-37
The Black Hills of South Dakota, a domal uplift in the Great Plains. Erosion of the central part exposes the core of Precambrian igneous and metamorphic rocks and the succession of Paleozoic and Mesozoic sediments that slope away from the center. [After *Geology of Soils* by C. B. Hunt. W. H. Freeman and Company. Copyright © 1972.]

(Fig. 20-37). Some geologists speculate that the doming of the Black Hills was caused by an upwelling of magma from deep within the crust.

Geologists have no ready mechanisms to account for most of these slow and broad epeirogenic movements, but hypotheses exist for some of them. The Fennoscandian uplift (Chapter 18) represents the slow upward recovery of the crust following the removal of the glacial load that had depressed it. The sediment-filled downwarp off the eastern coast of the United States may be evidence of the subsidence of the edge of the continent after it drifted apart from north Africa. This subsidence may represent contraction of the newly formed lithosphere as it cools in the spreading process, as mentioned in Chapter 19 (see Fig. 20-38). Some of the broad uplifts, such as the Tibetian Plateau, may be connected with plate collisions; in general, however, warping within continents and far from plate margins is still a puzzle.

Recent Movements

Up-and-down movement, or **warping,** of large regions is not restricted only to the geologic past; it is occurring at a measurable rate in our own time. For example, the city of Venice is slowly sinking into the Adriatic Sea at a rate of 4 millimeters per year. The process is mostly one of coastal down-warping, though subsidence due to the withdrawal of water and natural gas from the underlying sediments is, unfortunately, hastening this beautiful city's demise (Fig. 20-39).

Precision surveys across the United States have revealed the pattern of vertical movements shown in Figure 20-40. Large regions are sinking and rising at rates of 1 to 5 millimeters per year. If the pattern persists for the geologically short time of only a million years, much of New England and the Gulf Coast will have sunk to the sea floor, and plateaus a few kilometers in height will have grown in the Midwest. These seemingly slow

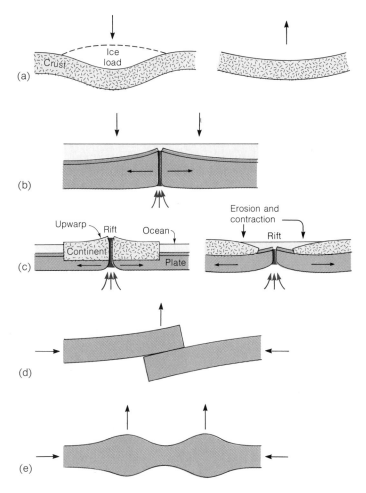

Figure 20-38
Some proposed mechanisms for vertical movements (not to scale). (a) The Glacial ice load buckles the crust; slow uplift follows removal of the ice. (b) Formation of lithosphere at mid-ocean ridges. Spreading sea floor subsides as the plate cools and contracts. (c) Rifting and upwarping due to heating, followed by separation of two segments of a continent; receding edges erode at the top, cool and contract within, forming subsiding continental margin. (d) Two plates collide; the overriding plate is uplifted. (e) Crust deforms by thickening due to horizontal forces.

Figure 20-39
Venice is slowly subsiding into the Adriatic Sea. Water level now reaches the base of the columns of this old building. The raised sidewalk, of recent construction, will be awash in several decades if subsidence continues. [Photo by Roberto Frassetto, National Research Council of Italy.]

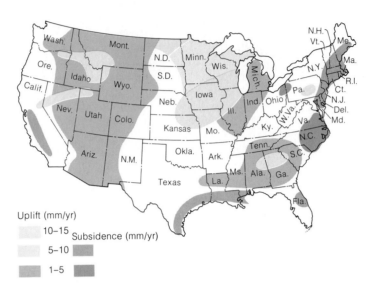

Figure 20-40
Crustal-movement map showing areas of present-day uplift and subsidence
in the United States. [Chart by S. P. Hand, National Oceanic and Atmospheric
Administration.]

rates of displacement, unnoticed by the inhabitants, are all that it takes to produce highlands and basins.

There is one type of crustal movement that the inhabitants of an affected area do take notice of: the sudden and large displacements along faults that produce large earthquakes. Over thousands or millions of years, the accumulated displacements due to earthquake movements may be of the same order as the gradual displacements depicted in Figure 20-40. The catastrophic nature and the large size of each impulse of displacement is what sets earthquake movements apart.

Earthquake displacements can be horizontal and vertical (Chapter 17) and, in the case of a truly great shock, can amount to as much as 10 or 15 meters. Horizontal displacements along the San Andreas fault at the time of the San Francisco earthquake of 1906 amounted to as much as 5 meters (Fig. 17-8). The earthquake of 1872 that devastated the Owens Valley just east of the Sierra Nevada in California produced a vertical displacement of 4 meters. At the rate of one such earthquake every few thousand years, this lofty mountain range could have been raised in but a few million years, even allowing for erosion.

The permanent displacements accompanying large earthquakes are not limited to the vicinity of the fault. They can affect immense areas. The great Alaskan earthquake of 1964 is a well-documented example of uplift and subsidence that look place in an area of more than 200,000 square kilometers. The maximum uplift amounted to 13 meters, and the greatest subsidence was 2 meters. All of this occurred in a few minutes as the Pacific plate slipped a few meters deeper into the mantle beneath southern Alaska. The displacements accompanying earthquakes, integrated over thousands or million of years, can be a major factor in regional deformation.

The broad, gradual uplifts described in the preceding pages almost always entail some slight undulating deformation of the rocks. Rocks that look flat-lying and horizontal prove to be slightly tilted and gently folded when careful surveys are made. Within the continent, in apparently flat terrain, there are also small faults along which very small amounts of movement have occurred. Thus folds and minor faults are widely distributed in most parts of the crust. In orogenic belts, strong folding and faulting are the major structural features.

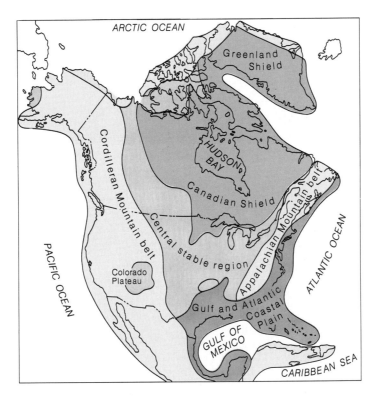

Figure 20-41
Major tectonic features of North America.

MAJOR REGIONAL STRUCTURES

Up to this point in the book, we have discussed the slow up-and-down movements of the crust, the deformation of layers by folding and faulting due to horizontal crustal forces, and the emplacement of plutons. These structural elements and the kinds of rocks they involve—in short, the sequence in space and time of magmatic, sedimentary, and deformational episodes—determine the geological character of a region. In this section, we shall discuss the broad regional characteristics of continents, using North America as an example.

The concepts of plate tectonics have given us new insight into the geology of continents. Convergent boundaries can leave records of deposits of material scraped off a subducting plate, fragments of continents or island arcs plastered onto a continent, magmatic belts, and fold and thrust belts. Divergent boundaries are evidenced by the orderly sequence of sediments deposited in geosynclines along a subsiding continental margin. Major orogenic belts like the Cordillera of western North and South America, the Appalachians of eastern North America, and the Caledonian

Mountains of Great Britain are now beginning to be understood in terms of plate boundary events over the past 700 million years. However, the older terrains, which typically occur in the interior of continents as large areas of Precambrian rocks, are even more difficult to interpret. Ancient episodes of metamorphism and erosion have removed much of the evidence, and the extent to which plate tectonics, as we know it today, was a major factor in Precambrian geology is not yet clear.

Perhaps because of the long, linear extent of plate boundaries and the characteristic events associated with the three types of boundaries, the geologic fabric of a continent is not random. Similar structures tend to occur in association so as to make up distinctive orogenic belts, stable interior platforms, marginal depressions, and so on. Moreover, the spatial relations of these regions seem to have a pattern, as we can see from Figures 20-41 and 20-42, in which the major physiographic and structural divisions of North America are indicated. A transcontinental section showing the major structural divisions schematically is shown in Figure 20-43.

Figure 20-42

Physiographic regions and provinces of the contiguous United States. [After *Natural Regions of the United States and Canada* by Charles B. Hunt. W. H. Freeman and Company. Copyright © 1974.]

1. *Superior Upland.* Hilly area of erosional topography on ancient crystalline rocks.
2. *Continental Shelf.* Shallow, sloping submarine plain of sedimentation.
3. *Coastal Plain.* Low, hilly to nearly flat terraced plains on soft sediments.
4. *Piedmond province.* Gentle to rough, hilly terrain on belted crystalline rocks becoming more hilly toward mountains.
5. *Blue Ridge Province.* Mountains of crystalline rock 900 to 1800 m (3000 to 6000 ft) high, mostly rounded summits.
6. *Valley and Ridge province.* Long mountain ridges and valleys eroded into strong and weak folded rock strata.
7. *St. Lawrence Valley.* Rolling lowlands with local rock hills.
8. *Appalachian Plateaus.* Generally steep-sided plateaus on sandstone bedrock, 900 to 1500 m (3000 to 5000 ft) high on the east side, declining gradually to the west.
9. *New England province.* Rolling, hilly, erosional topography on crystalline rocks in the southeastern part, changing to high mountainous country in the central and northern parts.
10. *Adirondack province.* Subdued mountains on ancient crystalline rocks rising to more than 1500 m.
11. *Interior Low Plateaus.* Low plateaus on stratified rocks.
12. *Central Lowland.* Mostly low, rolling landscape and nearly level plains. Most of area covered by veneer of glacial deposits, including ancient lake beds and hilly, lake-dotted moraines.
13. *Great Plains.* Broad river plains and low plateaus on weak, stratified sedimentary rocks. Rises toward Rocky Mountains, reaching altitudes above 1800 m at some places.
14. *Ozark Plateaus.* High, hilly landscape on stratified rocks.
15. *Ouachita province.* Ridges and valleys eroded on upturned, folded strata.
16. *Southern Rocky Mountains.* Complex mountains rising to more than 4300 m (14,000 ft).
17. *Wyoming Basin.* Elevated plains and plateaus on sedimentary strata.
18. *Middle Rocky Mountains.* Complex mountains with many intermontane basins and plains.
19. *Northern Rocky Mountains.* Rugged mountains with narrow intermontane basins.
20. *Columbia Plateau.* High rolling plateaus underlain by extensive lava flows; trenched by canyons.
21. *Colorado Plateau.* High plateaus on stratified rocks cut by deep canyons.
22. *Basin and Range province.* Mostly isolated ranges separated by wide desert plains. Many lakes, ancient lake beds, and alluvial fans.
23. *Cascade-Sierra Nevada Mountains.* The Sierra Nevada, in the southern part of the province, are high mountains eroded from crystalline rocks. The Cascades, in the northern part of the province, are high volcanic mountains.
24. *Pacific Border province.* Mostly very young, steep mountains; includes the extensive river plains in California.

[487]

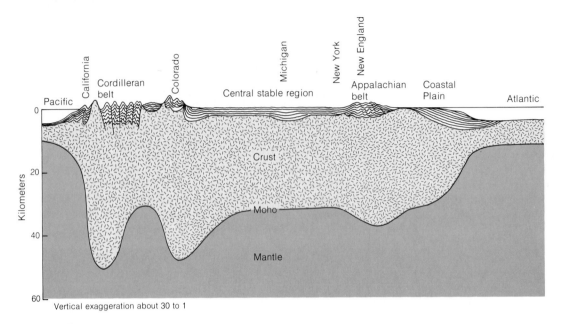

Figure 20-43
Schematic section across the United States, depicting major structural regions and
variations in crustal thickness.

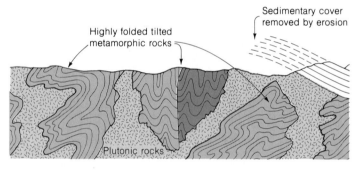

Figure 20-44
Idealized section in the Canadian Shield. The highly
deformed and metamorphosed rocks in this region
indicate that an intense orogenic episode took place
in Precambrian times, before stable conditions set in.

Stable Interior

The extensive, flat, stable interiors of continents,
which have been relatively undisturbed since
Precambrian time except for gentle warping, are
called **cratons.** The cratons typically include large
areas of exposed, very old crystalline basement
rocks called **shields,** surrounded by sediment-
covered platforms. The Canadian Shield is dominated by granitic and high-grade metamorphic
rocks, such as gneisses, which, together with
highly deformed and metamorphosed sediments
and volcanic rocks, imply a series of intense
mountain-making episodes in Precambrian time

before the stable conditions set in (Fig. 20-44).
The shield also includes some very old sediments
that were hardly touched by deformation and
metamorphism. This primitive block contains one
of the oldest and most complex records of geological history, much of it still unknown in detail. It
is famous for major deposits of iron, gold, copper,
and nickel. Other shields are found in Fennoscandia, Siberia, central Africa, Brazil, and Australia.

South of the Canadian Shield (see Figs. 20-41,
20-42, regions 12 and 13, 20-43) is the central
stable region, a sediment-covered platform. This
platform is in a sense a subsurface continuation of
the shield, for it contains Precambrian basement
rocks covered by a veneer of Paleozoic sediments
typically less than about 2 kilometers thick.
Within this region broad *sedimentary basins* are
defined by roughly oval areas where the sediments are somewhat thicker than the surrounding
platforms (Fig. 20-36). In parts of the Michigan
and Illinois basins sediments may be up to 2.5 or
3 kilometers thick. Some believe these basins to
be the result of rifting of the Precambrian surface
without spreading, but with later cooling and subsidence. The sediments, laid down under quiet
conditions, have remained unmetamorphosed
and only slightly deformed to this day. There is
abundant evidence of mild structural deformation, including shallow basins, dome, anticlines,
and synclines. The deformation is mostly in gentle tilts, folds, and faults with small displace-

Box 20-1   TRANSGRESSIONS AND REGRESSIONS OF THE SEA: A PLATE-TECTONIC EXPLANATION

The periodic advances and withdrawals of shallow seas, evidenced by the marine sediments of the platforms and shelves, have puzzled geologists for a long time. Do the transgressions and regressions represent vertical movements of the platforms and shelves, or do they result from changes in sea level? Certainly many local apparent changes of sea level can be explained by tectonic uplift or subsidence of the immediate region. Others, however, seem to be worldwide and controlled by sea-level changes. (See the discussion of eustatic sea-level changes in Chapter 11.) Glacial and interglacial periods can lead to eustatic sea-level changes of 100 meters, as occurred in the late Cenozoic Era (Chapter 9). There is no evidence, however, that a glacial-interglacial cycle was responsible for the extensive continental flooding of the late Mesozoic Era. A novel plate-tectonic explanation of sea-level changes has recently been proposed by J. D. Hays and W. C. Pitman III. We have seen in the preceding chapter that the new lithosphere that forms at a mid-ocean spreading ridge is initially hot and elevated. As it spreads from the ridge axis, the lithosphere cools and contracts, and the sea floor subsides. At times when sea-floor spreading is rapid, ocean ridge systems would be younger, hotter, and therefore substantially increased in volume, as the figure shows. Water would be displaced from the oceans, spilling onto the continents. When spreading slows down, the ocean ridge systems would be cooler

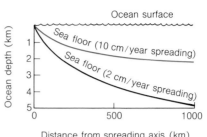

on the average and would occupy less volume; The seas would retreat from the continents back to the increased volume of the ocean basins. Support for this novel idea comes from the late Cretaceous transgression, which seems to correlate with an episode of rapid spreading between 110 and 85 million years ago. Similarly, the subsequent regression correlates with a reduction in spreading rates. Alternatively, changes in the total length of actively spreading ocean ridges might cause the transgressions and regressions of the sea. Many other changes of sea level are too rapid to be accounted for by this mechanism, but there is also no firm evidence of their correlation with glaciations. It may be that glaciations were more common than many think or that there is some other explanation for rapid transgressions and regressions.

---

ments. Some geologists think that these structures are related to local compression and tension as the basin subsided relative to the surrounding platform areas, rather than to later orogenic forces. One such structure, the Ozark dome, was an ancient structural high that was surrounded by and occasionally barely covered by mid-Paleozoic seas. The Russian platform, including the Don basin, which includes much of European Russia, is analogous to the North American platform.

The platform sediments were laid down under a variety of conditions. Nonmarine sediments testify to deposition on extensive alluvial plains, in occasional lakes, and particularly during Carboniferous times, in extensive swamps where peat beds accumulated. At other times, limestones, deltaic deposits, and marine sandstones and shales indicate extensive shallow inland seas, sometimes landlocked to form evaporite deposits. Some deposits of sandstone and shale imply the existence of low emergent lands near the margins of the craton; other sandstones and shale deposits

suggest the highlands elevated near continental margins, elevated perhaps by plate collisions. Much of the continent's uranium, coal, oil, and gas deposits are contained in the platform sediments. Paleogeographic maps for the Ordovician and Carboniferous periods (Fig. 19–16) show the distribution of ancient mountains, land areas, shallow seas, and deep seas as reconstructed from the sedimentary geological record and paleomagnetic data (see Box 20–1).

Orogenic Belts

Referring to Figures 20–41 and 20–42, we see that surrounding the great stable interior of the continent are **orogenic belts,** regions that were deformed by folding and faulting and were subjected to plutonism and metamorphism at various times in the Paleozoic, Mesozoic, and Cenozoic eras.

Orogenic (or folded) mountain belts usually show certain general features that enable us to

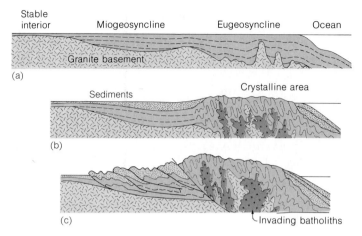

Figure 20-45
General features of an orogenic belt, showing its inferred development from geosyncline to deformed belt. (a) Geosynclinal stage. (b) After deformation of eugeosyncline and deposition of sediments eroded from the newly formed mountains. (c) After final deformation of miogeosyncline. According to the plate-tectonics hypothesis, stage (a) would follow plate separation and the opening of an ocean basin, with deposition on the receding margin of a continent. Stages (b) and (c) would occur with plate collision and the closing of an ocean basin. [After *The Evolution of North America* by P. B. King. Copyright © 1977 by Princeton University Press. Redrawn with permission of Princeton University Press.]

say something of how they were formed. Figure 20-45 reviews this important concept, which was discussed in Chapter 19 in terms of plate-tectonic theory. These deformed belts typically occur on the margins of stable shields in close connection with geosynclines. A geosyncline is a subsiding linear trough that earlier received a thick succession of layered sediments, sometimes more than 10,000 meters (30,000 feet) thick, many times the thin sedimentary veneer on the platform. Typically, two geosynclines are involved, an inner miogeosyncline with sandstone and limestone shelf deposits and an outer eugeosyncline with volcanic rock and a great thickness of such deep-water sediments as turbidites, shales, and pelagic limestones (see Chapter 11). Folded mountain chains seem almost always to be the result of a wave of deformation that begins in the eugeosyncline and later extends to the miogeosyncline. When plates collide, the lateral forces that develop compress the crust intensely—in some places by 20 percent or even more—as if the crust were caught in the jaws of a vise. In response to this squeezing, the overlying sedimentary cover in the eugeosyncline is deformed into great folds and thrust faults, generally trending parallel to the edge of the shield. Intrusions work their way into this compressed and contorted belt. New mountains are thrown up, reversing submarine slopes, so that erosion leads to episodes of rapid deposition of sediments toward the continent on

one side, both in arms of the sea and on land, as well as toward deeper water on the other side, as Figure 20-45 shows.

In a final stage, the miogeosyncline is deformed. The deforming forces are typically directed from the ocean basins toward the interior shield. This can be seen in the frequent transition from gentle folds adjacent to the shield to severe folding and thrusting and increasing metamorphism and magmatism farther from the shield. The deformed belts are asymmetrical in other ways as well—the folds lean or are overturned toward the shield, and the thrust sheets have been pushed in the same direction.

The fact that batholiths form the cores of many continental mountain ranges suggests that they have some connection with the mountain-making process. For example, the great mountain ranges that extend for thousands of kilometers along the western margins of the North American and South American plates have cores of contiguous batholiths. Huge areas of exposed Precambrian terrain in the now stable areas within continents are made up of successions of batholiths, and many of these seem to be roots of mountains that have long since been eroded. In Chapters 14 and 19, we discussed the idea that the formation of some batholiths is connected with subduction at the margin of colliding plates. It was hypothesized that reheating and melting of the descending plate may provide rising hot fluids as raw materi-

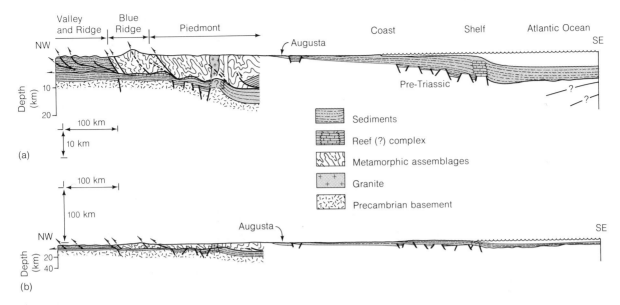

Figure 20-46
Geologic section of the southern Appalachians, the coastal plain, and continental shelf. Plate collisions in the Paleozoic thrust the Blue Ridge and Piedmont sheets over the sediments of an ancient North American shelf and deformed the Ridge and Valley province sediment. The modern shelf developed after the Triassic–Jurassic splitting of North America from Africa. In (a) the vertical scale is ten times the horizontal. In (b) the vertical and horizontal scales are the same, showing the thin overthrust sheets in true perspective. [After F. A. Cook, D. S. Albaugh, L. D. Brown, S. Kaufman, J. E. Oliver, and R. D. Hatcher, Jr., "Thin-Skinned Tectonics in the Crystalline Southern Appalachians," *Geology*, v. 7, pp. 563–567, 1979. Copyright © 1979 by Geological Society of America.]

als for batholith formation along the margins of the overriding plates (see Fig. 19-12). Crystalline rocks of mountain belts may also have formed elsewhere and then been displaced hundreds of kilometers as overthrust sheets.

Very often, following episodes of major folding and after erosion wears the mountains down, the belt is rejuvenated by later upwarping that exposes the old structures still remaining. Rejuvenated fold belts are exemplified by the modern topography of the Alps, the Urals, and the Appalachians.

The old, eroded Appalachian Mountains, a classical fold and thrust belt that extends along eastern North America from Newfoundland to Alabama, illustrates the general features of orogenic belts that enable us to reconstruct their history. Although the Appalachians are complex and the details vary greatly along their length, the development of the belt can be explained in the general terms of plate tectonics theory as consisting of two main stages: (1) Beginning in the late Precambrian, about 700 to 800 million years ago, North America split from Africa, an ancestral Atlantic opened, and geosynclines developed along the receding continental margins. (2) In the early Paleozoic, some 600 million years ago, the Atlantic began to close, and subduction along with volcan-

ism, folding, thrusting, and other convergent plate-boundary events continued through much of the era.

The present-day geologic features of the southern Appalachians are shown in Figure 20-46. The rock assemblages and structures, obtained from geological and geophysical studies, are the primary data available to reconstruct the history of the belt. From northwest to southeast this part of the Appalachians consists of a sequence of distinct geological zones:

1. *Valley and Ridge province.* Thick Paleozoic sediments were folded and thrust to the northwest by compressional forces from the southeast; deformation occurred in three orogenic episodes at the end of the Ordovician, at the end of the Devonian, and in Permian-Carboniferous time.
2. *Blue Ridge province.* These eroded mountains are composed largely of Precambrian and Cambrian crystalline rock, showing much metamorphism; deformed at the same times as the Valley and Ridge province, they were separated from it by a major thrust fault overriding to the northwest.
3. *Piedmont.* Precambrian and Paleozoic metamorphosed sedimentary and volcanic rocks

intruded by granite were eroded to low relief; the Piedmont is separated from the Blue Ridge by a major thrust fault, overriding to the northwest; volcanism began in the late Precambrian (about 700 million years ago) and continued into the Cambrian until about 500 million years ago; at least two episodes of deformation are evident, one at the end of the Devonian (about 450 million years ago), the other in early Carboniferous (about 350 million years ago).

4. *Coastal plain.* Relatively undisturbed sediments of Jurassic age and younger (less than 200 million years) are underlain by rocks similar to the Piedmont; the continental shelf is the offshore extension of the coastal plain.

Recent geological and seismic reflection results depicted in Figure 20–46 indicate that younger sediments of the Valley and Ridge province continue eastward and underlie the older crystalline and metamorphic rocks of the Blue Ridge and Piedmont!* It now appears that the Blue Ridge and Piedmont were not intruded and metamorphosed in place but were thrust as sheets over the continental-shelf sediments of ancestral North America. In this case accretion or continental growth during plate convergence occurred by the stacking of multiple thrust sheets.

Figure 20–47 shows a possible plate-tectonic reconstruction of southern Appalachian history consistent with these geological features. In late Precambrian time (Fig. 20–47a) a supercontinent split occurred, leaving what was to be North America on one side of the rift, Africa on the other, and the proto-Atlantic Ocean between. A continental fragment that would become the Blue Ridge and the Piedmont was separated from North America by a marginal sea. Continental-shelf sediments were deposited along the receding plate margins. In early Cambrian time (Fig. 20–47b) plate convergence began as the proto-Atlantic began to close. Evidence of the accompanying subduction and development of an island arc is seen today in the metamorphic sediments and volcanics of the Piedmont. Convergence continued from the Middle Cambrian through the Ordovician to early Silurian (Fig. 20–47c and d) when the continental fragment collided with North

America. The Blue Ridge and part of the Piedmont were thrust as sheets of crystalline rock over the younger North American shelf sediments. A second orogenic episode occurred near the close of the Devonian when the island arc collided with North America, pushing the Blue Ridge and Piedmont sheets farther west (Fig. 20–47e). The Atlantic closed completely in Carboniferous and Permian time with Africa slamming into North America. The extensive folding and thrusting in the Valley and Ridge province is associated with this culminating orogeny. The accumulated thrusting of all three deformational episodes may have pushed the Blue Ridge and Piedmont sheets more than 250 kilometers westward over the continental-shelf sediments of ancestral North America.

The modern North Atlantic opened some 200 million years ago in Triassic-Jurassic times when rifting occurred east of the Piedmont. Africa split away from North America, leaving a fragment behind that today is the basement under the coastal plain and shelf. The present-day shelf deposits began to develop on the receding margins (Fig. 20–47f). As a result of the multiple collisions and rifting, North America grew by the accretion of a continental fragment, an island arc, a piece of Africa, and a modern continental shelf.

This sequence of successive collisions with compression taken up by subduction of deeper layers and multiple thrusting and stacking of layers of the upper crust may also be important in other orogenies such as those that formed the Cordillera, the Alps, the Urals, the Caledonides, and the Himalayas. The Mauritanide mountain belt of western Africa appears to be the mirror image of the Appalachians as would be expected for the other side of a collision boundary.

Bounding the central stable region of North America on the south is a belt of folds and thrust faults not unlike the inner zone of the Appalachians. Its surface expression can be seen in the Ouachita Mountains of Arkansas and eastern Oklahoma and in the Marathon Mountains of western Texas. These mountains seem to be a continuation of the Appalachian belt, which curves west through Alabama, Mississippi, and Arkansas and then southwest to the Big Bend region of Texas.

On the west, the central stable region is bounded by a complex of orogenic zones of several types (refer to Fig. 20–42). This is the region of the North American Cordillera, a mountain belt extending from Alaska to Guatemala that contains some of the highest peaks on the continent. Across its middle section between San Francisco

---

*The seismic reflection experiments that traced the younger sediments under the crystalline Blue Ridge and Piedmont were the work of the Consortium for Continental Reflection Profiling (COCORP), a group of university, industry, and government scientists who have used petroleum exploration techniques to attack fundamental geological problems. An important practical result may be the discovery of an unexpected oil and gas province in the sediments below the overthrust sheets.

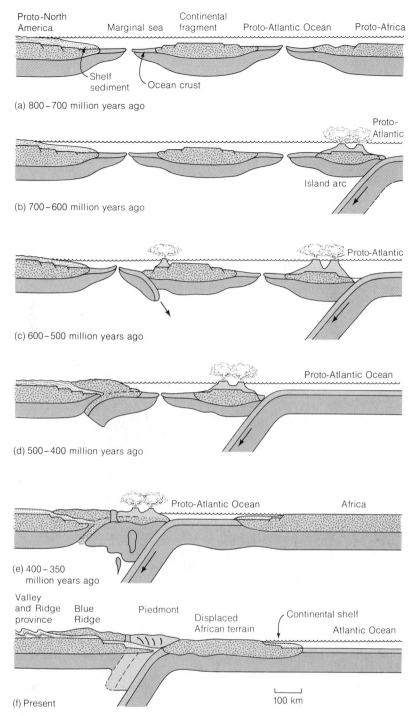

Figure 20-47
An interpretation of the Appalachian orogenic belt in terms of plate divergences and convergences from late Precambrian to the present. In the southern Appalachians successive collisions resulted in subduction and overthrusting of the continental margin of North America. [After F. A. Cook, D. S. Albaugh, L. D. Brown, S. Kaufman, J. E. Oliver, and R. D. Hatcher, Jr., "Thin-Skinned Tectonics in the Crystalline Southern Appalachians," *Geology,* v. 7, pp. 563–567, 1979. Copyright © 1979 by Geological Society of America.]

and Denver, the Cordilleran system is about 1600 kilometers (1000 miles) wide and includes several contrasting physiographic provinces: Region 24 of Figure 20–42, the Coast Ranges along the Pacific Ocean; the Great Valley of California; Region 23, the lofty Sierra Nevada; Region 22, the Basin and Range province (a region of faulted and tilted blocks forming many narrow mountain ranges and valleys extending from the California–Nevada border to western Utah); Region 21, the high table land of the Colorado Plateau; Region 16, the rugged Rocky Mountains, which end abruptly at the edge of the Great Plains on the stable interior (Region 13).

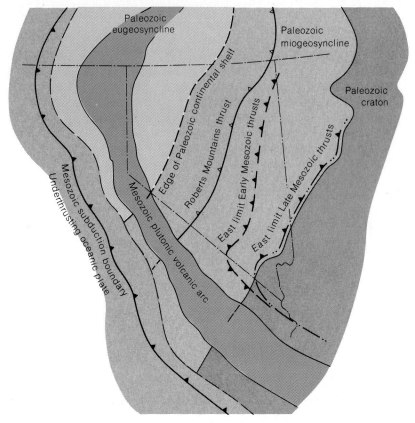

Figure 20-48
Summary tectonic sketch map showing the Paleozoic craton, miogeosyncline and eugeosyncline of the western United States. Also shown are the eastern limits of mid-Paleozoic (Roberts Mountains) and Mesozoic thrust sheets. The boundary of Mesozoic subduction of the Pacific Plate under the North American Plate is indicated. The Mesozoic magmatic belt associated with this phase of subduction crosscuts the Paleozoic geosynclinal and structural trends. [After B. C. Burchfiel and G. A. Davis, "Structural Framework and Evolution of the Southern Part of the Cordilleran Orogen, Western United States," *American Journal of Science*, v. 272, pp. 97–118, 1972.]

The Cordilleran system is topographically higher and more extensive than the Appalachians because its main orogeny was more recent, having occurred in the last half of Mesozoic and early Tertiary time. However, the form and height of the Cordillera we see today are not due to this major episode of folding, faulting, plutonism, and metamorphism. Rather, they are manifestations of more recent events that took place in Tertiary and Quaternary time and rejuvenated the mountains. At that time, for example, the Colorado Plateau and the central and southern Rockies attained much of their present height as a result of a broad regional uplift. Stream erosion was accelerated, the mountain topography sharpened, and the canyons deepened. Block faulting, which broke the crust into mountains and valleys in the Basin and Range province and tilted the Sierra Nevada block, occurred in Tertiary and Quaternary time. At that time, too, volcanism spread sheets of

ignimbrite widely over the southern parts of the Basin and Range province and covered large parts of the Pacific Northwest with lava. This culminating stage of uplift, block faulting, and volcanism served as a model for Figure 20-1. Geologists have read through the confusion of these recent events and reconstructed the orogenic history of the region.

The entire story of the Cordillera is a complicated one, with details that vary along its length. We tell it here in highly simplified form as it applies to the western United States. It is a story of the interaction of the Pacific plate and the North American plate over the past billion years. The interaction includes sedimentation in marginal geosynclines, episodic eastward subduction of the Pacific plate accompanied by volcanism, thrusting, folding, intrusion, and the juxtaposition of rock assemblages differing in age and origin.

From the nature and distribution of rocks, geol-

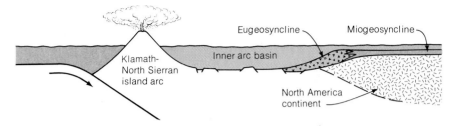

(a) Late Precambrian–Early Devonian

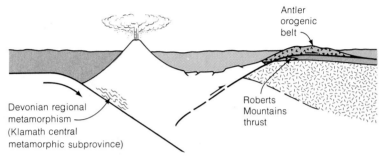

(b) Middle Devonian–Early Mississippian

Figure 20-49
(a) The relationship between miogeosynclinal shelf deposits and deep-sea eugeosynclinal deposits on the North American continental margin in early Paleozoic. Also shown is the inferred subduction of the ocean plate under an offshore volcanic island arc. (b) In late Devonian time the plate collision led to partial closing of the behind-the-arc basin, with an eugeosynclinal sheet thrust more than 100 km eastward onto the shelf, atop the miogeosynclinal section. [After B. C. Burchfiel and G. A. Davis, "Structural Framework and Evolution of the Southern Part of the Cordilleran Orogen, Western United States." *American Journal of Science,* v. 272, pp. 97–118, 1972.]

ogists can reconstruct the distributions of lands and seas of the past. From Cambrian to Devonian time, the main belt of the Cordillera existed as two undersea troughs. These geosynclines received voluminous deposits of sediments, forming a limestone, sandstone, and shale accumulation in the inner eastern miogeosynclinal shelf and a section of submarine volcanics, graywackes, shales, and cherts in the deep-water, outermost western eugeosynclinal belt. The eugeosynclinal depositional environment seems to include a continental slope and a deep-sea trench associated with island-arc subduction. Figure 20-48 shows the early Paleozoic northeast-trending boundaries of the craton, the miogeosynclinal shelf, and the deep-water eugeosyncline for the western United States. The edge of the continent in the early Paleozoic was the boundary between the two geosynclines. A sketch of a possible relationship between these belts is shown in Figure 20-49a.

At various times in the Paleozoic and Mesozoic, deformation accompanied by metamorphism and plutonism took place along much of the length of the eugeosyncline. In late Devonian time (Fig. 20-49b), perhaps due to the closure of the behind-

the-arc basin, eugeosynclinal deposits were thrust more than 100 kilometers eastward onto the shelf along the Roberts Mountain thrust (Figs. 20-48 and 20-49b). The thrust sheet contains a sequence of more than 10 kilometers of lower Paleozoic eugeosynclinal deposits, displaced atop a miogeosynclinal section of the same age. During the later Paleozoic the earlier offshore paleogeography was renewed only to be disturbed in the Mesozoic by the reactivation of eastward thrusting, much like the earlier episode, with eugeosynclinal assemblages thrust over and beyond the earlier thrust sheet onto the shelf (Fig. 20-48). A diagrammatic illustration of the successive episodes of Paleozoic and Mesozoic thrusting is shown in Figure 20-50.

The climactic Mesozoic and early Tertiary episodes of deformation ended geosynclinal conditions and refashioned much of the region into a land area. The Mesozoic-Tertiary orogenies are associated with the subduction of some 2000 kilometers of the Pacific Ocean plate eastward under the North American continent, influencing the geology as much as 1000 to 1500 kilometers inland (Figs. 19-13 and 20-51). From west to east the geo-

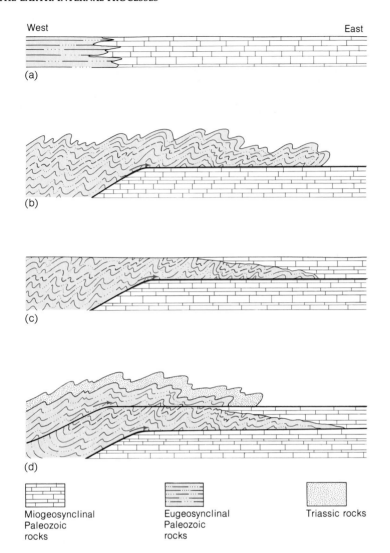

West   East

(a)

(b)

(c)

(d)

Miogeosynclinal
Paleozoic
rocks

Eugeosynclinal
Paleozoic
rocks

Triassic rocks

Figure 20-50
Structural history of north-central Nevada. (a) Deposition of eugeosynclinal and miogeosynclinal formations during early to mid-Paleozoic time.
(b) Eastward thrusting of eugeosynclinal over miogeosynclinal deposits in
mid-Paleozoic time. (c) Renewed miogeosynclinal deposition over thrust sheets
during Pennsylvanian and Permian time. (d) Renewal of eastward thrusting
during Mesozoic time. [After *The Evolution of North America* by P. B. King.
Copyright © 1977 by Princeton University Press. Redrawn with permission
of Princeton University Press.]

logical manifestations of this collision include: large-scale magmatism and metamorphism, seen today as extensive batholiths and metamorphic zones in the Klamath mountains of southwestern Oregon and northern California and in the Sierra Nevada of eastern California; east-directed folding and thrust faulting that deformed rocks of the Paleozoic miogeosyncline up to the edge of the craton from central Nevada eastward to the continental platform area along what is now the eastern edge of the Rocky Mountains; block faulting and volcanism in the Basin and Range province;

basement uplifts within the craton, some 1000 to 1500 kilometers east of the North American plate margin, that warped the relatively thin sediments of the continental platform into high, broad anticlines that have since been eroded to expose the basement granitic rocks. The Precambrian Pike's Peak granite of the Colorado Front Range is one of these. The deformations of the geosynclines and the intrusions of batholiths accompanying this great orogeny are preserved in the record of the rocks still remaining, as in Figure 20-32, a section across the Cordillera near the Canadian border.

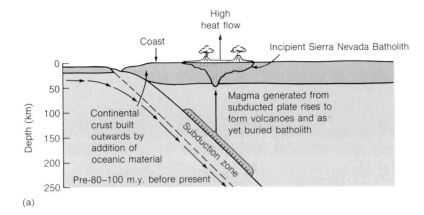

(a)

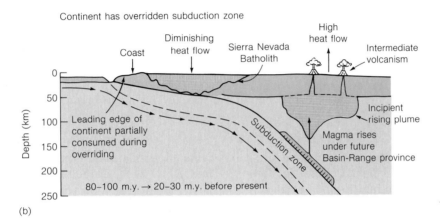

(b)

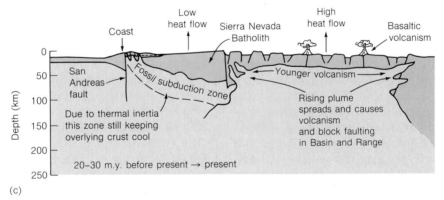

(c)

Figure 20-51
Geological effects of ocean-plate subduction under North America. (a) The continent grows by accretion of oceanic material; melting subducted plate is the source of the Sierra Nevada magmatic belt. (b) The continent overrides the subduction zone, and the region of magma generation shifts eastward under the future Basin and Range province. (c) The continent has overridden the spreading center, subduction stops, and the plate boundary changes to a transform fault. The heated crust under the Basin and Range province undergoes block faulting and volcanism [After T. L. Henyey and T. C. Lee, *Bulletin of the Geological Society of America*, v. 87, pp. 1179–1187, 1976.]

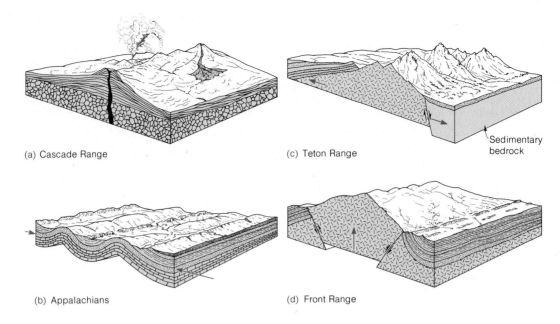

(a) Cascade Range

(c) Teton Range

Sedimentary bedrock

(b) Appalachians

(d) Front Range

Figure 20-52
Mountains vary in form and origin. (a) Mountains formed by volcanic action. (b) Mountains resulting from folded layers of rock. (c) Mountains formed from fault blocks. (d) Mountains originating in vertical uplift. [After U.S. Geological Survey.]

Well after the compression of the two great troughs, and mostly in late Tertiary time, the spectacular episode of **block faulting** occurred in a region extending southeast from southern Oregon to Mexico and including Nevada, western Utah, and parts of eastern California, Arizona, New Mexico, and western Texas. This latest structural imprinting, superposed on the earlier-deformed geosynclinal rocks, is responsible for the present-day features of the Basin and Range province. Thousands of high-angle faults sliced the crust into innumerable upheaved and downdropped blocks, forming hundreds of discontinuous, narrow mountain ranges and intervening, alluvium-filled basins separated by normal faults. Successive movements along the faults over millions of years caused the large vertical displacements evident today (Fig. 20-52c). Some of the faults have remained active through the Cenozoic. Unlike folded mountains (Fig. 20-52b), such as those of the Appalachians, which uplift primarily unmetamorphosed sedimentary geosynclinal deposits, the kinds of rocks and structures exposed in **fault-block mountains** vary greatly. Plutons, lavas and tuffs, metamorphic rocks, and folded geosynclinal deposits can be found in the different ranges. The intervening lowlands, once formed, accumulate sediments eroded from the adjacent mountains of upthrown blocks. (Fig. 20-53).

The Sierra Nevada fault block consists of granitic batholiths and tightly folded and metamorphosed beds, structures formed during the Late Jurassic–Early Cretaceous orogeny. The mountains of this earlier episode were worn down, leaving a deeply eroded remnant after the Mesozoic. About 10 million years ago, in Pliocene time, the block was pushed up on the east and tilted to the west. In 1872, the bounding fault on the east was the site of one of the greatest earthquakes in United States history. Many peaks of the Sierra Nevada, Mt. Whitney among them, stand more than 4000 meters above sea level.

Some other examples of fault-block mountains are the Wasatch Range of Utah, the Teton Range of Wyoming, and the tilted edges of the rift valleys of east Africa and the Dead Sea–Jordan Valley rift of Israel.

The present mountainous topography of the Rocky Mountains is due to epeirogeny that took place in Cenozoic time, after the earlier episodes of folding and deep erosion. The region was raised 1500 to 2100 meters (5000 to 7000 feet) in the past 15 or 20 million years, pushing Precambrian basement rocks and their veneer of later-deformed sediments above the level of their surroundings (Fig. 20-52d). Other examples of **upwarped mountains** are the Adirondacks, the Black Hills, the Labrador Highlands, and the mountains of Fennoscandia.

The Colorado Plateau, the classic showplace of geology discussed in Chapter 2, seems to be an island of the central stable region, cut off from the interior by the Rocky Mountain orogenic belt. Since the late Precambrian, it has been a stable shelf area—with no thick geosynclinal deposits and no major orogeny, with movements mainly

Figure 20-53
Fault-block mountains. The Sierra Nevada (skyline) viewed from the Panamint Mountains, California. Sierra fault scarp in distance. In the middle ground is the Argus range, a late Quaternary upfaulting of Plio-Pleistocene basalts. A small fault, or tectonic, valley, far side up, cuts the alluvium of Panamint Valley below the center of the picture. [Photo by W. B. Hamilton, U.S. Geological Survey.]

up and down, and with slight deformation in places into broad folds and basins. Volcanics and sedimentary debris from adjacent mountains filled the lowlands of the plateau in Cenozoic times.

The Coast Ranges of California lie between the Great Valley and the Pacific ocean. These mountains contain rocks younger than those to the east, great thicknesses of clastic rock, graywackes, shales, cherts, and pillow basalts that accumulated in a eugeosyncline seaward of the troughs discussed earlier. These deposits, which accumulated mainly after the climax of the orogenies to the east, show strong deformation in the form of a series of long strike-slip fault slices trending generally northwestward. Some of these faults existed as far back as the mid-Tertiary, and others, such as the famous San Andreas fault, are still active today.

Just how this chaotic terrain can be fitted into the framework of plate tectonics is not entirely clear, and many geologists are now occupied with this important queston. Apparently beginning some 30 million years ago in mid-Tertiary time the eugeosyncline and the mid-ocean ridge that produced the plate being subducted under North America were overridden by the continent. Subduction ceased, and the plate boundary changed from one of collision to transform faulting, creating the San Andreas systems of faults (Fig. 20–51).

Perhaps the most famous examples of orogenic belts are the Alps and the Himalayas. See Boxes 20–2 and 20–3.

Coastal Plains and Continental Shelf

The Atlantic **coastal plain** and the **continental shelf,** its offshore extension (see Chapters 5 and 19), began developing after the close of the Appalachian orogeny. The outer metamorphic zone of the Appalachians was block-faulted in Triassic

Box 20-2 THE ALPS

The Alps, another great orogenic belt, are a system of ranges extending for some 1000 kilometers (600 miles) in an arc running from the French-Italian Mediterranean coast through Italy, Switzerland, and Austria. The highest peak is Mont Blanc in southeastern France, with an elevation of 4810 meters (15,781 feet). Movement of the African plate against the Eurasian plate is the mechanism of the Alpine orogeny, which has been developing since early Mesozoic time.

In the development of the Alps, a whole series of geosynclines separated by anticlinal ridges received Mesozoic deposits. The southern Triassic geosyncline was strongly disturbed by faulting and folding—but without a major magmatic phase—and developed into the Lombardy Alps (see Fig. 20–32). The central geosyncline, a Jurassic one, now the Pennine Alps (see figure), was subjected to extreme folding, with recumbent folds piled up one atop the other; it also experienced

magmatic and strong metamorphic episodes. The northern geosyncline, which is now the Helvetides, is of Cretaceous age; it shows no magmatic phase but strong folding with thrust sheets developing from asymmetric anticlines, the sheets seemingly piled upon one another. In a later stage of compression and uplift, huge masses of sediments were detached from the rising belt, intricately folded and thrust into neighboring troughs in which Tertiary sediments were accumulating. The Alps have been termed unique in the grandeur of their scale and the complexity of their folding and overthrusting.

By Pliocene time, these complex structures were eroded to a low terrain. The present grand topography of the region reflects rather recent epeirogeny, most of the uplift and erosion having occurred prior to the Pleistocene glaciation.

Rugged topography of a young mountain belt. View over Bedretto Valley, Switzerland, to the front of the Pennine Alps on the left. The formations behind the front have glided a long distance from their original place of deposition to a position over the beds in the foreground. [Swissair.]

Box 20-3 THE COLLISION BETWEEN INDIA AND EURASIA

Some 40 to 60 million years ago the Indian subcontinent collided with Eurasia. After the collision, India slowed down but continued to drive northward into Eurasia, penetrating about 2000 kilometers. Since continental crust does not readily subduct, geologists must explain what happened to a piece of crust as wide as India and 2000 kilometers (1200 miles) long. Using earthquake data and satellite photographs, Peter Molnar and Paul Tapponier prepared this tectonic map and offered a startling proposal to account for the displaced crust. The Himalayas, the world's highest mountains, were formed from overthrust slices of the old north portion of India, stacked one on top of the other. This took up some of the compression. In Tibet,

horizontal compression found relief in vertical expansion; in this way the high Tibetan Plateau was formed. Compression by thrust faulting is the pattern of deformation in the Tien Shan mountains. These and other zones of compression could account for perhaps half of India's penetration into Eurasia. According to Molnar and Tapponier, the other half is made up by China's being pushed eastward and out of the way of India like toothpaste squeezed out of a tube. The movement took place along the enormously long strike-slip faults shown on the map. The mountains, plateaus, faults, and great earthquakes of Asia, thousands of kilometers from the suture, are thus influenced by the continuing collision of India and Eurasia.

After "The Collision Between India and Eurasia" by P. Molnar and P. Tapponier.
Copyright © 1977 by Scientific American, Inc. All rights reserved.

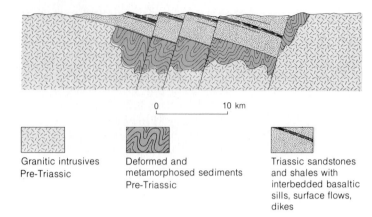

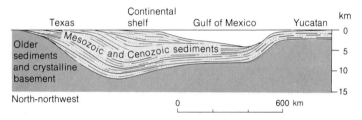

Granitic intrusives
Pre-Triassic

Deformed and
metamorphosed sediments
Pre-Triassic

Triassic sandstones
and shales with
interbedded basaltic
sills, surface flows,
dikes

Figure 20-54
Diagrammatic section of Triassic basins of Connecticut. Nonmarine sediments were trapped in basins formed by tilted fault blocks. Basaltic flows intruded and covered these deposits.

Figure 20-55
Schematic section across the Gulf of Mexico from Texas to Yucatan, showing inferred thicknesses of Mesozoic and Cenozoic sediments. Maximum thickness near the Texas coast may exceed 10 km. [After *The Evolution of North America* by P. B. King. Copyright © 1977 by Princeton University Press. Redrawn by permission of Princeton University Press.]

time, and a series of long, narrow, downdropped grabens developed along a belt running from South Carolina to Nova Scotia. This was the rifting that preceded the opening of the modern Atlantic Ocean. The basins that were formed trapped a thick series of red, clastic deposits, which were later intruded by basaltic sills and dikes. The Connecticut River valley and the Bay of Fundy are examples (Fig. 20-54).

Following this, in early Cretaceous time, the deeply eroded and beveled coastal plain and continental shelf began to subside and to receive sediments from the continent. Cretaceous and Tertiary sediments up to 5 kilometers thick filled the slowly subsiding trough, and even more material was dumped into the deeper water of the continental rise (Fig. 19-11). This living geosyncline, referred to in Chapter 19, is possibly a forerunner of some future orogeny.

The Gulf coastal plain and shelf are continuous extensions of the Atlantic ones, though interrupted in a way by Florida. The Mississippi, Rio Grande, and other rivers draining the interior of the continents have delivered sediments to fill a trough some 10 to 15 kilometers (40,00 to 50,000 feet) deep running parallel to the coast (Fig. 20-55). The Gulf coastal plain and shelf are a rich reservoir of petroleum and natural gas. The Atlantic shelf is now being actively explored for these and other resources.

DEFORMATIONAL FORCES

Ever since deformation of the crust was first recognized, geologists have speculated about the origin of the forces that shape the Earth's surface features. At one time, orthodox thought had it that the Earth was shrinking, and the crust was likened to the wrinkled skin of a drying apple. For a short time, recently, a few proposed that the Earth was expanding, in order to account for the fragmentation of continents and the opening of ocean basins. Convection currents in the mantle have been invoked at various times in the past—to move continents, to erode mountain roots, to downbuckle the crust.

In Chapter 19, a hypothesis was described that was startling in its simplicity: that geosynclinal-orogenic cycles are controlled by plate tectonics. This hypothesis is now generally accepted. However, the force that drives plate tectonics must still be specified. As we have noted, some form of mantle convection is usually invoked at this early stage in the development of the hypothesis. Upwarping, block faulting, and slow vertical movements of the stable interior are among the elements still to be integrated with the plate-tectonic concept. This integration may have to await the next generation of geologists and geophysicists.

SUMMARY

1. When rocks are subjected to forces, they deform. Laboratory studies show that rocks vary in strength. Some are ductile, others are brittle—depending on the kind of rock, the temperature, the surrounding pressure, the magnitude of the force, and how fast it is applied.

2. Among the geologic structures in rock formations that result from deformation are folds, domes, basins, joints, and faults. Geologists begin to deduce the deformational history of a region by fixing the ages of the formations using radioactive isotopes or fossils. The age of the deformation is then bracketed by finding a younger undeformed formation lying unconformably on an older deformed bed. Sediments that accumulate as a result of deformational movements also provide clues.

3. Forces acting within the crust can deform large regions. In some cases, the regional movements are simple up-and-down displacements (epeirogeny) without much deformation of the rock formations. In other cases, horizontal forces can produce extensive and complex folding and faulting (orogeny).

4. A continent typically is divided into large regions that have had different deformational histories. North America, for example, contains a large central stable region that has been relatively undisturbed since Precambrian time except for gentle vertical movements. Surrounding the stable interior are orogenic belts—the mountainous Cordillera and Appalachians, which were deformed by folding and faulting and were subjected to plutonism and metamorphism at various times in the Paleozoic, Mesozoic, and Cenozoic eras.

5. A typical orogenic episode is preceded by subsidence of marginal troughs in which sediments then accumulate. These geosynclines are subsequently deformed. Great folds and thrusts are generated, and metamorphism and intrusion by batholiths occur. The resulting belt is then uplifted and eroded. A culminating stage of uplift or block faulting that again raises the region accounts for many of the mountainous features we see today. An explanation of the geosynclinal-orogenic cycle in plate-tectonic terms would have the geosynclinal phase associated with fragmentation of a continent and plate separation. The deformation phase would be associated with plate collisions.

EXERCISES

1. Evidence of vertical crustal movements is frequently found in the geologic record. Give some examples of such evidence. Describe the geologic evolution of a region that leads to the formation of mesas, buttes, and tablelands bounded by high cliffs.

2. If the Greenland and Antarctic ice caps were to melt, how much would sea level rise? Assume that the ice caps average about 2 kilometers in thickness; the area of the Greenland ice cap and the Antarctic ice cap are 2×10^6 square kilometers and 10^7 square kilometers, respectively.

3. Draw a geological cross section that tells the following story. A series of marine sediments were deformed into folds and thrust faults. This was followed by erosion. Volcanic activity ensued, and lava flows covered the eroded surface. A final stage of block faulting broke the crust in numerous places.

4. Anticlines are upfolds and synclines are downfolds. Nevertheless, one often finds synclinal ridges and anticlinal valleys. Explain.

5. Summarize the stages of a typical orogenic episode in a series of sketches with legends.

6. Summarize the main features (orogenic belts, shields, platforms, coastal plains, shelves) of the major structural regions of a continent other than North America.

7. If a new mid-ocean ridge developed, with a spreading rate of 2 centimeters per year and a length of 10,000 square kilometers, how much would sea level rise? Assume that the cross section of the new ridge is the 2-centimeter-per-year profile shown in Box 20-1. Take the area of the world oceans as 350×10^6 square kilometers. If the spreading were faster, would the sea-level change be larger or smaller?

BIBLIOGRAPHY

Billings, M. P., *Structural Geology*, 3rd ed. Englewood Cliffs, N.J.: Prentice-Hall, 1972.

Burchfiel, B. C., "Structural Geology of the Earth's Exterior," *Proceedings of the National Academy of Sciences*, v. 76, pp. 4201–4207, 1979.

Cook, F. A., L. D. Brown, and J. E. Oliver, "The Southern Appalachians and the Growth of Continents," *Scientific American*, October 1980.

Dott, R. H., Jr., and R. L. Batten, *Evolution of the Earth*, 3rd ed. New York: McGraw-Hill, 1980.

Hills, E. S., *Elements of Structural Geology*, 2nd ed. New York: John Wiley & Sons. 1972.

King, P. B., *The Evolution of North America*, 3rd ed. Princeton, N.J.: Princeton University Press, 1977.

Wilson, J. Tuzo, ed., *Continents Adrift—Readings from Scientific American*. San Francisco: W. H. Freeman and Company, 1970.

21

The Planets: A Summary of Current Knowledge

The planets fall into two groups—the gaseous, giant, outer planets, which have retained their Sun-like composition, and the small, stony, Earth-like, inner planets. Space exploration has opened Earth's neighbors—the Moon, Mars, and Venus—for geological study, and they show surprising diversity. The Moon was geologically active in its early history, but it has been dead for three billion years. Mars, like Earth, is geologically youthful; it shows evidence of recent volcanism, of tectonism, and surprisingly, of erosion by water and wind, despite the absence of liquid water and the presence of only a very thin atmosphere. Venus, Earth's twin in many respects, is blanketed by a thick, heavy atmosphere of carbon dioxide, which causes oven-hot temperatures at the surface. Only when we know why the planets followed different evolutionary paths will we understand Earth's workings and environment and why life could originate here.

Human interest in the planets dates back to prehistoric times when the ancients noticed that a few of the brightest stars seemed to drift across the sky, not following the pattern of daily and annual motion of the fixed stars. It was the Greeks who called them *planetai,* or wanderers, and first mapped their paths; it was the Romans who named them after their gods: Mercury, Venus, Mars, Jupiter, and Saturn.

The invention of the telescope—first used to explore the skies by Galileo in 1609—opened the possibility for describing distinctive planetary features. Viewed through the telescope, the planets appeared to be disks, rather than points of light like the stars. They had individual features—Saturn with its rings; Jupiter with its alternating bands of yellow, blue, and brown clouds and its moons; Mars with its seasonally changing ice caps and its supposed canals; and, of course, our cratered Moon with its highlands and "seas." The asteroid belt and the outermost planets Uranus, Nepture, and Pluto were discovered in later centuries by astronomers searching the skies with

ever more powerful telescopes. As early as the nineteenth century, some geologists turned their attention to the Moon, seeking to explain its surface features in terms of Earth experience.

Since the first successful satellite in 1957, space exploration has quickened interest in the planets. Recent years have seen the birth of a new science, **planetology,** the comparative study of Earth and the other planets. This infant science has powerful new methods and tools: manned exploration of the Moon, automated and instrument-laden planetary satellites and probes, radar and radio telescopes exploring with radio waves rather than with light, and optical telescopes fitted with electronic devices to sense faint infrared and other invisible radiation as well as light.

In this chapter, we will review some of the recent discoveries of planetology. Perhaps the most intriguing aspect of the study of other planets is that we may one day arrive at a general theory of planetology that will account for both the origin of our solar system and the evolution of the individual planets with their differing features. Only in this way will we come really to understand our own Earth, its workings, its environment, and the reasons why life could originate here.

The framework upon which planetary scientists currently hang their hypotheses of planetary formation assumes the existence of a primitive solar nebula that condensed into planetary materials, followed by accretion into planets (Chapter 1). The planetary objects subsequently evolved under a variety of processes—bombardment by meteorites, igneous differentiation and formation of cores, crusts, and atmospheres, each planet following its unique course depending on its initial size, how fast it accreted (which determined its initial temperature), its composition, and its distance from the Sun.

SOME VITAL STATISTICS OF PLANETS

Our solar system is made up of the Sun, 9 planets, some 40 known satellites, and countless asteroids and comets occupying a region some eight billion miles across, a huge space yet a minuscule domain by cosmic standards. To bring planetary sizes and distances into perspective, one might think of the Sun as a baseball. A grain of sand about 30 feet away, then, would represent Earth; Jupiter would be a pea 150 feet away; Pluto, the outermost planet, would be another sand grain about 1200 feet distant; and the nearest star would be another baseball 1500 miles away. The layout of the planets, held in elliptical, approximately regularly spaced orbits by the Sun's gravitational attraction, is shown in Figure 1-2. Except for Pluto and Mercury, the planetary courses around the Sun lie approximately in the **ecliptic,** which is the name for Earth's orbital plane. Mercury's orbit is tilted 7°, and Pluto's is tilted an anomalous 17° from the ecliptic, as the figure shows. It has recently been discovered that Pluto's orbit comes quite close to Neptune at times, and that it actually lies inside Neptune's orbit over part of its path. This new result suggests that Pluto may be an escaped satellite of Neptune. The asteroid belt fills a conspicuous gap between Mars and Jupiter; it contains perhaps 100,000 small bodies in orbit around the Sun, an estimate obtained by viewing this region through large telescopes.

Table 21-1 lists the vital statistics of each planet—its dimensions, its mass and density, the lengths of its day and year, the pull of its gravity, and so on. "Vital" is an appropriate word: in this age of space exploration, planetary masses, dimensions, and positions are essential elements in planning the navigation of spacecraft. Planetary diameters are measured using telescopic observations or radar. To determine a planet's mass, one must see how its gravitational attraction affects the motion of celestial bodies that pass nearby, such as other planets, satellites, comets, asteroids, and nowadays, spacecraft. A planet's density (its mass divided by its volume) is a key factor in determining its composition, because the elements present, and the compounds they form, have characteristic densities, and the mean density of the planet depends on the details of the mix.

The planets—except for Pluto, about which little is known—can be separated into two distinct groups: the inner, small, high-density terrestrial planets and the outer, large, low-density giant planets, as Table 21-1 indicates. The other terrestrial planets resemble the Earth in density, and for this reason we expect them to have a similar composition—namely, a rocky ball mainly of magnesium silicates, with varying amounts of iron. The giant planets are more like the Sun in density and, therefore, we believe, in composition; they are light because, like the Sun, they are largely composed of hydrogen and helium, the lightest elements. In this way, the giant planets seem to have preserved much of the composition of the original solar nebula. Saturn is actually less dense than water—it would float if an ocean big enough could be found.

The Earth-like planets either never possessed the lighter elements or somehow lost them. Perhaps because of their smaller mass and weaker gravitational pull they couldn't hold these light gases. More likely, according to the chemical-condensation-sequence model described in Chapter 1, these elements did not condense from their gaseous state because of the high temperatures near the Sun and were expelled from the inner reaches of the solar system before the terrestrial planets formed.

The Earth, "with its atmosphere and oceans, its complex biosphere, its crust of relatively oxidized, silica-rich, sedimentary, igneous, and metamorphic rocks overlying [a magnesium silicate mantle and a core] of metallic iron, with its ice caps, deserts, forests, tundra, jungles, grasslands, fresh-water lakes, coal beds, oil deposits, volcanoes, fumaroles, factories, automobiles, plants, animals, magnetic field, ionosphere, mid-ocean ridges, convecting mantle, . . . is a system of stunning complexity." This quotation, essentially a one-paragraph summary of much in this book, is from a popular article by geochemist John S. Lewis (see the bibliography). It sets the scene for

Table 21-1
Vital statistics of the planets

Planet	Diameter (km)	Mass (Earth = 1)	Density (Water = 1)	Gravity (Earth = 1)	Number of satellites	Time for one rotation on axis (Earth hours or days)	Time for one revolution around Sun (Earth years)	Distance from Sun (10^6 km)	Distance from Sun (10^6 mi)	Composition of atmosphere
Terrestrial planets										
Mercury	4,835	0.055	5.69	0.38	0	59 days	0.241	57.7	36.8	none
Venus	12,194	0.815	5.16	0.89	0	243 days	0.616	107	66.9	CO_2
Earth	12,756	1.00	5.52	1.00	1	1.00 days	1.00	149	92.6	N_2, O_2
Mars	6,760	0.108	3.89	0.38	2	1.03 days	1.88	226	141	CO_2, N_2, Ar
Giant planets										
Jupiter	141,600	318.	1.25	2.64	15	9.83 hr	11.99	775	482	H_2, He
Saturn	120,800	95.1	0.62	1.17	15	10.23 hr	29.5	1421	883	H_2, He
Uranus	47,100	14.5	1.60	1.03	5	23.00 hr	84.0	2861	1777	H_2, He, CH_4
Neptune	44,600	17.0	2.21	1.50	2	22.00 hr	165.	4485	2787	H_2, He, CH_4
Pluto	14,000?	0.8?	4.2?	?	0	6.39 days	248.	5886	3658	?

the discussion here of the other, vastly different planets. It is essential to remember Earth's history as we consider her neighbors in the solar system.

THE MOON

Earth's satellite is exceptional, in that it is so large compared to its parent planet. For this reason, it is often regarded as a planet—a partner to the Earth—rather than a mere satellite. The Moon is the best-known planet next to Earth, thanks to its proximity and the programs of manned and unmanned lunar exploration that have been conducted by the United States and the Soviet Union.

Project Apollo

The signal to proceed with Project Apollo, the American program of manned lunar exploration, was given by President John F. Kennedy. It was basically a politically motivated decision made at a time when the superpowers were competing for world leadership. In those days, it was supposed that a demonstration of superior space technology was also a demonstration of a superior political-economic system. Regardless of their origin, the Apollo missions have paid off handsomely in terms of scientific discoveries. The explorations of the sea floor and of the Moon shared in opening the renaissance of major discoveries and new theories in the Earth sciences.

Lunar exploration provided photography of the Moon of unprecedented resolution; data from a network of emplaced observatories containing seismographs, magnetometers, and other geophysical tools; and, most important, Moon rocks. Although much (if not all) of what we have learned could have been achieved eventually with unmanned, automatic spacecraft, the value of having reasoning, trained human beings collecting samples and emplacing instruments should not be underestimated.*

Lunar Provinces

The view of the Moon in Figure 21-1 shows the major provinces of the Moon, together with locations of the landing sites of the exploratory missions. The side shown always faces the Earth because, for some not fully understood reason, the Moon rotates only once on its axis for each revolution around the Earth. Although the far side of the Moon has been photographed from lunar-orbiting spacecraft (Fig. 21-2), current technology is inadequate to let us land there. The lighter-colored, highly cratered highlands stand out from the large, dark basins, first called maria, or seas, by Galileo because they resembled oceans through his rudimentary telescope.

---

*The Apollo 17 mission included Dr. Harrison Schmitt, a petrologist trained at Caltech and Harvard. He was the first geologist to walk on the lunar surface. Earlier astronauts, who were flyers by profession, were nevertheless well versed in the rudiments of geology. Schmitt was elected to the U.S. Senate from New Mexico in 1976.

Lunar landing sites (white circles)

Apollo 11: Sea of Tranquility	Apollo 12: Ocean of Storms	Apollo 14: Fra Mauro
Apollo 15: Hadley–Apennine	Apollo 16: Descartes	Apollo 17: Taurus–Littrow

Figure 21-1
Photograph of lunar surface showing major features: dark maria covered by younger basal-
tic lava flows and light-colored highlands, whose heavy cratering testifies to their great age.
White circles show the Apollo landing sites. [Lick Observatory photograph.]

Figure 21-2
Apollo 8 view of the Moon over east limb. The right half of the photograph covers regions
not visible from Earth. The bright-rayed crater at top right is Bruno. For comparison with
the previous figure, note that the irregular mare in left center is Fecunditatis. Mare Crisium
is above it. [From National Aeronautics and Space Administration.]

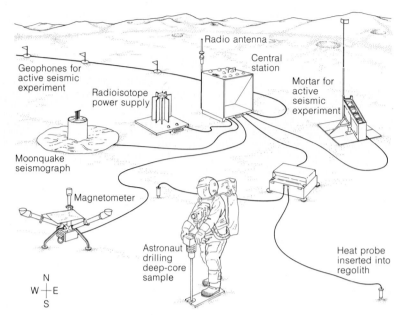

Figure 21-3
Alsep, the automatic observatory left behind on the surface of the Moon by the Apollo astronauts. Data from seismographs, magnetometers, heat-flow probes, and gas detectors will be transmitted back to Earth for many years. In the active seismic experiment, a mortar fires an explosive charge on signal from Earth to generate seismic waves in the lunar crust, which are picked up by the geophones. [From National Aeronautics and Space Administration.]

Surface Processes on the Moon

The Moon's gravity is only one-sixth that of Earth. It lacks the gravitational pull to retain an atmosphere as Earth does, and it lacks water. For these reasons, and also because it is tectonically almost dead, the surface features of the Moon are profoundly different from those of the Earth. As we have seen in earlier chapters, Earth's surface is molded by two opposing natural forces: air and water erode rocks and remove the debris to the oceans, whereas plate motions, with associated mountain-making and volcanism, continually renew the surface. As a result of this ongoing process of creation and destruction, most of the Earth's surface features are fresh ones—younger than 200 million years—formed in the last few percent of geological time. To be sure, many older relics, both rocks and structures, that date back 3 to 3.7 billion years can be found on continents, but these rocks cover only a small fraction of the Earth's surface.

Imagine the excitement among geologists when the first lunar rocks picked up by the Apollo astronauts proved to be 3.6 billion years old! This was no lucky find. According to radiometric dating, all Moon rocks examined so far were formed or recrystallized 3.1 to 4.6 billion years ago. Con-

sistent with these indications that the Moon has been geologically inactive for more than 3 billion years, the seismographs installed by the astronauts (Fig. 21-3) have indicated that the rate of seismic energy release by moonquakes is one billion times less than that of earthquakes. This negligible amount of moonquake activity clinches the idea that the Moon is now tectonically almost dead. We might also have reached the same conclusion from photographs that show none of the trademarks of plate tectonics—no globe-circling volcanic ridge-rifts, no transform faults, no extensive trenches with adjacent mountains—in short, no moving plates. Not only is the Moon now "dead," but when it was "alive," it had a different life style.

In the absence of erosion and orogeny as we know them on Earth, what does shape the lunar surface? What accounts for the principal structural features of the lunar surface—the nearly circular craters ranging in diameter from inches to hundreds of miles? The craters themselves, rimmed with rubble, often with rays (narrow, bright streaks of debris radiating outward), provide the clues (Figs. 21-2 and 21-4). The Moon's surface, unprotected by a blanket of air, has been riddled by meteorite impacts for the past 4.6 billion years. Coming in with velocities of as much

Figure 21-4
Apollo 16 mission to Descartes. This photograph was taken from the rim of a crater about 1 km in diameter and 230 m deep. The rim, 50 m high, is a flap of lunar formations blown out and overturned by the impact explosion. A meteorite about the size of a football field would have produced a crater this big. The rocks in the foreground are breccias. The undulating surface in the distance is the cratered, fine-grained lunar "soil," or regolith, with widely scattered pebbles and cobbles. [From U.S. Geological Survey and National Aeronautics and Space Administration.]

as 150,000 kilometers per hour (roughly 100,000 miles per hour), these celestial projectiles punch into the crust, vaporize, and explode as the energy of motion is converted almost instantaneously to heat. The blast excavates a crater, throwing out debris locally in the form of a surrounding rim, but also hurling ejecta to distant points. To give some idea of what an impact can do, a meteorite 3 meters (roughly 10 feet) in diameter and weighing 50 tons explodes with the energy of 10 kilotons of TNT and blasts out a crater 150 meters (roughly 500 feet) in diameter. A large meteorite weighing 10^6 tons explodes with as much energy as is released each year by earthquakes.

If the incoming body is big enough, a mare hundreds of miles in diameter can be excavated (see Figs. 21-1 and 21-2). In the smaller gravity field of the Moon, unimpeded by air, huge amounts of ejecta can be hurled for hundreds of kilometers, scarring the surface and littering it with soil and blocks of rock. Some of the ejecta fall back to the surface with enough speed to punch out secondary craters some distance from the primary crater.

A giant collision can throw up immense piles of rock—enough to form mountains. The most famous example is the great collision site known as Mare Imbrium, the largest circular basin on the moon. The massive Apennine Mountains, which tower some 5 kilometers (3 miles) above the floor of the basin on its southeast margin, were formed in this cataclysmic event, as were piles of rubble and many grooves, ridges, and secondary craters

Figure 21-5
Meteor Crater near Flagstaff, Arizona, is about the same size as the lunar crater in Figure 21-4. It was caused by the impact of a large meteorite in prehistoric times. Thousands of fragments of the meteorite have been found nearby. [From U.S. Geological Survey.]

Figure 21-6
Lunar landscape, showing the contrast between old and young craters. Old craters have rounded rims and filled bottoms as a result of erosion by micrometeorite bombardment. Young craters have sharp edges and bright rays emanating from them. [U.S. Geological Survey and National Aeronautics and Space Administration.]

radiating from the collision area. This region was the landing site for the Apollo 15 mission (landing site 15 in Fig. 21-1). The astronauts drove to the front of the Apennines in their lunar rover, collecting rocks and photographing the terrain.

Craters from meteorite impacts are known on Earth, but they are rare because the atmosphere slows all but the largest bodies and literally burns up (oxidizes) the smaller ones, almost all of them completely by the time they hit Earth. The bright light of shooting stars is produced by that burning. The evidence of impact of the larger objects that manage to get through is obliterated as a result of erosion by wind and water and burial by sediments. Craters and old buried structures believed to have been caused by impacts are called **astroblemes,** which means "star wounds" (Fig. 21-5). It has been estimated that on Earth one crater with a diameter greater than one kilometer is punched out by a meteorite every 50,000 years.

Craters on the Moon can be preserved for fantastically long times. The only processes working to erase them are a sort of sand blasting by small, high-velocity meteorite grains, called **micrometeorites,** blanketing by ejecta from other impacts, a direct hit by another meteorite, or burial by a lava flow. The contrast between a young and old crater side by side in Figure 21-6 shows that the eroding effect of micrometeorites tends to round off the sharp edges and to fill the bottoms of craters. The pitted surface of a Moon rock exposed to micrometeorite bombardment can be seen in Figure 21-7. The process of wearing down is extremely slow—perhaps 10 or 20 feet of abrasion in the 4.6 billion years of the lifetime of the Moon. A crater is more likely to be obliterated by burial under ejecta from nearby craters than by micrometeorite erosion. The larger the crater, however, the better its chances for survival. Craters larger than a few hundred feet might survive for billions of years.

The Regolith

The Moon has a soil of sorts that is called the **regolith.** It is not a soil in the conventional sense of the Earth's soil (see Chapter 4), which results from the breakdown of rocks mainly by the chemical and mechanical actions of water. The lunar regolith is simply the accumulated fragmental debris of billions of years of exposure to bombardment by meteorites, cosmic rays, and particles blown out from the Sun (the solar wind), as depicted in Figure 21-8. Its thickness generally ranges from a few meters to a few tens of meters. It accumulates

Figure 21-7
"Great Scott" rock, a vesicular mare basalt, collected during the Apollo 15 mission. Note the big, glass-lined "zap" pit caused by micrometeorite impact just left of center. [From U.S. Geological Survey and National Aeronautics and Space Administration.]

at the remarkably slow rate of about one millimeter per million years. Contrast this with the sediments of Earth's deep sea, which are laid down 5000 times as rapidly. Geochemists have also found a way to determine that the rate of turnover or "gardening" of the lunar soil due to the continuous bombardment is about 1 meter in 0.5 billion years. The lunar soil is particularly important because it is a mixture of materials thrown in from many different areas. One sample of soil, therefore, helps us to figure out what the rocks are like in other places on the Moon from which samples might not otherwise be collected. Figure 21-4 shows what the surface of the regolith looks like. Close inspection shows that, in addition to rock fragments, tiny glass spheres are an important component. Presumably, these are formed at the site of impacts, where melting is induced by the shock and the liquid rock is ejected as a fine spray. The droplets form glassy spherules if they freeze in flight, or they splatter rocks with a glassy crust if they remain molten until they land.

Lunar Stratigraphy

The mapping of geology from photographs is *photogeology.* Lunar geologists have used photographs to find the relative ages of different regions

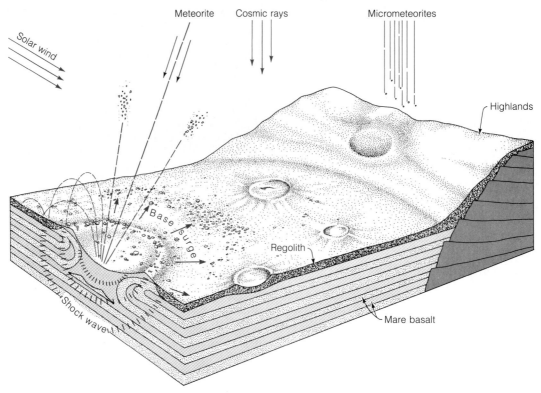

Figure 21-8
The original crust of the Moon, still present under the highlands, was broken up by large impacts. Afterward the basins were flooded by basaltic lava. Erosion of the surface, extremely slow compared to that on Earth, removes only about a millimeter in a million years. Bombardment by micrometeorites is believed to be the main cause. Larger meteorite impacts occur rarely, but they create craters and scatter ejecta over large distances. The regolith, up to a few tens of meters thick, is the accumulated, fragmental debris of billions of years of exposure to meteorite bombardment, cosmic rays, and the solar wind. [After "The Carbon Chemistry of the Moon" by G. Eglinton, J. R. Maxwell, and C. T. Pillinger. Copyright © 1973 by Scientific American, Inc. All rights reserved.]

of the Moon by counting the density of craters. The oldest terrain should show the highest crater count simply because it has been bombarded for the longest time. The densely cratered highlands, for example, in the south-central portion of the Moon (Fig. 21-1), are the oldest regions. In many places, the highlands are covered by ejecta blankets thrown out from the circular maria. This cover is punctured by more craters than the lava surface of the maria. Hence the impact that produced the ejecta blanket must be older than the volcanism that filled the maria with lava. The reader may wish to establish relative ages of the different features in Figure 21-2 by using this simple method of crater counting, as well as by looking for other evidence of "freshness," such as craters with sharp-edged features, or bright rays that have not yet disappeared due to erosion or other aging effects.

Stratigraphic mapping on the Moon—that is,

working out the chronological relations of the many lunar surface features—is quite different from similar procedures on the Earth. There are no fossils, and the events that define periods are quite different in nature and time scale from those on the Earth. Just as on Earth, however, the law of superposition—which places the youngest materials at the top of a stack of layers—is used in establishing relative ages of the principal formations: the original crust, lava flows and ejecta blankets of varying ages, and the regolith. Figure 21-9 shows one proposal for defining time periods on the Moon on the basis of crater counting, the law of superposition, and radiometric dating of returned lunar samples.

To be sure, the law of superposition holds on the Moon, but sometimes it must be inverted. Consider the case of material ejected from a crater, as in Figure 21-4. The deepest rocks encountered by the projectile become concentrated

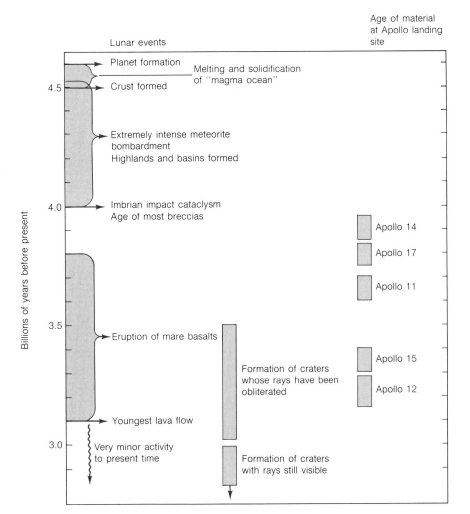

Figure 21-9
Lunar chronology and ages of material at Apollo landing sites. The Moon's early history was dominated by differentiation, impact bombardment, and mare volcanism. The Moon has been "geologically dead" for the last 3 billion years.

at the top of the ejecta that forms the rim. The stratigraphy is inverted, as if the formation were a flap folded back on itself. The shallowest ejecta are found on the surface, but farther out on the ejecta blanket. These features have been discovered on Earth in connection with underground nuclear explosions that were shallow enough to "blow out."

An unexpected and major discovery of lunar stratigraphy is the extreme intensity of the meteorite bombardment of the Moon and presumably the terrestrial planets in the first 600 million years following their formation. This may represent the last stage of "sweep up" of smaller asteroidal bodies by the larger planets. After about 4 billion years ago, the bombardment flux decreased rapidly. It may be that violent collisional events constituted the main process for shaping the surfaces of the young planets, obliterating most of their early history.

The Maria

Controversy raged for years prior to the Apollo program about the nature of the material in the maria. The lunar basins are not filled with sediments deposited in ancient seas, nor are they filled with dust abraded from the highlands, as some had supposed. Analysis of the returned samples from three maria (Tranquilitatis, Fecunditatis, and Procellarum) leaves no doubt about the answer: The maria are flooded by layer upon layer of basaltic lava, not unlike the basaltic flows of Iceland or the Columbia River Plateau discussed in Chapter 15. Apparently when the maria basins were plowed out by the impact of large meteorites early in the Moon's history, the partially molten interior was exposed, and lava subsequently flowed out to fill the depressions (Fig. 21-10). Mare Crisium and Mare Fecunditatis, both good examples of this, can be seen in the north-

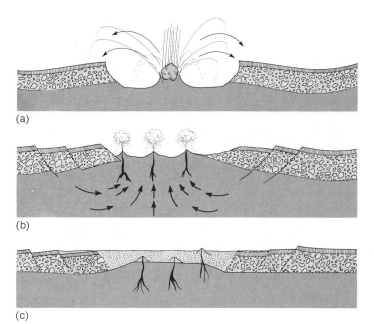

(a)

(b)

(c)

Figure 21-10
Model for the origin of the lunar maria. (a) The impact of a large meteorite excavates a basin, throwing material great distances. (b) The high temperatures generated by the impact cause volcanic activity. Alternatively, volcanism may be caused by fractures that reach partially molten rock beneath the lunar crust. (c) The basin fills with lava from the interior. [After National Aeronautics and Space Administration.]

western and western portions, respectively, of Figure 21-2. The Apollo 15 astronauts visited Hadley Rille, a long, narrow channel in Mare Imbrium, running parallel with the Apennine front. The astronauts could see and photograph ledges in the canyon wall that were some 60 meters thick. These are individual lava flows, layers that flowed one on top of the other in the process of filling the basin. Hadley Rille is thought by some to have been formed by the collapse of an underground lava tunnel whose roof caved in after the lava ceased flowing.

Lunar Rocks: What They Are and What They Reveal

Many hundreds of pounds of rocks have been returned to the Earth from the Moon. This precious cargo has been disseminated to hundreds of specialists all over the world. The United States and the Soviet Union have agreed to exchange lunar material and to cooperate in other ways in lunar exploration. NASA funding gave many investigators new laboratories in which to develop analyti-

cal equipment of unprecedented precision in order to ensure that all possible information be gleaned from the lunar samples. In this way, lunar exploration has inspired international cooperation and spurred innovations that might otherwise have been delayed for decades.

Differentiation occurred very early on the Moon, as evidenced by the ancient lunar crust. The Moon's crust turns out to be a relatively simple assemblage of rocks compared to that of Earth. The crust, as exposed in the highlands, is formed of anorthosite, an igneous rock consisting mostly of calcium-rich plagioclase feldspar. This material probably represents the first melting episode on the Moon, perhaps dating back 4.6 billion years, nearly to the beginning of the solar system. This early melting extended to a depth of several hundred kilometers, forming a "magma ocean." The crust formed as a result of cooling and fractional crystallizations of the magma ocean. Because anorthosite is rich in aluminum, a light element, and poor in iron, one of the heaviest elements, it is relatively light; its crystals floated to the top to form the primitive crust. This crust was later broken up by large impacts, following which the basins were flooded by iron-rich basalt (Fig. 21-9).

The lunar basalt began to flow out perhaps 4.1 billion years ago, but the most extensive mare flooding occurred between 3.8 and 3.1 billion years ago. It is startling that volcanic activity of Mt. St. Helens and Kilauea continue on Earth today, yet the youngest lava on the Moon is 3 billion years old! For some as yet unexplained reason, internal activity on the Moon mostly ceased at this point, as if the internal heat engine were shut off. The Moon simply stopped evolving (in the sense that its surface features have since changed relatively little) somewhat more than 3 billion years ago, leaving a "half-baked" planet with its ancient record still preserved. Four billion years ago, Earth's crust may have had a similar composition, but Earth continued on its evolutionary path to this very day, producing a complex and rich variety of rocks, an ocean, and an atmosphere, and initiating organic evolution. (Not a shred of evidence of biological life has been found on the lunar surface.)

In general terms, the Moon's basalt is similar to the commonest volcanic rock on Earth, yet lunar rocks differ in several essential features from their Earth counterparts. Materials that vaporize at relatively low temperatures—such as water, carbon, nitrogen, sulfur, chlorine, mercury, zinc, and lead—are depleted in Moon rocks. On the other hand, Moon rocks are enriched in certain refractory elements—that is, elements that are not easily melted, such as aluminum, titanium, and zirconium. The extreme dryness of the Moon rocks is striking. Even the driest Earth rocks contain some water bound in various physical and chemical ways, but water is absent from Moon rock. Astronauts can forget about trying to find water on the Moon to sustain supposed future settlements.

The big question is, at what stage did the Moon undergo this gain and loss of materials? Some specialists believe that these depletions and enrichments existed in the original matter that formed the Moon and that the Moon started off as chemically different from the Earth. The Moon may have formed in another part of the solar system of temperature high enough to drive off the volatile elements; subsequently, it was "captured" by the Earth. Alternatively, the Moon may have formed near the Earth, but because of its smaller size, may simply have captured different chemical elements than did Earth. Other hypotheses exist—for example, that the Moon formed by spinning off from the outer layers of Earth soon after Earth had differentiated. The debate about the Moon's origin will continue for some time.

A photograph of the Moon taken a few hundred million years ago would show essentially the same features as a modern picture—Figure 21-2, for example. Astronaut David Scott demonstrated this dramatically on the Apollo 15 mission when he picked up a rock at the foot of the Apennines that, as it turned out, had been lying there unmoved for 400 million years! A picture of the Earth taken 200 million years ago would show a surface completely different from the one the Earth shows today. The Atlantic would not yet have been formed and the map would be dominated by the supercontinent Pangaea. Incidentally, the youth of the Earth's surface features dispels the old speculation that the Moon was torn out of the Earth, leaving the Pacific Ocean as a scar. The Pacific is a recent feature of this rapidly changing planet—one that couldn't have existed in its present shape or form 4.6 billion years ago.

One other rock type should be mentioned because, although it has the same name as an Earth rock of similar appearance, it is peculiar to the Moon and extremely abundant there: the **lunar breccia.** Rocks of this type are cohesive agglomerates or clusters of rock fragments and fine regolith, apparently stuck together by the heat and pressure generated at the sites of meteorite impacts (Fig. 21-11). The rocks in the lunar highlands, the oldest regions of the Moon, are mainly breccias, testifying to the cataclysmic meteorite bombardment early in lunar history.

Figure 21-11
Lunar sample collected by the Apollo 15 astronauts near the Apennine Mountain front. The rock is a blocky, angular breccia. [From National Aeronautics and Space Administration.]

Magnetized Rocks and Mascons

Moon rocks show two other features, seemingly contradictory, that should be mentioned here. Lunar rocks are magnetized. This remanent magnetization in the returned samples cannot be isolated examples, because a magnetometer aboard a satellite launched into lunar orbit by the Apollo 15 astronauts shows that the whole lunar crust has a small residual magnetism. The simplest explanation is that the Moon once had its own planetary magnetic field. If so, the field existed at the times of formation of the magnetized rocks—that is, from the earliest moments in lunar history, when the primitive crust formed, to about 3.1 billion years ago, when the youngest mare basalts poured out. From our experience on Earth (Chapter 18), we surmise that planetary magnetic fields are caused by fluid convection in a molten-iron core, and many scientists have concluded that the Moon was hot enough to produce a small, molten-iron core in its early history. Some time after 3.1 billion years ago, the Moon supposedly cooled down enough for the iron to freeze into a solid, and the magnetic field died away. Thus, the magnetism of its rocks may imply that the Moon did have a hot interior, at least until about 3 billion years ago.

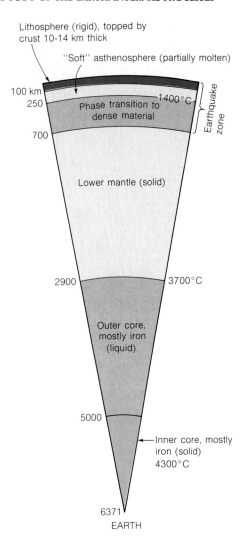

Lithosphere (rigid), topped by
crust 10-14 km thick

"Soft" asthenosphere (partially molten)

100 km
250

Phase transition to
dense material

1400°C

Earthquake zone

700

Lower mantle (solid)

2900

3700°C

Outer core,
mostly iron
(liquid)

5000

Inner core, mostly
iron (solid)
4300°C

6371
EARTH

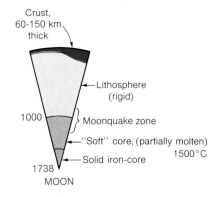

Crust,
60-150 km
thick

Lithosphere
(rigid)

1000

Moonquake zone

"Soft" core, (partially molten)
1500°C

Solid iron-core

1738
MOON

Figure 21-12
Comparison of the internal structures of
Earth and Moon. Note the thick, strong,
unbroken lithosphere on Moon in contrast to the thin lithosphere on Earth,
which breaks into plates moving over
the asthenosphere. Moonquakes are tiny
and infrequent and occur halfway to the
lunar center; earthquakes are a feature of
Earth's outer layers, where plate tectonics
takes place.

The word **mascons** is a contraction for mass
concentrations. Scientists at the Jet Propulsion
Laboratory discovered that spacecraft orbiting
the Moon were mysteriously pulled toward the
surface when they passed over certain locations.
These places with higher gravitational attraction
turned out to be the circular maria. Apparently,
massive bodies, or mascons, lie in or under the
circular basins, and they tug gravitationally on a
satellite each time it flies over. Aside from the
question of what the mascons might be, the most
fascinating question is how these heavy objects
could survive near the lunar surface without sinking into the interior in the 3 or 4 billion years
since the basins formed. Recall that, on Earth, a
surface load such as a continental glacier depresses the crust in only a few thousand years, effectively sinking it until it is supported buoyantly
(Chapter 18). The large positive gravity anomalies
of the mascons show that this has not yet occurred on the Moon, although 3 or 4 billion years
have elapsed. This means that an outer layer of

the Moon, perhaps 1000 kilometers thick (see Fig.
21-12), has been strong, rigid, and, therefore, cold
for 3 to 4 billion years, not warm, plastic, and
weak, as we might suppose from the evidence of
lunar volcanism and magnetized Moon rocks. As
to the nature of mascons, some scientists speculate that they are the buried remains of the massive meteorites whose impacts created the basins.
Most believe that when the primitive crust was
punctured, matter from the interior spurted
out—more than enough to fill the excavation—
and cooled to form a dense plug of mantle material where light crust had formerly been.

Moonquakes and the Lunar Interior

The Moon shows less seismic activity than the
Earth by a factor of about a billion, another indication of a geologically "dead" planet. Seismologists were surprised to discover, however, that
what tiny and infrequent moonquakes do occur

are concentrated around the time when the Moon is closest to the Earth—that is, when the tides of the Moon, pulled by Earth's gravity, are at a maximum. An old theory that tidal stresses can trigger quakes has been difficult to prove on Earth, but apparently it does work on the Moon. The moonquakes occur at the unusually great depths of 800 to 1100 kilometers, about halfway to the lunar center, near the boundary between a thick, strong lunar lithosphere and an underlying "soft," partially molten core (Fig. 21-12). Seismologists also tell us that the lunar crust is highly fractured in its outer portions, presumably because of the intense bombardment by meteorites.

What Does It All Mean?

It is a difficult task for the authors of a textbook to pull together this wealth of new and often contradictory material on the Moon when the experts themselves have not yet done so to everyone's satisfaction. Perhaps, in time, a concept with the simple elegance and synthesizing power of the concept of plate tectonics on Earth will emerge to unify the lunar data.

Whatever concept does emerge must begin with a Moon that formed 4.6 billion years ago from starting materials that were probably different chemically from those that formed the Earth. The Moon heated up rapidly, perhaps because of fast accretion or radioactivity. As a result, extensive melting occurred very early in lunar history, leading to the formation of a magma ocean. A primitive crust of anorthosite floated to the surface; the heavier mafic materials sank and became the sources of mare basalts that erupted later. The first 600 million years was a time of intense bombardment by meteorites; the ancient crust was heavily cratered, as evidenced by the brecciated relics in the highlands. Great impacts excavated mare basins, which later filled with basalts from the lunar interior, perhaps due to a second episode of partial melting of the outer portions of the Moon. A small iron core probably formed early in lunar history and started a magnetic field. The core solidified and the outer layers cooled rapidly about 3 billion years ago, forming a strong, one-plate lithosphere about ten times thicker than Earth's broken, many-plated lithosphere (Fig. 21-12). What the scientists must come up with is a thermal mechanism to account for a Moon initially hot enough to undergo igneous differentiation, forming crust and core, and then undergo a later stage of large-scale volcanism and fill the mare basins with basalt. They must explain how

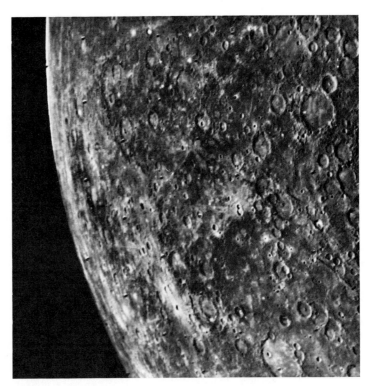

Figure 21-13
Mariner 10 photograph of the southwestern quadrant of Mercury taken from a distance of 200,000 km (120,000 mi) reveals features 5000 times smaller than could be seen by the best Earth-based telescopes. The largest craters seen in this picture are about 100 km (60 mi) in diameter. [From National Aeronautics and Space Administration.]

the Moon could cool fast enough to shut off volcanism abruptly and form a thick, strong lithosphere to support the mascons. The Moon has been essentially dead for 3 billion years. Its surface has been modified somewhat by a reduced amount of meteorite bombardment, but the large-scale ancient features have remained unchanged.

Moon geologists do not have all the answers at this time. They are working on materials that date back to the early days of the solar system, containing records of events no longer preserved on Earth. In a sense, they have Rosetta stones still waiting to be deciphered.

MERCURY

Before the 1974 Mariner 10 flight, Mercury appeared only as a fuzzy globe to Earth-based astronomers. The dramatic photographs and data radioed back by the space craft were startling in their indication that the planet has a Moon-like surface and an Earth-like interior (Figure 21-13).

Mercury is the densest of all the planets, indicating the presence of a large iron core (like that of Earth, only larger) that extends outward perhaps three-fourths of Mercury's radius. It has a lightweight silicate crust with heavily cratered regions like the lunar highlands; vast smooth plains occur, presumably of volcanic origin like the maria. There is a trace of an atmosphere, one-hundred billionth as dense as Earth's and mostly helium. Unexpectedly, Mercury has a weak but definite magnetic field, perhaps due to a "live" core dynamo, or the remanent field of an ancient core dynamo that has stopped working. Long, ancient scarps (up to several hundred kilometers in length) are seen on Mercury's surface. They have been interpreted as thrust faults due to contraction of the planet's radius by about 1 or 2 kilometers. Such global contraction would result from cooling.

Since Mercury is the planet closest to the Sun, one would expect its bulk constitution to be the richest in refractory elements, accounting for its large iron core. Like the Moon and presumably the other terrestrial planets, it underwent early chemical differentiation, with formation of crust and core, and received the same intense meteorite bombardment in the first half-billion years of its existence. The fact that the record of this ancient cataclysm is mostly preserved in the present surface implies that Mercury, like the Moon but unlike Earth, has been tectonically quiet for the past 2 to 3 billion years, and has not been subjected to atmosphere-related erosion.

Of all planets in our solar system, Mercury receives the most intense solar radiation. Surface temperatures reach 425°C (800°F) during the day. A night on Mercury lasts 88 Earth days, giving the surface time to cool down to −175°C (−280°F). This temperature range of 600°C is the widest in the solar system.

VENUS

Venus is closest to being Earth's twin in mass, diameter, density, gravity, and distance from the Sun (see Table 21-1). But here the similarity ends, according to the latest results from radar and automated space probes.

Venus differs profoundly from Earth in the nature of its atmosphere. It is covered by a thick blanket of carbon dioxide. More than 95 percent of the Venusian atmosphere is made up of this gas; the Earth's atmosphere, by comparison, is only 0.03 percent carbon dioxide. Nitrogen (3 percent), water vapor, and other gases are present but in extremely small amounts compared to the great quantities in the Earth's biosphere. The temperature at the surface is a searing 450°C (842°F), and atmospheric pressure has the crushing value of ninety times Earth's atmospheric pressure. The heavy, poisonous atmosphere, incredibly high temperatures, and clouds of corrosive sulfuric acid droplets would fit a description of hell. Certainly no organic life could exist on the surface.

What makes Venus hot is its insulating blanket of carbon dioxide. Solar energy is trapped by the atmosphere, which warms up as a result. This is entirely analogous to the much milder greenhouse effect on Earth (Chapters 1 and 12).

Venus is completely shrouded by dense clouds that hide its surface from view. However, the U.S. Pioneer Venus spacecraft placed in orbit around Venus in December 1978 mapped 93 percent of the planet with radar waves bounced off its surface. The radar mapper revealed a diversity of geological features, both similar to and different from those of other terrestrial planets. About 60 percent of Venus' surface consists of a planet-encompassing, relatively flat, rolling plain. Only 16 percent of the Venusian surface is in the form of low-lying basins—in contrast to the nearly two-thirds of Earth's surface that is ocean basins. Most of the remaining 24 percent of Venus is only one thousand or so meters higher than the plain, with about 8 percent in the form of true highlands, ranging to maximum altitudes of about 11 kilometers. Plateaus, mountain ranges, volcanoes, and volcanic terrains appear to be present, as well as great rift valleys as much as 5000 kilometers long. What appear to be large impact craters are scattered over the rolling plains. Satellite gravity data indicate that the mountains are made of light crustal rock and float on denser underlying mantle (that is, like mountains on Earth, they are isostatically compensated). Russia's Venera spacecraft, which landed on the surface, found lightweight crustal rocks whose chemistry suggested that they might be granite in one place and basalt in another.

Venus has no detectable magnetic field—perhaps because it has lacked a significant, fluid iron core.

What can we say from these preliminary data? Venus is a differentiated planet: it formed a crust and underwent extensive outgassing. A single huge continent covers most of the planet. Impact craters dating back to the heavy asteroidal bombardment of the first 1 or 2 billion years of the solar system appear to be preserved. These would be similar to the craters found on the ancient surfaces of single plate planets like Mercury and the Moon. However, such Earth-like features as plateaus and rifts are also present. These facts suggest that Venus evolved to form a single plate with a thicker, more buoyant crust than Earth, thus "squelching" plate tectonics. In a sense the global continent on Venus may have resisted subduction just as Earth's individual continents do.

The profound differences between Venus and Earth pose difficult questions: How could two planets, so similar in many respects, evolve so differently? Could it be that Venus, though Earth's neighbor, was just enough closer to the Sun so that it was too hot for water vapor to condense into oceans and carbon dioxide to be locked up in surface rocks? The absence of life prevented oxygen from becoming an important constituent of the atmosphere. It has been estimated that if Earth were only 7 percent, or 10 million kilometers (6 million miles), closer to the Sun, its climate would be similar to that of Venus.

MARS

Mars has been a planet favored by scientific speculators and science-fiction writers. Some have thought that Mars is like Earth, and that it has canals, a seasonally varying cover of vegetation, and even intelligent life. Others have thought it to be a tectonically dead planet like the Moon. Both of these extremes and other speculations as well have been laid to rest as a result of data radioed to Earth by the Mariner 9 spacecraft, which orbited Mars for 11 months in 1971 and 1972, and by the Viking project, which put two spacecraft into Martian orbits and landed two scientific packages on the surface in 1976. The unprecedented resolution of thousands of color photographs taken from orbit showed clearly features a few tens of meters across, in contrast with the 80-kilometer resolving ability of Earth-based telescopes, upon which so many of the early speculations depended. The landers were primarily designed to search for life, but they also contained high-resolution cameras, seismographs, analytical chemistry equipment, and instruments for meterological and other experiments.

Mars, the outermost of the terrestrial planets, travels in the next orbit to Earth's, farther from the Sun. Its diameter is a little over half that of the Earth, its mass about one-eighth, and its gravity is about two-fifths as strong. Mars has a density somewhat less than Earth's, but, when allowance is made for its smaller size and reduced internal pressures, its "uncompressed density" turns out to be the same as the Earth would have. This means that its overall composition is probably not much different from Earth's.

By coincidence, the Martian day and the Earth day are of nearly the same length, and the angle between the axis of rotation and the orbital plane is nearly the same for both planets. The Martian year is 687 Earth days. Mars has no large-scale magnetic field comparable to that of Earth today, but a small field cannot yet be ruled out.

The Viking data confirm that Mars has an atmosphere about 150 times thinner than Earth's. It is composed mostly of carbon dioxide (95 percent), nitrogen (2 to 3 percent), argon (1 to 2 percent), with lesser amounts of oxygen, water, and other gases. Water cannot now exist as a liquid at the surface; it would either evaporate or freeze. The surface temperatures are more extreme than those on Earth, ranging from about 25°C (roughly 80°F) at the equator to about −125°C (roughly −190°F) at the poles. At the Viking I landing site the daily temperature varied from −90°C to −10°C. Wind velocities of 270 kilometers per hour (roughly 170 miles per hour) can blow up in gigantic storms that cover the planet with dust for weeks at a time. One such dust storm blurred the view of the Mariner 9 television cameras for many weeks, delaying the transmission of pictures.

The portrait of Mars that has emerged from thousands of Mariner 9 and Viking photographs is one of a planet that has been shaped by a variety of processes. Among these are bombardment of the surface by meteorites, volcanism, tectonic activity, erosion by running water, action by wind, and sapping or undermining of ground layers by melting subsurface ice. Thus Mars is the product of processes that occur on the Moon and on Earth, but the unique mix on Mars has led to a distinctive history and character. The major geologic features of the Martian surface are summarized in the planetary map in Figure 21–14.

Preliminary Viking data indicate that most of the Martian surface is old. The crater density, an approximate age indicator, ranges from about one-tenth that of the lunar highlands to about the same density found there.

The contrast between the northern and southern hemispheres is striking. The southern hemisphere is mainly ancient cratered terrain, probably similar in age to the lunar highlands, 4 billion years old. The two large impact basins of the southern hemisphere, Hellas and Argyre, are ringed by mountains formed from basin ejecta. They are quite similar to Imbrium and other lunar

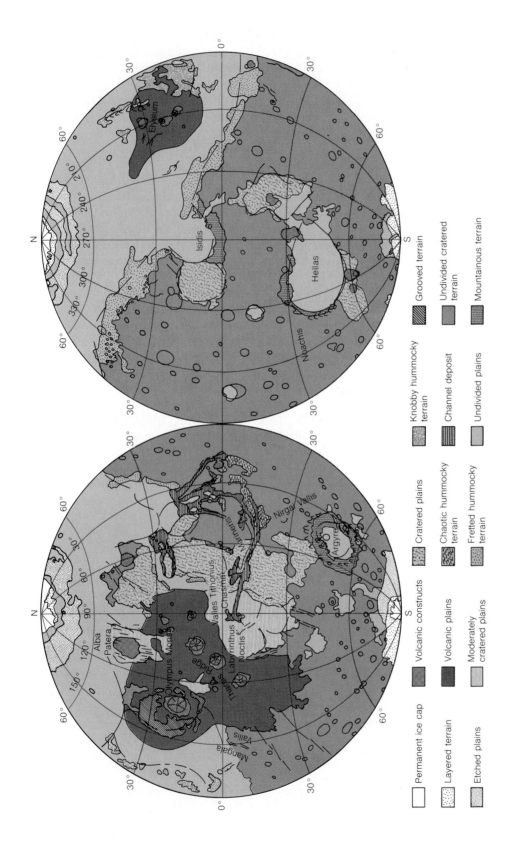

Figure 21-14
Geological features of Mars based on Mariner 9 pictures. Permanent ice caps are mostly water ice. The oldest and most cratered terrain occurs in the southern hemisphere. Large volcanoes and sparsely cratered volcanic plains predominate in the northern hemisphere. Channel deposits are flood features. The absence of mountain chains, transform faults, linear troughs and ridges, indicates a stable crust without plate tectonics. [From "Mars" by James B. Pollack. Copyright © 1975 by Scientific American, Inc. All rights reserved.]

Legend:
- Permanent ice cap
- Layered terrain
- Etched plains
- Volcanic constructs
- Volcanic plains
- Moderately cratered plains
- Cratered plains
- Chaotic hummocky terrain
- Fretted hummocky terrain
- Knobby hummocky terrain
- Channel deposit
- Undivided plains
- Grooved terrain
- Undivided cratered terrain
- Mountainous terrain

Labels on map: Elysium, Isidis, Hellas, Noachis, Alba Patera, Olympus Mons, Tharsis, Labyrinthus Noctis, Valles Marineris, Vallis Thonius, Chasma, Mangala Vallis, Nirgal Vallis, Argyre

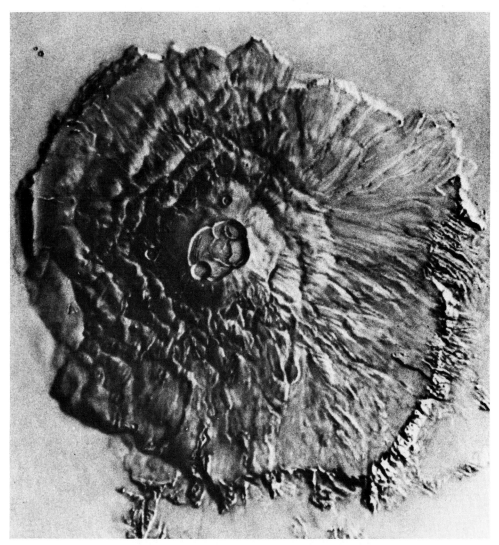

Figure 21-15
Viking 1 photograph of the great Martian volcano, Olympus Mons. The 27-km-high
(17 mi) mountain is about 600 km (370 mi) across at the base and would extend from
San Francisco to Los Angeles or from Boston to Baltimore. Note the complex, multiple vent, main crater, and the steep cliffs that drop off from the mountain's flanks
to the surrounding plain. [From U.S. Geological Survey and National Aeronautics
and Space Administration.]

mare basins but are more eroded. Great sand
dune fields are found in this area.

Mars has been volcanically active throughout
its history, the activity waning over time. The
northern hemisphere is the site of most of the
Martian volcanism. Volcanic structures on Mars
consist of enormous shield volcanoes (some probably still active), domes, cones, and large volcanic
plains. The shield volcano Olympus Mons—600
kilometers wide at the base, 27 kilometers high, and
with a summit caldera 80 kilometers wide (Fig.
21-15)—may be the largest volcano in the solar
system. On Earth volcanoes might last for a few
hundreds of thousands of years before plate mo-

tions carry them away from the source of magma.
In the absence of plate tectonics, Martian volcanoes sit over their sources and remain active for
long periods, possibly billions of years. This, together with a thick supporting lithosphere, accounts for their huge size. Martian lava plains
show flow features similar to those on the lunar
lava plains—wrinkle ridges and lobed escarpments at the fronts of individual lava flows. The
cratered plains are thought to be older volcanic
features. Most of the volcanics appear to be iron-
rich basalts.

The permanent ice caps, composed of water ice,
expand seasonally by the precipitation of carbon

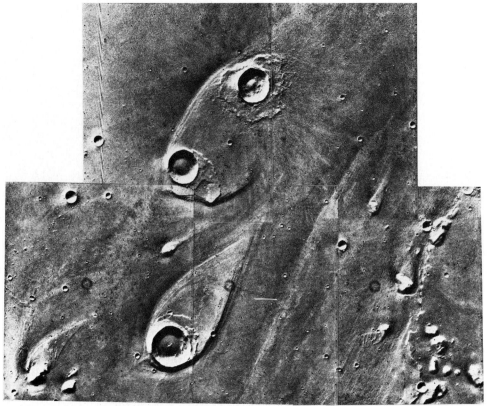

Figure 21-16
Some erosional features on Mars near the Viking 1 lander site photographed from a height of 1600 km (992 mi). The knobs and hummocks to the lower right are the erosional remnants of an old crater rim. The flow of water was from lower left to upper right. The teardrop-shaped islands formed when the water flowed around obstacles presented by an existing crater. Layering can be seen on the sides of the island. The crater density indicates that the floodplain surface is old and that the flooding occurred a long time ago. [From National Aeronautics and Space Administration.]

dioxide ice. The polar regions are ringed by extensive areas of layered deposits, sand dunes, and ice-free zones, suggesting a complicated climatic history.

The chaotic terrains (Fig. 21-14) are thought to have formed by sapping or melting of subsurface ice, resulting in the collapse of overlying layers and leaving an incoherent array of broken slabs of rock. Fretted terrain, which may have a similar origin, is lowland dotted with isolated mesas and bounded by cliffs of intricate geometry. Apparently, the cliffs retreat as a result of the melting of subsurface ice and the seeping of water from their bases. The cliffs are undermined, and landslides result.

Large areas show evidence of ancient catastrophic flooding without indications of ancient seas or accumulations of sediments. Among the features observed are narrow, winding channels with tributary systems and large, braided channels with bars and tear-drop-shaped islands (Fig.

21-16). Gullies can be seen, similar to drainage channels on Earth's desert hillsides.

But where is the water? The present CO_2 atmosphere has too low a pressure for liquid water to survive (it would vaporize), and the vapor content of the atmosphere is too small to account for the amount of liquid water that was evidently once present. The required water must exist as ice, or be trapped underground by permafrost or in aquifers, under present conditions. Moreover, about 5 to 20 times more CO_2 must have been present in the past for atmospheric pressures to be large enough to keep water in the liquid state. Somewhere on Mars there is probably a large hidden inventory of ice, both the H_2O variety and the CO_2 variety.

The polar ice caps contain some of the needed water. If the ice were some hundreds of meters thick, as some believe, and if all the ice melted at once, there would be enough water to cover the planet to a depth of 0.5 meter. Water and CO_2

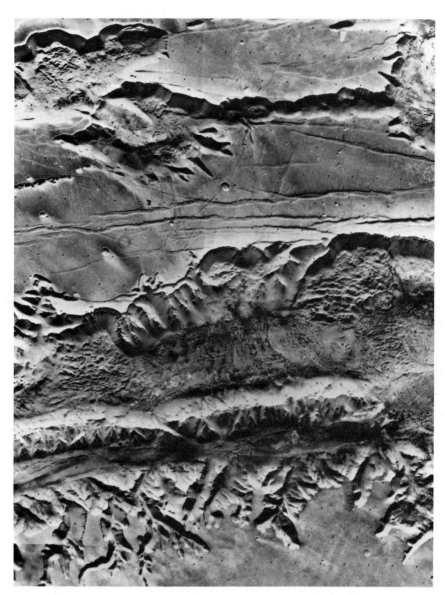

Figure 21-17
Viking 1 photomosaic of the enormous canyon Valles Marineris from a height of 4200 km
(2600 mi); north is to the top. The principal canyon crosses the bottom half of the picture
in an east–west direction. The northern wall of the main canyon shows several large landslides. The series of branching channels cut into the plateau on the bottom from the south
wall may have been formed by slow erosion as a result of the release of groundwater.
Other branches of the canyon are visible at the top. [From National Aeronautics and Space
Administration.]

could also be stored as ice in the Martian regolith,
which could be as much as a kilometer thick. During warm periods, perhaps triggered by volcanism
or climatic fluctuations due to slow periodic
changes in the Martian orbit, the frozen water and
CO_2 could be released. Alternatively, large impacts might release trapped underground water.
The collapse features with channels that seem to
drain them support the notion that water stored
underground in some form was released.

Among the most spectacular features are the
deep canyons and great cracklike rifts. One immense system of canyons, Valles Marineris, just
south of the equator, makes the Grand Canyon of
the Colorado look puny; it is about 2700 kilometers long, and in places 500 kilometers wide and 6
kilometers deep. These structures are tectonic in
origin, indicative of large-scale tensile (stretching)
forces, perhaps connected with uplift and volcanism elsewhere. Fluvial erosion and sapping have
also contributed to the morphology of the canyons (Fig. 21–17).

Unlike Earth, Mars shows no evidence of plate
movements. Mountain systems or linear troughs

Figure 21-18
This spectacular picture of the Martian landscape by the Viking 1 lander shows a dune field littered with boulders, a view remarkably similar to many seen in the deserts of Earth. The picture covers an angle of 100°, looking northeast at left and southeast at right. The sharp dune crests indicate the most recent wind storms capable of moving sand over the dunes blew in the general direction from upper left to lower right. Small deposits downwind of rocks also indicate this wind direction. The large boulder at the left is about 8 m (25 ft) from the spacecraft and measures about 1 m by 3 m (3 ft by 10 ft). The meteorology boom, which supports Viking's miniature weather station, cuts through the picture's center. The sun rose two hours before the picture was taken and is about 30° above the horizon near the center of the picture. [From National Aeronautics and Space Administration.]

forming under compression at convergent plate boundaries, linear ridge-rifts indicative of spreading boundaries, transform faults with large differential horizontal displacements—these are all lacking. The Viking seismograph has not detected any Mars quakes in the first few months of operation. Apparently, like the Moon, Mercury, and Venus, Mars is a single plate planet with a lithosphere too thick to break into plates.

Hundreds of color photographs have been taken at each of the Viking landing sites. The surface of Mars is a bright red color, as any amateur astronomer would have predicted. Both sites show a field of fine-grained, sandlike material littered with boulders. The Viking 1 landing site (22.4°N, 47.5°W) shows a dune field of fine-grained sand. Bedrock outcrops can be seen, and large numbers of rocks are strewn over the surface. Both vesicular and fine-grained rocks are in evidence, with indications of chemical weathering and wind-faceting (Fig. 21-18).

The Viking 2 landing site (47.9°N, 225.9°W) was chosen for its northerly location in the hope that more water would be present and the chance of finding life greater. The spacecraft landed in a boulder-strewn, flat, reddish desert cut by troughs that seem to form a polygonal network. Unlike the view from the first lander, photographs from Viking 2 show no outcrops or dunes. The blocks are highly vesicular, and they appear to be volcanic in origin but to have been transported as ejecta from a distant impact.

We see that Mars, though tectonically and volcanically active like Earth, evolved along a different course. Plate tectonics on Earth has renewed its features over many cycles of plate creation and destruction. Older surviving terrains on Earth have been profoundly altered by weathering, erosion and metamorphism. In contrast, weathering and erosion is extremely slow on Mars. Winds are inefficient because of the thin atmosphere; they serve only to transport particulate debris already on the surface. Running water at the surface has occurred rarely. A strong stable lithosphere has resisted plate tectonics. For these reasons the Martian surface reflects a rich accumulation of diverse geological events, which have taken place throughout its history and which have been preserved for Earthlings to untangle.

No signs of large organisms are apparent in the photographs at either landing site. Nevertheless, the environment is more hospitable to life than

had been expected (traces of nitrogen, oxygen, water vapor). Several experiments were designed to search for microscopic life. They were sensitive enough to detect signs of life in inhospitable soils from Antarctica and from barren Earth deserts. The tests made on Mars revealed a surprisingly active surface, subject to diverse interpretation. The consensus is that the activity is chemical in nature, caused by the presence of a strong oxidizer rather than by living things.

JUPITER

Jupiter is the best preserved sample of the early solar nebula and, with its satellites, may contain the most important clues to the origin of the solar system. It is the largest of the giant gas planets, with a volume 1300 times that of Earth. Like the other giant planets, it is composed predominantly of hydrogen, helium, ammonia, and methane with, possibly, a small rocky core. Jupiter has twice the mass of all of the other planets combined. Had it been slightly more massive, Jupiter might have attained internal temperatures as high as the ignition point for nuclear reactions, and it would have flamed as a star in its own right. One of the most successful space science experiments of recent years was the Voyager mission to Jupiter and Saturn. Voyager 1 and 2 swept by Jupiter and its four largest moons in 1979, returning 30,000 photos of unprecedented resolution (see Fig. 21–19) and an enormous amount of physical data.

Jupiter rotates very fast, once every 9.8 hours. As a result its clouds, which are largely frozen and liquid ammonia compounds, have been whipped into alternating colorful bands of jetlike winds that circle the planet at high speeds. Gigantic whirlwinds or cyclones, of which the Great Red Spot is an example, hover on the surface (Fig. 21–19). The Great Red Spot has persisted for centuries, perhaps because of its large size (Earth could easily fit inside it). Jupiter has a powerful magnetic field, which traps charged particles from the Sun, in the process sending out radio waves that were first detected by Earth-based radio telescopes in 1955. Surprisingly, Jupiter gives off twice as much heat as it receives from the Sun, probably heat generated by the continued gravitational contraction of the planet.

To geologists, interested primarily in solid materials, the four planet-sized moons of Jupiter provided the greatest excitement of the Voyager flyby. Like a miniature model of the solar system, Jupiter's moons decrease in density with distance—from rocky moons close to the planet to icy moons farther away (Fig. 21–20). The two outer large satellites, Callisto and Ganymede, have such low densities that half their bulk composition is probably water or water ice. Callisto is saturated with craters, so many that it may be the most cratered object in the solar system. It has been geologically inactive for so long that its thick, hard, icy surface preserves the record of the torrential bombardment that took place some 4 billion years ago and left imprints on the cratered highlands of Mercury, Mars, and Earth's moon as well. Ganymede has both old, cratered terrain like Callisto's and younger regions. This younger terrain is characterized by intricate patterns of par-

Figure 21-19
Voyager photos were assembled into this composite picture of Jupiter and its four planet-size moons. The alternating bands of east-west winds and cyclonic whirlwinds, including the Great Red Spot in the southern hemisphere, are seen on Jupiter's surface. Each of Jupiter's moons is different from its companions: reddish Io (upper left) is rough, splashed with bizarre colors, the most volcanically active body in the solar system; Europa (center) is smooth and white, its frozen ocean surface cut by a network of filled-in fractures; ice-covered Ganymede shows cratered terrain as well as younger regions grooved and fractured by tectonic activity; Callisto (lower right) is the most cratered body in the solar system, with a record of bombardment 4 billion years old preserved in its icy crust. The moons are not to scale but are in correct relative position. [Jet Propulsion Laboratory.]

allel troughs and ridges within blocks that seem to have been sheared and offset by fractures. Does this evidence of compression and shear imply a period of plate tectonics—with crustal plates made of ice?

Jupiter's second moon Europa is a rocky planet covered by a frozen ocean perhaps 100 kilometers (60 miles) thick. Europa has few craters, but its smooth, near-white surface is crisscrossed by bright and dark lines that resemble cracks in ice. Europa's icy crust may have remained warm and mobile enough to have flowed like a glacier, thereby erasing evidence of the ancient bombardment and covering any rugged topography on the rocky surface below.

Io is the moon of Jupiter most favored by geologists (Fig. 21-21). Its surface, bedecked in hues of red, orange, gold, silver, white and black, shows a variety of landforms: scarps, faults, calderas, mountains—and the most exciting discovery of all—active volcanoes! Io's uniquely colored, bi-

zarre surface is the product of volcanic eruptions that occur with such frequency and energy as to make Jupiter's innermost moon the most volcanically active body in the solar system. Not a single impact crater remains from the bombardment that scarred Callisto and Ganymede billions of years ago; volcanism has obliterated old features, leaving a young, continually changing surface on Io. The Voyager spacecraft found 10 erupting volcanoes, throwing plumes from 70 to 300 kilometers high, with ejection velocities ranging up to 3600 kilometers per hour (2200 miles per hour). Not only are Io's volcanoes more violent than those on Earth, they erupt sulfur, sulfur dioxide, and other sulfur compounds that are responsible for Io's exotic coloration. Scientists believe that Io's volcanoes draw their energy from tidal friction. As Io moves in its eccentric orbit around Jupiter, the planet's gravitational pull periodically stretches and distorts the satellite's crust, generating huge amounts of internal heat, just as a paper

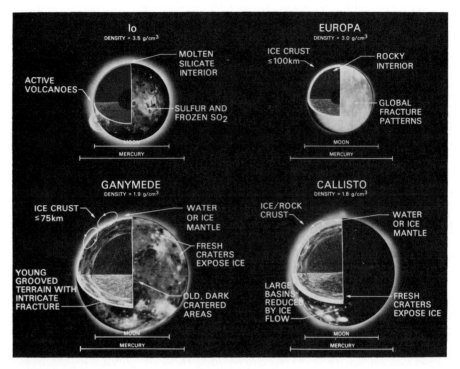

Figure 21-20
Schematic illustrations of the surface and interiors of Jupiter's largest moons.
The labels summarize physical data and current geological interpretations.
[Torrence Johnson, Jet Propulsion Laboratory.]

Figure 21-21
Voyager image of Io. An enormous volcanic eruption can be seen on the horizon silhouetted against dark space. The plume of solid ejecta was thrown up to an altitude of 160 km (100 mi). [Jet Propulsion Laboratory.]

Figure 21-22
Composite picture prepared by assembling images taken by Voyager 1 space-craft during its Saturn encounter in November 1980. Saturn and its rings and 6 of the planet's 15 known satellites are shown. Clockwise from the right: Tethys and crater-pocked Mimas in front of the planet; Enceladus in front of the rings; Dione in the forefront to the left; Rhea off the left edge of the rings; and smog-shrouded Titan in its distant orbit at the top. [Jet Propulsion Laboratory.]

clip becomes warm when it is bent back and forth.

In Jupiter's moons we are introduced to a new family of exotic worlds, each different from the others, and all different from anything we have ever seen before.

SATURN

After flying past Jupiter, Voyager 1 reached Saturn in November 1980. In the short space of just a few days, more was learned about this planet than has been learned in the 370 years since Galileo discovered it.

The second largest planet, Saturn turned out to be much like Jupiter in its turbulent complexity. It rotates rapidly and its surface shows colored, alternating bands of high-speed east-west winds and large cyclonic storms. Like Jupiter, Saturn has 15 moons, a magnetic field, and radiates more heat than it receives from the Sun (Fig. 21-22).

Although rings have recently been discovered circling Jupiter and Uranus, Saturn's immense ring system is unique in its extent. It forms a band

over 65,000 kilometers (40,000 miles) wide with a thickness of a few kilometers. The rings are composed of countless icy snowballs or ice-covered rocks ranging in size from a few microns to a meter. The rings are thought to be either fragments of a satellite broken up by tidal forces as it spiraled too close to Saturn or material that never accreted to form a satellite because a single body of any size could not exist so close to the planet.

Voyager's photos of Saturn's rings excited and disturbed planetary astronomers. Instead of being the three or four separate disks known from Earth observations, they turned out to be composed of as many as 1000 separate ringlets, like grooves in a phonograph record. Unexplained kinks, bulges, and radial "spokes" could be seen in what should have been uniform, spinning rings. It may be that the interaction of the ring particles with Saturn's magnetic field and its moons is responsible for these unexpected features.

Voyager swept past Saturn's inner moons and returned photographs of these previously unstudied smaller bodies (Figs. 21-22 and 21-23). These moons are composed mostly of water ice, and all but one show heavily cratered surfaces—evidence

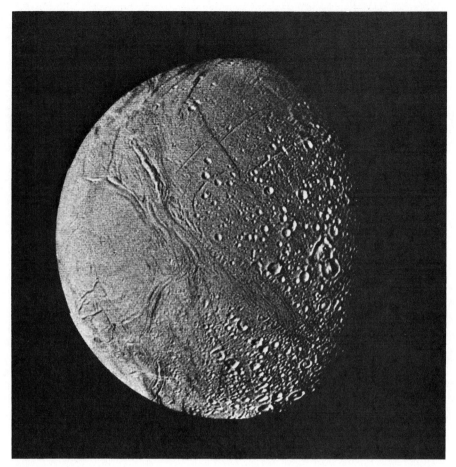

Figure 21-23
Image of Saturn's moon Enceladus made by Voyager 2 spacecraft in 1981 at a distance of 19,000 km (74,000 mi). Old cratered areas and relatively young uncratered terrain are present. The uncratered surface together with the grooves and linear features indicate that even though Enceladus is only 490 km (300 mi) in diameter, it has undergone relatively recent internal melting and crustal deformation. The largest crater is about 35 km (20 mi) across. [Jet Propulsion Laboratory.]

that this part of the solar system did not escape the heavy bombardment of 4 billion years ago.

Saturn's largest moon, Titan, is nearly twice the size of Earth's moon. It was of special interest to astronomers because it is the only moon in our solar system with a substantial atmosphere—one that unfortunately shrouds its surface with an opaque orange smog. Although Voyager could not photograph the surface, it discovered that like Earth's, Titan's atmosphere is mostly nitrogen, with small amounts of methane, ethane, acetylene, ethylene, and hydrogen cyanide. Atmospheric pressure is so high (one and a half times that on Earth) and its temperature so low that lakes of nitrogen and freezing hydrocarbons may be present. Some astronomers believe that Titan's atmosphere may be similar to Earth's early prebiotic atmosphere—in a state of deep freeze!

URANUS, NEPTUNE, AND PLUTO

Not much is known about Uranus and Neptune, the outermost of the giant planets, nor of Pluto, the ninth and most distant planet, beyond the data given in Table 21-1. Following the discovery of Uranus by telescope in 1781, the existence of Neptune and Pluto was predicted mathematically, because Uranus seemed to deviate from its predicted orbit, indicating that it might be influenced by the gravitational attraction of unknown planets. With the aid of these mathematical hints, Neptune was found in 1846 and Pluto in 1930. Uranus was only recently found to have as many as nine rings.

The most advanced Earth telescopes are now being trained on Uranus and Neptune, and the Voyager II spacecraft may be programmed to reach Uranus in 1986 and Neptune in 1989 after encountering Saturn in 1981. Before the end of the decade we are likely to learn more about these little-known outermost planets.

METEORITES

Meteorites are today believed to be broken fragments of asteroids and comets. These fragments are deflected into orbits that cross the path of Earth, which sweeps them up. Meteorites range in size from fine dust particles to bodies many miles in diameter, which blast out large craters or astroblemes when they impact the Moon, Mars, or Earth. Meteoritic dust falls almost continually on Earth. However, the larger, more spectacular meteorites that survived the fiery flight through the atmosphere were the only source of information about extraterrestrial materials before the lunar samples were collected.

Some of the asteroids may have been differentiated prior to breakup, with formation of an iron core, a stony mantle, and a crust. Other asteroids may have maintained their primitive mix of the original accretion products without undergoing chemical zonation. It may be that the fragmentation of the outer layers of the asteroids gave rise to the stony meteorites, the most abundant ones, made up of silicate minerals, mostly olivine and pyroxene. Deeper fragmentation, involving breakup of the core, could account for the stony-iron and the iron meteorites (Fig. 21–24). Some meteorites appear to be unheated, unmelted, undifferentiated primordial materials—containing actual samples of the condensates from the early solar nebula, as well as even older interstellar grains that were in transit through our solar system. These are receiving much attention now because of the unique information they contain about early solar-system and pre-solar-system processes that were not preserved on larger bodies.

Figure 21-24
Photograph of a piece of the Allende meteorite, thought to have formed early in the history of the solar system. The black crust on top is due to melting from frictional heating in flight through the atmosphere. [Smithsonian Astrophysical Observatory.]

Meteorites usually show formation ages of about 4.5 billion years, although breakup of their parent bodies may be as recent as 10 million years. The age of formation, determined by radiometric dating, was used to fix the time of formation of the solar system.

Meteorites are rare, prized objects. Recently large numbers were discovered in Antarctica, concentrated in certain parts of the ice cap. When the flowing ice encountered a barrier it could not pass, it evaporated, leaving a pile of meteorites that had fallen widely on the ice surface over the course of thousands of years. This was a bountiful discovery for meteorite scholars.

SUMMARY
1. The planets of the solar system fall in two distinct groups: the small, stony, inner Earth-like planets, and the giant, gaseous, outer planets, which are more Sun-like in their compositions.

2. The Moon was geologically active in its early history. Lunar rocks show the planet to have been differentiated and subjected to volcanism, but, for reasons not fully understood, the Moon became essentially inactive about 3 billion years ago. The absence of young rocks, the small number of moonquakes, and the great age of surface features confirm this conclusion. Without water or an atmosphere to foster erosion, and without the renewing forces of internal activity, the lunar surface has remained essentially

unchanged for billions of years. Its features were mostly shaped during its earlier active phase, in which meteorite impacts played a major role. The major formations on the Moon are its original anorthositic crust, now found primarily in the highlands, basaltic lava flows primarily filling the maria, ejecta blankets of varying ages, and the lunar regolith. The lunar lithosphere, which is much thicker and stronger than Earth's, is able to support mascons to resist plate tectonics.

3. Mercury is the smallest and densest planet. It has a large iron core and a silicate crust. Perhaps closeness to the Sun accounts for its lack of an atmosphere and the absence of the lighter, more volatile elements. Like the Moon, it has a heavily cratered surface, shows evidence of early differentiation, and has been geologically inactive for a long time.

4. Venus is similar to Earth in mass, size, density, and distance from the Sun, but there the resemblance ends. The surface is hidden by a thick and heavy blanket of carbon dioxide, which traps solar energy so that the surface temperature reaches 475°C. Radar data and photographs of the surface by the Venera spacecraft provide evidence of current geological activity and of ancient, cratered terrain. Venus appears to be a single-plate planet like Mercury and the Moon.

5. The Mariner 9 and Viking flights have revealed that Mars has been shaped by a variety of processes, including bombardment by meteorites, volcanism, tectonic activity, erosion by running water, action by wind, sapping of surface layers by melting subsurface ice. Most of the Martian surface is very old. The thin atmosphere is mostly carbon dioxide. The water, inferred from erosional features, may be stored as ice in the polar ice caps or in the regolith. There is no evidence of plate tectonics nor of any life.

6. Jupiter and Saturn are giant gas planets with turbulent and complex atmospheres displaying planet-circling bands of winds and large cyclonic disturbances. Each of the rocky and icy moons of these planets is geologically unique. Some show a predominance of ancient cratered terrains; others show evidence of tectonic activity. Active volcanism was discovered on Jupiter's moon Io. Saturn's moon Titan has a mainly nitrogen atmosphere with small amounts of hydrocarbons—much like Earth's early atmosphere, but in a state of deep freeze.

EXERCISES

1. The surface features of the Moon are profoundly different from those of Earth. Describe these differences, and explain why they occur.

2. The Earth and Moon are of comparable age, yet Earth's surface is youthful compared to the old surface features of the Moon. Most rocks on the surface of the Moon are about 3 billion years old or older, whereas most rocks on Earth's surface are younger than about 200 million years. Why?

3. What is the implication of the fact that mascons are found on the Moon and not on Earth?

4. Moon rocks and soil differ from Earth rocks. How? Why?

5. Contrast the geology of Mars and Earth, indicating similarities and differences.

6. If humans were forced to leave Earth because of some kind of catastrophe, where could they migrate to with the best chance of surviving? Why?

7. In size, mass, and distance from the Sun, Venus is nearly Earth's twin. Yet the two planets are radically different. Describe these differences, and hazard a guess as to why they exist.

8. Mars has been explored by an unmanned spacecraft that landed on the surface. The primary mission was to search for life. If you were the chief scientist, to which of the geological provinces of Mars would you send your spacecraft to maximize the chance of finding living forms? Why?

9. Why did plate tectonics occur on Earth and apparently not on the other terrestrial planets?

10. Jupiter's four moons decrease in density and increase in content of volatiles with distance from Jupiter, much as the planets change with distance from the Sun. Why?

BIBLIOGRAPHY

Brandt, J. C., and S. P. Maran, *New Horizons in Astronomy.* San Francisco: W. H. Freeman and Company, 1972.

Carr, M. H., "The Geology of Mars," *American Scientist,* v. 68, pp. 626–636, 1980.

Jupiter Issue, *Science,* v. 204, no. 4396, June 1, 1979. (The "Reports" section contains many articles about the Voyager-Jupiter mission.)

Lewis, J. S., "The Chemistry of the Solar System," *Scientific American,* March 1974.

Murray, Bruce, M. C. Malin, and Ronald Greeley, *Earthlike Planets.* San Francisco: W. H. Freeman and Company, 1981.

Pettingill, G. H., D. B. Campbell, and H. Masursky, "The Surface of Venus," *Scientific American,* August 1980.

Scientific American, *The Solar System.* San Francisco: W. H. Freeman and Company, September 1975. (Many articles on the planets and smaller bodies of the solar system.)

Simmons, G., *On the Moon with Apollo 17.* Washington, D.C.: U.S. Government Printing Office (Stock Number 3300–00470), 1972.

Snyder, C. W., "The Planet Mars as Seen at the End of the Viking Mission," *Journal of Geophysical Research,* v. 84, pp. 8487–8519, 1979.

Soderblom, L. A., "The Galilean Moons of Jupiter," *Scientific American,* January 1980.

22

Matter and Energy from the Earth

Throughout the evolution of civilization, humans have steadily broadened their use of the mineral and energy resources of the Earth. Since the eighteenth century, the abundance of fossil fuels has provided the industrial nations of the world with cheap energy, but we are faced today with a declining supply of oil. In the next few decades nations will need to conserve energy and arrange for the difficult transition from oil to the more plentiful fossil fuels and nuclear energy, while we await the development of clean, safe, inexhaustible energy sources sometime in the next century. In the next few decades, the geological hunt for new supplies of oil, gas, and coal will be coupled with exploration for new uranium ore reserves and a search for alternative energy supplies. High-grade sources of many vital metal ores and other useful minerals are becoming harder to find, and conservation, recycling, substitution, and exploitation of costlier, lower-grade deposits will become necessary in the decades ahead.

Unlike other living things, humans have learned to use the materials of the Earth to modify their environment extensively. We mine metal ores, coal, and salt. We drill for oil and gas. We strip away surface deposits on land and evaporate sea water to get dissolved metals. In our increasingly systematic search of the globe for new sources of the materials we depend upon, we use our geological knowledge of how known natural deposits are distributed in order to find more of them. At the same time, we rely on a plentiful supply of energy for concentrating, smelting, and refining the materials from their natural, impure forms to the final products that are machined or molded into various components of our technology. The source of the energy is itself in the Earth as oil, gas, coal, or uranium ore.

The metals that we use for machinery, the phosphate rock prepared for fertilizer, and the salt used in food never disappear. Some we return to

Earth as refuse. Some dissolve in sanitary drainage waters that find their way into groundwaters and into rivers and eventually into the ocean, just like the dissolved materials from weathering. In this sense, humans are hyperactive agents of erosion. Even though the used material is not really destroyed, much of it is so widely dispersed that it is recoverable only with a huge expenditure of energy. For this reason, without the systematic practice of conservation inordinate amounts of energy may be required to gather together and refine used materials for recycling (Fig. 22–1). Energy, unlike matter, is not recoverable. It is permanently lost to Earth. Regardless of which process we use, electrical, kinetic, or chemical, energy is converted to heat and irretrievably radiated to space.

Because the prospect of maintaining or increasing our supply of energy underlies all resource problems, we will first review the existing and

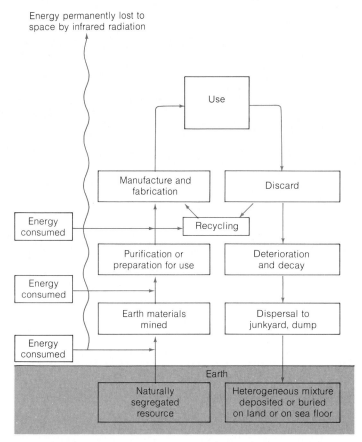

Figure 22–1
The cycle of mining, preparation, use, and discard of useful Earth materials is one in which a naturally segregated resource, such as an iron ore body, is mined, processed, and fabricated and then returned to Earth in a dispersed, heterogeneous mixture. Each step in deriving a useful object from the earth involves the consumption of energy.

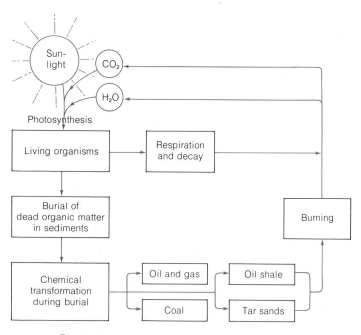

Figure 22–2
Photosynthesis produces organic matter that is buried, transformed, and so becomes a fossilized product of photosynthesis—a fossil fuel.

potential sources. Box 22–1 summarizes the various units used to measure energy.

ENERGY FROM FOSSIL FUELS

A wood fire, the extremely rapid oxidation of organic matter, is, in a superficial sense, deferred respiration. Throughout its lifetime, a tree respires, essentially as animals do, by slow, controlled oxidation of organic substances in cells. The decay of a tree following its death is another slow oxidation process, depending, like respiration, on the organic matter produced by photosynthesis. Thus we can look upon a piece of wood or any piece of plant matter as a photosynthetic product that can be returned, by respiration during life, by slow decay, or by burning, to the carbon dioxide and water from which it was made. If the wood was buried and transformed into coal (see Chapter 11) 250 million years ago and we burn it today, we are utilizing the energy stored

Box 22–1 MEASURING ENERGY*

Energy is used in a wide variety of forms, with different physical and thermal qualities and different capacities for mutual substitution. It is often convenient, however, to specify the quantity of energy in terms of a common unit. For this study, and most others undertaken in the English-speaking world, that unit is the British thermal unit, or Btu (the amount of energy required to raise the temperature of 1 pound of water 1°F from 39.2°F to 40.2°F). A barrel of crude oil, for example, contains about 5.8 million Btu. . . . When very large amounts of energy are discussed, it is convenient to use the unit quad, defined as one quadrillion (10^{15}) Btu.

The following table puts these quantities into perspective.

U.S. energy consumption in 1978

Energy source	Consumption		Conversion factor (values are equivalent to 1 quad)
	Standard units	Quads	
Coal[a]	623.5 million short tons	14.09	44.3 million short tons
Natural gas	19.41 trillion cubic feet	19.82	0.979 trillion cubic feet
Petroleum[b]	6838 million barrels	37.79	181 million barrels
Hydropower[c]	301.6 billion kilowatt-hours	3.15	95.7 billion kilowatt-hours
Nuclear power[c]	276.4 billion kilowatt-hours	2.98	92.9 billion kilowatt-hours
Geothermal and other[c,d]	3.3 billion kilowatt-hours	0.07	46.3 billion kilowatt-hours
Net imports of coke	5.0 million short tons	0.13	38.5 million short tons
Total[e]		78.01	

[a] Includes bituminous coal, lignite, and anthracite.
[b] Includes natural gas plant liquids and crude oil burned as fuel, as well as refined products.
[c] The conversions from kilowatt-hours to Btu's are necessarily arbitrary for these conversion technologies. The hydropower thermal conversion rates are the prevailing heat-rate factors at fossil-steam electric power plants. Those for nuclear power and geothermal energy represent the thermal conversion equivalent of the uranium and geothermal steam consumed at power plants. The heat content of 1 kilowatt-hour of electricity, regardless of the generation process, is 3413 Btu.
[d] Includes wood, refuse, and other organic matter burned to generate electricity.
[e] Details do not add to total due to rounding.
*Source: Reprinted from *Energy in Transition*, W. H. Freeman and Company, Copyright © 1980 by The National Academy of Sciences.

by photosynthesis from late Paleozoic sunlight. We are burning a "fossil." We can refer in the same way to all natural organic materials, ranging from coal to oil and gas, including oil shales, tar sands, and other geologic formations, as **fossil fuels** (Fig. 22–2).

The industrial revolution of the nineteenth century was powered by the energy from coal—in Britain, from the coal fields of England and Wales; in continental Europe, from the coal basins of western Germany and bordering countries; and in North America, from the Appalachian coal fields of Pennsylvania and West Virginia. As industrialization expanded, so did the hunger for coal. Geological exploration for this fuel spread over much of the world, from the Arctic deposits of Spitzbergen to Australia. Coal production climbed at an ever-accelerating pace.

Half a century after the first oil well in America was drilled in 1859, oil and gas were beginning to displace coal as the fuel of choice. Not only did they burn more cleanly, producing no ash, but they could be transported by pipeline as well as by rail or ship (Fig. 22–3). When oil exploration entered its heyday, the application of geology and geophysics to the search for prospects was combined with a rapidly developing innovative technology for drilling ever more deeply and rapidly. Understanding of the geologic conditions under which oil and gas are formed came quickly as data from oil exploration accumulated and inferences were drawn from the chemical composition of petroleum and natural gas.

PETROLEUM AND NATURAL GAS

Crude oil is made up of a great variety of **hydrocarbons**—compounds whose molecules are chains of carbon atoms with hydrogen atoms attached. Crude oil also contains some impurities, notably sulfur. The **paraffins,** a major subdivision of the hydrocarbons, are saturated compounds—that is, each carbon in the chain has attached to it all the hydrogens it can hold. The number of hydrogens in relation to the number of carbons is

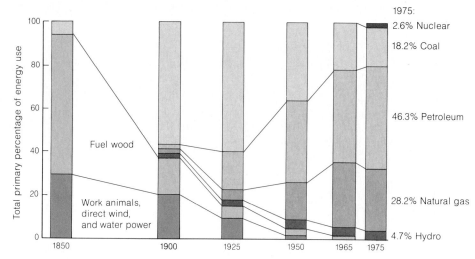

Figure 22-3
Changes in types of energy used from 1850 to 1975 as a percentage of total. Coal grew rapidly at the expense of wood, animal, wind, and water power to meet the needs of the industrial revolution. The explosive growth in the use of petroleum and natural gas and the reduced use of coal in the second and third quarters of the twentieth century reflect the cleanliness, convenience, and low cost of the liquid and gaseous fuels and the expansion of the transportation industry. The growth in the use of these fuels has now leveled off in response to the price increases of the past decade. Future needs will be met mainly by increased conservation to reduce energy growth rates, increased use of coal, increased use of nuclear fission energy in some countries, a gradual phase-in of solar and other renewable energy sources, and possibly energy derived from nuclear fusion in the next century. [Oak Ridge National Laboratory.]

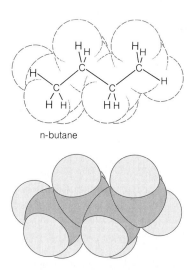

Figure 22-4
The structure of the hydrocarbon butane, a paraffin compound containing four carbon atoms and ten hydrogen atoms.

given by the general formula C_nH_{2n+2}. Methane, CH_4, or marsh gas, is the smallest and lightest member of the paraffin series. Methane and the other short-chain compounds are gases at room temperature and atmospheric pressure; examples are ethane, C_2H_6, and butane, C_4H_{10} (Fig. 22-4). Natural gas is a mixture of methane and other light paraffins. The mixture heptane, octane, and nonane, C_7H_{16} to C_9H_{20}, is what we know as gasoline. Lubricating oils are mixtures of still longer-chain paraffins.

How Oil and Gas Form

Small amounts of hydrocarbons and related compounds occur in organisms of all kinds, from algae to elephants. Other biological compounds can be transformed into hydrocarbons by organic chemical reactions in sediments. Petroleum, like coal, is a biological product—the organic debris of former life, buried, transformed, and preserved in sediments.

Petroleum forms when there is a favorable balance between the production of organic matter and its destruction by the scavenging of other organisms or by inorganic oxidation. This condition is met where productivity of organic matter is high, as in the coastal waters of the sea, where large numbers of organisms thrive, and where the supply of oxygen in bottom sediment layers is not enough to oxidize all organic matter. Many off-shore sedimentary basins on continental shelves—and perhaps also sedimentary accumulations on continental rises—satisfy these conditions; in such environments, and to a lesser degree in many others, small amounts of organic matter are buried and protected from oxidation. During the millions of years of burial, chemical reactions slowly transform some of the organic material into liquid and gaseous hydrocarbons. These fluids tend to be squeezed out of compacting muddy sediments into adjacent permeable beds, such as sandstones or porous limestones. The low density of oil and gas causes them to migrate to the highest place they can reach, where they float on top of the water in the pores of the formations.

The geologic environment that favors the large-scale accumulation of oil is a combination of structure and rock types that creates an impermeable barrier to upward migration—an **oil trap.** In order for the accumulation to qualify as an **oil field,** the oil-bearing formation must be permeable enough to allow the crude oil to be pumped out. The typical trap-forming structure is an anticline in which a permeable sandstone is overlain by an impermeable shale. The oil and gas accumulate at the highest part of the fold, the gas highest, the oil next, both floating on the water that saturates all permeable formations (Fig. 22-5). Other structures may trap oil in similar ways. For example, displacement at a fault may place a dipping permeable limestone bed opposite an impermeable shale, creating a trap for oil. A dipping sandstone bed may thin out against a shale and form a stratigraphic trap—one that is primarily the result of the original sedimentation pattern rather than of later structural deformation.

Geologists have mapped thousands of structural or stratigraphic traps, all conceivably favorable to oil accumulations, but only a fraction of them have proved to contain any oil or gas. The structures alone are not enough: the oil has to be available to migrate into them. There is a sequence of necessary geologic events that leads to oil accumulation, no one of which can be omitted, though the precise order of steps may vary.

1. Organic matter must be produced in some quantity in the environment.
2. Organic matter must be buried in the sediment before being destroyed by oxidation.
3. Some part of the organic matter must be transformed by slow chemical reactions to oil and gas.
4. The now-fluid organic matter in the sediment must migrate into nearby permeable beds.
5. The permeable beds must be deformed by folding or faulting to form a structural trap, or stratigraphic traps must be produced by conditions of sedimentation.
6. The oil-containing sediment must not be transformed into slate or higher-grade metamorphic rocks, or the organic matter will be transformed by heat into a nonhydrocarbon black carbonaceous material of no value.
7. Structural deformation in general must not be extremely severe. Intensely fractured rocks may render traps ineffective by causing leakage.
8. The beds must remain buried. If they are exposed to erosion, the oil may be dissipated.
9. The pore spaces of the permeable formations must not be so filled with swelling clays or mineral cements deposited by groundwaters that the permeability is too low to allow the oil to be pumped out of the formation.

With such a long list of requisites, one might wonder that there is as much oil as there is. Yet the probability of all those events happening in the right order is fairly high. A great quantity of organic matter *is* produced and buried in the sediments on and around continents, and the normal pattern *is* for that organic matter to be transformed slowly to oil. Stratigraphic and structural traps abound in most sedimentary basins. Although sediments are intensely deformed in mountain chains, the greater part of any continent consists of sediments of all ages that are only gently deformed and are subjected at most to mild warming by deep burial. Thus oil is not an unusual geologic product. Rather, we expect to find it where conditions are right, and the successful results of exploration during the past century have borne out those expectations.

The World Distribution of Oil and Gas

If you were to visit the exploration and research offices of any large oil company, you would find maps and reports of all the sedimentary rock terrains in which the company has operated, for where there is sediment, there may be oil. Thirty-one of the fifty United States produce commercially marketable oil, and some small occurrences are known in most of the others. Canada produces

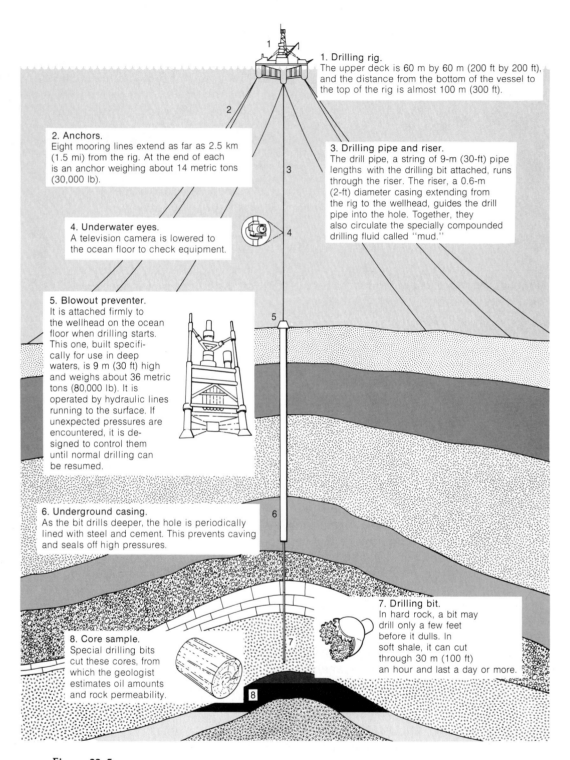

1. Drilling rig.
The upper deck is 60 m by 60 m (200 ft by 200 ft), and the distance from the bottom of the vessel to the top of the rig is almost 100 m (300 ft).

2. Anchors.
Eight mooring lines extend as far as 2.5 km (1.5 mi) from the rig. At the end of each is an anchor weighing about 14 metric tons (30,000 lb).

3. Drilling pipe and riser.
The drill pipe, a string of 9-m (30-ft) pipe lengths with the drilling bit attached, runs through the riser. The riser, a 0.6-m (2-ft) diameter casing extending from the rig to the wellhead, guides the drill pipe into the hole. Together, they also circulate the specially compounded drilling fluid called "mud."

4. Underwater eyes.
A television camera is lowered to the ocean floor to check equipment.

5. Blowout preventer.
It is attached firmly to the wellhead on the ocean floor when drilling starts. This one, built specifically for use in deep waters, is 9 m (30 ft) high and weighs about 36 metric tons (80,000 lb). It is operated by hydraulic lines running to the surface. If unexpected pressures are encountered, it is designed to control them until normal drilling can be resumed.

6. Underground casing.
As the bit drills deeper, the hole is periodically lined with steel and cement. This prevents caving and seals off high pressures.

7. Drilling bit.
In hard rock, a bit may drill only a few feet before it dulls. In soft shale, it can cut through 30 m (100 ft) an hour and last a day or more.

8. Core sample.
Special drilling bits cut these cores, from which the geologist estimates oil amounts and rock permeability.

Figure 22–5
Diagrammatic view of an offshore drilling rig. Natural gas (brown) and oil (black) are trapped in a fold with an impermeable layer above the permeable, oil producing formation. Oil floats above the groundwater line. With the exception of the special equipment in the water, the drill hole is drilled much as in land-based drilling. [Courtesy Exxon Corporation.]

Crude Oil World Total: 641.6 billion barrels

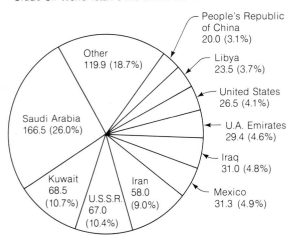

Figure 22-6
Estimate of world crude oil proved reserves (oil known to be recoverable). As of December 31, 1979, proved reserves were 641.6 billion barrels. Saudi Arabia with 166.5 billion barrels was the largest holder of reserves, followed by Kuwait, the U.S.S.R., and Iran. The U.S. ranked eighth with 26.5 billion barrels or 4.1 percent of the world total. Crude oil reserves listed here differ from data in Box 22-2 because natural gas liquids are not included in the latter and because the estimates come from different sources. [U.S. Department of Energy.]

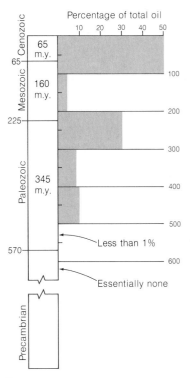

Figure 22-7
Percentage of the world's total past production and proven reserves of oil by age distribution in 100-million-year intervals. More than 80 percent of the world's known oil is found in rocks formed in the last 6 percent of geologic time.

oil from many of its provinces. Most of the countries of South America and Europe produce oil. Once considered oil-poor, Africa now produces a respectable quantity. Indonesia has many oil fields, and exploration of parts of Southeast Asia looks promising. This is not to say, however, that the world distribution is uniform; rather, it is most uneven. The two richest and most important oil-producing regions are the Middle East and the area between the Gulf of Mexico and the Caribbean. The Middle East includes about two-thirds of the known reserves of the world in the oil fields of Iran, Kuwait, Saudi Arabia, Iraq, and the Baku Region of the U.S.S.R. The highly productive Gulf Coast-Caribbean area includes the Louisiana-Texas province, Mexico, Colombia, Venezuela, and Trinidad (Fig. 22-6).

Why this uneven distribution? A political scientist once jokingly asked one of us, "Why is it that oil is always found in underdeveloped countries?" The distribution of oil, of course, is determined not by political boundaries but by geologic history. The age range of sediments from which oil is produced gives some idea of the reasons for the distribution. There is practically no oil in Precambrian rocks and very little in Cambrian rocks.

Rocks of the Triassic Period also contain very little.* Oil produced from rocks of these periods amounts to less than one percent of the world's total production. For all practical purposes, we can say that no oil is produced from Pleistocene rocks. But Cenozoic sediments, those deposited only in the last 70 million years of the Earth's 4.7 billion year history, account for about 60 percent of the total. Another large fraction, about 25 percent, comes from rocks of Mesozoic age, the next older era. Only about 15 percent comes from Paleozoic rocks (Fig. 22-7).

The correlation between geologic age and oil production is understandable in terms of the processes that destroy oil or allow it to disperse at the surface. The older the rock formation is, the more likely it has been eroded or involved in metamorphism and deformation that would eliminate the oil. The Cenozoic thus may represent the normal abundance of oil. Older oil-producing formations are those that have escaped the ravages of

---

*It is tempting to speculate that the lack of oil in Triassic sediments is related to a time of early rifting of Pangaea, when there was much less continental shelf and rise available in a single large continent.

time—exposure to destructive geologic events. If we were living in the Triassic period, we would probably find oil most abundant in late Paleozoic rocks.

Oil in the Middle East is concentrated largely in thick Cenozoic and Mesozoic sediments deposited in and adjacent to a major geosyncline situated along the southern margin of the Asian continent. Sedimentation has been accompanied by a variety of deformation structures and stratigraphic traps. There is no evidence of metamorphism, and relatively few of the sediments have been destroyed by erosion.

Continental Shelf Oil

Continental shelves are good places to explore for oil, as has been known since the general nature of their structure was worked out in the 1940s and 1950s. They easily meet the requisites for accumulation: the sediments are young, unmetamorphosed, mildly deformed, and little eroded. The shelves of the world have been geophysically combed for favorable structures by the major oil companies in anticipation of the large amounts of oil expected to be found there. Yet most shelf areas are still undeveloped. For example, less than 5 percent of the U.S. continental shelf has been leased by the government for commercial development. Drilling on continental shelves is not without its problems. A major one is the difficulty and expense of drilling in deep water. At this time, wells in offshore areas of Louisiana, Texas, and California easily can be drilled in water as much as several hundred meters deep. That depth can be increased to 2000 meters with known technology. The costs of drilling platforms with deep water capability are in the 100 million dollar range. The *Glomar Challenger*'s equipment, which has been used to drill the holes for the deep-sea drilling program, has served as a prototype for this technology.

Another major problem of offshore drilling is the prevention of oil pollution. The name "Santa Barbara" calls to mind the environmental damage that resulted from the accidental release of oil from an offshore drilling platform there in 1969. In 1979 as the result of a "blowout" of a well being drilled in the Gulf of Mexico off the Yucatan, as much as 100,000 barrels of petroleum per day escaped for many weeks before the well could be capped. Though it is difficult if not impossible to guarantee a "100 percent safe" drilling well, it *is* within the realm of possibility to devise technological systems and controls that would greatly reduce the chances of a serious accident. If there is adequate development and enforcement of safety procedures, offshore drilling need not pose an important threat to the oceans or nearby beaches. One analysis of the chances of oil spills on the East Coast of North America predicts that less oil would be spilled from drilling platforms and offshore pipelines than from fleets of oil tankers required to supply coastal states. (This assumes no major events such as that of Santa Barbara.) Flushing of oil tanks by ships at sea probably poses a more serious current danger. The important point is that there are large reserves of oil and gas under the sea bed that will eventually have to be drilled to satisfy the world's needs. We *can* tap those supplies without severe cost to the environment if precautions are taken. There is no doubt that we will be pushed to explore every conceivable oil resource as present reserves are depleted, whether our increasing demands are slowed or not.

Oil: The Exhaustible Supply

Inconceivable as it may seem to the landowner who sees a "gusher" spurting oil from the derrick on his property, that oil well will eventually run dry. What the world wants to know is when all the wells will run dry. Behind this loaded question are the facts of oil occurrence on Earth. Thirty years ago, many oil producers were optimistic; supplies seemed so immense, and so many scientific and engineering innovations held promise of tapping new sources, that no one needed to worry about the future. Since then world demand has accelerated rapidly, and a more sober analysis of the total quantity of oil remaining on Earth has changed the picture. Because the process of oil formation is so exceedingly slow, it is clear that we cannot rely on natural processes to replenish the oil we pump out of the ground at so rapid a rate. We are dealing with a fixed supply that steadily dwindles as we use it. Three significant questions that need to be answered are: how great is the total supply; how much of it have we already used; and at what rate do we expect to continue depleting the supply?

How does a petroleum geologist go about estimating the total amount of recoverable oil in the Earth? The individual oil well that identifies a new oil field gives only a hint of the ultimate size of the field. But from it the geologist learns how thick the oil-saturated porous beds are and how permeable they are, the permeability being a good guide to the ease of pumping the oil. As more

Box 22-2   NONRENEWABLE ENERGY RESERVES AND RESOURCES: WORLD AND UNITED STATES

These estimates were compiled from various recent studies. Data are presented in conventional units for each resource; the energy unit quad (10^{15} British thermal units) is also used to make comparisons easier. Although estimates are given to the nearest unit, differences between various estimates may be as much as 20 percent, or even more if lower-grade, high-priced, more speculative resources are considered.

	Proven resources currently recoverable	Estimated total resources reaining
Conventional crude oil (billion barrels)		
World	680 (3944 quads)	1696 (9839 quads)
U.S.	34 (201 quads)	115 (667 quads)
Natural gas (trillion cubic feet)		
World	2220 (2287 quads)	7423 (7572 quads)
U.S.	214 (221 quads)	695 (716 quads)
Coal (billions of short tons)		
World	716 (14,320 quads)	6755 (135,000 quads)
U.S.	219 (4,818 quads)	1410 (31,020 quads)
Syncrude from oil shale and tar sands (billions of barrels)		
World	265 (1540 quads)	2320 (13,500 quads)
U.S.	76.5 (444 quads)	1041 (6,038 quads)
*Uranium (thousands of short tons uranium oxide at costs up to $50/lb)		
World	2415 (966 quads)	4813 (1925 quads)
U.S.	600 (240 quads)	1843 (737 quads)

*If used in breeder reactors, energy content is increased by 60 to 100 times. Uranium resources do not include U.S.S.R. and China.

wells are drilled, the productive areas are outlined, and the volume of oil in the oil field becomes better known. Experience with older wells in the same geologic province gives some idea of how much oil might eventually be produced, for not all of the oil in the ground can be recovered, though improvements in recovery engineering are constantly being made.* The general abundance and distribution of known oil fields in the various kinds of sedimentary basins are guides to estimating existing reserves (Fig. 22-6 and Box 22-2). For example, exploration has been so thorough in the older oil provinces, such as those of Pennsylvania, Illinois, Oklahoma, and Texas, that there is little doubt that more than 90 percent of the oil there has already been discovered (though not all of it has been pumped). Such fields are **proven reserves,** those that have been geologically outlined and at least partially drilled. Estimates of proven reserves are constantly being updated as the balance changes between the increase from new discoveries and the decrease by production at pumping wells.

The next step is the geological estimate of the future exploration of as-yet-untapped oil fields.

---

*Oil companies are constantly seeking new engineering technologies to enhance the recovery of oil from existing oil fields. Water may be pumped into the periphery of an area to force more oil out of a central well. In order to loosen up very viscous oils and allow them to flow more easily, steam or CO_2 may be injected under pressure and then allowed to escape with the newly fluidized oil. All these methods have increased the proportion of oil that can be recovered, but we are still far from being able to get all of it.

With more than a hundred years experience in drilling oil and the increase in the geological knowledge of the world, geologists can make a good guess about how much oil is left to find. Geological mapping of the land areas of the world is essentially complete, at least on a general scale, and we know where all of the sedimentary basins are, both developed and undeveloped. The large discoveries of the past decade in Alaska were welcome but not exactly a great surprise, for geologists had known for years that the area was favorable for oil exploration.

The companion to geological mapping of exposed formations on land is geophysical exploration. Since the early 1920s the main practical use of seismology has been in oil-field exploration (see Box 17-1). Reflected seismic waves reveal subsurface structures and formation changes not apparent from surface mapping. This ability to "see" beneath the surface was utilized fully in the exploration of the continental shelves for structures favorable to oil accumulation. Newer, more sensitive seismic profiles sometimes show the presence of water-oil-gas interfaces in buried formations. The combination of geology and geophysics has made possible reliable estimates of all the oil that may eventually be discovered and produced on land, in the ocean, and under glacial ice—what we call **petroleum resources.** In spite of all of the major new discoveries in Alaska and elsewhere, the known petroleum resources of the world have not changed much in the past decade, increasing confidence in the estimates. Of course,

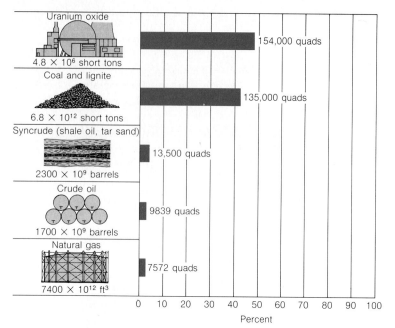

Figure 22-8
Remaining nonrenewable world energy resources (assuming nuclear breeder reactors) amount to 320,000 quads according to data in Box 22-2. Energy content is given in conventional units, percentage of total, and in quads. Coal and lignite resources, for example, amount to 6.8×10^{12} short tons, equivalent to 135,000 quads or 42 percent of the total.

we cannot completely rule out discoveries of a totally unpredictable kind, for example, in the deep water sediments of the continental slope and rise or below overthrust belts, but we would be unwise to count on them until they are found.

How Much Oil Is Left?

Recent estimates of the world's remaining oil resources are shown in Box 22-2 and Figure 22-8. Huge as the total, nearly two trillion barrels, may seem, projections for the future must be based on the rate at which we are now using oil, and the anticipated future rate. From past production figures we know that the rate of use has jumped with steadily expanding industrialization. It took a little more than 100 years of oil production to pump out and use the first half of all oil ever produced, but it took only another 10 years to pump the remaining half. At current world rates of usage (22 billion barrels per year), the remaining oil will be mostly depleted in some 80 years. However, with expanding world population, and the aspirations of both rich and poor countries for expanding economic growth, usage is likely to increase, shortening the depletion time, perhaps to about 60 years.

Estimates of this kind have some limitations. They do not, for example, take into account oil production from unconventional sources such as oil shale and tar sand. These resources exceed conventional crude oil resources (Box 22-2 and Fig. 22-8). Also, present methods of production leave more than half of the conventional oil in the ground. If the world economy could tolerate the higher price, much of this oil and the unconventional oil could be produced, providing liquid fuels into the second half of the next century. Predictions of future supply and demand are all the more uncertain because of the vulnerability of supply to political instabilities, the dependence of discovery and production on cost-price considerations, the control of prices by the OPEC cartel, and the sensitivity of demand to price.

In the United States, which consumes more than a fourth of the world's oil production, oil supplies 37 quads of energy per year, or 47 percent of the total annual energy consumption. In recent years this has amounted to about 17 million barrels of oil per day, of which about 7 million barrels per day were imported.* Most of the oil is used in transportation (~52 percent), followed by industry (~21 percent), residential and commercial (~18 percent), and electric power generation (~9 percent). Although some three-fourths of the world's conventional crude oil has not yet been exploited, the unexploited proportion in the U.S. is much less than that, perhaps only 20 percent, for the U.S. was one of the first countries to explore and produce oil systematically and efficiently. At current usage, U.S. proven crude oil reserves are equivalent to about 10 years of supply, and its resources to about 30 years of supply.

The world political and economic consequences of these facts of oil supply will continue to be a major factor in the next decades. Every student of international politics should ponder its implications. Yet oil does not tell all the story, even though we depend heavily on it today. There are enormous supplies of energy from other fossil fuels: coal, oil shale, and tar sands.

Natural Gas

Natural gas reserves and resources are comparable to those of crude oil (Box 22-2 and Fig. 22-8). However, the world reserve position in natural gas is favorable in that natural gas is a relative

---

*Conservation, in part due to increased oil prices and more efficient use of fuel, has produced a remarkable reduction in 1981 imports to 6 million barrels per day.

newcomer on the energy scene; it has been used on a large scale only in the U.S. and U.S.S.R. Because it is nonpolluting and easily transportable, natural gas is a premium fuel. It therefore should be conserved for use in residential heating and as chemical feedstock rather than for the generation of electricity or other uses for which the more plentiful coal or nuclear energy resources are available. In the U.S., some 20 quads of natural gas are used annually, most by industry ($\sim$39 percent), residential ($\sim$28 percent), electric power generation ($\sim$18 percent), and commercial ($\sim$13 percent). More than half of American homes and a great majority of commercial and industrial premises are connected to the gas distribution system. Resources should last about 35 years at current usage. However, resource estimates have been expanding in recent years; so the potential may be large for unconventional natural gas sources, produced from formations not exploited before, such as coal, tight (that is, relatively impermeable) sandstones, and shales.

COAL

Coal has been used as fuel for millennia, perhaps since early inhabitants of Wales used it for funeral pyres about 3000 or 4000 years ago. By the eighteenth century it was well established as a fuel for heating houses, and it became the mainstay of the industrial revolution in the nineteenth and early twentieth centuries. Coal was discovered in North America by several early expolorers; the first were Louis Joliet and Jacques Marquette, who found it in Illinois in 1673. Major production began in the 1800s, and increased steadily with demands brought on by the emerging steel industry, the development of the railroad steam engine, increased use for general heating, and finally the growth of electrical power. As of 1978, a total of about 162 billion short tons (147 billion metric tons) of all kinds of coal had been mined worldwide. The increase in the rate of use slackened in the twentieth century, largely because oil and gas supplanted coal for so many uses. There is now a resurgence in coal production prompted by the oil reserve and price picture.

It is far easier to estimate coal reserves than oil and gas reserves. Like other sedimentary layers, coal beds extend over large areas, hence they can easily be mapped from surface exposures and their subsurface extent inferred. There have been much mining and exploration drilling for coal; also, many wells drilled for oil have penetrated coal-bearing formations. Because coal-explora-

tion geologists have a good idea what kinds of sedimentary rock sections include coal and where to look for them (see Chapter 11), they are likely to be more concerned with thickness of the deposit and quality of coal than with its presence or absence. For all of these geological reasons, accurate estimates of the world's coal reserves are easier to make. At the same time, they are more complicated. Reserves are estimated with respect to ease of recovery and to the heat value of the coal. Commonly, detailed evaluations are made by thickness categories of the coal that might be mined, by the depth of rock overburden in the case of strip mine reserves, and by the quality of the coal. The last factor is usually displayed in terms of the rank, or metamorphic grade, of the coal. This ranges from lignite, the lowest stage, through the bituminous coals to the high-rank anthracites (see Chapter 11). The bituminous coals, the largest category, have been subdivided according to the relative proportion of volatile gases released from the coal when it is heated. The less the volatile content, the higher the heat value of the coal in general and, therefore, the higher the grade. Volatile content is also related to the suitability of a coal for manufacturing coke, a product important for iron ore smelting.

According to estimates (Box 22–2, and Fig. 22–8), about 6.75 trillion short tons of coal remain in the world. Thus we have so far used up only about 2.5 percent of the world's initial supply. United States resources are about 20 percent of the world total (Fig. 22–9). These figures include only coal deposits minable by present technology. Trillions of tons more coal exists, but the deposits are either so deep or in such thin beds that we cannot easily mine them by current methods. New technologies may allow the recovery of such coal in the future.

Technologies exist to convert coal to gaseous or liquid fuels comparable to those derived from petroleum today. The cost of synthetic oil from coal is higher than crude oil at today's prices, but with depletion of world reserves and political instability in the Middle East, the gap between the two is decreasing. A number of commercial scale synthetic oil and gas plants are now under construction in the U.S. It may well be that several million barrels per day of synthetic oil will be produced from coal in the U.S. before the turn of the century, and the potential for the next century is much larger.

The problems involved in the recovery and use of coal—problems that make it less desirable than oil or gas—still remain, whether the coal is burned or converted to liquid. Much coal contains

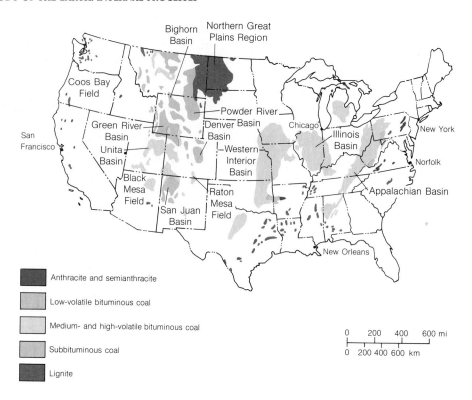

Figure 22-9
Coal fields of the contiguous United States. [U.S. Geological Survey.]

appreciable amounts of sulfur combined in the mineral pyrite (FeS_2), which vaporizes during combustion to liberate noxious sulfur oxides to the atmosphere. Acid rain, formed from the combination of these gases with rainwater, is becoming a serious problem in the northeastern U.S., Canada, and Scandinavia. Many lakes in these regions have become so acidic due to acid rain fallout that native fish no longer survive. Coal ash, which contains metal impurities in coal, some of which are toxic, amounts to several percent of the weight of the unburned coal and poses a significant disposal problem. Coal mining, particularly strip mining, can ravage the countryside. Underground mining accidents take the lives of miners each year, and many more suffer from "black lung" and other pulmonary diseases. Carbon dioxide released during combustion of fossil fuels may trigger climatic changes in the next century (Chapter 12). These human costs are as important to consider as the possible dangers of other energy sources such as nuclear reactors.

Except for the CO_2 problem, the technology for the "clean" combustion or liquefaction of coal, for restoring land ravaged by strip mining, and for reducing the danger to miners exists. It is expen-

sive and will add to the cost of coal, but these two drawbacks are unlikely to prevent the increased use of this important fuel, for it is one of the largest resources we have.

OIL SHALE AND TAR SANDS

Known for many years but practically untouched are large reserves of oil, called syncrude, that can be extracted from oil shale and tar sands. Oil shales are fine-grained sedimentary rocks containing a high proportion of solid organic matter mixed with the various minerals; some contain small amounts of liquid oil. This organic matter, called **kerogen,** has the same general origin as the other common organic materials, all ultimately derived from plant and animal matter. This complex and poorly understood solid mixture of carbon compounds has an extremely important property that is of practical value: if it is heated and the vapors distilled, oil can be recovered from the shale. Some oil shales can produce up to 150 gallons of oil per ton, but most yield about 25 to 50 gallons per ton.

Reserves of oil shales are known moderately

Figure 22-10
Typical cliffs in the principal oil-shale group of the Green River Formation.
Garfield County, Colorado. [Photo by D. E. Winchester, U.S. Geological Survey.]

well. Like coal, they occur as beds whose extent can be fairly reliably predicted. They are less abundant than coal, however, and have not been well explored in many parts of the world. Because many oil shales are too low in recoverable oil to make mining economical, the total practical resource is far less than the total amount of oil shale.

A little less than half of the world's supply of oil shale is in the Green River Shale of Eocene age in Wyoming, Utah, and Colorado (Fig. 22-10). Though pilot plants have been in operation for years, construction of the first commercial scale plant began in 1980. Like synthetic oil derived from coal, syncrude extracted from oil shale will be more expensive than crude oil in the near term, but the prices will probably converge with time. Disposal of immense tonnages of waste shale from the extraction process presents a serious problem, as does the need for large amounts of water. The combination of difficulty and expense of extraction and possible environmental problems makes it unlikely that syncrude will displace more than a few million barrels of crude oil per day in the next 15 to 20 years, but this is a significant amount.

A more unusual occurrence of organic matter usable as fuel is an asphalt or **tar sand.** Oil of a type similar to liquid petroleum can be recovered from these sands, and geologic evidence suggests that tar is a transformed product of a once-liquid petroleum. Some deposits of tar sands are oil pools that have "dried up" and become tarry by loss of the volatile hydrocarbon fractions; some are exposed at the surface by erosion. Others show signs of having been deposited as a combination of sand and oil, the oil having seeped from older formations. Some of the largest deposits, the Athabaska tar sands of Cretaceous age in Alberta, Canada, may be of this origin. The Athabaska deposit is a great resource that is now beginning to be exploited. Estimates are that 300 billion barrels of oil are recoverable from this one region. Other major deposits are known to exist in Venezuela. Tar sands are expensive to recover, and separation of the tar from the sand is difficult. As with oil shale, economic and environmentally safe recovery technology will have most to do with its future as an energy resource.

World syncrude resources derived from oil shale and tar sand are half again as great as all crude oil resources (Box 22-2, Fig. 22-8). Nearly

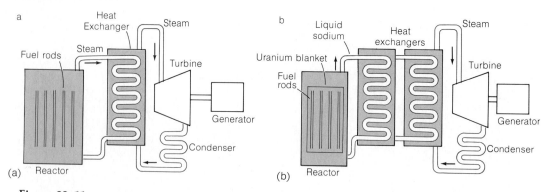

Figure 22-11
In a nuclear power plant (a) the fission of uranium-235 releases the energy to make steam, which then goes through the same cycle as in a fossil-fuel power plant. Under development are nuclear breeder reactors (b), in which surplus neutrons are captured by a blanket of nonfissile atoms of uranium-238 or thorium-232, which are transformed into fissile plutonium-239 or uranium-233. The heat of the reactor is removed by liquid sodium. [From "The Conversion of Energy" by C. M. Summers. Copyright © 1971 by Scientific American, Inc. All rights reserved.]

half the world's syncrude resources are in the U.S. This is a respectable quantity and clearly a significant future resource.

THE FUTURE OF FOSSIL FUELS

One of the currently popular occupations of geologists, economists, and others concerned with energy sources is to estimate our future dependence on fossil fuels. If crude oil and gas continue to be used as the major resource for satisfying the world's continually climbing energy appetite, the great bulk of the world's supply will be exhausted in half a century. Oil shale and tar sands will add some decades, but coal will become the predominant fossil fuel resource. It is reassuring to know that at modest annual energy growth rates, say three percent per year, coal and other fossil fuels can meet the world's energy needs for about 100 years, or even longer.

For the United States, the energy problem is acute, for we will have used up the bulk of our oil supply soon after the turn of the century. However, if our huge coal and oil shale resources can be developed expeditiously, our energy needs can be met totally from our own fossil fuel resources for over a hundred years.

These estimates do not take into account the possibility of meeting much of our energy needs through the development and use of alternate energy sources such as nuclear fission, nuclear fusion, solar, geothermal, and biomass. To the extent that these can be brought on stream, the pressure on our fossil fuel resources can be relaxed and their life extended.

NUCLEAR POWER

Nuclear physicists foresaw the possibility of peaceful uses of the gigantic amounts of energy stored in the atomic nucleus even before they were able first to liberate it in the form of an atomic bomb in 1944. Their predictions have started to come true as countries all over the world are building new nuclear reactors (Fig. 22-11). If its full potential is realized, nuclear energy can meet the world's energy needs for hundreds of years. However, reactor safety, radioactive waste disposal, and the possible diversion of nuclear fuel to nuclear weapons are political and technological problems that must be solved before we can light up the world with all the power it wants. One aspect of nuclear energy is most definitely in the geologic province: the question of reserves of nuclear fuel—in particular, uranium.

The older and still most widely used nuclear energy plants use the fission of Uranium-235 (^{235}U), the radioactive isotope whose nucleus spontaneously splits. When enough of this isotope is brought together in a carefully engineered assembly, called a pile, the controlled fission reaction proceeds steadily to produce a great quantity of energy, which is released as heat. Fission of one gram of ^{235}U liberates energy equivalent to about 2.7 metric tons of coal or 13.7 barrels of crude oil. The uranium is "used up" by this process, in the sense that it is converted to various fission products, other elements that cannot be used for energy production. An average nuclear power plant of this type, producing 1000 megawatts of electricity, uses up about 3 kilograms (6.6 pounds of ^{235}U each day. Uranium-235 constitutes only one atom out of every 141 of the average mixture of uranium isotopes mined (usually referred to as U_3O_8, or yellow cake), so the amount of U_3O_8 production needed to support a large number of reactors is large.

Uranium is present in very small amounts in the Earth's crust, constituting only 0.00016 percent of the average crustal rock. It is typically found as small quantities of the mineral uraninite (UO_2), frequently called by its common name, pitchblende, in granites and other felsic rocks and associated hydrothermal veins. Under near-surface groundwater conditions, this uranium may become oxidized and dissolved, transported in the groundwater, and later reprecipitated as uraninite if it is reduced by organic matter, which it may encounter in sedimentary rocks. The richest ores in the United States occur in the Triassic and Jurassic sedimentary rocks of the Colorado Plateau in western Colorado and adjacent parts of Utah, Arizona, Wyoming, and New Mexico. Rich Canadian deposits are found both in hydrothermal veins in the Great Bear Lake region of the Northwest Territories and Precambrian conglomerate sediments north of Lake Huron in Ontario.

In a 30-year useful lifetime, the average reactor will require 5600 tons of U_3O_8. Thus the estimated uranium resources of the U.S. (Box 22–2) are enough to support only about 300 such reactors. World resources could sustain some 900 reactors. At the present time it is estimated that some 150 to 200 reactors will be operating in the U.S. by the year 2000 (compared to the present 70), but much depends on political developments of the next decade. By early in the next century, uranium resources could limit the use of nuclear energy unless a reactor much more efficient in the use of nuclear fuel, such as the breeder reactor, is introduced.

The breeder reactor will make the picture brighter because it does not simply use up the ^{235}U, as do the fission reactors. In the breeder, ^{235}U fission is used to create new radioactive elements, such as plutonium-239 from uranium-238, and more nuclear fuel is produced than consumed. In addition, breeder reactors have a broader range of usable ores; the entire supply of natural uranium and thorium can be used instead of only ^{235}U. Breeder reactors can extend uranium resources by 60 to 100 times. Breeders are under active development in various countries, and some are already producing power. Technological problems remain, particularly in costs and in the important area of reactor safety, but there is reason to believe that the breeder can be made safe and efficient as a major supplier of power for the world in the next century.

Nevertheless, there remain some serious questions about the future of breeder reactors. The plutonium produced by the breeder is the same material that is used to manufacture nuclear bombs. A mere 25 kilograms (55 pounds) of plutonium is enough to make a big bomb, and each reactor might contain as much as 2.5 tons of this strategic material. Security of such a storehouse of plutonium in many widely scattered plants in many countries is not an easy problem to deal with. Finding ways to avoid widespread nuclear proliferation among many countries—and even among criminal groups—will be a central political issue in nuclear energy. The toxicity and radioactivity of reactor wastes are other areas of concern. They point to another issue that must be faced when we expand our nuclear power plant: how to dispose of these wastes (see Chapter 12). Although most scientists believe that geological containment in deep, stable, impermeable formations can be made to work, there is as yet no generally agreed upon plan for storage of high-level wastes for the hundreds or thousands of years required before they cease to be dangerous. No one wants to leave to our next generations the combined legacy of depleted energy resources and an unmanageable environmental hazard.

Beyond the breeder, nuclear physicists are looking to the possibility of building a fusion reactor, utilizing nuclear processes similar to those that fuel the Sun. The most promising approach—fusing the heavy hydrogen isotopes deuterium and tritium into helium with immense production of energy—requires achieving temperatures ten times those at the Sun's core and finding methods to contain the fuel at these temperatures. "Fuel" for fusion reactors is available: deuterium is abundant as a minor constituent of the water of the oceans, and tritium can be made in a reactor. The fusion reactor of this type awaits the application of science and engineering to build a workable reactor that will make energy so abundant and cheap that the world need not worry about an energy crisis. Unfortunately, there is no guarantee that fusion reactors can be made to work, but scientific progress is encouraging.

SOLAR ENERGY

Since all of our fossil energy sources come from the Sun anyway, why not convert its rays to energy? In principle, the Sun can provide us with all the energy we need, in all the forms we use: heat, electricity, and liquid or gaseous fuels. Solar energy is risk-free and nondepletable—the Sun will continue to shine at least for the next several billion years.

In the near term, the only form of solar energy for which the technology is available at costs nearly competitive to other sources is as heat for domestic space heating and hot water and for industrial and agricultural processes. Many homes and factories are beginning to use solar energy for these purposes, spurred by government tax credits and other incentives.

Electricity can be generated by solar energy in several ways: photovoltaic conversion (solar cells of the kind that power space satellites); solar thermal conversion, for example, concentrating sunlight to convert water to steam, which drives an electric generator; windpower. All of these systems can be built now and are being used for special purposes where costs are no impediment. However, solar-generated electricity is as yet too expensive for general use, for it costs several times as much as conventionally generated electricity. It must await technological advances to become competitive. Central solar electric power plants present significant environmental problems. A plant with 100 megawatt electric capacity (about 10 percent of the capacity of a nuclear power plant) located in the southwestern desert would require at least one square mile of land and might alter the local climate by significantly changing the solar radiation balance in the area.

Solar energy can be used to extract hydrogen from water by direct photochemical conversion. Hydrogen so produced could be used directly as a fuel or could be combined with carbon to synthesize hydrocarbon fuels. This approach has enormous potential but an economical technology has yet to be developed. Solar energy is stored in biomass, which can be obtained from municipal and agricultural wastes, or from crops grown for their energy content. Energy can be extracted by direct combustion or by conversion of biomass to gaseous or liquid fuels such as methane or alcohol. Using garbage as part of the biomass also puts waste material to good purpose. The technology for some of these processes is well developed, but the prices are not yet competitive without some form of government subsidy like those used for nuclear power. Energy farming presents ecological and ethical problems—the need for water, fertilizer, and the displacement of food crops. Hydroelectric power is a form of solar energy, delivering about 3 quads or about 4 percent of the U.S. annual energy consumption. It is clean and relatively riskless and cheap. Significant expansion of the present capacity would be resisted because it would involve the drowning of farmlands and wilderness areas in reservoirs behind dams.

How much solar energy can we depend on in the years ahead? Solar energy enthusiasts believe that, in the United States, some 20 quads per year (about half our present use of oil) might be supplied by the year 2010. Others think it more realistic to figure on less than 10 quads per year. All agree that under either scenario, important social benefits are realized: conservation, diversity of energy supply, and reduced imports. With adequate research and development, solar energy can probably become economically competitive and a major source of energy in the next century.

GEOTHERMAL POWER

In Chapter 13 we saw that Earth's internal heat, fueled by radioactivity, provides the energy for plate tectonics and continental drift, mountain-building and earthquakes. It can also do work for humankind.

The geothermal heat contained in the accessible part of the Earth's crust (that is, shallower than about 6 kilometers), at temperatures above 80°C is considered potentially producible. In the United States, this amounts to some 23,000 quads of energy. A significant fraction of this large quantity may eventually be used to generate electricity and to heat buildings. Only about one-fortieth of a quad per year of geothermal energy is currently produced in the United States, and perhaps twice this amount in the entire world. Geothermal energy will undoubtedly become increasingly useful in the years ahead as fuel prices rise.

By far the most abundant form of geothermal energy occurs at relatively low temperatures of 80° to 180°C, as a consequence of the normal increase in temperature with depth—the geothermal gradient, which averages about 25°C per kilometer. Water circulating through this heat reservoir, either naturally or artificially induced, can be used to extract heat for heating and cooling residential, commercial, and industrial space (Fig. 22–12). In France, more than 20,000 apartments are now heated by water drawn from the hydrothermal aquifer of the Paris Basin. Reykjavik, the capital of Iceland, is entirely heated by geothermal energy.

Geothermal reservoirs with temperatures above 180°C are useful for generating electricity. They occur primarily in regions of recent volcanism as

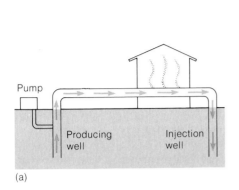

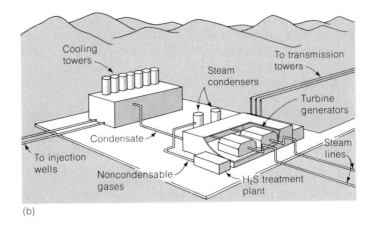

Figure 22-12
Applications of geothermal power. (a) Low-temperature hot water is used to heat
a building and then reinjected into the reservoir. (b) High-temperature steam drives
turbines to generate electricity. Toxic gases are removed before the cooled, condensed liquid is reinjected. [U.S. Geological Survey.]

hot, dry rock, natural hot water, or natural steam.
The latter two sources are limited to those few
areas where meteoric water circulates through
underground faults and fractures to reach deep
rocks heated by recent magmatic activity. Naturally occurring superheated water and steam are
highly prized resources for which geologists and
geophysicists are searching. The world's largest
supply of natural steam occurs at The Geysers, 120
kilometers (75 miles) north of San Francisco (Figs.
13-1 and 22-12b). Over 600 megawatts of electricity (about half the needs of San Francisco) is currently being generated there; this field, now in its
second decade of production, will probably expand to 1900 megawatts in the decade of the
1980s. Italian engineers first put the natural steam
vents at Larderello to work in 1904. These fields
are still producing 190 megawatts of electric
power.

Very hot, dry rocks present a more difficult
problem: the rocks must be fractured at depth to
permit the circulation of water, which must also
be provided. Experiments are underway to develop technologies for exploiting this resource.

An intriguing form of geothermal energy is the
so-called geopressured zone found primarily at
depths of three to five kilometers in deep sedimentary basins such as those on the Gulf Coast of
Texas and Louisiana. These zones contain hot salt
water (temperatures below 180°C), highly pressurized by the weight of the overlying sedimentary section. Although the temperatures are probably too low and the deposits too deep for
commercial exploitation, the geopressured zones
are also saturated with natural gas. The production of gas as a by-product could justify the devel-

opment of this resource. Experimental wells are
now being drilled.

Geothermal energy presents some environmental problems: subsidence can occur if hot water is
withdrawn without being replaced: geothermally
heated waters can be salty and toxic, presenting a
disposal problem if they are not reinjected; and
hot, dry rocks will require the investment of large
volumes of water to produce energy from the reservoir.

The contribution of geothermal energy to the
world's energy future is difficult to estimate. The
resource is in a sense nonrenewable, but it is so
large that its future will depend on the economics
of production. At the present time we know how
to utilize only naturally occurring hot water or
steam deposits. The technology to exploit some of
the largest types of geothermal reservoirs—normal geothermal gradient, hot, dry rock, geopressured zones—has not yet been commercially
demonstrated. Our guess is that in the near term
geothermal energy can make important local contributions where the proximity of the resource to
the user and the economics are favorable as they
are in San Francisco and Paris. Large-scale contributions of geothermal energy to the world energy
budget probably will not occur until well into the
next century, if ever.

ENERGY POLICY

In view of the diverse and plentiful energy resources described in the previous pages, the
reader may well wonder why the world faces an
energy crisis. The critical problem for the next

few decades is fluid fuels because world oil production will peak and begin to decline and world demand will increase in this time period. Shortages, the escalation of prices, and the political disruption of supply can shock the world economic and political order severely. High priority goals of energy policy during this period should therefore be to reduce demand by conserving energy, to encourage vigorous programs of exploration and more efficient production of conventional oil, to store oil in a strategic petroleum reserve for emergency use, and to develop synthetic liquid fuels from oil shale, tar sand, and the liquefaction of coal. During this period, we should adjust the distribution of our energy resources to the most appropriate end use: petroleum for transportation; natural gas for residential and industrial space heating; hydropower, coal, and nuclear fission for electric power generation. It is unwise to use petroleum or natural gas where more plentiful resources will do.

We have seen that each source of energy has risks and environmental problems. There is a very small but finite probability that nuclear reactors can have catastrophic accidents.* Coal presents both occupational and public health hazards. Climatic change with serious consequences may be induced by CO_2 released in the combustion of hydrocarbons. For these reasons, a balanced combination of coal and nuclear fission for electricity generation is all the more important, with high priority devoted to understanding and reducing the risk.

Although we can reduce the demand growth for energy in the decades ahead, economic growth is essential for world stability. This can be accommodated by only modest energy growth if efficiency is improved. In the near term of depleting oil resources, the obvious candidates to pace this growth are coal and nuclear fission: their resources are plentiful, and the technologies and costs are known. In the longer term, some combination of synthetic fuels, solar energy, geothermal energy, fusion, and other alternate energy sources can be used when the technologies and costs are better understood.

In a sense we are in a race against time—to develop the many options of indefinitely sustainable energy resources described earlier before the remaining oil and natural gas are depleted. The

transition to the era of energy sufficiency may be smooth or rough depending on our determination, technological skill, and our ability to solve the complex social and political problems. In our view, the required technological advances are achievable. Whether the sociopolitical problems will be resolved is less certain.

MINERALS AS ECONOMIC RESOURCES

Aside from agricultural products, just about everything we use in modern society comes from the ground, including all of the metals and their alloys and the thousands of products made from the chemicals refined from natural deposits. At various places in this book, we have mentioned the practical uses of different minerals and rocks. In this section, we shall survey a broad range of useful materials and discuss them in an economic context, because, at the moment that commercial utility enters the picture, matters of availability and price become important.

Just as fossil-fuel reserves are not limitless, neither are the economically recoverable deposits of many useful minerals. Thus the study of mineral reserves is as important to our subject as the study of their geologic occurrence. The distribution of minerals, economic aspects of their recovery, and estimates of reserves and resources are parts of the specialized field called **economic geology.** One of the major decisions an economic geologist must make is whether a mineral deposit has economic significance, and that assessment depends on a knowledge of modern economics as well as of how mineral resources are distributed geologically.

What Is a Mineral Deposit?

The chemical elements of the Earth's crust—the portion of the globe that is readily accessible to us for mining and drilling—are widely distributed in many different kinds of minerals, and those minerals are found in a great variety of rocks. Throughout this book, we have discussed examples of how nature homogenizes materials at some times and places and segregates them at others. In most places, a particular element will be found in amounts close to its average abundance in the crust—that is, it is homogenized with the other elements. Segregation occurs in a smaller number of geologic situations. The occurrences of elements in much higher abundance—those in which some geologic process has operated to seg-

---

*Although the accident at Three-Mile Island in 1979 was a major one, we were lucky that no serious threat to life and health developed, since the reactor could be shut down and the containment prevented the release of dangerous amounts of radioactivity.

regate much higher quantities of the element than normal—are the ones that interest us, because the richer the deposit, the cheaper it is to recover the resource, both in terms of energy and in terms of money. Rich deposits of metals are **ores;** the minerals containing these metals are **ore minerals.** Ore minerals include sulfides (the main group), oxides, and silicates. In addition, some metals, such as gold, are found in their **native** state—that is, uncombined with other elements. The average crustal abundances of a number of economically important elements, determined by evaluation of a great many chemical analyses of all rock types, are shown in Table 22-1.

Some of the elements such as aluminum, iron, and magnesium, are so abundant that any average crustal rock could conceivably be used as a raw material, though not necessarily economically. In contrast are the elements of low abundance, such as gold, platinum, and mercury, which are present in such small amounts in the average rock that enormous quantities of rock would have to be refined to recover even small amounts. *Mineral deposits of economic value are those in which an element occurs in much higher abundance than the average crustal rock, sufficiently high to make it economically worthwhile to mine.* Many of the most valuable mineral deposits are metal-ore deposits, and much of economic geology is concerned with them.

The **concentration factor**—that is, the ratio of the abundance of an element in a mineral deposit to its average abundance—is highly variable and depends on the particular element and its average crustal abundance (Table 22-2). Iron, one of the common elements of the crust, has an average abundance of 5.8 percent. A good iron ore contains 50 percent iron; thus its concentration factor is about 10. A less abundant metal, such as copper, which has a crustal abundance of 0.0058 percent, is concentrated by factors from 60 to 100 in its economic ores. Even more spectacular are the rarer elements, such as mercury or gold, which have concentration factors in the thousands. The crustal abundances of the elements are related to their atomic number and chemical affinities in a complex way, as has been described in Chapter 1. But the concentration factors, though generally increasing with decreasing crustal abundance, depend largely on the ways in which the metals are held in crystal structures and on their solubilities in various geologic solutions, such as groundwaters, hydrothermal solutions, or sea water.

Because the elements are so widely distributed in many common rocks, the issue of resources and reserves is dominated by costs of recovery. We

Table 22-1

Crustal abundance of economically important elements

Name	Chemical symbol	Atomic number	Crustal abundance (% by weight)
Aluminum	Al	13	8.00
Iron	Fe	26	5.8
Magnesium	Mg	12	2.77
Potassium	K	19	1.68
Titanium	Ti	22	0.86
Hydrogen	H	1	0.14
Phosphorus	P	15	0.101
Manganese	Mn	25	0.100
Fluorine	F	9	0.0460
Sulfur	S	16	0.030
Chlorine	Cl	17	0.019
Vanadium	V	23	0.017
Chromium	Cr	24	0.0096
Zinc	Zn	30	0.0082
Nickel	Ni	28	0.0072
Copper	Cu	29	0.0058
Cobalt	Co	27	0.0028
Lead	Pb	82	0.00010
Boron	B	5	0.0007
Beryllium	Be	4	0.00020
Arsenic	As	33	0.00020
Tin	Sn	50	0.00015
Molybdenum	Mb	42	0.00012
Uranium	U	92	0.00016
Tungsten	W	74	0.00010
Silver	Ag	47	0.000008
Mercury	Hg	80	0.000002
Platinum	Pt	78	0.0000005
Gold	Au	79	0.0000002

Table 22-2

Concentration factors of some economically important elements needed for profitable mining

Element	Crustal abundance (% by weight)	Concentration factor*
Aluminum	8.00	3–4
Iron	5.8	5–10
Copper	0.0058	80–100
Nickel	0.0072	150
Zinc	0.0082	300
Uranium	0.00016	1,200
Lead	0.00010	2,000
Gold	0.0000002	4,000
Mercury	0.000002	100,000

*Concentration factor = abundance in deposit/crustal abundance.
Sources: Data from B. J. Skinner, *Earth Resources,* Prentice-Hall, 1969; D. A. Brobst and W. P. Pratt, *Mineral Resources of the U.S.,* U.S. Geological Survey Prof. Paper 820, 1973.

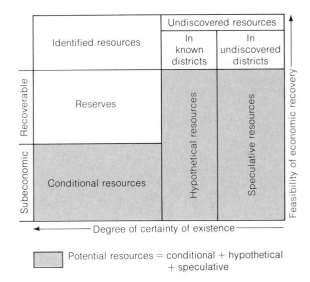

Potential resources = conditional + hypothetical + speculative

Figure 22-13
Categories of mineral resources that constitute our total primary resources. Identified resources consist of known deposits of reserves that are economically minable today and conditional resources whose locations are known but which are currently subeconomic. Undiscovered resources include hypothetical deposits that we may reasonably expect to find in known districts and the less certain, speculative resources. The latter may be unconventional deposits or discoveries in new provinces such as the sea floor. Shaded area contains the potential resources (conditional + hypothetical + speculative) which might become reserves by discovery, technological advance, or changes in economic conditions. [U.S. Geological Survey.]

could, theoretically, take almost any rock and extract both abundant and rare elements from it, given enough money and energy. So, from that point of view, it is clear that we will never "run out" of any vital elements. What is important is the exhaustion of the identified mineral deposits, the reserves that are economic to mine and purify. We can reasonably expect to add to the reserves by new discoveries—but at an uncertain rate in the face of increasing world demand. Once the deposits of the highest grade are mined out, we will be forced to rely on deposits of lower grades, that is, subeconomic resources whose concentration factors are lower (Fig. 22-13). It is not too soon to become concerned about mineral supplies and to start planning for the future, for only a few minerals are available domestically in quantities adequate to last for hundreds of years. Our dependence on imports is already great, sometimes because imports are cheaper, others because we lack the reserves. However, heavy dependence on imports for several strategically important metals such as cobalt, manganese, chromium, titanium,

and the platinum-group metals is a matter of national security concern (Fig. 22-14).

At some point in the history of mining a certain element, it may become economic to reuse (or "recycle") the element from worn-out manufactured materials that contain it. Gold and platinum, among other valuable metals, have been recycled for many years. A growing fraction of iron and aluminum in use is being recycled.

The decision whether or not to recycle is sometimes complex, one that can be made in a reasonable way only if all costs are calculated—not only the costs of mining, smelting, and transportation, but those of waste disposal and pollution control, both by the producer and the consumer. Because energy costs are rapidly rising, the cost pattern of mineral resources is bound to change drastically in the next decades. Smelting aluminum from its ore, for example, takes enormous quantities of electricity. It is estimated that the aluminum industry of the United States claimed three percent of the country's total power budget in 1965. For some materials, recovery from lower-grade ores will require such high expenditures of energy that it will be more economical to substitute other materials. In the future, we may well become familiar with costs of materials expressed in terms of units of energy per ton, as a measure on a par with dollars per ton.

The economic geologist, mindful of these costs, naturally seeks to minimize them, if possible, by finding new high-grade deposits rather than automatically resorting to the use of previously known lower-grade ones. The tools used for exploration are the subjects we have explored in this book. Understanding of Earth processes and Earth history, knowledge of how to use geophysics and geochemistry in exploration, and practical training in recognizing and mapping rocks and minerals in the field are brought together in the search. The prospector for mineral resources must be familiar with the geology of igneous, sedimentary, and metamorphic rocks, for mineral deposits are found in all three. But the prospector becomes more specialized than most geologists, always looking for the special situation in which nature has segregated a purer product from the surrounding rock masses.

GEOLOGY OF MINERAL DEPOSITS

A mineral segregation is the product of so many different kinds of geologic processes that we would need to restate most of this book to catalogue them. To give some idea of the origins of

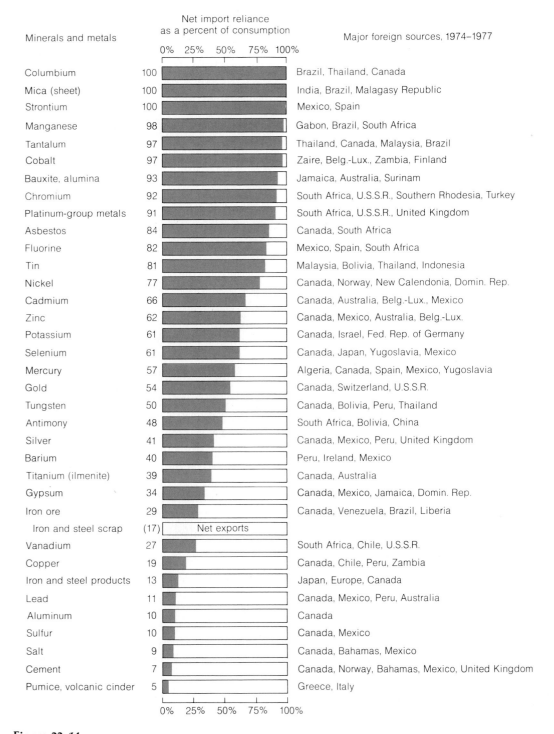

Minerals and metals	Net import reliance as a percent of consumption	Major foreign sources, 1974–1977
Columbium	100	Brazil, Thailand, Canada
Mica (sheet)	100	India, Brazil, Malagasy Republic
Strontium	100	Mexico, Spain
Manganese	98	Gabon, Brazil, South Africa
Tantalum	97	Thailand, Canada, Malaysia, Brazil
Cobalt	97	Zaire, Belg.-Lux., Zambia, Finland
Bauxite, alumina	93	Jamaica, Australia, Surinam
Chromium	92	South Africa, U.S.S.R., Southern Rhodesia, Turkey
Platinum-group metals	91	South Africa, U.S.S.R., United Kingdom
Asbestos	84	Canada, South Africa
Fluorine	82	Mexico, Spain, South Africa
Tin	81	Malaysia, Bolivia, Thailand, Indonesia
Nickel	77	Canada, Norway, New Calendonia, Domin. Rep.
Cadmium	66	Canada, Australia, Belg.-Lux., Mexico
Zinc	62	Canada, Mexico, Australia, Belg.-Lux.
Potassium	61	Canada, Israel, Fed. Rep. of Germany
Selenium	61	Canada, Japan, Yugoslavia, Mexico
Mercury	57	Algeria, Canada, Spain, Mexico, Yugoslavia
Gold	54	Canada, Switzerland, U.S.S.R.
Tungsten	50	Canada, Bolivia, Peru, Thailand
Antimony	48	South Africa, Bolivia, China
Silver	41	Canada, Mexico, Peru, United Kingdom
Barium	40	Peru, Ireland, Mexico
Titanium (ilmenite)	39	Canada, Australia
Gypsum	34	Canada, Mexico, Jamaica, Domin. Rep.
Iron ore	29	Canada, Venezuela, Brazil, Liberia
Iron and steel scrap	(17) Net exports	
Vanadium	27	South Africa, Chile, U.S.S.R.
Copper	19	Canada, Chile, Peru, Zambia
Iron and steel products	13	Japan, Europe, Canada
Lead	11	Canada, Mexico, Peru, Australia
Aluminum	10	Canada
Sulfur	10	Canada, Mexico
Salt	9	Canada, Bahamas, Mexico
Cement	7	Canada, Norway, Bahamas, Mexico, United Kingdom
Pumice, volcanic cinder	5	Greece, Italy

Figure 22–14
U.S. net import reliance on selected minerals and metals as a percentage of consumption in 1978. [From Status of Mineral Industries for 1979, U.S. Bureau of Mines.]

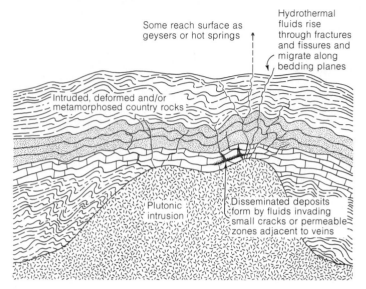

Some reach surface as geysers or hot springs

Hydrothermal fluids rise through fractures and fissures and migrate along bedding planes

Intruded, deformed and/or metamorphosed country rocks

Plutonic intrusion

Disseminated deposits form by fluids invading small cracks or permeable zones adjacent to veins

Figure 22–15
Many ore deposits are found in hydrothermal veins formed by hot solutions rising from magmas. As the solutions cool and react with surrounding rocks, they may precipitate ore minerals together with quartz, calcite, or other common vein-filling minerals.

mineral deposits, we shall describe some of the main geological occurrences of a few elements of major economic importance.

Hydrothermal Deposits

Many of the most useful ore deposits are found in hydrothermal veins—that is, in joints and fractures filled with minerals precipitated from heated waters, usually those emanating from an igneous intrusion (see Chapters 14 and 16). Some of the ores are found in the veins themselves; others are found in the rock adjacent to the veins that has been altered by heating and infiltration by the vein-forming solutions, a process similar, in some ways, to contact metamorphism (Fig. 22–15).

Most hydrothermal deposits are thought to be magmatic in origin because most are associated with igneous intrusions; in some places, several deposits occur around a single intrusion, and the veins are more abundant nearer the rocks of the intrusive bodies. The minerals found in hydrothermal veins are often similar to those found in volcanic and contact-metamorphic rocks, especially the sulfides of various metals: iron (FeS_2, pyrite), lead (PbS, galena), zinc (ZnS, sphalerite), mercury (HgS, cinnabar), and copper (CuS, covellite; CuS_2, chalcocite; and $CuFeS_2$, chalcopyrite). Only some magmas are closely associated with ore deposits. Some of these appear to have been

particularly enriched in one or a group of metallic elements and so have generated hydrothermal veins rich in the sulfides of those metals. Other hydrothermal deposits do not seem to be clearly associated with any intrusion; many economic geologists believe that these are the result of groundwater circulating through areas of abnormal heating related to an episode of regional metamorphism.

Where hydrothermal solutions reach the surface, they become hot springs and geysers, many of which precipitate minerals—including lead, zinc, and mercury ores—as they cool. At depth, such fluids deposit their metallic ores as they traverse the surrounding country rock and cool. Geochemists studying the ore-forming process have concluded that the ability of water to dissolve and later precipitate many of the metallic elements is dependent on the high temperatures and pressures found at depth in the crust and on chemical interactions with surrounding rocks. The ability of hot waters to carry significant quantities of metals in solution results partly from changes in the properties of the water itself and partly from interactions between the different substances in solution that occur at high concentrations, temperatures, and pressures and that promote greater solubility.

In 1963, a most remarkable oceanographic find was made in the Red Sea. Several deep basins were discovered at the bottom that were filled with brines dramatically hotter and much more concentrated in dissolved salts than the overlying normal Red Sea water. The bottom sediments of those basins are rich in precipitates of oxides and sulfides of copper, zinc, lead, iron, and other metals. Apparently, these hot brines are escaping from the sea floor after travelling upward through subsurface formations below the Red Sea—which lies in a rift-valley area, a young zone of sea-floor spreading. Since then, coring by the deep-sea drilling program has led to the discovery of other areas whose sediments are abnormally rich in such metal ores; one core brought up from sediments of the continental rise about 550 kilometers (359 miles) southeast of New York City contained a small vein of native copper.

One of the most important geological events in many decades was the discovery of hot springs on the East Pacific Rise spreading center in 1979 (see Chapter 13 and Fig. 13–9). Sea water circulating through fractures along the rift is heated to temperatures of several hundred degrees centigrade by contact with magma or hot rocks deep in the crust. The hot waters ascend by convection and vent through the sea floor. When the hot solutions, laden with dissolved metals leached from the hot rocks, reach the cooler upper crust and near-freezing ocean bottom waters, the dissolved minerals precipitate. In this manner, enormous quantities of sulfide ores rich in zinc, copper, iron, and other metals are being deposited along mid-ocean spreading centers. Collision of ocean plates occasionally exposes on land ancient fragments of oceanic lithosphere, known as ophiolites (Chapter 19). The rich copper, lead, and zinc deposits in the ophiolites of the Apennines, Cyprus, the Philippines, and elsewhere undoubtedly owe their origin to the same process of hydrothermal circulation along mid-ocean rifts, witnessed "live" for the first time by geologists from their deep-diving submarine.

A great many of the major deposits of sulfide ores of hydrothermal origin, or associated with volcanism or plutonism, are found on modern or ancient plate subduction-collision boundaries, including those of the Andes, the Coast Ranges, the eastern Mediterranean to Pakistan, the Philippine Islands, and Japan. A current hypothesis proposes that these deposits represent a second stage of the ore-forming process—the partial melting of subducted oceanic sediments and crust containing minerals previously concentrated by hydrothermal circulation at a spreading center. The metals "boil off" one at a time as the plate descends into increasingly hot regions of the mantle and rise into the overriding plate along with magma. Perhaps the iron, copper, molybdenum, lead, zinc, tin, and gold found along convergent plate boundaries originate this way (Fig. 22–16). This two-stage process of ore formation has been called the *geostill*.

How well these ideas will serve the economic geologist in the exploration for new high-grade ores and mineral deposits remains to be seen. The mountain belts they suggest for exploration are, for the most part, those that have already been singled out for exploration on the basis of earlier theories. How much undersea exploration may be stimulated is an open question—depending in part on the development of efficient marine technology for deep-sea mining. In any case, the most difficult part of mineral prospecting is pinpointing the generally small ore bodies, most of which seem to be distributed sparsely in an irregular pattern along structurally favorable zones. Zeroing in on them is still mostly a matter of geological common sense and good luck, but newer methods of geophysical and geochemical prospecting show promise of being very helpful in this respect.

Once a hydrothermal deposit is found—per-

haps from an outcrop, perhaps from a drill hole—the geologist and mining engineer must attempt to predict the extent of the deposit, both as a guide to future mining and as the basis for estimating reserves. Such predictions are based on the characteristic shapes of ore bodies. **Veins** are bodies of rock that have been deposited along joints, cracks, faults, dikes, bedding planes, or other zones of structural weakness that offered entry for the infiltrating hot waters. Like most igneous dikes, most veins are tabular—that is, thin in comparison to their length and width; others, sometimes called **lodes,** may be very thick, and some may even be pencil-shaped. Some are uniform in thickness; others pinch and swell in an irregular manner. Many veins are easily distinguished from the surrounding rock because of their distinctly different mineral composition and texture; others may merge imperceptibly with the surrounding rock in an extensive transition zone of rock alteration, and so may not be clearly distinguishable as veins in the field. Some veins contain cavities lined with crystals and crusts, remnants of gas pockets trapped in the fractures through which the hydrothermal solutions percolated.

Important hydrothermal deposits that are dispersed through much larger volumes of rock than vein deposits are called **disseminated deposits.** In igneous and sedimentary rocks alike, the dissemination takes place along abundant cracks and fractures; in sedimentary rocks, it may also take place along zones of higher permeability produced by variations in the lithology of the rocks. One important type of disseminated deposit is exemplified by the copper porphyry deposits of the southwestern United States and Chile. These deposits occur in a great number of tiny fractures in porphyritic felsic intrusives (granitic rocks with large feldspar or quartz crystals in a finer-grained matrix) and country rocks surrounding the higher parts of the plutons. Some unknown process associated with the intrusion or its aftermath broke the rocks into millions of pieces, and hydrothermal solutions penetrated and recemented the rocks with tiny veins and pore fillings. Such widespread dispersal of the ore deposit produces a low-grade but very large resource of many millions of tons of ore, which may be mined economically by large-scale mining methods. The most common copper mineral in the porphyry is chalcopyrite ($CuFeS_2$).

Sedimentary rocks are the hosts of disseminated hydrothermal deposits in the lead-zinc province of the upper Mississippi Valley, which extends from southwestern Wisconsin to Kansas

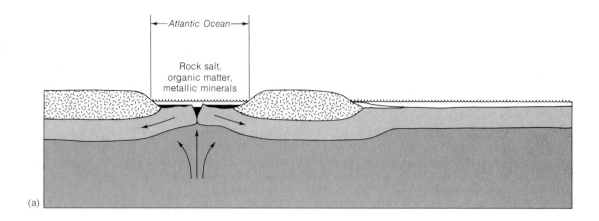

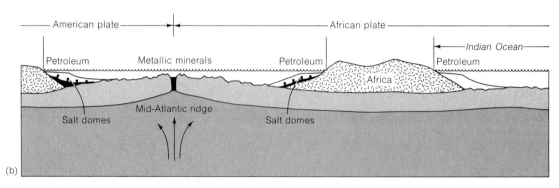

Figure 22-16
Role of plate boundaries in the accumulation of mineral deposits. (a) The ancestral continent of Pangaea rifts; South America and Africa drift apart. Thick layers of rock salt and organic matter accumulate on the receding margins. As a result of convection in fractures along the mid-ocean spreading center, sea water becomes an ore-forming hydrothermal solution, leaching metals from the hot rocks deep in the crust, ascending, and precipitating metallic ores on the sea floor and in the sediments and upper crust. (b) As sea-floor spreading continues in the Atlantic, salt domes originating in the thick layers of rock salt rise through the sediments of the continental margin, trapping oil and gas that are generated from the organic matter preserved

and Oklahoma. These ores are not associated with any known intrusion, and the evidence from the chemical and isotopic compositions of remnants of ore-depositing fluid encased in tiny cavities in quartz crystals, called **fluid inclusions,** is that the solutions were relatively low in temperature, no more than 150°C. The origin of such solutions is unknown; speculation centers on possible localized heat sources in the lower part of the lithospheric plate. The major minerals of these deposits are lead sulfide (galena, PbS) and zinc sulfide (sphalerite, ZnS). They have been deposited as **replacements** of calcite and dolomite in the surrounding limestones—that is, the solutions have dissolved some carbonate and replaced it with an equal volume of new crystals of sulfide.

Igneous and Contact-Metamorphic Deposits

The most important igneous deposits are found as segregations of ore minerals near the bases of layered intrusives, which are formed as minerals crystallizing from a melt settle to the floor of a magma chamber to form distinct layers (see Chapters 14 and 16). One of the largest ore bodies of this type ever found is at Sudbury, Ontario. It is a large mafic intrusive lopolith containing great quantities of layered nickel, copper, and iron sulfides near its base. These sulfide deposits are believed to have formed from crystallization of a dense sulfide-rich liquid that separated from the rest of the cooling magma and sank to the bottom of the chamber before congealing. Important plat-

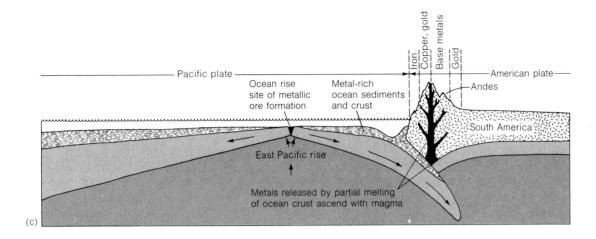

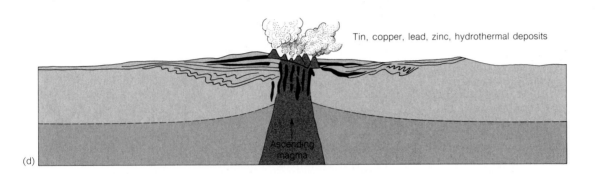

in the sediments. (c) Pacific Ocean sediment and crust, enriched in metallic ores by hydrothermal ore deposition along East Pacific rise, are partially melted when the Pacific plate is subducted. The metals ascend with magma to form the metal-bearing provinces of the Andes. (d) Hot spots, or plumes of rising magma, can occur within plates. Intracontinental hot spots can be the sites of deposits containing such metals as tin, lead, copper, and zinc. [After "Plate Tectonics and Mineral Resources" by P. A. Rona. Copyright © 1973 by Scientific American, Inc. All rights reserved.]

inum and chromium deposits have been found in layered intrusives in South Africa and Montana. One of the most valuable minerals, diamond, occurs in ultramafic rocks called **kimberlites** that extend to the surface from deep in the crust and upper mantle, where the extremely high pressures needed for their formation are found. These rocks are in the form of narrow pipes, and the mechanism of their eruption to the surface is a matter of controversy.

Contact metamorphism is the transformation of the mineral composition and texture of country rock along zones of contact with igneous intrusives (Chapter 16). In this process, minerals of commercial value—such as garnet and emery (corundum), which are used for abrasives—may

be formed. Some of the most important deposits of this type are iron ores, such as the small but high-grade deposit of magnetite (Fe_3O_4) near Cornwall, Pennsylvania, where a diabase has intruded and metamorphosed shale. In this, as in most contact-metamorphic deposits, the ore is the result of an interaction of the invaded rock with permeating material from the hot intrusion, not just an alteration produced only by heat and pressure without exchange of material.

Sedimentary Mineral Deposits

Chemical and mechanical segregations of many economically important minerals are the ordinary

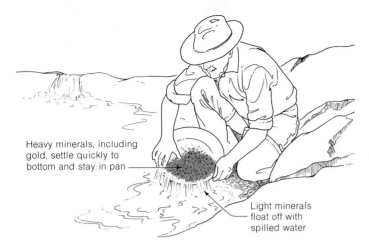

Heavy minerals, including gold, settle quickly to bottom and stay in pan

Light minerals float off with spilled water

Figure 22-17
The gold pan, an old prospecting tool that remains useful.

result of sedimentary processes. Limestones, separated out as chemical precipitates mainly by organisms, are used for agricultural lime, cement, and building stone. Pure sands, abraded and winnowed by waves and currents so that all materials other than quartz are removed, are the raw materials for glassmaking. Coarse sand and gravel, suitable for construction purposes, have been abundantly distributed in many areas of the northern United States and southern Canada by the Pleistocene glaciations; these materials are also widely distributed in channels and former channels of many rivers. Clays of high purity produced by prolonged weathering are used for pottery and ceramics, for both home and industrial use. Evaporite deposits of gypsum, separated from sea water by fractional crystallization, are used for plaster, and sodium and potassium salts from evaporites have varied uses from table salt to fertilizer. Phosphate rocks—marine shales and limestones enriched in phosphate by the chemical action of deep-sea waters—are the major raw materials of the world's fertilizer industry.

Sedimentary ore deposits are some of the world's most important sources of copper, iron, and other metals. They are chemical precipitates formed in sedimentary environments to which large quantities of the metals were transported in solution. Some of the important sedimentary copper ores, such as those of the Permian Kupferschiefer (German for copper shale) beds of Germany, are possibly the products of interaction between metal-rich sulfides derived from hot brines of hydrothermal origin with bottom sediments.

The major iron ores of the Earth have been found in Precambrian sedimentary rocks. It is now thought that the low oxygen content of the Earth's atmosphere at that time (see Chapter 12) allowed great quantities of iron to be transported in its soluble, reduced (ferrous) form into broad, shallow marine environments where it was oxidized to its insoluble (ferric) form and precipitated. In many of these basins, it was deposited in thin layers alternating with layers of chert. Such iron ores of alternating silica and hematite have been called banded iron ores. Post-Cambrian iron ores were formed in relatively small, restricted marine basins at times when abundant iron was being leached from the land surface by deep weathering. One of the many hypotheses is that the iron was transported in reduced form from land to these basins by marine seepage of groundwaters that were sufficiently acid and poor in oxygen to dissolve large quantities of iron from soils and subsurface formations. In the marine environment, the iron combines with clays or oxidizes and precipitates.

Possibly the most publicized (and romanticized) type of mineral prospecting is "panning" for gold, in which the gold seeker shakes a flat pan of river sediment in hopes of turning up the glint of a nugget. Many rich deposits of gold, diamonds, and such heavy-metal ore minerals as magnetite and chromite (Fe_2CrO_4) are found in **placers,** deposits concentrated by the mechanical sorting action of currents. Ore minerals are some of the most common heavy minerals—those much denser than the abundant quartz and feldspar of most sand—that are concentrated by current action in some places in streams or along beaches. Because the heavy minerals settle more quickly out of the current than the lighter quartz and feldspar, they tend to concentrate in accumulations on river bottoms and bars, where the current is strong enough to keep the lighter minerals suspended and in transport. The same kind of concentration is induced by waves, which preferentially deposit heavy minerals on the beach or on shallow offshore bars. The gold panner does the same thing by shaking the water-filled pan and washing out the lighter minerals. Because of the recent high prices for gold, gold panning is undergoing a revival (Fig. 22-17).

Some placers can be followed upstream to the location of the original mineral deposit from which they were eroded. Erosion of the Mother Lode, an extensive gold-bearing vein system lying along the western flanks of the Sierra Nevada batholith, produced the placers that were discovered in 1848 and led to the California gold rush. The placers were discovered first, then their source. That was also the sequence in the Kimberley diamond mines of South Africa two decades later.

This brief summary of the geology of mineral deposits barely touches on the great diversity of geologic situations in which various minerals of value are found. Some minerals or ores are found mainly or only in one kind of deposit; others are found in many different situations. Table 22–3 shows the geologic occurrence of some of the principal kinds of mineral deposits.

FINDING NEW MINERAL DEPOSITS

As the Earth's human population grows larger at an ever increasing rate and people all over the world demand higher standards of living (usually stated in terms of more food, goods, and materials), the requirements for mineral resources shoot upward. The total dollar value of all mineral resources, including fuels, produced in the United States has grown from less than 5 billion dollars in 1925 to more than 77 billion dollars in 1976. This fifteenfold growth in 51 years (which includes some inflationary increases because of the lowering of the value of the dollar in that time period) took place in a highly industrialized society that had already built a huge technological capability and whose population grew by only 150 percent in the same period. The rate of increase in countries that are rapidly developing their industries is even faster, and all countries have aspirations to speed their growth even further (Fig. 22–18).

In the United States the rate at which mineral resources are being used is much greater than the rate of population growth. This is shown by the increase in per capita consumption of almost all metals and other rock and mineral products. This trend shows no signs of diminishing: all segments of society demand more and new kinds of manufactured products and a volume of agricultural commodities that is highly dependent on the fertilizer industry, and those with lower incomes seek a greater share in the material wealth of their society.

Unequal sharing in the exploitation of mineral resources is an important fact on an international scale. North America, with less than a tenth of the world's population, consumes almost three-quarters of the world's production of aluminum, while Asia and Africa, with about two-thirds of the

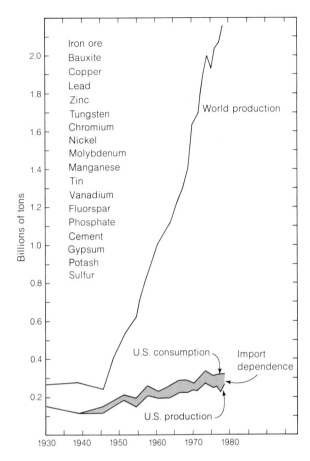

Figure 22–18
World production and U.S. production and consumption of minerals listed. Following World War II, the growth in the economies of many nations led to an enormous increase in world production. In the same period, U.S. dependence on imports grew. [Compiled from published sources by E. N. Cameron.]

world's population, use a little over 5 percent of the production. The same extreme imbalance is true of other materials as well. We are now seeing how international relations can be deeply affected by struggles over the control of resources, as some nations have nationalized, or demanded a greater share in the profits of, oil or mining companies owned by corporations based in North America or Europe. In late 1973, some nations of the Middle East began to restrict exports of oil as a political response to Arab-Israeli hostilities, and the ensuing oil embargo signaled the overwhelming importance of oil to our economy. Shortlived as it was, it led to shortages of gasoline that woke people up—at least temporarily—to the realities of our dependence on oil imports. The same could happen with other resources, though it might not affect us on the same scale as an oil embargo.

Table 22–3
Principal types of economic mineral deposit

Mineral deposit	Typical minerals	Geological occurrence	Uses	Major deposits Remarks
METALS PRESENT IN MAJOR AMOUNTS IN EARTH'S CRUST				
Iron	Hematite, Fe_2O_3 Magnetite, Fe_3O_4 Limonite, $FeO(OH)$	Sedimentary banded iron formation Contact metamorphic Magmatic segregation Sedimentary bog iron ore	Manufactured materials, construction, etc.	Mesabi, Minn.; Cornwall, Pa.; Kiruna, Sweden Resources immense; economics determines exploitation.
Aluminum	Gibbsite, $Al(OH)_3$ Diaspore, $AlO(OH)$	Bauxite: residual soils formed by deep chemical weathering	Lightweight manufactured materials	Jamaica Resources great, but expensive to smelt.
Magnesium	Dolomite, $CaMg(CO_3)_2$ Magnesite, $MgCO_3$	Dissolved in sea water Hydrothermal veins, limestones	Lightweight alloy metal, insulators, chemical raw material	Most extracted from sea water; unlimited supply.
Titanium	Ilmenite, $FeTiO_3$ Rutile, TiO_2	Magmatic segregations Placers	High-temperature alloys; paint pigment	Allard Lake, Quebec; Kerala, India Reserves large in relation to demand.
Chromium	Chromite, $(Mg,Fe)_2CrO_4$	Magmatic segregations of mafic and ultramafic rocks	Steel alloys	Bushveldt, S. Africa Extensive reserves in a number of large deposits.
Manganese	Pyrolusite, MnO_2	Chemical sedimentary deposits, residual weathering deposits, sea-floor nodules	Essential to steel making	Ukraine, U.S.S.R. World's land resources moderate, but sea-floor deposits immense.
METALS PRESENT IN MINOR AMOUNTS IN EARTH'S CRUST				
Copper	Covellite, CuS Chalcocite, Cu_2S Digenite, Cu_9S_5 Chalcopyrite, $CuFeS_2$ Bornite, Cu_5FeS_4	Porphyry copper deposits Hydrothermal veins Contact metamorphic Sedimentary deposits in shales (Kupferschiefer type)	Electrical wire and other products	Bingham Canyon, Utah: Kuperschiefer: Germany; Poland
Lead	Galena, PbS	Hydrothermal (replacement) Contact metamorphic Sedimentary deposits (Kupferschiefer type)	Storage batteries, gasoline additive (tetraethyl lead)	Mississippi Valley: Broken Hill, Australia Large resources, Many lower-grade deposits.
Zinc	Sphalerite, ZnS	Same as lead	Alloy metal	Same as lead
Nickel	Pentlandite, $(Ni, Fe)_9S_8$ Garnierite, $Ni_3Si_2O_5(OH)_4$	Magmatic segregations Residual weathering deposits	Alloy metal	Sudbury, Ontario High grade ores limited; large resources of low-grade ores; also in sea-floor Mn nodules.
Silver	Argentite, Ag_2S In solid solution in copper, lead, and zinc sulfides	Hydrothermal veins with lead, zinc, and copper	Photographic chemicals; electrical equipment	Most produced as by-product of copper, lead, and zinc recovery.
Mercury	Cinnabar, HgS	Hydrothermal veins	Electrical equipment, pharmaceuticals	Almadén, Spain Few high-grade deposits with limited reserves.
Platinum	Native metal	Magmatic segregations (mafic rocks) Placers	Chemical industry; electrical; alloying metal	Bushveldt, S. Africa Large reserves in relation to demand.
Gold	Native metal	Hydrothermal veins Placers	Coinage; dentistry, jewelry	Witwatersrand, S. Africa Reserves concentrated in a few larger deposits.

Table 22-3 (Continued)

Mineral deposit	Typical minerals	Geological occurrence	Uses	Major deposits Remarks
		NONMETALS		
Salt	Halite, NaCl	Evaporite deposits Salt domes	Food; chemicals	Resources unlimited; economics determines exploitation
Phosphate rock	Apatite, $Ca_5(PO_4)_3OH$	Marine phosphatic sedimentary rock Residual concentrations of nodules	Fertilizer	*Florida* High-grade deposits limited but extensive resources of low-grade deposits.
Sulfur	Native sulfur Sulfide ore minerals	Caprock of salt domes (main source) Hydrothermal and sedimentary sulfides	Fertilizer manufacture; chemical industry	*Texas; Louisiana; Sicily.* Native sulfur reserves limited but immense resources of sulfides.
Potassium	Sylvite, KCl Carnallite, $KCl \cdot MgCl_2 \cdot 6H_2O$	Evaporite deposits	Fertilizer	*Carlsbad, New Mexico* Great resources of rich deposits.
Diamond	Diamond, C	Kimberlite pipes Placers	Industrial abrasives	*Kimberly, S. Africa* Synthetic diamond now commercially available.
Gypsum	Gypsum, $CaSO_4 \cdot 2H_2O$ Anhydrite, $CaSO_4$	Evaporite deposits	Plaster	Immense resources widely distributed.
Limestone	Calcite, $CaCO_3$ Dolomite, $CaMg(CO_3)_2$	Sedimentary carbonate rocks	Building stone; Agricultural lime; cement	Widely distributed; transportation a major cost.
Clay	Kaolinite $Al_2Si_2O_5(OH)_4$ Smectite* Illite*	Residual weathering deposits; sedimentary clays and shales	Ceramics: china, electrical; structural tile	Many large pure deposits; immense reserves of all grades.
Asbestos	Chrysotile, $Mg_3Si_2O_5(OH)_4$	Ultramafic rocks altered and hydrated in near-surface crustal zones	Nonflammable fibers and products	*Southeastern Quebec* Limited high-grade reserves but great low-grade reserves

*Formula highly variable; a hydrous aluminum silicate with other cations, such as Na^+, K^+, Ca^{2+}, Mg^{2+}.

A hard fact of life is that an equal per capita sharing of the world's resources would not bring everyone to a "satisfactory" level of consumption—if that could ever be defined—certainly not to a level of consumption anywhere near those of the affluent countries of Europe or North America. How can all of the world's peoples enjoy the benefits of the Earth's resources, benefits that require a highly technological society to utilize them in support of a great population? One of the needs is a great increase in the discovery and exploitation of mineral and energy resources and the development of new kinds of materials to substitute for scarce minerals. In a sense, we are all dependent on the skill of the economic geologist and the materials scientist. However, their skills will be of no avail if population growth continues undiminished.

The prospect for future discovery of major resources is neither rosy nor bleak. It is not rosy because mineral resources are finite and much of the Earth's land surface has already been explored geologically to some extent. It is not very probable that ore bodies of the magnitude of the Sudbury nickel deposit will be discovered in outcrop. In addition, most known major mineral deposits are at least moderately well mapped, so there is little chance of finding appreciable new extensions of these deposits. Even if large quantities of reserves were found, the cost of mining lower-grade deposits added to the costs of reducing environmental pollution resulting from the expanded mining and processing operations could limit their development.

On the other hand, we can extend our supplies by increasing the efficiency of recovery, by recycling, and by substituting an abundant material for a scarce one. Moreover, there are regions of Asia, Africa, South America, Australia and Antarctica that have not been subjected to intensive scrutiny by prospectors. These regions, as well as parts of North America and Europe, have "blind ground," that is, extensive areas covered by deep soil and recent sediment. These areas await prospecting methods capable of detecting ore bodies below the cover. About half of the U.S. falls in this category. The oceans, as we have seen, represent an immense new mineral province of possibly huge potential if we can develop the technology to mine the sea floor. The new ideas about plate boundaries and mineral deposits may help focus mineral exploration more successfully.

The methods of experimental geophysics have been applied with great success to many resource-exploration campaigns. Most notable has been the use of seismic prospecting in the hunt for geologic structures favorable for oil and gas accumulation. Magnetic, electrical (mainly electrical-conductivity), and gravity mapping have likewise been used in the service of mineral exploration, both to reveal structures and to locate specific ore bodies. Airborne surveys of this type have been effective in locating good prospects, particularly in country difficult to cover on the ground, such as great areas of northern Canada and other polar terrain. Radioactivity is of great importance in prospecting for uranium ore. In the late 1940s and 1950s, no prospector for uranium on the Colorado Plateau was well equipped without a radioactive-particle counter (usually a Geiger counter) to detect small quantities of radiation from uranium-235.

The application of geochemistry to mineral prospecting is a little less direct. The main object is to see if unusual concentrations of minor elements, which show up in some places in surface or subsurface waters, in soils, or in vegetation, indicate a buried mineral deposit. The amounts of an element present in a sample may be so small that only the most sensitive laboratory instruments can detect them, yet their variation may be sufficient to draw maps that locate the source of the elements in a mineral deposit. The elements may have been distributed to surface materials by small leakages from the original vein-depositing fluids of hydrothermal deposits, by weathering processes and soil formation of near-surface rock masses, or by circulating ground waters. The trace-element content (and sometimes even the appearance) of vegetation may be a guide to small amounts of particular elements, diagnostic of a nearby ore body.

The advent of satellites that photograph the surface as they orbit Earth has opened a new era of geological mapping. Photographs in different colors are transmitted to receiving stations in many countries. Manipulating these images can produce composite pictures that bring out fires, flooding, volcanic eruptions, different types of vegetation, and many different rock and soil types. New ore bodies have already been discovered by this surveying of the surface from a great distance, called *remote sensing*. These methods have given the geologist bigger and better eyes for more effective surveillance of the Earth and seven-league boots with which to cover more territory.

Most of the minerals that we may recover from the ocean will come either from the water or from the sea floor, which has been studied by a variety of remote-sensing devices, such as echo-sounding instruments. The ocean is an immense storehouse of some easily extractable elements. For example, almost all magnesium is now recovered from evaporated sea water. For an element present in sea water, it is ease of recoverability, rather than abundance, that makes the source attractive, for, as Tables 22–1 and 22–4 show, the concentration of almost every element is greater in crustal rocks than in sea water. Because silver is 900 times more concentrated in the average rock than in sea water, the cost of recovering it from the sea would have to be at least 900 times less than the cost of recovering it from rock in order for sea-water extraction to be economical. These figures suggest that, for most metals, it will be a long time before we "mine" sea water.

The element bromine, however, is a different story, as can be seen from Table 22–4: it is much more plentiful in sea water than in average crustal rocks. Bromine has been economically recovered from sea water for many years, and only recently has it been extracted more profitably from concentrated saline groundwaters. These brines (see Chapter 6) are a potential source for many other materials as well, for there are indications that some heavy trace metals may be concentrated in them.

Sodium chloride has been recovered from sea water probably since prehistoric times. In a great many coastal regions of the world, the separation of salt from sea water is an economic process. One of the best-known processing plants is in San Francisco Bay, where a series of evaporating basins is used to extract the pure salt. Although the idea of plate tectonics may be a stimulus to further mineral exploration in the sea, there have long been mining operations near shore that extended from land under the ocean floor. Because the continental borderlands have the same range of rocks as the nearby continental areas, they may contain the same range of mineral deposits as the

Table 22-4
Concentrations of the elements in seawater and their values in elemental or combined form (The elements are listed in order of abundance; those in italic type are in concentrations valued at $1 or more per million gallons of seawater.)

Element	Concentration ($lb/10^6$ gal)*	Value ($$/10^6$ gal)	
		as	
Chlorine	166,000	NaCl	924
Sodium	92,000	Na_2CO_3	378
Magnesium	11,800	Mg	4,130
Sulfur	7,750	S	101
Calcium	3,500	$CaCl_2$	150
Potassium	3,300	K_2O (equiv)	91
Bromine	570	Br_2	190
Carbon	250	Graphite	8×10^{-5}
Strontium	70	$SrCO_3$	2
Boron	40	H_3BO_3	3
Fluorine	11	CaF_2	0.35
Nitrogen	4	NH_4NO_3	1
Lithium	1.5	Li_2CO_3	36
Rubidium	1.0	Rb	125
Phosphorus	0.6	$CaHPO_4$	0.08
Iodine	0.5	I_2	1
Zinc	0.09	Zn	0.013
Iron	0.09	Fe_2O_3	0.001
Aluminum	0.09	Al	0.04
Molybdenum	0.09	Mo	0.004
Tin	0.03	Sn	0.05
Copper	0.03	Cu	0.01
Arsenic	0.03	As_2O_3	0.002
Uranium	0.03	U_3O_8	0.3
Nickel	0.02	Ni	0.02
Manganese	0.02	Mn	0.006
Titanium	0.009	TiO_2	0.003
Silver	0.003	Ag	0.02
Chromium	0.0004	Cr_2O_3	0.00001
Thorium	0.0004	ThO_2	0.0009
Lead	0.0003	Pb	0.00004
Mercury	0.0003	Hg	0.002
Gold	0.00004	Au	0.02
Radium	9×10^{-10}	Ra (in salts)	0.002

*One $lb/10^6$ gal is approximately 0.000012 percent.
Source: From *Resources and Man*, National Academy of Sciences—National Research Council. W. H. Freeman and Company, 1969.

Figure 22-19
Manganese nodules dredged from the ocean floor helped to demonstrate the feasibility of mining the seabed for them. Deposits are estimated to be in the trillions of tons. [Photo by B. J. Nixon/Deepsea Ventures; Courtesy National Science Foundation.]

land. The metal that seems likely to be the first candidate for deep-sea mining is manganese. In Chapter 11 we pointed out the widespread occurrence on the sea floor of manganese nodules, spheroidal aggregates of manganese, iron, and other metal oxides that range in size from tiny encrustations weighing less than a gram to large masses of several hundred kilograms. The majority of the nodules are a few centimeters in diameter. In the past few years, interest in the possibility of economic recovery of the deposits has been strong, both because of the gradual depletion of high-grade manganese ores on land and because nodules are rich in many other valuable metals, such as copper, nickel, and cobalt. The average nodule contains more than 20 percent manganese, 6 percent iron, and about 1 percent each of copper and nickel—a highly enriched ore by any standard (Fig. 22-19).

The economic recovery of manganese nodules awaits the development of a technology of seafloor surface mining that will be nonpolluting and cheap enough to match the current costs of recovery of known ores from land deposits. Mining operations have been carried out on an experimental basis since 1970 by American and Japanese companies. The prospects for mining the sea floor are good—and make more urgent—an international agreement on economic exploitation of the sea. In the summer of 1973, a major convocation of nations began to work on the thorny international legal problems, all of which have political overtones. Meetings have continued since then, but a stable international agreement on the law of the sea seems to be some time off. Ultimately, some resolution of the question of ownership must be found before the resources of the sea can be used.

Even with advanced technology, conservation, and recycling, it is clear that exponential growth of population and the demand for minerals cannot be sustained indefinitely on a finite planet with finite resources. At present, population doubles every 35 years, and mineral production doubles about every 20 years (Fig. 22–18). If these growth rates continue, even with the very optimistic assumption that ultimate recoverable resources exceed known reserves 100-fold, about half the commonly used metals will be exhausted in less than 150 years. Well before that time, by the year 2000, we will be unable to feed the world population with presently known methods of farming. It is mind-boggling to calculate that sometime in the next century human mining and agricultural activities will become the dominant factor in modifying Earth's crust, exceeding the crust-mass production at mid-ocean ridges!

The world clearly faces major readjustments in the decades ahead, and it is not too soon to start working out equitable, humane, and lasting solutions. How well we use our planet will depend on how well we understand how it works and the extent to which the people of the world cooperate intelligently in the use, regulation, and control of its resources. We have only one Earth. To live on it we must appreciate it better.

SUMMARY

1. Matter of the Earth is theoretically recyclable, but energy is permanently consumed when we use its resources. Energy from fossil fuels—coal, oil, and gas—has been the mainstay of the world's industrialization.

2. Oil and gas form by the preservation and alteration of organic matter deposited in sediments that chemically transform part of the organics into liquid and gaseous hydrocarbons, compounds of carbon and hydrogen. Oil and gas accumulate in structural and stratigraphic traps that confine the fluids with impermeable barriers. Oil and gas are found in many of the buried sedimentary rocks of the Earth that have not been too compacted, deformed, or metamorphosed to allow preservation of the hydrocarbons. The geographic distribution of oil and gas is related to geologic age of sedimentary rocks accumulated in major geosynclines. Middle East oil is accumulated in thick Cenozoic and Mesozoic sediments in a major geosyncline.

3. Coal, formed by the compaction and chemical alteration of swamp vegetation, is present as huge resources in sedimentary rocks. We have used only 2.5 percent of the world's minable coal resources. Coal mining and pollution caused by coal burning are risky to human life, but increased use of coal in the next decades seems certain for electric power generation and for conversion to liquid and gaseous fuels. Reserves of oil shale and tar sands are great and may be extensively exploited in the next few decades as oil costs rise and technology improves. These fossil fuels will last well into the next century, when alternate energy sources should be available. However, combustion of fossil fuels releases CO_2 which could trigger worldwide climatic change in the next century.

4. Nuclear power from controlled fission reactors can be used as a major energy source but only if the public can be assured of its safety. Known high-grade reserves of uranium-235 ore will not be adequate to support major use of conventional nuclear power plants for much more than a few decades. Thus fast breeder reactors will be needed early in the next century. The use of nuclear energy could serve to extend our fossil fuel resources.

5. Alternative energy sources are water power, solar energy, geothermal power, and nuclear fusion, none of which has any immediate prospect of being a major answer to world energy needs. Solar energy and fusion could become major energy sources in the next century.

6. The goal of energy policy should be to guide the nations of the world through the transition from oil to the more plentiful fossil fuels and thus to buy time in order to develop alternate, safe, and unlimited energy sources to replace fossil fuels in the next

century. The resources and technology exist to accomplish this peacefully and without economic dislocation if the sociopolitical problems can be worked out.

7. Mineral deposits of economic value are those in which an element occurs in much higher abundance than in the average crustal rock—high enough to make the deposit economically worthwhile to mine. When high-grade ores or other economic deposits are depleted, we can move to recover elements from less concentrated ores, though at greater cost. In many cases it may become profitable to recycle or substitute.

8. Hydrothermal veins, which are some of the most important ore deposits, are formed by hot water emanating from igneous intrusions or by circulating groundwater or sea water in areas of high heat flow. Many of these deposits may be related to boundaries of lithospheric plates. They may occur in vein or lode deposits or in such disseminated deposits as the copper-porphyry type.

9. Igneous and contact-metamorphic deposits occur as segregations of layered intrusives, such as copper and nickel deposits of lopoliths, or in contact-metamorphic aureoles.

10. Sedimentary deposits include many of the chemical and mechanical segregations of such ordinary rocks as limestone, sand, and gravel, and evaporite salt deposits. Sedimentary ores of copper and iron have formed in special sedimentary environments, the iron ores chiefly in Precambrian times. Placers are current-laid deposits rich in gold or other heavy minerals.

11. Finding new mineral deposits is vital to support an increasingly industrialized world civilization. Prospects for finding new resources, based on geological, geophysical, and geochemical prospecting, are good. The sea represents a largely untapped resource.

EXERCISES

1. What sedimentary environments favor the formation of sediments containing organic matter that might later be transformed into petroleum?

2. Do you think the mid-Atlantic ridge might be a good prospect for oil drilling, assuming that we can invent the technology to drill there? Why?

3. Given two permeable, anticlinal sandstones, one Ordovician, the other Miocene, which would you choose to drill first in the hope of finding large reserves of oil and gas?

4. Which of the following factors are most important in estimating the future supply of oil and gas: (a) rate of oil accumulation, (b) rate of natural seepage of oil, (c) rate of pumping of oil from known reserves, (d) rate of discovery of new reserves, (e) the total amount of oil now present in the Earth?

5. In terms of use and economic recovery, rank according to relative importance all of the different

forms of fossil fuels today, and explain how their ranking might differ at the end of the next century.

6. Contrast the risks and benefits of nuclear fission and coal combustion as energy sources.

7. How would you use knowledge of the distribution of plate boundaries to make a map showing the most likely areas of Earth to investigate for geothermal power?

8. In what important respects do uranium reserves differ from fossil fuel reserves?

9. What evidence might you marshal to show that a particular ore deposit was formed by hydrothermal solutions emanating from an igneous intrusion?

10. Should we stockpile oil and strategic minerals against future shortages caused by political disruptions? What are the costs and benefits?

BIBLIOGRAPHY

Atwood, G., "The Strip-Mining of Western Coal," *Scientific American,* June 1974.

Bonatti, E., "The Origin of Metal Deposits in the Oceanic Lithosphere," *Scientific American,* February 1978.

Brobst, D. A., and W. P. Pratt. *United States Mineral Resources.* Washington, D.C.: U.S. Government Printing Office, 1973.

Energy Information Administration, U.S. Department of Energy, Annual Report to Congress (3 vols.).

Washington, D.C.: U.S. Government Printing Office, 1979.

International Institute for Applied Systems Analysis, *Energy in a Finite World* (2 vols.). Cambridge, Mass.: Ballinger Publishing Company, 1981.

Levorsen, A. I., *Geology of Petroleum,* 2nd ed. San Francisco: W. H. Freeman and Company, 1967.

National Academy of Sciences, *Energy in Transition, 1985–2010.* San Francisco: W. H. Freeman and Company, 1980.

Park, C. F., Jr., and R. A. MacDiarmid, *Ore Deposits.* San Francisco: W. H. Freeman and Company, 1975.

Rona, P. A., "Plate Tectonics and Mineral Resources," *Scientific American,* July 1973.

Skinner, B. J., *Earth Resources,* 2nd ed. Englewood Cliffs, N.J.: Prentice-Hall, 1976.

Skinner, B. J., and K. K. Turekian, *Man and the Ocean.* Englewood Cliffs, N.J.: Prentice-Hall, 1976.

APPENDIXES

APPENDIX I

Conversion Factors: Metric-English

LENGTH	
1 centimeter	0.3937 inch
1 inch	2.5400 centimeters
1 meter	3.2808 feet
1 foot	0.3048 meter
1 meter	1.0936 yards
1 yard	0.9144 meter
1 kilometer	0.6214 mile (statute)
1 kilometer	3281 feet
1 mile (statute)	1.6093 kilometers
1 mile (nautical)	1.8531 kilometers
1 fathom	6 feet
1 fathom	1.8288 meters
1 angstrom	10^{-8} centimeter
1 micrometer (formerly micron)	0.0001 centimeter

VELOCITY	
1 kilometer/hour	27.78 centimeters/second
1 mile/hour	17.60 inches/second

AREA	
1 square centimeter	0.1550 square inch
1 square inch	6.452 square centimeters
1 square meter	10.764 square feet
1 square meter	1.1960 square yards
1 square foot	0.0929 square meter
1 square kilometer	0.3861 square mile
1 square mile	2.590 square kilometers
1 acre (U.S.)	4840 square yards

VOLUME	
1 cubic centimeter	0.0610 cubic inch
1 cubic inch	16.3872 cubic centimeters
1 cubic meter	35.314 cubic feet
1 cubic foot	0.02832 cubic meter
1 cubic meter	1.3079 cubic yards
1 cubic yard	0.7646 cubic meter
1 liter	1000 cubic centimeters
1 liter	1.0567 quarts (U.S. liquid)
1 liter	33.815 ounces (U.S. fluid)
1 gallon (U.S. liquid)	3.7853 liters

MASS

1 gram	0.03527 ounce (Avoirdupois)
1 ounce (Avoirdupois)	28.3495 grams
1 gram	0.03215 ounce (Troy)
1 kilogram	2.20462 pounds (Avoirdupois)
1 pound (Avoirdupois)	0.45359 kilogram

DENSITY

1 gram/cubic centimeter	62.4280 pounds/cubic foot

PRESSURE

1 kilogram/square centimeter	0.96784 atmosphere
1 kilogram/square centimeter	0.98067 bar
1 kilogram/square centimeter	14.2233 pounds/square inch
1 bar	0.98692 atmosphere
1 foot of water	0.03048 kilogram/square centimeter
1 kilometer of granite	265 kilograms/square centimeter (approx.)

ENERGY

1 erg	2.39006×10^{-8} calorie (gram)
1 erg	9.48451×10^{-11} BTU
1 erg	10^{-7} joule
Explosion equivalent to 1000 tons of TNT	4×10^{19} ergs

POWER

1 watt	10^7 ergs/second
1 watt	0.001341 horsepower (U.S.)
1 watt	0.05688 BTU/minute
1 watt	0.73756 foot pound/second

Prefix Names of Multiples and Submultiples of Units

Prefix	Abbreviation	Factor by which unit is multiplied
giga	g	10^9
mega	M	10^6
kilo	k	10^3
hecto	h	10^2
deka	da	10
deci	d	10^{-1}
centi	c	10^{-2}
milli	m	10^{-3}
micro	μ	10^{-6}
nano	n	10^{-9}
pico	p	10^{-12}

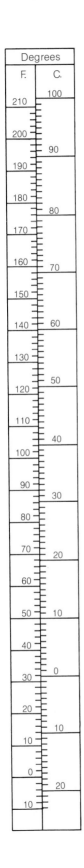

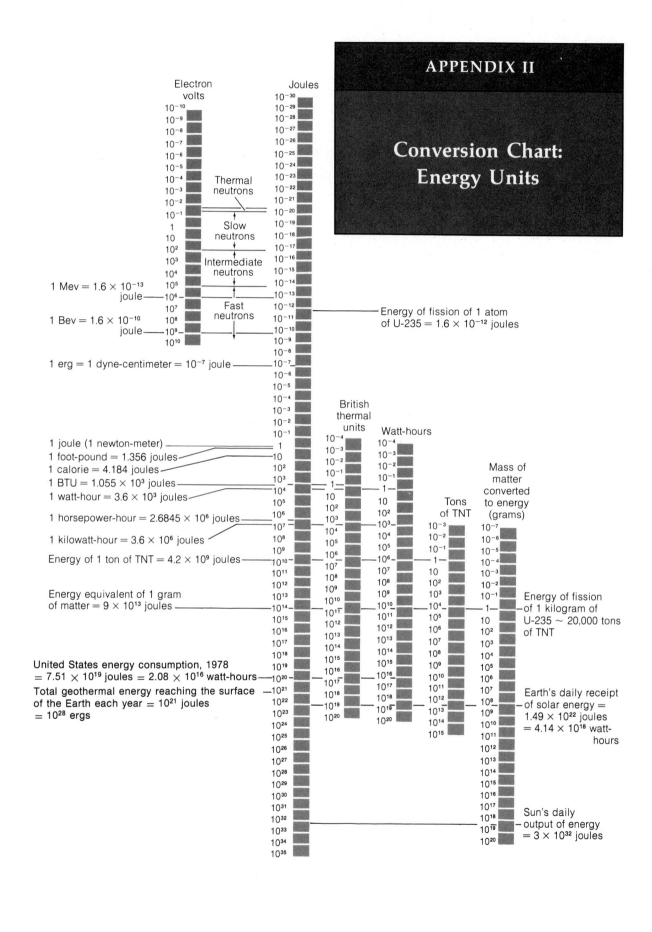

APPENDIX II

Conversion Chart: Energy Units

Electron volts

Joules

Thermal neutrons

Slow neutrons

Intermediate neutrons

$1 \text{ Mev} = 1.6 \times 10^{-13}$ joule

Fast neutrons

$1 \text{ Bev} = 1.6 \times 10^{-10}$ joule

$1 \text{ erg} = 1 \text{ dyne-centimeter} = 10^{-7}$ joule

Energy of fission of 1 atom of U-235 $= 1.6 \times 10^{-12}$ joules

British thermal units

Watt-hours

Mass of matter converted to energy (grams)

1 joule (1 newton-meter)
1 foot-pound $= 1.356$ joules
1 calorie $= 4.184$ joules
1 BTU $= 1.055 \times 10^3$ joules
1 watt-hour $= 3.6 \times 10^3$ joules

Tons of TNT

1 horsepower-hour $= 2.6845 \times 10^6$ joules

1 kilowatt-hour $= 3.6 \times 10^6$ joules

Energy of 1 ton of TNT $= 4.2 \times 10^9$ joules

Energy of fission of 1 kilogram of U-235 $\sim$ 20,000 tons of TNT

Energy equivalent of 1 gram of matter $= 9 \times 10^{13}$ joules

United States energy consumption, 1978
$= 7.51 \times 10^{19}$ joules $= 2.08 \times 10^{16}$ watt-hours
Total geothermal energy reaching the surface of the Earth each year $= 10^{21}$ joules
$= 10^{28}$ ergs

Earth's daily receipt of solar energy $=$ 1.49×10^{22} joules $= 4.14 \times 10^{18}$ watt-hours

Sun's daily output of energy $= 3 \times 10^{32}$ joules

Numerical Data Pertaining to Earth

Equatorial radius:	6378 kilometers
Polar radius:	6357 kilometers
Radius of sphere with Earth's volume:	6371 kilometers
Volume:	1.083×10^{27} cubic centimeters
Surface area:	5.1×10^{18} square centimeters
Percent surface area of oceans:	71
Percent surface area of land:	29
Average elevation of lands:	623 meters
Average depth of oceans:	3.8 kilometers
Mass:	5.976×10^{27} grams
Density:	5.517 grams/cubic centimeter
Gravity at equator:	978.032 centimeters/second/second
Mass of atmosphere:	5.1×10^{21} grams
Mass of ice:	$25–30 \times 10^{21}$ grams
Mass of oceans:	1.4×10^{24} grams
Mass of crust:	2.5×10^{25} grams
Mass of mantle:	4.05×10^{27} grams
Mass of core:	1.90×10^{27} grams
Mean distance to Sun:	1.496×10^{8} kilometers
Rotational velocity:	7.292×10^{-5} radians/second (40,000 km/day linear velocity at equator)
Average velocity around Sun:	29.77 kilometers/second
Ratio—mass of Sun/mass of Earth:	3.329×10^{5}
Ratio—mass of Earth/mass of Moon:	81.303

APPENDIX IV

Properties of the Most Common Minerals of the Earth's Crust

	Mineral or group name	Varieties and chemical composition	Form, diagnostic characters	Cleavage, fracture	Color	Hardness
LIGHT COLORED MINERALS. **VERY ABUNDANT** IN EARTH'S CRUST IN **ALL MAJOR** ROCK TYPES	FELDSPAR (FRAMEWORK SILICATES)	*POTASSIUM FELDSPARS* [KAlSi$_3$O$_8$] Sanidine Orthoclase Microcline	Cleavable coarsely crystalline or finely granular masses. Isolated crystals or grains in rocks, most commonly not showing crystal faces.	Two at right angles, one perfect and one good; pearly luster on perfect cleavage.	White to gray, frequently pink or yellowish; some green.	6
		PLAGIOCLASE FELDSPARS [NaAlSi$_3$O$_8$] Albite [CaAl$_2$Si$_2$O$_8$] Anorthite		Two at nearly right angles; one perfect, one good. Fine parallel striations on perfect cleavage.	White to gray, less commonly greenish or yellowish.	
	QUARTZ (FRAMEWORK SILICATES)	SiO$_2$	Single crystals or masses of 6-sided prismatic crystals. Also formless crystals and grains or finely granular or massive.	Very poor or nondetectable; conchoidal fracture.	Colorless, usually transparent; also slightly colored smoky gray, pink, yellow.	7
	MICA (SHEET SILICATES)	*MUSCOVITE* [KAl$_3$Si$_3$O$_{10}$(OH)$_2$]	Thin, disc-shaped crystals, some with hexagonal outlines. Dispersed or aggregates.	One perfect; splittable into very thin, flexible, transparent sheets.	Colorless; slight gray or green in thick pieces.	2–2½
DARK COLORED MINERALS **ABUNDANT** IN MANY KINDS OF **IGNEOUS** AND **METAMORPHIC** ROCKS		*BIOTITE* [K(Mg,Fe)$_3$AlSi$_3$O$_{10}$(OH)$_2$]	Irregular, foliated masses; scaly aggregates.	One perfect; splittable into thin, flexible sheets.	Black to dark brown. Translucent to opaque.	2½–3
		CHLORITE [(Mg,Fe)$_5$(Al,Fe)$_2$Si$_3$O$_{10}$(OH)$_8$]	Foliated masses or aggregates of small scales.	One perfect: thin sheets flexible but not elastic.	Various shades of green.	2–2½
	AMPHIBOLE (DOUBLE CHAINS)	*TREMOLITE–ACTINOLITE* [Ca$_2$(Mg,Fe)$_5$Si$_8$O$_{22}$(OH)$_2$]	Long, prismatic crystals, usually 6-sided. Commonly in fibrous masses or irregular aggregates.	Two good cleavage directions at 56° and 124° angles.	Pale to deep green. Pure tremolite white, vitreous luster.	5–6
		HORNBLENDE [Complex Ca,Na,Mg,Fe, Al silicate]			Dark green to black.	

Mineral	Subclass	Name / Composition	Crystal form & occurrence	Cleavage / fracture	Color	Hardness
PYROXENE	SINGLE CHAINS	ENSTATITE–HYPERSTHENE [(Mg,Fe)$_2$Si$_2$O$_6$]	Prismatic crystals, either 4- or 8-sided. Granular masses and scattered grains.	Two good cleavage directions at about 90°.	Green and brown to grayish or greenish white.	5–6
		DIOPSIDE [(Ca,Mg)$_2$Si$_2$O$_6$]			Light to dark green.	
		AUGITE [Complex Ca,Na,Mg,Fe, Al silicate]			Very dark green to black.	
OLIVINE	ISOLATED TETRAHEDRA	[(Mg,Fe)$_2$SiO$_4$]	Granular masses and disseminated small grains.	Conchoidal fracture.	Olive to grayish green and brown.	6½–7
GARNET		[Ca,Mg,Fe,Al silicate]	Isometric crystals, well-formed or rounded; high specific gravity, 3.5–4.3.	Conchoidal and irregular fracture.	Red and brown, less commonly pale colors.	6½–7½
CALCITE	CARBONATES	CaCO$_3$	Coarsely to finely crystalline in beds, veins, and other aggregates. Cleavage faces may show in coarser masses. Calcite effervesces rapidly, dolomite slowly, only in powders.	Three perfect cleavages, at oblique angles; splits to rhombohedral cleavage pieces.	Colorless, transparent to translucent; variously colored by impurities.	3
DOLOMITE		CaMg(CO$_3$)$_2$				3½–4
CLAY MINERALS	HYDROUS ALUMINO-SILICATES	KAOLINITE [Al$_2$Si$_2$O$_5$(OH)$_4$] ILLITE [similar to Muscovite + Mg,Fe] SMECTITE [Complex Ca,Na,Mg,Fe Al silicate + H$_2$O]	Earthy masses in soils; bedded; in association with other clays, iron oxides, or carbonates. Plastic when wet; montmorillonite swells when wet.	Earthy, irregular.	White to light gray and buff; also gray to dark gray, greenish gray, and brownish depending on impurities and associated minerals.	1½–2½
GYPSUM	SULFATES	CaSO$_4$·2H$_2$O	Granular, earthy, or finely crystalline masses. Tabular crystals.	One perfect, splitting to fairly thin slabs or sheets. Two other good cleavages.	Colorless to white. Transparent to translucent.	2
ANHYDRITE		CaSO$_4$	Massive or crystalline aggregates in beds and veins.	One perfect, one nearly perfect, one good; at right angles.	Colorless, some tinged with blue.	3–3½
HALITE		NaCl	Granular masses in beds. Some cubic crystals. Salty taste.	Three excellent cleavages at right angles.	Colorless, transparent to translucent.	2½
OPAL–CHALCEDONY		SiO$_2$ [Opal is an amorphous variety; chalcedony is a formless microcrystalline quartz.]	Beds in siliceous sediments and chert; in veins or banded aggregates.	Conchoidal fracture.	Colorless or white when pure, but tinged with various colors by impurities in bands, especially in agates.	5–6½

LIGHT COLORED, TYPICALLY AS **ABUNDANT** CONSTITUENTS OF **SEDIMENTS** AND **SEDIMENTARY ROCKS**

Mineral or group name		Varieties and chemical composition	Form, diagnostic characters	Cleavage, fracture	Color	Hardness
DARK MINERALS **COMMON** IN MANY ROCK TYPES	**MAGNETITE** (IRON OXIDES)	Fe_3O_4	Magnetic. Disseminated grains, granular masses; occasional octahedral isometric crystals. High specific gravity: 5.2.	Conchoidal or irregular fracture.	Black, metallic luster.	6
	HEMATITE (IRON OXIDES)	Fe_2O_3	Earthy to dense masses, some with rounded forms, some granular or foliated. High specific gravity: 4.9–5.3.	None; uneven, sometimes splintery fracture.	Reddish-brown to black.	$5\frac{1}{2}$–$6\frac{1}{2}$
	"LIMONITE" (IRON OXIDES)	$HFeO_2$ [GOETHITE is the major mineral of the mixture called "limonite," a field term.]	Earthy masses, massive bodies or encrustations, irregular layers. High specific gravity: 3.3–4.7.	One excellent in the rare crystals; usually an earthy fracture.	Yellowish-brown to dark brown and black.	5–$5\frac{1}{2}$
LIGHT COLORED MINERALS, MAINLY IN **IGNEOUS** AND **METAMORPHIC** ROCKS AS **COMMON** OR **MINOR** CONSTITUENTS	**KYANITE** (ALUMINOSILICATES)	Al_2SiO_5	Long, bladed or tabular crystals or aggregates.	One perfect and one poor, parallel to length of crystals.	White to light-colored or pale blue.	5 parallel to crystal length 7 across crystals
	SILLIMANITE (ALUMINOSILICATES)	Al_2SiO_5	Long, slender crystals or fibrous, felted masses.	One perfect parallel to length, not usually seen.	Colorless, gray to white.	6–7
	ANDALUSITE (ALUMINOSILICATES)	Al_2SiO_5	Coarse, nearly square prismatic crystals, some with symmetrically arranged impurities.	One distinct; irregular fracture.	Red, reddish-brown, olive-green	$7\frac{1}{2}$
	FELDSPATHOIDS (ALKALI SILICATES)	NEPHELINE [(Na,K)AlSiO$_4$]	Compact masses or as embedded grains, rarely as small prismatic crystals.	One distinct. Irregular fracture.	Colorless, white, light gray. Gray-greenish in masses, with greasy luster.	$5\frac{1}{2}$–6
	(ALKALI SILICATES)	LEUCITE [KAlSi$_2$O$_6$]	Trapezohedral crystals embedded in volcanic rocks.	One very imperfect.	White to gray.	$5\frac{1}{2}$–6
	SERPENTINE (MAGNESIUM SILICATES)	$Mg_6Si_4O_{10}(OH)_8$	Fibrous (asbestos) or platy masses.	Splintery fracture.	Green; some yellowish, brownish, or gray. Waxy or greasy luster in massive habit; silky luster in fibrous habit.	4–6
	TALC (MAGNESIUM SILICATES)	$Mg_3Si_4O_{10}(OH)_2$	Foliated or compact masses or aggregates.	One perfect, making thin flakes or scales. Soapy feel.	White to pale green. Pearly or greasy luster.	1

Group	Mineral	Class	Formula	Occurrence / Form & Specific Gravity	Cleavage / Fracture	Color / Luster	Hardness
DARK COLORED MINERALS COMMON IN METAMORPHIC ROCKS	CORUNDUM	OXIDE	Al_2O_3	Some rounded, barrel-shaped crystals; most often as disseminated grains or granular (emery) masses.	Irregular fracture.	Usually brown, pink, or blue. Emery black. Gem stone varieties: ruby, sapphire.	9
	EPIDOTE	SILICATES	$Ca_2(Al,Fe)Al_2Si_3O_{12}(OH)$	Aggregates of long prismatic crystals, granular or compact masses, embedded grains.	One good, one poor at greater than right angles. Conchoidal and irregular fracture.	Green, yellow-green, gray, some varieties dark brown to black.	6–7
	STAUROLITE		$Fe_2Al_9Si_4O_{22}(O,OH)_2$	Short prismatic crystals, some cross-shaped, usually coarser than matrix of rock.	One poor.	Brown, reddish, or dark brown to black.	7
METALLIC LUSTER, COMMON IN MANY ROCK TYPES, ABUNDANT IN VEINS	PYRITE	SULFIDES	FeS_2	Granular masses or well-formed cubic crystals in veins and beds or disseminated. Specific gravity high: 4.9–5.2.	Uneven fracture.	Pale brass-yellow.	6–6½
	GALENA		PbS	Granular masses in veins and disseminated. Some cubic crystals. Specific gravity very high: 7.3–7.6.	Three perfect cleavages at mutual right angles, giving cubic cleavage fragments.	Silver-gray.	2½
	SPHALERITE		ZnS	Granular masses or compact crystalline aggregates. Specific gravity high: 3.9–4.1.	Six perfect cleavages at 60° to one another.	White to green, brown, and black. Resinous to submetallic luster.	3½–4
	CHALCOPYRITE		$CuFeS_2$	Granular or compact masses; disseminated crystals. Specific gravity: 4.1–4.3.	Uneven fracture.	Brassy to golden-yellow.	3½–4
	CHALCOCITE		Cu_2S	Fine-grained masses. Specific gravity: 5.5–5.8.	Conchoidal fracture.	Lead-gray to black. May tarnish green or blue.	2½–3
MINERALS FOUND IN MINOR AMOUNTS IN A VARIETY OF ROCK TYPES AND IN VEINS OR PLACERS	RUTILE	TITANIUM OXIDES	TiO_2	Slender to prismatic crystals; granular masses. Specific gravity: 4.25.	One distinct, one less distinct. Conchoidal fracture.	Reddish-brown, some yellowish, violet, or black.	6–6½
	ILMENITE		$FeTiO_3$	Compact masses, embedded grains, detrital grains in sand. Specific gravity: 4.79.	Conchoidal fracture	Iron-black metallic to submetallic luster.	5–6
	ZEOLITES	SILICATES	Complex hydrous silicates, many varieties of minerals, including analcime, natrolite, phillipsite, heulandite, and chabazite.	Well-formed radiating crystals in cavities in volcanics, in veins, and hot springs. Also as fine-grained and earthy bedded deposits.	One perfect for most.	Colorless, white, some pinkish.	4–5

[577]

Topographic and Geologic Maps

A map is a quantitative representation of the spatial distribution of some attribute or property of the Earth. It is a kind of graph in which the axes are lines of latitude and longitude and the positions of points on the surface (or beneath it) are plotted in relation to those axes or some other established reference. Geologists have frequent need of showing the configuration and nature of the geological materials at or near the surface in a meaningful way, so that they can construct a three-dimensional mental picture of the geology from this two-dimensional graph. Once the nature of maps becomes familiar, the map reader can become practiced at deducing much of the geologic structure and history of an area.

The use of topographic and geologic maps has spread widely throughout our culture. To the more traditional users of such maps—the geologists and surveyors—have been added city planners, industrial zoning commissions, and a large number of the public seeking recreational areas for hiking, camping, fishing, and other activities. In 1976 the U.S. Geological Survey distributed nearly 10 million copies of its 40,000 published topographic maps. Maps are a necessity for all kinds of geological and mineral resource studies, as well as for studies of groundwater, flood control, soil management, and such environmental concerns as land-use planning, which involves the location of highways, industrial areas, oil and gas pipelines, and recreational areas. Each may use maps with different emphasis for some specific need, but all follow the same general conventions.

TOPOGRAPHIC MAPS

The beginning of a map is the choice of a suitable graphical framework to use for plotting the position of points on the globe onto a flat piece of paper. Since a curved surface cannot be made to lie flat without distortion, there will necessarily be some adjustments in order to approximate the true size and shape of an area, such as a lake, when it is mapped. Though there are a large number of ways of doing this, most topographic and geologic maps are made by means of **projections,** the term used for the method of transferring a three-dimensional surface onto a two-dimensional one, just as a camera transfers what it "sees" onto photographic film. The projections used for most maps strike a compromise between distortion of shape and maintaining true relative sizes of areas. The deviation from true size and shape is small for maps of relatively small areas of the globe, such as a county or a state because the surface of a small area of this huge globe is for all practical purposes flat.

Because the size of the area covered, and thus the amount of detail that can be shown, is always important, we use the concept of *scale*—that is, the relation of a distance (or area) on the map to the true distance on the Earth. This is simply done by stating a ratio, such as 1:24,000, which indicates that a distance of one unit on the map represents a distance of 24,000 such units on the Earth. It does not matter what the units are: a map of scale 1:24,000 is the same whether we use metric or English systems. The scale can be thought of in any convenient units: 1 inch = 2000 feet, or 1 meter = 24 kilometers, or 10 centimeters = 2.4 kilometers. For convenience, maps have a graphic scale, usually at the bottom margin, in which a distance such as one kilometer or one mile, usually with subdivisions, is shown as it would appear on the map. Common scales for detailed topographic and geologic maps are the 1:24,000, used by the U.S. Geological Survey for most modern maps, or a scale that is somewhat smaller (because the ratio is smaller) of 1:62,500, roughly an inch to a mile, which was used for many of the older maps. The scale used for regional maps covering much larger areas is 1:250,000. Scales of 1:1,000,000 are used for aeronautical charts. In 1976 the U.S. Geological Survey introduced the first of a new series of 1:100,000 all-metric topographic maps on which graphic scales show both kilometers and miles.

On most maps, natural and constructed features of the surface are represented by conventional symbols. Those used by the U.S. Geological Survey are typical. Rivers, lakes, and oceans are shown in blue. Topography is shown in brown. Constructed features are shown in black, with main highways and urban areas in red. Green shaded areas show wooded land. Some special symbols may be shown on the explanation, or **legend,** of the map, which is usually displayed along the bottom margin. Most complex to represent are the topographic elevations of the surface, usually shown on North American maps by contours (see Chapter 5 for discussion of various ways of showing topography). Special maps are sometimes prepared to show environmental variables, such as the distribution of slopes of various steepness.

Topographic maps used to be made entirely by geologists or surveyors in the field. They first established a major network of points accurately located by surveying instruments with respect to latitude, longitude, and elevation. This major network, still used as a framework for most maps, was gradually extended from coast to coast across the continent. Within specific areas to be mapped, smaller networks of surveyed points were established and tied to the major network. Between points the locations were drawn in by sight by practiced topographers: surveyors, topographic engineers, or geologists. One can still see such surveyors operating at new highway or construction sites where high precision is needed. Since the advent of the airplane and the development of aerial photography in the 1930s and 1940s, most topographic maps are made by *photogrammetry,*

the science of making measurements of position and elevation from aerial photographs. Because maps so prepared are checked on the ground and tied to the major surveyed network, they are the most accurate and precise maps available.

GEOLOGIC MAPS

Geologic maps are a representation of the distribution of rocks and other geologic materials of different lithologies and ages over the Earth's surface or below it. The geologist perceives the Earth not only in its surface expression of topography and patterns of land and water but in terms of its pattern of subsurface structures, stratigraphic sequences, igneous intrusions, unconformities, and other geometric relationships of rocks. Just as an anatomist can visualize the muscles and bones beneath the skin, so can a geologist visualize details of the Earth's subsurface. What the map looks like is very much a product of the geologic ideas—the concepts of origin of rocks and structure—that the geologist uses. An area mapped a hundred years ago might look somewhat different today as remapped. For example, some older maps of metamorphosed sedimentary rock terrains show relations of one rock type to another only in terms of lithology. The pattern is a complex and disorganized array of schists, gneisses, and other rock types. The same region, mapped later by a geologist who conceptualized the nature of the original sedimentary rocks by "looking through" the metamorphism, appears as a simpler, more organized pattern of a deformed sedimentary rock sequence with a superimposed pattern of different grades of metamorphism. So even though the rocks themselves have not changed, the way we look at them, and therefore the way we map them, has changed over the years.

Detailed geologic maps are normally constructed on a topographic map base. This serves the useful purpose of making it easy to locate geologic structures with respect to surface features of the Earth. It is also important because topography is so often related to the nature of the underlying rocks and their structures. Because it contains so much more information than a topographic map alone, a geologic map is the most valuable for many of the purposes for which maps are used.

Geologic maps are ordinarily made by a geologist who roams over the area and notes the kinds of rocks, sediments, and soils and their structural and stratigraphic relationships. In modern times

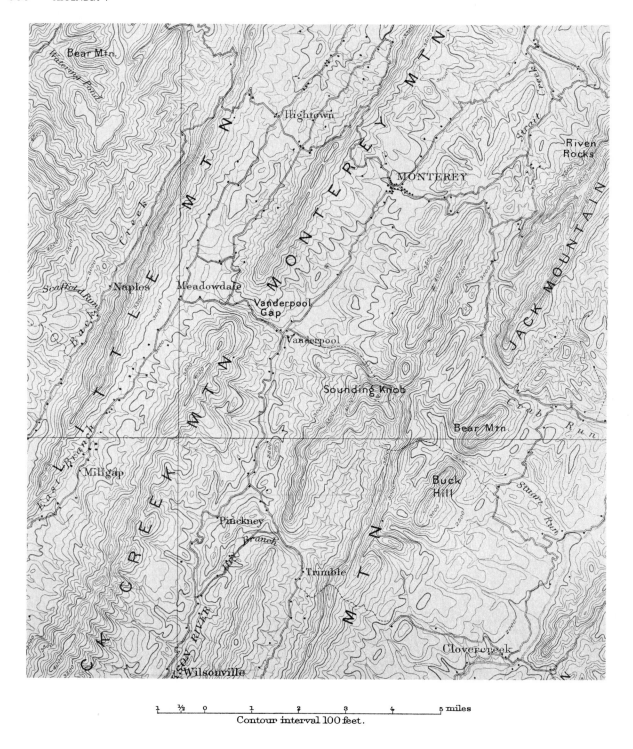

Contour interval 100 feet.

Topographic map (above) and geologic map with cross sections (facing page) of folded sedimentary rocks in the Valley and Ridge province of the Appalachian Mountains. Contours show the pronounced trends of valleys and ridges that reflect the parallel folds. The ridges have developed along the formations that are resistant to erosion, some at the crests of anticlines, such as Jack Mountain north of Crab Run, and others along the flanks of folds, such as Little Mountain. The valleys are in the easily eroded formations, some in synclines, such as Jackson River, some on anticlines, such as East Branch, and some in the flanks of folds,

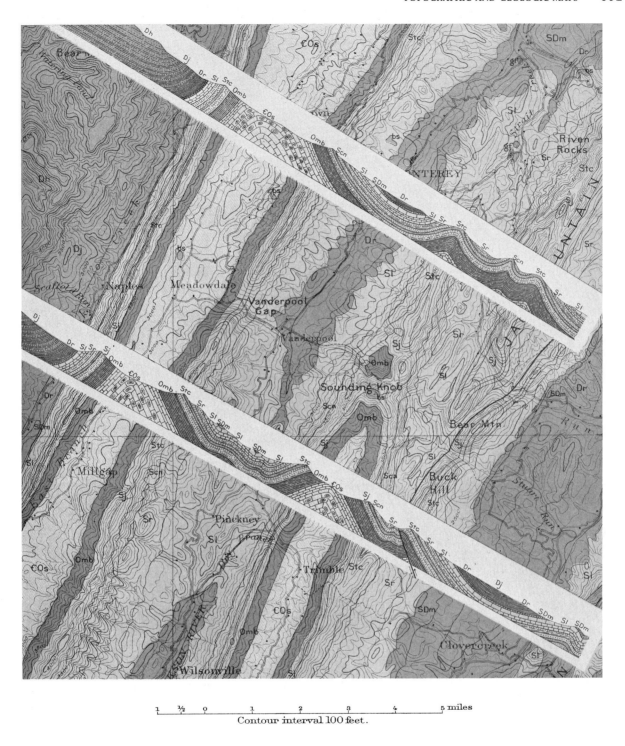

1 ½ 0   1   2   3   4   5 miles

Contour interval 100 feet.

such as Back Creek. On the geologic map the pattern of anticlines and synclines can be read from the positions of formations of different age, such as at East Branch, where the oldest rocks, Cambrian and Ordovician formations (ЄOs), are at the surface bordered on both sides by younger formations of Ordovician and Silurian age (Omb, Stc, Sj, and others). The cross sections make these relationships clearer and add some detail. [From U.S. Geological Survey.]

this is supplemented or even entirely supplanted by remote sensing by aerial photography or geophysical instruments. Remote, inaccessible regions, such as those in some polar or desert regions, may be mapped almost entirely by this method, with the geologist ground checking in scattered places. The mapping of the Moon is an extreme example of this approach. Mars is being mapped with no ground check at all except for the area immediately surrounding the landing site of Viking. The best and most accurate maps, however, are those made by traditional means, geologists covering the ground on foot to see most if not all of the outcrops; they are of course helped enormously by automobile, sometimes by helicopter, and, in some places, by horse or donkey.

The mapping proceeds by the following steps:

1. *Principal observations.* Description of outcrop location, lithology, age, fossil content, and structural attitude as measured by dip and strike, direction of fault movement, fold axes, etc. (see Chapter 20). Plotting observations on work map.

2. *First Integration.* Conceptualizing the spatial relationship of one outcrop to another by stratigraphic correlation of rocks of the same age, facies, degree of metamorphism and deformation. Grouping of mappable rock units into formations. Drawing of lines on the primitive geologic map of inferred connections where formations are hidden. Compilation of the complete or composite stratigraphic sequences, ages of deformational or igneous intrusive events.

3. *Synthesizing the Map.* Visualizing the larger pattern of geologic relationships and constructing the map, together with geologic cross sections made both to help geologists in their thinking and to illustrate more detail and inference from the map.

In the later stages, analyses of rock composition, radiometric determinations of absolute age, and such geophysical information as seismic, gravity, and magnetic data are incorporated in the map. The geologist further draws on the geologic literature or personal experience of the geology of nearby and similar kinds of regions. The final result is the finished geologic map, a codified mass of information displayed in a form in which anyone familiar with geology can quickly read the nature of the Earth's crust in the area and a good deal of its geologic history.

Geologic maps are of many kinds. The most common shows the bedrock geology and gives a picture of what the land would look like if all soil were stripped away. Surficial geological maps, on the other hand, emphasize the nature of soils, unconsolidated river sediment, sand dunes, and whatever other materials, including outcrops, that appear at the surface. A special kind of surficial geological map is used for environmental hazards. One kind shows areas of high-angle or unsupported slopes that are likely to slump or slide (see Chapter 5). Tectonic maps show the disposition of large groups of rocks and their structural relationships to each other. Paleogeologic maps show a geologic map of a former land surface now buried beneath an unconformity. Whatever the geologic purpose, there is a map that can be made to show the relevant data. There is no question that the map is at one time both the best device for geological research into the origin of the distribution of important geologic characteristics over the Earth and the best way to illustrate the patterns discovered from such research.

Glossary

Words in *italic* are defined elsewhere in the glossary.
Specific minerals are defined in Appendix IV.

Aa: A blocky and fragmented form of lava occurring in flows with fissured and angular surfaces.

Ablation zone: The lower part of a glacier, where annual water loss exceeds snow accumulation.

Abyssal hill: A low, rounded submarine hill with a relief of 100 to 200 meters, common in deep ocean basins.

Abyssal plain: A flat, sediment-covered province of the sea floor that slopes at less than 1:1000.

Accumulation zone: The upper part of a glacier, where annual snowfall exceeds melting and evaporation.

A-horizon: The uppermost layer of a soil, containing organic material and leached minerals.

Algal mat: A layered growth of algae observed in fossils and in present-day tidal zones associated with carbonate sedimentation.

Alluvial fan: A low, cone-shaped deposit of terrestrial sediment formed where a stream undergoes an abrupt reduction in slope.

Alluvium: An unconsolidated terrestrial sediment composed of sorted or unsorted sand, gravel, and clay deposited by water.

Alpha (α) particle: A helium nucleus, emitted from a larger nucleus undergoing *radioactive decay*.

Amphibolite: A metamorphic rock containing mostly amphibole and plagioclase feldspar.

Andesite: A volcanic rock type intermediate in composition between *rhyolite* and *basalt*.

Angle of repose: The steepest slope angle at which a particular sediment will lie without cascading down.

Angstrom: A unit of length equal to 10^{-10} meter.

Angular momentum: The product of a body's angular velocity, or rotation rate, and its moment of inertia, which reflects the intensity of rotation of the body.

Angular unconformity: An unconformity in which the bedding planes of the rocks above and below are not parallel.

Anion: Any negatively charged ion; the opposite of *cation*.

Anisotropic: Any material in which physical properties (for example, light transmission or seismic wave velocity) vary quantitatively with the direction in which they are measured.

Anorthosite: A plutonic rock composed mainly of calcium-rich plagioclase feldspar.

Anthracite: The most highly metamorphosed form of coal, containing 92 to 98 percent fixed carbon. It is black, hard, and glassy.

Anticline: A fold, usually from 100 meters to 300 kilometers in width, that is convex upward with the oldest strata at the center.

Antiroot: An accumulation of higher-density material in the sub-oceanic crust that compensates for the low density of sea water.

Aphanite: Generic term for rocks in which individual grains are not visible to the naked eye.

Aphanitic texture: In igneous rocks, having a grain size that is so uniformly small that crystals are invisible to the naked eye.

Aquiclude: An impermeable stratum that acts as a barrier to the flow of groundwater.

Aquifer: A permeable formation that stores and transmits groundwater in sufficient quantity to supply wells.

Arkose: A variety of sandstone containing abundant feldspar and quartz, frequently in angular, poorly sorted grains.

Arroyo: A steep-sided and flat-bottomed gully in an arid region that is occupied by a stream only intermittently, after rains.

Artesian well: A well that penetrates an *aquiclude* to reach an *aquifer* containing water under pressure. Thus water in the well rises above the surrounding water table.

Aseismic region: One that is relatively free of earthquakes. (Actually, all areas show some seismicity over a sufficiently long interval.)

Aseismic ridge: A linear submarine ridge characterized by the absence of seismic activity distinguished from a *mid-oceanic ridge*, which is seismically active.

Ash: See *Volcanic ash*.

Asthenosphere: The layer below the *lithosphere* that is marked by low seismic wave velocities and high *seismic attenuation.* The asthenosphere is a soft layer, probably partially molten.

Astrobleme: A circular erosional feature that has been ascribed to the impact of a meteorite or comet.

Asymmetrical fold: A fold that is inclined to one side. The dips of the two limbs are unequal.

Asymmetrical ripple: A ripple whose cross section is asymmetric, with a gentle slope on the upcurrent side and a steeper face on the downcurrent side.

Atmosphere (atm): A unit of pressure equal to 101,325 newtons per square meter, or about 14.7 pounds per square inch.

Atoll: A continuous or broken circle of coral reef and low coral islands surrounding a central lagoon.

Atomic number: The number of protons in the nucleus of an atom.

Atomic weight: The average weight of one atom of an element, relative to a standard weight of 12 for the carbon-12 isotope.

Augen gneiss: A gneiss containing *phenocrysts* or *porphyroblasts* that have been deformed into eye-shaped grains or grain clusters.

Aureole: The area surrounding an intrusion that has been affected by contact metamorphism.

Axial plane: In folds, the plane that most nearly separates two symmetrical limbs. In a simple *anticline,* it is vertical; in complex folding, it is perpendicular to the direction of compression.

Axis (fold): Within each stratum involved in a fold, the axis connecting all the points in the center of the fold, from which both limbs bend.

Axis of symmetry: An imaginary line about which an object (for example, a crystal) may be rotated $\frac{1}{2}$, $\frac{1}{3}$, $\frac{1}{4}$, $\frac{1}{6}$, or any other simple fraction of a turn without changing its appearance. The denominator corresponds to the order of the axis.

Backwash: The return flow of water down a beach after a wave has broken.

Banded iron ore: A sediment consisting of layers of chert alternating with bands of ferric iron oxides (hematite and limonite) in valuable concentrations.

Bankfull stage: The height of water in a stream that just corresponds to the level of the surrounding floodplain.

Bar: A unit of pressure equal to 10^6 dynes per square centimeter; approximately one atmosphere.

Bar (stream): An accumulation of sediment, usually sandy, that forms at the border or in the channels of streams or offshore from a beach.

Barchan: A crescent-shaped sand dune moving across a clean surface with its convex face upwind and its concave *slip face* downwind.

Bar-finger sand: An elongated lens of sand deposited during the growth of a distributary in a delta. The bar at the distributary mouth is the growing segment of the bar finger.

Barrier island: A long, narrow island parallel to the shore, composed of sand, and built by wave action.

Basalt: A fine-grained, dark, mafic igneous rock composed largely of plagioclase feldspar and pyroxene.

Base level: The level below which a stream cannot erode: usually sea level, sometimes locally the level of a lake or resistant formation.

Basement: The oldest rocks recognized in a given area, a complex of metamorphic and igneous rocks that underlies all the sedimentary formations. Usually Precambrian or Paleozoic in age.

Basic rock: Any igneous rock containing mafic minerals rich in iron and magnesium, but containing no quartz and little sodium-rich plagioclase feldspar. (See preferred term, *mafic rock.*)

Basin: In *tectonics,* a circular, synclinelike depression of strata. In sedimentology, the site of accumulation of a large thickness of *sediments.*

Batholith: A great irregular mass of coarse-grained igneous rock with an exposed surface of more than 100 square kilometers, which has either intruded the country rock or been derived from it through metamorphism.

Bathymetry: The study and mapping of sea-floor topography.

Bauxite: A rock composed primarily of hydrous aluminum oxides and formed by weathering in tropical areas with good drainage; a major ore of aluminum.

Bedding: A characteristic of sedimentary rocks in which parallel planar surfaces separating different grain sizes or compositions indicate successive depositional surfaces formed at the time of sedimentation.

Bed load: The sediment that a stream moves along the bottom of its channel by rolling and bouncing (*saltation*).

Beta (β) particle: An electron emitted with high energy and velocity from a nucleus undergoing *radioactive decay.*

B-horizon: The intermediate layer in a soil, below the A-horizon, consisting of clays and oxide minerals. Also called the zone of accumulation.

Bicarbonate ion: The anion group $HCO_3{}^-$.

Biochemical precipitate: A sediment, especially of limestone or iron minerals, formed from elements extracted from sea water by living organisms.

Bioturbation: Reworking of existing sediments by organisms.

Bituminous coal: A soft coal formed by an intermediate degree of metamorphism and containing 15 to 20 percent *volatiles.* The most common grade of coal.

Block fault: A structure formed when the crust is divided into blocks of different elevation by a set of *normal faults.*

Blowout: A shallow circular or elliptical depression in sand or dry soil formed by wind erosion. (See also *deflation.*)

Blueschist: A metamorphic rock formed under conditions of high pressure (in excess of 5000 bar) and relatively low temperature, often containing the blue minerals glaucophane (an amphibole) and kyanite.

Bolson: In arid regions, a basin filled with *alluvium* and intermittent *playa* lakes and having no outlet.

Bomb: See *Volcanic bomb.*

Bottomset bed: A flat-lying bed of fine sediment deposited in front of a *delta* and then buried by continued delta growth.

Bouguer correction: The correction of a measured gravity value by an amount theoretically calculated to compensate for the mass of known topography around the station.

Bowen's reaction series: A simple schematic description of the order in which different minerals crystallize during the cooling and progressive crystallization of a *magma*.

Braided stream: A stream so choked with *sediment* that it divides and recombines numerous times, forming many small and meandering channels.

Breaker: An ocean wave that becomes so steep on encountering shallow water that it collapses turbulently.

Breccia: A clastic rock composed mainly of large angular fragments.

Breeder reactor: A planned nuclear reactor that would use the high-energy particles created in fission to create more fissionable fuel from stable uranium or thorium isotopes.

Brine: Sea water whose salinity has been increased by evaporation, or groundwater with an unusually high concentration of salts.

Brittle: Describes a material that breaks abruptly when its *elastic limit* is reached; the opposite of *ductile*.

Butte: A steep-sided and flat-topped hill formed by erosion of flat-lying strata where remnants of a resistant layer protect softer rocks underneath.

Calcium carbonate compensation depth: The depth in the oceans below which the solution rate of $CaCO_3$ becomes so great that no carbonate organisms or sediments are preserved on the sea floor.

Caldera: A large, circular depression in a volcanic terrain, typically originating in collapse, explosion, or erosion.

Canyon: A very large, deep valley with precipitous walls formed mainly by stream downcutting.

Capacity (stream): The amount of sediment and detritus a stream can transport past any point in a given time.

Carbonate ion: The anion group CO_3^{2}.

Carbonate platform: A submarine or intertidal shelf whose elevation is maintained by active shallow-water carbonate deposition.

Carbonate rock: A rock composed of carbonate minerals, especially *limestone* and *dolomite*.

Carbon-14 activity: The number of atoms of the isotope of carbon with atomic weight 14 that decay radioactively in a unit time.

Carbonic acid: The weak acid H_2CO_3, formed by the dissolution of CO_2 in water.

Cataclastic rock: A breccia or powdered rock formed by crushing and shearing during tectonic movements.

Cation: Any ion with a positive electric charge.

Central vent: The largest vent of a volcano, situated at the center of its cone.

Chalk: The lithified equivalent of nannoplankton ooze.

Chemical bonding: The process by which two or more atoms of the same or different elements combine to form a molecule.

Chemical differentiation: The formation of more than one igneous rock composition from a common magma as a result of crystals settling out as they form, thus changing the composition of the remaining melt.

Chemical sediment: One that is formed at or near its place of deposition by chemical precipitation, usually from sea water.

Chemical weathering: The total set of all chemical reactions that act on rock exposed to water and atmosphere and so change its minerals to more stable forms.

Chert: A sedimentary form of amorphous or extremely fine-grained silica, partially hydrous; commonly a chemical sediment.

C-horizon: The lowest layer of a soil, consisting of fragments of rock and their chemically weathered products.

Cinder cone: A steep, conical hill built up about a volcanic vent and composed of coarse pyroclasts expelled from the vent by escaping gases.

Cirque: The head of a glacial valley, usually with the form of one half of an inverted cone. The upper edges have the steepest slopes, approaching the vertical, and the base may be flat or hollowed out and occupied by a small lake or pond.

Clastic rock: A sedimentary rock formed from particles (clasts) that were mechanically transported.

Clay: Any of a number of hydrous aluminosilicate minerals with sheetlike crystal structure, formed by weathering and hydration of other silicates; also, any mineral fragments smaller than $1/256$ mm.

Cleavage: (1) Of minerals: the tendency of a crystal to break along certain preferred planes in the crystal lattice; also, the geometric pattern of such breakage. (2) Of rocks: the tendency of a rock to break along certain planes; usually the result of *preferred orientation* of the minerals in the rock.

Coal: The metamorphic product of stratified plant remains. It contains more than 50 percent carbon compounds and burns readily.

Coastal plain: A low plain of little relief adjacent to the ocean and covered with gently dipping sediments.

Coccoliths: A type of marine algae with calcareous shells; also called coccolithophorids.

Coefficient of thermal expansion: The increase in volume of a material with increasing temperature relative to the original volume.

Collision hypothesis: The theory that the material forming the planets was separated from the solar mass by an interstellar collision or near miss.

Columnar jointing: The division of an igneous rock body into prismatic columns by cracks produced by thermal contraction on cooling.

Compaction: The decrease in volume and porosity of a sediment caused by burial.

Compensation (gravity): The mechanism by which segments of the crust rise or sink to equilibrium positions, depending upon the mass and density of the rocks above and below a certain depth, called the *compensation depth*.

Compensation depth: In relation to gravity, the critical depth in the above definition. (See also *calcium carbonate compensation depth*.)

Competence (rock): The ability of a stratum to withstand deformation without fracturing or changing in thickness.

Competence (stream): A measure of the largest particle a stream is able to transport, not the total volume.

Composite cone: The volcanic cone of a *stratovolcano*, composed of both cinders and lava flows.

Concentration factor: The ratio of the abundance of an element in a mineral deposit to its average abundance in the crust.

Concordant contact: The planar contact of an intrusion that follows the bedding of the country rock.

Concordant stratification: Sedimentary layering in which successive bedding planes are parallel.

Conduction: See *heat conduction.*

Cone: See *volcanic cone.*

Confined water reservoir: A body of groundwater surrounded by impermeable strata.

Conformable succession: A sequence of sedimentary rocks that indicates continuous deposition with no erosion over a geologically long period.

Conglomerate: A sedimentary rock, a significant fraction of which is composed of rounded pebbles and boulders; the lithified equivalent of *gravel.*

Conservation of angular momentum: The physical law that the total angular momentum of a system of isolated bodies cannot change without outside interference.

Contact metamorphism: Mineralogical and textural changes and deformation of rock resulting from the heat and pressure of an igneous intrusion in the near vicinity.

Continental divide: An imaginary line connecting high points across a continent and dividing regions whose streams drain into one ocean from regions that drain into another.

Continental drift: The horizontal displacement or rotation of continents relative to one another.

Continental glacier: A continuous, thick glacier covering more than 50,000 square kilometers and moving independently of minor topographic features.

Continental rise: A broad and gently sloping ramp that rises from an abyssal plain to the *continental slope* at a rate of less than 1:40.

Continental shelf: The gently sloping submerged edge of a continent, extending commonly to a depth of about 200 meters or the edge of the continental slope.

Continental shelf deposits: Sediments laid down in a tectonically quiet *syncline* at a passive continental margin.

Continental slope: The region of steep slopes between the continental shelf and continental rise.

Continuous reaction series: A reaction series in which the same mineral crystallizes throughout the range of temperatures in question, but in which there is a gradual change in the chemical composition of the mineral with changing temperature.

Contour map: A map that shows topography by means of contour lines. Each contour line connects points of equal elevation, and the elevation interval between lines is constant.

Convection: A mechanism of heat transfer in a flowing material in which hot material from the bottom rises because of its lesser density while cool surface material sinks.

Convection cell: A single closed flow circuit of rising warm material and sinking cold material.

Convergence zone: A band along which moving plates collide and area is lost either by shortening and crustal thickening or by subduction and destruction of crust. The site of volcanism, earthquakes, trenches, and mountain-building.

Coordination number: In a mineral, the number of ions bonded to a given ion of opposite charge.

Coral: Organisms, mostly colonial, with carbonate skel-etons that live in sea water at sufficiently shallow depths to receive light from the surface.

Cordillera: If capitalized, the continuous mountain system extending from Alaska to extreme South America and ranging up to 1500 kilometers in width. If not, any similar chain of parallel mountain ranges.

Core: The central part of the Earth below a depth of 2900 kilometers. It is thought to be composed of iron and nickel and to be molten on the outside with a central solid inner core.

Coriolis effect: An apparent force that a moving object feels, tending to the right in the Northern Hemisphere and to the left in the Southern. It is due to the Earth's rotation, which requires that objects near the equator have a greater eastward velocity than those nearer the poles.

Country rock: The rock into which an igneous rock intrudes or a mineral deposit is emplaced.

Covalent bond: A bond between atoms in which outer electrons are shared.

Crater: An abrupt circular depression formed by extrusion of volcanic material and its deposition in a surrounding rim or by explosive ejection of matter upon meteorite impact.

Craton: A portion of a continent that has not been subjected to major deformation for a prolonged time, typically since Precambrian or early Paleozoic time.

Crevasse: Any large vertical crack in the surface of a glacier or snowfield.

Cross-bedding: Inclined beds of depositional origin in a sedimentary rock. Formed by currents of wind or water in the direction which the bed slopes downward.

Cross-cutting: Having *discordant contacts* with the *country rock.*

Cross section: A drawing showing the features that would be exposed by a vertical cut through a structure.

Crust: The outermost layer of the *lithosphere* consisting of relatively light, low-melting materials. The continental crust consists largely of *granite* and granodiorite; the oceanic crust is mostly *basalt.*

Crystal: A form of matter in which the atoms, ions, or molecules are arranged regularly in all directions to form a regular, repeating network.

Crystal face: A planar growth surface of a crystal.

Crystallinity: The proportion of crystal (as opposed to *glass*) in an igneous rock.

Cuesta: A ridge with one steep and one gentle face formed by the outcrop and slower erosion of a resistant, gently dipping bed.

Curie point: The temperature above which a given mineral cannot retain any permanent magnetization.

Dacite: Volcanic equivalent of *granodiorite.*

Datum plane: An artificially established, well-surveyed horizontal plane against which elevations, depths, tides, etc. are measured (for example, *mean sea level*).

Daughter element: Also "daughter product." An element that occurs in a rock as the end-product of the radioactive decay of another element.

Debris-avalanche: A fast downhill *mass movement* of soil and rock.

Decay constant: A constant, different for each radioactive isotope, which indicates how rapidly that isotope breaks down spontaneously by *radioactive decay*. (See also *half-life*.)

Declination: At any place on Earth, the angle between the magnetic and rotational poles.

Deflation: The removal of clay and dust from dry soil by strong winds.

Delta: A body of sediment deposited in an ocean or lake at the mouth of a stream.

Delta kame: A deposit having the form of a steep, flat-topped hill, left at the front of a retreating continental glacier.

Dendritic drainage: A stream system that branches irregularly and resembles a branching tree in plan.

Density: The mass per unit volume of a substance, commonly expressed in grams per cubic centimeter.

Density current: A current that flows on the bottom of a sea or lake because entering water is more dense due to temperature or suspended sediments.

Deposition: A general term for the accumulation of sediments by either physical or chemical sedimentation.

Depositional remanent magnetization: A weak magnetization created in sedimentary rocks by the rotation of magnetic crystals into line with the ambient field during settling.

Depth recorder: See *echo-sounder*.

Desert pavement: A residual deposit produced by continued deflation, which removes the fine grains of a soil and leaves a surface covered with close-packed cobbles.

Detrital sediment: A sediment deposited by a physical process.

Desert varnish: A dark coating commonly found on the surface of rock in the desert. It consists of clays, iron oxides, and magnesium oxides, produced during weathering.

Diagenesis: The physical and chemical change undergone by a sediment during *lithification* and *compaction*, excluding *erosion* and *metamorphism*.

Diatom: A one-celled plant that has a siliceous framework and grows in oceans and lakes.

Diatomite: A siliceous chertlike sediment formed from the hard parts of diatoms.

Diatom ooze: A fine, muddy sediment consisting of the hard parts of diatoms.

Diatreme: A volcanic vent filled with breccia by the explosive escape of gases.

Differentiated planet: One that is chemically zoned because heavy materials have sunk to the center and light materials have accumulated in a crust.

Diffraction pattern: The pattern formed by the reflection of x rays from the internal planes of atoms when a crystal is exposed to a beam of x radiation.

Dike: A roughly planar body of intrusive igneous rock that has *discordant contacts* with the surrounding rock.

Dike swarm: A group of dikes emanating from a common magma chamber.

Diorite: A plutonic rock with composition intermediate between *granite* and *gabbro*; the intrusive equivalent of *andesite*.

Dip: The maximum angle by which a stratum or other planar feature deviates from the horizontal. The angle is measured in a plane perpendicular to the strike.

Dip needle: A type of compass that measures the inclination of the Earth's magnetic field from the horizontal.

Dip-slip fault: A fault in which the relative displacement is along the direction of dip of the fault plane; either a normal or a reverse fault.

Discharge: The rate of water movement through a stream, measured in units of volume per unit time.

Disconformity: A stratigraphic boundary where bedding planes of sediments on either side of the boundary are parallel, but where there is evidence that a geologically significant time span is not represented in the sedimentary record—that is, that sedimentation has not been continuous.

Discontinuity: See *seismic discontinuity*.

Discontinuous reaction series: A *reaction series* in which the end members have different crystal structures (are distinct mineral phases).

Discordant contact: A contact that cuts across *bedding* or *foliation* planes, such as the contact between a *dike* and the *country rock*.

Disintegration constant: See *decay constant*.

Disseminated deposit: A deposit of ore in which the metal is distributed in small amounts throughout the rock, not concentrated in veins.

Distributary: A smaller branch of a large stream that receives water from the main channel; the opposite of a tributary.

Divergence zone: A belt along which plates move apart and new crust and lithosphere are created: the site of mid-ocean ridges, shallow-focus earthquakes, and volcanism.

Divide: A ridge of high ground separating two drainage basins emptied by different streams.

Dolomite: A carbonate mineral with the general formula $CaMg(CO_3)_2$; also, a sedimentary rock composed primarily of this mineral.

Dome: In structural geology, a round or elliptical upwarp of strata resembling a short anticline.

Drainage basin: A region of land surrounded by divides and crossed by streams that eventually converge to one river or lake.

Drift (continental): See *continental drift*.

Drift (geomorphology): A pileup of sand and sediment to the leeward (downcurrent) side of an obstacle.

Drift (glacial): A collective term for all the rock, sand, and clay that is deposited by a glacier either as *till* or as *outwash*.

Dripstone: See *stalactite, stalagmite*.

Drowned valley: A river valley flooded by the sea when sea level rises or the land subsides.

Drumlin: A smooth, streamlined hill composed of *till*.

Dry wash: An intermittent stream bed in an *arroyo* or *canyon* that carries water only briefly after a rain.

Ductile material: A material that is capable of undergoing considerable change in shape without breaking.

Ductile rock: A rock that can withstand 5 to 10 percent *strain* without fracturing after its elastic limit has been reached.

Dune: An elongated mound of sand formed by wind or water.

Dunite: An *ultramafic rock* composed almost entirely of the mineral olivine.

Duricrust: A hard surface crust made up of the products of chemical weathering in the desert; it can consist of silica, calcium carbonate, iron oxides, and other materials.

Earthflow: A detachment of soil and broken rock and its subsequent downslope movement at slow or moderate rates in a stream- or tonguelike form.

Earthquake: The violent oscillatory motion of the ground caused by the passage of seismic waves radiating from a fault along which sudden movement has taken place.

Ebb tide: The part of the tide cycle during which the water level is falling.

Echo-sounder: An oceanographic instrument that emits sound pulses into the water and measures its depth by the time elapsed before they return.

Ecliptic: The plane that contains the Earth's orbit around the Sun.

Eclogite: An extremely high-pressure metamorphic rock containing garnet and pyroxene.

Ecology: The study of the interactions among different species of organisms and their relationships with the physical environment.

Economic geology: The study of and location of ores, fossil fuels, and other useful materials that occur in sufficient concentration to be marketable in a particular economy.

Elastic rebound theory: A theory of fault movement and earthquake generation that holds that faults remain locked while strain energy accumulates in the rock, and then suddenly slip and release this energy.

Electron: A negatively charged atomic particle with a mass of 9.1×10^{-28} gram and negative charge of 1.6×10^{-19} coulomb. The position of an electron about an atomic nucleus is not fixed, but is described by a probability statement.

Electron microprobe: An instrument that bombards a minute sample with electrons to determine its chemical composition from the resulting x radiation.

Electron sharing: See *covalent bond.*

Electron structure of elements: The arrangement of electrons into shells surrounding the nucleus. This arrangement determines the electrical characteristics of an atom, and in part, controls its bonding properties.

Elevation: The vertical height of one point on the Earth above a given *datum plane,* usually sea level.

Elliptical orbit: An orbit with the shape of a geometrical ellipse. All orbits are elliptical or hyperbolic, with the Sun occupying one focus.

Emission spectroscopy: The identification of elements by the characteristic wavelengths of light that they emit when heated. A prism separates the spectrum into characteristic bright lines.

Environment of deposition: A geographically limited area where sediments are preserved; characterized by its landforms, relative energy of currents, and chemical equilibria.

Eolian: Pertaining to or deposited by wind.

Eon: The largest division of geologic time, embracing several *eras,* (for example, the Phanerozoic, 600 m.y. ago to present); also any span of one billion years.

Epeirogeny: Large-scale, primarily vertical movement of the crust. It is characteristically so gradual that rocks are little folded and faulted.

Epicenter: The point on the Earth's surface directly above the focus or hypocenter of an earthquake.

Epoch: One subdivision of a geologic period, often chosen to correspond to a stratigraphic series. Also used for a division of time corresponding to a paleomagnetic interval.

Equilibrium: A stable state of a system. Once equilibrium has been reached, no further net change in the chemical or physical state of materials in the system, or in their proportions, will occur without some external interference (such as a change in temperature or pressure or addition of some component).

Era: A geological division including several periods, but smaller than an eon. Commonly recognized eras are Precambrian, Paleozoic, Mesozoic, and Cenozoic.

Erosion: The set of all processes by which soil and rock are loosened and moved downhill or downwind.

Esker: A glacial deposit in the form of a continuous, winding ridge, formed from the deposits of a stream flowing beneath the ice.

Estuary: A body of water along a coastline, open to the ocean but diluted by fresh water.

Eugeosyncline: The seaward part of a *geosyncline;* characterized by clastic sediments and volcanism.

Eustatic change: Sea-level changes that affect the whole Earth.

Eutrophication: A superabundance of algal life in a body of water; caused by an unusual influx of nitrate, phosphate, or other nutrients.

Evaporite: A chemical sedimentary rock consisting of minerals precipitated by evaporating waters; especially salt and gypsum.

Excess volatiles: Chemical constituents of the Earth's crust, including water, carbon dioxide, and chloride, that are more abundant than would be expected on the basis of their abundances in igneous rocks. Their discovery led to the suggestion that these materials were released from deep within the Earth through volcanism.

Exfoliation: A physical weathering process in which sheets of rock are fractured and detached from an *outcrop.*

Exobiology: The study of life outside the Earth.

Experimental petrology: Laboratory study of the response of rocks and minerals to different physical and chemical conditions, often using special apparatus to simulate the high pressures and temperatures found deep in the crust and upper mantle.

Extrusive rock: One formed from *lava* or from other volcanic material spewed out onto the surface of the Earth.

Facies: The set of characteristics of a sedimentary rock that indicates its particular *environment of deposition* and that distinguishes it from other facies in the same rock unit. Also, for a metamorphic rock, the particular range of pressure and temperature under which metamorphism occurred.

Fathometer: See *echo-sounder.*

Fault: A planar or gently curved fracture in the Earth's crust across which there has been relative displacement.

Fault-block mountain: A mountain or range formed as a horst when it was elevated (or as the surrounding region sank) between two normal faults.

Fault plane: The plane that best approximates the fracture surface of a fault.

Faunal succession: The sequence of life forms, especially as recorded by the fossil remains in a stratigraphic sequence.

Feedback: Any system in which a process controls itself. See also *negative feedback process* and *positive feedback process.*

Feldspar: General term for a group of aluminosilicate minerals containing sodium, calcium, or potassium and having a framework structure. Feldspars are the most common minerals in the Earth's crust.

Felsic: An adjective used to describe a light-colored igneous rock that is poor in iron and magnesium and contains abundant feldspars and quartz.

Ferric iron: Iron with a $+3$ *valence* (Fe^{3+}).

Ferrous iron: Iron with a $+2$ *valence* (Fe^{2+}).

Fetch: The distance that wind blows over the water in producing waves.

Field relations: The pattern of contacts, faults, intrusions, and other boundaries where rock formations meet, from which the field geologist infers the chronology and history of an area.

Fining-upward alluvial cycle: A characteristic sequence of sedimentary rocks in which coarse-grained material at the bottom grades into finer and finer material toward the top. Such sequences are termed cyclic because they tend to recur at intervals, as in the deposition of *turbidites* or the migration of a meandering stream.

Fiord: A former glacial valley with steep walls and a U-shaped profile, now occupied by the sea.

Firn: Old, dense, compacted snow.

First motion: On a seismogram, the direction of ground motion at the beginning of the arrival of a *P*-wave. Upward ground motion indicates a compression; downward motion, a dilation.

Fissility: A breaking property of rock, characterized by a tendency to fracture along parallel planes resulting in tabular fragments.

Fission tracks: Scars on the order of 10^{-3} centimeters in length produced in minerals as a result of the release of energetic atomic particles during radioactive decay of contained unstable atoms. ^{238}U, for example, may leave fission tracks in mica minerals upon decaying. Fission tracks can be used in dating rocks.

Flood basalt: A plateau basalt extending many kilometers in flat, layered flows originating in fissure eruptions.

Floodplain: A level plain of stratified alluvium on either side of a stream; submerged during floods and built up by silt and sand carried out of the main channel.

Flood tide: The part of the tide cycle during which the water is rising or leveling off at high water.

Flow cleavage: In a metamorphic rock, the parallel arrangement of all planar or linear crystals as a result of rock flowage during metamorphism.

Fluid inclusion: A small body of fluid that is entrapped in a crystal and has the same composition as the fluid from which the crystal formed.

Flume: A laboratory model of stream flow and sedimentation consisting of a rectangular channel filled with sediment and running water.

Focus (earthquake): The point at which the rupture occurs; synonymous with *hypocenter.*

Fold: A bent or warped stratum or sequence of strata, which was originally horizontal, or nearly so, and was subsequently deformed.

Fold axis: See *axis.*

Fold belt: Synonym of *orogenic belt.*

Foliation: Any planar set of minerals or banding of mineral concentrations, including cleavage, found in a metamorphic rock.

Foraminifer: An order of oceanic protozoa most of which have shells composed of calcite.

Foraminiferal ooze: A calcareous sediment composed of the shells of dead foraminifera.

Foreset bed: One of the inclined beds found in *cross-bedding;* also an inclined bed deposited on the outer front of a *delta.*

Formation: The basic unit for the naming of rocks in lithostratigraphy: a set of rocks that are or once were horizontally continuous, that share some distinctive feature of lithology, and that are large enough to be mapped.

Fossil: An impression, cast, outline, track, or body part of an animal or plant that is preserved in rock after the original organic material is transformed or removed.

Fossil fuel: A general term for combustible geologic deposits of carbon in reduced (organic) form and of biological origin, including coal, oil, natural gas, oil shales, and tar sands.

Fossil magnetism: See *paleomagnetism.*

Fractional crystallization: The separation of a cooling magma into components by successive formation of and removal of crystals at progressively lower temperatures.

Fractionation: See *magmatic differentiation.*

Fracture (mineralogy): The irregular breaking of a crystal along a surface not parallel to a crystal face. The nature of the surface texture is considered diagnostic.

Fracture cleavage: A set of closely spaced, parallel joints in a rock resulting from deformation and metamorphism.

Free-air correction: In gravity, the addition to a measured value of gravity of an amount calculated theoretically to correct for the effect of elevation only.

Free oscillation: The "ringing" or periodic deformation of the whole Earth at characteristic low frequencies after a major earthquake.

Friction breccia: A breccia formed in a fault zone or volcanic pipe by the relative motion of two rock bodies.

Fringing reef: A coral reef that is directly attached to a landmass not composed of coral.

Fumarole: A small vent in the ground from which emerge volcanic gases and heated groundwater, but no lava.

Gabbro: A black, coarse-grained, intrusive igneous rock, composed of calcic feldspars and pyroxene. The intrusive equivalent of basalt.

Geochronology: The science of absolute dating and relative dating of geologic formations and events, primarily through the measurement of the abundances of radioactive *parent* and *daughter* elements in minerals and rocks.

Geologic cycle: The sequence through which rock material passes in going from its sedimentary form, through diagenesis and deformation of sedimentary rock, then through metamorphism and eventual melting and magma formation, then through volcanism and plutonism to igneous rock formation, and finally through erosion to form new sediments.

Geomorphic cycle: An idealized model of erosion wherein a plain is uplifted epeirogenically, then dissected by rapid streams (youth), then rounded by downslope movements into a landscape of steep hills (maturity), and finally reduced to a new *peneplain* at sea level (old age).

Geomorphology: The science of surface landforms and their interpretation on the basis of geology and climate.

Geosyncline: A major downwarp in the Earth's crust, usually more than 1000 kilometers in length, in which sediments accumulate to thicknesses of many kilometers. The sediments may eventually be deformed and metamorphosed during a mountain-building episode.

Geotherm: A curve on a temperature-pressure or temperature-depth graph that describes how temperature in the Earth changes with depth. Different tectonic provinces are characterized by more or less rapid increases of temperature with depth. Alternatively, a curving surface within the Earth along which the temperature is constant (geoisotherm).

Geothermal power: Power generated by utilizing the heat energy of the crust, especially in volcanic regions.

Geyser: A hot spring that forcibly ejects hot water and steam into the air. The heat is thought to result from the contact of groundwater with magma bodies.

Giant planet: One of the large planets located beyond the orbit of Mars, as opposed to the terrestrial planets.

Glacial rebound: Epeirogenic uplift of the crust that takes place after the retreat of a continental glacier, in response to earlier subsidence under the weight of the ice.

Glacial striations: Scratches left on bedrock and boulders by overriding ice and showing the direction of glacial motion.

Glacial valley: A valley occupied or formerly occupied by a glacier, typically with a U-shaped profile.

Glacier: A mass of ice and surficial snow that persists throughout the year and flows downhill under its own weight. The size range is from 100 meters to 10,000 kilometers.

Glacier surge: A period of unusually rapid movement of a glacier, sometimes lasting more than one year.

Glass: A rock formed when magma is cooled too rapidly (quenched) to allow crystal growth.

Glassiness: The content or extent of glass in an igneous rock.

Gneiss: A coarse-grained regional metamorphic rock that shows compositional banding and parallel alignment of minerals.

Gondwanaland: A hypothetical supercontinent comprising approximately the present continents of the Southern Hemisphere.

Graben: A downthrown block between two normal faults of parallel strike but converging dips; hence a tensional feature. (See also *horst*.)

Graded bedding: A bed in which the coarsest particles are concentrated at the bottom and grade gradually upward into fine silt, the whole bed having been deposited by a waning current.

Graded stream: A stream whose smooth profile is unbroken by resistant ledges, lakes, or waterfalls, and which exactly maintains the velocity required to carry the sediment provided to it.

Granite: A coarse-grained, intrusive igneous rock composed of quartz, orthoclase feldspar, sodium-rich plagioclase feldspar, and micas. Also sometimes a metamorphic product.

Granitization: The formation of metamorphic granite from other rocks by recrystallization with or without complete melting.

Granodiorite: *Plutonic* rock similar to *granite* in composition, except that *plagioclase feldspar* is present in greater abundance than orthoclase feldspar.

Granular snow: Snow that has been metamorphosed into small granules of ice.

Granulite: A regional metamorphic rock with coarse interlocking grains, generally formed under conditions of relatively high pressure and temperature.

Gravel: The coarsest of alluvial sediments, containing mostly particles larger than 2 millimeters and including cobbles and boulders.

Gravimeter: An instrument for measuring the force of gravity at a given point on the Earth.

Gravity anomaly: The value of gravity left after subtracting from a gravity measurement the reference value based on latitude and any free-air and Bouguer corrections.

Gravity survey: The measurement of gravity at regularly spaced grid points with repetitions to control instrument drift.

Graywacke: A poorly sorted sandstone containing abundant feldspar and rock fragments, often in a clay-rich matrix.

Greenhouse effect: The heating of the atmosphere by the absorption of infrared energy reemitted by the Earth as it receives light energy in the visible band from the Sun.

Greenschist: A metamorphic *schist* containing chlorite and epidote (which are green) and formed by low-temperature, low-pressure metamorphism.

Ground moraine: A glacial deposit of till with no marked relief, interpreted as having been transported at the base of the ice.

Groundwater: The mass of water in the ground below the *phreatic zone,* occupying the total pore space in the rock and moving slowly downhill where *permeability* allows.

Gully: A small steep-sided valley or erosional channel one meter to about ten meters across.

Guyot: A flat-topped submarine mountain or *seamount.*

Gyre: The circular rotation of the waters of each major sea, driven by prevailing winds and the *coriolis effect.*

Half-life: The time required for half of a sample of a given radioactive isotope to decay. The half-life of an isotope is inversely related to its *decay constant.*

Hanging valley: A former glacial tributary valley that enters a larger glacial valley above its base, high up on the valley wall.

Hard coal: See *anthracite.*

Hard water: Water that contains enough dissolved calcium and magnesium to cause a carbonate scale to form when the water is boiled or to prevent the sudsing of soap.

Harmonic tremor: Seismic vibrations that emanate from a volcanic tube as lava flows through it.

Heat conduction: The transfer of the vibrational energy of atoms and molecules, which constitutes heat energy, by the mechanism of atomic or molecular impact.

Heat engine: A device that transfers heat from a place of high temperature to a place of lower temperature and does mechanical work in the process.

Heat flow: The rate at which heat escapes at the Earth's surface, related to the nature of the surface rocks and the rate at which heat is supplied to the crust from below.

Hill: A natural land elevation, usually less than 300 meters (1000 feet) above its surroundings, with a rounded outline. The distinction between hill and mountain depends on the locality.

Hogback: A formation similar to a *cuesta* in that it is a ridge formed by slower erosion of hard strata, but having two steep, equally inclined slopes.

Hornfels: A high-temperature, low-pressure metamorphic rock of uniform grain size showing no foliation. Usually formed by contact metamorphism.

Horst: An elongate, elevated block of crust forming a ridge or plateau, typically bounded by parallel, outward-dipping normal faults.

Hot spot: The surface expression of a mantle *plume.*

Hot spring: A spring whose waters are above both human body and soil temperature as a result of *plutonism* at depth.

Humus: The decayed part of the organic matter in a soil.

Hydration: A chemical reaction, usually in weathering, that adds water or OH^- to a mineral structure.

Hydrocarbon: An organic chemical compound made up of carbon and hydrogen atoms arranged in chains or rings.

Hydrologic cycle: The cyclical movement of water from the ocean to the atmosphere, through rain to the surface, through runoff and groundwater to streams, and back to the sea.

Hydrology: The science of that part of the *hydrologic cycle* between rain and return to the sea; the study of water on and within the land.

Hydrothermal activity: Any process involving high-temperature groundwater, especially the alteration and emplacement of minerals and the formation of *hot springs* and *geysers.*

Hydrothermal vein: A cluster of minerals precipitated by hydrothermal activity in a rock cavity.

Hypocenter: See *focus (earthquake).*

Hypsometric diagram: A graph that shows in any way the relative amounts of the Earth's surface at different elevations with regard to sea level.

Igneous rock: A rock formed by the solidification of a magma.

Ignimbrite: An igneous rock formed by the *lithification* of *volcanic ash* and *volcanic breccia.*

Inclination: The angle between a line in the Earth's magnetic field and the horizontal plane; also a synonym for *dip.*

Index mineral: A mineral that forms or is stable over a limited range of pressures and temperatures; useful in determining the conditions under which a rock has been metamorphosed.

Index of refraction: The ratio of the speed of light in a vacuum to the speed in a material; this ratio determines the amount that light is refracted as it passes into a crystal.

Infiltration: The movement of groundwater or hydrothermal water into rock or soil through joints and pores.

Interfacial angle: The angle between two faces of a crystal. The law of constancy of interfacial angles states that, for any particular pair of crystal faces of a given mineral, this angle is constant, for it arises from the characteristic crystal symmetry of the mineral.

Interior drainage: A system of streams that converges in a closed basin and evaporates without reaching the sea.

Intermontane basin: A basin between mountain ranges, often formed over a graben.

Intrusion: Igneous rock body that has forced its way in a molten state into surrounding *country rock.*

Intrusive rock: Igneous rock that is interpreted as a former intrusion from its *cross-cutting* contacts, chilled margins, or other *field relations.*

Ion: An atom or group of atoms that has gained or lost electrons and so has a net electric charge.

Ionic bond: A bond formed between atoms by electrostatic attraction between oppositely charged ions.

Ionic radius: Assuming that the shape of an ion approximates a sphere, the distance from the center of an ion to its edge.

Iron catastrophe: An event inferred to have taken place early in the history of the Earth, when the Earth had heated up enough to cause iron to melt. The dense drops of molten iron would have coalesced and fallen to the center of the Earth to form the *core;* this process would have converted a great deal of gravitational potential energy to more heat, causing widespread melting in the Earth.

Iron formation: A sedimentary rock containing much iron, usually more than 15 percent, as sulfide, oxide, hydroxide, or carbonate; a low-grade ore of iron.

Island arc: A linear or arcuate chain of volcanic islands formed at a convergent plate boundary. The island arc is formed in the overriding plate from rising melt derived from the subducted plate and from the asthenosphere above that plate.

Isochron: A line connecting points of equal age.

Isograd: A line or curved surface connecting rocks that have undergone an equivalent degree of metamorphism.

Isostasy: The mechanism whereby areas of the crust rise or subside until the mass of their topography is buoyantly supported or compensated by the thickness of crust below, which "floats" on the denser mantle. The theory contends that continents and mountains are supported by low-density crustal "roots."

Isotope: One of several forms of one element, all having the same number of protons in the nucleus, but differing in their number of neutrons and thus in atomic weight.

Isotope geology: The study of the relative abundances of isotopes in rocks to determine their ages (see *geochronology*) or conditions of formation.

Isotropic substance: One in which the magnitude of a physical property, such as transmission velocity of light, depends upon crystallographic direction.

Joint: A large and relatively planar fracture in a rock across which there is no relative displacement of the two sides.

Juvenile gas: Gases that come to the surface for the first time from the deep interior.

Kame: A ridgelike or hilly local glacial deposit of coarse alluvium formed as a delta at the glacier front by meltwater streams.

Karst topography: An irregular topography characterized by *sinkholes*, caverns, and lack of surface streams; formed in humid regions because an underlying carbonate formation has been riddled with underground drainage channels that capture the surface streams.

Kerogen: A mixture of organic substances found in many fine-grained sedimentary rocks and a major constituent of oil shale.

Kettle: A small hollow or depression formed in glacial deposits when *outwash* was deposited around a residual block of ice that later melted.

Kilobar: A unit of pressure equal to 1000 *bars*.

Kimberlite: A *peridotite* that contains garnet and olivine and is found in volcanic pipes.

Laccolith: A *sill*-like igneous intrusion that forces apart two strata and forms a round, lens-shaped body many times wider than it is thick.

Lahar: A mudflow of unconsolidated *volcanic ash, dust, breccia,* and *boulders* mixed with rain or the water of a lake displaced by a lava flow.

Laminar flow: A flow regime in which particle paths are straight or gently curved and parallel.

Landslide: The rapid downslope movement of soil and rock material, often lubricated by groundwater, over a basal shear zone; also the tongue of stationary material deposited by such an event.

Lapilli: Fragments of volcanic rock formed when magma is ejected into the air by expanding gases. The fragments range from sand to cobble size.

Lateral moraine: A moraine formed along the side of a valley glacier and composed of rock scraped off or fallen from the valley sides.

Laterite: A distinctive soil formed in humid regions, characterized by high alumina and iron oxide content, and produced by rapid weathering of feldspar minerals.

Laurasia: Hypothetical supercontinent comprising approximately the present continents of the Northern Hemisphere.

Lava: Magma or molten rock that has reached the surface.

Lava fountain: A fluid pillar of lava spewed into the air at regular intervals as a result of gas and fluid pressure that build up in the crust below.

Lava tube: A hollow tunnel formed when the outside of a lava flow cools and solidifies and the molten material passing through it is drained away.

Leaching: The removal of materials from a soil by dissolution in water moving downward in the ground.

Leaky transform fault: *Transform fault* along which some plate separation and volcanism occur.

Left-lateral fault: A *strike-slip fault* on which the displacement of the more distant block is to the left when viewed from either side.

Levee: A ridge along a stream bank, formed by deposits left when floodwater slows on leaving the channel; also an artificial barrier to floods built in the same form.

Lignite: A very soft coallike material formed by the burial of peat.

Limb (fold): The relatively planar part of a fold or of two adjacent folds (for example, the steeply dipping part of a stratum between an anticline and syncline).

Limestone: A sedimentary rock composed mainly of calcium carbonate ($CaCO_3$), usually as the mineral calcite.

Lineation: Any linear arrangement of features found in a rock.

Lithic arenite: A sediment composed of sand-size particles, many of which are rock fragments rather than mineral grains.

Lithification: The processes that convert a sediment into a sedimentary rock.

Lithology: The systematic description of rocks, in terms of mineral composition and texture.

Lithosphere: The outer, rigid shell of Earth, situated above the *asthenosphere* and containing the *crust*, continents, and *plates*.

Lode: An unusually large vein or set of veins containing ore minerals.

Loess: An unstratified, wind-deposited, dusty sediment rich in clay minerals.

Longitudinal dune: A long dune parallel to the direction of the prevailing wind.

Longitudinal profile: A cross section of a stream from its mouth to its head, showing elevation versus distance to the mouth.

Longshore current: A current that flows parallel to the shoreline; the summed longshore components of water motion of waves that break obliquely with respect to the shore.

Longshore drift: The movement of sediment along a beach by *swash* and *backwash* of waves that approach the shore obliquely.

Lopolith: A large *laccolith* that is bowl-shaped and depressed in the center, possibly by subsidence of an emptied magma chamber beneath the intrusion.

Lowland: Region of little topographic relief, and at an elevation that is low relative to surrounding regions.

Low-velocity zone: A region in the Earth, especially a planar layer, that has lower seismic-wave velocities than the region immediately above it.

Lunar breccia: The lithified equivalent of the lunar *regolith*.

Luster: The general quality of the shine of a mineral surface, described by such subjective terms as dull, glassy, metallic.

M discontinuity: See *Mohorovičić discontinuity.*

Mafic mineral: A dark-colored mineral rich in iron and magnesium, especially pyroxene, amphibole, or olivine.

Magma: Molten rock material that forms igneous rocks upon cooling. Magma that reaches the surface is *lava.*

Magma chamber: A magma-filled cavity within the lithosphere.

Magmatic differentiation: The process by which a uniform parent magma may lead to rocks of a variety of compositions. Magmatic differentiation occurs because different mineral phases crystallize from a melt at different temperatures.

Magmatic water: Water that is dissolved in a magma or that is derived from such water.

Magnetic anomaly: The value of the local magnetic field remaining after the subtraction of the dipole portion of the Earth's field.

Magnetic epoch: A geologically long period during which the Earth's magnetic field was of predominantly one polarity; the epochs immediately before and after a given epoch would by definition be characterized by a field of opposite polarity.

Magnetic events: Geologically short periods within *magnetic epochs* during which the field had a reversed polarity.

Magnetic north pole: (1) The point where the Earth's surface intersects the axis of the dipole that best approximates the Earth's field. (2) The point where the Earth's magnetic field dips vertically downward.

Magnetic reversal: A change of the Earth's magnetic field to the opposite polarity.

Magnetic stratigraphy: The study and correlation of polarity epochs and events in the history of the Earth's magnetic field as contained in magnetic rocks.

Magnetometer: An instrument for measuring either one orthogonal component or the entire intensity of the Earth's magnetic field at various points.

Magnitude: A measure of earthquake size, determined by taking the common logarithm (base 10) of the largest ground motion observed during the arrival of a *P* wave or seismic surface wave and applying a standard correction for distance to the epicenter.

Manganese nodule: A small, rounded or elongate concentrically banded concretion found on the deep ocean floor that may contain as much as 20 percent manganese and smaller amounts of iron, copper, and nickel oxides and hydroxides.

Mantle: The main bulk of the Earth, between the crust and core, ranging from depths of about 40 to 2900 kilometers. It is composed of dense mafic silicates and divided into concentric layers by phase changes that are caused by the increase in pressure with depth.

Marble: The metamorphosed equivalent of *limestone* or other *carbonate rock.*

Mare: A dark, low-lying lunar plain, filled to an undetermined depth with mafic volcanic rocks. Plural: maria.

Mascon: A mass concentration located below the lunar surface; detected as a positive gravity anomaly.

Massive rock: A rock that is little or not at all broken by joints, cracks, foliation, or bedding, tending to present a homogeneous appearance.

Mass movement: A downhill movement of soil or fractured rock under the force of gravity.

Mass spectrometer: An instrument for separating ions of different mass but equal charge (mainly isotopes in geology) and measuring their relative quantities.

Matrix: Relatively fine-grained material in which coarser fragments or crystals are embedded; also called groundmass.

Maturity: A stage in the *geomorphic cycle* in which maximum relief and well-developed drainage are both present.

Meander: Broad, semicircular curves in a stream that develop as the stream erodes the outer bank of a curve and deposits sediment against the inner bank.

Mean sea level: The average height of the sea between high and low tides.

Mechanical weathering: The set of all physical processes by which an *outcrop* is broken up into small particles.

Medial moraine: A long strip of rock debris carried on or within a glacier resulting from the convergence of lateral moraines where two glaciers join.

Medical geology: The application of geological science to problems of health, especially those relating to mineral sources of toxic or nutritious elements and natural dispersal of toxic pollutants.

Mélange: A formation consisting of a heterogeneous mixture of rock materials on a mappable scale. Fragments of diverse composition, size, and texture have been mixed and consolidated by tremendous deformational pressure.

Mesa: A flat-topped, steep-sided upland topped by a resistant formation; larger than a butte.

Mesosphere: The lower *mantle.*

Metamorphic grade: The relative intensity of metamorphism; that is, low-grade metamorphism implies low pressures and temperatures, and so on.

Metamorphic rock: A rock whose original mineralogy, texture, or composition has been changed due to the effects of pressure, temperature, or the gain or loss of chemical components.

Metamorphism: The changes of mineralogy and texture imposed on a rock by pressure and temperature in the Earth's interior.

Metastable: Describes a system that appears to be at *equilibrium,* in that no changes are observed to be taking place in it, but which we know is not in its most stable configuration for the given conditions. Metastable states may be preserved when the reactions required to bring the system into its stable state take place too slowly to be noticed on the time scale of our observations.

Metazoan: Multicellular life forms that first evolved in late Precambrian time.

Meteoric water: Rainwater, snow, hail, and sleet.

Meteorite: A stony or metallic object from interplanetary space that penetrates the atmosphere to impact on the surface.

Mica: General term for a group of hydrous aluminosilicate minerals with a sheetlike crystal structure, common in igneous and metamorphic rocks.

Micrometeorite: A meteorite less than one millimeter in diameter.

Microseism: A weak vibration of the ground detectable by seismographs that is caused by waves, wind, or human activity, but not by an earthquake.

Mid-ocean ridge: A major elevated linear feature of the sea floor consisting of many small, slightly offset segments, with a total length of 200 to 20,000 kilometers. Characteristic type of plate boundary occurring in a *divergence zone;* a site where two *plates* are being pulled apart and new oceanic *lithosphere* is being created.

Migmatite: A rock with both igneous and metamorphic characteristics that shows large crystal and laminar flow structures. Probably formed metamorphically in the presence of water and without melting.

Mineral: A naturally occurring, solid, inorganic element or compound, with a definite composition or range of compositions, usually possessing a regular internal crystalline structure.

Mineralogy: The study of mineral composition, structure, appearance, stability, occurrence, and associations.

Miogeosyncline: A *geosyncline* that is situated near a *craton* and receives chemical and well-sorted clastic sediments from the continent.

Moho: See *Mohorovičić discontinuity.*

Mohorovičić discontinuity: The boundary between *crust* and mantle, marked by a rapid increase in seismic wave velocity to more than eight kilometers per second. Depth: 5 to 45 kilometers. Abbreviated *Moho* or *M discontinuity.*

Mohs scale of hardness: An empirical, ascending scale of mineral hardness. See Table 3–4.

Monadnock: An isolated *hill* or *mountain* rising above a peneplain.

Monocline: The S-shaped fold connecting two horizontal parts of the same stratum at different elevations. Its central limb is usually not overturned.

Moraine: A glacial deposit of *till* left at the margin of an ice sheet. (See also *ground moraine, lateral moraine, medial moraine,* and *terminal moraine.*)

Mountain: A steep-sided topographic elevation larger than a *hill;* also a single prominence forming part of a ridge or mountain range.

Mudflow: A mass movement of material finer than sand, lubricated with large amounts of water.

Mudstone: The lithified equivalent of mud, a fine-grained sedimentary rock similar to shale but more massive.

m.y.: Abbreviation for "million years."

Mylonite: A very fine lithified fault breccia commonly found in major thrust faults and produced by shearing and rolling during fault movement.

Nannofossil ooze: A deep-sea sediment composed largely of nannofossils, that is, organisms less than about 60 micrometers in diameter.

Native metal: A natural deposit of a metallic element in pure metallic form, not combined with other elements.

Natural levee: See *levee.*

Neap tide: A tide cycle of unusually small amplitude, which occurs twice monthly when the lunar and solar tides are opposed—that is, when the gravitational pull of the Sun is at right angles to that of the Moon.

Nebula: An immense, diffuse body of interstellar gas and dust that has not condensed into a star.

Nebular hypothesis: A theory of the formation of the planets that holds that a rotating nebula contracted and was then torn into fragments by centrifugal forces, with planets condensing from the fragments.

Negative feedback process: A process in which the products that are produced suppress the production of more product.

Neutron: An electrically neutral elementary particle in the atomic nucleus having the mass of one proton.

Neutron-activation analysis: A method of identifying isotopes of an element by bombarding them with neutrons and observing the characteristic radioactive decay products emitted.

Nonpenetrative property: One that does not occur throughout the body of a rock, but only in isolated patches, or systematically associated with certain features such as bedding planes. Fracture cleavage is an example of a nonpenetrative feature. (Compare *penetrative.*)

Normal fault: A *dip-slip fault* in which the block above the fault has moved downward relative to the block below.

Nuclear geology: See *isotope geology.*

Nuée ardente: A "glowing cloud" of hot volcanic ash, dust, and gas that moves rapidly downhill as a density current in the atmosphere.

Obduction: A process occurring during plate collision, whereby a piece of the subducted plate is broken off and pushed up onto the overriding plate; this mechanism probably explains why we find blocks of ophiolite on the continents.

Oblique-slip fault: A fault that combines some *strike-slip* motion with some *dip-slip* motion.

Obsidian: Dark volcanic glass of *felsic* composition.

Octahedral coordination: The packing of six ions around an ion of opposite charge to form an octahedron.

Oil field: An underground accumulation of oil and gas concentrated beneath an impermeable trap, preventing its upward escape.

Oil shale: A dark-colored shale containing organic material that can be crushed and heated to liberate gaseous hydrocarbons.

Oil trap: See *trap.*

Old age: A stage in the *geomorphic cycle,* characterized by formation of a *peneplain* near sea level.

Oolite: A spherical *carbonate* particle precipitated from warm ocean water on a *carbonate platform;* also, a rock composed of such particles.

Ophiolite suite: An assemblage of *mafic* and *ultramafic* igneous rocks with deep-sea sediments supposedly associated with *divergence zones* and the sea-floor environment.

Optical mineralogy: The identification and study of minerals on the basis of their optical properties, as observed with a *petrographic microscope.*

Orbit: The elliptical or hyperbolic path traced by a planet or meteorite or satellite in the presence of a more massive body.

Order: See *stream order.*

Ore: A natural deposit in which a valuable metallic element occurs in high enough concentration to make mining economically feasible.

Ore mineral: The mineral of an ore that contains the useful element.

Organic sediment: Carbon-rich deposits, such as coal, gas, and oil, formed by the decay of once-living material after burial.

Original horizontality, principle of: The proposition of

Steno, that all sedimentary bedding is horizontal at the time of deposition.

Orogenic belt: A linear region, often a former *geosyncline*, that has been subjected to folding and other deformation in a mountain-building episode.

Orogeny: The tectonic process in which large areas are folded, thrust-faulted, metamorphosed, and subjected to plutonism. The cycle ends with uplift and the formation of mountains.

Oscillation ripple: A ripple with a symmetrical cross section and a sharp peak formed by waves.

Outcrop: A segment of bedrock exposed to the atmosphere.

Outgassing: The release of *juvenile gases* to the atmosphere and oceans by volcanism.

Outwash: A sediment deposited by meltwater streams emanating from a glacier.

Overturned fold: A fold in which a limb has tilted past vertical so that the older strata are uppermost.

Oxbow lake: A long, broad, crescent-shaped lake formed when a stream abandons a *meander* and takes a new course.

Oxidation: A chemical reaction in which electrons are lost from an atom and its charge becomes more positive.

Oxidized element: An element occurring in the more positively charged of two common ionic forms (for example, the Fe^{3+} form of iron, as opposed to Fe^{2+}).

Pahoehoe: A *basaltic* lava flow with a glassy, smooth, and undulating, or ropy, surface.

Paleoclimate: The average state or typical conditions of climate during some past geologic period.

Paleocurrent map: A map of depositional currents that have been inferred from *cross-bedding, ripples,* or other sedimentary structures.

Paleogeographic map: A map showing the surface landforms and coastline of an area at some time in the geologic past.

Paleogeography: The pattern of physical features of the Earth's surface at a particular time in the past. These changing configurations are the subject matter of historical geology.

Paleomagnetism: The science of the reconstruction of the Earth's ancient magnetic field and the positions of the continents from the evidence of *remanent magnetization* in ancient rocks.

Paleontology: The science of fossils of ancient life forms, and their evolution.

Paleowind: A prevailing wind direction in an area, inferred from dune structure or the distribution of volcanic ash for one particular time in geologic history.

Pangaea: Supercontinent that coalesced in the latest Paleozoic and was comprised of all present continents. Breakup of Pangaea began in Mesozoic time as inferred from paleomagnetic and other data.

Panthalassa: The single, immense world ocean that existed at the time of the Pangaean supercontinent.

Paraffin: Hydrocarbons that contain chains of carbon and hydrogen atoms with the general formula C_nH_{2n+2}.

Parent element: An element that is transformed by *radioactive decay* to a different (*daughter*) element.

Partial melt: The product produced during *partial melting.*

Partial melting: A process in which heating causes a mass of rock to become partially molten. Partial melting occurs because the minerals that compose a rock melt at different temperatures.

Peat: A marsh or swamp deposit of water-soaked plant remains containing more than 50 percent carbon.

Pedalfer: A common soil type in humid regions, characterized by an abundance of iron oxides and clay minerals deposited in the *B-horizon* by leaching.

Pediment: A planar, sloping rock surface forming a ramp up to the front of a mountain range in an arid region. It may be covered locally by thin *alluvium.*

Pedocal: A common soil type of arid regions, characterized by accumulation of calcium carbonate in the *A-horizon.*

Pegmatite: An igneous rock with extremely large grains, more than a centimeter in diameter. It may be of any composition but most frequently is granitic.

Pelagic sediment: Deep-sea sediments composed of fine-grained detritus that slowly settles from surface waters. Common constituents are *clay, radiolarian ooze,* and *foraminiferal ooze.*

Peléan eruption: A volcanic eruption accompanied by great explosions and emanations of hot gas and *nuées ardentes.* Named for Mont Pelée, Martinique.

Peneplain: A hypothetical extensive area of low elevation and relief reduced to near sea level by a long period of erosion and representing the end product of the ideal *geomorphic cycle.*

Penetrative property: A property that occurs throughout a rock, as in *slaty cleavage.*

Perched groundwater: An isolated body of groundwater that is perched above and separated from the main water table by an aquiclude.

Peridotite: A coarse-grained mafic igneous rock composed of olivine with small amounts of pyroxene and amphibole.

Period (geologic): The most commonly used unit of geologic time, representing one subdivision of an *era.*

Period (wave): The time interval between the arrival of successive crests in a homogeneous wave train; the period is the inverse of the frequency of a cyclic event.

Permafrost: A permanently frozen aggregate of ice and soil occurring in very cold regions.

Permeability: The ability of a formation to transmit groundwater or other fluids through pores and cracks.

Petrographic microscope: An optical microscope designed for examining thin sections of rocks and minerals using polarized light.

Petrography: The study of rocks using the *petrographic microscope.*

Petroleum resources: The total quantity of natural oil in the Earth's crust that it may be possible to extract.

Petrology: The study of the composition, structure, and origin of rocks.

Phanerite: Generic term for rocks in which individual grains are visible to the naked eye.

Phaneritic texture: A rock texture in which individual crystals are visible to the unaided eye.

Phanerozoic: Originally, the expanse of time during which life was known to exist. Now, broadly refers to the time of complex life; that is, from just before the Cambrian to the Recent.

Phase change: (1) The transformation of an element or compound from solid to liquid, gaseous to solid, etc. (2) The transformation of a solid mineral to a different solid form of different structure and density.

Phenocryst: A large crystal surrounded by a finer *matrix* in a *porphyry*.

Phosphate sediment: A sediment composed largely of calcium phosphate, usually as a variety of the mineral apatite and largely in the form of concretions and nodules.

Photogeology: The study of geologic features as exposed in aerial photographs of Earth or other planets.

Photolysis: The chemical breakdown of water into hydrogen and oxygen by solar radiation in the upper atmosphere.

Photosynthesis: The conversion of water, carbon dioxide, and the energy of sunlight into oxygen and organic compounds such as sugars by plants containing chlorophyll.

Phreatic eruption: A volcanic eruption of mud and debris caused by the expansion of steam formed when magma comes in contact with confined groundwater.

Phreatic zone: The zone of soil and rock in which pores are completely filled with groundwater. Also called the saturated zone.

Physical sedimentation: The deposition of *clastic* particles derived by erosion.

Pillow lava: A type of lava formed underwater, in which many small pillow-shaped tongues break through the chilled surface and quickly solidify, leading to a rock formation resembling a pile of sandbags.

Pipe: See *volcanic pipe*.

Piracy: See *stream piracy*.

Placer: A *detrital* sedimentary deposit of a valuable mineral or native metal in unusually high concentration, usually segregated because of its greater density.

Plagioclase feldspar: One of the most common minerals in crustal rocks. Plagioclase feldspar ranges in composition from albite ($NaAlSi_3O_8$) to anorthite ($CaAl_2Si_2O_8$).

Plain: An area of low relief, without major hills, valleys, or ridges; not necessarily at a low elevation.

Plane of symmetry: An imaginary plane passing through an object in such an orientation that each feature of the object on one side of the plane has a mirror image on the other side.

Planetary evolution: The process by which a differentiated planet is formed.

Planetesimal: A body of rock in space, smaller than a planet and attracted by gravity to other planetesimals in the formation of a planet.

Planetology: The science of the distribution, composition, and origin of matter in the planets of the solar system.

Plastic deformation: Deformation that proceeds to large strains at constant stress without fracturing.

Plate: One of the dozen or more segments of the *lithosphere* that are internally rigid and move independently over the interior, meeting in *convergence zones* and separating at *divergence zones*.

Plateau: An extensive upland region at high elevation with respect to its surroundings.

Plate tectonics: The theory and study of plate formation, movement, interactions, and destruction; the attempt to explain seismicity, volcanism, mountain-building, and paleomagnetic evidence in terms of plate motions.

Playa: The flat floor of a closed basin in an arid region. It may be occupied by an intermittent lake.

Plume: Hypothetical rising jet of hot, partially molten material, perhaps emanating from the mantle, and believed by some to be responsible for intraplate volcanism.

Plunging fold: A fold whose axis is not horizontal but dips. Thus progressively younger strata are found at the center of the fold as one travels along the direction of plunge, and the geologic map pattern is one of nested V-shaped outcrops of formations.

Pluton: A large igneous intrusion, formed at depth in the crust.

Plutonic: Pertaining to igneous activity at depth.

Plutonism: (1) The formation and worldwide distribution of plutons. (2) The eighteenth- and nineteenth-century theory that the Earth was at one time entirely molten.

Point bar: A deposit of sediment on the inner bank of a *meander* that forms because the stream velocity is lower against the inner bank.

Polarity epoch: See *magnetic epoch*.

Polar wandering: An interpretation given to the observation that the position of the magnetic north pole, as inferred from the thermoremanent magnetization of rocks, varies with time as a result of plate movements.

Pole of spreading: An imaginary point on the Earth's surface that represents the emergence of an imaginary axis passing through the Earth's center and about which one plate moves relative to another; thus, for each pair of plates, there is a unique pole.

Polymorph: One of two or more alternative possible structures for a single chemical compound (for example, the minerals calcite and aragonite are polymorphs of calcium carbonate).

Pool: The deep part of an undulating stream bed.

Porosity: The percentage of the total volume of a rock that is pore space (not occupied by mineral grains).

Porphyroblast: A large crystal in a finer-grained *matrix* in a metamorphic rock; analogous to a *phenocryst* in an igneous rock.

Porphyry: An igneous rock containing abundant *phenocrysts*.

Positive-feedback process: A process in which the products that are produced stimulate the production of more product.

Potable water: Water that is agreeable to the taste and not dangerous to the health.

Pothole: A hemispherical hole in the bedrock of a stream bed, formed by abrasion of small pebbles and cobbles in a strong current.

ppm: Abbreviation for "parts per million."

Preferred orientation: Any deviation from randomness in the distribution of the crystallographic or grain shape axes of minerals in a rock; produced by flow in igneous rocks, directional stress during recrystallization of a metamorphic rock, or depositional currents during sedimentation.

Products: The compounds, elements, or ions that are

produced in a chemical reaction. The starting materials are referred to as *reactants*.

Prograde metamorphism: Metamorphic reactions that occur as a result of increasing pressure and temperature.

Projection: A method of representing a three-dimensional object on a two-dimensional surface.

Proton: An elementary particle found in the atomic nucleus with a positive charge of 1.602×10^{-19} coulomb and a mass of 1836 electrons; one H^+ ion.

Proto-sun: A large cloud of dust and gas gradually coalescing into a star under the force of gravity.

Protozoan: Simple, unicellular life forms that first evolved during the Precambrian.

Proven reserves: Deposits of fossil fuels whose location and extent are known, as opposed to potential but unproved ("discovered") deposits.

Pumice: A form of volcanic glass, usually of *felsic* composition, so filled with *vesicles* that it resembles a sponge and has very low density.

P wave: The primary or fastest wave traveling away from a seismic event through the solid rock, and consisting of a train of compressions and dilations of the material.

Pyroclastic rock: A rock formed by the accumulation of fragments of volcanic rock scattered by volcanic explosions.

Pyroclastic texture: The unsorted, angular texture of the fragments in a *pyroclastic rock*.

Pyroxene granulite: A coarse-grained regional metamorphic rock containing pyroxene; formed at high temperatures and pressures deep in the crust.

Quartz arenite: A *sandstone* containing very little except pure quartz grains and cementing material.

Quartzite: (1) A very hard, clean, white metamorphic rock formed from a *quartz arenite* sandstone. (2) A quartz arenite containing so much cement that it resembles (1).

Quartzose sandstone: (1) A *quartz arenite*. (2) A clean quartz sandstone, less pure than a quartz arenite, that may contain a moderate amount of other *detrital* minerals and/or calcite cement.

Radial drainage: A system of streams running in a radial pattern away from the center of a circular elevation, such as a volcano or dome.

Radial mode: A type of *free oscillation* in which the whole earth periodically (with a period of 20 minutes) increases and decreases in radius.

Radiative transfer: One mechanism for the movement of heat, in which it takes the form of long-wavelength infrared radiation.

Radioactive decay: The spontaneous breakdown of certain kinds of atomic nuclei into one or more nuclei of different elements, involving the release of energy and subatomic particles.

Radiolarian: A class of one-celled marine animals with siliceous skeletons that have existed in the ocean throughout the Phanerozoic Eon.

Radiolarian ooze: A siliceous deep-sea sediment composed largely of the skeletons of *radiolaria*.

Radiolarite: The lithified sedimentary rock formed from radiolarian ooze.

Radiometric dating: The method of obtaining ages of geological materials by measuring the relative abundances of radioactive parent and daughter isotopes in them.

Radius ratio: The ratio of the radius of an *anion* to the radius of a *cation*. This ratio determines the *coordination number*.

Rank: Referring to coal, the degree to which the coal has undergone metamorphism.

Ray: A linear landform of the lunar surface emanating from a large crater and extending as much as 100 kilometers outward, probably consisting of fine ejecta thrown out by the impact of a meteorite.

Reactants: The chemical constituents that participate in a chemical reaction to produce *products*.

Reaction series: A series of chemical reactions occurring in a cooling magma by which a mineral formed at high temperature becomes unstable in the melt and reacts to form another mineral.

Recharge: In hydrology, the replenishment of groundwater by infiltration of *meteoric water* through the soil.

Recrystallization: The growth of new mineral grains in a rock at the expense of old grains, which supply the material.

Rectangular drainage: A system of streams in which each straight segment of each stream takes one of two characteristic perpendicular directions. The streams are usually following perpendicular sets of joints.

Recumbent fold: An *overturned* fold with both limbs nearly horizontal.

Reduced element: Of elements that have two different ions of different charge, the one with the less positive (more negative) charge is said to be reduced (for example, Fe^{2+} as opposed to Fe^{3+}).

Refraction (wave): The departure of a wave from its original direction of travel at the interface with a material of different index of refraction (light) or seismic wave velocity. (See also *seismic refraction*.)

Regional metamorphism: Metamorphism occurring over a wide area and caused by deep burial or strong tectonic forces of the Earth.

Regolith: Any solid material lying on top of bedrock. Includes soil, alluvium, and rock fragments weathered from the bedrock.

Regression: A drop in sea level that causes an area of the Earth to be uncovered by sea water, ending marine deposition.

Relative humidity: The amount of water vapor in the air, expressed as a percentage of the total amount of water vapor that the air could hold at that temperature if saturated.

Relief: The maximum regional difference in elevation.

Remanent magnetization: See *thermoremanent magnetization*, and *depositional remanent magnetization*.

Remote sensing: The study of surface conditions and materials of the Earth or other bodies by means of photography, spectroscopy, or radar observations from airplanes or satellites.

Replacement deposit: A deposit of ore minerals by hydrothermal solutions that have first dissolved the original mineral to form a small cavity.

Reserves: See *Proven reserves*.

Residence time: The amount of a given element or compound in a given body (especially the sea or atmosphere) divided by the average removal rate, assuming the system is in a *steady state*.

Respiration: The chemical process by which animals and plants convert their food into energy, through oxidation of carbohydrates; the process consumes oxygen and produces carbon dioxide.

Retrograde metamorphism: A situation in which a rock that has been metamorphosed to a fairly high grade is remetamorphosed by a later thermal event of somewhat lower intensity, so that lower-grade metamorphic mineralogy is superimposed on higher-grade mineral assemblages.

Reverse fault: A *dip-slip* fault in which the upper block, above the fault plane, moves up and over the lower block, so that older strata are placed over younger ones.

Reversible process: One that can proceed in either of two opposite directions, depending on the balance of the forces controlling it.

Rhyolite: The fine-grained volcanic or extrusive equivalent of granite, light brown to gray and compact.

Richter magnitude scale: See *magnitude*.

Ridge: See *mid-ocean ridge*.

Riffle: A shallow stretch of an undulating stream bed.

Rift valley: A fault trough formed in a *divergence zone* or other area of tension.

Right-lateral fault: A *strike-slip fault* on which the displacement of the more distant block is to the right when viewed from either side.

Ring dike: A dike in the form of a segment of a cone or cylinder, having an arcuate outcrop.

Rip current: A current that flows strongly away from the sea shore through gaps in the surf zone at intervals along the shoreline.

Ripple: A very small dune of sand or silt whose long dimension is formed at right angles to the current.

River order: See *stream order*.

Rock: A solid, cohesive aggregate of grains of one or more *minerals*.

Rock cycle: The geologic cycle, with emphasis on the rocks produced; sedimentary rocks are metamorphosed to metamorphic rocks, or melted to create igneous rocks, and all rocks may be uplifted and eroded to make sediments, which lithify to sedimentary rocks.

Rock flour: A glacial sediment of extremely fine (silt- and clay-size) ground rock formed by abrasion of rocks at the base of the glacier.

Rock glacier: A glacierlike mass of rock fragments or *talus* with interstitial ice that moves downhill under the force of gravity.

Rockslide: A landslide involving mainly large blocks of detached bedrock with little or no soil or sand.

Rounding: The degree to which the edges and corners of a particle become worn and rounded as a result of abrasion during transportation. Expressed as angular, subrounded, well-rounded, etc.

Runoff: The amount of rain water directly leaving an area in surface drainage, as opposed to the amount that seeps out as groundwater.

Rupture strength: The greatest *stress* that a material can sustain at one atmosphere confining pressure without fracturing.

Sabkha: A characteristic boundary between a desert and an ocean, at which crusts of *evaporites* form above the high-water line.

Saltation: The movement of sand or fine sediment by short jumps above the ground or stream bed under the influence of a current too weak to keep it permanently suspended.

Sandblasting: A physical weathering process in which rock is eroded by the impact of sand grains carried by the wind, frequently leading to *ventifact* formation of pebbles and cobbles.

Sandstone: A *detrital* rock composed of grains from 1/16 to 2 millimeters in diameter, dominated in most sandstones by quartz, feldspar, and rock fragments, bound together by a cement of silica, carbonate, or other minerals or a *matrix* of clay minerals.

Saturated zone: See *phreatic zone*.

Schist: A metamorphic rock characterized by strong *foliation* or *schistosity*.

Schistosity: The parallel arrangement of sheety or prismatic minerals like micas and amphiboles resulting from nonhydrostatic stress in metamorphism.

Scoria: Congealed lava, usually of mafic composition, with a large number of *vesicles* formed by gases coming out of solution.

Sea-floor spreading: The mechanism by which new sea floor crust is created at ridges in *divergence zones* and adjacent *plates* are moved apart to make room. This process may continue at a few centimeters per year through many geologic periods.

Seamount: An isolated tall mountain on the sea floor that may extend more than 1 kilometer from base to peak. (See also *guyot*.)

Secular variation: Slow changes in the orientation of the Earth's magnetic field that appear to be long lasting and internal in origin as opposed to rapid fluctuations, which are external in origin.

Sediment: Any of a number of materials deposited at Earth's surface by physical agents (such as wind, water, and ice), chemical agents (precipitation from oceans, lakes, and rivers), or biological agents (organisms, living or dead).

Sedimentary environment: See *environment of deposition*.

Sedimentary rock: A rock formed by the accumulation and cementation of mineral grains transported by wind, water, or ice to the site of deposition or by chemical precipitation at the depositional site.

Sedimentary structure: Any structure of a sedimentary or weakly metamorphosed rock that was formed at the time of deposition; includes bedding, cross-bedding, graded bedding, ripples, scour marks, mudcracks.

Sedimentation: The process of deposition of mineral grains or precipitates in beds or other accumulations.

Seif dune: A *longitudinal dune* that shows the sculpturing effect of cross-winds not parallel to its axis.

Seismic discontinuity: A surface within the Earth across which *P-wave* or *S-wave* velocities change rapidly, usually by more than ± 0.2 kilometer per second.

Seismicity: The worldwide or local distribution of earthquakes in space and time; a general term for the number of earthquakes in a unit of time.

Seismic profile: The data collected from a set of seismographs arranged in a straight line with an artificial seismic source; especially the times of *P-wave* arrivals.

Seismic reflection: A mode of seismic prospecting in which the *seismic profile* is examined for waves that have reflected from near-horizontal strata below the surface.

Seismic refraction: A mode of seismic prospecting in which the *seismic profile* is examined for waves that have been refracted upward from seismic discontinuities below the profile. Greater depths may be reached than through *seismic reflection*.

Seismic shadow: Region on the far side of the Earth's surface from an earthquake, not reached by *P waves* from that earthquake because they have been deflected at the surface of the core.

Seismic stratigraphy: The use of methods of seismic exploration to resolve sedimentary interfaces at depth and thus facilitate construction of stratigraphic sections in the absence of outcrop data.

Seismic surface wave: A seismic wave that follows the Earth's surface only, with a speed less than that of *S waves*.

Seismic transition zone: A seismic discontinuity, found in all parts of the Earth, at which the velocity increases rapidly with depth; especially the one at 400 to 700 kilometers.

Seismic travel time: See *travel-time curve*.

Seismograph: An instrument for magnifying and recording the motions of the Earth's surface that are caused by seismic waves.

Seismology: The study of earthquakes, seismic waves, and their propagation through the Earth.

Self-exciting dynamo: A system involving a rotating electrical conductor that creates and sustains its own magnetic field.

Series: A set of rocks formed in one area during a geologic *epoch*.

Settling velocity: The rate at which a sedimentary particle of a given size falls through water or air.

Shale: A very fine grained *detrital* sedimentary rock composed of silt and clay which tends to part along *bedding* planes.

Shard (volcanic): An angular fragment of volcanic glass.

Sheeting: See *exfoliation*.

Sheet wash: A flow of rainwater that covers the entire ground surface with a thin film and is not concentrated into streams.

Shield: A large region of stable, ancient basement rocks within a continent.

Shield volcano: A large, broad *volcanic cone* with very gentle slopes built up by nonviscous basalt lavas.

Shooting flow: A very fast form of water flow with a high velocity induced by steep slopes in the flow bed, typically developed in rapids.

Shoreline: The straight to sinuous, smooth to irregular interface between land and sea.

Silicate: Any of a vast class of minerals containing silicon and oxygen and constructed from the tetrahedral group $(SiO_4)^{4-}$. The bulk of the crust is composed of silicate minerals.

Silicic rock: An igneous rock containing more than two-thirds silicon-oxygen tetrahedra by weight, usually as quartz and feldspar (for example, *granite*).

Sill: A horizontal tabular intrusion with *concordant* contacts.

Sill (topographic): A small submarine ridge that nearly separates or restricts water flow between two adjacent bodies of water.

Sinkhole: A small, steep depression caused in *karst topography* by the dissolution and collapse of subterranean caverns in *carbonate* formations.

Sinking current: A downward movement of sea or lake water that has become denser through cooling or increased salinity or that sinks because of an offshore wind piling up water against the shore.

Sinuosity: The length of the channel of a stream divided by the straight-line distance between its ends.

Slate: The metamorphic equivalent of *shale;* a hard, gray, red, green, or black fine-grained rock with *slaty cleavage*.

Slaty cleavage: A *foliation* consisting of the parallel arrangement of sheety metamorphic minerals (see *schistosity*) at an angle to bedding planes, and related to deformational structures, along which the rock tends to part.

Slickensides: Parallel grooves, ramps, and scratches on one or both of the inside faces of a fault plane, showing the direction of slip.

Slip (fault): The motion of one face of a fault relative to the other.

Slip face: The steep downwind face of a dune on which sand is deposited in cross-beds at the angle of repose.

Slope wash: The motion of water and sediment down a slope by the mechanism of *sheet wash*.

Slump: A collapse structure in unlithified sediments, caused by gravity sliding or collapse under the weight of later sediments.

Snowfield: An area of permanent snow that is not moving or whose movement is not visible.

Soft coal: See *bituminous coal*.

Soft water: Water that is free of calcium and magnesium *carbonates* and other dissolved materials of *hard water*.

Soil: The surface accumulation of sand, clay, and humus that compose the *regolith,* but excluding the larger fragments of unweathered rock.

Soil creep: The imperceptible downhill flow of soil under the force of gravity. It is a shear flow with velocity decreasing downward and occurs even on gentle slopes.

Solar wind: The flow of ionized gas (plasma) from the Sun toward the Earth, which interacts with the Earth's magnetic field.

Solid-solution series: A series of minerals of identical structure that can contain a mixture of two elements over a range of ratios (for example, plagioclase feldspars).

Solidus: A curve on a pressure versus temperature graph representing the beginning of melting of a rock or mineral.

Solifluction: The soil creep of material saturated with water and/or ice; most common in polar regions.

Solubility: The mass of a substance that can be dissolved in a certain amount of solvent if chemical *equilibrium* is attained.

Solute: The substance dissolved in a *solvent*.

Solution: A homogeneous mixture of a *solute* in a *solvent*.

Solvent: A medium, usually liquid, in which other substances can be dissolved.

Sorting: A measure of the homogeneity of the sizes of particles in a sediment or sedimentary rock.

Spatter cone: A conical deposit of cooled *lava* congealed around a volcanic vent that disgorges mostly gas with occasional globs of molten rock.

Specific gravity: The ratio of the *density* of a given substance to the density of water.

Spheroidal mode: A type of *free oscillation* in which the whole Earth periodically (period of 53 minutes) lengthens along one axis while shortening along a perpendicular axis, and then lengthens along the former short axis while shortening along the former long axis.

Spheroidal weathering: The formation of spherical residual inner cores by the weathering of boulders.

Spit: A long ridge of sand deposited by longshore current and drift where the coast takes an abrupt inward turn. It is attached to land at the upstream end.

Splay deposit: A small *delta* deposited on a *floodplain* when the stream breaches a *levee* during a flood.

Spring tide: A tide cycle of unusually large amplitude that occurs twice monthly when the lunar and solar tides are in phase. (Compare *neap tide*.)

Stage (stream): The elevation of the water level of a stream measured against some constant reference.

Stalactite: An icicle- or toothlike deposit of calcite or aragonite hanging from the roof of a cave. It is deposited by evaporation and precipitation from solutions seeping through limestone.

Stalagmite: An inverted icicle-shaped deposit that builds up on a cave floor beneath a *stalactite* and is formed by the same process as a stalactite.

Steady-state: An adjective describing a system that is in a stable dynamic state in which inputs balance outputs.

Steepness (wave): The ratio of the height of a wave to its wavelength.

Stock (volcanic): An intrusion with the characteristics of a *batholith* but less than 100 square kilometers in area.

Stoping: The process by which an intruding igneous mass makes room for itself in the *country rock* by breaking rock fragments off the walls and "absorbing" them when they sink.

Strain: A quantity describing the exact deformation of each point in a body. Roughly, the change in a dimension or volume divided by the original dimension or volume.

Strain seismograph: An instrument that measures changes of *strain* in surface rocks to detect seismic waves.

Stratification: A structure of sedimentary rocks, indicating recognizable parallel beds of considerable lateral extent.

Stratigraphic sequence: A set of deposited beds that reflects the geologic history of a region.

Stratigraphic time scale: See *time scale*.

Stratigraphy: The science of description, correlation, and classification of strata in sedimentary rocks, including the interpretation of the depositional environments of those strata.

Stratovolcano: A volcanic *cone* consisting of both *lava* and *pyroclastic* rocks.

Stratum: A single sedimentary rock unit with a distinct set of physical or mineralogical characteristics or fossils such that it can readily be distinguished from beds above and below. Plural: strata.

Streak: The fine deposit of mineral dust left on an abrasive surface when a mineral is scraped across it; especially the characteristic color of the dust.

Streak plate: A ceramic abrasive surface for streak tests.

Streaming flow: A tranquil flow slower than *shooting flow*.

Streamline: A curved line representing the successive positions of a particle in a flow as time passes.

Stream order: The hierarchical number of a stream segment in *dendritic drainage*. The smallest tributary streams have order one, and at each junction of streams of equal order the order of the subsequent segment is one higher.

Stream piracy: The erosion of a divide between two streams by the more competent, leading to the capture of all or part of the drainage of the slower stream by the faster.

Stress: A quantity describing the forces acting on each part of a body in units of force per unit area.

Striation: See *glacial striation*.

Strike: The angle between true North and the horizontal line contained in any planar feature (inclined bed, dike, fault plane, etc); also the geographic direction of this horizontal line.

Strike-slip fault: A fault whose relative displacement is purely horizontal. (See also *right-* and *left-lateral fault*).

Stromatolite: A fossil form representing the growth habit of an algal mat: concentric spherules, stacked hemisphere, or flat sheets of calcium carbonate and trapped silt encountered in limestones.

Subbituminous coal: A form of coal produced at pressures and temperatures greater than those that produce *lignite*, but less than those required for *bituminous coal*.

Subduction zone: A dipping planar zone descending away from a *trench* and defined by high *seismicity*; interpreted as the shear zone between a sinking oceanic plate and an overriding plate. (See also *convergence zone*.)

Sublimation: A phase change between the solid and gaseous states without passing through the liquid state.

Submarine canyon: An underwater canyon in the continental shelf.

Subsidence: A gentle epeirogenic movement where a broad area of the crust sinks without appreciable deformation.

Superposed stream: A stream that flows through resistant formations because its course was established at a higher level on uniform rocks before down-cutting began.

Superposition, law of: The principle stated by Steno that, except in extremely deformed strata, a bed that overlies another bed is always the younger.

Supersaturation: The unstable state of a solution that contains more solute than its solubility allows.

Surf: The breaking or tumbling forward of water waves as they approach the shore.

Surface wave: See *seismic surface wave*.

Surf zone: An offshore belt along which the waves collapse into breakers as they approach the shore.

Surge: See *glacier surge*.

Suspended load: The fine sediment kept suspended in a stream because the *settling velocity* is lower than the upward velocity of eddies.

Swash: The landward rush of water from a breaking wave up the slope of the beach.

S wave: The secondary seismic wave, which travels more slowly than the *P wave*, and consists of elastic vibrations transverse to the direction of travel. *S* waves cannot penetrate a liquid.

Swell: An oceanic water wave with a wavelength on the order of 30 meters or more and a height of perhaps 2 meters or less that may travel great distances from its source.

Symbiosis: The interaction of two mutually supporting species who do not compete with or prey upon each other.

Symmetrical fold: One whose limbs *dip* at the same angle.

Syncline: A large fold whose limbs are higher than its center; a fold with the youngest strata in the center. (Compare *anticline*.)

System (stratigraphy): A stratigraphic unit larger than a *series* consisting of all the rocks deposited in one *period* of an era.

Table land: A large elevated region with a surface of relatively low relief. (See also *plateau*.)

Talus: A deposit of large angular fragments of physically weathered bedrock, usually at the base of a cliff or steep slope.

Tar sand: A *sandstone* containing the densest asphaltic components of petroleum, the end product of evaporation of *volatile* components or of some thickening process.

Tectonics: The study of the movements and deformation of the crust on a large scale, including *epeirogeny*, *metamorphism*, folding, faulting, and *plate tectonics*.

Terminal moraine: A sinuous ridge of unsorted glacial *till* deposited by a glacier at the line of its farthest advance.

Terrestrial planet: A planet similar in size and composition to the Earth; especially Mars, Earth, Venus, and Mercury.

Terrestrial sediment: A deposit of sediment that accumulated above sea level in lakes, *alluvial fans*, *floodplains*, *moraines*, etc., regardless of its present elevation.

Tetrahedron: A geometric form of four equal triangular sides, or an arrangement of four atoms around a fifth at the corners of such a form, as in $(SiO_4)^{-4}$.

Tetrapod: A vertebrate animal with four legs.

Texture (rock): The rock characteristics of grain or crystal size, size variability, rounding or angularity, and preferred orientation.

Thalweg: A sinuous imaginary line following the deepest part of a stream.

Thermal conductivity: A measure of a rock's capacity for heat conduction.

Thermal expansion: The property of increasing in volume as a result of an increase in internal temperature.

Thermonuclear reaction: A reaction in which atomic nuclei fuse into new elements with a large release of heat, especially one that is self-sustaining; occasionally used to include fission reactions as well.

Thermoremanent magnetization: A permanent magnetization acquired by igneous rocks as they cool through the *Curie point* in the presence of the Earth's magnetic field.

Thrust fault: A *reverse fault* in which the *dip* of the fault plane is relatively shallow.

Tidal current: A horizontal displacement of ocean water under the gravitational influence of Sun and Moon, causing the water to pile up against the coast at high tide and move outward at low tide.

Tidal flat: A broad, flat region of muddy or sandy sediment, covered and uncovered in each tidal cycle.

Till: An unconsolidated sediment containing all sizes of fragments from clay to boulders deposited by glacial action, usually unbedded.

Tillite: The lithified equivalent of *till*.

Time scale: The division of geologic history into *eras*, *periods*, and *epochs* accomplished through *stratigraphy* and *paleontology*.

Topographic map: See *contour map*; also a schematic drawing of prominent landforms indicated by conventionalized symbols, such as hachures or contours.

Topography: The shape of the Earth's surface, above and below sea level; the set of landforms in a region; the distribution of elevations.

Topset bed: A horizontal sedimentary bed formed at the top of a *delta* and overlying the *foreset beds*.

Toroidal mode: A type of *free oscillation* in which the whole Earth periodically (period of 44 minutes) twists in opposite directions on opposite sides of a great circle.

Trace element: An element that appears in minerals in a concentration of less than 1 percent (often less than 0.001 percent).

Tranquil flow: See *streaming flow*.

Transform fault: A *strike-slip fault* connecting the ends of an offset in a *mid-ocean ridge*. Some pairs of plates slide past each other along transform faults.

Transgression: A rise in sea level relative to the land that causes areas to be submerged and marine deposition to begin in that region.

Transition elements: Elements of atomic number 21 through 30, 39 through 48, and 72 through 80, and the actinide series elements.

Transition zone: See *seismic transition zone*.

Transpiration: The removal of water from the ground into plants, ultimately to be evaporated into the atmosphere by them.

Transverse dune: A dune that has its axis transverse to the prevailing winds or to a current. The upwind or upcurrent side has a gentle slope, and the downwind side lies at the *angle of repose*.

Trap (oil): A sedimentary or tectonic structure that impedes the upward movement of oil and gas and allows it to collect beneath the barrier.

Travel-time curve: A curve on a graph of travel time versus distance for the arrival of seismic waves from distant events. Each type of seismic wave has its own curve.

Travertine: A terrestrial deposit of limestone formed in caves and around hot springs where cooling carbonate-saturated groundwater is exposed to the air.

Trellis drainage: A system of streams in which tributaries tend to lie in parallel valleys formed in steeply dipping beds in folded belts.

Trench: A long and narrow deep trough in the sea floor; interpreted as marking the line along which a plate bends down into a *subduction zone.*

Trilobites: Extinct class of arthropods that ranged from Cambrian to Permian. Related to crustaceans.

Triple junction: A point that is common to three plates and that must also be the meeting place of three boundary features, such as *divergence zones, convergence zones,* or *transform faults.*

Trough: The depression between two *ripples.* Troughs are wider than the ripples that are juxtaposed against them.

Tsunami: A large destructive wave caused by sea-floor movements in an earthquake.

Tufa: See *travertine.* Not related to *tuff.*

Tuff: A consolidated rock composed of *pyroclastic* fragments and fine *ash.* If particles are melted slightly together from their own heat, it is a "welded tuff."

Turbidite: The sedimentary deposit of a *turbidity current,* typically showing *graded bedding* and sedimentary structures on the undersides of the sandstones.

Turbidity current: A mass of mixed water and sediment that flows downhill along the bottom of an ocean or lake because it is denser than the surrounding water. It may reach high speeds and erode rapidly. (See also *density current.*)

Turbulent flow: A high-velocity flow in which streamlines are neither parallel nor straight but curled into small tight eddies. (Compare *laminar flow.*)

Ultramafic rock: An igneous rock consisting mainly of *mafic* minerals, containing less than ten percent feldspar. Includes *dunite, peridotite, amphibolite,* and pyroxenite.

Unconformity: A surface that separates two strata. It represents an interval of time in which deposition stopped, erosion removed some sediments and rock, and then deposition resumed. (See also *angular unconformity.*)

Unconsolidated material: Nonlithified sediment that has no mineral cement or *matrix* binding its grains.

Uniformitarianism, principle of: The concept that the processes that have shaped the Earth through geologic time are the same as those observable today.

Unsaturated zone: See *vadose zone.*

Uplift: A broad and gentle epeirogenic increase in the elevation of a region without a *eustatic change* in sea level.

Upwarped mountains: Mountains elevated by *uplift* of broad regions without faulting.

Upwelling current: The upward movement of cold bottom water in the sea that occurs when wind or currents displace the lighter surface water.

U-shaped valley: A deep valley with steep upper walls that grade into a flat floor, usually eroded by a glacier.

Vadose zone: The region in the ground between the surface and the *water table* in which pores are not filled with water. Also called the unsaturated zone.

Valence electron: An electron of the outermost shell of an atom; one of those most active in bonding.

Valley glacier: A glacier that is smaller than a *continental glacier* or an icecap, and that flows mainly along well-defined valleys, many with tributaries.

Van der Waals bond: A bond much weaker than the *ionic* or *covalent,* which bonds atoms by small electrostatic attraction.

Varve: A thin layer of sediment grading upward from coarse to fine and light to dark, found in a lake bed and representing one year's deposition.

Vector: A mathematical element that has a direction and magnitude, but no fixed position. Examples are force and gravity.

Vein: A deposit of foreign minerals within a rock fracture or *joint.*

Ventifact: A rock that exhibits the effects of *sandblasting* or "snowblasting" on its surfaces, which become flat with sharp edges in between.

Vertical exaggeration: The ratio of the horizontal scale (for example, 100,000:1) to the vertical scale (for example, 500:1) in an illustration.

Vesicle: A cavity in an igneous rock that was formerly occupied by a bubble of escaping gas.

Viscosity: A measure of resistance to flow in a liquid.

Volatiles: Gaseous materials that are readily lost from a system if not confined; also, substances such as water and carbon dioxide that are loosely bound into a mineral structure and that can escape from a rock if the mineral breaks down during metamorphism.

Volcanic ash: A volcanic sediment of rock fragments, usually glass, less than 4 millimeters in diameter that is formed when escaping gases force out a fine spray of *magma.*

Volcanic ash fall: A deposit of volcanic ash resting where it was dropped by eruptions and winds.

Volcanic ash flow: A mixture of volcanic ash and gases that moves downhill as a *density current* in the atmosphere.

Volcanic block: A pyroclastic rock fragment ranging from about fist- to car-size.

Volcanic bomb: A pyroclastic rock fragment that shows the effects of cooling in flight in its streamlined or "bread-crust" surface.

Volcanic breccia: A pyroclastic rock in which all fragments are more than two millimeters in diameter.

Volcanic cone: The deposit of lava and pyroclastic materials that has settled close to the volcano's central vent.

Volcanic dome: A rounded accumulation around a volcanic vent of congealed lava too viscous to flow away quickly; hence usually *rhyolite* lava.

Volcanic dust: See *volcanic ash.*

Volcanic ejecta blanket: A collective term for all the *pyroclastic* rocks deposited around a volcano, especially by a volcanic explosion.

Volcanic emanations: Gases, especially steam, emitted from a vent or released from *lava.*

Volcanic pipe: The vertical chamber along which magma and gas ascend to the surface; also, a formation of igneous rock that cooled in a pipe and remains after the erosion of the volcano.

Volcano: Any opening through the crust that has allowed *magma* to reach the surface, including the deposits immediately surrounding this vent.

V-shaped valley: A valley whose walls have a more-or-less uniform slope from top to bottom, usually formed by stream erosion.

Wadi: A steep-sided valley containing an intermittent stream in an arid region.

Warping: In *tectonics*, refers to the gentle, regional bending of the crust, which occurs in epeirogenic movements.

Water mass: A mass of water that fills part of an ocean or lake and is distinguished by its uniform physical and chemical properties, such as temperature and salinity.

Water table: A gently curved surface below the ground at which the *vadose zone* ends and the *phreatic zone* begins; the level to which a well would fill with water.

Wave-cut terrace: A level surface formed by wave erosion of coastal bedrock to the bottom of the turbulent *breaker* zone. May appear above sea level if uplifted.

Wavelength: The distance between two successive peaks, or between troughs, of a cyclic propagating disturbance.

Wave refraction: The bending of water waves as they encounter different depths and bottom conditions, or of other waves as they pass from one medium to another of different properties.

Wave steepness: The maximum height or amplitude of a wave divided by its wavelength.

Weathering: The set of all processes that decay and break up bedrock, by a combination of physical fracturing or chemical decomposition.

Welded tuff: See *tuff* and *ignimbrite*.

Xenolith: A piece of *country rock* found engulfed in an intrusion.

X-ray diffraction: In mineralogy, the process of identifying mineral structures by exposing crystals to x rays and studying the resulting diffraction pattern.

X-ray fluorescence: A technique useful in the determination of trace-element composition in minerals. The mineral is bombarded by x rays, inducing secondary emission of radiation that is characteristic of the trace elements present.

Youth (geomorphology): A stage in the *geomorphic cycle* in which a landscape has just been uplifted and is beginning to be dissected by canyons cut by young streams.

Zeolite: A class of silicates containing H_2O in cavities within the crystal structure. Formed by alteration at low temperature and pressure of other silicates, often volcanic glass.

Zoned crystal: A single crystal of one mineral that has a different chemical composition in its inner and outer parts; formed from minerals belonging to a *solid-solution series*, and caused by the changing concentration of elements in a cooling magma that results from crystals settling out or by crystallization so rapid that the composition of the mineral cannot adjust fast enough so that all parts of the mineral achieve the *equilibrium* composition for each temperature.

Zone of accumulation (soil): See *B-horizon*.

Zone of aeration: See *vadose zone*.

Index

Page numbers in **boldface** refer to a definition;
page numbers followed by an asterisk indicate an illustration.
The appendixes and glossary are not covered by this index.